Lecture Notes in Mechanical Engineering

Lecture Notes in Mechanical Engineering (LNME) publishes the latest developments in Mechanical Engineering—quickly, informally and with high quality. Original research or contributions reported in proceedings and post-proceedings represents the core of LNME. Volumes published in LNME embrace all aspects, subfields and new challenges of mechanical engineering.

To submit a proposal or request further information, please contact the Springer Editor of your location:

Europe, USA, Africa: Leontina Di Cecco at Leontina.dicecco@springer.com
China: Ella Zhang at ella.zhang@cn.springernature.com
India, Rest of Asia, Australia, New Zealand: Swati Meherishi at swati.meherishi@springer.com

Topics in the series include:

- Engineering Design
- Machinery and Machine Elements
- Mechanical Structures and Stress Analysis
- Automotive Engineering
- Engine Technology
- Aerospace Technology and Astronautics
- Nanotechnology and Microengineering
- Control, Robotics, Mechatronics
- MEMS
- Theoretical and Applied Mechanics
- Dynamical Systems, Control
- Fluid Mechanics
- Engineering Thermodynamics, Heat and Mass Transfer
- Manufacturing Engineering and Smart Manufacturing
- Precision Engineering, Instrumentation, Measurement
- Materials Engineering
- Tribology and Surface Technology

Indexed by SCOPUS, EI Compendex, and INSPEC.

All books published in the series are evaluated by Web of Science for the Conference Proceedings Citation Index (CPCI).

To submit a proposal for a monograph, please check our Springer Tracts in Mechanical Engineering at https://link.springer.com/bookseries/11693.

Martin Manns
Editor

New Paradigms for Anticipated Uncertainty

Proceedings of the 10th Changeable, Agile, Reconfigurable and Virtual Production Conference (CARV 2025) and the 12th World Mass Customization and Personalization Conference (MCPC 2025), Siegen, Germany, September 2025

Springer

Editor
Martin Manns
Department of Mechanical Engineering
PROTECH - Institute for Production Technology
University of Siegen
Siegen, Nordrhein-Westfalen, Germany

ISSN 2195-4356 ISSN 2195-4364 (electronic)
Lecture Notes in Mechanical Engineering
ISBN 978-3-032-16888-7 ISBN 978-3-032-16889-4 (eBook)
https://doi.org/10.1007/978-3-032-16889-4

This Springer imprint is published by the registered company Springer Nature Switzerland AG
The registered company address is: Gewerbestrasse 11, 6330 Cham, Switzerland

Organization

Conferences Chair CARV 2025 - MCPC 2025

Martin Manns, University of Siegen, Siegen, Germany

Local Organizing Committee

Peter Burggräf, Germany
Bernd Engel, Germany
Karsten Kluth, Germany
Martin Manns, Germany
Ulrich Stache, Germany
Tadele Belay Tuli, Germany
Florian Schreiber, Germany
Peter Frohn-Sörensen, Germany
Michael Jonek, Germany
Daniel Klein, Germany

Honorary Co-chairs CARV 2025

Michael Zaeh, Germany
Waguih ElMaraghy, Canada

Honorary Co-chairs MCPC 2025

Kjeld Nielsen, University of Aalborg, Denmark
Frank Piller, RWTH Aachen University, Germany

International Scientific Committee CARV 2025

Ann-Louise Andersen, Denmark
Jan Aurich, Germany
Ahmed Azab, Canada
Peter Burggräf, Germany
Emanuele Carpanzano, Switzerland
George Chryssolouris, Greece
Waguih ElMaraghy, Canada
Bernd Engel, Germany
Michael Freitag, Germany
Norbert Gronau, Germany
Karsten Kluth, Germany
Gisela Lanza, Germany
Luc Laperrière, Canada
Martin Manns, Germany
Frank Mantwill, Germany
Aki Mikkola, Finland
Niels H. Mortensen, Denmark
Dimitris Mourtzis, Greece
Khumbulani Mpofu, South Africa
Giovanni Perrone, Italy
Ulrich Stache, Germany
Kristina Säfsten, Sweden
Sebastian Thiede, Netherlands
Tullio Tolio, Italy
Kirsten Tracht, Germany
Jill Urbanic, Canada
Jozsef Vancza, Hungary
Michael F. Zaeh, Germany

International Scientific Committee MCPC 2025

Ann-Louise Andersen, Denmark
Jocelyn Bellemare, Canada

Paul Blazek, Austria
Claudio Boër, Switzerland
Marco Bortolini, Italy
Salvador Fabrizio, Spain
Francesco Galizia, Italy
Paul Gembarski, Germany
Stephan Hankammer, Germany
Lars Hvam, Denmark
Niels H. Mortensen, Denmark
Kjeld Nielsen, Denmark
Frank Piller, Germany
Herwig Schreiner, Austria
Lars Skjelstad, Norway
Kristina Säfsten, Sweden
Maria Thomassen, Norway

Reviewers

Adane Kassa Shikur
Ahmed Azab
Aki Mikkola
Alfred Hypki
Ali Al-Yacoub
Alina Peter
Anna-Lena Dreisbach
Annika Raatz
Apostolos Aslanidis
Aydin Ünlü
Christopher Kuhnhen
Clara Fischer
Daniel Nebeling
Dawid Stade
Dirk Schreiber
Emanuele Carpanzano
Emmanouil Bakopoulos
Fabian Kost
Fazel Ansari
Felix Carros
Fengyun Shao
Florian Schreiber
Francesco Gabriele Galizia
Frank Mantwill
Frank Piller

Giacinto Barresi
Gisela Lanza
Herwig Schreiner
Jana Marciniak
Janis Sprenger
Jannis Middendorf
Jill Urbanic
Jocelyn Bellemare
Jonas Knoche
Jonas Reuter
Jorge Morales
Jozsef Vancza
Julian Schützenberger
Karl Hribernik
Karsten Kluth
Kirsten Tracht
Kjeld Nielsen
Kosmas Alexopoulos
Kristina Säfsten
Lars Skjelstad
Lili Wu
Luc Laperriere
Marco Fries
Maria Thomassen
Martin Manns
Maximilian Schmidt
Michael Freitag
Michael Geueke
Michael Jonek
Michael Schiller
Michael Zaeh
Muhammad Ali Kaleem
Natacha Kuete Meli
Nathanael L. Baisa
Nicolas Meusel
Nicole Goeckel
Niels Henrik Mortensen
Norbert Gronau
Paul Blazek
Paul Christoph Gembarski
Peter Frohn-Sörensen
Ralf Polzin
Rasmus Andersen
Raza Saeed
René Sauer

Salvador Fabrizio
Sebastian Thiede
Sisay Chala
Sotiris Makris
Stefano Baraldo
Syed Muhammad Raza
Tadele Belay Tuli
Tamara Reinicke
Theresa Breckle
Thomas Ludwig
Thomas Trautner
Tobias Girresser
Tobias Kotzian
Ulrich Stache
Verena Kiesler
Nasser Gyagenda

Preface

In recent years, the manufacturing industry and its supply chains have been confronted with an unprecedented level of global disruptions, significantly increasing the degree of uncertainty faced by industry and society. While these challenges test the resilience of industrial systems, they also offer a powerful impetus to rethink established paradigms and unlock opportunities for innovation. The theme of this year's CARV - MCPC 2025 Joint Conference, "New Paradigms for Anticipated Uncertainty," reflects this imperative. The conference aims to explore innovative technical and organizational approaches that enable production systems and supply chains to become more robust, agile, and resilient in the face of uncertainty.

As a contribution to respond to these challenges and changes, academia and industry are focusing their research on the definition of methods and solutions for the design and management of flexible, agile, changeable, and reconfigurable production systems on the one hand and on innovative and sustainable techniques to best manage mass customization on the other. For all these reasons, the theme of the 10th Changeable, Agile, Reconfigurable, and Virtual Production Conference (CARV 2025) and the 12th World Mass Customization and Personalization Conference (MCPC 2025) is "New Paradigms for Anticipated Uncertainty." These international conferences are hosted for the first time by the University of Siegen, Siegen, Germany. The Local Organizing Committee for CARV 2025 and MCPC 2025 is affiliated with the PROTECH - Institute for Production Technology, Department of Mechanical Engineering at the University of Siegen. Following a recent tradition, the two conferences are co-organized and co-located for the third time. While CARV's traditional topics focus on advanced production systems, MCPC explores the mass customization theme; hence, they are two sides of the same coin.

CARV 2025 and MCPC 2025 comprise keynote talks, oral presentations, invited sessions, and industrial workshops, covering theory and applications, methodologies, and case histories in the dynamic field of resilient production, smart manufacturing, manufacturing system design and customized products within the Industry 4.0 framework, which serves as an enabler for achieving goals related to mass customization management and the design of dynamic and sustainable supply chains. CARV 2025 and MCPC 2025 provide an excellent platform for presenting and discussing new data and concepts, fostering knowledge exchange and the generation and sharing of innovative ideas.

We look forward to an excellent meeting point for scientists from diverse backgrounds and countries worldwide, as well as industrial practitioners from multiple sectors and areas, to share new and promising results in the field of advanced manufacturing systems design and mass customization management.

This book of proceedings includes 83 papers accepted for presentation at CARV 2025 and MCPC 2025. All papers underwent and passed a two-step peer-review process involving independent reviewers from the conferences' communities. Eight relevant topics are covered and featured in this book:

- Resilient production
- Digital business models, smart products and services
- Changeable, reconfigurable and flexible manufacturing
- Supply chains and production strategies
- Production planning, operation, and control
- Smart automation
- Collaborative robotics
- Additive manufacturing and smart tools

We hope our readers, from academia and industry, will enjoy and find inspiration in reading this book contributing to progress in this scientific field.

Enjoy your CARV 2025 and MCPC 2025 Conferences!

Siegen, Germany
September 2025

Univ.-Prof. Dr.-Ing. Martin Manns
The CARV 2025 and MCPC 2025
Conferences Chair

Acknowledgements

The 10th Changeable, Agile, Reconfigurable, and Virtual Production Conference (CARV 2025) and the 12th World Mass Customization and Personalization Conference (MCPC 2025) are jointly organized by the PROTECH - Institute for Production Technology, Department of Mechanical Engineering at the University of Siegen. The chair for Production Automation and Assembly (FAMS) at the University of Siegen is responsible for administrative coordination and secretarial duties. The CARV 2025 and MCPC 2025 conferences are sponsored by the German Research Foundation (DFG). The CARV 2025 Conference is sponsored by the International Academy for Production Engineering (CIRP). We thank both organizations for their relevant support.

The editor and the Local Organizing Committee would like to express their gratitude to all the conference delegates and to the authors for their attendance, the high-quality contributions, and cooperation in preparing the manuscripts. We also express our gratitude to the reviewers and the International Scientific Committees for their support in the manuscript evaluation process, providing expert knowledge and valuable comments. Furthermore, a deep sense of gratitude goes to the Honorary Co-chairs, Prof. Waguih ElMaraghy, Prof. Michael Zaeh, Prof. Frank Piller, and Prof. Kjeld Nielsen, for inspiring us, for their collaboration, and for the continuous support that they provided to the Local Organizing Committee in managing the Conferences.

Contents

Resilient Production

A Review of Challenges and Strategies Towards Integrating Sustainability and Due Diligence in the Buyer–Supplier Relationship ... 3
Lasse Snogdal Sørensen, Magnus Sommer Klampe, Rasmus Andersen, Brian Vejrum Wæhrens, and Ann-Louise Andersen

Adaptive Simulation Modeling of Intertwined Supply Networks with Demand Uncertainty in Disruptions ... 13
Mohaddeseh Roshan, Jessica Olivares-Aguila, and Waguih ElMaraghy

Decision Support Model for Component Remanufacturing ... 23
Emma B. Worup, Ann-Louise Andersen, Thomas D. Brunoe, and Kjeld Nielsen

Resilience Performance Assessment in Metal Forming Machines: Bayesian Networks-Based Methodology ... 35
Adane Kassa, Ulrich Stache, and Martin Manns

Resilience Assessment in Steel Manufacturing: Evaluating Disruption Scenarios Using the Penalty of Change Methodology ... 47
Emmanouil Bakopoulos, Panagiotis Mavrothalassitis, Vasilis Siatras, Sotiris Makris, and Kosmas Alexopoulos

Enhancing Resilience in Smart Manufacturing: A Method for Connected Machine Networks ... 57
Elena Urkia, Elena Montejo, Emmanouil Bakopoulos, and Kosmas Alexopoulos

Resilient Manufacturing Through Data Spaces: Leveraging IDS and Gaia-X in a Compute-on-Demand Environment ... 67
Emmanouil Bakopoulos, Konstantinos Sipsas, and Kosmas Alexopoulos

Analysis of Human Action Variability for Reconfiguring Tooling Systems ... 77
Syed Muhammad Raza, Tadele Belay Tuli, and Martin Manns

Time-Domain Analysis of Human Motion for Reconfiguring Machine Tools ... 87
Syed Muhammad Raza, Tadele Belay Tuli, and Martin Manns

A Digital Twin and Dataspace Framework for Resilient and Reconfigurable Manufacturing ... 97
Syed Muhammad Raza, Maximilian Schmidt, Adane Kassa Shikur, Bernd Engel, and Martin Manns

Fault Detection and Reconfiguration on Resource Level with Fuzzy-Logic and Case-Based Reasoning ... 105
Maximilian Schmidt and Bernd Engel

Dynamic Resilience Assessment in Smart Make-to-Order Job Shops with Attention-Based LSTM ... 115
Adane Kassa, Elena Urkia, Elena Montejo, and Martin Manns

Digital Business Models, Smart Products and Services

A Unified Framework for Intelligent Product Configuration Using Knowledge Graphs and Bayesian Networks ... 131
Stefan Berlik and Mohammad Seidpisheh

Managing Variety in the Era of Mass Customization: A Decisional Algorithm for Product Platform Design ... 143
Ludovica Diletta Naldi, Riccardo Venturi, Francesco Gabriele Galizia, Marco Bortolini, and Matteo Gabellini

Engaging Consumer Uncertainty and Creativity During the Mass Customization Co-design Experience ... 153
Frances Turner, Marie Watts, and Nikola Suzic

Algorithmic and Knowledge-Based Modeling of Bus- and Sectional-Modular Designs: A Seating Layout Case Study ... 165
Paul Christoph Gembarski, Alejandro Coronas Firchow, and Shakiran Sandran

Conceptualizing Compound Features: On Resource- and Capability Adaption in Sustainable Ecosystems ... 175
Paul Christoph Gembarski and Friedemann Kammler

Concept for the Sustainability-Oriented Variant Management of Technical Products Supported by Machine Learning ... 185
Nikolai Kelbel, Alexander Keuper, and Günther Schuh

Bringing Innovation Systems to Life: Collaborative Upskilling and Customized Connectivity ... 197
Paul Blazek and Marton Liszka

Model-Based Configuration as a Tool for Discovering Design Knowledge: A Skate Park Case-Study ... 209
Paul Christoph Gembarski, Leo Duchan, Julian Schröder, and Julius Vahle

AI-Driven Transformation of Online Product Configurators: Enhancing User Interaction in Mass Customization ... 219
Paul Blazek and Clarissa Streichsbier

Changeable, Reconfigurable and Flexible Manufacturing

Classification of Contact Points in Car Body Production to Enable a Mechanization Assessment ... 231
Julian Schützenberger, Claus-Dieter Reiniger, and Martin Manns

Design Principles for Front-End Radical Manufacturing Innovation: Insights from Pharmaceutical Manufacturing ... 241
Rasmus Andersen, Casper Schou, Ole Madsen, Astrid H. Lassen, Martin B. Jensen, Tristan Schwörer, Bjørn C. Dueholm, and Oline V. Olesen

Systematic Literature Review to Derive Basic Elements for Model-Based Product Development ... 253
Gereon C. Bönsch, Alexander Keuper, and Günther Schuh

Scalable Cell for the Production of Sheet Metal Car Body Components ... 261
Michael Schiller, Peter Frohn-Sörensen, Michael Jonek, Walid Elleuch, Marco Fries, Ireneus Wior, Izel Gediz, Anja Elser, Christopher Heftrich, Tony Joost, Meik Ebbinghaus, Marvin Engel, Dario Winterberg, Michael Keckeisen, Martin Manns, and Bernd Engel

The Learning Factory Approach: From Knowledge Transfer to Knowledge Creation ... 275
Lasse Christiansen, Jonas Frendrup, Maria S. S. Larsen, and Astrid H. Lassen

A Concept for Reconfigurable Gripping Systems for Flexible Disassembly Systems ... 283
Marc-André Weismüller, Oliver Petrovic, and Christian Brecher

Economic Justification of Reconfigurable Manufacturing Systems: Insights from the Swedish Automotive Industry ... 293
Gary Linnéusson, Filip Skärin, Itxaso Andoaga Ayestaran, and Carin Rösiö

Design of a Flexible Tool System for the Scalable Production of Secondary Forming Elements ... 305
Michael Schiller, Peter Frohn-Sörensen, Martin Freitag, Martin Manns, and Bernd Engel

Design and Experimental Validation of a Soft, Shape-Programmable Multi-fold Gripper ... 317
Florian Schreiber and Martin Manns

Development of a Software Tool for the Company-Individual and Change-Specific Support of Processes in Manufacturing Change Management ... 327
Jan-Philipp Rammo, Clément Rouvelle Romegoux, Moritz Goeldner, and Michael F. Zaeh

Supply Chains and Production Strategies

Balancing Between Efficiency Value and Service Value in Delivering Customized Solutions ... 339
Junsong He and Miia Martinsuo

Data-Driven Order Management for Built-to-Stock Products in Multi-variant Series Production ... 349
Jonas Strecker, Simon Duerr, and Stephan Daurer

Emotional Durability Through Mass Customization in the Fashion and Apparel Industry ... 359
Daniella Besse, Jocelyn Bellemare, Claudia Déméné, and Marie-Eve Faust

Exploring the Flexibility and Impact of Hybrid Production Models in Mass Customization Within the Fashion and Apparel Industry ... 371
Anne Léger and Jocelyn Bellemare

Spare Parts Supply Capabilities and Equipment Coverage: A Systematic, Data-Driven Portfolio Assessment Approach ... 383
Simon Klarskov Didriksen, Kristoffer Wernblad Sigsgaard, Christian Brunbjerg Jespersen, and Niels Henrik Mortensen

Bridging the Gap: A Conceptual Framework for Aligning Internal Capabilities and Strategic Goals for Technology Integration in Manufacturing Companies ... 393
Louise Paaske Jørgensen, Mikkel Østergaard, Søren Odgaard, Zsigmond Csaba Dósa, Vita Jutinskaitė, and Torben Tambo

Battery Systems Logistics in Automotive Supply Chains: Success Factors and Best Practices for Efficiency and Sustainability ... 407
Henrik Uebach

Towards Intelligent Supply Chain Reconfiguration: A Framework Integrating Dynamic Knowledge Graph and AI-Driven Optimization ... 417
Chaouki Saidi, Nadia Hamani, Mounir Benaissa, Benjamin Rolf, Tobias Reggelin, and Sebastian Lang

The Effect of Learning Factories on Manufacturing Innovation in SMEs ... 427
Lasse Christiansen, Jonas Frendrup, and Astrid H. Lassen

Process Design for Digitally Assisted Location Selection for Sustainable Factories ... 437
Günther Schuh, Esben Schukat, Siyuan Wang, Felix Paßlick, and Carsten Engeln

Empirical Study on Asset Fingerprinting with Natural Markers to Improve Supply Chain Traceability Using Digital Product Passports ... 447
Marco Buecheler, Maximilian Hentsch, Frithjof Dorka, Grant Richards, and Daniel Palm

Production Planning, Operation and Control

Optimizing Offer Management: Reuse Strategies for Efficient Production System Planning ... 459
Max Eichenwald and Rainer Müller

Conception of an Automated Graph-Based Assembly System Development Process for a Balancing Scooter ... 469
Timo Schuchter, Theresa Breckle, Ralf Stetter, Markus Till, Patricia Derksen, and Stephan Rudolph

Proposing a Multi-domain and Life-Cycle Based Analysis Process for Service Architecture Design: A Case Study from a Danish Third-Party Logistics Company ... 479
Anton Seistrup Hermann, Frederik Holm Nielsen, and Niels Henrik Mortensen

Carbon Footprint Reduction of Queueing Systems in Manufacturing ... 493
Sebastian Schötz and Philipp Gölzer

Data Acquisition Pipeline for the Generation of a Digital Twin for Large Compliant Thin-Walled Structures ... 505
Guillaume Tréheux, Dominik Schäfer, Alexander Moriz, Lars Köttner, Tobias Weber, Amon Göppert, and Robert H. Schmitt

Layout Optimization of Robotized Production Systems Using a Mass-Spring-Damper-Model in a Brownfield Example 515
Ramez Awad and Katharina Barbu

Cost Allocation in Modular Product Development: Reviewing Existing Practices and Identifying Opportunities 525
Morten Nørgaard, Jakob Meinertz Grønvald, and Niels Henrik Mortensen

Generating Assembly Instructions by Demonstration—System Implementation and Evaluation 535
Dennis Keiser, Christoph Petzoldt, Dario Niermann, Burak Vur, and Michael Freitag

Development of a Methodology for the Semi-automated Determination of Optimized Workplace Design in Manual Assembly 547
Fabian Adler, Sebastian Flierl, and Rainer Müller

Optimizing Finite State Machines for Process Time Analysis in Industrial Production 557
Dan Eisenkrämer and Bernd Lüdemann-Ravit

Resistance Spot Welding: Quantitative Assessment of Its Impact on Cycle Time and Robotic Assembly Line Balancing 567
Dawid Stade and Martin Manns

A Dynamic Digital Twin Modelling Environment Integrating Maintenance Policy in RMS Operations' Management 577
Marco Bortolini, Francesco Gabriele Galizia, Ludovica Diletta Naldi, Giulia Cardelli, Michele Micci, Andrea Sanfilippo, and Alberto Regattieri

Design of a Decision Logic of a Hybrid Matrix Assembly for Electric Vehicle Production 587
Stefanie Dechant and Hans-Christian Möhring

Improving the Robustness of Reconfigurable Assembly Systems with an Agent-Based Control Architecture 597
Jasper Wilhelm, Artem Schurig, Aaron Heuermann, Burak Vur, and Michael Freitag

Smart Automation

Identification of Barriers and Opportunities for Adoption of Machine Vision for Small- and Medium-Sized Enterprises: A Stakeholder Investigation 609
Mikkel Graugaard Antonsen, Lasse Christiansen, and Christian Black Jørgensen

Comparison of Feature Extraction Methods for Time Series Data in Fault Classification of Screw Connections ... 619
Michelle Henkies, Nikolai West, and Jochen Deuse

A 3D Scanning-Based Method for Surface Defect Detection in Galvanized Scaffolding ... 629
Fengyun Shao, Tadele Belay Tuli, Florian Schreiber, and Martin Manns

Anomaly Detection in Industrial Robotic Assembly with Variational Autoencoders ... 639
Niklas Grambow, Lisa-Marie Fenner, Dingyuan Wan, Jörg Krüger, and Kevin Haninger

A Force Controlled Pose Adaptation Method for Robot Based Defect Removal ... 649
Fengyun Shao, Tadele Belay Tuli, Florian Schreiber, and Martin Manns

Rebound Reasoning: Enhancing Quantized Language Models Through Iterative Prompting as a Service for Production Systems ... 657
David Golchinfar, Daryoush Vaziri, Darius Hennekeuser, and Dirk Schreiber

Leveraging LLM-Based Reasoning for Human-Robot Cooperative Disassembly ... 667
Daniel Gusenburger, Marco Giangreco, Attique Bashir, and Rainer Mueller

Leveraging LLM for Assembly Instructions Using Controlled Natural Language ... 677
Michael Jonek and Martin Manns

Human Motion Modelling for Collaborative Handling of Rigid Objects ... 689
Raza Saeed, Tadele Belay Tuli, and Martin Manns

AI-Driven Analysis of 2D Technical Drawings for Agile and Reconfigurable Manufacturing Systems ... 699
Marco Fries and Thomas Ludwig

Combining Virtual Commissioning and Discrete-Event Simulation in Digital Manufacturing: A Literature Review and Future Research Directions ... 713
Martin Birtic, Enrique Ruiz Zúñiga, and Anna Syberfeldt

Implementing AI-Driven Augmented Reality: Insights from an Industrial Use Case Implementation ... 725
Katharina Schmenn, Lukas Baeck, Peter Burggräf, Carl René Sauer, Alexander Becher, and Maximilian Lutz

Ergonomics-Oriented Parameter Analysis Based on Digital Human Models with a Focus on Manual Work 735
Aydin Ünlü, Pascal Petri, and Karsten Kluth

Analysis of Cooperative Activities Between Two People Including Motion and Muscle Activity Measurements 743
Aydin Ünlü, Saeed Raza, Martin Manns, and Karsten Kluth

Modular Cognitive Robotics: Enabling Flexible Handling and Inspection in Automated Manufacturing 751
Martin Naumann, Leutrim Gjakova, Rico Löser, and Martin Dix

Timing Analysis of Low-Cost Edge Vision for Object Detection for Smart Factory Systems 763
Christopher Tofaute, Tadele Belay Tuli, Florian Schreiber, and Martin Manns

Collaborative Robotics

Robust Automated Unscrewing with Cost-Effective Robot Control System 775
Sara Menetrey, Manuel Möckel, Holger Schlegel, Matthias Rehm, and Martin Dix

Role-Based Human-Robot Interaction for Symbiotic Collaboration 787
Tadele Belay Tuli, Raza Saeed, and Martin Manns

Scenario-Based Collaborative Robot Evaluation (SCORE) Framework 795
Ali Al-Yacoub, Alina Peter, Dennis Guck, Felix Fleschhut, and Michael Keckeisen

Energy Consumption in Robotics: A Simplified Modeling Approach 805
Lisa Lokstein, Kevin Haninger, Valentyn Petrichenko, Sophie Matthews, Miriam Schleipen, and Gregor Thiele

A Methodology for Applying HRC on Remanufacturing Workstations: A Use Case from Household Appliances 815
Dionisis Vallianatos, Dimitrios Kaliakatsos-Georgopoulos, Dimosthenis Dimosthenopoulos, Christos Gkournelos, George Michalos, and Sotiris Makris

Additive Manufacturing and Smart Tools

Benchmarking Additive Manufactured Silicone Actuators: A Trajectory and Bending Analysis 829
Florian Schreiber, Lukas Gugel, Philipp Vogelgsang, Jörg Franke, and Martin Manns

Fabrication of Cranioplasty Implants from Sheet Metal Plates in a Rapid Tooling Approach 839
Peter Frohn-Sörensen and Michael Schiller

A Prototype System Integration Solution for Limp Fabric Handling Utilizing Reconfigurable and Scalable Grippers 849
Morteza Alebooyeh, Jill Urbanic, and Thanh Dat Vo

A Framework for Medium Volume Production Strategies for Directed Energy Deposition Additive Manufacturing 859
Jill Urbanic

Investigation of the Long-Term Durability of Soft Pneumatic Actuators 869
Florian Schreiber and Martin Manns

Optimized Data Preprocessing for Design of Passive Compliance via Solid Generation from G-Code 879
Richard Matthias Hartisch, Danny Huang, Patrick Kahl, Linus Schneider, and Jörg Krüger

Resilient Production

A Review of Challenges and Strategies Towards Integrating Sustainability and Due Diligence in the Buyer–Supplier Relationship

Lasse Snogdal Sørensen, Magnus Sommer Klampe, Rasmus Andersen, Brian Vejrum Wæhrens, and Ann-Louise Andersen

Abstract Global supply chains account for 80% of international trade but pose significant human rights and environmental risks. The EU Corporate Sustainability Due Diligence Directive and similar regulations impose new compliance requirements, challenging companies to integrate sustainability, due diligence, and supplier relationship management. This paper reviews the challenges of complying with these regulations, including supply chain dilemmas related to regulations, complexity, and transparency. It explores how risk-based and collaborative approaches and leveraging technologies can promote effective sustainability due diligence in global supply chains. This study contributes to understanding how large firms can navigate evolving supply chain regulations to ensure compliant, sustainable practices by addressing challenges and mitigating strategies.

Keywords Due diligence · Supply chain management · Sustainability · Literature review · CSDDD

1 Introduction

Global supply chains drive economic growth [1] but simultaneously pose risks to human rights and the environment. This complexity makes monitoring and enforcing ethical practices across dispersed supply chains increasingly difficult [2], leading to regulations requiring large companies to ensure compliance within their supply chains [3]. The EU's Corporate Sustainability Due Diligence Directive (CSDDD) marks a significant shift, imposing transnational accountability with penalties of up to 5% of companies' global turnover [4]. By targeting large companies, the CSDDD leverages their influence to drive industry-wide responsibility in supply chains [5].

L. S. Sørensen · M. S. Klampe · R. Andersen (✉) · B. V. Wæhrens · A.-L. Andersen
Department of Materials and Production, Aalborg University, Aalborg, Denmark
e-mail: rasmus@mp.aau.dk

M. Manns (ed.), *New Paradigms for Anticipated Uncertainty*, Lecture Notes in Mechanical Engineering, https://doi.org/10.1007/978-3-032-16889-4_1

Supply chain regulations, including the EU's CSDDD and Germany's Supply Chain Due Diligence Act, aim to enhance corporate accountability, transparency, and protection against environmental and human rights violations. The CSDDD, in particular, fundamentally shifts corporate responsibility by integrating environmental and human rights into companies' operational and strategic planning within the EU market [6]. It mandates due diligence across supply chains, encompassing subsidiaries and indirect relationships [4]. Despite their intent, these regulations present substantial practical challenges for companies in managing buyer–supplier relationships and maintaining compliance.

Despite increasing regulatory pressure, supply chain management research typically focuses on sustainability, due diligence, or supplier relationships as distinct management disciplines or as pairwise intersections, with little focus on their full integration. This presents a critical theoretical and practical gap essential for regulatory compliance and sustainability of global supply chains [7, 8]. As its primary contribution, this study investigates current literature for insights into the barriers and challenges companies are facing in transitioning to more responsible and sustainable supply chain practices as well as strategies for how companies can integrate these disciplines to enhance such practices. The study is guided by the following research questions:

1. What are the main challenges to complying with current and future legislation?
2. How can large companies effectively conduct due diligence in their supplier networks to ensure sustainable human rights and environmental practices?

The remainder of this paper is structured as follows. Section 2 briefly describes the methodology. Sections 3 and 4 focus on inherent dilemmas and challenges to sustainability due diligence in global supply chains. Section 5 reviews strategies to accommodate these challenges. Lastly, Sect. 6 presents concluding remarks.

2 Methodology

Literature reviews are ways of collecting and synthesizing published research and can take on more or less systematic approaches [9]. This study has adopted a narrative review approach, and have sought to improve the quality of the review by adopting principles from more structured and systematic review methods [10]. Specifically, this study has used keyword-based searches in Elsevier's Scopus and subsequent snowballing to identify relevant studies for inclusion. Keywords relevant to the three domains of (1) supply chain management, (2) sustainability, and (3) due diligence formed initial search strings, and search delimiters were applied to search only title, abstract, and keywords, as well as English language studies. Subsequently, backward and forward searches [11] were applied to relevant studies. Pertinent additional keywords, identified through the snowballing process, were added to the initial search string to investigate further studies related to specific topics, such as "risk-based approaches."

3 Dilemmas of Sustainability and Due Diligence in a Buyer–Supplier Relationship

Several dilemmas related to integrating sustainability and due diligence in a buyer–supplier relationship exist, especially in relation to the CSDDDs requirements.

Firstly, most supply chain acts, CSDDD included, only apply to a defined scope of companies, creating a division between companies that are legally required to ensure sustainable practices within their supplier networks and those with no such legal obligations [4]. Consequently, companies in the scope of the legal obligations are constrained by a lack of direct control over their supply chain partners [12]. This, in turn, drives large companies to rely on traditional buyer–supplier dynamics, which, in combination with the sustainability angle, poses new dilemmas [12]. These complexities challenge companies to balance conventional operational methods with the demands of more sustainable practices.

In 2021, the EU had 30.1 million companies, with 99.8% being micro, small, or medium-sized [13]. The CSDDD, applying only to companies with > 1000 employees [4], affects only a small fraction of European companies directly. This creates compliance challenges for large companies as they, besides complying with the CSDDD, most also influence and support their smaller supply chain partners [4]. It is, therefore, crucial for companies in scope of the CSDDD, to analyze their relationships with suppliers, focusing on evaluating the tensions between trust and opportunism [7]. Understanding this dynamic is essential, as it highlights the complexities of achieving a higher degree of sustainability within buyer–supplier relationships. Furthermore, it significantly impacts the purchasing companies' efforts to conduct due diligence on their suppliers as required by the CSDDD [1].

Another contradictory situation exists between supply chain transparency and long-term performance on sustainability parameters [14]. For suppliers, enhancing supply chain transparency is often associated with short-term reputational risks regarding human rights and environmental issues in the supply chain. This will give the suppliers a tendency to act in a more opportunist manner to avoid the potential consequences of increased supply chain transparency [15]. This can hinder long-term improvements in the supply chain's environmental and social dimensions, which is a challenge for the buying companies to address properly [16].

4 Challenges and Barriers to Sustainability Due Diligence of Suppliers

Despite increasing attention from both policymakers and industry on supply chain due diligence, examples of companies excelling in human rights and environmental due diligence are still relatively uncommon [17]. This is mainly due to some inherent challenges and barriers [18].

Regulatory Fragmentation: Industry-leading companies within this area have for a long time been pushing for a common legal framework to ensure uniformity and accountability of sustainability-related standards across supply chains [19]. This push aims to create a level field where companies operate with shared guidelines, fostering sustainable practices. However, ambiguous rules hindered progress, keeping efforts to improve human rights and environmental due diligence in supply chains scattered and fragmented globally [20]. In this context, the CSDDD marks a significant enhancement as it harmonizes the legal requirements for companies operating within the EU market. This will lay the necessary groundwork for the industry to promote sustainable business practices within global supply chains more effectively [6].

Complex Global Supply Chains: Despite existing legal frameworks, intensified globalization, shorter product lifecycles, and technological advancements immensely complicate supply chain due diligence. These factors result in supplier networks stretching across regions with varying standards and higher supplier turnover, which in turn make compliance especially difficult for large companies managing vast supplier networks [8]. Moreover, EU production networks exhibit a "small world" structure [21], meaning firms are highly interconnected. Hurt et al. [21] further estimate that 8.5% of EU companies have direct links to firms engaged in child or forced labor, rising to 82.4% at the second tier and over 99.1% at the third tier. This highlights the difficulty of ensuring due diligence, particularly for large firms at the network's center.

High Transaction Costs: Coordinating with numerous suppliers, auditing them, and ensuring data quality and consistency can result in significant transaction costs for the involved companies [18].

Power Dynamics: The efficiency with which large companies can engage in due diligence practices with their suppliers is significantly influenced by the power dynamics in the supply chain [14], making it essential to find an appropriate strategy to handle these challenges.

Geographical and Cultural Distances: Conducting human rights and environmental due diligence is further complicated by the geographical and cultural distance due to the global nature of modern supply chains, making effective communication, compliance, and monitoring more challenging [8]. This diversity introduces additional layers of complexity in understanding local regulations, labor practices, and environmental standards, which are critical for ensuring sustainability and ethical practices across the entire supply chain [18].

Incomplete Information: Vital details concerning, e.g., the origin of materials or labor practices [22], is often missing during supplier selection, highlighting large companies' challenges when incorporating sustainability and due diligence into buyer–supplier relationships.

Uncertainty: Investing resources into due diligence practices carries risks, particularly when the future direction of regulations, market trends, and technological advancements remains uncertain [23].

Market Competition: Companies with high ambitions (a "strong buyer") risk spill-over effects due to the phenomenon known as "free-riding" [24]. When different companies share the same supplier, a company with less focus on responsibility and sustainability (a "weak buyer") might gain from the efforts of a more diligent and sustainable company without investing as much (i.e., "free riding"), resulting in the strong buyer bearing the cost of the weak buyer [24].

5 Strategies for Addressing Sustainability Due Diligence of Suppliers

Addressing the highlighted challenges necessitates a nuanced strategy that includes collaborative approaches, risk-based approaches, and leveraging available technologies.

5.1 Collaborative Approaches

Improving sustainability and due diligence within global supply chains necessitates a collaborative approach recognized in the literature as well as by recent regulations (see, e.g., CSDDD [4] and OECD Due Diligence Guidance for Sustainable Business Conduct [25]. This involves engaging with suppliers, non-governmental organizations (NGOs), competitors, and other third-party actors.

Organizations should foster a collaborative supply chain culture, as their reputation is linked closely to suppliers' environmental and human rights practices. Shifting from transactional to long-term strategic partnerships, particularly concerning sustainability, enhances resilience, fosters innovation, improves reputation [15], and supports sustainable growth. Collaborative strategies should adapt to the supplier's historical context: newer relationships thrive on balanced trust, fostering high collaboration through proactive conflict resolution and shared sustainability goals. In contrast, mature relationships leverages established trust to manage conflicts, thereby continuing collaborative sustainability efforts [7]. Engaging closely with tier-one suppliers via regular monitoring, training, incentives, and common sustainability goals can accommodate passivity of lower-tier suppliers. This approach enables sustainability practices to cascade efficiently throughout the supply chain by leveraging tier-one suppliers' deeper operational knowledge of their downstream partners [26].

Viewing NGOs and other third-party actors as collaborators rather than adversaries promotes more effective global supply chain management concerning sustainability outcomes [8]. Specifically, NGOs can aid in training local suppliers, promoting sustainable business practices, and increasing supply chain transparency through reliable monitoring and reporting [8]. Cao et al. [24] found that competitors can serve as allies in managing the sustainable practices of shared suppliers. Despite the risk of overlapping due diligence efforts, collaboration among competitors can mitigate risks, influence industry standards, and enhance customer value through shared resources and expertise. Strategic alliances with competitors can provide mutual benefits while supporting industry-wide sustainability [24].

Companies that are more advanced in integrating environmental considerations into the supply chain are moving towards joint initiatives for measurements and incentives for sharing the benefits, demonstrating the importance of collaborative approaches when integrating sustainability and due diligence into buyer–supplier relationships [16].

5.2 Risk Management-Based Approaches

Conducting due diligence within their supplier network is an immense and challenging undertaking for large companies. The CSDDD [4] and the OECD Guidelines for Due Diligence [25], highlight that due diligence efforts should be risk-based, driving companies to prioritize and address potential risks, thereby ensuring the appropriate allocation of efforts and resources [27], and effective management of the complexities of global supply chains.

The OECD framework, also referenced in the CSDDD, offers six universally recognized steps and practical guidelines, from policy integration to remediation [4]. Additionally, adapting risk management standards such as ISO 31000:2018 and 31010:2019 can be advantageous when adapted to company-specific needs as they provide the foundation for establishing a consistent risk assessment and evaluation process [28].

Hurt et al. [21] describe two approaches to risk-based due diligence monitoring, each carrying their benefits. "Node-based" due diligence assesses compliance and risks of each company separately. On the other hand, "link-based" due diligence concentrates on the relationships or "links" between companies, examining their transactions and interactions. Hurt et al. [21] note that node-based due diligence monitoring would reduce required due diligence assessments by a factor of 30 while maintaining compliance with the CSDDD.

5.3 Leveraging Technologies

Integrating technology into buyer–supplier relationships and supply chains will become crucial for achieving sustainability and effective due diligence in the future. Existing literature highlights the multifaceted role of technologies such as blockchain and artificial intelligence (AI) applications to achieve enhanced transparency, efficiency, and overall sustainability within supply chains [29].

Blockchain technology can enhance supply chain sustainability by reducing costs, increasing transparency [30], and ensuring regulatory compliance [31]. It provides all parties access to the same information, enabling material traceability and verification of ethical sourcing. This transparency supports tracking products, automating payments, and assigning persistent records to suppliers, simplifying due diligence [30, 31].

Digital product passports (DPPs) also contribute to supply chain sustainability by improving transparency for stakeholders, thereby facilitating due diligence and regulatory compliance [32]. Adopting distributed ledger technologies, such as blockchain, is essential for ensuring data accessibility in DPPs [33].

AI integration in supply chains enables predictive analytics and optimization, enhancing efficiency, sustainability, and responsiveness. It allows real-time adaptation and improved decision-making, aligning with Industry 4.0 objectives [29]. AI can automate data analysis [34], aiding due diligence efforts by identifying patterns and risks. This approach helps address potential sustainability risks before they escalate.

6 Concluding Remarks

Sustainability and due diligence in the buyer–supplier relationship are becoming increasingly relevant, as evidenced by, e.g., the EU's CSDDD. However, this presents new challenges and opportunities for companies in global supply chains, including dilemmas related to buyer–supplier dynamics, transparency, and compliance efforts. In addition, companies face challenges, including regulatory fragmentation, complex supply chains, and high transaction costs. To address these challenges, companies should adopt collaborative, risk-based, and technology-focused strategies. It is found that closer engagement with suppliers and third parties, adopting risk management frameworks and standards, and leveraging AI and blockchain technologies can promote sustainability and due diligence in supply chains. An integrated approach to managing sustainability due diligence in buyer–supplier relations can enable companies to navigate an evolving regulatory landscape and mitigate environmental and social risks while promoting sustainable supply chain practices.

Future research should focus on investigating practical applications and developing frameworks to assist practitioners in operationalizing these integrated strategies. This need is exemplified by the challenge of balancing conventional methods and more sustainable practices, where the presence of persistent transactional supplier relationships forces companies to decide whether a high-risk supplier should be developed or replaced—an issue requiring further study. In developing such frameworks, the challenges, barriers, and integrated strategies found in this study should be addressed.

References

1. Bird RC, Soundararajan V (2020) From suspicion to sustainability in global supply chains. Tex AM Law Rev 7:383–418
2. Wiedmann T, Lenzen M (2018) Environmental and social footprints of international trade. Nat Geosci 11:314–321
3. European Commission. Directorate General for Justice and Consumers., British Institute of International and Comparative Law., Civic Consulting., LSE (2020) Study on due diligence requirements through the supply chain: final report. Publications Office, LU
4. Directive (EU) 2024/1760 (2024) Directive (EU) 2024/1760 of the European Parliament and of the Council of 13 June 2024 on corporate sustainability due diligence and amending Directive (EU) 2019/1937 and Regulation (EU) 2023/2859. Off J Eur Union, pp 1–58
5. Ventura L (2022) Supply chain management and sustainability: the new boundaries of the firm. Unif Law Rev 26:599–634
6. Velluti S (2024) Labour standards in global garment supply chains and the proposed EU corporate sustainability due diligence directive. Eur Labour Law J 15:822–850
7. Blome C, Paulraj A, Preuss L, Roehrich JK (2023) Trust and opportunism as paradoxical tension: implications for achieving sustainability in buyer-supplier relationships. Ind Mark Manag 108:94–107
8. Koberg E, Longoni A (2019) A systematic review of sustainable supply chain management in global supply chains. J Clean Prod 207:1084–1098
9. Snyder H (2019) Literature review as a research methodology: an overview and guidelines. J Bus Res 104:333–339
10. Ferrari R (2015) Writing narrative style literature reviews. Med Writ 24:230–235
11. Levy YJ, Ellis T (2006) A systems approach to conduct an effective literature review in support of information systems research. Inf Sci Int J. Emerg Transdiscipl 9:181–212
12. Saloranta J, Hurmerinta-Haanpää A (2022) Proactive contract theory in the context of corporate sustainability due diligence. J Strateg Contract Negot 6:221–236
13. Eurostat: International trade in goods by type of good. https://ec.europa.eu/eurostat/statistics-explained/index.php?title=International_trade_in_goods_by_type_of_good. Accessed 05 Feb 2025
14. Gardner TA, Benzie M, Börner J, Dawkins E, Fick S, Garrett R, Godar J, Grimard A, Lake S, Larsen RK, Mardas N, McDermott CL, Meyfroidt P, Osbeck M, Persson M, Sembres T, Suavet C, Strassburg B, Trevisan A, West C, Wolvekamp P (2019) Transparency and sustainability in global commodity supply chains. World Dev 121:163–177
15. Seuring S, Müller M (2008) From a literature review to a conceptual framework for sustainable supply chain management. J Clean Prod 16:1699–1710
16. Murfield ML, Tate WL (2017) Buyer and supplier perspectives on environmental initiatives: potential implications for supply chain relationships. Int J Logist Manag 28:1319–1350
17. Smit L, Holly G, McCorquodale R, Neely S (2020) Human rights due diligence in global supply chains: evidence of corporate practices to inform a legal standard. Int J Hum Rights 25:945–973
18. Patchell J (2018) Can the implications of the GHG Protocol's scope 3 standard be realized? J Clean Prod 185:941–958
19. Business and Human Rights Resource Centre (2024) A fair and sustainable future for people and planet
20. Schilling-Vacaflor A, Lenschow A (2023) Hardening foreign corporate accountability through mandatory due diligence in the European Union? New trends and persisting challenges. Regul Gov 17:677–693
21. Hurt J, Ledebur K, Meyer B, Friesenbichler K, Gerschberger M, Thurner S, Klimek P (2023) Supply chain due diligence risk assessment for the EU: a network approach to estimate expected effectiveness of the planned EU directive. http://arxiv.org/abs/2311.15971
22. Büyüközkan G, Çifçi G (2011) A novel fuzzy multi-criteria decision framework for sustainable supplier selection with incomplete information. Comput Ind 62:164–174

23. Brandenburg M, Warasthe R, Seuring S (2023) Impact pathways: Improving supply chain sustainability by due diligence acts? Insights from a German case. Int J Oper Prod Manag 44:859–874
24. Cao Y, Li Q, Shen B, Wang Y (2023) Buyer collaboration in managing supplier responsibility with ESG due diligence effort spillover and fairness concerns. Transp Res Part E Logist Transp Rev 180:103333
25. OECD (2018) OECD due diligence guidance for responsible business conduct. OECD
26. Villena VH, Gioia DA (2018) On the riskiness of lower-tier suppliers: managing sustainability in supply networks. J Oper Manag 64:65–87
27. Tummala R, Schoenherr T (2011) Assessing and managing risks using the Supply Chain Risk Management Process (SCRMP). Supply Chain Manage Int J 16:474–483
28. De Oliveira UR, Marins FAS, Rocha HM, Salomon VAP (2017) The ISO 31000 standard in supply chain risk management. J Clean Prod 151:616–633
29. Moorthy S, Balakrishanan S, Nithish K, Rohith CMR (2022) Blockchain and artificial intelligence feasibility, implementation and sustainability in supply chain. In: 2022 8th international conference on advanced computing and communication systems (ICACCS), pp. 2112–2116. IEEE, Coimbatore, India
30. Radmanesh S-A, Haji A, Fatahi Valilai O (2023) Blockchain-based architecture for a sustainable supply chain in cloud architecture. Sustainability 15:9072
31. Saberi S, Kouhizadeh M, Sarkis J, Shen L (2019) Blockchain technology and its relationships to sustainable supply chain management. Int J Prod Res 57:2117–2135
32. Jensen SF, Kristensen JH, Adamsen S, Christensen A, Waehrens BV (2023) Digital product passports for a circular economy: data needs for product life cycle decision-making. Sustain Prod Consum 37:242–255
33. Chaudhuri A, Wæhrens BV, Treiblmaier H, Jensen SF (2024) Impact pathways: digital product passport for embedding circularity in electronics supply chains. Int J Oper Prod Manage
34. Baryannis G, Validi S, Dani S, Antoniou G (2019) Supply chain risk management and artificial intelligence: state of the art and future research directions. Int J Prod Res 57:2179–2202

[illegible]

Adaptive Simulation Modeling of Intertwined Supply Networks with Demand Uncertainty in Disruptions

Mohaddeseh Roshan, Jessica Olivares-Aguila, and Waguih ElMaraghy

Abstract Despite the growing conceptual emphasis on Intertwined Supply Networks (ISNs), there is still a lack of simulation-based studies that quantitatively assess their operational adaptability under disruptions and uncertainty. This study develops a discrete-event dynamic simulation model using Python for an intertwined pharmaceutical-food supply network, to evaluate the effectiveness of adaptive sourcing and dynamic re-routing under three scenarios: baseline, supplier disruption, and transportation disruption. The model tracks operational Key Performance Indicators (KPIs) such as backlog, service level, and transportation cost. Comparative analysis of all three scenarios show that adaptive mechanisms reduce backlog by 61% and transportation costs by 14.2%, demonstrating substantial improvements in resilience and efficiency. The approach offers valuable insights for designing more resilient and adaptable supply networks under uncertainty.

Keywords Intertwined supply network · Adaptability · Resilience · Demand uncertainty · Simulation modeling

1 Introduction

In contrast to the studies in traditional supply chain (SC) with limited interaction between industry sectors, ISNs represent a significant evolution by interconnecting multiple SCs from distinct industries, facilitating shared resources, coordinated operations, and integrated response strategies [1]. This interconnectedness enhances the

M. Roshan · W. ElMaraghy (✉)
Intelligent Manufacturing Systems (IMS) Centre, University of Windsor, Windsor, ON, Canada
e-mail: wem@uwindsor.ca

M. Roshan
e-mail: heidarpm@uwindsor.ca

J. Olivares-Aguila
Shannon School of Business, Cape Breton University, Sydney, NS, Canada
e-mail: jessica_olivares@cbu.ca

M. Manns (ed.), *New Paradigms for Anticipated Uncertainty*, Lecture Notes in Mechanical Engineering, https://doi.org/10.1007/978-3-032-16889-4_2

adaptability and resilience of supply networks, which is critical in today's highly dynamic environment. Adaptability within ISNs refers to their ability to rapidly adjust the network to mitigate disruptions [2]. ISNs achieve enhanced flexibility through adaptive mechanisms for faster recovery and minimized disruption costs [3]. SC disruptions present substantial challenges to the networks [4, 5]. Linear SCs typically struggle to adapt swiftly due to their inflexible structure. Another critical factor influencing SC performance is integrating demand uncertainty into the analysis and design of supply networks, which is vital for maintaining service levels, controlling costs, and ensuring efficient resource utilization [6].

The primary aim of this research is to investigate how intertwining SCs across different industries, combined with adaptive mechanisms, enhances supply network performance during disruptions. Specifically, this study fills a gap in analyzing operational adaptability in ISNs by evaluating performance outcomes under supplier and transportation disruptions through simulation modeling.

2 Literature Review

In the field of Supply chain (SC) risk management, Ivanov [7] addressed disruption risks specific to pandemics, noting their rapid expansion and unpredictability. Further, Ivanov and Dolgui [1] investigated the differences between resilience and viability and proposed a framework to ensure long-term viability. Ivanov [2] expanded on these concepts by introducing the Viable Supply Chain (VSC) Model, which integrates resilience, sustainability, and survivability to cope with crises. SC risk management has been investigated from different perspectives, such as utilizing system dynamics [8], developing goal programming models [9], developing a framework for assessing SC robustness and complexity [4], and using mathematical models to identify disaster-resilient configurations [10]. Ivanov [5] focused on disruption recovery, particularly how SCs manage post-disruption backlogs and delays. Hosseini et al. [11], along with studies by Bode and Macdonald [12] and Dubey et al. [13], identify these as key resilience factors. Roshan et al. [6] proposed an agile pharmaceutical SC model addressing social responsibility. Inman and Green [14] and Niemann et al. [15] identified agility as crucial in improving performance amidst uncertainty. Baah et al. [16] emphasized information sharing as vital for fostering agility and enhancing SC performance.

In the field of ISNs, Roshan et al. [17] focused on designing sustainable ISNs under uncertainty. Gorbaneva and Ougolnitsky [18] utilized a game-theoretic framework to address sustainability within ISNs. Wang and Yao [3] developed a redundancy optimization model to balance network resilience and cost efficiency during disruptions. Echefaj et al. [19] offered practical insights through discrete event simulations. Feizabadi et al. [20] examined adaptive behaviors, specifically jury-rigging, during a crisis. Ivanov [21] focused on how disruptions propagate through ISNs, known as the ripple effect. Ivanov [2] proposed the VSC model, integrating agility, resilience, and sustainability. Ivanov and Dolgui [1] expanded the concept of ISN viability, emphasizing the network's capacity to survive extraordinary disruptions through dynamic adaptability.

Traditional SC studies have explored resilience and adaptability, yet few examine these concepts in operationally adaptive, multi-echelon ISNs. This research addresses that gap by modeling adaptive sourcing and re-routing under disruptions. The lack of studies incorporating demand uncertainty and evaluating recovery using managerial KPIs marks a critical gap this work aims to fill.

3 Problem Statement

Modern supply chains are more vulnerable to disruptions like transportation delays and supplier failures. Traditional linear supply chains frequently lack the ability necessary to adapt in these kinds of circumstances. By connecting several supply chains from various industries, ISNs provide increased resilience through shared facilities and interconnected flows. However, there is still a gap in investigating how ISNs with adaptive mechanisms can function in the face of dynamic disruptions, especially under uncertainty. This study addresses this gap by exploring the effectiveness of combining adaptive mechanism and dynamic re-routing within ISN framework in improving network resilience. To illustrate on this problem, a case study is conducted consisting of two interconnected SCs, a pharmaceutical factory and a food factory, explained in detail in Sect. 4.1. The methodology of this research involves a discrete-event predefined rule-based simulation approach, which is a powerful and flexible tool for studying complex scenarios. A discrete-event simulation model, developed in Python and executed in the Google Colab environment, is conducted over a 30-day period, to assess the designed ISN's response under two disruption scenarios: supplier shortage and transportation bottleneck. Demand uncertainty is incorporated using stochastic distributions. The simulation evaluates how adaptive mechanisms impact service levels, backlogs, inventory availability, and transportation costs (Fig. 1).

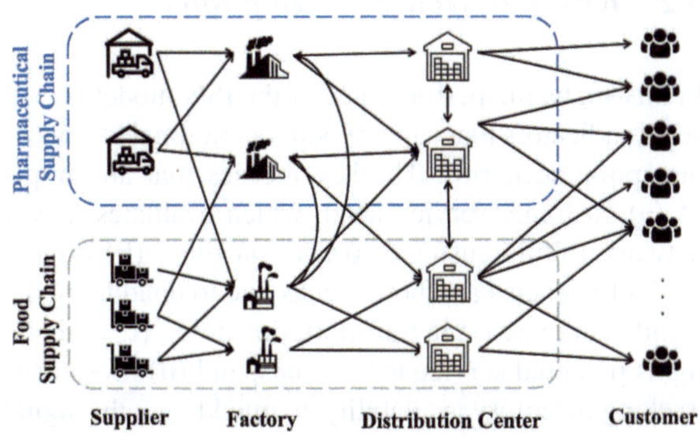

Fig. 1 The intertwined supply network structure

Table 1 Customer demand and distribution center capacity

Products	Hyaluronic acid	Glycerol	Processed food	Fermented product
Demand	$\sim U(90, 100)$	$\sim U(50, 60)$	$\sim U(45, 55)$	$\sim U(60, 70)$
DCs	DC1	DC2	DC3	DC4
Capacity	2500 units	1900 units	2800 units	2200 units

4 Simulation Model

4.1 Simulation Model Logic and Input Data

The supply network in this study represents an ISN that consists of two interconnected SCs: a pharmaceutical SC and a food SC. Four key echelons are integrated into this ISN: four suppliers (three primary and one backup), two factories (one pharmaceutical factory and one food factory), four Distribution Centers (DCs), and ten customers. The data used in this case study is synthetically generated. Pharmaceutical factories produce two types of products: hyaluronic acid and glycerol, which is a byproduct used in food production. Food factories produce two types of food: processed food and a fermented product. The latest requires glycerol from pharmaceutical factories. Suppliers provide raw materials to factories. Finished products are then distributed from factories to DCs and ultimately to customers. The simulation model follows a discrete-event logic where system state updates are triggered by events such as daily demand generation, shipment execution, disruption onset, and adaptive response. The demand of each customer for each product follows a uniform distribution, as presented in Table 1. Also, the capacity of each DC is provided in Table 1. The transportation cost is considered \$1.5/unit/mile, the inventory holding cost is \$5/unit, the DC operating cost is \$30 K per year, the factory production cost is \$50 K per year, and the penalty per unit of unmet demand is \$10.

4.2 KPI and Model Assumptions

In this study, the performance of the ISN model is evaluated using a comprehensive set of indicators that capture both operational efficiency and resilience under various disruption scenarios. The key metrics that are employed in this research consist of **(a)** customer service level, which evaluates how well and effectively a supply network fulfills customer orders on time, **(b)** transportation costs, which signals the inefficiencies and bottlenecks due to unaddressed disruptions, and consequently significant increase in transportation costs, **(c)** unmet demand/backlog, which highlights potential service level issues, and **(d)** inventory level of all DCs, which helps tracking inventory availability, to quickly get the signal of potential shortages due to disruption.

The assumptions of this study are as follows:

- The network involves a shared reverse flow, where one of the pharmaceutical factory's products is a byproduct used as an input by the food factory.
- Backlog processing is managed at the distribution center level.
- The model uses rule-based adaptive mechanisms for disruption response. Each factory has a predefined backup supplier, and each DC–customer pair has a main and backup route.
- Baseline operations assume that each network flow functions at 85% of its original capacity, and transportation disruptions create bottlenecks that increase delays and costs, leading to re-routing strategies.
- Transportation costs are assumed to be directly proportional to the distance traveled and the volume of goods transported.

4.3 Scenarios

The proposed dynamic model simulates a one-month period, considering three scenarios to assess the adaptability of the ISN under various disruptions. Table 2 shows the experimental setup for simulation scenarios. The baseline scenario (SC1) represents normal operations where all flows operate at 85% of their original capacity. The results will be the benchmark for comparing the model's performance during disruptions. In the supplier disruption scenario (SC2), one supplier experiences a shortage due to an external disruption that reduces the capacity of flows carrying raw materials. The transportation disruption scenario (SC3) models a transportation bottleneck, which results in cost increases and delays in shipment.

Table 2 Scenarios' characteristics

Scenarios	Incidents	Experimental setup
SC1	Baseline scenario	• No disruption, and business operate as usual
SC2	Supplier shortage	• The disruption occurs on day 5 • Supplier 2 faces a shortage due to external disruption • Supplier 4, the backup supplier, should be activated
SC3	Transportation bottleneck	• The disruption occurs on day 15 • Immediate 80% increase in total transportation costs due to disruption in links: DC4 to Customer 8, DC3 to Customer 3, and DC1 to Customer 2

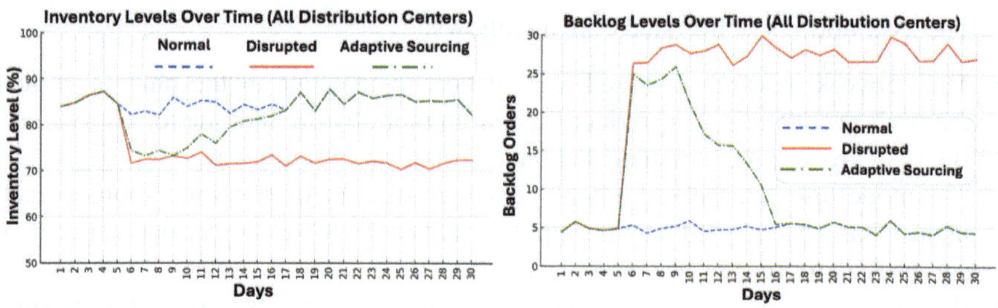

Fig. 2 Experiment results: supplier disruption (SC2)

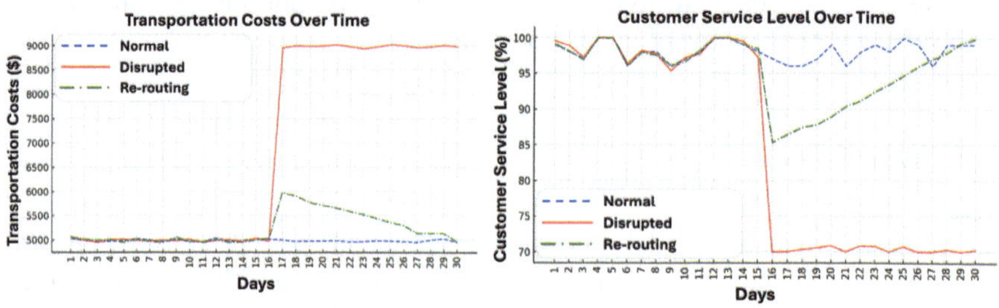

Fig. 3 Experiment results: transportation disruption (SC3)

5 Experiments, Results, and Analysis

5.1 Baseline: Normal Scenario (SC1)

In SC1, the ISN operates normally, with no disruption. The results are displayed in Figs. 2 and 3, along with the disrupted scenarios, to allow clear comparison. Results in SC1 show small variations in KPIs, which is normal due to the uncertainty of the demand parameter. Furthermore, transportation costs can fluctuate slightly, even in a stable environment, due to changes in shipment consolidation or slight route variations. Small backlog orders, stable inventory level, high service level (around 85%), and transportation cost show that the ISN is working seamlessly and is profitable.

5.2 ISN Under Supply Disruption (SC2)

In SC2, primary supplier 2 faces an external disruption and experiences a shortage on day 5. The results are visualized in Fig. 2. In the disrupted scenario, inventory levels drop to approximately 72% on day 6 and remain low for the remainder of the period. Additionally, backlog levels rose sharply to approximately 28 orders as supply shortages prevented meeting customer demand. Conversely, the adaptive sourcing scenario demonstrates a backup supplier's effectiveness in stabilizing the SC. From days 6 to 8, the inventory remains low, indicating a brief period of constrained supply while the network activates the backup supplier. However, after day 8, inventory levels begin a rapid recovery, increasing steadily until day 16, at which point they closely align with the normal scenario. Moreover, backlog levels also rise following the disruption, reaching 24 orders between days 6 and 8, but once the backup supplier becomes active, the backlog level starts to decline rapidly, which shows the increasing availability of inventory as replenishments flow into the system. By day 16, backlog levels have nearly returned to normal. Results demonstrate the success of adaptive sourcing in restoring SC functionality.

5.3 *ISN Under Transportation Disruption (SC3)*

In the third scenario, on day 15, a transportation bottleneck happens for the routes DC4 to Customer 8, DC3 to Customer 3, and DC1 to Customer 2, which causes delays and increased transportation costs. Subsequently, the model dynamically re-routes shipments through alternative paths, and the results are represented in Fig. 3. As observed in Fig. 3, the disrupted scenario shows a prolonged decrease in service performance, with the service level stabilizing at 70% for the remainder of the simulation. Furthermore, transportation costs surge by about $2000 and remain elevated for the remainder of the simulation period. Inversely, the adaptive ISN reaches 85% at the beginning instead of 70% and then gradually improves over time, eventually returning to 100% as the alternative transportation routes compensate for the initial disruption. Besides, the re-routing scenario gradually reduces transportation costs, indicating a progressive recovery as the system stabilizes.

5.4 *Comparative Analysis of All Three Scenarios*

Table 3 shows the differences in transportation costs and backlogs in all three scenarios. In stable ISN, the results show the lowest total transportation cost ($149,950) and the smallest backlog (147 units). Without an adaptive mechanism, the disrupted ISN resulted in the highest transportation costs ($209,450) and a substantial increase in backlog orders (729 units). However, applying an adaptive mechanism within the disrupted ISN resulted in a 14.2% decrease in the total transportation costs and about a 61% decrease in the amount of unmet demand. These outcomes validate the effectiveness of using adaptive mechanisms to mitigate the impact of disruption.

Table 3 Comparative analysis results

	Normal ISN	Disrupted ISN	Adaptive ISN
Total transportation cost	$149,950	$209,450	$179,700
Total backlog (rounded)	147	729	285

6 Managerial Insights

ISNs include shared resources, integrated operational strategies, and coordinated response efforts across all echelons, and this interconnected structure enables the system to be more resilient and adaptive compared to a traditional linear SC view. Additionally, the adaptive mechanisms utilized in this research validate critical benefits in minimizing disruption impacts. In the scenario of supplier disruption, adaptive sourcing quickly activated a backup supplier and prevented the system from having significant inventory shortages and reduced backlog by approximately 61%. Similarly, in the transportation disruption scenario, the transportation paths were rapidly adjusted using a dynamic re-routing mechanism, and reduced transportation costs by 14.2%. In conclusion, the comparative analysis demonstrates that integrating adaptive mechanisms within the ISN framework significantly enhances supply chain resilience and performance. Consequently, managers should strategically prioritize the adoption of adaptive capabilities and foster interconnected supply networks to ensure robust operational performance and greater responsiveness to disruptions.

7 Conclusion

This research presents a simulation-based case study that investigates an ISN of pharmaceutical and food SCs under disruption and demand uncertainty. Three scenarios are examined: a baseline/normal scenario, a supplier disruption scenario, and a transportation disruption scenario. The simulation results demonstrated that the incorporation of adaptive mechanisms significantly enhanced network resilience and adaptability. Under the supplier disruption scenario, adaptive sourcing rapidly activated backup suppliers, effectively restoring inventory levels and reducing backlog. Similarly, dynamic re-routing during the transportation disruption scenario successfully maintained higher customer service levels and substantially lowered transportation costs. Future research should address more complex type of uncertainties such as correlated disruptions and cascading failures to better reflect real-world ISN dynamics, incorporate empirical data sources to validate the simulation, and expand the disruption logic to improve model realism and practical relevance.

References

1. Ivanov D, Dolgui A (2020) Viability of intertwined supply networks: extending the supply chain resilience angles towards survivability. A position paper motivated by COVID-19 outbreak. Int J Prod Res 58(10):2904–2915
2. Ivanov D (2022) Viable supply chain model: integrating agility, resilience and sustainability perspectives—lessons from and thinking beyond the COVID-19 pandemic. Ann Oper Res 319(1):1411–1431
3. Wang M, Yao J (2023) Intertwined supply network design under facility and transportation disruption from the viability perspective. Int J Prod Res 61(8):2513–2543
4. Aguila JO, ElMaraghy W (2018) Structural complexity and robustness of supply chain networks based on product architecture. Int J Prod Res 56(20):6701–6718
5. Ivanov D (2019) Disruption tails and revival policies: a simulation analysis of supply chain design and production-ordering systems in the recovery and post-disruption periods. Comput Ind Eng 127:558–570
6. Roshan M, Tavakkoli-Moghaddam R, Rahimi Y (2019) A two-stage approach to agile pharmaceutical supply chain management with product substitutability in crises. Comput Chem Eng 127:200–217
7. Ivanov D (2020) Predicting the impacts of epidemic outbreaks on global supply chains: a simulation-based analysis on the coronavirus outbreak (COVID-19/SARS-CoV-2) case. Transp Res Part E Logist Transp Rev 136:101922
8. Aguila JO, ElMaraghy W (2019) Supply chain resilience and structure: an evaluation framework. Proc Manuf 28:43–50
9. Aguila JO, ElMaraghy W, ElMaraghy H (2019) Impact of risk attitudes on the concurrent design of supply chains and product architectures. Proc CIRP 81:974–979
10. Aguila JO, ElMaraghy W (2018) Simultaneous global supply chain and product architecture design considering natural hazard exposure and geographical facility location. Proc CIRP 72:533–538
11. Hosseini S, Ivanov D, Dolgui A (2019) Review of quantitative methods for supply chain resilience analysis. Transp Res Part E Logist Transp Rev 125:285–307
12. Bode C, Macdonald JR (2017) Stages of supply chain disruption response: direct, constraining, and mediating factors for impact mitigation. Decis Sci 48(5):836–874
13. Dubey R, Altay N, Gunasekaran A, Blome C, Papadopoulos T, Childe SJ (2018) Supply chain agility, adaptability and alignment: empirical evidence from the Indian auto components industry. Int J Oper Prod Manag 38(1):129–148
14. Inman RA, Green KW (2021) Environmental uncertainty and supply chain performance: the effect of agility. J Manuf Technol Manag 33(2):239–258
15. Niemann J, Breschan D, Schlegel A, Putz M (2019) Method for optimizing supply chain flexibility in the production of electrified powertrains. In: 8th international conference on industrial technology and management (ICITM 2019). IEEE, pp 32–36
16. Baah C, Agyeman DO, Acquah ISK, Agyabeng-Mensah Y, Afum E, Issau K, Faibil D (2021) Effect of information sharing in supply chains: understanding the roles of supply chain visibility, agility, collaboration on supply chain performance. Benchmarking Int J 29(2):434–455
17. Roshan M, Olivares-Aguila J, ElMaraghy W (2025) Design of sustainable intertwined supply networks. Proc CIRP 134:855–860
18. Gorbaneva O, Ougolnitsky G (2022) Sustainability of intertwined supply networks: a game-theoretic approach. Games 13(3):35
19. Echefaj K, Charkaoui A, Cherrafi A, Ivanov D (2024) Design of resilient and viable sourcing strategies in intertwined circular supply networks. Ann Oper Res 1–40
20. Feizabadi J, Gligor DM, Choi TY (2023) Examining the resiliency of intertwined supply networks: a jury-rigging perspective. Int J Prod Res 61(8):2432–2451
21. Ivanov D (2023) Collaborative emergency adaptation for ripple effect mitigation in intertwined supply networks. Ann Oper Res 1–17

Decision Support Model for Component Remanufacturing

Emma B. Worup, Ann-Louise Andersen, Thomas D. Brunoe, and Kjeld Nielsen

Abstract Remanufacturing is widely regarded as an end-of-life strategy that can improve sustainability and resilience. Remanufacturing restores components to their original functionality, potentially allowing them to reintegrate into standard inventory procedures alongside non-remanufactured components. By going through the literature on design for remanufacturing, this study identifies critical parameters for assessing the ability to remanufacture components, including the ability to disassemble, clean, inspect, repair, upgrade, and test, as well as the cost/resource efficiency of these processes. Based on these parameters, a decision support model is proposed to evaluate whether the components are suitable for remanufacturing. An illustrative example of a circulating pump demonstrates the practicality of the model, distinguishing components suitable for remanufacturing from those better suited for recycling. The findings highlight the potential of integrating a decision support model for end-of-life strategy selection and how such a model can enhance sustainability and resilience.

Keywords Remanufacturing · End-of-life · Component remanufacturing · Circular economy · Circularity · Sustainability · Resilience

1 Introduction

The "Emission Gap report 2024: No more hot air… please!" emphasizes that greenhouse gas emissions from the industry must be reduced. As of 2023, the industry accounts for 20 percent of the total emissions. Moreover, with an increase of 1.3 percent in emissions from 2022 to 2023, there is an urgent need for a shift in industry practices [1]. Therefore, companies must begin to assess the different sustainability practices that could help reduce their environmental impact. Remanufacturing is an approach that has gained attention for its effectiveness in recovering the value of products, as it requires products to be restored to at least their original functionality. The

E. B. Worup (✉) · A.-L. Andersen · T. D. Brunoe · K. Nielsen
Department of Materials and Production, Aalborg University, Aalborg, Denmark
e-mail: ebw@mp.aau.dk

M. Manns (ed.), *New Paradigms for Anticipated Uncertainty*, Lecture Notes in Mechanical Engineering, https://doi.org/10.1007/978-3-032-16889-4_3

potential of remanufacturing goes beyond sustainability and should be investigated with resilience in mind [2]. As companies need to be able to adjust to fluctuations and disruptions in supply and demand, a possible solution is to implement take-back mechanisms to retrieve products, remanufacture their components, and store these for future use [2]. This approach could enhance resilience against supply disruptions by ensuring an alternative supply of critical components. Despite the promising potential of remanufacturing, its widespread adoption in the manufacturing industry remains limited. While there are numerous successful examples of remanufacturing within the automotive industry [3], broader implementation across various industries is lacking [4]. This likely stems from companies' uncertainty about which products and components are suitable for remanufacturing, where to begin or the potential benefits of remanufacturing. The existing research has focused largely on the design of products to facilitate remanufacturing and the strategic selection of appropriate end-of-life strategies [5–7]. Nevertheless, the remanufacturing of components at the individual part level remains largely unexplored, and there is a lack of methods and models designed to support companies in identifying which specific components are suitable for remanufacturing. The importance of remanufacturing at the component level is twofold. Firstly, it allows for the exploitation of potential waste streams, promoting sustainability through, for example, enabling after-sales service using remanufactured components. Secondly, components are often used in multiple products, so focusing on remanufacturing components can create a secondary supply of goods, thereby enhancing resilience [2]. Accurately identifying components suitable for remanufacturing reduces the need for extensive inspection and testing of unsuitable components. Additionally, incorporating remanufacturing practices at the component level ensures an alternative flow of goods, which can act as a buffer against supply chain disruptions, such as resource shortages.

The grading and sorting process is one of the critical points for choosing the appropriate end-of-life strategy. Figuring out what end-of-life strategy is suitable for the components should be decided on multiple levels. At the general level, a decision must be made about whether components should be included in an end-of-life strategy. This decision, which applies to a component type rather than an individual product, should be made during development, launch, or any stage of the product's market life cycle, and can be revisited as needed. The other decision is on the specific product level, when the products have been used in the market, reached the end-of-life, and returned. This decision is related to the actual condition of the products and components returned to the company. In this paper, the focus is exclusively on the general level.

The purpose of this paper is to identify appropriate parameters for determining the suitability of individual components for remanufacturing, where a component is defined as "*a part that combines with other parts to form something bigger*" [8]. Based on the identified parameters, the paper proposes a decision support model to identify components suitable for remanufacturing. While the model does not predict the condition of each component, it aims to provide insights into the ability to use remanufacturing on the components based on the components' characteristics. The paper is structured as follows: Sect. 2 presents the literature review, and Sect. 3

presents the proposed decision support model for assessing if the components are suitable for remanufacturing. Finally, the conclusion and future research are presented in Sect. 4.

2 Literature Review

To understand the information required about components to assess their suitability for remanufacturing, a literature review was conducted. The review aims to identify and synthesize existing knowledge on end-of-life strategies, remanufacturing criteria, design for remanufacturing (DfRem) parameters, and disassembly considerations. The search strategy was designed to capture the key parameters and criteria relevant to the various end-of-life strategies, with a particular focus on remanufacturing. The literature review addresses the end-of-life strategies, key design aspects for remanufacturing, and the processes involved. This is achieved using the following search strings: [("remanufacturing" AND "criteria" AND "end-of-life") OR ("remanufacturing" AND "component reuse")] and [("design for remanufacturing" OR "DfRem") OR ("design for disassembly")].

Remanufacturing is often regarded as the most effective end-of-life strategy, since the products must return to "as-good-as-new" or "better-than-new" quality, consequently offering both environmental and economic benefits [4]. Several studies have explored methods and models to assess and optimize remanufacturing of products and other end-of-life strategies [5, 6, 9–12]. End-of-life strategies include reuse, refurbish, remanufacture, resynthesize, cannibalize, recover, and recycle [5]. Research predominantly underscores the importance of quality and cost considerations in evaluating the effectiveness and suitability of these strategies [5, 10, 13]. Additionally, assessing damage and remaining life is essential for determining the value recovery option in the processes. However, the research primarily emphasizes strategic decision-making based on economic and technical factors, often overlooking market demand, social trends, and legislative pressures [6, 7].

Design for remanufacturing (DfRem) has been widely researched, with numerous studies contributing to developing techniques for designing products for remanufacturing. The key design parameters include the ability to disassemble, clean, inspect, upgrade (update), repair (remanufacture), test, and reassemble the products [9, 14–16]. One of the most significant contributions to DfRem was made by Sundin and Lindahl, who developed the RemPro matrix. The matrix maps how the different design aspects correspond to the remanufacturing processes. Sundin and Lindahl's work continues to be frequently cited in the literature as the foundation for understanding the relation between the remanufacturing processes and design parameters. Different methodologies have been used to examine the topic, ranging from literature reviews to case studies, offering insights into how the different design aspects should be applied and how companies can utilize them [3, 17]. Some studies extend the scope of DfRem by considering factors such as the cost of remanufacturing, market

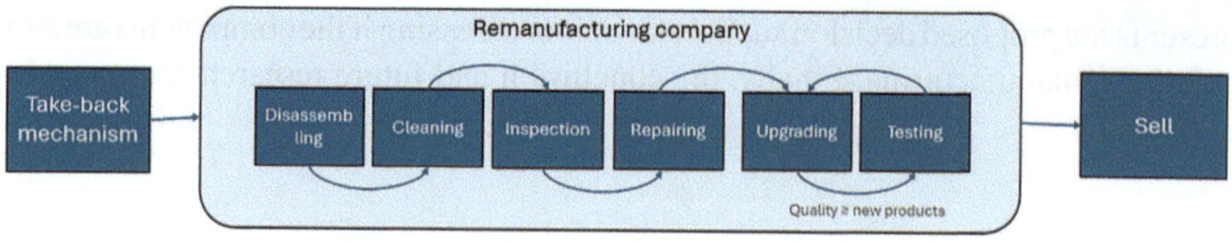

Fig. 1 The typical remanufacturing processes

demand, and the need for effective reverse logistics channels [9, 15], whilst some also highlight modularity as a key enabler of design for remanufacturing [3, 17, 18].

Much of the research references Sundin and Lindahl's description of the remanufacturing processes. However, other authors have proposed various sequences and specific procedures that products must undergo. Sundin and Lindahl outline the processes of *inspection, cleaning, storage, disassembly, remanufacturing, reassembly, and testing* [9, 16]. Although some authors offer more detailed descriptions or propose alternative sequences, there is a consensus in the literature around the following core processes: disassembly, cleaning, inspection, repair (including remanufacturing and reprocessing), upgrading, reassembly, and testing. These steps are illustrated in Fig. 1. For remanufacturing to be feasible, companies must implement a take-back system and secure customers for the remanufactured products [3].

Based on the literature on end-of-life strategies, design for remanufacturing, and the remanufacturing processes, it is possible to compare the different parameters and establish criteria for the components. The information from the literature review is used in the following sections to develop a model, enabling companies to identify components suitable for remanufacturing.

3 Proposed Model

The literature identified the key design parameters for components that are suitable for remanufacturing, including ease of disassembly, cleaning, inspection, upgrading, repair, and testing. These parameters align well with those derived from the remanufacturing process. This implies that, for a component to be suitable for remanufacturing, it must be able to undergo each of these processes, if necessary. This is further enabled by the components possessing the identified design characteristics. Additionally, given that remanufacturing is a sustainable practice, it is essential to include a parameter concerning the resources and costs involved in the processes [18–20]. Therefore, components must also undergo a cost/benefit analysis to ensure no process results in excessive resource use or costs. The conceptualization of each of the remanufacturing characteristics can be seen in Table 1.

Table 1 Conceptualization of remanufacturing characteristics

Ability to be disassembled	[18, 19, 21, 22]
Component removal must not damage other components Simplify de-manufacturing for minimal time and cost Allow recovery of components and materials Use appropriate fastening methods for easy disassembly	
Ability to be cleaned	[3, 14, 18, 19]
Components must be cleanable for inspection The cleaning profile depends on the contamination profile and use case Components must withstand the cleaning process and chemicals Minimize geometric shapes and features that could trap contaminants	
Ability to be inspected	[15–18]
Component conditions are unpredictable and depend on environments and usage Components must be inspectable to identify quality and restorability issues Ease of inspection is measured by the time required Longer inspection time indicates more complexity	
Ability to be repaired	[9, 14, 17, 18]
Components must be repairable and withstand repair activities Corrosion can make restoration challenging Repairability is evaluated by the accessibility of high-value parts Component complexity influences the ease of repairs	
Ability to be upgraded	[3, 12, 17, 18]
Not all components need upgrades, but those that do must be upgradable Facilitated by processing options or replacing worn-out parts Upgradability is enhanced by easily accessible and marked setting adjustments	
Ability to be tested	[14, 18, 21]
Components must be testable to verify performance meets standards Testability features include the ability to withstand testing Presence of features that identify performance metrics	
Cost and resource efficiency	[18–20]
Components must undergo remanufacturing without excessive resource use or costs Components requiring costly and environmentally harmful processes, like chemical cleaning, may not be suitable for remanufacturing	

3.1 Decision Support Model

Based on the identified parameters derived from analyzing remanufacturing processes and design for remanufacturing, a model has been developed to support the decision-making process in determining which components can be remanufactured. This model can be seen in Fig. 2. The model also considers other end-of-life strategies for the components. Prioritizing these strategies is essential. Following the principle of the end-of-life strategies, reuse is the second priority after reduce, due to its minimal resource requirements. Remanufacturing is prioritized next, as it restores components to their original quality and warranty, thereby extending their lifespan. Refurbishment is the next priority, as it does not require the same warranty as a new product and, consequently, does not extend the component's lifespan as effectively. Finally, recycling is the last priority, as it requires more resources and reduces components to raw materials [7]. The model does not incorporate reuse, as this decision must be made after the products are taken back. The potential for reuse depends on the component's quality, which is largely influenced by the product's usage and cannot be assessed until it is returned. However, during the remanufacturing process, this can be determined during the inspection phase, potentially requiring only minor cleaning, repair, and upgrading [17].

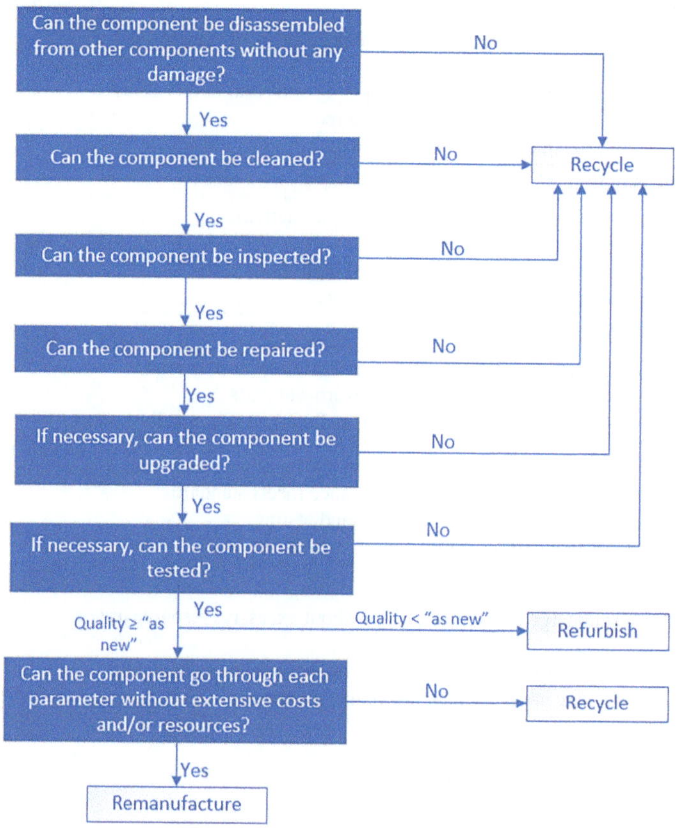

Fig. 2 Decision support model for component end-of-life strategy selection

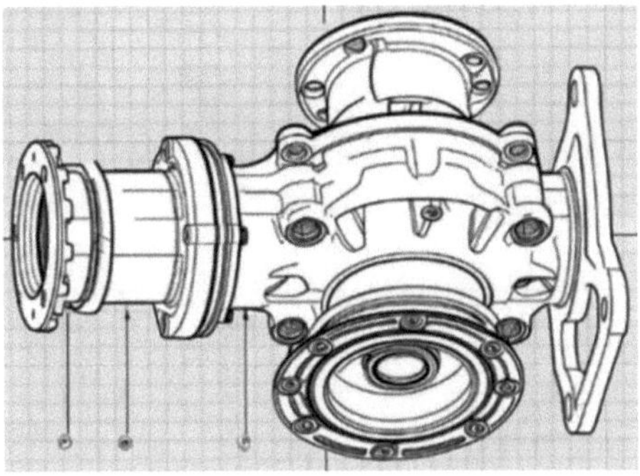

Fig. 3 A simplified view of a circulating pump generated using an AI tool

3.2 Illustrative Example Using a Circulating Pump

An illustrative example based on a circulating pump is presented to understand the model's practical application. To simplify the pump's complexity, not all of its components are considered. Figure 3 illustrates the circulating pump. The circulating pump consists of a pump casing, impeller, backplate, bearing, rotor and shaft, seal, rotor can, stator, and windings. The pump casing has the inlet and the outlet for the water. Within the pump casing, the impeller is placed in the volute channel. The impeller rotates and imparts a centrifugal force that forces the water into and out of the pump. Behind the impeller, the backplate is placed as a barrier to keep the water within the pump [23].

The possibility of using remanufacturing is assessed by going through each of the parameters to identify whether the component can be remanufactured. This process can be seen in Fig. 4.

The pump casing is an example of a component that can be remanufactured. It can be disassembled because it is joined together with the other components using bolts and screws. It can be cleaned using standard methods such as pressure washing or chemical agents. Visual inspection can identify wear, corrosion, damage, and cracks. Minor repairs, including patching small cracks, replacing worn seals, or re-machining surfaces, can be performed. It can be tested to ensure the pump casing is leak-free and maintains its structural integrity. The remanufacturing processes for the pump casing do not require excessive costs or resources. However, the condition of returned components can influence this assessment, potentially necessitating more intensive chemical cleaning. Three components of the circulating pump, the seal, rotor, and stator with windings are unsuitable for remanufacturing due to high cleaning demands and repair challenges. The seal is irreparable and costly to replace, rotor resurfacing is more expensive than replacement, and disassembling the stator is impractical due to soldered wire separation.

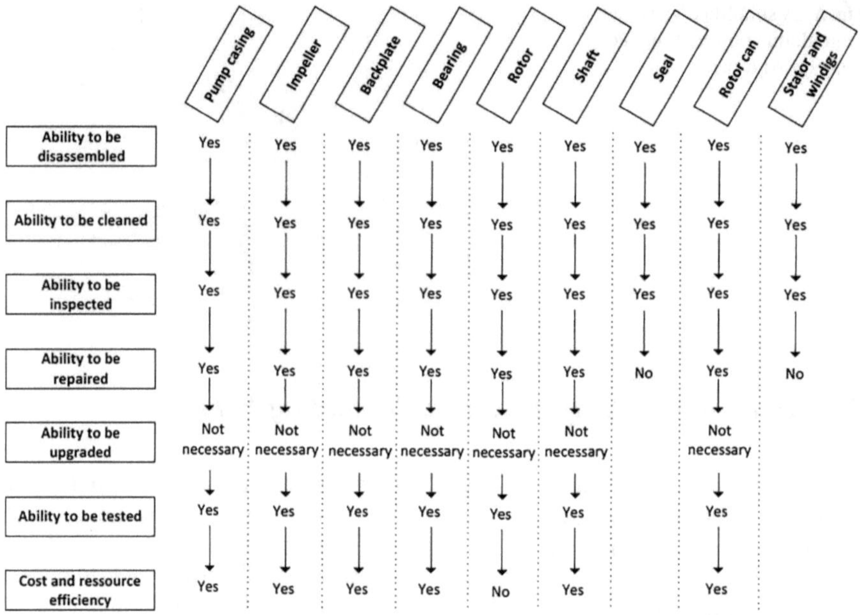

Fig. 4 An overview of the model used on the components from the circulating pump

4 Discussion

The proposed model serves as a tool for identifying components suitable for remanufacturing, thereby supporting the selection of products for take-back operations. However, it is important to acknowledge that grading components after their return remains a critical step towards implementing successful remanufacturing. This highlights the need for companies to implement reliable methods for assessing the condition of returned components. Besides finding components suitable for remanufacturing, the model also helps identify parts or products that need redesign to support reuse, refurbishment, or remanufacturing. A key strategic advantage of focusing on component-level remanufacturing is its potential to establish alternative supply flows. Being able to remanufacture components can improve the resilience of a company during supply chain disruptions [2]. While the model provides a clear overview of components with remanufacturing potential, further research is needed to determine how production systems should be configured to support these processes effectively. Moreover, additional investigation is required to assess whether the identified components can be reintegrated into new products or incorporated into more extensively remanufactured assemblies.

The model supports academic research, providing a clear basis for better decision-making in product recovery and circular manufacturing. Nevertheless, to fully validate the model's practical applicability, further empirical research is required. Specifically, theory-confirming case studies across diverse industrial contexts are necessary to test the model's robustness and to identify potential limitations or contextual constraints. Such cross-sectoral validation is critical for uncovering the potential operational differences and ensuring the model's adaptability and generalizability.

5 Conclusion

A decision support model for identifying components suitable for remanufacturing was developed by analyzing key parameters from remanufacturing and design literature. The parameters for assessing whether components are suitable for remanufacturing are the ability to be disassembled, cleaned, inspected, repaired, upgraded, tested, and the cost/resource efficiency of these processes. The illustrative example using a circulating pump demonstrates the model's practical application, effectively distinguishing components suitable for remanufacturing from those better suited for other end-of-life strategies. The model enables companies to identify components suitable for remanufacturing, providing the opportunity to enter the remanufacturing market. This allows companies to differentiate themselves by offering sustainable and cost-effective solutions for environmentally conscious customers. Furthermore, the model can enhance sustainability by identifying products for take-back programs. The main objective for companies should be to maximize the number of components that meet the remanufacturing criteria identified in this paper. To fully leverage the model's potential, companies can further use it to identify components with remanufacturing potential, even if they initially do not meet all criteria. By redesigning these components to fulfill the necessary criteria, companies can drive innovations and improvements in product design and manufacturing processes.

Furthermore, the structured approach for assessing if components can be remanufactured can help increase the supply chain resilience, by reintegrating the remanufactured components into the standard inventory procedures. The remanufacturing of components can help companies achieve stability in their supply chains, by reducing the dependence on volatile material supplies affected by disruptions such as geopolitical tensions, natural disasters, etc. By securing an alternative supply of components, companies can mitigate the reliance on suppliers. Implementing the model at the component level ensures that these components can be used across multiple products, rather than being limited to specific items. This flexibility allows resources to be allocated to where they are most needed, further enhancing resilience. Further research should investigate the applicability of the decision support model combined with remanufacturing processes, providing further insights into the potential of remanufacturing as a strategy to enhance sustainability and resilience.

References

1. United Nations Environment Programme et al (2024) Emissions Gap Report 2024: No more hot air … please! With a massive gap between rhetoric and reality, countries draft new climate commitments. UNEP. https://doi.org/10.59117/20.500.11822/46404
2. Bag S, Gupta S, Foropon C (2019) Examining the role of dynamic remanufacturing capability on supply chain resilience in circular economy. Manag Decis 57(4):863–885. https://doi.org/10.1108/MD-07-2018-0724
3. Geist H, Balle F (2024) Remanufactured products, components, and their materials: a circularity engineering focused empirical status quo analysis. Sustain Prod Consum 45:525–537. https://doi.org/10.1016/j.spc.2024.02.003
4. Ferreira C, Goncalves G (2021) A systematic review on life extension strategies in industry: the case of remanufacturing and refurbishing. In: 2021 26th IEEE international conference on emerging technologies and factory automation (ETFA). IEEE, Vasteras, pp 1–8. https://doi.org/10.1109/ETFA45728.2021.9613610
5. Alamerew YA, Brissaud D (2019) Circular economy assessment tool for end of life product recovery strategies. J Remanuf 9(3):169–185. https://doi.org/10.1007/s13243-018-0064-8
6. Jiang Z, Wang H, Zhang H, Mendis G, Sutherland JW (2019) Value recovery options portfolio optimization for remanufacturing end of life product. J Clean Prod 210:419–431. https://doi.org/10.1016/j.jclepro.2018.10.316
7. Sihvonen S, Ritola T (2015) Conceptualizing ReX for aggregating end-of-life strategies in product development. Proc CIRP 29:639–644. https://doi.org/10.1016/j.procir.2015.01.026
8. COMPONENT | English meaning—Cambridge Dictionary. https://dictionary.cambridge.org/dictionary/english/component. Accessed 28 Jan 2025
9. Ahlstedt E, Sundin E (2023) Assessing product suitability for remanufacturing—a case study of a handheld battery-driven assembly tool. Proc CIRP 116:582–587. https://doi.org/10.1016/j.procir.2023.02.098
10. Gao Y, Feng Y, Wang Q, Zheng H, Tan J (2018) A multi-objective decision making approach for dealing with uncertainty in EOL product recovery. J Clean Prod 204:712–725. https://doi.org/10.1016/j.jclepro.2018.09.080
11. Mangun D, Thurston DL (2002) Incorporating component reuse, remanufacture, and recycle into product portfolio design. IEEE Trans Eng Manag 49(4):479–490. https://doi.org/10.1109/TEM.2002.807292
12. Schuh G, Schmitz S, Schopen M, Hermann A (2023) Concept for maturity assessment of remanufacturing ability in production process development. Proc CIRP 120:846–851. https://doi.org/10.1016/j.procir.2023.09.086
13. Wang W, Mo DY, Wang Y, Tseng MM (2019) Assessing the cost structure of component reuse in a product family for remanufacturing. J Intell Manuf 30(2):575–587. https://doi.org/10.1007/s10845-016-1267-1
14. Ijomah WL, McMahon CA, Hammond GP, Newman ST (2007) Development of design for remanufacturing guidelines to support sustainable manufacturing. Robot Comput Integr Manuf 23(6):712–719. https://doi.org/10.1016/j.rcim.2007.02.017
15. Cai Y, Ke C, Ji Q (2024) A fuzzy decision-making method for green design for remanufacturability. Processes 12(5):911. https://doi.org/10.3390/pr12050911
16. Sundin E (2018) Circular economy and design for remanufacturing. In: Charter M (ed) Designing for the circular economy, 1st ed. Routledge, Abingdon, Oxon; New York, NY, pp 186–199. https://doi.org/10.4324/9781315113067-18
17. Wahjudi D, Yamin IY (2024) Product design evaluation of refrigerators to facilitate remanufacturing process. In: Fera M, Caterino M, Macchiaroli R, Pham DT (eds) Advances in remanufacturing. Lecture Notes in Mechanical Engineering. Springer Nature Switzerland, Cham, pp 459–467. https://doi.org/10.1007/978-3-031-52649-7_36
18. Hilton B, Thurston M (2019) Design for remanufacturing. In: Nasr N (ed) Remanufacturing in the circular economy, 1st ed. Wiley, pp 137–168. https://doi.org/10.1002/9781119664383.ch6

19. Abuzied H, Senbel H, Awad M, Abbas A (2020) A review of advances in design for disassembly with active disassembly applications. Eng Sci Technol Int J 23(3):618–624. https://doi.org/10.1016/j.jestch.2019.07.003
20. Ijomah WL, McMahon CA, Hammond GP, Newman ST (2007) Development of robust design-for-remanufacturing guidelines to further the aims of sustainable development. Int J Prod Res 45(18–19):4513–4536. https://doi.org/10.1080/00207540701450138
21. Geda MW, Kwong CK, Jiang H (2019) Fastening method selection with simultaneous consideration of product assembly and disassembly from a remanufacturing perspective. Int J Adv Manuf Technol 101(5–8):1481–1493. https://doi.org/10.1007/s00170-018-3027-1
22. Vanegas P et al (2018) Ease of disassembly of products to support circular economy strategies. Resour Conserv Recycl 135:323–334. https://doi.org/10.1016/j.resconrec.2017.06.022
23. Brown RN (2005) Centrifugal compressors. In: Compressors. Elsevier, pp 166–261. https://doi.org/10.1016/B978-075067545-1/50007-9

Resilience Performance Assessment in Metal Forming Machines: Bayesian Networks-Based Methodology

Adane Kassa, Ulrich Stache, and Martin Manns

Abstract In recent years, research on supply chain disruptions and internal failures has increased due to their significant impacts on extended downtime, reduced revenue, and diminished customer trust. Traditional reliability and risk management approaches primarily focus on failure prevention, whereas resilience extends beyond by incorporating prediction and adaptive strategies. Current methods for measuring resilience rely on classical reliability metrics, such as mean time between failures (MTBF) and mean time to repair (MTTR). While useful, these metrics fail to capture the multi-dimensional aspects of resilience, specifically the ability to prevent, predict, and adaptively respond to breakdowns. This limitation leaves production systems exposed to critical and unanticipated failures, making it difficult for practitioners to prioritize investments in predictive maintenance technologies or design effective resilience-enhancing strategies. To address this gap, this study develops a quantitative, data-driven resilience assessment framework tailored for production machinery. The framework employs Bayesian Networks (BNs) to model probabilistic dependencies among machine components, failure modes, and resilience strategies, enabling a holistic evaluation of machine resilience performance. A case study on a compound forming machine demonstrates the practical application of the methodology, highlighting its ability to identify critical failure modes and support decision-making for fault detection and human-in-the-loop assistance systems. By facilitating data-driven decisions and targeted retrofitting with emerging technologies, this approach enhances production resilience, reduces downtime, and provides valuable insights for both industrial practitioners and researchers.

Keywords Bayesian networks · Resilience quantification · Predictive maintenance · Resilience score · Artificial intelligence

---

A. Kassa (✉) · U. Stache · M. Manns
Mechanical Engineering Department, PROTECH: Institute for Production Technology, Universität Siegen, Siegen, Germany
e-mail: adane.kassa@aait.edu.et

M. Manns (ed.), *New Paradigms for Anticipated Uncertainty*, Lecture Notes in Mechanical Engineering, https://doi.org/10.1007/978-3-032-16889-4_4

1 Introduction

Modern production systems face growing challenges from supply chain (SC) disruptions and internal failures, particularly machine breakdowns. These failures can severely disrupt production continuity, leading to increased downtime, reduced revenue, and diminished customer trust. Such disruptions often ripple through the entire SC, as exemplified by the 2017 BMW incident. A shortage of steering gears, originating from machine breakdowns at an Italian second-tier supplier, cascaded through the SC. This caused Bosch, BMW's first-tier supplier, to miss delivery schedules, ultimately halting production at BMW's assembly plants worldwide [1]. This incident underscores the need for resilience strategies to contain such disruptions at the production level. Resilience extends beyond classical risk and reliability approaches, which focus primarily on preventing breakdowns. Instead, resilience encompasses the holistic ability to prevent, predict, and adaptively respond to disruptions [2].

Advances in Industry 4.0 technologies, such as smart sensors, cyber-physical systems, and artificial intelligence (AI), offer a transformative paradigm shift. These advancements enable a move from reactive maintenance to intelligent, data-driven resilience enhancement by providing the tools necessary to monitor, predict, and adapt to failures in real-time [2–4]. Consequently, there is growing interest in retrofitting existing machinery with these technologies to accurately predict component failures, detect faults, and assist operators in real-time breakdown resolution.

However, to effectively guide such retrofitting, it is essential to first assess the resilience performance of machines, particularly identifying vulnerable components where preventive, predictive, or adaptive capacities are limited. Despite the urgency of this need, no established methodology in the existing literature that comprehensively assesses the resilience performance of production machinery. Current approaches to managing breakdowns are primarily limited to classical reliability engineering metrics, such as mean time between failure (MTBF) and mean time to repair (MTTR). While useful, these metrics fail to capture the multi-dimensional aspects of resilience, specifically the ability to prevent, predict, and adaptively respond to breakdowns.

The absence of such a methodology has tangible, practical consequences: production systems remain exposed to critical, unanticipated failures due to unrecognized weaknesses in their internal structure. Moreover, without clear resilience indicators, it is difficult for practitioners to prioritize investments in predictive technologies or design effective maintenance strategies. Given that resilience is a complex, multi-dimensional phenomenon, a more holistic assessment framework is required to effectively evaluate the resilience performance of machinery.

To address this critical research gap, this study develops a quantitative, data-driven resilience assessment framework tailored for production machinery. It proposes a Bayesian Network (BN)-based methodology to evaluate the resilience of a compound forming machine, identify critical failure modes, and support decision-making for fault detection and human-in-the-loop assistance systems. BNs are particularly well-suited for this task because they can:

- Capture complex probabilistic dependencies among machine components, failure modes, and resilience factors.
- Perform inference from effect to causes and back to the root cause.
- Integrate both expert knowledge and operational data.
- Dynamically update resilience estimates as new evidence becomes available [5–8].

Our approach operationalizes resilience across three key dimensions: prevention, prediction, and restoration. It uses systematically collected operational and maintenance data to generate a machine-specific resilience score. This dynamic and modular framework not only provides insights into current system vulnerabilities but also supports continuous resilience monitoring over time.

By bridging the gap between traditional reliability analysis and modern adaptive technologies, the proposed approach contributes to the emerging field of operational resilience in smart manufacturing. It offers a scalable and transferable methodology applicable across diverse industrial contexts, making it a valuable tool for advancing resilience engineering in practice. The following research questions guide our investigation:

- RQ1: How can a comprehensive, probabilistic metric for machine resilience, integrating key dimensions, be quantitatively defined and operationalized?
- RQ2: How can a BN model be effectively constructed and validated using operational data to quantify the resilience and identify critical vulnerabilities of a production machine?
- RQ3: How can insights from the BN model inform data-driven retrofitting strategies and targeted interventions to enhance machinery resilience?

2 Related Works on Quantitative Resilience Assessment

In recent years, as the need to manage disruptions in systems has grown, resilience has become a critical indicator for evaluating system performance across fields, such as disaster prevention, engineering, electrical power networks [9], ecological, and production systems [10, 11]. Quantitative resilience assessment methodologies can be classified into three categories: optimization, numerical, and data-driven approaches [10, 12, 13].

Optimization-based methods utilize mathematical models to analyze the influence of system structure on resilience and to identify the optimal configuration for resilience. For instance, Li and Zhao [14] proposed a model for analyzing supply chain (SC) resilience. The model's objective function was defined based on resilience cells with self-adaptive and self-recovery abilities. Numerical experiments were conducted to explore the relationship between risk factors and system resilience.

Numerical analysis methods utilize probabilistic or deterministic performance indicators to evaluate and compare the resilience metrics of different systems or alternative actions. These involve constructing mathematical models to represent system behaviors and analyzing them to gain insights into the underlying dynamics influencing resilience. Alexopoulos et al. [15] introduced a resilience metric using the penalty of change (POC) method. The method was applied to an industrial scenario to evaluate the resilience of two alternative production systems, additive manufacturing (3D printing) and injection molding, during the COVID-19 disrupted production.

Data-driven methods deploy historical data or simulations to assess system resilience. Yin et al. [10] proposed Bayesian networks to evaluate the resilience of urban rail transit systems based on data collected from disruption incidents. They developed a resilience matrix, a function of severity and recovery time, to evaluate and compare the rail systems in Beijing. While major disruptions are rare, and data collection can be challenging, even for complex systems with frequent disruptions, advances in cyber systems now allow for more accurate and precise data collection.

A review of the literature across these methodological categories reveals a significant research gap. While resilience has been extensively examined in domains such as SCs, transportation, and critical infrastructure systems, there is a notable absence of established methodologies specifically targeting the resilience assessment of production machinery. This gap exists despite the frequent internal breakdowns and the increasing availability of relevant operational and maintenance data. Furthermore, most existing research on handling breakdowns primarily focuses on classical engineering reliability metrics, such as mean time between failure (MTBF), which measures the average time a system operates before failing, and mean time to repair (MTTR), which measures the average time it takes to repair a failed system. While useful, these metrics fall short in capturing the complex, multi-dimensional, and interdependent nature of resilience, particularly the abilities to prevent, monitor, predict, and adaptively respond to failures.

Therefore, there is a clear need to conceptualize and define resilience at the machine level and to quantitatively assess it. Such an assessment would support decision-making in production and maintenance planning by incorporating resilience perspectives and enabling the retrofitting of critical components using emerging technologies that enhance resilience.

3 Resilience in Production Machinery

The concept of systems resilience lacks a single, universal definition due to the varied nature of disruptions, which differ in frequency, severity, and origin. Furthermore, a system's ability to cope with disruptions depends on its unique capabilities, resources, and design. In the context of production machinery, we define resilience as the machine's ability to manage internal breakdowns through preventive, predictive, and restorative strategies.

From this perspective, a resilient production machinery system, along with its operation and maintenance personnel, can:

- Prevent random breakdowns through planned maintenance (proactive) measures,
- Predict failures before they occur using skilled analysis or integrated sensing and actuation systems, and
- Restore operations efficiently after unexpected breakdowns without compromising quality and production schedules or supply chain commitments.

Based on these capabilities, we propose measuring resilience as the union of successful maintenance strategies: prevention, prediction, and restoration. A system is considered resilient if at least one of these strategies is effective during a breakdown incident (event).

Conceptually, resilience (denoted by ψ) can be represented as the sum of the likelihood of successful maintenance strategies, as shown in Eq. (1). Prediction comes into play when prevention fails, and restoration is required when both prevention and prediction are ineffective. We assume that each strategy is attempted in sequence and that their outcomes are mutually exclusive during each incident, meaning only one strategy is considered responsible for managing a given failure. Under this assumption, the sum of the probabilities of the strategies equals 1.

To quantitatively operationalize this concept, we define a new probabilistic resilience metric as:

$$\Psi = P(\beta) + P(\delta|\beta = 0) + P(\theta|\beta = 0 \text{ and } \delta = 0). \tag{1}$$

where:

- Ψ represents the total probability of resilience,
- β is the probability of successful prevention,
- $P(\delta|\beta = 0)$ is the probability of successful prediction given that prevention has failed, and
- $P(\theta|\beta = 0 \text{ and } \delta = 0)$ is the probability of successful restoration given that both prevention and prediction have failed.

This metric provides a quantitative framework for assessing the resilience of production machinery systems by evaluating the effectiveness of maintenance strategies in managing breakdowns.

4 Bayesian Networks

Bayesian networks (BNs) are transparent, white-box artificial intelligence (AI) techniques originating from the field of explainable AI. They serve as a formal method for analyzing decision strategies under uncertainty. A BN is a directed acyclic graph (DAG), denoted as G = (V, E), where V represents a set of vertices (random variables), and E represents a set of directed edges (arcs or links). The arcs indicate direct conditional dependencies between variables. For example, if there is a directed edge from node X_i to X_j, then X_j is conditionally dependent on X_i, meaning the occurrence of X_j is influenced by X_i [16].

Each node in a BN is associated with a conditional probability table (CPT) [16, 17]. One of the key advantages of BNs is their ability to represent causal relationships and dependencies among variables through conditional probabilities. As a result, they have been widely used for risk assessment and resilience quantification [16, 18].

In this study, a BN is employed to model conditional independence among the three dimensions, prevention, prediction, and restoration, of resilience based on our definition. It captures probabilistic dependencies among the machine components, failure modes, and resilience factors while quantitatively evaluating the resilience of a compound forming machine.

5 Implementation and Case Study

5.1 Case Study Description

The proposed probabilistic resilience quantification framework is demonstrated using a compound forming machine in an internationally active company specializing in the production of deep-drawn parts and metallic tubular components.

Due to frequent random breakdowns, management aims to strategically integrate sensors beneath functional tool surfaces to monitor and analyze process-relevant parameters. This data will be processed using AI, which will compare real-time sensor inputs with a process-related knowledge base that links past faults to successful corrective actions. Based on this analysis, the AI system will derive optimal adjustment measures to minimize downtime and prevent rejects.

Additionally, management wants to enhance operator skills in preventing, predicting, and adaptively restoring breakdowns. To support this, the resilience performance of the machine must be assessed to identify critical components, determine which functional parts require fault detection systems, and develop targeted operator training programs.

A systematic data collection process was initiated in 2022, utilizing machine operation and maintenance logbooks to record breakdown incidents and repair activities. The dataset, covering the period from January 1, 2022, to March 13, 2025, serves as the foundation for the resilience assessment.

5.2 Modeling and Quantification

To assess and quantify the resilience of the compound forming machine, we employed a step-by-step research methodology (Fig. 1). Our approach centers on a Bayesian Network (BN) model, built using a combination of expert knowledge and operational data. We used the open-source Python library pgmpy [19] to construct and train the BN network. The model's structure was developed based on domain expertise. At the heart of this network is the "Resilience" node (leaf or target node), which quantifies the machine's capacity to withstand and recover from failures. The three key aspects of resilience, prevention, prediction, and restoration, are each represented as intermediate nodes, helping to evaluate how well the machine can avoid breakdowns, predict or anticipate issues, and recover once a failure has occurred. Table 1 provides an overview of the basic variables and their possible states.

The three core dimensions of resilience, prevention, prediction, and restoration, are explicitly modeled as intermediate nodes. These nodes capture the machine's capacity to prevent, anticipate potential breakdowns, and restore functionality effectively after breakdowns. The basic nodes and their corresponding states are presented in Table 1.

Data Collection and Encoding: During data collection, we applied clear criteria for encoding the resilience dimensions. For instance, the prevention dimension is marked as "1" when a component is replaced or adjusted during planned maintenance, and as "0" when no such preventive action is taken. Prediction: A value of 1 is assigned when failures not addressed by the prevention strategy are detected by sensors or operators

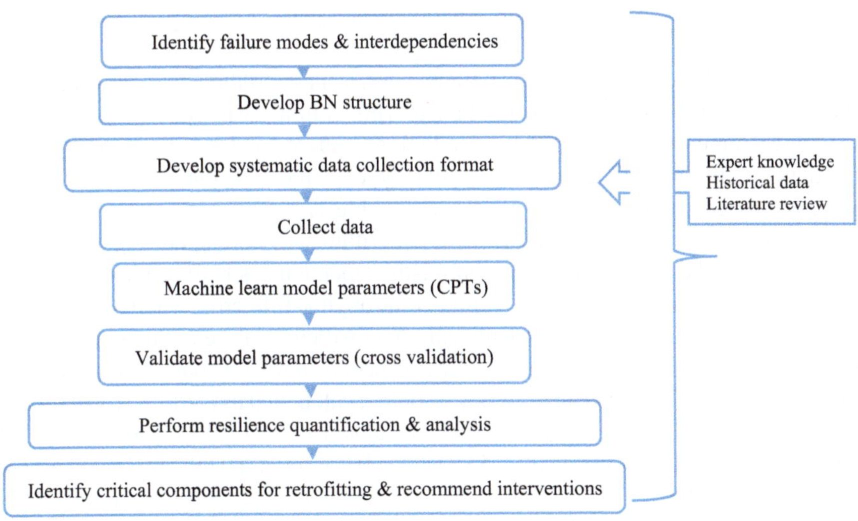

Fig. 1 Research methodology

Table 1 State of basic nodes for BN

Node	State
Resilience score	1: Resilient, 0: Not resilient
Prevention	1: Yes, 0: No
Prediction	1: Yes, 0: No, 2: None
Restoration	1: Effective, 0: Ineffective, 2: None
Forming Machine	2: Planned Maintenance, 1: Predicted repair, 0: Breakdown
Component (e.g. Die)	1: Failed, 0: Functional
Subcomponent (e.g. forming Die)	1: Failed, 0: Functional
Failure mode	1: occur, 0: Not occur

before a breakdown occurs. If the issue is already mitigated by the prevention strategy, a value of 2 is assigned.

Restoration: A value of 1 is assigned when a random failure, not mitigated by previous strategies, is repaired without compromising production schedules or supply deliveries, and no defective products enter the production lot. Since restoration is assessed post-failure, it is evaluated based on these criteria after the incident is resolved.

A value of 0 is assigned when the restoration strategy is ineffective, while a value of 2 is assigned when the failure is mitigated by the prevention or prediction strategies.

Machine State: A value of 2 is assigned when the machine is interrupted due to planned maintenance (prevention), a value of 1 when a fault is detected by operators or sensors, and a value of 0 when a breakdown occurs, resulting in downtime or defective production.

Based on the defined resilience framework, a new column was introduced in the original dataset to represent resilience. This column was mapped to the three resilience dimensions (prevention, prediction, and restoration) according to the resilience definition. To ensure compatibility with the BN, the numerical values were converted into node states, as shown in Table 1.

The Expectation–Maximization (EM) algorithm was employed to estimate the parameters of the BN, capturing probabilistic dependencies between nodes. Model performance and robustness were validated through k-fold cross-validation (with $k = 8$), ensuring its reliability in quantifying the resilience of the compound forming machine. Table 2 presents the results in the form of a confusion matrix. Based on this matrix, the model achieved an accuracy of 0.967, calculated as $(2811 + 1308)/(2811 + 1308 + 86 + 55)$; a precision of 0.9703, calculated as $2811/(2811 + 86)$; a recall of 0.9808, calculated as $2811/(2811 + 55)$; and an F1 score of 0.9755, computed as $2 \times (0.9703 \times 0.9808)/(0.9703 + 0.9808)$. These results further confirm the effectiveness and predictive accuracy of the proposed BN-based resilience assessment model.

Table 2 Confusion matrix

Actual	Predicted	
	Resilient	Not resilient
Resilient	2811 (TP)	55 (FN)
Not resilient	86 (FP)	1308 (TN)

Note: *TP* true positive, *FP* false positive, *FN* false negative, *TN* true negative

5.3 Results and Discussion

The resilience of the compound forming machine was evaluated using the BN based on the proposed metric. The overall resilience of the machine is found to be 68%, as represented in the leaf node of the BN model. This implies that a 68% probability that the machine can avoid breakdowns through planned maintenance, anomaly detection systems, and operator interventions, while also being effectively restored without compromising quality, production, or supply chain objectives. Conversely, in 32% of cases, resilience strategies fail, affecting performance.

Examining the prevention node, 32% of failures are mitigated through planned inspections, repairs, adjustments, and scheduled replacements. Among the remaining 68% of failures not addressed by preventive strategies, approximately 18% (0.68*(19/(19 + 52) are detected and mitigated by fault detection systems and operator interventions before machine breakdown. Additionally, around 18% (0.68*0.52*(32/(32 + 29)) of machine breakdowns are effectively restored without compromising production quality and supply chain commitments.

Certain machine components and subcomponents exhibit higher failure probabilities. The objective of this analysis is to identify failures that are least likely to be prevented, predicted, or effectively restored. Forward inference was applied by propagating evidence from parent nodes to child nodes, revealing the failure modes, subcomponents, and most critical component failures. Findings indicate that failures in the forming punch (forming section) and forming die (die section), primarily due to wear and misalignment, are the most critical. These failures are difficult to prevent, predict, and recover effectively. For example, when the "Failed" state in the forming die node is set to 100%, the probability of the Die section being in a "Failed" state rises to 90%. This also impacts the overall machine state, where the probabilities for Planned Maintenance, Anticipated Failure Detection, and Breakdown shift to 31%, 18%, and 51%, respectively. This indicates:

- 31% of forming die failures are addressed through planned maintenance.
- 18% are detected before machine breakdown.
- 51% lead to breakdowns, of which only 32% are effectively restored without impacting production and supply chain plans.

Based on these findings, the forming punch and die sections require retrofitting to improve resilience. There is a critical need to develop a smart anomaly detection and operator assistance system to sense deviations in forming parameters, assist in real-time adjustments, and prevent catastrophic failures.

To address this, a fault detection and human assistance system is proposed, enabling operators to identify and mitigate failures through reconfiguration measures before actual breakdowns. This system leverages fuzzy case-based reasoning (fuzzy CBR), which learns from expert experience and employs fuzzy logic to represent case variables as degrees of agreement or probability distributions.

Current experimental efforts focus on embedding fiber optic sensors to monitor and report critical parameters during the drawing, ironing, and embossing stages of the forming process. Additionally, motion capture technology is being used to record experienced operators' actions when adjusting misaligned parts, correcting dimensions, or replacing failed components. This data will be used to develop a knowledge base for the fuzzy CBR system, improving real-time decision-making and adaptive intervention capabilities.

6 Conclusion and Future Work

The objective of evaluating resilience in compound press machine to identify critical components for retrofitting with emerging technologies has been successfully achieved. This was accomplished by conceptualizing and defining resilience at the machine level, developing a novel quantitative metric, and creating a comprehensive evaluation model using Bayesian networks (BNs). The proposed framework has proven effective in assessing a compound forming machine, quantifying overall resilience, identifying critical component failures, and highlighting areas where resilience can be enhanced through retrofitting, leveraging AI and sensing and actuating technologies. Although the framework was demonstrated in the context of a compound forming machine in the metal forming industry, it is not limited to this specific application. The approach is generalizable and can be adapted to other industries and machinery by customizing the model structure according to machine-specific failure modes and component interdependencies. This adaptability makes the framework applicable to a wide range of industrial settings where machine behavior data is available.

As a novel approach to modeling and quantitatively assessing machine-level resilience, some limitations exist. First, the data collection process in the case study was manual due to the absence of established automated data acquisition systems. This manual recording can introduce errors or inconsistencies. Future work should aim to incorporate automated data collection mechanisms using smart sensors and IoT devices to improve data reliability and granularity. Second, the current definition and operationalization of resilience assume mutually exclusive maintenance strategies, resilience dimensions, and failure events. However, in practice, these events may overlap, and simultaneous failures across multiple components can occur. Future studies could extend the framework to account for overlapping and concurrent failure dynamics, potentially by exploring alternative modeling techniques or enhancements to BNs.

Moreover, future research could expand the BN model by incorporating exogenous factors that influence machine health and downtime, such as variations in material grade, quality deviations, or operator behavior. Time-series AI models, such as Long Short-Term Memory (LSTM) recurrent neural networks, could also be employed to predict the remaining useful life of machines and their components. This would support proactive maintenance decisions based on historical failure trends, strengthening preventive strategies. This would also allow for the identification of hidden patterns and underlying drivers of resilience, enabling continuous improvement of production system robustness.

Acknowledgements This research has been funded by the Horizon Europe FLEX4RES project with Grant Agreement Number 101091903.

References

1. Hosseini S, Ivanov D (2019) A new resilience measure for supply networks with the ripple effect considerations: a Bayesian network approach. Ann Oper Res. https://doi.org/10.1007/s10479-019-03350-8
2. Spieske A, Birkel H (2021) Improving supply chain resilience through industry 4.0: a systematic literature review under the impressions of the COVID-19 pandemic. Comput Ind Eng 158(June):107452. https://doi.org/10.1016/j.cie.2021.107452
3. Tortorella G, Fogliatto FS, Gao S, Chan TK (2022) Contributions of Industry 4.0 to supply chain resilience. Int J Logist Manag 33(2):547–566. https://doi.org/10.1108/IJLM-12-2020-0494
4. Kassa A, Kitaw D, Stache U, Beshah B, Degefu G (2023) Artificial intelligence techniques for enhancing supply chain resilience: a systematic literature review, holistic framework, and future research. Comput Ind Eng 186(October):109714. https://doi.org/10.1016/j.cie.2023.109714
5. Ojha R, Ghadge A, Tiwari MK, Bititci US (2018) Bayesian network modelling for supply chain risk propagation. Int J Prod Res 56(17):5795–5819. https://doi.org/10.1080/00207543.2018.1467059
6. Hosseini S, Barker K (2016) A Bayesian network model for resilience-based supplier selection. Int J Prod Econ 180:68–87. https://doi.org/10.1016/j.ijpe.2016.07.007
7. Hosseini S, Ivanov D, Dolgui A (2019) Review of quantitative methods for supply chain resilience analysis. Transp Res Part E Logist Transp Rev 125(December 2018):285–307. https://doi.org/10.1016/j.tre.2019.03.001
8. Hosseini S, Ivanov D (2020) Bayesian networks for supply chain risk, resilience and ripple effect analysis: a literature review. Expert Syst Appl 161:113649. https://doi.org/10.1016/j.eswa.2020.113649
9. Yodo N, Wang P, Zhou Z (2017) Predictive resilience analysis of complex systems using dynamic Bayesian networks. IEEE Trans Reliab 66(3):1–10
10. Yin J, Ren X, Liu R, Tang T, Su S (2021) Quantitative analysis for resilience-based urban rail systems: a hybrid knowledge-based and data-driven approach. Reliab Eng Syst Saf 219(November):2022
11. Holling CS (1973) Resilience and stability of ecological systems. Annu Rev Ecol Syst 4:1–23
12. Henry D, Ramirez-marquez JE (2012) Generic metrics and quantitative approaches for system resilience as a function of time. Reliab Eng Syst Saf 99:114–122. https://doi.org/10.1016/j.ress.2011.09.002

13. Hosseini S, Barker K, Ramirez-Marquez JE (2016) A review of definitions and measures of system resilience. Reliab Eng Syst Saf 145:47–61. https://doi.org/10.1016/j.ress.2015.08.006
14. Li Y, Zhao L (2010) Analyzing deformation of supply chain resilient system based on cell resilience model. Lecture notes on Computer Science (including Subseries Lecture Notes on Artificial Intelligence. Lecture Notes Bioinformatics), vol 6328 LNCS, no PART 1, pp 26–35. https://doi.org/10.1007/978-3-642-15621-2_4
15. Alexopoulos K, Anagiannis I, Nikolakis N, Chryssolouris G (2022) A quantitative approach to resilience in manufacturing systems. Int J Prod Res 60(24):7178–7193. https://doi.org/10.1080/00207543.2021.2018519
16. Hosseini S, Ivanov D, Blackhurst J (2022) Conceptualization and measurement of supply chain resilience in an open-system context. IEEE Trans Eng Manag 69(6):3111–3126. https://doi.org/10.1109/TEM.2020.3026465
17. Jun HB, Kim D (2017) A Bayesian network-based approach for fault analysis. Expert Syst Appl 81:332–348. https://doi.org/10.1016/j.eswa.2017.03.056
18. Hosseini S, Al Khaled A, Sarder MD (2016) A general framework for assessing system resilience using Bayesian networks: a case study of sulfuric acid manufacturer. J Manuf Syst 41:211–227. https://doi.org/10.1016/j.jmsy.2016.09.006
19. Ankan A, Textor J (2024) Pgmpy: a Python toolkit for Bayesian networks. J Mach Learn Res 25:1–8

Resilience Assessment in Steel Manufacturing: Evaluating Disruption Scenarios Using the Penalty of Change Methodology

Emmanouil Bakopoulos, Panagiotis Mavrothalassitis, Vasilis Siatras, Sotiris Makris, and Kosmas Alexopoulos

Abstract As manufacturing systems have become more complex, uncertainties and disruptions have increased, where resilience has become essential for manufacturers, in order to ensure production continuity and quick adaptations to disruptions such as equipment failures, supply chain issues, or the introduction of new products. This work presents an approach of assessing resilience in steel manufacturing, using the Penalty of Change (PoC) methodology. Combining PoC with a heuristic scheduling algorithm, the goal is to simulate and evaluate the impact of disruption. Resilience is quantified by measuring the degree of production deviation caused by disruptions such as machinery breakdowns and unplanned orders with high priority. Through a case study of a steel manufacturing plant, the proposed framework offers a clear methodology for assessing resilience and identifying areas for improvement. This study emphasizes the importance of resilience in manufacturing, offering a practical tool for decision-makers to optimize production planning and enhance operational robustness under certain conditions.

Keywords Resilience assessment · Penalty of change · Disruption Scenarios · Resilient manufacturing

1 Introduction

Manufacturing systems are becoming increasingly complex, driven by the need for higher efficiency, flexibility, resilience and competitiveness [1]. However, this complexity also amplifies the impact of disruptions, such as equipment failures,

E. Bakopoulos · P. Mavrothalassitis · V. Siatras · S. Makris · K. Alexopoulos (✉)
Laboratory for Manufacturing Systems and Automation (LMS), Department of Mechanical Engineering and Aeronautics, University of Patras, Rio-Patras, Greece
e-mail: kalexopoulos@dind.uoa.gr; alexokos@lms.mech.upatras.gr

K. Alexopoulos
Digital Industry Technologies, National and Kapodistrian University of Athens, Psahna, Greece

M. Manns (ed.), *New Paradigms for Anticipated Uncertainty*, Lecture Notes in Mechanical Engineering, https://doi.org/10.1007/978-3-032-16889-4_5

raw material shortages, and demand fluctuations. In 2024, supply chain disruptions contributed to an estimated 1.6 trillion USD in missed revenue [2]. EUROFER reports that the EU steel market has faced four recessions in five years, with a 6% drop in apparent steel consumption in 2023 and a projected further decline of 2.3% in 2024—highlighting the industry's vulnerability to disruptions and the need for robust resilience assessment methods [3]. These figures underscore the critical need for enterprises to develop methodologies that assess and enhance a production system's ability to quickly adapt and maintain operational continuity in the face of unexpected disruptions. This work introduces a structured methodology for evaluating the resilience of production systems using the Penalty of Change (PoC) metric [4]. Based on simulated disruption scenarios (including machine breakdowns and urgent high-priority orders derived from historical data in a steel manufacturing plant) the approach quantifies deviations from baseline production schedules. The methodology enables systematic analysis of plan adaptability and responsiveness. Results from a steel industry case study demonstrate that the PoC metric effectively captures trade-offs in production planning under uncertainty, enabling the identification of more resilient and stable scheduling strategies. The study addresses the following research question: *How effectively does PoC differentiate schedule resilience in steel-manufacturing simulations under machine breakdowns and high-priority orders?*

2 Related Work

Resilient manufacturing systems are increasingly recognized as essential due to global disruptions such as the COVID-19 pandemic. Resilient manufacturing is defined as the ability of manufacturing systems to efficiently quickly mitigate any external or internal disruption. Lee et al. [5] laid foundational concepts for resilient manufacturing by emphasizing agility and system robustness. Additionally, Magodi et al. [6] conducted a systematic literature review to identify key design, operational, and technological enablers of resilience. Several quantitative frameworks have since emerged. Reichsthaler et al. [7] proposed the ORSE-framework (Optimization of Resilience, Sustainability and Efficiency), a multi-criteria decision-making approach that jointly assesses resilience, sustainability, and efficiency. Molenda et al. [8] introduced the Resilience Balanced Scorecard, extending the traditional scorecard to capture resilience at multiple organizational levels. Aruväli et al. [9] analyzed existing quantitative metrics for human-centered cyber-physical production systems (CPPS) and highlighted the Penalty of Change (PoC) as a promising resilience indicator. The PoC metric was first introduced by Chryssolouris and Lee [4] to evaluate manufacturing flexibility and later adapted by Alexopoulos et al. [10] for resilience assessment, providing a quantitative measure of disruption impact and guide improvement strategies. De Marchi et al. [11] developed a holistic resilience assessment model combining qualitative and quantitative indicators, while Faruquee et al. [12] proposed a typology of supply chain resilience distinguishing proactive from reactive capabilities. Wu et al. [13] and Caputo et al. [14] contributed hybrid scheduling frameworks to dynamically respond to uncertainty in real time. Design-oriented strategies have also gained traction. Ihlenfeldt et al. [15] proposed resilience-by-design principles for integrated production systems. Bakopoulos et al. [16] digital twins implemented by Asset Administration Sheel (AAS) technology and data spaces framework that facilitates real-time data sharing and system reconfiguration to improve value chain resilience. Despite these advances, many methods remain qualitative or context-specific, limiting their scalability and comparability. PoC and scorecard-based approaches also lack practical integration into operations. Scorecards give static views; our scenario-weighted PoC captures disruption timing, so the resilience signal is more actionable. This paper addresses these gaps by operationalizing the PoC metric in a simulation-driven setting to evaluate production plan robustness under various disruptions, offering a probabilistic, interpretable resilience score that supports real-time scheduling decisions.

3 Methodology

The approach used to evaluate resilience is based on the PoC methodology [10], a metric that quantifies the cost or impact of deviating from an original production plan when a disruption occurs. PoC offers operational-level insights derived directly from production data and simulations. This methodology allows for a structured assessment of how deviations from a baseline production plan affect resource utilization, makespan, and overall operational efficiency. By systematically measuring these effects, PoC provides valuable insights into the system's ability to recover from unexpected disruptions. The PoC measurement aims to support the development of strategies to enhance resilience and maintain production stability under varying conditions. In detail, PoC methodology is presented by Eq. (1) below:

$$PoC = \sum_{i=1}^{D} Pn(X_i) \cdot \Pr(X_i) \tag{1}$$

where D is the number of potential changes, X_i the i-th potential change, $Pn(X_i)$ is the penalty of the i-th potential change, and $Pr(X_i)$ is the probability of the i-th potential change to occur. To compute PoC, it is important to define the penalty function, the potential change X_i itself, the number of the potential changes D, and also the probability of each potential change to occur. To evaluate the resilience of a manufacturing system, disruption scenarios are simulated, where each disruption scenario is represented by a set of potential changes that could impact production stability. These disruptions, such as machinery breakdowns and unplanned high-priority orders, introduce deviations from a baseline production plan, requiring quick

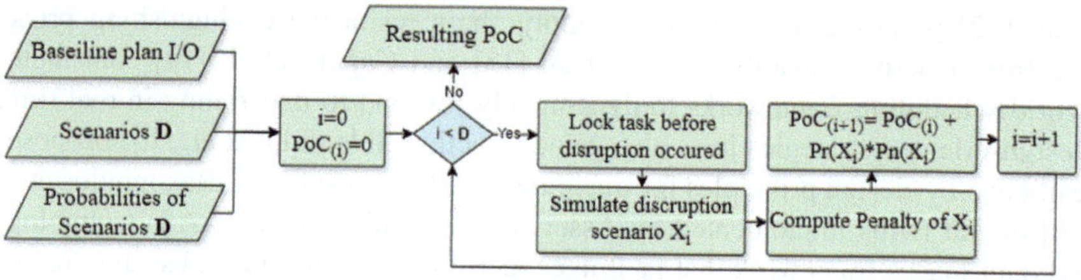

Fig. 1 Resilience assessment utilizing PoC methodology

adjustments to minimize operational delays. Each potential change is characterized by a specific start and end time, allowing for a structured analysis of its effects.

The PoC methodology is used to quantify the impact of disruption scenarios by comparing key performance indicators of the baseline plan to those of the adjusted plan under each scenario, as illustrated in Fig. 1. For instance, if a critical rolling mill machine fails at 10:00 AM and remains out of service for four hours, this unexpected change reduces machine utilization and extends the production makespan. By applying PoC, the deviation from the planned schedule is measured, providing proactive insights into how resilient the system is to such disruptions. This assessment supports the implementation of targeted strategies, such as predictive maintenance or dynamic rescheduling, to enhance adaptability and minimize production losses.

4 Experimental Results

4.1 Case Study

The case study focuses on a steel manufacturing group that produces a diverse range of steel products, including wire rod coils, rebar coils, and straight rebar, which are used across multiple industrial sectors. The group's production infrastructure includes enterprise-level operations, warehouses, and multiple manufacturing plants. For the purposes of this study, the resilience assessment is conducted at the factory level, targeting one specific manufacturing plant within the group. The manufacturing plant operates in an environment vulnerable to uncertainties and disruptions, which pose risks to the production schedule. Such disruptions can lead to schedule deviations, increased operational costs, and reduced ability to meet customer requirements. To assess resilience, historical data on disruptions from a one-year period was collected and analyzed from the companies Enterprise Resource Planning (ERP). Two representative disruption scenarios were selected to study in this work (i) machine breakdown affecting production continuity and requiring rapid rescheduling, and (ii) arrival of a new high-priority customer order, requiring real-time adjustment of the production plan to meet urgent demand. These scenarios serve as test cases to evaluate the robustness and adaptability of production schedules using PoC methodology.

4.2 Implementation

The PoC metric is designed to quantify the impact of disruptions, providing a structured method for evaluating the resilience of a manufacturing system. PoC methodology was selected for this study due to its compatibility with discrete-event simulation tools and its capacity to reflect order-specific and time-sensitive disruptions in production environments. By simulating disruption scenarios, PoC enables comparisons across multiple conditions, making it well-suited for industrial applications where flexibility and responsiveness are critical. This section focuses on how PoC is operationalized in the case study. The metric measures the deviation in system performance, specifically makespan, between a baseline (planned) schedule and disrupted scenarios. The penalty reflects the increase in the adjusted makespan caused by disruptions such as machine breakdowns or urgent order insertions. Since multiple disruption scenarios (D) are considered, each scenario has different probability of occurrence. The formal computation of PoC in our implementation is as follows:

$$PoC = \sum_{i=1}^{D} \Pr(X_i)\left(\max_{j \in J(X_i)} C_j - \max_{k \in J(Baseline)} C_k\right) \tag{2}$$

- Where $\Pr(X_i)$ is the probability of occurrence for disruption scenario X_i, ensuring that $\sum_{i=1}^{D} \Pr(X_i) = 1$
- $J(X_i)$ is the set of jobs in the schedule under disruption scenario X_i
- $J(Baseline)$ is the set of jobs in the original schedule
- C_j is the completion time of job j and C_k is the completion time of job k
- $\max_{j \in J(X_i)} C_j$ represents the adjusted makespan under disruption scenario X_i
- $\max_{k \in J(Baseline)} C_k$ represents the makespan of the original schedule

By integrating the probabilities of each disruption scenario, PoC provides a risk-adjusted resilience score. For instance, frequent machine breakdowns may have higher weights, while rare but severe disruptions, like high-priority customer orders, may carry lower probabilities but significant penalties. These probabilities can be estimated from ERP-derived historical data, expert assessments, or predictive analytics, allowing organizations to tailor resilience strategies to their risk profile. For example, Fig. 2 illustrates three disruption scenarios evaluated in this study. To simulate the production adjustments and calculate makespan deviations, we used a heuristic production scheduling tool that solves the resource allocation problem [17, 18].

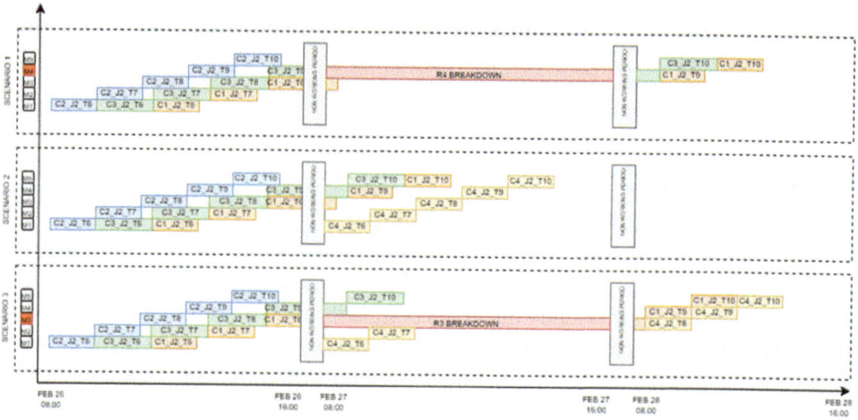

Fig. 2 Disruption scenarios: (i) machine breakdown, (ii) arrival of a new high-priority customer order, (iii) combination

4.3 Experimental Results

The experimental workload consists of three customer orders; each composed of five tasks. These tasks are executed across five machines (R1 to R5), each operating on a fixed shift schedule from 08:00 to 16:00. The hours between 16:00 and 08:00 are non-working periods during which no tasks are processed. The aim of the experiment is to evaluate the resilience of a baseline production schedule by simulating disruptive events such as machine breakdowns and the insertion of urgent customer orders. Six disruption scenarios were designed based on historical disruption data extracted from the company's ERP system, covering a one-year operational period. In Fig. 2 is shown three discussion scenarios. In Scenario 1, a one-shift breakdown affects Resource 5 (R5). Scenario 2 adds an urgent customer order. Scenario 3 combines both a breakdown in Resource 3 (R3) and an unexpected new order. These scenarios were further validated through interviews with operations managers to ensure their practical relevance. The selected scenarios include two involving machine breakdowns, lasting one working shift, two introducing unexpected high-priority customer orders, and two combining both types of disruptions. These represent the most frequent and impactful causes of schedule instability observed in the plant: machine failures, urgent demand, and material delays. Although the experimental case is simplified to three orders and five tasks to ensure clarity and tractability, it preserves industrial relevance and provides sufficient complexity for meaningful scenario comparisons.

In Fig. 3 is shown the Gantt charts of the baseline production schedules used for comparison. PoC values, computed for each schedule, offer a quantitative measure of robustness by indicating the impact of disruptions on the original schedule. A PoC of zero means that disruptions had no impact, while higher values indicate more significant deviations from the baseline makespan. Additionally, Fig. 4 shows

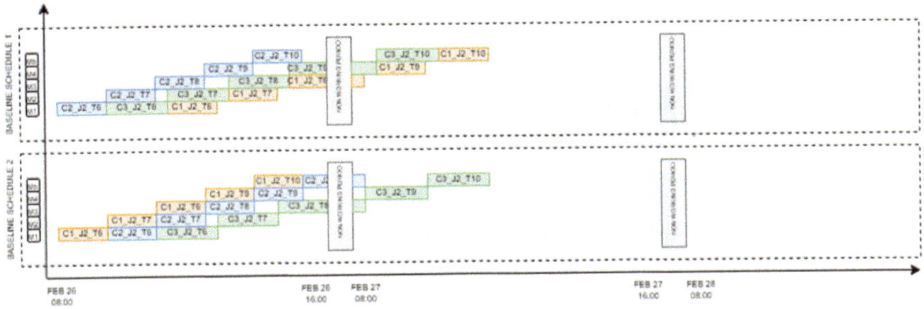

Fig. 3 Production schedules—baseline plans

the adjusted makespans for each scenario. The baseline schedule reflects normal conditions without disruption.

Table 1 presents the PoC values and the standard deviation across the six disruption scenarios for the two evaluated production schedules, which were generated by simulating the scenarios using the heuristic production scheduling tool. Although both schedules achieve the same makespan, Schedule 1 records a PoC of 28.7 with a standard deviation (StD) across the six disruption scenarios of 33.5, whereas Schedule 2 records a PoC of 66.5 with a StD of 86.6, an increase of 132%. The higher PoC and volatility (StD) indicate that Schedule 2 is both more sensitive to disruptions under machine-breakdown and rush-order disturbances. Hence PoC uncovers resilience gaps that conventional time-based metrics leave unseen, guiding planners toward the more robust Schedule 1.

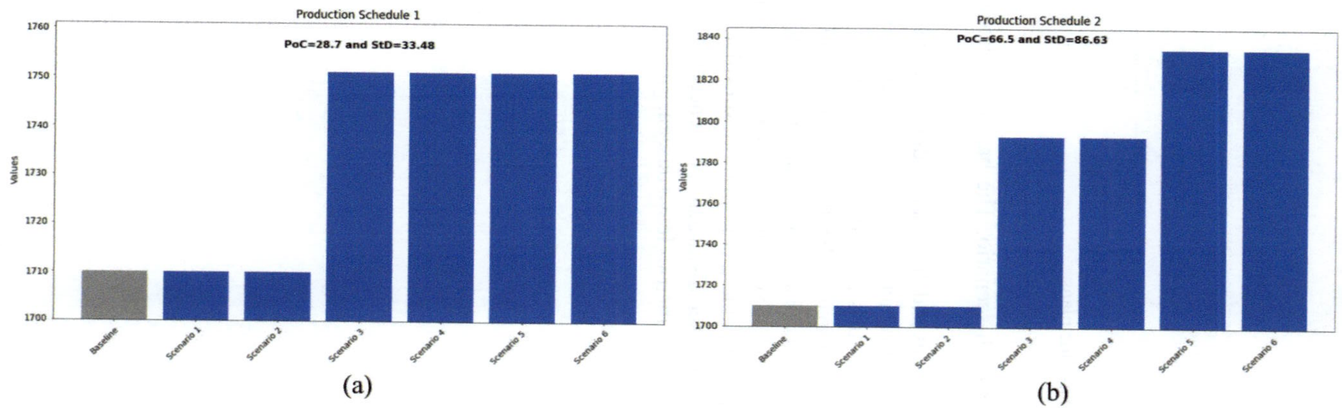

Fig. 4 Resulting makespan per production schedule: **a** schedule 1 and **b** schedule 2

Table 1 Experimental results

No. of production schedule	Penalty of Change (PoC)	Standard deviation of disruption scenarios
1	28.7	33.48
2	66.5	86.63

5 Conclusion

This work presents a structured approach to assessing manufacturing resilience by leveraging the Penalty of Change (PoC) methodology to quantify deviations in production schedules under various disruption scenarios. By employing a heuristic algorithm to simulate disruptions, the study provides a systematic means to evaluate how different events affect production performance. The results show that the proposed resilience assessment methodology can serve as a practical decision-support tool, enabling planners to identify production plans that are inherently more robust to common disruptions. While the case study focuses on a simplified industrial use case involving a limited number of orders, tasks, and machines, it retains realistic elements and serves as a proof of concept. This study uses three orders, five machines, fixed repair times, and no supply variability, which means PoC may misrate resilience in more complex plants. Future work will scale the methodology to more complex environments, disruptions, and industrial sectors. Integrating real-time decision-making and optimization will enhance responsiveness. A comparative analysis with other resilience assessment methods will also be carried out to benchmark effectiveness, adaptability, and suitability across different manufacturing contexts.

Acknowledgements This work is under the framework of Horizon Europe Flex4Res project. This project has received funding from the European Union's Horizon research and innovation program under grant agreement No. 101091903. The dissemination of results herein reflects only the authors' view, and the Commission is not responsible for any use that may be made of the information it contains.

References

1. Chryssolouris G, Alexopoulos K, Arkouli Z (2023) Artificial intelligence in manufacturing systems. In: A perspective on artificial intelligence in manufacturing. Springer, Cham, pp 79–135
2. Statista. Losses from supply chain disruptions by industry 2020. https://www.statista.com/statistics/1259124/loss-supply-chain-disruption-industry. Accessed 27 Feb 2025
3. EUROFER. Economic and Steel Market Outlook 2025–2026 (First Quarter). https://www.eurofer.eu/publications/economic-market-outlook/economic-and-steel-market-outlook-2025-2026-first-quarter. Accessed 30 May 2025
4. Chryssolouris G, Lee M (1992) An assessment of flexibility in manufacturing systems. Manuf Rev 5(2):105–116

5. Lee J, Siahpour S, Jia X, Brown P (2022) Introduction to resilient manufacturing systems. Manuf Lett 32:24–27
6. Magodi Y, Daniyan I, Mpofu K, Ale F (2024) Towards the development of resilient manufacturing systems: a systematic literature review. In: 2024 International conference on science, engineering and business for driving sustainable development goals (SEB4SDG), pp 1–8. IEEE
7. Reichsthaler L, Toth D, Szaller Á, Sihn W (2023) A multi-criteria approach for assessing resilience, sustainability and efficiency measures in manufacturing companies. Proc CIRP 120:547–552
8. Molenda P, Groneberg H, Schötz S, Döpper F (2023) Resilience Balanced Scorecard: Measuring resilience of manufacturing companies at multiple levels. Procedia CIRP 120:189–194
9. Aruväli T, De Marchi M, Rauch E (2023) Analysis of quantitative metrics for assessing resilience of human-centered CPPS workstations. Sci Rep 13(1):2914
10. Alexopoulos K, Anagiannis I, Nikolakis N, Chryssolouris G (2022) A quantitative approach to resilience in manufacturing systems. Int J Prod Res 60(24):7178–7193
11. De Marchi M, Friedrich F, Riedl M, Zadek H, Rauch E (2023) Development of a resilience assessment model for manufacturing enterprises. Sustainability 15(24):16947
12. Faruquee M, Paulraj A, Irawan CA (2024) A typology of supply chain resilience: recognising the multi-capability nature of proactive and reactive contexts. Prod Plan Control 35(12):1503–1523
13. Wu LH, Chen X, Chen XD, Chen QX (2009) The research on proactive-reactive scheduling framework based on real-time manufacturing information. In: Materials science forum, vol 626, pp 789–794. Trans Tech Publications Ltd
14. Caputo AC, Donati L, Salini P (2023) Estimating resilience of manufacturing plants to physical disruptions: model and application. Int J Prod Econ 266:109037
15. Ihlenfeldt S, Wunderlich T, Süße M, Hellmich A, Schenke CC, Wenzel K, Mater S (2021) Increasing resilience of production systems by integrated design. Appl Sci 11(18):8457
16. Bakopoulos E, Sipsas K, Nikolakis N, Alexopoulos K (2024) A digital twin and data spaces framework towards resilient manufacturing value chains. IFAC-PapersOnLine 58(19):163–168
17. Siatras V, Bakopoulos E, Mavrothalassitis P, Nikolakis N, Alexopoulos K (2024) Production scheduling based on a multi-agent system and digital twin: a bicycle industry case. Information 15(6):337
18. Bakopoulos E, Siatras V, Mavrothalassitis P, Nikolakis N, Alexopoulos K (2024) Digital-twin-enabled framework for AI agents in production scheduling. In: Artificial intelligence in manufacturing. Springer, Cham, pp 147–179

Enhancing Resilience in Smart Manufacturing: A Method for Connected Machine Networks

Elena Urkia, Elena Montejo, Emmanouil Bakopoulos, and Kosmas Alexopoulos

Abstract Resilience in production environments is defined as the system's ability to avoid, adapt, and rapidly recover from disruptive events, attaining normal operational conditions with minimal time and financial investment. Current methods lack the capability to quantify risks based on real-time data from production devices, creating a critical gap. This article presents a novel method aimed at enhancing resilience by integrating data from micro and meso levels within precision machining production plants. By utilizing real-time monitoring data, from machines and production processes, the proposed approach is based on the assessment of risks of failure of machines and early detection of anomalies in machining processes, to support decision-making. Leveraging the Industrial Internet of Things (IIoT), the method collects and analyses real-time machines and machining process data for risk assessment and early anomaly detection. The proposed method minimizes the impact of unexpected failures by incorporating failure risk calculations into plant-level decision-making. Validation of the method was conducted using a milling machine in a controlled laboratory setting, demonstrating the significance of real-time data and health assessment in developing resilient industrial systems. Preliminary results indicate strong potential for industrial application, with the next step being implemented in a production plant.

Keywords Resilience assessment · IIoT · Smart manufacturing · Health diagnosis · Decision-making

E. Urkia (✉) · E. Montejo
ICT and Automation Department, IDEKO Research Centre, BRTA, Elgoibar, Spain
e-mail: eurkia@ideko.es

E. Bakopoulos · K. Alexopoulos
Laboratory for Manufacturing Systems and Automation (LMS), Department of Mechanical Engineering and Aeronautics, University of Patras, Rio-Patras, Greece

M. Manns (ed.), *New Paradigms for Anticipated Uncertainty*, Lecture Notes in Mechanical Engineering, https://doi.org/10.1007/978-3-032-16889-4_6

1 Introduction

Systems resilience, as defined in [1], encompasses both the robustness of a system to withstand perturbations that may disrupt operations and the speed at which it can recover its capacity following such events. In recent years, the scientific community has intensified efforts to enhance the resilience of production systems in response to global disruptions such as pandemics, armed conflicts, and geopolitical tensions. A central strategy in this context involves diversifying supply chains to mitigate regional dependencies and reduce geopolitical risks. Simulation models, as demonstrated in [2], have proven effective in analyzing and forecasting supply chain behavior under stress.

Resilience in manufacturing is addressed at three hierarchical levels: the micro level, which focuses on individual machines and processes within a plant; the meso level, which considers the entire manufacturing facility or interconnected systems; and the macro level, which encompasses the broader supply chain and external influences. At the micro level, resilience is supported through advanced monitoring systems, Predictive Maintenance (PdM), and real-time analytics aimed at preventing failures and optimizing machine-level planning. The meso level emphasizes strategies such as production line optimization, increased operational flexibility, and inter-company collaboration through eco-industrial parks, where resources and by-products are shared. At the macro level, resilience strategies include supplier diversification, the development of alternative logistics routes, and policy advocacy to reduce geopolitical exposure and ensure supply chain continuity, as discussed in [3].

Concurrently, industrial digitalization has advanced significantly. Modern manufacturing systems feature interconnected machines equipped with intelligent monitoring capabilities. The increasing volume and complexity of data presents new challenges in information management. Enhancing resilience under these conditions requires the integration of data sources and the acceleration of decision-making processes. PdM has emerged as a key topic in smart manufacturing and digital transformation, with numerous recent studies exploring data-driven approaches. The integration of Artificial Intelligence (AI) into operational workflows further supports resilience by reducing manual effort, shortening analysis time, and providing objective recommendations for maintenance and investment decisions.

This work aims to integrate these concepts by leveraging machine digitalization to collect real-time data on machine condition and behavior during production. This information is used to detect early signs of failure, assess the associated risk, and inform planning systems to prevent disruptions or mitigate their impact. As a result, the proposed approach contributes to resilience enhancement at both the micro and meso levels.

The remainder of this paper is structured as follows: Sect. 2 presents a literature review, Sect. 3 describes the proposed method for resilience improvement using digitized plant data, Sect. 4 details the case study, and Sect. 5 provides the conclusions and discussion.

2 Literature Review

In the managerial literature, as noted in [4], resilience has been extensively examined as an organizational challenge with increasing research interest in recent years. According to [2] many approaches focus on resilience measurement, with simulation models commonly used to analyse and predict supply chain behaviour. Core strategies include supplier diversification, the development of alternative logistics routes, and policy advocacy to mitigate geopolitical risks and promote supply chain stability.

Recent studies have deepened the understanding of resilience in manufacturing contexts. Probabilistic modelling at the plant level is emphasized in [5], introducing the "Penalty of Change" metric to quantify resilience and proposing a framework for production network reconfiguration. Diagnostic tools for identifying value stream risks are presented in [6], validated through real-world applications and oriented toward value chain design. A broader view is adopted in [3], exploring the role of digital manufacturing—particularly Cyber-Physical Production Systems (CPPS)—in bolstering resilience, and call for further integration of smart technologies into resilience strategies.

Additional contributions come from [7–9], which extend resilience research to encompass efficiency evaluation, plant recovery modelling, and supply chain assessment, respectively. In [7], dynamic efficiency metrics grounded in mathematical modelling for process industries are proposed. A refined framework for estimating plant recovery and economic loss is offered in [8], supporting disruption preparedness. A fuzzy index system for evaluating supply chain resilience is introduced in [9].

Despite these advances, a significant gap remains, current methodologies rarely connect machine-level disruptions with higher-level resilience strategies, limiting their effectiveness in dynamic manufacturing settings. While recent work in [10] introduces resilience-aware monitoring systems that leverage historical data and event logs to simulate shop-floor scenarios, such systems are primarily reactive. They fall short of implementing proactive, predictive strategies that anticipate disruptions using AI-driven models.

Moreover, PdM has emerged as a pivotal component of smart manufacturing, aiming to reduce unplanned downtime, extend asset life, and optimize maintenance activities. Unlike traditional reactive or scheduled maintenance strategies, PdM utilizes sensor data, condition monitoring, and machine learning to predict failures before they occur. As highlighted in surveys [11, 12], data-driven PdM approaches enabled by IIoT systems are gaining momentum. The integration of AI further enhances predictive accuracy and supports autonomous decision-making, contributing to reduced diagnostic time, improved planning, and greater operational reliability.

Nevertheless, while resilience strategies often focus on supply chain management or planning algorithms, and PdM has seen widespread application, our review reveals a critical gap: the absence of methods that combine real-time machine data with high-level planning systems to enhance proactive resilience in manufacturing.

This paper addresses that gap by proposing an integrated framework that links machine-level PdM data with planning and decision-making systems, thereby strengthening manufacturing resilience against disruptions in an Industry 4.0 context.

3 Method for Resilience Improvement Using Digitized Plant Data

In response to the gap identified in the scientific literature, this paper develops a method that leverages IIoT technologies to enhance resilience at the plant level. By digitalizing machines, real-time data on their condition and behavior during production is collected. This data enables early detection of problems or potential failures and supports timely decision-making.

The method focuses on two main aspects: quantification of the risk of failure in critical machine components and anomaly detection in machining processes. A milling machine is used as the reference system for the development and implementation of the method. Although a Failure Mode and Effects Analysis (FMEA) is not conducted within this study, the selection of the spindle, axes, and counterweight as the critical components of the machine is based on prior work where an FMEA was performed on the same type of milling machine. Fingerprint Routines (FR), as defined in [13] as predefined test cycles in no-load conditions, are applied to evaluate the health status of these components. The results are used to assess the risk of failure and are integrated into the company's planning tool, facilitating the scheduling of predictive maintenance (PdM) activities. Additionally, machining processes are continuously monitored, and AI is employed to detect anomalies, which are promptly communicated to users to support rapid decision-making.

3.1 Risk Quantification Method

To quantify the risk of failure, the monitoring system collects data to assess the condition of the three components identified as most critical in the FMEA: the spindle, the axes, and the counterweight. Their condition is evaluated through FRs, which are executed on a monthly basis under no-load conditions.

During each cycle, several Key Performance Indicators (KPIs) are measured and analyzed. Based on these indicators, a risk level R is assigned to each component on a scale from one to four, where one indicates no risk and four indicates imminent failure (Fig. 1).

The risk of failure of a machine, MR, is calculated based on the risks of its critical components, each weighted by the Mean Time To Repair (MTTR) [14]:

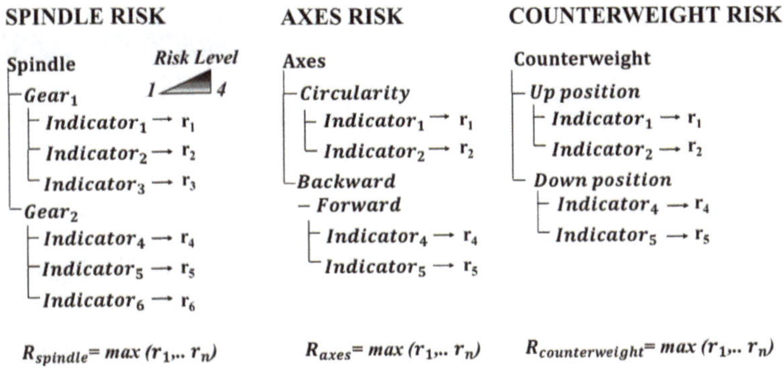

Fig. 1 Risk evaluation in critical components

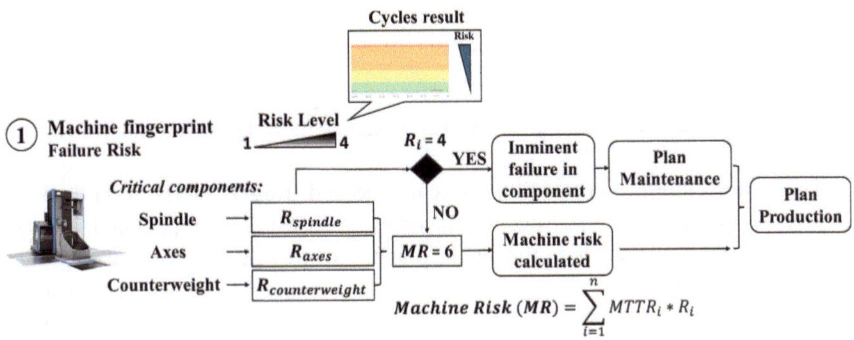

Fig. 2 Machine fingerprint and planning method

$$MR = MTTR_{spindle} * R_{spindle} + MTTR_{axes} * R_{axes} + MTTR_{counterweight} * R_{counterweight} \quad (1)$$

Equation 1 shows machine risk calculation.

These values are considered for computing the planning. If any of the risk values reach four, indicating imminent failure, an alert is sent to the planner to schedule maintenance and arrange alternative fabrication routes. Otherwise, the overall machine risk value is used as input for production planning, allowing the selection of fabrication routes that minimize the risk of failure (Fig. 2).

3.2 *Process Anomaly Detection*

While the risk of failure estimation targets critical machine components, this section focuses on process-level anomaly detection as a complementary aspect of overall risk assessment. Machine learning models are trained on repetitive machining processes

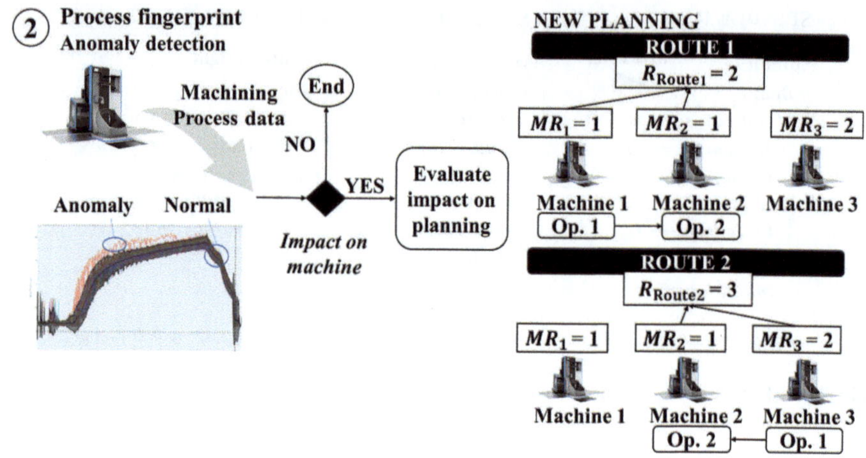

Fig. 3 Process fingerprint and planning method

using novelty detection, an unsupervised approach to identify deviations from normal behavior.

The system monitors operations in real time and triggers alerts upon detecting anomalies. When an anomaly occurs, the operator assesses its impact and, if necessary, reports the expected downtime to the planner, who evaluates its effect on production and may generate an alternative plan to mitigate disruptions, as schematized in Fig. 3. The overall risk level R of the route is a cumulative value computed from the individual risks of the machines involved in the route.

4 Case Study

The method was tested in a laboratory setting as a preliminary step before deployment in a real job shop specialized in precision part machining. This environment is characterized by long, unattended milling operations and flexible routing. Machines are equipped with sensors and data acquisition systems connected to the planning system. The goal of the test was to validate the method under controlled conditions that reflect real-world scenarios of the job shop.

4.1 Implementation

The demonstration was carried out on a horizontal and vertical multifunction machining center with a mobile column at Ideko's Digital Grinding Innovation Hub (DGIH). The machine is equipped with accelerometers for performance monitoring.

Additionally, historical data from actual machining operations in the job shop was used to develop and test the process anomaly detection logic, leveraging signals collected from a production machine operating under real conditions.

4.2 *Risk Quantification Method*

The method was tested in two stages. In the first stage, a subset of signals was selected from the outputs of FRs, based on expert knowledge and historical data linking specific signal patterns to degradation. Once selected, FRs were systematically executed to monitor these signals under both normal and controlled degradation conditions. From the observed variations, threshold values were defined to quantify different levels of risk, enabling the system to detect when a signal indicates a potential deterioration event.

In a second stage, the health condition of certain components was forced to deteriorate to observe how the system calculates risk and estimates the impact on planning. The system successfully detected the increased risk and triggered an alert. An anomaly was identified when one or more of the selected signals exceeded their predefined threshold values, which had been established during the first stage.

4.3 *Process Anomaly Detection*

In the process domain, data from several machining operations was gathered at the job shop. A total of 46,281 normal instances were used to train three novelty detection algorithms—One-Class Support Vector Machine (OCSVM), Local Outlier Factor (LOF), and k-Nearest Neighbors (KNN)—each modeling the standard behavior of the system (Fig. 4). To evaluate the models, three anomalous scenarios were manually injected, simulating realistic deviations such as altered operation sequences, abnormal power peaks, and extended durations. Detection performance was assessed by verifying whether each algorithm flagged these anomalies as outliers. Figure 5 illustrates this by plotting signal values from various production orders over time, where the red line highlights an anomalous instance that clearly deviates from the normal operational profile.

5 Conclusion and Discussion

The proposed method makes a substantial contribution to the advancement of resilient manufacturing systems by integrating digitized machine monitoring, predictive maintenance, and process anomaly detection. Through the acquisition and analysis of real-time data from critical machine components and machining operations,

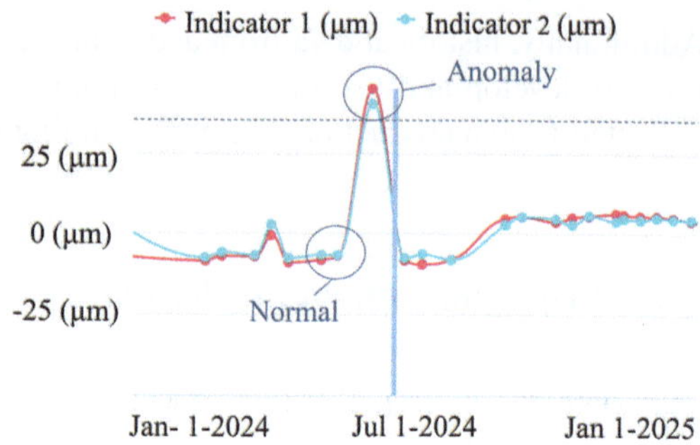

Fig. 4 Machine fingerprint analysis with an anomaly peak

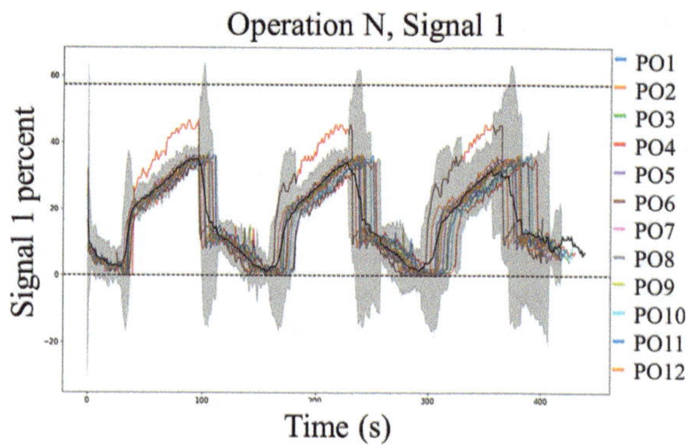

Fig. 5 Process fingerprint analysis with anomaly in production order (PO4)

the approach enables dynamic risk assessment and supports informed, proactive planning decisions.

Laboratory validation confirmed the method's effectiveness in early fault detection. Initial validation of the risk quantification method showed that the system effectively detected increased risk under controlled degradation, confirming its potential for early fault detection and decision support. In the process domain, all three injected anomalies were successfully identified by the implemented algorithms, demonstrating their applicability for detecting deviations in real-world machining scenarios. Full deployment in a job shop environment is expected to improve machine availability, reduce operational risk, and enhance overall plant performance.

Furthermore, certain elements of the method open promising avenues for future research, particularly the application of AI technologies to automatically establish context-sensitive thresholds for both Fingerprint Routine (FR) indicators and process monitoring parameters. These thresholds could adapt to varying working conditions in actual manufacturing settings, further increasing system autonomy and robustness.

The initial validation underscores the feasibility and potential impact of the method. Future work will focus on full-scale implementation, long-term performance assessment, and integration with adaptive scheduling algorithms suited for dynamic and variable production environments.

Acknowledgements This project has received funding from the European Union's Horizon research and innovation program in the Flex4Res Project under grant agreement No. 101091903. The dissemination of results herein reflects only the authors' view, and the Commission is not responsible for any use that may be made of the information it contains.

Declaration of Generative AI and AI-Assisted Technologies in the Writing Process During the preparation of this work the authors used ChatGPT and DeepL to improve readability and language. After using these tools, the authors reviewed and edited the content as needed and take full responsibility for the content of the publication.

References

1. Caputo AC, Paolacci F (2017) A method to estimate process plant seismic resilience. In: Conference: Proceedings of the ASME 2017 pressure vessels & piping conference at Waikoloa, Hawaii, United States
2. Alexopoulos K, Nikolakis N, Chryssolouris G (2020) Digital twin-driven supervised machine learning for the development of artificial intelligence applications in manufacturing. Int J Comput Integr Manuf 33(5):429–439
3. Fowler D, Gregory Epiphaniou M, Higgins M, Maple C (2023) Aspects of resilience for smart manufacturing systems. Strateg Change 32(6):167–237. Special Issue: Resilience, Sustainability and Identity in the Digital Age
4. Burnard K, Bhamra R (2011) Organisational resilience: development of a conceptual framework for organisational response. Int J Prod Res 49(18):5581–5599
5. Alexopoulos K, Anagiannis I, Nikolakis N, Chryssolouris G (2022) A quantitative approach to resilience in manufacturing systems. J Prod Res 24–60
6. Steinmeyer M, Bentz D (2024) Identification of risks in value streams for resilience assessment diagnostic models. Proc CIRP 130:319–324
7. Namvar H, Bamdad S (2021) Performance evaluation of process industries resilience: risk-based with a network approach. J Loss Prev Process Indus
8. Caputo AC, Donati L, Salini P (2023) Estimating resilience of manufacturing plants to physical disruptions: model and application. Int J Prod Econ
9. Rajesh R (2019) A fuzzy approach to analyzing the level of resilience in manufacturing supply chains. Sustain Prod Consump
10. Aruvali T, Marchi MD, Rauch E (2024) Monitoring system architecture for long-term resilience in manufacturing. AIP Conf Proc 2989(1)
11. Carvalho TP, Soares FAAMN, Vita R, Francisco RP, Basto JP, Alcalá SG (2019) A systematic literature review of machine learning methods applied to predictive maintenance. Comput Ind Eng 137:106024
12. Lei Y, Li N, Guo L, Li N, Yan T, Lin J (2020) Machinery health prognostics: a systematic review from data acquisition to RUL prediction. Mech Syst Signal Process 104761
13. Fogliazza G, Arvedi C, Spoto C, Trappa L, Garghetti F, Grasso M, Colosimo BM (2021) Fingerprint analysis for machine tool health condition monitoring. IFAC-PapersOnLine 54:1212–1217
14. Laplante PA (ed) (2023) Mean time to repair (MTTR). In: Encyclopedia of computer science and engineering. Springer, Cham

Resilient Manufacturing Through Data Spaces: Leveraging IDS and Gaia-X in a Compute-on-Demand Environment

Emmanouil Bakopoulos, Konstantinos Sipsas, and Kosmas Alexopoulos

Abstract Resilience is increasingly critical in modern manufacturing, especially within complex supply chains. Information sharing based on data sovereignty principles enables informed, timely decisions, improving supply chain resilience. Data space technologies, like International Data Spaces (IDS) and Gaia-X, support secure, multi-stakeholder data sharing. This work integrates IDS and Gaia-X data spaces into a data spaces framework for resilience (DSF4R). The framework facilitates secure data exchange using a compute on demand environment, ensuring data remains secure within the provider's system while enabling external services to process it. This eliminates the need for data transfer, ensuring privacy and compliance. Applied in a real-world industrial use case from the steel industry sector, it demonstrates enhanced resilience and broader applicability.

Keywords Resilience manufacturing framework · Compute on demand · International data space · Gaia-X data space · Reconfiguration

1 Introduction

The global manufacturing sector has undergone significant transformation in recent decades, driven by globalization, automation, and digital technologies [1]. While these changes have enhanced efficiency and innovation, they have also introduced

E. Bakopoulos (✉) · K. Alexopoulos
Laboratory for Manufacturing Systems and Automation (LMS), Department of Mechanical Engineering and Aeronautics, University of Patras, Rio-Patras, Greece
e-mail: bakopoulos@lms.mech.upatras.gr

K. Alexopoulos
e-mail: kalexopoulos@dind.uoa.gr; alexokos@lms.mech.upatras.gr

K. Sipsas
Netcompany-Intrasoft, Luxembourg City, Luxembourg

K. Alexopoulos
Digital Industry Technologies, National and Kapodistrian University of Athens, Psachna, Greece

M. Manns (ed.), *New Paradigms for Anticipated Uncertainty*, Lecture Notes in Mechanical Engineering, https://doi.org/10.1007/978-3-032-16889-4_7

new complexities and vulnerabilities within supply chains. In this evolving landscape, resilience has become a critical requirement [2]. Modern manufacturing systems must not only withstand disruptions but also rapidly adapt to changing conditions, ensuring continuity and performance [3]. Resilience in manufacturing refers to the capacity of a system to maintain or quickly recover its desired functionality in the face of internal or external disturbances [4]. This includes the ability to reconfigure operations, reroute supply chains, and make data-driven decisions in real time. A key enabler for such resilience is trusted, sovereign, and secure data exchange between stakeholders. Data spaces, such as the International Data Spaces (IDS) and Gaia-X, provide the necessary infrastructure for secure, interoperable, and standardized data sharing across organizational boundaries [5, 6]. This paper addresses the following research question: *How can an integrated data space architecture combining IDS and Gaia-X support resilient manufacturing through a compute-on-demand environment?*

To answer the research question, this work proposes a Data Spaces Framework for Resilience (DSF4R), which leverages the architectural strengths of both IDS and Gaia-X to facilitate a compute-on-demand service execution without requiring sensitive data to leave its original environment. Conceptually, the use of data spaces improves resilience by reducing the coordination burden during disruptions. When supply chain stakeholders are connected through a standardized and trusted data-sharing environment, response times can be significantly shortened. For instance, a raw material supplier can securely communicate stock delays through the data space infrastructure, enabling downstream partners to reconfigure plans proactively and efficiently. The DSF4R is validated through a case study in the steel manufacturing sector, demonstrating its practical value in improving the resilience and adaptability of manufacturing systems. The framework demonstrates the feasibility of integrating two data space technologies in a compute-on-demand environment, providing a foundation for more flexible and robust industrial ecosystems.

2 Literature Review

Resilience in manufacturing is critical for the long-term sustainability of production systems, especially in uncertain and volatile environments. It involves the capacity to prepare for, absorb, recover from, and adapt to disruptions [3, 7, 8]. Alexopoulos et al. [3] propose a quantitative framework to assess manufacturing system resilience using the Penalty of Change (PoC) methodology, allowing for the comparison of alternative plans based on disruption impact. Zhang and Van Luttervelt [4] introduced the foundational concept of resilient manufacturing systems, focusing on system-level robustness. Liu and Lee [8] highlight the importance of integration in third-party logistics for enhancing service resilience. Ivanov and Dolgui [7] contribute operational research methods for managing ripple effects in supply chains during crises. Conceptual frameworks further refine resilience understanding. Dubey et al. [9] emphasize agility, adaptability, and alignment in supply chains, while Hohenstein et al. [10] outline a four-phase model of resilience (readiness, response, recovery, and growth). Srinivasan and Swink [11] argue that visibility and flexibility are essential complements to analytics in disruption handling. Chiffoleau et al. [12] illustrates resilience through the coexistence of supply chains in urban food systems. Azadegan et al. [13] empirically assesses how firms navigate supply chain disruptions, stressing business continuity. Fowler et al. [14] propose a framework for resilience at macro (network and supply chain), meso (plant and production system), and micro (machine and devices) levels. Romero et al. [15] link production innovation to resilience by emphasizing the integration of digital tools and reconfigurable systems.

To enable operational resilience, secure and governed data sharing across organizations has become a priority. Data spaces, particularly the International Data Spaces (IDS) and Gaia-X, are emerging as foundational frameworks that enable trusted, interoperable, and sovereign data collaboration. Guo et al. [5] defines key technologies and architecture for Industrial Data Spaces, proposing how smart manufacturing systems can leverage them for data-driven resilience. Bakopoulos et al. [6] extend this by integrating Digital Twins with data spaces to support reconfiguration decisions. Otto and Jarke [16], and Otto [17] articulate the evolution of IDS and its role as a multi-sided platform for policy-enforced data exchange. Otto [18] and the DSSC Glossary [19] describe the governance and identity components required for trustworthy data spaces. Gaia-X complements IDS by extending trust to the infrastructure level. Otto et al. [20] describe Gaia-X architecture components, such as Self-Descriptions and Compliance Services, which enforce usage policies. Farmakis et al. [21] explore how data spaces can support circular and resilient information systems by linking service orchestration to data governance. Despite this progress, the joint integration of IDS and Gaia-X remains underexplored. Most contributions treat them as parallel ecosystems, without demonstrating their combined application in real-world industrial settings. Existing studies rarely describe compute-on-demand environments or cross-framework orchestration mechanisms. This paper addresses this critical gap by introducing the Data Spaces Framework for Resilience (DSF4R), by integrating IDS and Gaia-X within a compute-on-demand architecture, enabling secure, policy-compliant service execution on both internal (e.g., ERP, AAS) and external data.

3 Methodology

This work proposes a Data Spaces Framework for Resilience (DSF4R) that enables secure, sovereign, and efficient collaboration among manufacturing stakeholders by integrating two data space technologies. The DSF4R establishes interoperability between the key architectural components of IDS and Gaia-X, where central to this framework is the concept of a compute-on-demand environment, where services are executed on data that remains within the data provider's infrastructure. However, services are not limited to internal data only, but they can also access, and process external data assets retrieved via the IDS data space. The objective of the framework is to allow services using data that may originate from both within the company and from external supply chain partners, without compromising data sovereignty or exposing sensitive assets. The architecture is built upon the IDS Reference Architecture Model (IDS-RAM) for secure data exchange, while utilizing Gaia-X components to enable compliance verification within a federated, trusted environment. The IDS enables organizations to share data under defined policies and usage constraints, enforced via IDS Connectors and supported by core services such as the IDS Broker, which is a service for publishing and searching metadata of Connectors and resources between IDS participants. The communication between an IDS Connector and an IDS Broker is based on the same principles as any other Connector-to-Connector communication within the IDS architecture. Moreover, the IDS Clearing House is an implementation that stores data space transactions in an encrypted and immutable record. Furthermore, the Clearing House component ensures transparency and accountability by logging usage data, supporting clearing and settlement based on usage contracts, enabling billing services, and validating usage claims in line with predefined governance policies. Furthermore, the Gaia-X data space enables federated trust through the use of verifiable credentials, identity management, and compliance verification. Participants and services in the DSF4R ecosystem are described using Gaia-X Self-Descriptions and are published in Federated Catalogs, allowing for structured discovery and assessment. Gaia-X Core Services, including the Notary, Registry, and Compliance Services, collectively referred to as Gaia-X Clearing Houses (GXDCH), validate whether data exchange and service execution conform to established policies. This combination of IDS governance and Gaia-X trust services ensures that services are only executed when both technical and organizational compliance is verified, across participants. This trust and certification layer is particularly important in heterogeneous environments where participants span different organizational and regulatory contexts (Fig. 1).

The compute-on-demand environment serves as the bridge between the two data space technologies, where services can access and process data originating both from internal systems (i.e., Asset Administration Shell middleware, ERP, local databases, or other data sources) and from assets retrieved via the IDS data space. This approach functions as the execution layer for third-party or internal services. Unlike traditional centralized processing, this model ensures that services are executed in environments governed by the data holder or a certified, trusted provider. The compute-on-demand environment can be deployed either locally within the organization's infrastructure or externally in trusted cloud platforms. By enabling seamless execution on hybrid data (i.e., internal and external), the compute-on-demand environment bridges the IDS and Gaia-X data spaces and makes their integration actionable.

4 Case Study

This case study examines the application of the DSF4R framework within a steel manufacturing group operating fifteen facilities across multiple countries. The company produces a wide range of steel products for sectors such as construction

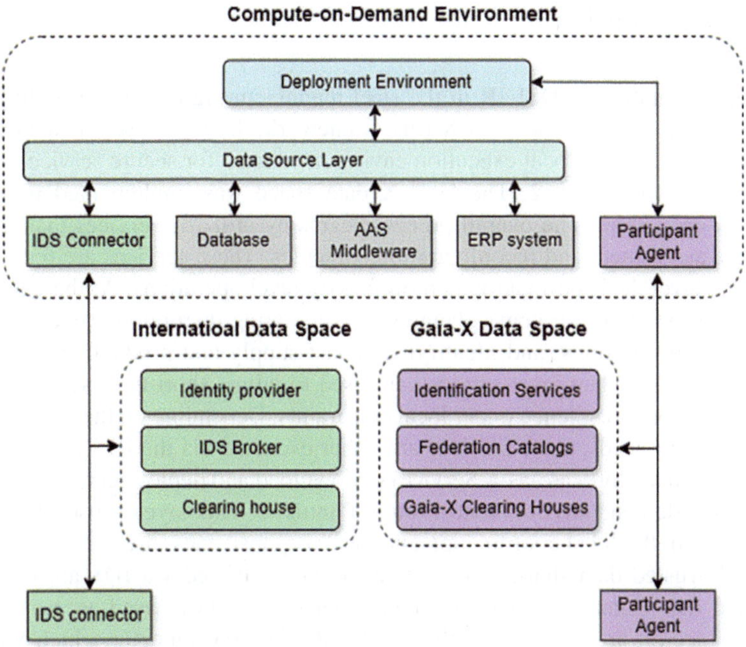

Fig. 1 Data spaces framework for resilience

and automotive, and its operations are exposed to frequent disruptions caused by market volatility, logistics bottlenecks, and evolving regulatory constraints. This industrial setting was selected due to its complex, multi-site structure and high sensitivity to both internal and external disruptions. The organization relies on centralized data collection through an ERP system but planning and scheduling are performed manually and independently at each plant. This fragmentation introduces delays and limits the company's ability to respond effectively to disruptions. At the same time, the highly confidential nature of its production and demand data restricts the use of conventional external services. The DSF4R framework provides a solution by allowing secure and governed access to external supply chain data via IDS, while enabling the use of advanced services (such as scheduling and resilience assessments) to be executed within a local compute-on-demand environment. This ensures that sensitive data remains within the steel manufacturing group's internal network and are not externally exposed.

5 Implementation

The implementation of DSF4R in the steel manufacturing case study is based on the interaction between the Gaia-X trust framework, the IDS-based data exchange infrastructure, and the local execution environment used for secure service deployment, as shown in Fig. 2. The Gaia-X data space was implemented using the Pontus-X ecosystem. The overall process begins by utilizing services that support resilience assessment and reconfiguration strategies. These services are focused on enterprise-level decision-making and on factory-level operations. At the enterprise level, services enable strategic planning by simulating disruption scenarios, evaluating recovery options, and identifying robust supply and production configurations. In parallel, plant-level services are used to adjust short-term schedules and assess operational resilience under local constraints. Depending on the data location and processing needs, these services are either executed via the IDS infrastructure (when offered as Software-as-a-Service) or download and deploy services in a local compute-on-demand environment through Pontus-X. This layered use of services ensures that both global coordination and local responsiveness are supported within a unified, trusted data sharing ecosystem. Services offered via IDS are deployed to trusted IDS Connectors and executed under strict policy-based access control. These connectors are integrated with the FIWARE TRUE Connector, which enforces data usage policies, while all data transactions are logged by the IDS Clearing House to ensure auditability and compliance. Identity management within the IDS ecosystem is handled by the Fraunhofer AISEC omejdn-server, which issues self-signed X.509 certificates. These certificates authenticate participants and enable secure data exchange between connectors while preserving data sovereignty. In parallel, services requiring localized execution or access to a combination of internal and external data are delivered through the Pontus-X platform. Pontus-X provides compute-on-demand environments, where service configuration files and metadata are transferred and securely executed close to the data. Execution is orchestrated using Ocean Protocol components such as the Ocean Enterprise Connector, while identity and compliance are governed by Gaia-X principles. Participant and service identities are represented using verifiable credentials and Self-Descriptions, which are validated by Gaia-X Core Services, including the Gaia-X Clearing Houses (GXDCH), to ensure conformance with established trust and usage policies.

The Master Production Scheduler (MPS) is deployed as an IDS service and generates a supply chain plan by accessing centralized ERP data. This plan is then assessed by a Supply Chain Resilience Assessment service, also on IDS, using the Penalty of Change (PoC) methodology [3]. After validation, the plan is segmented and distributed to the group's production facilities through the IDS data space. Each facility computes a local production schedule tailored to its constraints and availability using a Production planning and Production scheduling services downloaded and deployed in a local environment via Nautilus, a TypeScript wrapper for Ocean.js, and executed as Docker containers within secure compute-on-demand environments near the data. A Factory-Level Resilience Assessment is performed after schedule

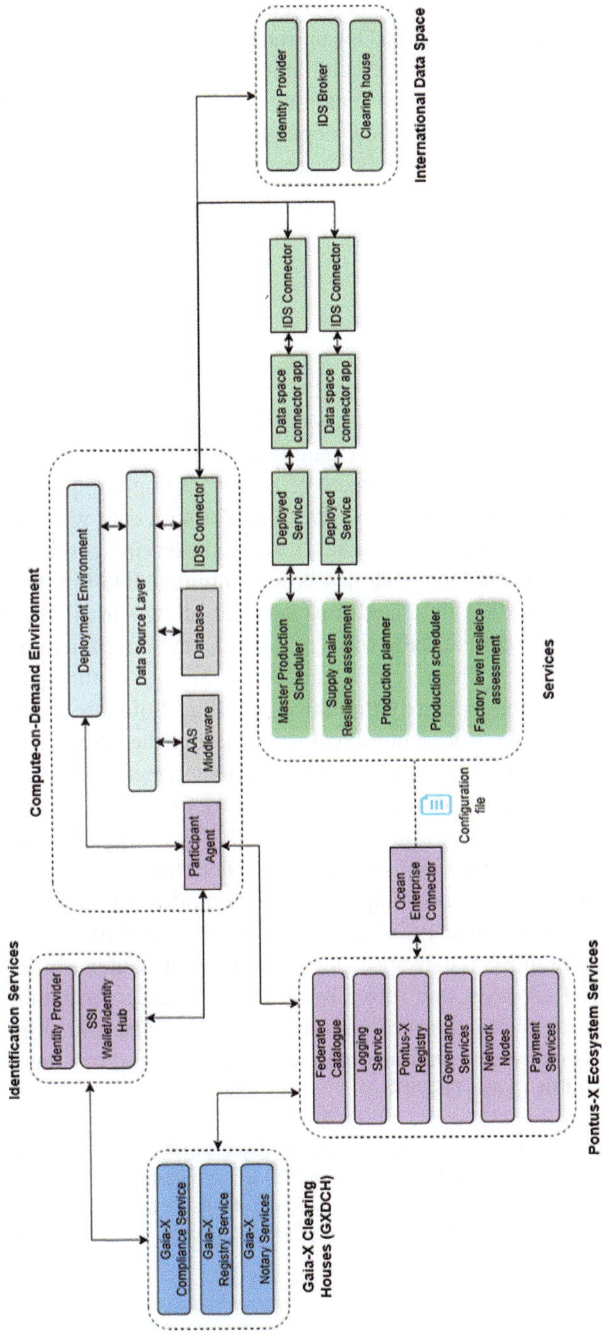

Fig. 2 DSF4R implementation—steel manufacturing industry

creation, using the same PoC methodology to consider local risks such as equipment failure or material shortages. These services are also executed within Pontus-X environments with governance by the Ocean Enterprise Connector. By coordinating planning, simulation, and distributed execution through a secure data-sharing infrastructure, DSF4R enhances resilience by enabling rapid reconfiguration, secure cross-organizational data exchange, and local deployment of services without compromising data sovereignty. The case study illustrates how combining IDS and Gaia-X technologies supports sovereign collaboration and operational adaptability across manufacturing networks.

6 Conclusion

This work presented the Data Spaces Framework for Resilience (DSF4R), which integrates IDS and Gaia-X to support resilient manufacturing through a secure, sovereign, and interoperable data-sharing infrastructure. By leveraging compute-on-demand environments, DSF4R enables the execution of services on data without requiring its transfer across organizational boundaries, thereby preserving confidentiality and compliance. The framework demonstrates how combining IDS's policy-enforced data exchange with Gaia-X's federated trust services allows distributed manufacturing actors to access external data, run simulations, and coordinate actions in a secure and verifiable way. The research question is addressed through a layered deployment of services that operate either through IDS as Software-as-a-Service or via compute-on-demand environments downloaded through the Pontus-X ecosystem. The framework was validated through its architectural implementation in a real-world steel manufacturing use case, showcasing its capacity to coordinate cross-facility scheduling. This implementation highlights the feasibility and practical applicability of DSF4R across heterogeneous industrial systems. Nevertheless, limitations persist, including the challenge of harmonizing diverse data models and enforcing unified trust and identity policies across jurisdictions. Future work will focus on expanding DSF4R to integrate additional data space technologies (e.g., Catena-X), applying it to broader industrial contexts, and formalizing its resilience contributions through quantitative evaluation. While the current study emphasizes infrastructure-level integration and operational orchestration, future research will explore measuring the qualitative impact of utilizing DSF4R.

Acknowledgements This work is under the framework of Horizon Europe Flex4Res project. This project has received funding from the European Union's Horizon research and innovation program under grant agreement No 101091903. The dissemination of results herein reflects only the authors' view, and the Commission is not responsible for any use that may be made of the information it contains.

References

1. Chryssolouris G, Alexopoulos K, Arkouli Z (2023) A perspective on artificial intelligence in manufacturing, vol 436. Springer, Cham, pp 1–135
2. Woods DD (2015) Four concepts for resilience and the implications for the future of resilience engineering. Reliab Eng Syst Saf 141:5–9
3. Alexopoulos K, Anagiannis I, Nikolakis N, Chryssolouris G (2022) A quantitative approach to resilience in manufacturing systems. Int J Prod Res 60(24):7178–7193
4. Zhang WJ, Van Luttervelt CA (2011) Toward a resilient manufacturing system. CIRP Ann 60(1):469–472
5. Guo J, Cheng Y, Wang D, Tao F, Pickl S (2023) Industrial dataspace for smart manufacturing: connotation, key technologies, and framework. Int J Prod Res 61(12):3868–3883
6. Bakopoulos E, Sipsas K, Nikolakis N, Alexopoulos K (2024) A digital twin and data spaces framework towards resilient manufacturing value chains. IFAC-PapersOnLine 58(19):163–168
7. Ivanov D, Dolgui A (2021) OR-methods for coping with the ripple effect in supply chains during COVID-19 pandemic: managerial insights and research implications. Int J Prod Econ 232:107921
8. Liu CL, Lee MY (2018) Integration, supply chain resilience, and service performance in third-party logistics providers. Int J Logist Manage 29(1):5–21
9. Dubey R, Altay N, Gunasekaran A, Blome C, Papadopoulos T, Childe SJ (2018) Supply chain agility, adaptability and alignment: empirical evidence from the Indian auto components industry. Int J Oper Prod Manage 38(1):129–148
10. Hohenstein NO, Feisel E, Hartmann E, Giunipero L (2015) Research on the phenomenon of supply chain resilience: a systematic review and paths for further investigation. Int J Phys Distrib Logist Manage 45(1/2):90–117
11. Srinivasan R, Swink M (2018) An investigation of visibility and flexibility as complements to supply chain analytics: an organizational information processing theory perspective. Prod Oper Manage 27(10):1849–1867
12. Chiffoleau Y, Brit AC, Monnier M, Akermann G, Lenormand M, Saucède F (2020) Coexistence of supply chains in a city's food supply: a factor for resilience? Rev Agric Food Environ Stud 101:391–414
13. Azadegan A, Mellat Parast M, Lucianetti L, Nishant R, Blackhurst J (2020) Supply chain disruptions and business continuity: an empirical assessment. Decis Sci 51(1):38–73
14. Fowler DS, Epiphaniou G, Higgins MD, Maple C (2023) Aspects of resilience for smart manufacturing systems. Strat Chang 32(6):183–193
15. Romero D, Stahre J, Larsson L, Rönnbäck AÖ (2021) Building manufacturing resilience through production innovation. In: 2021 IEEE international conference on engineering, technology and innovation (ICE/ITMC). IEEE, pp 1–9
16. Otto B, Jarke M (2019) Designing a multi-sided data platform: findings from the international data spaces case. Electron Mark 29(4):561–580
17. Otto B (2022) The evolution of data spaces. In: Designing data spaces: the ecosystem approach to competitive advantage, pp 3–15. Springer, Cham
18. Otto B (2022) A federated infrastructure for European data spaces. Commun ACM 65(4):44–45
19. Data Spaces Support Center: DSSC Glossary, Version 2.0 (2023). https://dssc.eu/space/Glossary/176553985/DSSC+Glossary. Accessed 22 June 2025
20. Otto B, Rubina A, Eitel A, Teuscher A, Schleimer AM, Lange C, Stingl D et al (2021) GAIA-X and IDS. International Data Spaces Association
21. Farmakis T, Koukopoulos A, Zois G, Mourtos I, Lounis S, Kalaboukas K (2024) Developing a circular and resilient information system: a design science approach. In: IFIP international conference on advances in production management systems, pp 64–79. Springer, Cham

Analysis of Human Action Variability for Reconfiguring Tooling Systems

Syed Muhammad Raza, Tadele Belay Tuli, and Martin Manns

Abstract Reconfiguration of machine tooling components in industrial production systems is highly dependent on the expertise of human operators. This expertise is developed over time through the experience of skilled workers. A significant challenge is the absence of a predefined set of actions and their sequences. Moreover, due to variability in user behavior, structured tasks are often performed with diverse patterns and sequences that differ across users and contexts. This work analyzes the variability in reconfiguration patterns caused by differences in action sequences of an operator, by employing an action prediction model that annotates action and sequence semantically. The findings are evaluated based on experts' action duration and sequence. Results highlight the potential to identify skillsets for reconfiguring machine tooling in forming processes use-case, paving a way for systems to enhance their productivity on the shopfloor.

Keywords Machine tooling reconfiguration · Human motion capture · Eye gaze tracker · Human action · Operator 5.0

1 Introduction

In the present industrial landscape, production systems are desired to be adaptable and flexible in order to address global turbulences and rapid market demands [1]. In response to the growing demand for manufacturing systems capable of swiftly adapting to dynamic market conditions and evolving product requirements, the concept of reconfigurable manufacturing systems (RMS) has emerged during the final decade of the twentieth century [2, 3]. An RMS is defined by its capability to rapidly adjust its hardware, software, or both in response to shifting market demands. This adaptability is strengthened by key characteristics such as diagnosability, convertibility, flexibility, integrability, and modularity [4].

S. M. Raza (✉) · T. B. Tuli · M. Manns
PROTECH—Institut für Produktionstechnik, Universität Siegen, Siegen, Germany
e-mail: syed.raza@uni-siegen.de

M. Manns (ed.), *New Paradigms for Anticipated Uncertainty*, Lecture Notes in Mechanical Engineering, https://doi.org/10.1007/978-3-032-16889-4_8

Reconfiguration can be initiated by both internal and external disruptions, including machine failures, variability in task times, queuing delays, product rejections, rework processes, fluctuations in demand or pricing, resource scarcity, and broader economic factors. Since each disruption requires a customized response, identifying the appropriate reconfiguration steps tailored to the specific triggering event is critical for effective system adaptation [5].

The reconfiguration of machine tooling components within industrial production systems needs specialized operator expertise, which is typically accumulated over time through the practical knowledge of skilled personnel [6]. It is frequently observed that standardized reconfiguration procedures and their sequential execution are often lacking thus resulting in variations across different operators [7].

2 State of the Art

Numerous studies are undertaken to investigate human-centric manufacturing operations, examining worker interactions, behavioral patterns, operational sequences, and procedural actions in both assembly processes and machine tooling reconfiguration tasks [8, 9]. The variability observed in human actions and operational sequences during assembly tasks and machine tooling reconfiguration is influenced by multiple factors, including task complexity and the adaptive capabilities of both human operators and manufacturing systems [10]. This variability is established primarily in temporal differences in execution time and kinematic variations in physical actions or movements [11]. Such fluctuations may arise from differences in operator skill levels, cognitive processing of complex tasks, or the flexible nature of reconfigurable equipment [12, 13].

Through sensor fusion technique varied intentions of operators are identified and activity sequences are predicted by integrating body, finger, and eye trackers data [14]. Moreover, with multimodal sensor fusion approach reliability of captured data to recognize varied actions in assembly manufacturing context is enhanced [15]. Complementing this, action variability through an automated clustering method based on probabilistic modeling of spatial motion distributions are addressed and computational approaches to characterize temporal fluctuations in work sequences and spatial variations in movement patterns in industrial settings are presented [16].

Influence of temporal variation of operators on the overall performance of the assembly system is assessed by employing discrete event simulation model of an assembly system [17]. Further, within the context of automotive production systems, action variability in reach motions during assembly operations are investigated and gait patterns in manufacturing settings are assessed providing insights into human motion dynamics and variations [18, 19].

In previous work, operational variability through human action sequence prediction for unstructured reconfiguration tasks are examined in the context of reconfiguration of machine tool components [7]. Findings reveal significant variations in action duration, frequency, and sequencing patterns, which are influenced by both task-specific requirements and individual operator factors including intention and attention levels. It highlights the inherent behavioral variability in human–machine interactions during complex reconfiguration processes.

3 Objective

This study investigates variations in operator actions during structured tasks associated with reconfiguration of machine tooling components. The research employs a semantic action prediction model to analyze variability in reconfiguration patterns arising from differences in operator action sequences. Through quantitative and qualitative evaluation of duration and frequency, the work assesses the impact of variations on manufacturing performance.

Acknowledging action-sequence variability and integrating it with assistive technologies may optimize reconfiguration processes in human–machine collaborative environments.

4 Method

The proposed methodology for the analysis of human action variability during the reconfiguration of machine tooling components includes:

- Motion capture using Inertial Measurement Units (IMUs) and preprocessing
- Semantic definition of actions required for reconfiguring tooling systems used in forming processes
- Semantic labeling and manual keyframe annotation
- Filtering, frequency counting, and keyframe extraction
- Computation of duration of action and overall mean for each action class
- Duration and frequency/count analysis using both qualitative (box plot) and quantitative (table) measures
- Identification of a skillset using radar chart.

5 Experimental Setup

Motion capture is performed with an XSENS Awinda IMU sensor array operating at 60 Hz sampling frequency. Seventeen inertial measurement units are strategically positioned on key anatomical landmarks, including the pelvis, upper and lower extremities, hands, feet, thoracic region, and cephalic segment. This configuration yielded a comprehensive dataset of approximately 778,626 motion-capture frames.

Table 1 Action classes mean frame length

Actionclasses	Semantics	Mean (μ)			Count or frequency		
		01	02	03	01	02	03
ET	Eject tool	122	120	120	1	10	26
IT	Insert tool	100	179	191	1	10	28
PB	Push button	494	412	385	2	19	53
PiT	Pick tool	189	1329	418	1	9	27
PlTW	Place tool	240	153	148	1	10	26
RWA	Rotate wheel anticlockwise	80	59	57	1	10	26
RWC	Rotate wheel clockwise	166	119	113	1	6	19
SD	Screw die	436	410	269	1	9	20
TS	Tighten screw with wrench	425	5614	2948	1	6	28
USD	Unscrew die	N/A	8222	1125	N/A	2	7
UTS	Untighten screw with wrench	N/A	339	12,906	N/A	1	27

Visual attention patterns are recorded using Pupil Invisible wearable eye-tracking technology to capture operator intentions during task execution. A Panasonic HC-X1 4K camcorder provided complementary scene documentation. Post-processing involved three specialized software platforms: manual annotation and motion data analysis with MVN Analyze, gaze pattern processing via Pupil Invisible Companion, and scene analysis through VLC Media Player.

The experimental protocol is in a controlled laboratory environment featuring 50 reconfiguration trials of a Zwick Roell Z250 materials testing machine, producing 414,740 motion frames. Through systematic analysis, eleven discrete action primitives are identified and semantically categorized in Table 1.

6 Results

The results from the test scenarios, achieved by capturing motions and intentions of an expert in three session datasets, sample 01, 02 and 03, are depicted in Table 1 and Fig. 1. Each session dataset contains 9–12 action classes that vary in duration and frequency. The sample 01 is recorded with a single take whereas samples 02 and 03 are recorded with multiple takes of the same reconfiguration task. According to Table 1, samples 02 and 03 show that the time duration for USD, UTS, and TS actions is significantly higher than for other actions since it requires finger movements, in tightening and untightening the screw, in addition to the hand and wrist movements that are out of scope of this work. Thus, for the sake of visualization, these actions are excluded in Fig. 1. The RWA action exhibits the minimum time duration because the rotation of yoke wheel in counterclockwise is followed once the hydraulics is released, thus requiring less effort in loosening the machine tool. The SD action exhibits maximum time duration due to the complexity of screwing die inside the tool, since it requires precision for the placement of screw and the wrench. Apparently, sample 01 shows the significance of the action duration in the absence of USD, UTS, and TS. The analysis given in Table 1. In this study, the primary focus is on the mean values of the action classes across the datasets of human motion capture, as well as the skillset, which are collected at different times but involve similar actions. The count of occurrences for each action class in the respective datasets significantly shows the temporal and frequency variability of human actions in each take.

Moreover, the action classes such as ET, IT, PB, PiT, PlTW, RWA, RWC, SD, and TS are consistently present across all datasets. This is due to the structured reconfiguration strategy applied, as recommended by the expert. In this section, unlike previous work presented in [7], which found it difficult to find similarities in motion sequences, current work focuses on structured action sequences. This consistency allows for a comparative analysis of their mean values and counts across different samples.

There is a noticeable variation in the mean values of the action classes across the datasets. For instance, the mean value for PiT in Sample 02 is significantly higher

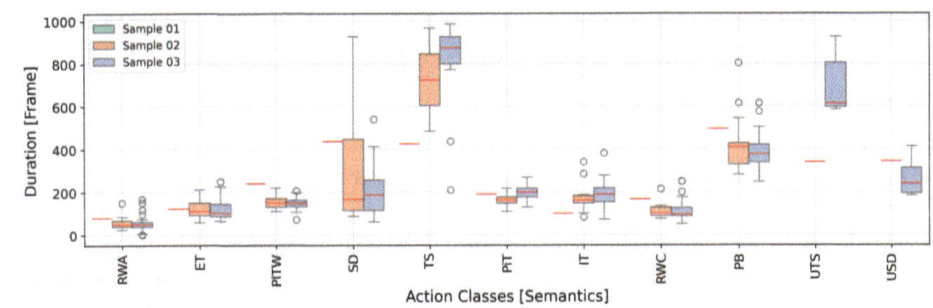

Sequence Pattern	Frequency
PB → RWA → ET	33
TS → PiT → IT	32
PiT → IT → PB	32
ET → PlTW → UTS	27
SD → TS → PiT	24
IT → PB → RWC	19
RWC → PB → RWA	17
PB → RWC → PB	16
PlTW → UTS → SD	14

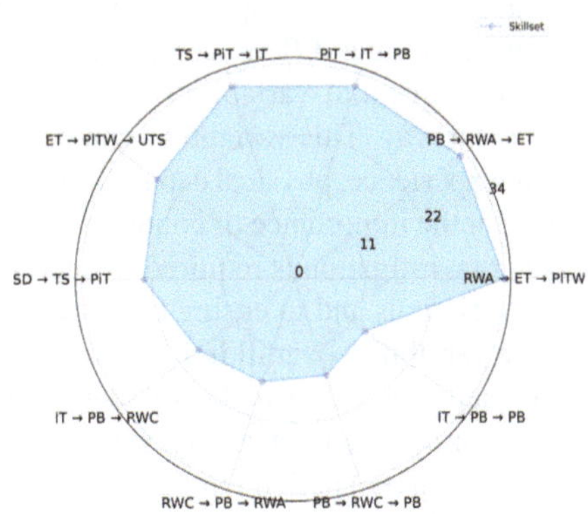

Fig. 1 Variability of human action duration (top) and overall sequence radar chart (bottom-right) supplemented with a sequence pattern and frequency table for common sequence patterns repeated more than ten (bottom-left) during the reconfiguration operation of machine tooling components on the shop floor

(1329) compared to the other datasets, indicating a potential outlier or a different context of measurement. It is due to the placement of machine components in different positions on the workbench, adding the overall time to reach and pick the tool. The action classes USD and UTS are unique to Samples 02 and 03, with no corresponding entries in Sample 01. This uniqueness suggests that these action classes were either not observed or not recorded in Sample 01. The count of occurrences for each action class varies significantly across the datasets. For example, PB has a count of 2 in Sample 01 but 53 in Sample 03, indicating a broader observation or recording in the latter datasets. It is due to two reasons. The push buttons are pressed to release and restore the hydraulics. Since sample 01 is recorded for a single take, the push button action is recorded only twice. Also, in sample 02 and sample 03, the frequency of push button action is irregular. It is because the amount of hydraulics to release or restore varies thus pressing multiple times at some point reaching the desired goal.

High-frequency sequences (e.g., PB → RWA → ET, RWA → ET → PlTW) depicted in Fig. 1 suggest strong procedural structures within the session. The lower frequency for some sequences (e.g., PB → RWC → PB, RWC → PB → RWA) shows action sequential variability that may result from workers diverging more at certain transitions, possibly due to task complexity, flexibility in methods, or errors.

7 Discussion

The comparative analysis of action classes across multiple datasets reveals significant variations in mean values and frequencies of the actions. The results show that human factors play a significant role in the assembly process, affecting both the speed and accuracy of assembly tasks. For example, placement of machine component facing upwards or sidewards impact the cycle times, leading to temporal variability in human actions. Also, the differences in mean values of action classes across the datasets indicate sequential variability in how an operator may perform the same task in a different way. This variability can be attributed to factors such as individual skill levels, experience, physical capabilities, and worker situations. These differences also highlight the importance of context and methodology in data collection and analysis. Further investigation is required to understand the underlying factors contributing to these variations and to ensure consistency in future studies.

The variability overall in human actions during the assembly of machine tooling components is a critical factor that influences the efficiency and effectiveness of reconfigurable manufacturing systems (RMS). This study highlights several key aspects of human action temporal and sequential variability and its implications for the assembly process, particularly in the context of reconfiguring machine tooling components. For instance, how distant the component is placed on the workbench from the operator increases the reach time affecting efficiency.

Reconfigurable machine tooling components are designed to adapt to changing production requirements. The ability to quickly reconfigure these components is essential for maintaining productivity in a dynamic manufacturing environment.

However, the variability in human actions overall can pose challenges to the reconfiguration process. Ensuring adequate training for operators and standardized that assembly procedures can mitigate these challenges.

The study of operator behavior is crucial for optimizing assembly performance while performing reconfiguration. Digital human models (DHMs) and AI-assisted monitoring systems can help in understanding and mitigating the impact of human behavioral variability. By simulating operator behavior and continuously monitoring assembly operations, it is possible to identify bottlenecks and optimize the assembly process. This approach can lead to more consistent and efficient assembly operations, reducing the impact of human variability overall.

Further research is required to explore the integration of advanced technologies such as machine learning and computer vision in monitoring and optimizing human actions during reconfiguration. These technologies can provide real-time feedback and support to operators, enhancing their performance and reducing variability overall. Additionally, the development of more intuitive and user-friendly interfaces can contribute to improved reconfiguration efficiency.

8 Conclusion and Outlook

The temporal, frequency and sequential variability in human actions in reconfiguring tooling systems of machines is complex thus requires a multifaceted approach. By leveraging digital human models, and AI-assisted monitoring, it is possible to enhance the efficiency and consistency of assembly operations. Future research should focus on exploring new methods to integrate these technologies to support operators in a reconfigurable manufacturing environment and analyzing intentional deviations during reconfiguration tasks.

Acknowledgements The authors would like to acknowledge financial support from Horizon Europe within FLEX4RES project (Grant Agreement Number 101091903).

References

1. Alexopoulos K, Weber M, Trautner T, Manns M, Nikolakis N, Weigold M, Engel B (2023) An industrial data-spaces framework for resilient manufacturing value chains. Procedia CIRP 116:299–304
2. Koren Y (2020) The emergence of reconfigurable manufacturing systems (RMSs). In: Benyoucef L (ed) Reconfigurable manufacturing systems: from design to implementation. Springer International Publishing, Cham, pp 1–9
3. Koren Y (2022) The invention of reconfigurable manufacturing systems—a survey of RMS early patents. Manuf Lett 33:51–53
4. Koren Y (2010) The global manufacturing revolution. Wiley
5. Caesar B, Tilbury DM, Barton K, Fay A (2022) Requirements for reconfiguration management for manufacturing systems. IFAC-PapersOnLine 55(10):3214–3219

6. Choudhari SJ, Singh SA, Kumar AS, Desai KA (2022) Machine setup abnormality detection using machine vision and deep learning. In: Volume 1: additive manufacturing; biomanufacturing; life cycle engineering; manufacturing equipment and automation; nano/micro/meso manufacturing. ASME 2022 17th international manufacturing science and engineering conference, West Lafayette, Indiana, USA. 27.06.2022–01.07.2022. American Society of Mechanical Engineers
7. Raza SM, Tuli TB, Manns M (2024) Human action sequence prediction for (re)configuring machine tools. Procedia CIRP 130:1170–1175
8. Fabio G, Margherita P, Luca Z, Marcello P (2019) An automatic procedure based on virtual ergonomic analysis to promote human-centric manufacturing. Procedia Manuf 38:488–496
9. Krugh M, Garimella RS, Baburaj A, Wescoat E, Mears L (2020) Closed loop feedback mechanism effect pilot investigation on manual assembly time and process variation. Procedia Manuf 48:95–104
10. Torres Y, Nadeau S, Landau K (2021) Classification and quantification of human error in manufacturing: a case study in complex manual assembly. Appl Sci 11(2):749
11. Butturi M, Lolli F, Menini C (2020) Balancing of manual reconfigurable assembly systems with learning and forgetting effects. In: Proceedings of the international conference on innovative intelligent industrial production and logistics, Budapest, Hungary. 02.11.2020–04.11.2020. SCITEPRESS—Science and Technology Publications, pp 124–131
12. Bagnara S, Tartaglia R, Albolino S, Alexander T, Fujita Y (eds) (2019) Proceedings of the 20th congress of the international ergonomics association (IEA 2018). Springer International Publishing, Cham
13. Gilles MA, Gaudez C, Savin J, Remy A, Remy O, Wild P (2019) How age and pace of work affect movement variability during repetitive assembly tasks. In: Bagnara S, Tartaglia R, Albolino S, Alexander T, Fujita Y (eds) Proceedings of the 20th congress of the international ergonomics association (IEA 2018), vol 826. Springer International Publishing, Cham, pp 16–22
14. Manns M, Tuli TB, Schreiber F (2021) Identifying human intention during assembly operations using wearable motion capturing systems including eye focus. Procedia CIRP 104:924–929
15. Al-Amin M, Tao W, Doell D, Lingard R, Yin Z, Leu MC, Qin R (2019) Action recognition in manufacturing assembly using multimodal sensor fusion. Procedia Manuf 39:158–167
16. Tuli TB, Manns M (2023) Explainable human activity recognition based on probabilistic spatial partitions for symbiotic workplaces. Int J Comput Integr Manuf 36(12):1783–1800
17. Peças P, Semeano J (2019) Industrial-based time performance of assembly cell operators: impact assessment characterization by discrete-event simulation. Int J Math Eng Manage Sci 5(1):29–44
18. Agethen P, Gaisbauer F, Manns M, Link M, Rukzio E (2018) Towards realistic walk path simulation of single subjects. In: Proceedings of the 11th annual international conference on motion, interaction, and games. MIG '18: motion, interaction and games, Limassol Cyprus. 08.11.2018–10.11.2018. ACM, New York, NY, USA, pp. 1–10
19. Manns M, Fischer K, Du H, Slusallek P, Alexopoulos K (2018) A new approach to plan manual assembly. Int J Comput Integr Manuf 31(9):907–920

Time-Domain Analysis of Human Motion for Reconfiguring Machine Tools

Syed Muhammad Raza, Tadele Belay Tuli, and Martin Manns

Abstract Traditionally, in industrial production systems, the reconfiguration of machinery relies on the expertise of human operators. This expertise is accumulated over time as the worker gains experience. The actions and intentions of experienced operators can be recorded with motion capture technologies such as body and eye-gaze trackers. This data can potentially support inexperienced or less proficient workers in acquiring knowledge to perform similar tasks by simulating the actions of the experts. This work presents a method to capture and analyze the motions of an expert operator while reconfiguring machine tools of the forming machine. The findings reveal that actions associated with reconfiguring machine tools for forming press machine can be recognized and segmented by applying time-domain analysis on the motion data. The results show potential of developing a knowledge-based assistance system that encapsulates the expertise of operators and supports novice operators to replicate learned actions effectively in order to enhance productivity on the shop floor.

Keywords Machine tool reconfiguration · Motion capture · Eye gaze tracker · Human action recognition · Operator 5.0

1 Introduction

The evolving market demands, reduced production batch sizes, and increasing product diversity compel contemporary manufacturing systems to be reconfigurable and support flexible and resilient production processes [1]. By rapid machine adjustments, reconfiguration minimizes downtime and improves production efficiency, thus enhancing capability of manufacturing system to maintain continuous production even during disruptions [2].

The higher number of components in a machine increases its overall complexity, which can contribute to a greater risk of failures and defects, potentially resulting

S. M. Raza (✉) · T. B. Tuli · M. Manns
PROTECH—Institut für Produktionstechnik, Universität Siegen, Siegen, Germany
e-mail: syed.raza@uni-siegen.de

M. Manns (ed.), *New Paradigms for Anticipated Uncertainty*, Lecture Notes in Mechanical Engineering, https://doi.org/10.1007/978-3-032-16889-4_9

in increased downtime [3]. Machine-level reconfiguration is a complex process that requires a certain level of expertise and specialized skill sets, typically acquired through hands-on experience. The reconfiguration process varies depending on the specific operation and the proficiency of an expert, thus establishing a standardized procedure remains a challenge [4]. Inefficient reconfiguration leads to an increase in non-value-adding activities and compromises the operational resilience of the manufacturing systems [5]. Maintaining a knowledge database, comprising of actions and intentions of an expert, is essential to simulate and guide a novice operator to perform similar reconfiguration tasks on the shop floor [6].

2 State of the Art

The concept of reconfigurability in manufacturing systems was initiated in the last decade of the twentieth century in a response to the need for manufacturing systems that could rapidly adapt to changing market conditions and product demands [7, 8]. Defined by Koren [9], the key feature of a reconfigurable manufacturing system is its ability to quickly adjust software, hardware, or both components in response to market changes. This reconfigurability is characterized by diagnosability, convertibility, flexibility, integrability, and modularity [9].

Reconfiguration can be triggered from internal as well as external disturbances, such as machine breakdown, variable task times, changes in demand, product prices, scarcity of resources, and economic factors etc. Each type of disruption requires a tailored solution, therefore identifying the necessary steps to respond to a specific reconfiguration trigger is crucial [4].

Various reconfiguration strategies are proposed, and studies are conducted to explore their application in the development of different types of reconfigurable machines, including assembly machines, machining tools, and inspection systems. However, challenges in reconfiguring machine tools arise mostly due to system design [10]. Virtual interactive prototyping approach can support design engineers in decision making by providing an appropriate reconfiguration method for reconfigurable machine tools [11].

Modular research platforms, such as simulator software, user interface and human operator physiology sensors, are found to be the key components to design, test and understand human interactions with the machine tools [12]. Several studies analyzed and showed how humans perform reconfiguration tasks related to assembly or disassembly based on human motion behaviors in industrial and laboratory environments [13, 14]. Sensor fusion techniques allow identification of operators' intentions and prediction of activity sequences by integrating data from body, finger, and eye trackers [15, 16]. Furthermore, multimodal sensor fusion enhances the reliability of captured data to recognize human action in assembly manufacturing [17].

3 Objective

Based on the complexity of reconfiguring machine tools, this study proposes a method to capture and characterize expert competencies involved in reconfiguring machine tools with a particular focus on forming technologies. The research evaluates if the duration of the actions can be modelled with Gaussian distribution. This method can be advantageous in maintaining expert knowledge database that may facilitate effective knowledge transfer from experts to novice operators, assist shopfloor operators to identify optimal reconfiguration strategies and execute reconfiguration tasks accordingly.

4 Method

The proposed method follows a four-step structured process, as illustrated in Fig. 1. Each step incrementally builds toward a fully developed method for capturing and analyzing expert motions for the tasks associated with machine tool reconfiguration within forming technologies.

The first step involves selecting appropriate technologies to capture various aspects of expert behavior in task specific context. The technologies include (i) Inertial Measurement Unit (IMU) sensors to record body movements and gestures to recognize actions related to reconfiguration tasks, (ii) Gaze-tracking sensors to infer the intention of the experts while reconfiguring the machine tool, and (iii) Video camera to document the complete task environment and interactions, providing contextual understanding of the reconfiguration workflow.

The selected technologies are integrated through a sensor fusion approach. The multimodal data streams from IMU sensors, gaze sensors, and video recordings are synchronized and combined to provide holistic representation of user activity, intention, and task context. This integration ensures temporal alignment and supports accurate interpretation of the reconfiguration process.

Following the integration of sensing technologies, expert operators are recorded while performing various reconfiguration tasks on forming machines. The empirical

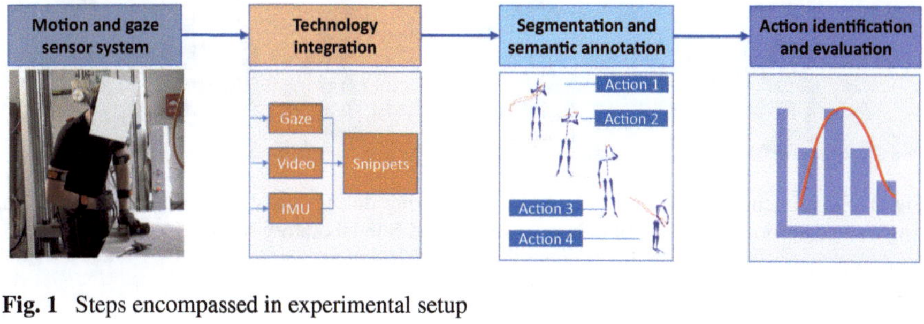

Fig. 1 Steps encompassed in experimental setup

data provides procedural information for the reconfiguration tasks through motion trajectories, gaze patterns, and environmental video recordings of an expert.

The collected data is manually annotated and segmented to identify meaningful actions, decision points, and task sequences. Further, the data is modelled to fine tune the annotation and segmentation to extract implicit knowledge related to reconfiguration activities.

The annotated and segmented data acquired is evaluated by a time-based model. The model analyzes the temporal distribution of actions that may support in finding dependencies between task segments, assessing efficiency, identifying bottlenecks, optimizing the reconfiguration strategies and validating its reproducibility for the development of knowledge-base and operator guiding system.

5 Experimental Setup

Figure 2 illustrates the experimental setup for capturing human actions using XSENS Awinda IMU sensors at 60 Hz. Seventeen sensors, anchored at the hips, are distributed across major body segments including the pelvis, limbs, feet, hands, sternum, and head. In total, approximately 778,626 data points are recorded.

Eye-tracking data is collected using Pupil Invisible eye-tracking glasses to gain deeper insights into the intention of the expert during task execution. A Panasonic HC-X1 video camera records the entire scene. Video data is processed with VLC Player, eye-tracking data with Pupil Invisible Companion, and motion data with MVN Analyze software.

Data is collected in two settings. In the industrial environment, an expert with ten years of experience reconfigures a machine tool for a multi-stage progressive forming press, yielding 363,886 frames over 55 repetitions (see Fig. 2a). In the laboratory, 414,740 frames are captured over 50 repetitions during reconfiguration of a Zwick Roell Z250 universal testing machine (see Fig. 2b).

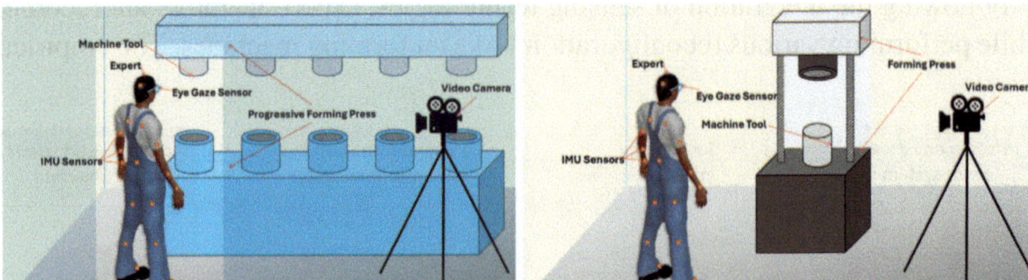

Fig. 2 Experimental setup for human motion capturing during press tool reconfiguration for a single stage in forming press **a** in industrial setting and **b** in laboratory setting

6 Result

The proposed experimental setup identifies 7 distinct actions during the reconfiguration of one of the stages of multi-stage forming press machine. Combining these actions forms a procedure necessary to reconfigure the machine tools of a forming press. The recognized actions are semantically defined as insert tool (IT), place tool on workbench (PlTW), screw the die inside tool (SD), untighten screw (UTS) using an Allen key, untighten screw manually (UTSM), tighten screw manually (TSM), and tighten screw with the Allen key (TS).

The statistical analysis shown in Fig. 3 reveals that simpler operations, such as tool insertion (IT) and placement (PlTW), exhibit low temporal variability across participants. In contrast, complex mechanical operations like screwing the die (SD) and untightening screws (UTS) show significant variability, with UTS operations sometimes resulting in extreme delays (max: 56,093 frames). These findings suggest that certain mechanical steps present substantial operator-dependent challenges, potentially serving as performance bottlenecks during machine reconfiguration tasks. In Fig. 4, fine-grained actions such as tightening (TS) and untightening screws (UTS) displayed greater temporal variability compared to untighten screw manually (UTSM) and tighten screw manually (TSM). While detailed screw operations fluctuated based on tool handling or task complexity, the associated body motions are performed with higher consistency, suggesting a higher degree of procedural learning for macro-movements.

In order to test the normality of the Gaussian fitted action durations, the D'Agostino and Pearson normality test [18] is applied to two datasets (lab and industrial environment). As shown in Table 1, seven out of eight actions deviate from a

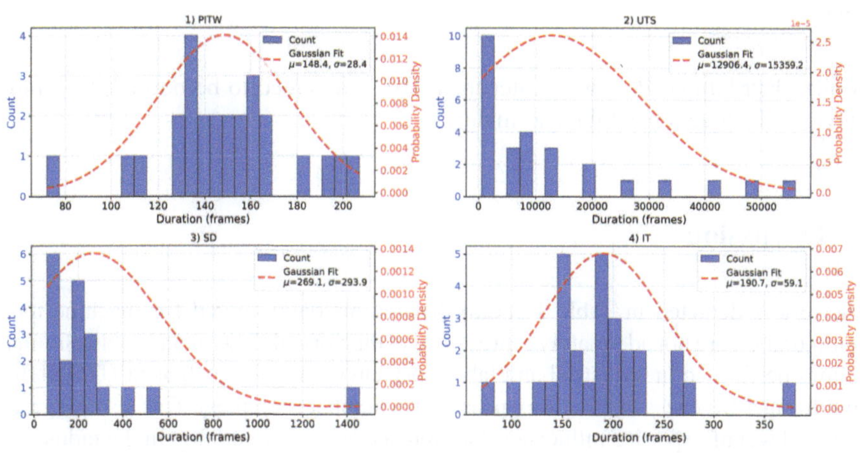

Fig. 3 Test at lab environment for a non-experienced participant action duration: **a** place tool on workbench (PlTW), **b** untighten screw (UTS) using Allen key, **c** screw the die inside tool (SD) and **d** insert tool (IT)

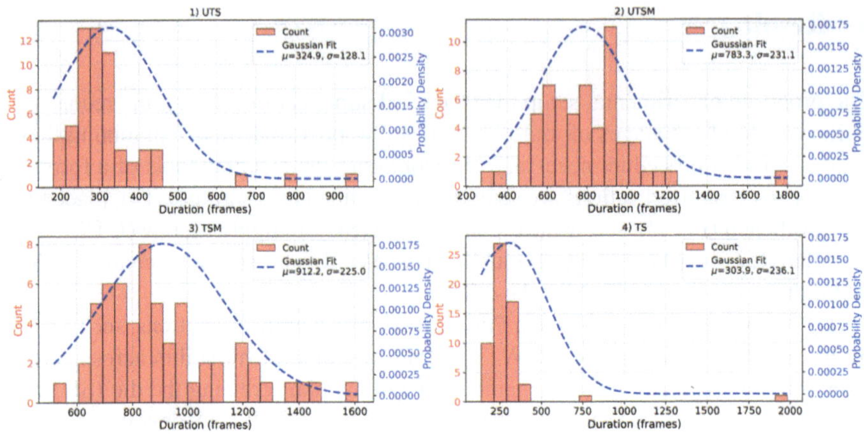

Fig. 4 Test at industrial environment for experienced expert's action duration: **a** untighten screw (UTS) using an Allen key, **b** untighten screw manually (USM), **c** tighten screw manually (TSM), and tighten screw with the Allen key (TS)

Table 1 Table normality test for Gaussian fitted action duration

Industrial environment				Lab environment			
Action	μ	σ	*p*-value	Action	μ	σ	*p*-value
UTS	324.90	128.09	0.000	PITW	148.38	28.35	0.369*
UTSM	783.28	231.05	0.000	UTS	12,906.37	15,359.17	0.002
TSM	912.20	224.97	0.003	SD	269.15	293.91	0.000
TS	303.86	236.10	0.000	IT	190.68	59.13	0.007

*Normal

normal distribution. *P*-value greater than 0.05 is assumed to be normal, therefore only the PITW action exhibits normality.

7 Discussion

The results depicted in Table 1 reveal that the actions required for machine tool configuration are not adequately represented using normal Gaussian models. Similar actions performed in industrial and laboratory environments, such as at UTS, show significant differences in duration (measured in number of frames). This also suggests that the level of expertise influences duration, as the reconfiguration in the industrial setting (Fig. 4) has been conducted by an expert with over 10 years of experience, in contrast to the lab-based reconfiguration (Fig. 3). Furthermore, the results indicate that uncertainties arising from the condition of the machine tool can make the process unpredictable. For example, some screws may be challenging to loosen due to dirt or oil buildup. Additionally, the style of tool usage differs between novice and expert workers, further complicating the modeling of actions using Gaussian distributions. Nevertheless, the method presented in this study demonstrates the feasibility of systematically capturing and analyzing expert actions involved in the reconfiguration of forming machine tools. Implementation of the proposed setup in both industrial and laboratory settings shows generalizability across different forming machines along with its reproducibility for both structured and unstructured tasks cf. [16]. However, further research is required to extend the time model and validate the applicability of this method across different machine types and task complexities to confirm its versatility and adaptability.

Normal analysis reveals that reconfiguration of machine tools associated with the forming process is complex, therefore operators with higher level of expertise are required for effective reconfiguration. This may lead to a necessity of AI driven human assistance system to support operators in performing reconfiguration tasks on the shopfloor. It opens pathways for integrating expert knowledge into digital platforms, such as virtual training environments and decision support systems.

Integration of motion capture through IMU sensors, intention tracking via gaze sensor, and contextual observation through video recordings provides a multimodal representation of expert behavior. Further, annotation and segmentation of task sequences allow the identification of distinct actions and strategies executed by experienced operators. However, the manual annotation and segmentation process is subject to human interpretation, which introduces potential irregularity in the segmentation and labeling of tasks, affecting the consistency of action identification across different annotators or scenarios. Integration of semi-automated or AI-assisted annotation tools may enhance scalability and reduce subjectivity.

8 Conclusion and Outlook

Overall, the proposed method identifies actions and sequences to perform complex reconfiguration tasks associated with forming machine tools. These actions and sequences provide a foundation to maintain a knowledge database that can transfer expert knowledge to novice operators through a real-time operator assistance system for reconfiguring machine tools, which is intended to be developed in future work.

Acknowledgements The authors would like to acknowledge financial support from Horizon Europe within the FLEX4RES project (Grant Agreement Number 101091903).

References

1. Alexopoulos K, Weber M, Trautner T et al (2023) An industrial data-spaces framework for resilient manufacturing value chains. Procedia CIRP 116:299–304. https://doi.org/10.1016/j.procir.2023.02.051
2. Bruccoleri M, Pasek ZJ, Koren Y (2006) Operation management in reconfigurable manufacturing systems: reconfiguration for error handling. Int J Prod Econ 100:87–100. https://doi.org/10.1016/j.ijpe.2004.10.009
3. Samy SN, ElMaraghy H (2010) A model for measuring products assembly complexity. Int J Comput Integr Manuf 23:1015–1027. https://doi.org/10.1080/0951192X.2010.511652
4. Caesar B, Tilbury DM, Barton K et al (2022) Requirements for reconfiguration management for manufacturing systems. IFAC-PapersOnLine 55:3214–3219. https://doi.org/10.1016/j.ifacol.2022.10.147
5. Gu X, Jin X, Ni J et al (2015) Manufacturing system design for resilience. Procedia CIRP 36:135–140. https://doi.org/10.1016/j.procir.2015.02.075
6. Hoerner L, Schamberger M, Bodendorf F (2023) Using tacit expert knowledge to support shop-floor operators through a knowledge-based assistance system. Comput Support Coop Work 32:55–91. https://doi.org/10.1007/s10606-022-09445-4
7. Koren Y (2022) The invention of reconfigurable manufacturing systems—a survey of RMS early patents. Manuf Lett 33:51–53. https://doi.org/10.1016/j.mfglet.2022.07.007
8. Koren Y (2020) The emergence of reconfigurable manufacturing systems (RMSs). In: Benyoucef L (ed) Reconfigurable manufacturing systems: from design to implementation. Springer International Publishing, Cham, pp 1–9
9. Koren Y (2010) The global manufacturing revolution. Wiley
10. Gadalla M, Xue D (2017) Recent advances in research on reconfigurable machine tools: a literature review. Int J Prod Res 55:1440–1454. https://doi.org/10.1080/00207543.2016.1237795
11. Marin FB, Constantin IC, Marinescu V et al (2015) Machine tool configuration and analysis on the base of virtual interactive prototyping. Int J Interact Des Manuf 9:77–82. https://doi.org/10.1007/s12008-014-0222-7
12. Hatlem LA, Chen J, Dybvik H et al (2020) A modular research platform—proof-of-concept of a flexible experiment setup developed for rapid testing of simulators, UIs and human physiology sensors. Procedia CIRP 91:407–414. https://doi.org/10.1016/j.procir.2020.02.193
13. Manns M, Fischer K, Du H et al (2018) A new approach to plan manual assembly. Int J Comput Integr Manuf 31:907–920. https://doi.org/10.1080/0951192X.2018.1466396
14. Agethen P, Gaisbauer F, Manns M et al (2018) Towards realistic walk path simulation of single subjects. In: Charalambous P, Chrysanthou Y, Jones B et al (eds) Proceedings of the 11th annual international conference on motion, interaction, and games. ACM, New York, NY, USA, pp 1–10
15. Manns M, Tuli TB, Schreiber F (2021) Identifying human intention during assembly operations using wearable motion capturing systems including eye focus. Procedia CIRP 104:924–929. https://doi.org/10.1016/j.procir.2021.11.155
16. Raza SM, Tuli TB, Manns M (2024) Human action sequence prediction for (re)configuring machine tools. Procedia CIRP 130:1170–1175. https://doi.org/10.1016/j.procir.2024.10.223
17. Al-Amin M, Tao W, Doell D et al (2019) Action recognition in manufacturing assembly using multimodal sensor fusion. Procedia Manuf 39:158–167. https://doi.org/10.1016/j.promfg.2020.01.288
18. Breton MD, Devore MD, Brown DE (2008) A tool for systematically comparing the power of tests for normality. J Stat Comput Simul 78:623–638. https://doi.org/10.1080/00949650701377984

A Digital Twin and Dataspace Framework for Resilient and Reconfigurable Manufacturing

Syed Muhammad Raza, Maximilian Schmidt, Adane Kassa Shikur, Bernd Engel, and Martin Manns

Abstract Uncertainties and disruptions across value chains, from supply chains to shopfloor resources, significantly impact production systems. Addressing these challenges requires manufacturing systems that are both resilient and reconfigurable to maintain robust and flexible operations. This work presents a conceptual framework to improve the resilience and reconfiguration capabilities of manufacturing systems at the machine level. The resilience quantification and reconfiguration support are enabled by dataspaces and the Asset Administration Shell (AAS). The goal is to facilitate shopfloor workers in measuring resilience and implementing reconfiguration strategies effectively in response to disruptions. This framework is designed with the focus on progressive forming press machines potentially supporting digital manufacturing. The findings reveal the possibility for production systems to be adaptable and robust. Future work will focus on implementation and evaluation of the proposed framework and scaling it to other manufacturing processes.

Keywords Digital manufacturing · Digital twin (DT) · Human assistance system (HAS) · Industry 5.0 · Operator 5.0

1 Introduction

Contemporary industries face significant challenges in delivering high quality products at accelerated production rates while minimizing non value adding activities [1]. Dealing with growing complexity and shorter lifecycles of products, and variability in customer demands stipulate the industries to be more flexible and robust [2, 3]. Global disruptions, such as the COVID-19 pandemic and natural disasters, including flooding in Central and Eastern Europe, have further emphasized the necessity for manufacturing systems to respond swiftly to changing conditions and volatile market demands [4, 5]. Therefore, the manufacturing sector needs to be resilient and reconfigurable to address rapid fluctuations in product demand across complete value chain

S. M. Raza (✉) · M. Schmidt · A. K. Shikur · B. Engel · M. Manns
PROTECH—Institut für Produktionstechnik, Universität Siegen, Siegen, Germany
e-mail: syed.raza@uni-siegen.de

M. Manns (ed.), *New Paradigms for Anticipated Uncertainty*, Lecture Notes in Mechanical Engineering, https://doi.org/10.1007/978-3-032-16889-4_10

[6]. Changes in raw materials may require reconfiguration at the machine level to ensure production continuity and sustain product quality.

Flex4Res project aims to enhance flexibility and reconfiguration in manufacturing by enabling seamless integration and real-time data exchange across the entire value chain through asset administration shell (AAS) and dataspaces, fostering resilient and adaptable industry in dynamic environments [7, 8].

2 State of the Art

Resilience is generally known as the adeptness of a system to withstand by absorbing changes of state variables, driving variables and parameters. However it is conceptualized differently in different domains [9]. In the manufacturing domain, resilience is defined as the capability of a manufacturing system or process to withstand disruptions, fluctuations, and unanticipated events while sustaining or regaining its desired functionality and performance swiftly with negligible additional expenses [10, 11]. Particularly for the machine level, resilience may be interpreted as the ability of individual resources, such as machines, workers and processes, to withstand, adapt to, and recover from disruptions within a manufacturing environment so as to ensure continuity in production without compromising overall shopfloor performance.

Sensors allow real-time monitoring of key operational parameters and detecting faults at an early stage which may enhance resilience at machine level [12]. Incorporating sensors within machine tools support operational resilience by continuous condition monitoring of the tool [13]. Laser metal deposition (LMD) is an additive manufacturing technology enables sensor integration inside existing machine tools closely beneath functional surfaces [14–16].

Reconfiguration is an attribute of manufacturing system that allows adjustment of manufacturing elements in order to address environmental and technological changes and demand fluctuations [17]. Six main characteristics that govern reconfigurability of manufacturing system are modularity, integrability, customization, convertibility, scalability and diagnosability [18]. Machine level reconfiguration refers to the capability of manufacturing machines to modify their structure and functionality in order to adapt evolving production requirements [19]. Machine level reconfiguration enhances the operational flexibility of manufacturing systems which is crucial to cope with uncertainties and change with minimal penalty of time, effort, cost and performance [20].

Both the initiatives, AAS and dataspaces have progressed as standardized frameworks and conceptual models that facilitate digital transformation across various industries while aligning with the objectives of the Industry 4.0 revolution [21, 22]. The AAS serves as the digital representation of physical assets, facilitating interoperability and lifecycle management within the industrial digitalized ecosystem, enabling service exchange and aiming to standardize digital twins (DTs) for industrial assets while adhering safety and security [23]. The AASX Package Explorer serves as an enabler for modeling and managing AAS with a graphical interface for creating, editing, and validating AAS data structures [22]. It supports the integration of DTs by allowing users to create and embed submodels in compliance with Industrial Digital Twin Association (IDTA) specifications.

Similarly, dataspaces enable organizations to create an ecosystem for securely sharing and managing data while maintaining sovereignty by retaining control over access and usage rights, thereby fostering a trustworthy data exchange environment [24]. Gaia-X is a European initiative that establishes a federated and secure data infrastructure, enabling sovereign data exchange and interoperability across digital services and platforms. Pontus-X is Gaia-X compliant ecosystem for the exchange of data, software, and infrastructure services securely while ensuring data sovereignty and the protection of intellectual property (IP) [25].

A framework implementing digital twin (DT) through the integration of international data spaces (IDS) and the AAS, along with resilience assessment and reconfiguration services developed using service oriented architecture (SOA) approach is proposed in previous work [10]. The framework primarily addresses the supply chain and enterprise levels of manufacturing. In contrast, the present work focuses on the machine level of manufacturing systems.

3 Objective

Based on the FLEX4RES approach [8], this work presents a conceptual framework to improve resilience and support reconfiguration within the manufacturing sector at the machine level. The framework encompasses the integration of resilience assessment and reconfiguration toolboxes through dataspaces, while a DT is implemented through the AAS.

4 Conceptual Framework

The proposed conceptual framework is depicted in Fig. 1. The framework comprises resilience and reconfiguration toolboxes. Resilience services will be triggered to evaluate system resilience either manually or automatically following the reconfiguration of machine tooling components. The resilience toolbox will play a supervisory role by measuring performance indicators continuously at certain intervals and prompting reconfiguration actions as needed. In response, the reconfiguration toolbox will determine the root cause, identify reconfiguration strategy, and support in performing the necessary reconfiguration tasks to preserve operational continuity. After the completion of reconfiguration, resilience will be reassessed to confirm system robustness. A workflow diagram for the complete setup is illustrated in Fig. 2.

Resilience will be quantitatively assessed using a probabilistic metric, formulated as the sum of the probability associated with successful prevention, prediction, and

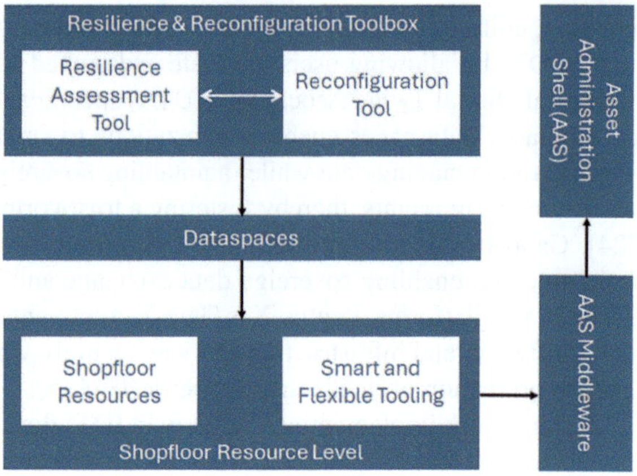

Fig. 1 System overview diagram

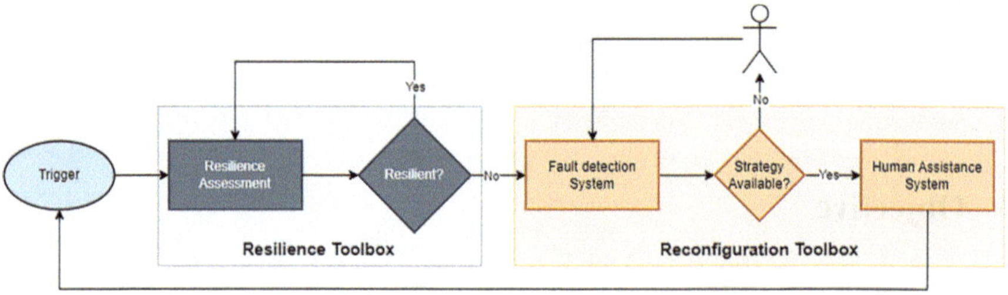

Fig. 2 Flow diagram of resilience assessment and reconfiguration toolbox for machine level

restoration mechanisms, each scored as one and zero upon success and failure respectively. To model and evaluate resilience at the machine level, a Bayesian network will be employed, enabling the representation of causal relationships and uncertainty inherent in machine-level behaviors and responses.

Reconfiguration toolbox consists of two components. Fault detection system based on fuzzy case-based reasoning (Fuzzy-CBR) will be developed to support resilience driven decision making. Sensor and machine data from the shop floor and resilience indicators with linguistic variables (e.g., low, medium and high) will be linked to Fuzzy logics. The fuzzy sets and their associated membership functions will help manage uncertainty by allowing variables to partially belong to multiple linguistic categories (e.g., a variable might be 46% medium). These linguistic variables will then be used to formulate rules by comparing current situation with the case repository. The fault detection system will present the operator with the most relevant case built upon the CBR cycle and propose a reconfiguration strategy.

Operators on the shop floor will be supported by a human assistance system during the reconfiguration process. The human assistance system will comprise of a knowledge database accumulated from the skilled experts by capturing human intentions, actions and their sequences through gaze and inertial measurement unit (IMU) sensors. The knowledge database will assist novice operators in performing reconfiguration tasks through a user interface. The case will be revised, or a new case will be added to the knowledge base by an expert if a novice operator identifies the proposed reconfiguration ineffective. This iterative process will enhance the ability of a reconfiguration toolbox to detect faults and recommend suitable reconfiguration strategies, while continuously improving its decision-making capabilities over time.

Resilience and reconfiguration toolboxes will be deployed within a dataspace environment as Software-as-a-Service (SaaS) solutions specifically accessible at the machine level of the shopfloor for localized adaptability and real-time interaction with physical assets while maintaining the data sovereignty.

Flexible tooling combined with the integration of sensors at the machine level will enable direct data acquisition from individual machines, thereby enhancing the implementation of resilience and reconfiguration measures at the machine level. Further, to provide a semantic and interoperable interface between the physical shopfloor resources and the digital environment, AAS will serve as DT for the physical assets via an AAS middleware, which will facilitate secure and standardized communication across the system components.

5 Implementation Scenario

The proposed conceptual framework is projected for implementation and evaluation within a steel manufacturing industry producing sheet metal components for the automotive and heating industry. The manufacturing process involves multi-stage progressive forming machines to gradually form complex parts. Variations in critical machine components such as gas spring pressure, tool misalignment, and tool wear may directly influence the product quality. To facilitate resilience assessment of machine components through real-time monitoring of tool condition, LMD is employed to embed sensors inside machine tools [14–16]. Further, machine data is modeled using the AASX Package Explorer in accordance with the IDTA Submodel for Time Series Data (IDTA 02008-1-1).

Historical machine failure data from the shopfloor is used to identify root causes and suggest reconfiguration strategies. This data allows training the fuzzy CBR fault detection system, a key part of the reconfiguration toolbox. The system processes sensor data from the smart tool, machine operational data, and resilience assessment feedback to characterize current conditions. These inputs are matched against a case base, with new cases formed by associating input parameters with fuzzy rules and corresponding membership functions. The case with the highest alignment is then forwarded to the human assistance system to support reconfiguration.

To develop the human assistance system, a sensor fusion approach is employed to capture expert behavior during the execution of reconfiguration tasks [26]. IMU sensors capture expert movements, gaze sensor records intention whereas a video camera documents the full scenario for contextual analysis. A dedicated tool is developed to automatically annotate and segment the collected data, allowing for the analysis of action duration, frequency, and sequential patterns [27]. This data contributes to the development of a knowledge database that serves as a reference for performing reconfiguration tasks. Moreover, a user interface is designed to assist novice operators in executing reconfiguration procedures on a forming press machine.

The resilience assessment and reconfiguration toolboxes are offered as digital services, deployed locally on the shopfloor via Pontus-X.

6 Expected Outcomes, Implications and Limitations

The proposed framework, which integrates resilience assessment and reconfiguration services via dataspace at the machine level, demonstrates potential to enhance the operational resilience and reconfiguration of machine tooling components within manufacturing environments. Embedding sensors within forming tools near functional surfaces with LMD will offer a foundation for continuous condition monitoring of critical parameters essential to assess machine resilience. However, further investigations will be needed to ensure reliable and consistent data acquisition from the sensors embedded inside machine tooling components.

Quantifying resilience through metrics focusing on prevention, prediction, and restoration will enhance the ability of a system to proactively detect and respond to disruptions. The integration of fuzzy CBR based fault detection system will support early identification of machine faults and supports decision making by proposing reconfiguration strategies accordingly. These actions may contribute to reduced unplanned downtime and improve overall reliability by enabling timely reconfiguration. While the approach addresses machine failures, the underlying focus is on process adaptability through reconfiguration of machines, positioning resilience as the ability of system to maintain operational continuity under disruption. Consecutively, integration with a human assistance system may enhance system responsiveness and support real-time adaptability. However, to sustain system performance, it is essential to maintain high quality historical data, a comprehensive case repository, and accurate input data.

The human assistance system is a part of reconfiguration toolbox, aimed at guiding novice operators in performing reconfiguration tasks. Since it is crucial for knowledge base to capture expert actions precisely, gaze sensors capture intention IMU sensors will be selected to capture expert actions due to their ability to record motions without being affected from occlusions in the manufacturing environment. Operator workshops are proposed to enhance adaptability and usability of the selected technology.

Implementing the IDTA Time Series Submodel to digitally represent machine data can offer operators increased visibility into machine status with improved interoperability across system components. Further, hosting resilience and reconfiguration services on the Pontus-X will allow standardized access while maintaining control over sensitive operational data within organizational infrastructure.

Finally, preliminary validation within controlled laboratory environment is recommended prior to implementing the proposed framework in industrial environments to address anticipated challenges with respect to system integration.

7 Conclusion and Outlook

A conceptual framework to enhance resilience and support reconfiguration at the machine level is proposed. Resilience and reconfiguration toolboxes are proposed to be deployed as services on the shopfloor through dataspace while machine assets are proposed to be visualized through AAS. Future work will focus on implementing the framework within a progressive sheet metal forming use-case, with an aim of scaling the framework across diverse manufacturing processes and investigating its acceptance and usability on the shopfloor by human centric approach.

Acknowledgements The authors would like to acknowledge financial support from Horizon Europe within the FLEX4RES project (Grant Agreement Number 101091903).

References

1. Mourtzis D, Angelopoulos J, Panopoulos N (2021) Robust engineering for the design of resilient manufacturing systems. Appl Sci 11(7):3067
2. Fenta EW, Tsegaye AA, Abere AE, Tefera GT (2025) Opportunities in flexible manufacturing systems in the near future. Glob J Flex Syst Manage 26(2):247–267
3. Tolio T, Urgo M, Váncza J (2011) Robust production control against propagation of disruptions. CIRP Ann 60(1):489–492
4. Reiter K, Knittel N, Bachner G, Hochrainer-Stigler S (2022) Barriers and ways forward to climate risk management against indirect effects of natural disasters: a case study on flood risk in Austria. Clim Risk Manage 36:100431
5. Sharma M, Luthra S, Joshi S, Joshi H (2022) Challenges to agile project management during COVID-19 pandemic: an emerging economy perspective. Oper Manage Res 15(1–2):461–474
6. Acevedo Amaya MR, Ortega-Jimenez CH (2024) Revolutionizing supply chains: unleashing reconfigurability and flexibility through industry 4.0 integration. In: Proceedings of the 22nd LACCEI international multi-conference for engineering, education and technology (LACCEI 2024): sustainable engineering for a diverse, equitable, and inclusive future at the service of education, research, and industry for a society 5.0, 17–19 July 2024. Latin American and Caribbean Consortium of Engineering Institutions
7. Data Spaces for Flexible Production Lines and Supply Chains for Resilient Manufacturing (2022)
8. Alexopoulos K, Weber M, Trautner T, Manns M, Nikolakis N, Weigold M, Engel B (2023) An industrial data-spaces framework for resilient manufacturing value chains. Procedia CIRP 116:299–304
9. Xiao L, Cao H (2017) Organizational resilience: the theoretical model and research implication. ITM Web Conf 12:4021
10. Bakopoulos E, Sipsas K, Nikolakis N, Alexopoulos K (2024) A digital twin and data spaces framework towards resilient manufacturing value chains. IFAC-PapersOnLine 58(19):163–168

11. Zhang WJ, van Luttervelt CA (2011) Toward a resilient manufacturing system. CIRP Ann 60(1):469–472
12. Albano M, Lino Ferreira L, Di Orio G, Maló P, Webers G, Jantunen E, Gabilondo I, Viguera M, Papa G (2020) Advanced sensor-based maintenance in real-world exemplary cases. Automatika 61(4):537–553
13. Costa S, Silva F, Campilho R, Pereira T (2020) Guidelines for machine tool sensing and smart manufacturing integration. Procedia Manuf 51:251–257
14. Manns M, Raza SM, Morez D, Schreiber F, Engel B (2024) Embedding optical fiber with laser metal deposition. Prog Addit Manuf 9(6):2013–2016
15. Raza SM, Li Z, Choubey B, Manns M (2025) Embedding electronics inside metal with laser metal deposition. Procedia CIRP 134:301–306
16. Raza SM, Manns M, Choubey B (2024) Laser metal deposition based embedding of optical fibers. Procedia CIRP 130:1739–1744
17. Maganha I, Silva C, Ferreira LMD (2018) Understanding reconfigurability of manufacturing systems: an empirical analysis. J Manuf Syst 48:120–130
18. Prasad D, Jayswal SC (2019) Reconfigurable manufacturing system—a new class of manufacturing system. Manage Prod Eng Rev
19. Goyal KK, Jain PK, Jain M (2011) A novel approach to measure machine reconfigurability in reconfigurable manufacturing system. Ann DAAAM Proc 959–960
20. Pérez Pérez M, Serrano Bedia AM, López Fernández MC (2016) A review of manufacturing flexibility: systematising the concept. Int J Prod Res 54(10):3133–3148
21. Atzori M, Ciaramella A, Diamantini C, Di Martino B, Distefano S, Facchinetti T, Montecchiani F, Nocera A, Ruffo G, Trasarti R (2023) Dataspaces: concepts, architectures and initiatives. In: CEUR workshop proceedings, vol 3606
22. van Erp T, Pedersen FB, Larsen NPL, Lund RB (2023) Industrial digital twin in industry 4.0: enabling service exchange between assets in manufacturing systems. In: Kohl H, Seliger G, Dietrich F (eds) Manufacturing driving circular economy. Springer International Publishing, Cham, pp 567–575
23. Kohl H, Seliger G, Dietrich F (eds) (2023) Manufacturing driving circular economy. Springer International Publishing, Cham
24. Quix C (2025) Toward advanced query processing in dataspaces. In: International workshop on data warehousing and OLAP
25. Gast F, Berchtenbreiter V, Dumss S, Gehrer R, Weißenbach P, Weigold M, Grafinger M (2024) Automatic publication of data to data- and service ecosystems from the shopfloor. Procedia CIRP 130:1589–1597
26. Tuli TB, Manns M (2023) Explainable human activity recognition based on probabilistic spatial partitions for symbiotic workplaces. Int J Comput Integr Manuf 36(12):1783–1800
27. Raza SM, Tuli TB, Manns M (2024) Human action sequence prediction for (re)configuring machine tools. Procedia CIRP 130:1170–1175

Fault Detection and Reconfiguration on Resource Level with Fuzzy-Logic and Case-Based Reasoning

Maximilian Schmidt and Bernd Engel

Abstract Uncertainties and disruptions across value chains, spanning from supply chain to shopfloor at resource level, highly affect production. To ensure robust and flexible production, resilient and reconfigurable manufacturing systems are required. The work presents a service for the resource level, in this case multi-stage forming tools, which recognize fault situations and suggest actions for reconfiguration to the operator in order to react appropriately to the situation. The methods behind the service are fuzzy logic and case-based reasoning. The utilisation of machine and sensor data, in conjunction with expert knowledge, which will be employed in the construction of membership functions, fuzzy rules and case-based-reasoning (CBR), facilitates the early detection, identification and removal of faults in the production process. The performance of the service is quantified using a benchmark. In addition, an outlook is given on the effects of the applied method on robustness and flexibility of the manufacturing system.

Keywords Fuzzy logic · Case-based reasoning · Decision-making · Reconfiguration

1 Introduction

The rapid onset of the Covid-19 pandemic precipitated a swift and pervasive disruption to industrial production. Consequently, bottlenecks have emerged in the availability of labour and material supplies, and product demand has changed. In the mechanical- and automotive engineering sectors, as well as in metalworking industries, 77% of companies reported significant production losses [1]. These uncertainties and disruptions have underscored the necessity for resilient and reconfigurable systems at levels of industrial production [2]. The reconfiguration of production processes at the micro level is a complex task that requires a significant degree of

M. Schmidt (✉) · B. Engel
Forming Technology Siegen (UTS), University of Siegen, Breite Straße 11, 57076 Siegen, Germany
e-mail: maximilian.schmidt@uni-siegen.de

M. Manns (ed.), *New Paradigms for Anticipated Uncertainty*, Lecture Notes in Mechanical Engineering, https://doi.org/10.1007/978-3-032-16889-4_11

expertise in the specific process. This knowledge is known to increase with experience and is found to be strongly dependent on the individual and the complexity of the task [3].

In the context of metal forming, the adjustment of tools represents a complex process whose duration and procedure depend on the individual experience and skills of the worker [4]. This experience is also required for deep cup-shaped products made from cold-rolled sheet, which are produced by multiple deep drawing stages. The deep drawing process is a widely used forming process for manufacturing sheet metal components. The process involves the placement of a sheet on the die, followed by the application of a defined force by the blank holder [5]. Recognizing the need for reconfiguration (e.g. the detection of faults) and the reconfiguration itself of such a system is a complex undertaking. The proposed approach combines the linguistic interpretability of fuzzy logic with the systematic knowledge preservation enabled by case-based reasoning. By using transparent, human-readable rules with a structured case base, the method delivers clear, trustworthy decision rationales while capturing and retaining expert experience for future reconfigurations. This study investigates whether a fuzzy-logic and case-based-reasoning system can detect fault scenarios in deep-drawing with a fulfilment-score > 50% and recommend reconfigurations that strengthen equipment resilience.

2 Objective

This study investigates the applicability of fuzzy logic and case-based reasoning, following the classical CBR cycle [6], in the identification of faults and the assistance of decision making for the reconfiguration of deep drawing processes. The utilization of case generation facilitates the preservation of knowledge possessed by experienced workers, thereby ensuring its availability to less trained workers. Furthermore, the interpretation of the sensor data creates an additional flow of information that supports the experienced worker in finding faults. The objective of this study is threefold: firstly, to demonstrate the applicability of fuzzy sets for the interpretation of sensor data from deep drawing tools; secondly, to demonstrate the creation of cases; and thirdly, to assess the transmission of measures for reconfiguration.

3 Methodology

The procedure for fault detection and decision-making at the resource level using case-based reasoning following the retrieve–reuse–revise–retain cycle [6] and fuzzy logic is structured as follows. Initially, cases that trigger reconfiguration actions are defined. Based on experimental trials, membership functions for the sensor input data are derived, and associated fuzzy rules are formulated, which builds the initial case base. Each case also includes a recommended reconfiguration action to support decision-making.

3.1 *Deep Drawing Process and Reconfiguration Options*

For the demonstrator, five scenarios are presented in Table 1, which results in either a change in process stability or an impact on the quality of the cup (see Fig. 1b) produced. Every case presented is linked with the status of the process, which describes the possible fault, and the measure for reconfiguration.

Table 1 Overview of fault cases and corresponding reconfiguration measures

Case name	Adjustment	Status/reconfiguration need
Normal	None	Working/no reconfiguration
Geometric deviation	Insert a 2.5 mm thick underlay to raise the forming punch	Risk for deviation part quality/yes
Low blank holder force	Reduce the blank holder force to 4 kN	Failing part ejection/yes
Low lubrication	Omission of lubrication, both for the workpiece and the tooling	Increased tool wear/yes
Workpiece misalignment	Insertion of workpiece 1 mm off centric to the tool	Wrong workpiece placement/yes

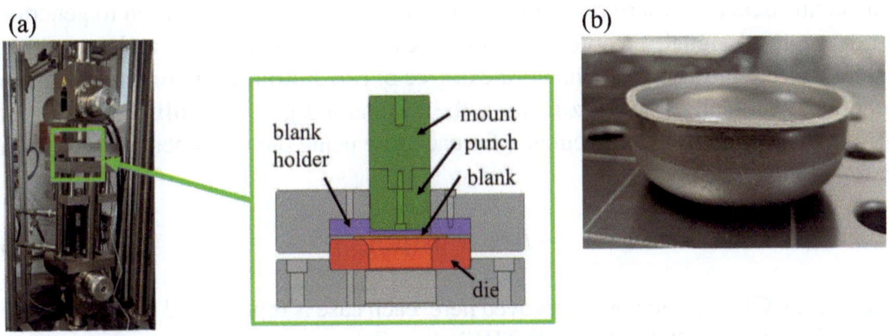

Fig. 1 **a** View of the deep drawing tool setup and a cut view of the used tooling with the blank holder (violet), forming punch (green), inserted workpiece (orange) and the die (red), **b** deep drawn cup

3.2 Fuzzy Logic and Membership Function

This study employs fuzzy logic as an alternative to binary classification for interpreting sensor input data. Fuzzy logic introduces the concept that a value can simultaneously partially belong to and partially not belong to a set at the same time [7]. The membership function μ(x) assigns a degree of membership to a variable x, as described by Eq. (1):

$$\mu = \{(x_1, \mu(x_1)), (x_2, \mu(x_2)), \ldots, (x_n, \mu(x_n))\} \tag{1}$$

Membership functions can take various shapes, including triangular, trapezoidal, and bell-shaped forms. The choice of function depends on expert experience or can be derived through machine learning techniques. These functions are visualized as curves, where the x-axis represents the input values and the y-axis the corresponding membership levels (see Fig. 3).

3.3 Fuzzy Case Based Reasoning for Decision Making

Another application of fuzzy logic in forming processes is the use for process control. In such systems, fuzzy logic processes fuzzy variables derived from precise measurement data to support control strategies (cf. Engel [8]). The characteristic fuzziness arises during the assignment of input variables to their corresponding membership functions (see Eq. 1). Unlike conventional fuzzy controllers, such as Mamdani or min-max controllers [9], where rule fulfilment is followed by defuzzification to generate a crisp control signal, the present study does not aim to produce a sharp output value. Instead, the system evaluates the degree of rule fulfillment purely for decision support. Specifically, after fuzzification, the minimum degree of fulfillment across all membership conditions is calculated for each rule using the min-operator, as shown in Eq. (2):

$$F_i = \mathrm{MIN}(\mu_{ij}) \tag{2}$$

In the fuzzy CBR framework employed here, each case is represented by a fuzzy rule of the form *IF A is B AND C is D THEN E is F*, accompanied by a corresponding reconfiguration recommendation. The full collection of these fuzzy rules constitutes the case base, which serves as the core of the CBR system. The integration of fuzzy logic into classical CBR transforms binary case structures into flexible, continuous models. This enhances the system's ability to handle uncertainty in sensor data and provides more flexible and robust support for decision-making.

4 Experimental Setup

The deep drawing demonstrator in this study was previously utilised set up by Geueke [10]. The demonstrator is mounted on the Universal Testing Machine, designated Zwick/Roell 250 (see Fig. 1a).

The stroke of the demonstrator is controlled via the testing machine. The blank holder is actuated by an external hydraulic system. The signals from the pressure, force and displacement sensors to monitor the deep drawing process is recorded by using the Q.Station X from Gantner Instruments. The tool was designed for 2.75 mm thick blanks made of DC05 (cold-rolled steel) to produce the cup, shown in Fig. 2b. The forming punch has a diameter of 60 mm and a radius of 11 mm, while the die has an internal diameter of 66.7 mm. The blank holder force was set at 16 kN to prevent the formation of creases and to ensure the successful removal of the deep drawn cup from forming punch by the blank holder. In the 'low blank holder force' case, the blank holder force is reduced to 4 kN (see Table 1). All tests were conducted at a constant velocity of 4 mm/s and a stroke of 41 mm. The lubricant employed was Tim-Form GFO 50 at a concentration of 20 g/mm^2. In the Low Lubrication case, the lubricant was not applied to the blank. The results of this study are compared to the normal state in Table 1, which details the other cases and the deliberate changes that were introduced to them. To ensure statistical validity, each case was subjected to 12 independent tests. To rectify the changes in the system, a task list for reconfiguration was determined by an expert and entered for the respective case as reconfiguration task.

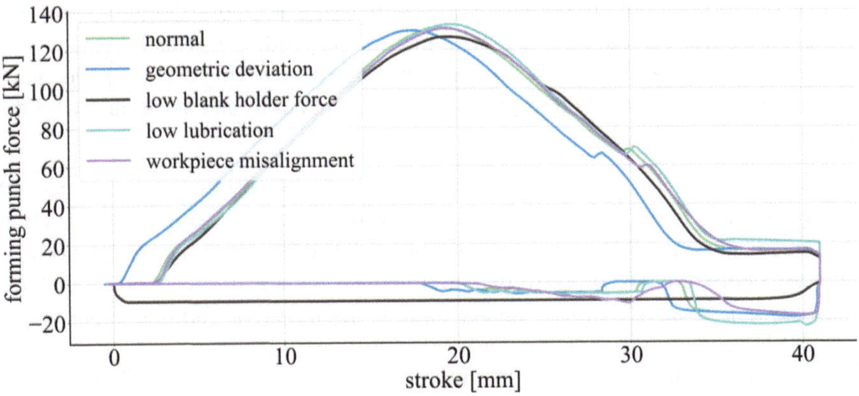

Fig. 2 Measurement of forming punch force over stroke for s single run per case

5 Results

The results of the deep drawing experiments demonstrate that the force–displacement curves within each case exhibit a certain degree of variability. This variability is illustrated in Fig. 2, where six measurement curves per case are plotted, showing the forming punch force as a function of stroke. The curves highlight clear distinctions between the cases. For example, in the "Geometric Deviation" case, an early rise in force is observed, which is attributed to the insertion of an oversized punch.

To construct the case base according to the methodology in Sect. 3.3, characteristic variables were first identified, which enabled the derivation of the membership functions as outlined in Sect. 3.2. Preliminary experiments under constant process parameters revealed low variability (coefficient of variation $\leq 3\%$), indicating that 12 independent runs per case are sufficient to ensure reliable and reproducible results. These variables were extracted from the measurement data of the initial 35 experimental runs (six per case) and include: the begin of forming in mm, the maximum force on the punch in kN, the friction force on the punch after the forming is complete in kN, the initial blank holder force in kN, the average force during retraction between 20 and 10 mm in kN and the position of the second local maximum of the force curve in mm. Trapezoidal membership functions were used to delineate clear regions of uncertainty and to provide plateau zones for enhanced stability, thereby ensuring a balance between sensitivity and robustness. For each characteristic variable, the mean and standard deviation (σ) were calculated from the first six runs per case. The boundaries of the linguistic terms were then set to $\mu \pm 2\sigma$, since this interval covers approximately 95% of the observed measurements, thus providing a good compromise between sensitivity to outliers and robustness against noise. Figure 3 presents the resulting membership functions with their corresponding linguistic terms.

The following fuzzy rules were defined by an expert, having prior knowledge of the experimental setup, based on the membership functions and associated with the five cases from Table 1:

1. Case normal: IF maximum punch force is normal AND backward force is normal AND blank holder force is normal AND forming begin is normal AND friction force is normal AND peak position is normal THEN process status is working
2. Case geometric deviation: IF maximum punch force is normal AND backward force is normal AND blank holder force is normal AND forming begin is late THEN geometric deviation in tooling is negative
3. Case geometric deviation: IF maximum punch force is normal AND backward force is normal AND blank holder force is normal AND forming begin is early THEN geometric deviation in tooling is positive
4. Case workpiece ejection: IF backward force is high AND blank holder force is low AND forming begin is normal AND peak position is no Peak THEN workpiece ejection is failing
5. Case low lubrication: IF blank holder force is normal AND forming begin is normal AND friction force is high AND peak position is late THEN lubrication level is low

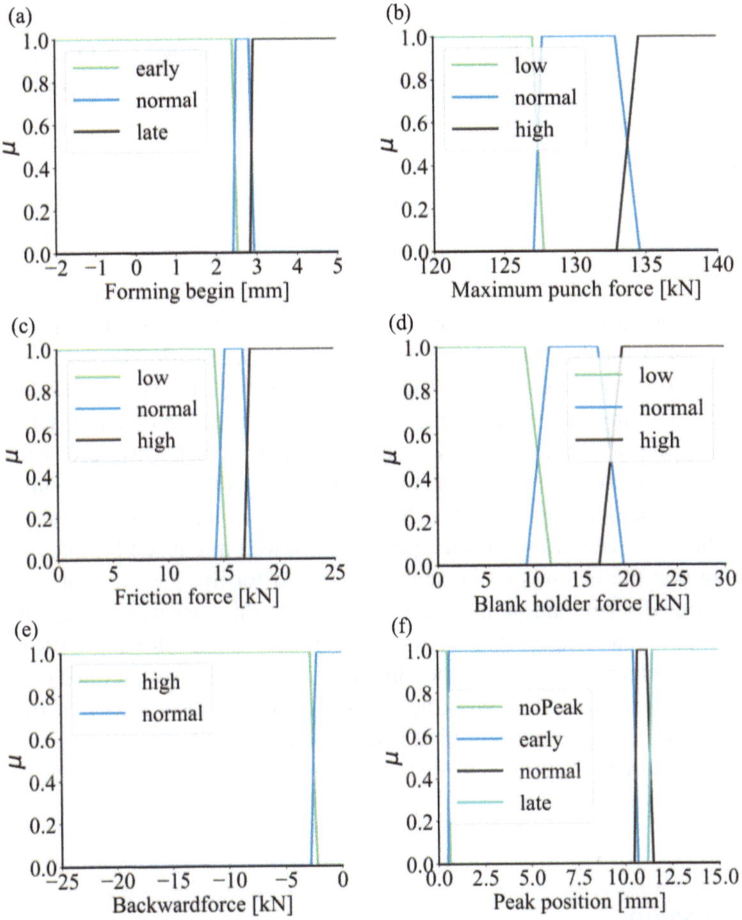

Fig. 3 Membership functions for **a** forming begin, **b** maximum punch force, **c** friction force, **d** blank holder force, **e** backward force and **f** peak position

6. Case workpiece alignment: IF maximum punch force is normal AND backward force is normal AND blank holder force is normal AND peak position is late THEN alignment of workpiece is misaligned.

Following the establishment of the case base, an additional six experimental runs per case were conducted and used as input for the Fuzzy-CBR system. Table 2 summarizes the hit rate, indicating the proportion of correctly identified cases, the average degree of rule fulfilment and the F1-Score to capture the balance between precision and recall for each scenario.

In all test runs for the cases “Normal”, “Geometric Deviation”, and “Low Blank Holder Force”, the system correctly identified the scenario, and the corresponding

Table 2 Results of Fuzzy-CBR in fault detection and decision-making

Case name	Hit rate (%)	Average fulfilment (%)	F1-score (–)
Normal	100	100	0,923
Geometric deviation	100	88.02	1
Low blank holder force	100	100	1
Low lubrication	66.6	53.72	0.8
Workpiece misalignment	83.3	83.3	0.91

reconfiguration tasks were accurately proposed. The Fuzzy-CBR achieved F1-Scores of 0.923 for the Normal case, 1.00 for both Geometric Deviation and Low Blank Holder Force, 0.80 for Low Lubrication, and 0.91 for Workpiece Misalignment, demonstrating robust classification performance across all scenarios. However, for the "Geometric Deviation" case, although identification was successful in all instances, a lower average degree of fulfilment indicates the presence of uncertainty in some of the underlying variables. Nevertheless, this uncertainty did not lead to misclassification. The "Low Lubrication" case showed a reduced hit rate of 66.6%, suggesting that the features in the sensor data used for this case do not offer sufficiently unique discrimination. In the instances where this case was misclassified, the system suggested a different case than the expert decision. Following CBR methodology (cf. Kolodner [6]), the corresponding rule can be iteratively refined to improve future performance. For the "Workpiece Misalignment" case, 83.3% of runs were correctly identified. One misclassified instance was found to be a statistical outlier, where the measured values lay outside the expected range defined by the membership functions.

6 Discussion

The findings of this study highlight the potential of integrating fuzzy logic with case-based reasoning to enable real-time decision support in complex forming processes. The system demonstrated high accuracy in detecting faults such as geometric deviations and insufficient blank holder force, thereby validating the effectiveness of fuzzy membership functions in addressing uncertainty in sensor data interpretation.

Nevertheless, the reduced hit rate in the "Low Lubrication" scenario suggests that certain process conditions may not produce uniquely distinguishable sensor patterns. This limitation may stem from the current design of the membership functions or the selection of sensor metrics, which may not be sufficiently sensitive to lubrication-related anomalies. Future developments could address this challenge by incorporating additional sensor (e.g., for temperature) or by applying supervised learning methods to improve rule definition and fault discrimination.

The performance in the "Workpiece Misalignment" case further underscores the system's sensitivity to localized deviations in force characteristics. Although the hit rate was relatively high, the occurrence of a statistical outlier that was misclassified points to the need for better handling of rare or marginal cases. This could involve expanding the training dataset.

A key advantage of the proposed approach is its interpretability: the use of linguistic terms in the fuzzy rules enables operators to comprehend the rationale behind decision-making processes. This transparency can foster trust in the system and supports its practical deployment in industrial settings. The current case base and fuzzy rules were elicited from a single domain expert and limited types of sensors, which may introduce operator dependency, subjective bias and reduced fault-coverage due to unmonitored process variables. To mitigate this, ongoing work should include a multi-expert knowledge elicitation protocol, and the development of semi-automated rule-refinement methods (e.g., clustering-based rule adaptation) to improve robustness and generalizability across different users.

In industrial practice, frequent product or material changes increase the likelihood of fault errors, which are difficult to detect without fault detection. The fuzzy-CBR system presented in this study detect the occurrence faults early and prevent them from arising by giving reconfiguration advise, which mitigates the impact of learned errors, enhancing process resilience and supporting robust production with fewer defective parts, even under variable conditions.

In addition, the methodology facilitates the systematic capture and reuse of expert knowledge through the case base. This not only supports consistent decision-making across varying levels of operator expertise but also promotes resilience and adaptability in reconfigurable manufacturing environments.

7 Conclusion

This study presents a methodology that integrates fuzzy logic with case-based reasoning (CBR) to support decision-making for reconfiguration at the resource level in forming processes. By interpreting sensor data through fuzzy membership functions and utilizing expert knowledge within a structured case base, the system enables early fault detection and the recommendation of appropriate corrective measures.

The method was experimentally validated in a deep drawing process and demonstrated high accuracy in identifying well-defined fault scenarios, such as geometric deviations and insufficient blank holder force. The approach proved to be robust in handling data uncertainty and effective in standardizing decision-making by embedding experiential knowledge into the system.

A key strength of the methodology lies in its transparency and interpretability. By representing rules in linguistic form, the system supports user comprehension and fosters trust in automated recommendations. This facilitates deployment in environments with varying levels of operator expertise and contributes to consistent and adaptive process control. Future work will explore the system's usability in extended multi-stage processes with a larger data set and its performance under real industrial conditions. A comparative analysis with alternative approaches for decision-making (e.g. data-based AI). Of particular interest is how operators interact with the system over time, including whether it enhances their confidence, consistency, and decision quality in complex fault situations.

Acknowledgments The authors would like to acknowledge the financial support by the Horizon Europe FLEX4RES project with Grant Agreement Number 101091903.

References

1. Lerch C, Jäger A, Horvat D (2021) Resilient in der Krise. In: Conference 2016, LNCS, vol 9999. Springer, Heidelberg, pp 1–13 (2016)
2. Alexopoulos K, Anagiannis I, Nikolakis N, Chryssolouris G (2022) A quantitative approach to resilience in manufacturing systems. Int J Prod Res 60(24):7178–7193. https://doi.org/10.1080/00207543.2021.2018519
3. Caesar B, Tilbury DM, Barton K, Fay A (2022) Requirements for reconfiguration management for manufacturing systems. IFAC-PapersOnLine 55(10):3214–3219. https://doi.org/10.1016/j.ifacol.2022.10.147
4. Essig P, Liewald M, Bolay C, Schubert T (2019) Digital process support in toolmaking by using optical metrology. IOP Conf Ser: Mater Sci Eng 651(1):12026. https://doi.org/10.1088/1757-899X/651/1/012026
5. Lange K (1990) Umformtechnik. Springer, Berlin, Heidelberg
6. Kolodner J (1995) Promoting transfer through case-based reasoning: rituals and practices in the learning by design classroom and evidence of transfer. Morgan Kaufmann, San Mateo, CA
7. Kruse R, Borgelt C, Braune C, Klawonn F, Moewes C, Steinbrecher M (2015) Computational intelligence: eine methodische Einführung in künstliche neuronale Netze, evolutionäre Algorithmen, Fuzzy-Systeme und Bayes-Netze. 2., überarbeitete und erweiterte Auflage. Springer Vieweg (Computational intelligence), Wiesbaden
8. Engel B (1996) Verfahrensstrategie zum Innenhochdruck-Umformen. In: Als Ms. gedr. Shaker (Berichte aus Produktion und Umformtechnik), Aachen
9. Schulz G, Graf C (2013) Regelungstechnik 2: Oldenbourg Wissenschaftsverlag Verlag
10. Geueke M, Steinheimer R, Frohn-Sörensen P, Schiller M, Engel B (2024) Sustainable tool technology: wood-based forming tools for deep drawing of sheet metal. IOP Conf Ser: Mater Sci Eng 1307(1):12018. https://doi.org/10.1088/1757-899X/1307/1/012018

Dynamic Resilience Assessment in Smart Make-to-Order Job Shops with Attention-Based LSTM

Adane Kassa, Elena Urkia, Elena Montejo, and Martin Manns

Abstract Modern make-to-order (MTO) job shops are vulnerable to disruptions, ranging from machine failures to supply chain fluctuations, threatening their ability to meet delivery commitments. This study introduces a framework designed to enhance operational resilience by integrating the Relative Lateness Proportion (RLP) to quantify delay propagation, along with the Active Period Percentage (APP) for monitoring productive capacity. These metrics are combined with an advanced Attention-based Long Short-Term Memory (A-LSTM) prediction model. Validated through extensive testing in a real-world precision manufacturing facility, the proposed solution demonstrates significant advantages over conventional methods, including Autoregressive Integrated Moving Average (ARIMA), vanilla Recurrent Neural Network (RNN), and basic LSTM architecture. The framework enables the early identification of vulnerable workstations, accurate prediction of disruption patterns, and data-driven interventions to maintain production schedules. Its performance is attributed to integrating operational metrics with machine learning, where the attention mechanism effectively prioritizes critical production data. The proposed approach provides manufacturers with a practical and scalable method to improving delivery reliability in complex, high-variability production environments.

Keywords Make-to-order · Dynamic resilience · LSTM · Predictive maintenance · Smart manufacturing · Deep learning

---

A. Kassa (✉) · M. Manns
PROTECH—Institut für Produktionstechnik, Universität Siegen, Siegen, Germany
e-mail: adane.kassa@aait.edu.et
URL: https://protech.mb.uni-siegen.de/fams/

E. Urkia · E. Montejo
ICT and Automation Department, IDEKO—Basque Research and Technology Alliance (BRTA), Elgoibar, Spain

M. Manns (ed.), *New Paradigms for Anticipated Uncertainty*, Lecture Notes in Mechanical Engineering, https://doi.org/10.1007/978-3-032-16889-4_12

1 Introduction

The manufacturing sector's transition from make-to-stock (MTS) to make-to-order (MTO) production has redefined job shop operations, driven by increasing demand for customized, low-volume manufacturing [1]. These MTO job shops, distinguished by their functional layouts, shared resource pools, and variable process routings, demonstrate superior flexibility for handling product variety but face unique scheduling complexities not encountered in flow production systems. While this adaptability allows effective responses to both internal disruptions (machine breakdowns, quality defects) and external variability (supply chain disruptions, demand shifts), it simultaneously introduces systemic challenges for maintaining delivery reliability. The dynamic environment of MTO job shops generates constantly evolving production schedules, nonlinear workflow patterns, and queue-dependent processing times that create stochastic bottlenecks. When combined with progressive equipment degradation and growing supply chain instabilities, these factors produce complex disruption cascades that are especially difficult to manage in resource-constrained job shop environments.

Traditional risk management approaches, with their reliance on static prevention controls, are inadequate for such dynamic operational contexts. Modern manufacturing demands integrated resilience strategies that combine three key capabilities: real-time monitoring, data-driven disruption prediction, and adaptive response mechanisms.

Advances in Industry 4.0 technologies have significantly contributed to this goal. Cyber-physical systems equipped with IoT sensors and AI-driven analytics now enable real-time equipment monitoring, predictive maintenance, and dynamic recovery processes [2]. These technological developments provide the infrastructure for building smart and resilient MTO production systems.

Despite these technological advances, notable research gaps remain in resilience assessment for smart MTO job shops. Current studies frequently lack high-resolution operational data capturing complete job flow dynamics and workstation state transitions. Moreover, prevailing assessment methods remain predominantly reactive, focusing on post-disruption mitigation rather than proactive vulnerability identification. Most critically, conventional approaches using static historical data cannot accommodate the continuous evolution of manufacturing systems, where equipment performance, workforce skills, and product requirements change dynamically.

This is particularly critical in job shops serving industries like aerospace and energy, where custom-made products are often subject to strict deadlines, and delays can lead to contractual penalties and reputational damage. In the real-world industrial case studied, a high-precision MTO job shop serving both the aerospace and energy sectors, observations revealed that even small disturbances at upstream workstations frequently propagated downstream, leading to system-wide lateness. The dynamic interactions between job routing, resource contention, and machine availability created cascading effects that traditional scheduling tools could not predict or mitigate. These operational realities clearly highlight the need for a more robust and forward-looking approach to production resilience.

This research addresses these limitations by developing and validating a dynamic resilience assessment framework tailored to MTO job shops. The framework integrates three key components: the Relative Lateness Proportion metric, which tracks how delays spread across workstations; the Active Period Percentage, which captures productive capacity retention during disruptions; and an Attention-based Long Short-Term Memory multivariate time-series model that enables real-time resilience prediction from operational data streams. Together, these components form a proactive, data-driven framework that empowers operators to detect emerging vulnerabilities and initiate corrective actions before schedule adherence is jeopardized.

2 State of the Art in Resilience Assessment

Recent years have witnessed growing recognition of resilience as a critical performance indicator across diverse systems, including power networks, ecological systems, and production environments [3, 4]. Contemporary quantitative resilience assessment methodologies have evolved into three principal approaches: optimization-based, numerical analysis, and data-driven methods, each offering distinct advantages for system evaluation.

Optimization-based methods utilize mathematical models to analyze the influence of system structure on resilience and to identify the optimal configuration for resilience. Yonghong and Lindu [5] propose a model for analyzing supply chain resilience. The model's objective function is defined based on resilience cells with self-adaptive and self-recovery abilities. Numerical experiments are conducted to explore the relationship between risk factors and system resilience.

Numerical analysis methods provide another approach, evaluating and comparing resilience performance through sophisticated mathematical modeling of system behaviors. Zobel et al. [6] propose a resilience metric that calculates the total system loss during a certain time interval, where the total system loss is regarded as the quality degradation of system service function and is measured by the triangular area associated with the system service function. While offering theoretical rigor and precise quantification capabilities, such methods typically demand extensive system characterization and may lack the flexibility required for rapidly changing production environments. Their strength lies in scenario comparison rather than dynamic adaptation.

Data-driven methods have emerged as a powerful alternative, leveraging historical data and simulations for more adaptive resilience assessment. Jiateng et al. [4] propose Bayesian networks to evaluate the resilience of urban rail transit systems based on data collected from disruption incidents. They develop a resilience matrix, a function of severity and recovery time, to evaluate and compare the rails systems in Beijing.

Data-driven methods demonstrate effectiveness in handling complex, interdependent systems, outperforming the other methods in capturing real-world operational complexities.

Despite these methodological advances, significant gaps persist in addressing resilience challenges within smart manufacturing systems, particularly in make-to-order (MTO) job shop environments. Such systems face frequent disruptions, from internal equipment failures to external supply chain fluctuations, yet current research often overlooks comprehensive data collection on job order flows and real-time production state dynamics, which are critical for accurate resilience assessment. This data gap undermines the practical effectiveness of existing approaches.

Moreover, the field relies heavily on reactive methodologies, such as post-disruption optimization (firefighting) or static historical performance analyses. These methods are increasingly inadequate for modern production environments, where machine conditions, workforce capabilities, and product complexity continuously evolve. To address these challenges, dynamic manufacturing systems require more proactive and adaptive assessment techniques [7, 8].

To address these limitations, this study proposes a dynamic resilience assessment framework combining three core components: Relative Lateness Proportion for capturing disruption propagation, Active Period Percentage for quantifying capacity retention during disruptions, and an Attention-based Long Short-Term Memory (A-LSTM) model for real-time resilience prediction using multivariate operational data. This integrated framework aims to offer a more holistic, forward-looking approach to resilience in complex manufacturing systems.

3 Proposed Methodology

3.1 *Relative Lateness Proportion in Assessing Resilience*

In MTO job shops, lateness, the deviation between actual and planned processing times, serves as a critical performance metric. It reflects disruptions to the production schedule, such as machine breakdowns, urgent job insertions, or stochastic variability, which propagate through workstations and jeopardize on-time delivery [7, 9, 10].

Relative lateness quantifies the discrepancy between target and actual throughput times, offering a dynamic measure of system vulnerability. Under ideal conditions, production adheres to the planned schedule. However, real-world disturbances introduce lateness, which cascades across workstations, complicating mitigation efforts. To enable proactive intervention, a predictive metric is needed to identify workstations with elevated lateness propagation risks.

Mathematical Formulation

(a) Machine-level relative lateness (RL):

$$\mathrm{RL}_{\mathrm{m}}^{\mathrm{t}} = \sum_{\mathrm{j}=1}^{\mathrm{k}} \left(\mathrm{RL}_{\mathrm{j}}^{\mathrm{t}}\right) \tag{1}$$

(b) Workstation-Level Relative Lateness (RL):

$$\mathrm{RL}_{\mathrm{w}}^{\mathrm{t}} = \sum_{\mathrm{m}=1}^{\mathrm{M}} \sum_{\mathrm{j}=1}^{\mathrm{k}} \left(\mathrm{RL}_{\mathrm{j}}^{\mathrm{t}}\right) \tag{2}$$

(c) Relative Lateness Proportion (RLP):

$$\mathrm{RLP}_{\mathrm{w}}^{\mathrm{t}} = \frac{RL_{\mathrm{w}}^{\mathrm{t}}}{\sum_{\mathrm{w}=1}^{\mathrm{W}} \left(\mathrm{RL}_{\mathrm{w}}^{\mathrm{t}}\right)} \tag{3}$$

where:

- $\mathrm{RL}_{\mathrm{j}}^{\mathrm{t}}$: Relative lateness of job j in time window t.
- $\mathrm{RL}_{\mathrm{m}}^{\mathrm{t}}$: Total relative lateness of machine m in t.
- $\mathrm{RL}_{\mathrm{w}}^{\mathrm{t}}$: Total relative lateness of workstation w in t.
- $\mathrm{RLP}_{\mathrm{w}}^{\mathrm{t}}$: Relative lateness proportion of workstation w in t.
- k: Jobs completed by machine m in t.
- M: Number of machines in workstation w.
- W: Total workstations in the production line.
- t: Time window (8-hour shift in this case).

Workstations with high RLP values contribute disproportionately to total system delays, signaling reduced resilience. Such stations act as bottlenecks, amplifying the risk of delay propagation. Conversely, resilient workstations localize disruptions through dynamic adjustments, like job rescheduling, prioritization, or preemption, preserving overall performance.

The RLP metric enables a structured approach to disruption management. First, workstations are ranked by their RLP values to identify critical hotspots requiring intervention. For workstations with elevated RLP, operators can deploy targeted corrective measures, such as rescheduling jobs across parallel machines, prioritizing critical orders, or implementing preemption strategies in single-machine stations. Further diagnostics, such as analyzing the active period percentage or other predictive factors, help uncover root causes of non-resilient behavior (e.g., chronic machine downtime or unbalanced workloads). This systematic process ensures timely reconfiguration and sustains production stability.

3.2 Active Period Percentage

The Active Period Percentage (APP) is a key metric for identifying dynamic bottlenecks in job shop production systems [11]. It quantifies machine utilization by measuring the proportion of uninterrupted operation time within a given observation window. APP is calculated as the total active time across all machines in a workstation, divided by the product of the number of machines and the total time window:

$$\mathrm{APP_{wt}} = \frac{\sum_{m=1}^{M} \text{Active time (t)}}{\mathrm{M} \times \text{Total time window}} \tag{4}$$

Machine states are classified to distinguish active periods, processing, breakdowns, interruptions, and setups, from inactive ones such as preventive maintenance and idle time. APP relies on accurate state tracking via event logs, enabling real-time identification of underutilized machines, performance comparison across workstations, and early bottleneck detection. It also highlights operational issues like prolonged inactivity or excessive state switching. In our framework, APP functions as the key predictive input for the multivariate RLP time-series model, supporting dynamic resilience analysis.

3.3 Attention-Based Long Short-Term Memory for RLP Prediction

Long Short-Term Memory (LSTM) networks, a specialized type of recurrent neural network (RNN), excel at capturing long-term dependencies in sequential data. Through memory cells and gating mechanisms, LSTMs overcome the vanishing gradient problem of traditional RNNs, making them ideal for time-series forecasting. In production systems, they effectively model temporal dynamics, such as machine performance degradation and bottleneck formation [12]. Attention-based LSTM (A-LSTM) further enhances multivariate time-series analysis by integrating LSTM's temporal modeling with an attention mechanism that dynamically weights the most relevant input features. This approach improves both prediction speed and accuracy [13].

To benchmark the performance of our proposed A-LSTM model, we included three comparative models: ARIMA, Vanilla RNN, and standard LSTM. ARIMA (Autoregressive Integrated Moving Average) is a classical statistical model well-regarded for capturing linear temporal dependencies, providing a strong and interpretable baseline for time-series prediction [14]. The Vanilla RNN, a foundational deep learning model for sequential data, struggles with learning long-term dependencies due to vanishing gradients. Standard LSTM improves upon Vanilla RNN by addressing this limitation through gating mechanisms that capture both short- and long-term dependencies in time-series data.

In this study, an A-LSTM network models the temporal dependencies and dynamic evolution of RLP. The proposed multivariate time-series framework predicts RLP for workstations using job flow dynamics and machine state data. The LSTM architecture, integrated with a soft attention mechanism, prioritizes critical input sequence segments to boost prediction accuracy. By capturing both short-term fluctuations and long-term trends in APP and RLP, the model predicts future RLP values.

4 Implementation and Case Study

4.1 Case Study Description and Data Generation

We demonstrate our proposed approach within a large make-to-order (MTO) job shop specializing in precision components for demanding industries like aerospace, energy, and machinery. This facility features nine specialized workstations, each equipped with multiple parallel or duplicate machines. Specifically, the machine count per station is: 3D measuring (1), painting (2), surface grinding (3), gantry milling (2), 4–5-axis machining center (3), multifunction milling (1), spot welding and boring (2), travelling column milling (3), and coordinate grinding (2). This configuration, depicted in Fig. 1, enables the shop to support high-mix, low-volume production, where each job follows a distinct routing sequence through the workstations.

The study leveraged IoT-enabled sensors to collect two years of real-world operational data, capturing both machine states and job flow dynamics. Machine state transitions were recorded with second-by-second precision, allowing detailed tracking of equipment status changes. Simultaneously, RFID-tagged work orders provided accurate timestamps as jobs entered and exited each workstation, creating a comprehensive record of production flow. To strengthen the dataset's predictive capabilities, we extended the observation period to five years through discrete-event simulation using Python's SimPy library. From this comprehensive data, time-series metrics per shift, including Relative Lateness Proportion (RLP) and Active Period Percentage (APP), were derived for analysis.

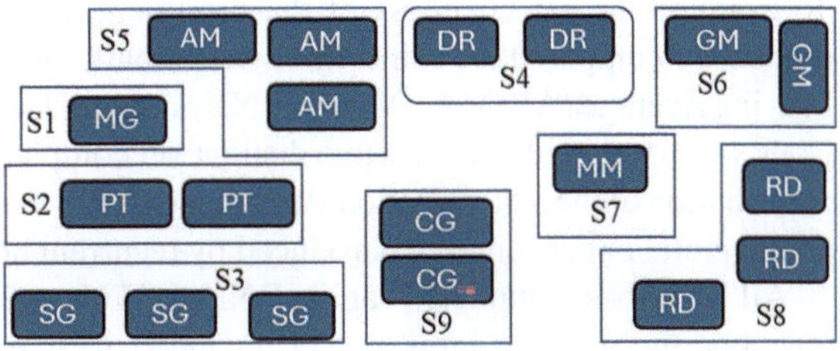

Fig. 1 Systemic layout of the MTO shop

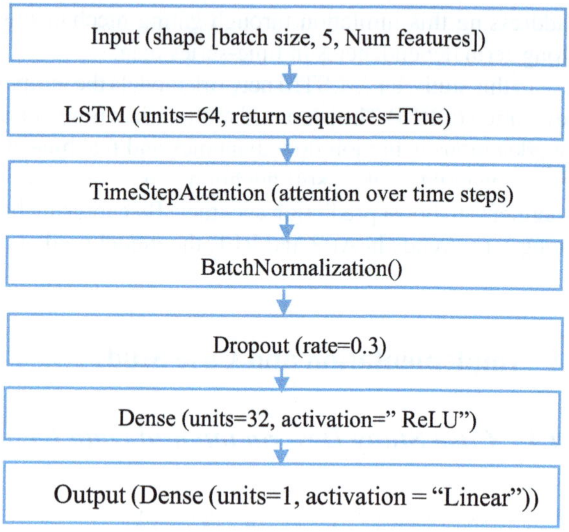

Fig. 2 Model architecture

4.2 Model Development and Training

We developed an Attention-based LSTM (A-LSTM) model to enhance temporal dependency learning by integrating an attention mechanism with a standard LSTM network. The model uses Active Period Percentage (APP) and Relative Lateness Proportion (RLP) as input features to predict future RLP values. We applied extensive feature engineering to the dataset to extract meaningful patterns, trends, and inter-variable relationships, ensuring the model could robustly learn complex temporal dynamics.

Bayesian Optimization determined the optimal set of hyperparameters for our model. This resulted in a configuration of 750 epochs, a batch size of 5, 64 LSTM units, a learning rate of 0.0023, a sequence length of 5, 32 dense units, and a dropout rate of 0.3. Figure 2 illustrates the complete architecture of the model.

4.3 Results

To evaluate the effectiveness of the proposed Attention-based LSTM (A-LSTM) model, we compared it against three benchmark models: Autoregressive Integrated Moving Average (ARIMA), Vanilla RNN, and standard LSTM. These model comparisons were designed to assess two distinct yet complementary advancements: (1) the contribution of the Active Period Percentage (APP) as an auxiliary feature, and (2) the architectural enhancements introduced by recurrent neural network (RNN) variants.

All RNN-based models (Vanilla RNN, LSTM, and A-LSTM) were trained using the same multivariate input, which included APP, ensuring a consistent basis for

Table 1 The performance of A-LSTM and baseline models

Metric	ARIMA (4,1,1)	Vanilla RNN	LSTM	A-LSTM
MAE $\times 10^{-2}$	35.1	22.10	22.2	14.46
RMSE $\times 10^{-2}$	36.2	25.7	23.7	16.4
R^2	0.47	0.71	0.73	0.89

comparing their architectural performance. In contrast, ARIMA, being a univariate statistical model, was trained solely on the Relative Lateness Proportion (RLP) time series, as it cannot integrate auxiliary features like APP.

Model performance was evaluated using three key metrics: Mean Absolute Error (MAE), Root Mean Squared Error (RMSE), and R-squared (R^2). Results for the entire MTO job shop are presented in Appendix, while detailed performance metrics for Workstation 3 are summarized in Table 1. To demonstrate the effectiveness of the proposed approach, a comprehensive analysis of Workstation 3 is provided in the Discussion section.

4.4 Discussion

The ARIMA model showed the weakest performance among the evaluated models, with an RMSE of 0.362, MAE of 0.351, and R^2 of 0.47. Its limited capacity to model complex temporal patterns and long-term dependencies resulted in reduced predictive accuracy compared to neural network-based approaches. In addition, as a univariate statistical model, ARIMA does not incorporate operational features such as APP, limiting its ability to capture complex production dynamics.

The Vanilla RNN offered improvements over ARIMA, achieving an RMSE of 0.257, MAE of 0.221, and R^2 of 0.71. This performance gain is primarily due to the inclusion of APP as an additional feature, which enriches the model's understanding of machine utilization patterns. While it captures sequential dependencies more effectively, it still struggles with modeling long-range relationships, which impacts prediction accuracy in dynamic production settings.

The LSTM model provided further enhancements, with an RMSE of 0.237, MAE of 0.222, and R^2 of 0.73. Its architecture effectively addresses vanishing gradients and models long-term dependencies, making it well-suited for environments where historical performance influences future states. Although its MAE is comparable to that of the Vanilla RNN, LSTM generally yields more stable and interpretable outputs over extended time periods.

Finally, the Attention-based LSTM (A-LSTM) model outperformed all other models, achieving an MAE of 0.145, RMSE of 0.164, and R^2 of 0.89. This represents a 34% reduction in MAE compared to the standard LSTM, highlighting its superior accuracy and consistency. Such precision is particularly crucial in production dynamics forecasting, where even minor deviations can disrupt operations. The significant performance gains from A-LSTM can be attributed to two key factors:

1. Inclusion of APP as an auxiliary feature: This universally enhanced all neural network models relative to the ARIMA baseline.
2. Architectural advantage of the attention mechanism: This enables the model to selectively focus on the most informative time steps, thereby refining temporal pattern recognition and substantially improving prediction accuracy.

In summary, the A-LSTM model demonstrated the highest accuracy across all evaluation metrics, confirming the effectiveness of integrating attention mechanisms and operational metrics for forecasting dynamic resilience in complex production systems.

5 Conclusion and Outlook

This study presents an innovative dynamic resilience assessment framework integrating relative lateness proportion (RLP) and active period percentage (APP) metrics with an Attention-based LSTM (A-LSTM) predictive model. Implementation in a precision Make-to-Order (MTO) job shop demonstrates the framework's effectiveness in quantifying and predicting production system resilience. The A-LSTM model outperforms conventional approaches, delivering actionable insights for proactive disruption management. The framework's predictive capability for RLP values advances production analytics, enabling early identification of vulnerable workstations before delays escalate. Combining operational metrics with machine learning establishes a robust foundation for data-driven decision-making in complex manufacturing environments. Future work should enhance the framework by incorporating digital twins for real-time reconfiguration and advanced diagnostics for bottleneck analysis. Expanding the scope to include supply chain and workforce resilience will further evolve the system into a comprehensive resilience management platform suited for high-stakes, time-sensitive production environments.

Acknowledgments This research has been funded by the Horizon Europe FLEX4RES project with Grant Agreement Number 101091903.

Appendix: Model Evaluation Results for All Workstations

Workstation	Model	Metric		
		MAE	RMSE	R^2
WS1	ARIMA (4,1,1)	25.6	35	0.42
	Vanilla RNN	22.1	33.9	0.61
	LSTM	20.2	30.3	0.83
	A-LSTM	16.2	32	0.97
WS2	ARIMA (4,1,1)	33.	29.6	0.57
	Vanilla RNN	26.1	26.8	0.61
	LSTM	22.6	25	0.52
	A-LSTM	1.6	22.8	0.79
WS3	ARIMA (4,1,1)	35.1	36.2	0.47
	Vanilla RNN	22.1	25.7	0.71
	LSTM	22.2	23.7	0.73
	A-LSTM	14.46	16.4	0.89
WS4	ARIMA (4,1,1)	45.2	37.2	0.6
	Vanilla RNN	26.1	35.7	0.7
	LSTM	24.25	32.1	0.74
	A-LSTM	23.47	29.3	0.78
WS5	ARIMA (4,1,1)	45.54	34.6	0.57
	Vanilla RNN	29.4	35.56	0.66
	LSTM	30.34	25	0.74
	A-LSTM	25.46	26.4	0.99
WS6	ARIMA (4,1,1)	33.87	26.5	0.55
	Vanilla RNN	30	25.8	0.74
	LSTM	28.24	24.1	0.75
	A-LSTM	27.02	22.5	0.92
WS7	ARIMA (4,1,1)	29.14	40.1	0.61
	Vanilla RNN	28.4	35.7	0.72
	LSTM	27.9	37.8	0.73
	A-LSTM	26.8	30.3	0.8
WS8	ARIMA (4,1,1)	36.4	36.7	0.53
	Vanilla RNN	39	35.2	0.65
	LSTM	30.2	29.7	0.6
	A-LSTM	30	26.4	0.72
WS9	ARIMA (4,1,1)	36.8	30.25	0.78
	Vanilla RNN	32.1	25.2	0.81

(continued)

(continued)

Workstation	Model	Metric		
		MAE	RMSE	R^2
	LSTM	31.8	23.4	0.81
	A-LSTM	30.6	20.6	0.76

References

1. Gansterer M (2015) Aggregate planning and forecasting in make-to-order production systems. Int J Prod Econ 170:521–528. https://doi.org/10.1016/j.ijpe.2015.06.001
2. Kassa A, Kitaw D, Stache U, Beshah B, Degefu G (2023) Artificial intelligence techniques for enhancing supply chain resilience: a systematic literature review, holistic framework, and future research. Comput Ind Eng 186, no. October:109714. https://doi.org/10.1016/j.cie.2023.109714
3. Yodo N, Wang P, Zhou Z (2017) Predictive resilience analysis of complex systems using dynamic Bayesian networks 66(3):1–10
4. Yin J, Ren X, Liu R, Tang T, Su S (2021) Quantitative analysis for resilience-based urban rail systems: a hybrid knowledge-based and data-driven approach. Reliab Eng Syst Saf 219(November):2022
5. Y. Li and L. Zhao, "Analyzing deformation of supply chain resilient system based on cell resilience model," Lect Notes Comput Sci (including Subser Lect Notes Artif Intell Lect Notes Bioinformatics), vol. 6328 LNCS, no. PART 1, pp. 26–35, 2010, doi:https://doi.org/10.1007/978-3-642-15621-2_4
6. Zobel CW, Khansa L (2014) Characterizing multi-event disaster resilience. Comput Oper Res 42:83–94. https://doi.org/10.1016/j.cor.2011.09.024
7. Wenzel A, Moschref A, Beers L, Gehlhoff F, Fay A, Nyhuis P (2024) Framework for logistics-oriented risk assessment by integrating resilience analysis. IFAC-PapersOnLine 58(27):1505–1510. https://doi.org/10.1016/j.procir.2024.10.274
8. Zhang D, Xie M, Yan H, Liu Q (2021) Resilience dynamics modeling and control for a reconfigurable electronic assembly line under spatio-temporal disruptions. J Manuf Syst 60(January):852–863. https://doi.org/10.1016/j.jmsy.2021.01.011
9. Scholz-Reiter B, Windt K, Liu H (2011) Modelling dynamic bottlenecks in production networks 24. https://doi.org/10.1080/0951192X.2010.511655
10. A. Mütze, T. Hiller, and P. Nyhuis, Lateness in production systems—in a nutshell: How to determine the causes of lateness at work systems? Adv Ind Manuf Eng 8(October 2023):100132. 2024, doi:10.1016/j.aime.2023.100132
11. Roser C (2002) Shifting bottleneck detection. In: Proceedings of the 2002 winter simulation conference
12. X. Lai, H. Shui, and J. Ni, A two-layer long short-term memory network for bottleneck prediction in multi-job manufacturing systems. In: ASME 2018 13th International manufacturing science and engineering conference MSEC 2018, vol 3, no. January 2023, 2018, doi:https://doi.org/10.1115/MSEC2018-6678
13. Kumar I, Tripathi BK, Singh A (2023) Attention-based LSTM network-assisted time series forecasting models for petroleum production. Eng Appl Artif Intell 123(May):106440. https://doi.org/10.1016/j.engappai.2023.106440
14. Ma X, Tao Z, Wang Y, Yu H, Wang Y (2015) Long short-term memory neural network for traffic speed prediction using remote microwave sensor data. Transp Res Part C Emerg Technol 54:187–197. https://doi.org/10.1016/j.trc.2015.03.014

Digital Business Models, Smart Products and Services

A Unified Framework for Intelligent Product Configuration Using Knowledge Graphs and Bayesian Networks

Stefan Berlik and Mohammad Seidpisheh

Abstract The increasing complexity of product configurations demands intelligent systems that effectively integrate customer requirements, dependencies, and uncertainties. This paper introduces a unified framework combining Knowledge Graphs (KGs) and Bayesian Networks (BNs) to enhance the efficiency and adaptability of product configuration processes. KGs provide a semantic foundation for product information, ensuring interoperability and explicit relationship modeling. BNs enhance this through probabilistic reasoning, allowing the system to manage uncertainties and dynamically generate optimal configurations. The integration of deterministic, rule-based reasoning from ontologies with the probabilistic nature of BNs automates suggestions, predicts user preferences, and reduces complexity. This framework streamlines user interactions through intelligent form pre-filling and contextually relevant suggestions, even under uncertainty. By employing an ontology-based representation of BNs, the components fit seamlessly into the KG, creating a cohesive and unified framework that balances scalability and user-centric design to address modern configuration challenges.

Keywords Product configuration · Knowledge graphs · Ontology · Bayesian networks

1 Introduction

As customizable products become more complex, intelligent configuration systems must effectively integrate customer preferences, constraints, and uncertainties. This growing complexity necessitates more sophisticated reasoning frameworks that can maintain both flexibility and efficiency in decision making. Traditional rule-based configuration systems often struggle to adapt to these dynamic aspects, limiting their effectiveness in real-world applications. Over time, product configuration sys-

S. Berlik (✉) · M. Seidpisheh
Faculty of Engineering and Mathematics, Hochschule Bielefeld, University of Applied Sciences and Arts, Bielefeld, Germany
e-mail: stefan.berlik@hsbi.de

M. Manns (ed.), *New Paradigms for Anticipated Uncertainty*, Lecture Notes in Mechanical Engineering, https://doi.org/10.1007/978-3-032-16889-4_13

tems have evolved to balance expressive knowledge representation with computational efficiency. Early rule-based systems such as R1/XCON [1] are tightly coupled with structural and policy knowledge, posing challenges for maintenance. Subsequent approaches, such as Constraint Satisfaction Problem (CSP) and Dynamic CSP (DCSP), addressed dynamic variable integration [2], but lacked high-level modeling abstractions. Ontology-driven methods, such as OWL (Web Ontology Language) and SWRL (Semantic Web Rule Language) [3], enable structured reasoning and knowledge reuse, but struggle to handle uncertainty. To address this limitation, probabilistic reasoning frameworks such as BNs and Multi-Entity Bayesian Networks (MEBN) [4] have been integrated into ontologies. PR-OWL [5–7] extends OWL with Bayesian semantics, enabling probabilistic inference in OWL. Although ontology-based configuration offers a structured and semantic approach, it lacks mechanisms to manage uncertainty in product customization. By integrating PR-OWL and MEBN with OWL and KGs, a scalable and intelligent configuration framework can be developed to support probabilistic reasoning within complex product models.

Although traditional systems rely on rigid rule-based logic, this paper explores the integration of KGs and BNs to develop a scalable and adaptable product configuration framework. KGs organize product knowledge by explicitly defining relationships and dependencies, ensuring consistency and interoperability across various domains. Meanwhile, BNs introduce probabilistic reasoning, allowing the system to infer missing information and assess uncertainties in decision-making. Integrating these technologies results in a more flexible and intelligent configuration system, improving user interactions and optimizing decision-making efficiency.

The proposed approach combines deterministic reasoning from ontologies with probabilistic inference from PR-OWL and MEBN to address the uncertainties inherent in product customization. This integration enhances decision-making efficiency, dynamically generates optimal configurations, and streamlines user interactions by pre-filling forms and offering relevant recommendations based on real-time data. Furthermore, the framework is designed to be scalable, adapting to evolving product configurations and changing user preferences without requiring extensive manual updates. By incorporating probabilistic assessments, the system improves the accuracy of the recommendation, making the configuration process more efficient and user-friendly.

The remainder of this paper introduces ontology-based configuration and probabilistic reasoning, elaborates on the proposed framework, presents an implementation case study, and concludes with key findings and directions for future research.

2 Ontology Based Configuration

Ontology-based configuration provides a structured and semantic approach to managing complex product configurations, ensuring consistency and scalability through formalized knowledge representation. However, real-world customization scenarios often involve incomplete information, evolving user preferences, and inherent uncertainties that deterministic methods alone cannot effectively handle. To address these challenges, this section first examines the deterministic foundations of ontology-based configuration, followed by an exploration of probabilistic reasoning frameworks that enhance adaptability and decision-making in intelligent configuration systems.

2.1 Deterministic Framework

As product configurations become increasingly complex, intelligent systems must effectively integrate customer requirements, dependencies, and constraints while maintaining logical consistency. A structured framework is essential to organize information about components, their attributes, and their interrelationships in a machine-readable format. Ontologies provide a semantic foundation for product configuration by not only storing data but also defining formal relationships and rules that govern component interactions.

Unlike traditional data storage methods, ontologies enable dynamic and flexible knowledge representation, eliminating the need to manually define every possible component combination. As integral components of KGs, ontologies establish formal vocabularies and logical constraints that support automated reasoning and consistency across product domains. This structured approach facilitates knowledge reusability and sharing, reducing the effort required to develop and maintain configuration models while ensuring adaptability to evolving requirements.

Within this deterministic framework, rule-based systems enforce hard constraints, ensuring that configurations adhere to predefined logical relationships. By leveraging deterministic reasoning, ontology-based Configuration systems can autonomously evaluate constraint satisfaction, infer missing dependencies, and maintain logical coherence, thereby enhancing the efficiency, scalability, and reliability of intelligent product customization.

2.2 Probabilistic Framework

While deterministic frameworks provide structured reasoning and enforce logical consistency, they struggle to accommodate real-world product configurations. These challenges necessitate the integration of probabilistic reasoning to enhance adaptability and inference capabilities in configuration systems. MEBN, a framework that combines Bayesian inference with First-Order Logic (FOL), complement ontology-based reasoning by introducing probabilistic dependencies, enabling configuration systems to dynamically refine recommendations in response to uncertainty and missing data. MEBN assigns a probability distribution over interpretations of a first-order logical theory, ensuring a structured yet flexible approach to handling uncertainty.

A key representation for integrating uncertainty into ontologies is PR-OWL 2, an enhanced version of PR-OWL, which extends OWL with constructs to represent MEBN theories [8]. In MEBN, probabilistic knowledge is structured into Modular Fragments (MFrags), which represent coherent and reusable probabilistic relationships between entities. These MFrags are organized within an MEBN Theory (MTheory), a top-level structure that ensures a consistent joint probability distribution, dynamically adapting to new information and evolving data. MFrags, as the building blocks of a probabilistic ontology, encapsulate probabilistic dependencies and enable context-aware inference in complex reasoning tasks. Furthermore, Situation-Specific Bayesian Networks (SSBNs) refine inference by constructing Bayesian models tailored to specific queries, enhancing the system's ability to handle uncertain and incomplete configurations.

For example, while deterministic reasoning enforces strict rules such as "if a door has a sliding operation type, it must have a continuous hinge", probabilistic modeling enhances flexibility by suggesting likely operational types based on historical data and observed user preferences. Integration of PR-OWL 2 and MEBN facilitates scalable probabilistic reasoning, allowing configuration systems to predict user preferences, infer missing details, and generate optimal recommendations while maintaining logical consistency within an ontology-based framework. This approach improves decision-making by enabling systems to manage ambiguity, predict customer preferences, and generate personalized recommendations, ultimately enhancing user experience and configuration efficiency.

3 Proposed Framework

As modern product configuration grows increasingly complex, systems must efficiently manage evolving customer requirements and data uncertainty. For example, a customer may specify a door's color and material while leaving insulation or security features unspecified, causing deterministic systems to either fail or make arbitrary assumptions. To address this, probabilistic reasoning with BNs and MEBN provides a flexible, data-driven approach to decision-making under uncertainty. The integration of KGs with probabilistic models enables configuration systems to infer missing information by analyzing historical data, contextual knowledge, and predefined constraints. This approach reduces user effort, enhances decision support, and continuously improvement over time by learning from past configurations, making the system more adaptive and intelligent.

In our proposed ontology-based configuration system, we model two types of relationships within a unified product knowledge framework. First, deterministic relationships are captured using an ontology-based knowledge graph, which defines strict rules and logical constraints between product components. These relationships ensure structural consistency and rule-based validation. Second, we incorporate probabilistic relationships between components using BNs, allowing the system to handle uncertainty, missing data, and user preferences that evolve over time.

These reasoning paradigms are unified through PR-OWL 2, a probabilistic extension of OWL that embeds MEBN into the ontology. This integration allows deterministic rules and probabilistic dependencies to coexist and interact within a single knowledge model (see [7]).

Ontologies, expressed in OWL, define a structured, machine-interpretable vocabulary of product components, constraints, and relationships. These serve as the deterministic backbone for configuration reasoning. However, deterministic approaches alone cannot effectively manage real-world uncertainties, such as incomplete user inputs, dynamic preferences, or variability in component availability.

To address this, we incorporate PR-OWL 2, an extension of OWL designed for uncertainty modeling, which enables ontologies to represent and reason about probabilistic dependencies. PR-OWL 2 builds on the MEBN formalism, where knowledge is decomposed into MFrags–modular fragments that encapsulate random variables and their probabilistic relationships. These MFrags are composed into an MTheory, a coherent model ensuring a globally consistent joint probability distribution. This formalization allows the system to:

- Represent probabilistic rules within the same framework as deterministic ones.
- Instantiate SSBNs at runtime, tailored to specific user inputs or queries.
- Seamlessly map ontology entities (classes and properties) to MEBN random variables, enabling joint reasoning across symbolic and probabilistic domains.

In deterministic systems, strict rules enforce one fixed outcome–for example, always recommending Steel as the material regardless of door type. In contrast, probabilistic reasoning can predict which material is most likely preferred by the user, even if the input is incomplete. For instance, if a user selects a Folding door but does not specify the material, the system might recommend Aluminum as the most likely option based on past configurations and learned preferences. This integration ensures that configuration systems are not only consistent and reusable (via ontologies), but also adaptive and uncertainty-aware (via BNs). By embedding probabilistic capabilities directly within the ontological structure, our framework supports scalable, real-time reasoning that aligns with evolving customer needs and product constraints–paving the way for automated knowledge discovery and dynamic configuration optimization.

The integration of KGs and BNs addresses several critical challenges in modern product configuration:

Continuous Improvement Over Time Unlike deterministic systems, probabilistic models continuously adapt by learning from past configurations. Bayesian inference dynamically updates probabilities based on real-world observations, ensuring more accurate and personalized recommendations over time. This adaptability facilitates the system align with changing customer preferences, market trends, and manufacturing constraints.

Managing Complexity As product configurations become more complex, traditional systems struggle to scale efficiently. The integration of BNs with KGs supports scalable and modular reasoning, enabling dynamic incorporation of new product components and constraints without extensive manual intervention. By breaking down configurations into probabilistic dependencies, the system can manage highly complex relationships without performance bottlenecks.

Handling Missing Information Incomplete customer inputs necessitate intelligent inference mechanisms rather than rigid rule enforcement. Probabilistic reasoning enables context-aware recommendations, filling in missing details based on historical data. For example, if a user selects a operational type of door (sliding, folding), the system can suggest likely material choices (wood or steel) without explicit input.

Incorporating Uncertainty Configuration decisions must account for uncertainties, including fluctuations in material availability and ambiguous customer preferences. Probabilistic ontologies represent confidence levels in recommendations, allowing the system to suggest feasible alternatives instead of rejecting incomplete configurations due to hard constraints.

Enhanced Decision-Making Probabilistic methods enhance decision-making by ranking configurations according to their likelihood of success. This enables the system to prioritize the most compatible and feasible options, reducing decision fatigue and helping both customers and manufacturers make informed choices.

Enhanced User Experience As product configurations grow more complex, users often experience decision inefficiency due to overwhelming choices. Integrating BNs with KGs enables intelligent pre-filling, dynamically suggesting and ranking configuration options based on posterior probability. This reduces cognitive load, minimizes manual input, and streamlines decision-making, ensuring a more intuitive and efficient user experience.

Probabilistic reasoning enhances modern product configuration by enabling continuous learning, robust decision-making, and seamless user interaction. By combining deterministic reasoning in KGs with probabilistic inference from BNs, the framework ensures adaptability, efficiency, and enhanced user experience, making it a powerful solution for intelligent product configuration.

4 Implementation and Case Study

Implementing the proposed intelligent product configuration framework follows a structured approach that integrates KGs and BNs to enhance reasoning under uncertainty. The key stages of the implementation are as follows:

Ontology Development for Product Configuration A domain-specific ontology for door configuration is developed using OWL, containing essential concepts, relationships, and constraints integral to the configuration process. The ontology defines core entities such as door components (e.g., material, operational type, hinges) and their relationships. The ontology structure is illustrated in Fig. 1.

Integration of instance data to form a KG Instances of door components and configurations are added to the ontology, transforming it into a KG. This step involves defining real-world instances (individuals) that populate the ontology and represent possible configurations.

Embedding deterministic rules with SWRL SWRL rules are incorporated into the ontology to enforce deterministic constraints. These rules ensure that specific configurations adhere to predefined logical constraints, such as requiring that a heavy door material necessitates a reinforced hinge.

Extending ontology with probabilistic properties using PR-OWL 2 Probabilistic reasoning is introduced by extending the ontology with PR-OWL 2, enabling the representation of random variables and uncertain dependencies. In door configuration, PR-OWL 2 models uncertain dependencies, such as operational type and material. Bayesian properties, such as conditional dependencies between operational type and material, are encoded within the ontology, see Fig. 1.

Definition of MTheory and MFrags for probabilistic reasoning MEBN modeling is utilized to define an MTheory including MFrags. The MFrags include operational type and material random variables and their distributions, which are interconnected probabilistically, as visualized in Fig. 2. It illustrates the structure of a Modular Bayesian Network (MTheory) defined using PR-OWL 2 for the domain of door configuration. The Material_MF MFrag models probabilistic dependencies between the operational type random variable (OperationalTypeRV) and the material random variable (MaterialRV). Both are resident nodes (shown as yellow rectangles) that capture conditional dependencies, specifically, MaterialRV is conditionally dependent on OperationalTypeRV. This structure enables probabilistic inference of likely materials based on selected operational types, supporting a more flexible and adaptive configuration process under uncertainty. Context nodes (depicted as green polygons) represent boolean conditions that must be satisfied for the associated probabilistic distribution to be applicable. In this context, they assert that the ordinary variables op and mat are of types OperationalTypeRV and MaterialRV, respectively, thereby ensuring semantically valid type assignments during the inference process.

Implementation of probabilistic reasoning using UnBBayes API [9] Reasoning is implemented using the UnBBayes API, facilitating probabilistic inference over the defined ontology. The system allows querying probabilistic relationships, enabling predictions based on incomplete customer inputs.

Comparison of deterministic and probabilistic reasoning in a simple configuration system A comparative evaluation is performed between traditional deterministic reasoning and probabilistic reasoning (MEBN), as visualized in Fig. 3. In a scenario where a customer specifies an operational type but not the material, the system provides a recommendation based on:

- Deterministic Approach: Enforcing strict rules, suggesting a single material based on predefined logical constraints.
- Probabilistic Approach: Computing posterior distribution based on likelihoods and historical data and uncertain dependencies, ranking the most suitable material

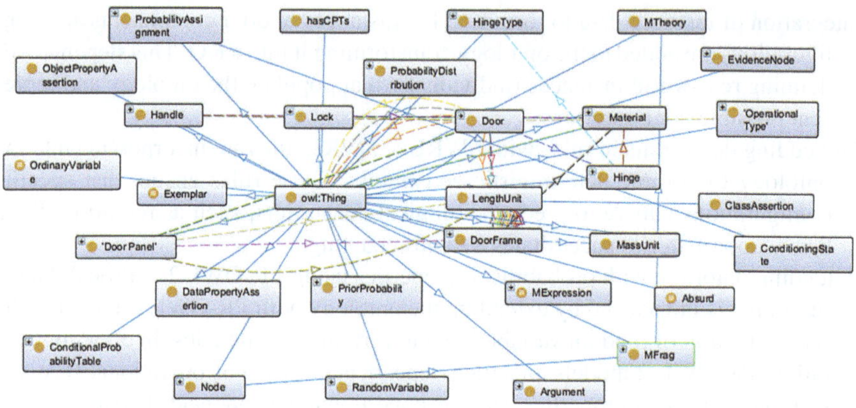

Fig. 1 The ontology defines core components such as Door, Material, and Hinge, capturing their hierarchical structure and logical constraints. It supports deterministic reasoning through formal relationships and SWRL rules. In addition, the ontology is extended with probabilistic constructs using PR-OWL 2 to model uncertain dependencies (e.g., between operational type and material), forming the foundation for Bayesian reasoning within the configuration system

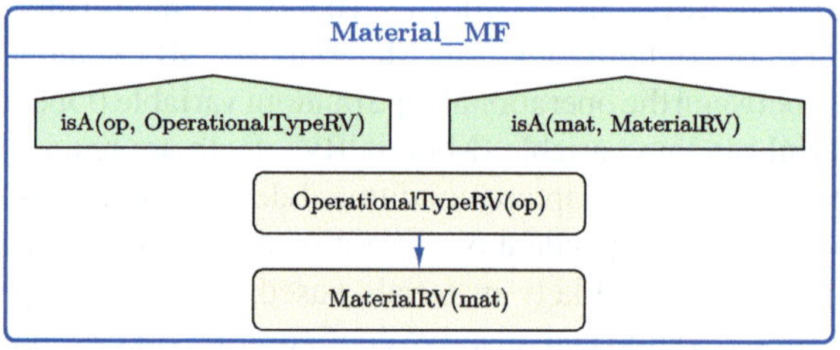

Fig. 2 Modeling the dependency between Material and Operational Type random variables in door configurations with PR-OWL 2

options.

Integration of both results in a robust and flexible configuration system that can handle both hard constraints and probabilistic dependencies, enhancing user experience by accommodating uncertainty and missing information.

In the configuration scenario illustrated in Fig. 3, the system compares material recommendations for three operational door types–Swinging, Sliding, and Folding–using both deterministic and probabilistic reasoning approaches.

The deterministic reasoning method, implemented via ontology and SWRL rules, enforces predefined logical constraints to ensure structural validity. It identifies Steel as the only valid material across all operational types due to strict rule enforcement. While this guarantees that configurations meet hard constraints, it also limits flexibility. Deterministic models do not account for uncertainty or incomplete inputs;

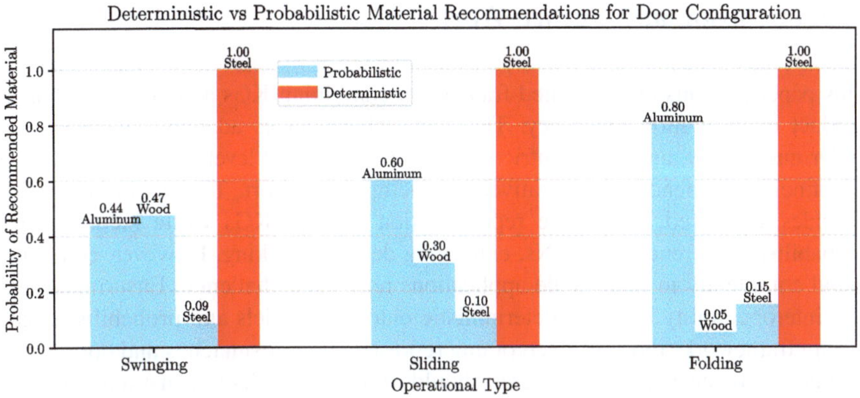

Fig. 3 Comparison of probabilistic vs deterministic reasoning for three different operational types (Swinging, Sliding, and Folding). While deterministic reasoning suggests a single fixed material based on rules, probabilistic reasoning ranks materials by likelihood, handling incomplete inputs and uncertainty. This enables more adaptive and user-friendly configuration support

they provide a single feasible solution but cannot prioritize alternatives or adapt to evolving user needs.

By contrast, the probabilistic reasoning method, powered by BNs within the PR-OWL 2 framework, captures the uncertainty inherent in real-world configurations. This method assigns likelihoods to different material options based on prior knowledge, historical usage data, and expert-defined probabilistic dependencies. For example, in the case of a folding door, the probabilistic model determines that Aluminum has the highest likelihood (0.80) of being suitable, followed by Steel and Wood with lower probabilities. These values reflect learned preferences and contextual patterns rather than fixed rules.

This probabilistic approach supports ranking and intelligent default selection, providing users with the most probable and contextually appropriate choices. It also allows the system to infer missing values when the user input is incomplete, such as selecting an operational type without specifying a material. Furthermore, by continually updating its probabilities based on new data, the system becomes more adaptive over time.

Overall, Fig. 3 shows how integrating deterministic knowledge from the ontology with probabilistic dependencies captured by BNs enhances the configurator's reasoning. It enables more nuanced, data-driven decision support while maintaining logical consistency, offering a significant improvement over traditional rule-based systems in terms of flexibility, user guidance, and scalability.

5 Conclusion and Future Work

This paper presents an integrated framework combining KGs and BNs to enhance product configuration efficiency. The approach enhances adaptability, reasoning under uncertainty, and user-centric recommendations and leverages historical data to refine recommendations iteratively, reducing user effort. The reasoning model provides structured knowledge representation through KGs while incorporating probabilistic inference with BNs, enhancing decision-making. However, computational complexity in large-scale applications remains a challenge. Ensuring seamless interoperability between deterministic ontology models and probabilistic reasoning frameworks requires overcoming representation mismatches and optimizing inference efficiency. Future research should refine probabilistic constraints in KGs and optimize runtime performance. Additionally, integrating probabilistic inference with semantic web technologies will enhance scalability for industrial applications. Owing to the generality of the proposed framework, in future work the integration of dynamic BNs and semantic models might also be explored in domains such as product development, quality management, and maintenance. Investigating these areas will provide deeper insights into the role of ontologies, KGs, and BNs across the entire product life-cycle and strengthen the foundation for intelligent and adaptive decision support systems.

References

1. Barker VE, O'Connor DE (1989) Expert systems for configuration at digital: Xcon and beyond. Commun ACM 32(3):298–318
2. Mittal S, Falkenhainer B (1990) Dynamic constraint satisfaction problems. In: Proceedings of the AAAI conference, pp 25–32
3. Yang D, Dong M, Miao R (2009) Development of a product configuration system with ontology-based approach. Expert Syst Appl 36:4399–4411
4. Laskey KB (2008) Mebn: a language for first-order bayesian knowledge bases. Artif Intell 172(2):140–178
5. da Costa PCG, Laskey KB (2005) Pr-owl: A bayesian ontology language for the semantic web. In: Proceedings of the international workshop on uncertainty reasoning for the semantic web, pp 23–33
6. Costa PCG, Laskey KB, Hughes TR (2013) A reference architecture for probabilistic ontology development. In: Proceedings of the 10th international conference on information fusion, pp 1–8
7. Carvalho RN, Laskey KB, Costa PC (2017) Pr-owl – a language for defining probabilistic ontologies. Int J Approx Reason 91:56–79
8. Carvalho RN, Laskey KB, Costa PCG (2013) Pr-owl 2.0 – bridging the gap to owl semantics. In: Uncertainty reasoning for the semantic web II. Springer, Berlin, Heidelberg, pp 1–18 (2013)
9. Matsumoto S, Carvalho RN, Ladeira MB, Costa PCG, Santos LL, Silva D, Onishi M, Machado EL (2011) Unbbayes : a java framework for probabilistic models in AI

Managing Variety in the Era of Mass Customization: A Decisional Algorithm for Product Platform Design

Ludovica Diletta Naldi, Riccardo Venturi, Francesco Gabriele Galizia, Marco Bortolini, and Matteo Gabellini

Abstract In recent years, companies have been struggling to best cope with the increasing request for customized and personalized products that characterizes mass customization paradigm. The efficient management of product variety requires effective solutions to make companies able to deliver such products in a short time and at a high-quality rate. The Delayed Product Differentiation (DPD) is rising as one of the most effective strategies to manage mass customization, implemented in practice using product platforms. Platforms are intermediate products formed by the most common components within a product family and managed through a make-to-stock strategy. After the arrival of the customer order, platforms are transformed into the final variants through assembly/disassembly customization tasks, delaying the final product differentiation point. In such a scenario, this paper proposes an innovative decisional algorithm for product platform design and selection to reduce the overall effort for platform customization into final variants. The application of the procedure to a reference industrial case study showcases its relevance and validity in managing high product variety.

Keywords Mass customization · Product platform · Variety management

1 Introduction and Paper Positioning

Mass customization (MC) is a business strategy that combines the efficiency of mass production with flexibility to meet individual customer preferences. This approach has seen significant growth due to advanced technologies, manufacturing processes, and consumer demand for personalized products. However, managing an increasing variety of products, both in terms of design and configuration, remains a major challenge. Product platforms offer a solution, enabling companies to create product variants through shared components, thus reducing management complexity. These

L. D. Naldi · R. Venturi · F. G. Galizia (✉) · M. Bortolini · M. Gabellini
Alma Mater Studiorum—University of Bologna, Bologna, Italy
e-mail: francesco.galizia3@unibo.it

M. Manns (ed.), *New Paradigms for Anticipated Uncertainty*, Lecture Notes in Mechanical Engineering, https://doi.org/10.1007/978-3-032-16889-4_14

platforms are produced following a Make to Stock (MTS) strategy and then reconfigured into the final variants through a Make to Order (MTO) approach, achieving economies of scale while maintaining customization [1, 2]. MC is particularly relevant in industries such as automotive, electronics, and fashion, where customers ask for personalized products. However, the growing variety increases challenges in production and inventory management. Existing literature proposes various strategies and models for product platform design and implementation, including optimization methods. For instance, Ben-Arieh et al. [4] proposed a model for designing multiple platforms to minimize production costs. Hanafy and ElMaraghy [5] introduced a model for modular product platform design, while Xiong et al. [6] addressed product modularity and supply chain management through a Stackelberg game approach. Other studies developed non-linear bi-objective models to optimize platform design and inventory management [1, 7]. While optimization models are effective, they are computationally intensive, making heuristic methods necessary for practical industrial applications. Studies by Hanafy and ElMaraghy [8] and Galizia et al. [9] developed decision-support systems using heuristic algorithms for platform design. Additionally, Longo et al. [10] and Bortolini et al. [2] proposed two-stage platform design models to manage MC in industrial contexts, applying these methods to real case studies. However, heuristic methods often simplify production constraints, lacking the ability to consider technological precedencies and customization activities' impact to produce the variants' demand in the creation of efficient platforms platforms. According to this background, this paper presents an innovative decisional algorithm for product platform design and selection with the aim of reducing the global effort customizing the platform into final variants. Based on the research object, the paper attempts to answer the following research question (RQ): *How can a decisional algorithm optimize platform design to reduce customization effort?* The proposed algorithm is validated through an industrial case study, showcasing its relevance for the industrial context.

2 The Decisional Algorithm

The increasing variety in production mixes characterizing the current industrial scenario requires innovative strategies to manage the resulting high manufacturing complexity. The adoption of a Delayed Product Differentiation (DPD) strategy, taking advantage of the positive sides of MTO and MTS, is of help in reducing the manufacturing complexity delaying the customization activities through the introduction of product platforms. The proposed decisional algorithm generates product families and associated product platforms trying to reduce the global effort for platform customization into the final variants. The customization effort considers the number of assembly/disassembly customization tasks required to reconfigure the platform structure deriving the associated variants. The algorithm lies on the following assumptions, validated by the standard literature and industrial practice [2, 9, 11]: the production mix is known, as well as the bill of materials (BOMs) and assembly cycle constraints of each product variant; the manufacturing is made through assembly tasks and the cycle is expressed as a linear sequence of components to be assembled; each product family is associated with a single product platform from which derive all the product variants in the family; customization of product platforms is realized by assembly and disassembly of components to/from platforms; a product variant can be obtained from a product platform if the variant assembly cycle shares the platform's technological cycle from the beginning.

The following notations are introduced.

Indices

Component	$c, d = 1, \ldots, C$
Product variant	$v = 1, \ldots, V$
Product family	$f = 1, \ldots, V$
Product platform	$p = 1, \ldots, P$

Parameters

n Number of components shared by the product variants from the beginning of their assembly cycles

$$BOM_{cv}\begin{cases} 1 \text{if variant } v \text{ includes component} c \text{in its BOM} \\ 0 \text{otherwise} \end{cases}$$

D_v Demand of variant v [pcs/month].

The proposed algorithm is divided into two consecutive steps. Step I requires the BOM and the assembly precedence relationships among the components for each product variant. In output, this first step defines product families based on the similarity of the assembly cycles of the variants and made of at least two variants. This last consideration is required to avoid the situation where a single variant is derived from a platform. These families are used as input to step II to further design the product platform for each family minimizing the customization effort to derive the associated variants according to the Platform Reconfiguration Index (PRI) introduced by [9]. The flowchart of the algorithm is proposed in Fig. 1.

A parameter n is introduced, with $n = 1, \ldots, C$, to represent the level of similarity of the assembly cycle of product variants in the same family.

Step I starts with setting the value of n and grouping the product variants sharing the same initial n components of their assembly cycle in product families. As a fixed value of n corresponds to a specific and univocal set of f different product families, by varying the n parameter different sets of product families are obtained. Small values of n are normally associated with large families of variants sharing a small initial part of their assembly cycle, while higher values of n are associated with smaller families of variants having more similar assembly cycles. To design the related product platform, each product family has to be composed of at least two variants, so its feasibility is checked. If a single family is composed of a single

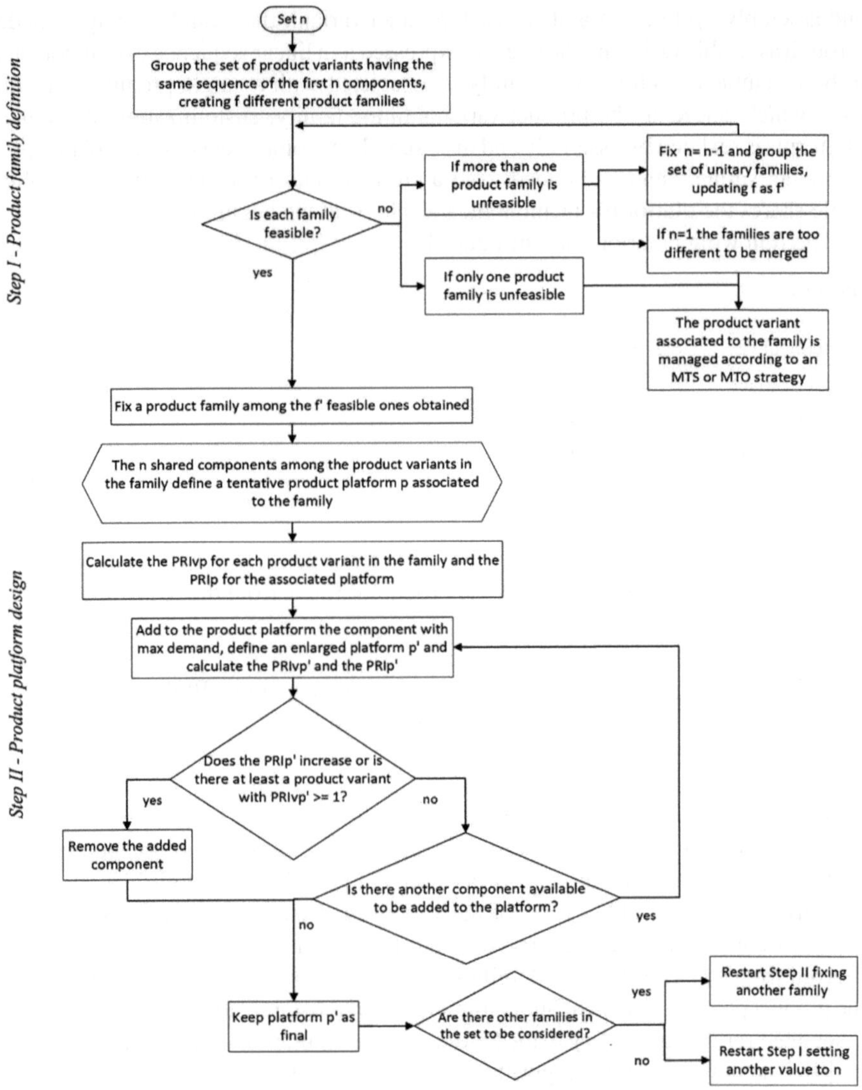

Fig. 1 Flowchart of the decisional algorithm proposed

variant, this variant is considered too different from the others, requiring ad-hoc production methodologies such as Assembly to Order (ATO) and MTS according to its characteristics and demand. If more than one family is composed of a single variant, the value of the n parameter is decreased by one trying to group these unitary families to get feasible ones. If n is already set to 1, the variants in the families do not share even the first component and they are considered too dissimilar to be obtained by DPD.

The set of f' feasible product families obtained from the step I is given in input to step II to provide the design and assessment of the related product platforms. Fixed a family among the f' feasible ones, step II starts defining a tentative product platform associated with the family. This tentative platform includes the n, or less if the family is obtained through merging of unfeasible ones, shared components among the variants in the family. According to this identified tentative platform and to the variants' assembly cycles, the values of the following variables are computed.

$$a_{pcv}\begin{cases}1 \text{ if component } c \text{has to be assembled to platform } p \text{ to derive variant} v\\ 0 \text{ otherwise}\end{cases}$$

$$d_{pcv}\begin{cases}1 \text{ if component } c \text{ has to be disassembled from platform } p \text{ to derive variant } v\\ 0 \text{ otherwise}\end{cases}$$

$$x_{vp}\begin{cases}1 \text{ if variant } v \text{ is associated to platform } p\\ 0 \text{ otherwise}\end{cases}$$

The platform considered is then evaluated according to the PRI index, that represents the effort required to reconfigure the product platform into a specific variant. The PRI is calculated for each variant v that is derived from the platform p, i.e., PRI_{pv}, comparing the number of assembly and disassembly tasks required for the platform customization with the total number of assembly tasks required to produce the variant from scratch, e.g., using an ATO approach, hence assembling the total number of components of the variant. A global platform reconfiguration effort, i.e., PRI_p, is calculated as a weighted average over the associated variants' demand. The mathematical formulation of PRI_{pv} and PRI_p is expressed by Eqs. [1] and [2].

$$PRI_{vp} = \frac{\sum_{c=1}^{C} a_{pcv} + \sum_{c=1}^{C} d_{pcv}}{\sum_{c=1}^{C} BOM_{cv}} v = 1, \ldots, V \tag{1}$$

$$PRI_p = \frac{\sum_{v=1}^{V} PRI_{vp} * D_v * x_{vp}}{\sum_{v=1}^{V} D_v * x_{vp}} p = 1, \ldots, P \tag{2}$$

According to the value of PRI_{vp} three cases are considered:

- $PRI_{vp} = 0$, no effort is required for platform p customization as the structure of p coincides with the assembly cycle of the variant v;
- $0 < PRI_{vp} < 1$, platform p shares some components with variant v but requires some assembly/disassembly customization tasks to derive the variant;
- $PRI_{vp} \geq 1$, the effort to produce variant v through platform p customization is greater than the effort to assembly v using an ATO strategy so the use of the platform p should be avoided.

The values of PRIvp and PRIp are used throughout the algorithm to assess the platform structure. The algorithm evaluates whether to enlarge the platform by adding a new component, prioritizing variants with higher market demand. If the added component helps serve high-demand variants, it is included, forming an enlarged platform p′. The PRI values are updated accordingly to assess the impact on reconfiguration effort. If any PRIvp' > 1, or if the average PRIvp' increases compared to PRIvp, the platform is considered inefficient, and the component is removed. Otherwise, further components may be added. This process continues until stopping conditions are met, defining the final platform for each product family. Step II is repeated for each family f′ from Step I. The algorithm can be rerun with a different n value to generate alternative product families and platforms. The decisional algorithm is implemented in Microsoft Excel through VBA, enabling the resolution of real industrial case studies over an Intel® Core™ i7-3770 CPU @ 3.40 GHz and 16.0 GB RAM workstation with an almost null computational time.

3 Industrial Application

This section applies the proposed decisional algorithm to an Italian industrial case study, representative of a SME facing a medium–high product variety of standard products. In particular, the company produces 38 different plastic valves involving 93 different components and is investigating the possibility of implementing a DPD strategy by introducing appropriate product platforms, to face the production and management complexity. For the sake of brevity, the BOMs and assembly cycles of the production mix are omitted, and they can be found in [2].

According to the algorithm structure, step I defines a set of feasible product families for each fixed value of the parameter n. The value of n is set in the range from 1 to 8, grouping the product variants having from a single shared initial component up to 8 in different families and identifying 8 main scenarios. Since the maximum number of shared components among the variants is 8, higher values for n are not considered. Each different set of families is given as input to step II, that provides the final platforms minimizing the reconfiguration effort to derive their associated variants.

Two metrics are used to analyze and compare the platform's results for the different scenarios. A total demand for a platform is derived as the sum of the demand of the associated variants and used to compute a mean PRI for the n scenario, i.e., PRI_n. It is calculated as a weighted average of all the PRI_p of the platform related to each n over the platform's demand, and it is used to map the mean platform reconfiguration effort for a single scenario, as in Fig. 2. The second metric in Fig. 3., i.e., "*Variety*", represents the number of different items required to be stocked in the n scenario. It considers the number of different platforms designed and the number of possible variants detected in step I where the DPD strategy cannot be applied, assuming they are managed according to an MTS strategy, due to industrial considerations.

According to the graphs, as the value of common components among the variants associated to a platform n is set from 1 to 8, the mean reconfiguration effort for the platforms corresponding to the n-scenario decreases while the number of different stocked items increases. More generalized and broad platforms, including a low

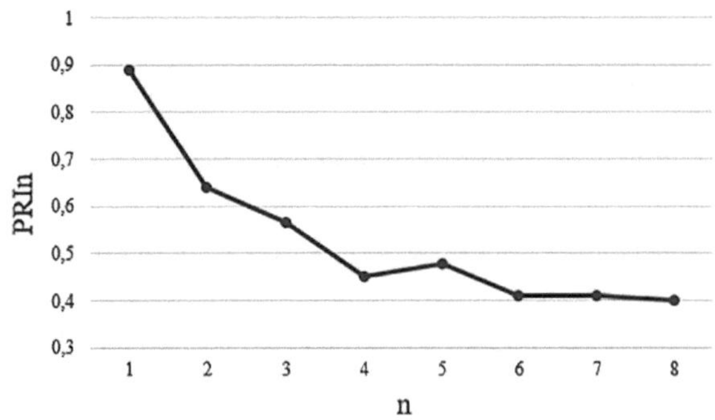

Fig. 2 Trend of PRI_n as function of the n value

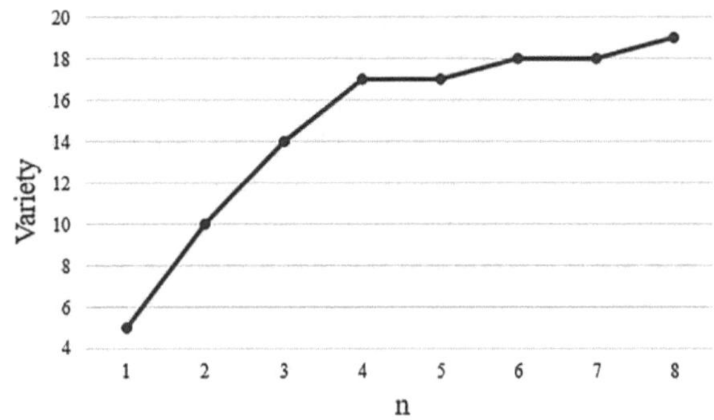

Fig. 3 Trend of variety as function of the n value

number of shared components among variants, require many assembly tasks to derive the associated variants, while more specific platforms require fewer customization tasks and corresponding reconfiguration effort. As an example, the first scenario is associated to $PRI_1 = 0.8892$ while the last one corresponds to $PRI_8 = 0.3994$. On the contrary, using broad and small platforms allows companies to stock a low number of different items from which derive the entire production mix, while fixing a high value of n leads to the definition of a high number of different families and associated specific platforms. Thus, the *Variety* index ranges from 5 to 19 at the extremes of the second graph, considering 18 different platforms to stock and a product variant to produce using an MTS when $n = 8$.

A reference example of the outcome of the algorithm when $n = 3$ is presented in Table 1. Step I defines 12 different families, 10 of which are associated with variants sharing n initial components while the last 2 are obtained by grouping previously infeasible families decreasing the value of n. Two product variants, i.e., V19 and V27, are left out from the grouping and it is assumed they are produced through an MTS. Step II enlarges the tentative platform associated with each family, designing the final platform structure in Table 1. Such a scenario is characterized by a value of the *Variety* index equal to 14 and an average PRI_3 equal to 0.5638.

Table 1 Outcome of the algorithm for $n = 3$

Family/platform ID	Associated variants	Included components	PRI_p
F/PL1	V2, V3, V5, V6	C2–C6–C7–C10–C16	0.6009
F/PL2	V4, V7, V10	C1–C2–C4–C7–C8	0.6825
F/PL3	V8, V9, V11, V12	C2–C7–C10–C13–C15–C25–C27–C29–C30	0.5767
F/PL4	V13, V16	C13–C34–C35–C36–C37–C38–C39–C40–C41	0.0491
F/PL5	V14, V15, V17, V18	C13–C40–C41–C42–C43	0.6487
F/PL6	V23, V34, V37	C10–C15–C16–C17	0.6804
F/PL7	V24, V35, V38	C10–C15–C25–C26	0.6941
F/PL8	V25, V26, V29, V30	C10–C15–C78–C79–C80–C81–C82–C83–C84	0.2327
F/PL9	V28, V31, V32	C78–C79–C84–C86–C87	0.4977
F/PL10	V20, V21	C10–C50–C52	0.6729
F/PL11	V22, V33	C1–C3	0.8091
F/PL12	V1, V36	C1	0.9016

The scenarios represent different possible platform configurations for the case study, leaving the final decision on which scenario or single platform structure to adopt according to the company marketing and production considerations.

4 Conclusions and Future Research

In today's industrial landscape, marked by growing product variety, strategies like Delayed Product Differentiation (DPD) help reduce manufacturing complexity. This paper introduces a decisional algorithm for defining product families and designing product platforms, using a reconfiguration index to assess customization effort. Applied to a real case in the plastic valve sector, the algorithm proved effective in managing variety and streamlining production. These results suggest its potential for broader application, especially in industries facing high product variability. Future work will focus on validating scalability and integration with digital tools for automated decision-making.

References

1. Naldi LD, Galizia FG, Bortolini M (2025) Balancing storage cost and customization time in product platform design: a bi-objective optimization model. Int J Adv Manuf Technol 1–14
2. Bortolini M, Calabrese F, Galizia FG, Regattieri A (2023) A two-step methodology for product platform design and assessment in high-variety manufacturing. Int J Adv Manuf Technol 126(9):3923–3948
3. Pine BJ, Victor B, Boynton AC (1993) Making mass customization work. Harv Bus Rev 71(5):108–111
4. Ben-Arieh D, Easton T, Choubey AM (2009) Solving the multiple platforms configuration problem. Int J Prod Res 47(7):1969–1988
5. Hanafy M, ElMaraghy H (2015) A modular product multi-platform configuration model. Int J Comput Integr Manuf 28(9):999–1014
6. Xiong Y, Du G, Jiao RJ (2018) Modular product platforming with supply chain postponement decisions by leader-follower interactive optimization. Int J Prod Econ 205:272–286
7. Moussa M, ElMaraghy H (2022) Multi-period additive/subtractive product platform design and inventory management. Int J Prod Res 60(24):7262–7280
8. Hanafy M, ElMaraghy H (2015) Developing assembly line layout for delayed product differentiation using phylogenetic networks. Int J Prod Res 53(9):2633–2651
9. Galizia FG, ElMaraghy H, Bortolini M, Mora C (2020) Product platforms design, selection and customisation in high-variety manufacturing. Int J Prod Res 58(3):893–911
10. Longo F, Padovano A, Cimmino B, Pinto P (2021) Towards a mass customization in the fashion industry: an evolutionary decision aid model for apparel product platform design and optimization. Comput Ind Eng 162:107742
11. Naldi LD, Galizia FG, Bortolini M (2025) Balancing storage cost and customization time in product platform design: a bi-objective optimization model. Int J Adv Manuf Technol 136:4933–4946

References

Engaging Consumer Uncertainty and Creativity During the Mass Customization Co-design Experience

Frances Turner, Marie Watts, and Nikola Suzic

Abstract Uncertainty is an enduring facet of our world, shaping individual experiences and societal dynamics, including business processes and strategies. While challenging, uncertainty also serves to catalyze creativity, a foundational concept of mass customization (MC) co-design toolkits (i.e., product configurators). The present research explores whether the MC experience could be deliberately designed to foster uncertainty to nudge a variety of consumer types to embrace unknowns. This promotion of uncertainty may seem counterintuitive due to uncertainty's role in fostering complexity which acts as a significant barrier to delivering value-laden MC experiences. However, the interaction of uncertainty and complexity presents an opportunity to enhance consumers' creative engagement. The MC co-design toolkit, therefore, holds a potential role in empowering individuals struggling with the ambiguity of uncertain outcomes through cultivating confidence during any challenges of the MC experience. This approach envisions a toolkit to facilitate collaboration between consumers who thrive in uncertain environments and those whose risk aversion inhibits their inherent creative potential. This interactivity can yield intangible relational benefits, including self-realization and enhanced individual and societal well-being. Moreover, optimizing the MC toolkit to embrace creative uncertainty could offer providers unique observational insights into managing the unpredictability inherent in consumer-driven creativity. By leveraging concepts of Knightian uncertainty and behavioral economics, particularly ecological and bounded rationality, the present research evaluates the transformative potential of designing MC experiences to harmonize creativity and uncertainty.

F. Turner (✉)
School of Business, Ithaca College, Ithaca, NY, USA
e-mail: fturner@ithaca.edu

M. Watts
San Francisco, CA, USA
e-mail: lmw2@cornell.edu

N. Suzic
Department of Industrial Engineering, University of Trento, Trento, Italy
e-mail: nikola.suzic@unitn.it

M. Manns (ed.), *New Paradigms for Anticipated Uncertainty*, Lecture Notes in Mechanical Engineering, https://doi.org/10.1007/978-3-032-16889-4_15

Keywords Mass customization · Product configurator · Creativity · Knightian uncertainty · Bounded rationality · Ecological rationality · Co-design toolkit

1 Introduction

The Latin phrase "Dubito ergo creo," translating to "I doubt therefore I create," is utilized by Beghetto [1] in the title of his keynote address at the 2021 Creativity Conference. This phrase aptly introduces the central argument of this paper, which posits that the design of the MC co-design toolkit to promote uncertainty can engage consumer creativity, thereby enhancing the perceived value of the MC experience. Through the integration of social, learning, and feedback elements within the co-design configurator, consumers with a creative inclination can be encouraged to collaborate with or assist those who are averse to ambiguity, enabling them to embrace their inherent human creativity. In doing so, the MC co-design experience contributes to individual, community, and societal well-being.

Uncertainty is a persistent aspect of our world, influencing individual experiences and societal dynamics, including business processes and strategies. While it presents challenges, uncertainty also acts as a catalyst for creativity [1] which is a fundamental concept of mass customization (MC) co-design toolkits (i.e., product configurators). This research investigates whether the MC experience can be intentionally designed to foster uncertainty, encouraging a diverse range of consumer types to embrace the unknown. Promoting uncertainty may appear counterintuitive given its relationship to complexity [2] since the latter is a significant barrier to delivering value-laden MC experiences. The interaction between uncertainty and complexity offers an opportunity to enhance consumers' creative engagement. Consequently, the MC co-design toolkit may empower individuals who struggle with ambiguity by cultivating confidence during the challenges of the MC experience.

The paper is structured as follows: Sect. 2 explores uncertainty and related concepts. Section 3 examines the nature of creativity. Section 4 considers the interplay between uncertainty and creativity and how it fosters collaboration between consumers who thrive creatively in uncertain environments and those who do not. Section 5 describes how the MC co-design toolkit can be configured to promote the benefits that uncertainty and creativity provide in delivering value-laden consumer customization experiences. Section 6 proposes the contributions that consumers' creative uncertainty might yield for MC providers. In Sect. 7, the evaluation is concluded, noting limitations and offering recommendations for future research.

2 Uncertainty

Kozyreva and Hertzwig [3] observe a significant "interdisciplinary rise in interest in uncertainty and its kindred topics of risk, complexity, and unpredictability" (p. 1518). Uncertainty embodies both negative and positive implications. It can evoke pessimism or optimism, discomfort, or a sense of potential [4]. The dual nature of uncertainty is influenced by cognitive and affective processes inherent in human beings [5]. Despite numerous definitions, a conceptual consensus on uncertainty remains elusive [3, 6]. Knight [7] characterized uncertainty as an outcome with an immeasurable probability, a scenario where forming future expectations becomes impossible [8]. Taylor [9] described uncertainty as the inability to predict future outcomes and consequences of a decision. Dequech [10] attributes uncertainty to a lack of knowledge about the future, stemming from the complexity inherent in decision-making environments. Rumsfeld [11] coined the term "unknown unknowns" to describe uncertainty as "things ... we don't know we don't know" (p. 6). Booker and Ross [12] define uncertainty as "that which is not precisely known" (p. 669).

Desrochers and Outreville [13] differentiate between risk, uncertainty, and ambiguity, suggesting that uncertain situations lack prior information on loss probability. Beghetto [4] describes uncertainty as "a state of doubt...of not knowing" (p. 1), advocating that the benefits of engaging with uncertainty surpass the costs of avoiding it. Dibiasi and Iselin [8] define uncertainty as "ignorance of future events, representing a situation in which it is no longer possible to form expectations about the future" (p. 2). Mousavi and Gigerenzer [6] propose that uncertainty arises when options, consequences, and their probabilities are unknown. Kurdoglu et al. [14] assert that "extremely uncertain" environments (p. 4) are those that are subjectively novel, lacking prior experience, where information searches yield no quantitative or qualitative data, and cannot be navigated with heuristics. Hovious [5] identifies three information theory definitions of uncertainty: gaps in understanding, cognition that triggers negative emotions, and noise or entropy in communication, where higher levels indicate greater complexity.

Ambiguity, a lesser degree of uncertainty, is vagueness [13] or doubt "about probability, created by missing information that is relevant and could be known" [15], p. 330]. Decision-makers prefer definitive, certain outcomes over vague or ambiguous ones, with those confident in their competence more inclined to act and take risks [16]. Conversely, individuals who question their competence or confidence in their knowledge compared to others tend to exhibit ambiguity aversion, i.e., reluctance to act in the face of uncertain outcomes [16].

3 Knightian Uncertainty, Bounded Rationality, and Ecological Rationality

In the domains of entrepreneurship and behavioral economics, Knightian uncertainty posits that profit is the reward for individuals who demonstrate fearlessness, take risks, and innovate by venturing into the unknown to create new offerings [6, 7]. Recognizing Knight's [7] impact on the scholarship of judgment and decision-making, [3] observe, "He argued that in situations of risk, probabilistic information is either available a priori or can be gathered statistically, but in situations of uncertainty people have no choice but to rely on subjective estimates. These estimates are formed under conditions of unmeasurable uncertainty around unique events and situations for which no probabilistic information yet exists" (p. 1519). Simon's [17] concept of bounded rationality addresses uncertainty as a condition where cognitive constraints and environmental complexities hinder an individual's ability to assess options and predict outcomes. His conceptualization of satisficing acknowledges that individuals adapt to manage situations with unknown or complex outcomes and cognitive costs by settling for satisfactory outcomes rather than pursuing optimal ones [3, 18].

Mousavi and Gigerenzer [6] emphasize that "under uncertainty, optimization is by definition infeasible, either because the full set of options and consequences or their probabilities cannot be known in advance or because the problem is computationally intractable" (p. 364). Ecological rationality conceptualizes uncertainty as an element of an ecosystem, described as "the organism-environment system" [3, p. 1532]. Uncertainty is contingent upon the nature and complexity of the environment, as well as the individual's cognition, knowledge, and experience. "The ecological approach to rationality, exemplified in theories such as … [fast-and-frugal heuristics] posits that different decision-making procedures are rational in various decision environments" [19, p. 1]. Essentially, ecological rationality is situational, allowing decision-makers to develop and employ cognitive and affective shortcuts, or heuristics, that enable them to adapt and navigate the "uncertainty [that] comprises both environmental unpredictability and uncertainties that stem from the mind's boundaries" [3, p. 1529]. Kurdoglu et al. [14, 20] argue that a different adaptive strategy, eristics, is better when dealing with extreme uncertainty. Eristics are "action-triggering short-cuts that draw on hedonic urges" [20, p. 763] and as such make no consideration for decision consequences. Further, [14, 20] assert that this type of decision making reduces the individual's anxiety, allowing for creation of more novel, radical solutions.

4 Creativity and Uncertainty

No singularly agreed upon definition of creativity exists, but several scholars conceptualize it as the production of novel and task-relevant outcomes [21]. Florida [22] describes creativity as "multifaceted and multidimensional" (p. 15). In work on the measurement of creativity and the relationship between firms' products, consumer behavior and willingness to pay (WTP), Horn and Salvendy [23] define product creativity as "the originality and appropriateness of a product that elicits a positive affect compatible with the consumer or judge" (p. 155). Csikszentmihalyi's [24] systems-based view posits that for creativity to exist, interaction must occur between person, society, and a specific realm, or environment, or culture. Walia [25] describes creativity as the act of generating new ideas, products, or solutions that are both novel and useful, and emerges from the interaction between individual cognition, the social environment, and perceived environmental disequilibrium. Runco and Chand [26] define creative thinking as a process leading to insights, solutions, and adaptive ideas. Their model highlights the interplay of motivational, cognitive, and affective processes, including judgment, ideation, problem finding, knowledge, and motivation. Mehta and Dahl [27] identify four key factors influencing consumer creativity: cognitive ability, motivation, social personality, and environment. Intrinsic motivation is crucial for creativity, whereas extrinsic motivation focused solely on rewards, grades, and scores diminishes creativity [28, 29].

Fox and Weber [16] demonstrate that increased information can lead to greater uncertainty and reduced willingness to take risks, even if the information is not directly relevant to the outcome. Shulman et al. [30] note that uncertainty-reducing information provided before a purchase decision can increase decision reversals. Mueller et al. [31] found that individuals hold ambivalent attitudes towards creativity, with an implicit bias against it under conditions of uncertainty, which can hinder the recognition of creative ideas. Their study revealed that participants experiencing uncertainty exhibited bias against creativity when compared to practicality, differing significantly from those in a baseline condition. Participants with low tolerance for uncertainty rated creative ideas as less creative than those with high tolerance, with implicit bias mediating this relationship; but the authors acknowledge that these findings contradict existing research, suggesting their methodology may have reduced social desirability bias [31].

Rios et al. [32] assert that uncertainty does not invariably negatively impact creativity. Beghetto [4] explores the relationship between uncertainty and creativity, positing that creativity cannot exist without uncertainty, which is inherent in all creative endeavors. The author suggests that encounters with uncertainty can generate doubt, leading to "new possibilities" [4, p. 5] for creative action. Recognizing these opportunities helps resolve doubt in the short term, inspiring continued exploration and engagement in creative activities. Furthermore, belief in one's agency or control influences both creative action and the capacity to think and act creatively in uncertain situations.

Faraji-Rad and Pham [33] affirm the interplay between affect, self, and uncertainty. They emphasize the critical role of one's "feelings, moods, and emotions" in relation to the self (p. 3). Uncertainty poses a threat to the sense of self, prompting a "reliance on inputs that are closer to the self and have high subjective validity" [33, p. 1]. Rios et al. [32] suggest creative action mitigates self-uncertainty and restores self-clarity. For individualists, who define themselves by their distinctiveness, creative expression serves as an effective coping mechanism for uncertainty. Conversely, for conformists who do not define themselves by unique characteristics, creative expression may be less relevant and potentially incongruent with their self-concept [34]. This indicates that self-uncertainty and individualism, rather than merely influencing creative performance through effort, motivate individuals to become genuinely engaged and absorbed in tasks perceived as creative [32, 34].

5 Creativity and MC

In describing creativity's ascendancy, [22] foretold its economic significance, ubiquity, and the transition of the world from "creative economy to creative society" (p. 17). Hearn [35] defines creative economy as "the creation, capture, and consumption of intangible value through the application of creative, technological and innovation know-how" (p. 2). The MC co-design toolkit facilitates creative experiences for consumers and fosters a collaborative relationship with the firm. Participation in MC co-design yields both tangible and intangible benefits, including opportunities for individual creative expression. Consumers derive enjoyment from the creativity involved in co-designing new products, appreciate the empowerment it offers, and value the social interaction and public recognition associated with co-production [36].

As noted by [37], the "joy of performing a creative or artistic act" (p. 413) is integral to the MC co-design process. The MC toolkit encourages users to perceive the outcome of the experience as uniquely their own. This perception is manifested in "strong feelings of pride" [38, p. 323], "creative achievement value" [39, p. 505], and the "I designed it myself" effect [40]. Franke et al. [41] demonstrated that providing feedback on the uniqueness of a co-designed offering enhances its perceived value and WTP. MC studies illustrate the significant value individuals place on self-designed objects, stemming from the belief that they are the originators of these creations [40]. By informing them that their design is one-of-a-kind, MC consumers perceive greater value in their customized creations [42]. Consumers' creative engagement increases their WTP for self-created products, and when the MC configurator is optimally designed, the full value of MC is captured through psychological ownership of the co-designed MC offerings [40]. Furthermore, [43] found that framing customization as a creative process enhances consumer perception of the co-creation experience, thereby boosting their WTP more effectively than when options are framed as addressing objective needs.

6 The MC Co-design Toolkit: Cultivator of Creative Collaboration and Confidence

The interplay between uncertainty and complexity offers an opportunity to enhance consumers' creative engagement. The mass customization (MC) experience can be intentionally designed to foster uncertainty, thereby encouraging a diverse range of consumer types to embrace the unknown. The MC co-design toolkit, therefore, has the potential to empower individuals who struggle with the ambiguity of uncertain outcomes by cultivating confidence during the challenges of the MC experience. This approach envisions a toolkit that facilitates collaboration between consumers who thrive in uncertain environments and those whose aversion to risk or ambiguity inhibits their creative potential.

Scholars propose methods to harness the opportunities presented by uncertainty, suggesting interventions that may be integrated into MC co-design toolkits. Moreau et al. [44] discovered that, in the absence of time constraints, input limitations during generative tasks stimulate consumer creativity and the development of more innovative solutions. Beghetto [4] advocates structured uncertainty to foster creative learning, which emerges from the formation of new and meaningful thoughts and actions, leading to creative contributions that enhance the lives and learning of others (p. 4). Structuring customized options through creative attributes improves consumer perception of an offering as customized, particularly when these choices align with consumer preferences [43]. This perception diminishes when objective needs are presented as choices. The authors describe this decision framing of choices as a creative process termed the "illusion of increased customization." This results from an enhanced sense of creative achievement, which leads to a higher perception of value and customization, as evidenced by willingness to pay. [43] assert that presenting each design choice on an individual page elevates perceived customization.

Zaggl et al. [45] identified that consumers opt for hedonic features in product options to express individuality, while utilitarian features are selected to satisfy the need for conformity. They propose mass customization (MC) toolkits should offer a combination of these product attributes by "providing a broad range of options related to hedonic attributes to allow for differentiation ... [and] ... limit[ing] the choice set for utilitarian product attributes" (p. 205). Conformity reduces the cognitive complexity of choice for consumers, as utilitarian product options mitigate risk and uncertainty. Conversely, the affective need for uniqueness diminishes uncertainty, as emotions such as joy and delight drive the hedonic pursuit of adventure and novel, creative solutions. MC scholars have determined that enhancing perceived control and enjoyment is fundamental to MC configurator design [46]. Hovious [5] supports this notion, suggesting that promoting joyous uncertainty can inspire individuals to address the negativity associated with information uncertainty. Achieving this requires agency, the self-efficacy known to foster well-being and the ability to cope with information uncertainty. Since agency is not a fixed trait, it can be nurtured in individuals. MC toolkits can serve as a conduit and space to connect creatively inclined and uncertainty averse consumers.

Haase et al. [47] propose interventions that incorporate self-assessment measures of creativity and training programs aimed at developing creative self-efficacy, which refers to individuals' beliefs in their ability to produce creative outcomes. Richter et al. [48] emphasize the importance of the interaction between self-efficacy and access to others' knowledge and skills within social environments, showing that "shared team informational resources" enhance "more positive ... creative self-efficacy and individual creativity" (p. 1282). This finding is promising for structuring MC experiences to foster agency in individuals who are averse to uncertainty by connecting them with more creatively inclined peers. As previously noted, creativity is not a fixed trait, rather, everyone possesses creative potential, and creative skills can be cultivated.

Nohutlu et al. [49] argue that intrinsic and extrinsic motivations affect consumer participation and engagement in co-creation via online communities with the combination of intrinsic and internalized extrinsic motivation—learning, skills development, and needs satisfaction—enhances consumers' willingness to engage in online community co-creation. Amabile and Pillimer [28] recommend the development of training programs and interventions for enhancing creative performance and promoting intrinsic motivation and creativity in organizational settings, such as providing autonomy support and fostering a sense of "harmonious passion" (pp. 8–9). On both individual and team levels, [28] assert that creative behavior is supported by environments that encourage interpersonal congruence, paradoxical thinking, moral flexibility and provide opportunities for social interaction, feedback, and autonomy. Both [49] and [28] concur on the significance of the social environment in fostering creativity and underscore intervention tactics and programs should reduce the emphasis on extrinsic motivators, such as evaluations and rewards, as they tend to diminish intrinsic motivation and creativity.

Tansley [50] introduced the concept of the ecosystem, emphasizing the inseparable, dynamic nature of continuous exchanges and relationships between organisms and their environments, shaped by interactions and mutual adjustments. Participants' roles shift over time and as systems and environments evolve [51] these interconnected networks adapt continuously to changing conditions [52]. MC toolkits that are designed to empower consumers to co-design in concert with their social networks, can serve to create webs of interaction among users, firms, and platforms. They may be seen as creating mini ecosystems, fostering collaboration and iterative feedback loops that drive innovation [37].

MC scholars have demonstrated that integrating access to social networks during the MC experience is beneficial to consumers. In the co-design toolkit's key role as the enabler of relational value, it is important to design robust collaborative social elements into MC configurators [53–55]. Shared ideas, designs, and experiences facilitated by peer input, user communities, visual feedback, and real-time interaction enhance MC consumers' perceived control, enjoyment, and sense of accomplishment while mitigating complexity [40, 56–59]. These all promote development of problem-solving skills, nurturing the individual's creative and self-efficacy equipping her to manage uncertainty [47, 48, 57]. By leveraging collective intelligence to refine offerings [60], the shared value creation enabled by an MC ecosystem makes customer participation a catalyst for ongoing adaptation and competitive advantage [61].

7 Conclusion

Optimizing the MC toolkit with features that enhance consumer creativity, encourage Knightian uncertainty, and prompt ecological rationality may provide firms with distinctive observational insights to manage the unpredictability inherent in consumer-driven creativity. Rather than succumb to operational rigidity that may lead to stagnant business practices and financial challenges, MC embraces differentiated consumer needs and preferences spanning individual proclivities for creativity. Ignoring these dynamics is akin to neglecting the lessons of natural ecosystems, where limited genetic diversity restricts adaptive evolutionary processes that may lead to species collapse. Successful and sustainable ecosystems regularly confront challenges and continuously adapt to the uncertain unknown unknowns.

Asymptotic risk necessitates establishing a boundary between the business imperative to reduce or mitigate risk and the consumer's desire for security, stability, surprise, and delight, all derived from uncertainty. Indeed, there exists a competitive tension between offering too little to highly creative consumers and too much to individuals who avoid ambiguity or uncertainty. Nevertheless, the interaction and collaboration of firms and diverse consumer types are essential, requiring an ecosystem that mimics nature's variety and adaptive essence to thrive.

This research should delve further into the relationship between task complexity, creativity, and uncertainty, and how consumers and providers can manage these via MC toolkit design. Further, the work would benefit from deeper inquiry into the effects of individual differences, as well as how to embed extant technology into MC configurator features that promote interaction between consumers of varying degrees of creative expression. Designing MC co-design toolkits to facilitate creative uncertainty represents the transformative potential of MC experiences that harmonize collaborative creativity and unknown unknowns to enhance the well-being of individuals, providers, communities, and society.

References

1. Beghetto RA (2021) There is no creativity without uncertainty: Dubito Ergo Creo. Journal of Creativity 31:100005
2. Winkler JA (2016) Embracing complexity and uncertainty. Ann Am Assoc Geogr 106(6):1418–1433

3. Kozyreva A, Hertwig R (2021) The interpretation of uncertainty in ecological rationality. Synthese 198(2):1517–1547
4. Beghetto RA (2020) Uncertainty. In: The Palgrave encyclopedia of the possible. Springer International Publishing, Cham, pp. 1691–1697
5. Hovious AS (2022) Finding joy in uncertainty. Libr Trends 70(4):472–489
6. Mousavi S, Gigerenzer G (2017) Heuristics are tools for uncertainty. Homo Oeconomicus 34:361–379
7. Knight, F. H.: Risk, uncertainty, and profit, Vol. 31, Houghton Mifflin (1921).
8. Dibiasi A, Iselin D (2021) Measuring knightian uncertainty. Empir Econ 61(4):2113–2141
9. Taylor JW (1974) The role of risk in consumer behavior: A comprehensive and operational theory of risk taking in consumer behavior. J Mark 38(2):54–60
10. Dequech D (2001) Bounded rationality, institutions, and uncertainty. J Econ Issues 35(4):911–929
11. Rumsfeld, D.: Defense.gov News Transcript: DoD News Briefing – Secretary Rumsfeld and Gen. Myers. United States Department of Defense. February 12, 2002. Archived from the original on April 6, 2016. https://en.wikipedia.org/wiki/There_are_unknown_unknowns, last accessed 2025/3/14.
12. Booker JM, Ross TJ (2011) An evolution of uncertainty assessment and quantification. Scientia Iranica 18(3):669–676
13. Desrochers, J., Outreville, J. F.: Uncertainty, ambiguity, and risk taking: An experimental investigation of consumer behavior and demand for insurance. ICER Working Paper Series 10/2013. https://ssrn.com/abstract=2363870 or https://doi.org/10.2139/ssrn.2363870 (2013).
14. Kurdoglu RS, Jekel M, Ateş NY (2023) Eristic reasoning: Adaptation to extreme uncertainty. Front Psychol 14:1004031
15. Camerer C, Weber M (1992) Recent developments in modeling preferences: Uncertainty and ambiguity. J Risk Uncertain 5:325–370
16. Fox CR, Weber M (2002) Ambiguity aversion, comparative ignorance, and decision context. Organ Behav Hum Decis Process 88(1):476–498
17. Simon HA (1972) Theories of bounded rationality. Decision and Organization 1(1):161–176
18. Simon HA (1956) Rational choice and the structure of the environment. Psychol Rev 63(2):129–138
19. Filipović N (2025) Material foundations of ecological rationality. Synthese 205(1):110. https://doi.org/10.1007/s11229-025-04939-2
20. Kurdoglu RS, Ates NY, Lerner DA (2022) Decision-making under extreme uncertainty: Eristic rather than heuristic. Int J Entrep Behav Res 29(3):763–782
21. Kaufman, J. C., Glăveanu, V. P.: An overview of creativity theories. In Creativity: An introduction, pp. 17–30, Cambridge University Press (2021).
22. Florida, R.: The rise of the creative class, Basic Books (2003).
23. Horn D, Salvendy G (2006) Consumer-based assessment of product creativity: A review and reappraisal. Human Factors and Ergonomics in Manufacturing & Service Industries 16(2):155–175
24. Csikszentmihalyi M (1988) Society, culture, and person: A systems view of creativity. In: Sternberg RJ (ed) The nature of creativity. Cambridge University Press, New York, pp 325–339
25. Walia C (2019) A dynamic definition of creativity. Creat Res J 31(3):237–247
26. Runco MA, Chand I (1995) Cognition and creativity. Educ Psychol Rev 7(3):243–267
27. Mehta R, Dahl DW (2019) Creativity: past, present, and future. Cons Psychol Rev 2(1):30–49
28. Amabile TM, Pillemer J (2012) Perspectives on the social psychology of creativity. The Journal of Creative Behavior 46(1):3–15
29. Hennessey BA, Amabile TM (2010) Creativity. Ann Rev Psychol 61:569–598
30. Shulman JD, Cunha M Jr, Saint Clair JK (2015) Consumer uncertainty and purchase decision reversals: Theory and evidence. Mark Sci 34(4):590–605
31. Mueller JS, Melwani S, Goncalo JA (2012) The bias against creativity: Why people desire but reject creative ideas. Psychol Sci 23(1):13–17

32. Rios K, Markman KD, Schroeder J, Dyczewski EA (2014) A (creative) portrait of the uncertain individual: Self-uncertainty and individualism enhance creative generation. Pers Soc Psychol Bull 40(8):1050–1062
33. Faraji-Rad A, Pham MT (2017) Uncertainty increases the reliance on affect in decisions. Journal of Consumer Research 44(1):1–21
34. Ng AK (2003) A cultural model of creative and conforming behavior. Creat Res J 15(2–3):223–233
35. Hearn, G.: The future of creative work: Creativity and digital disruption. In The Future of Creative Work, pp. 1–12, Edward Elgar Publishing (2020).
36. Dellaert BG (2019) The consumer production journey: marketing to consumers as co-producers in the sharing economy. J Acad Mark Sci 47(2):238–254
37. Franke N, Piller FT (2004) Value creation by toolkits for user innovation and design: The case of the watch market. J Prod Innov Manag 21(6):401–415
38. Schreier M (2006) The value increment of mass-customized products: An empirical assessment. Journal of Consumer Behavior 5(4):317–327
39. Merle A, Chandon JL, Roux E, Alizon F (2010) Perceived value of the mass customized product and mass customization experience for individual consumers. Prod Oper Manag 19(5):503–514
40. Franke N, Schreier M, Kaiser U (2010) The "I designed it myself" effect in mass customization. Manage Sci 56(1):125–140
41. Franke N, Burghartz P, de Bellis E (2019) Being and staying the only one: Creating value through uniqueness in mass customization. Adv Consum Res 47:559–560
42. Krause F, Görgen J, de Bellis E, Franke N, Burghartz P, Klanner IM, Häubl G (2023) One-of-a-kind products: Leveraging strict uniqueness in mass customization. Int J Res Mark 40(4):823–840
43. Moon, A., Bos, M. W.: The illusion of increased customization: Framing choices as a creative process increases perceived customization. In Proceedings of the 2024 CHI Conference on Human Factors in Computing Systems, pp. 1–12, (2024).
44. Moreau CP, Dahl DW (2005) Designing the solution: The impact of constraints on consumers' creativity. Journal of Consumer Research 32(1):13–22
45. Zaggl MA, Hagenmaier MA, Raasch C (2019) The choice between uniqueness and conformity in mass customization. R&D Management 49(2):204–221
46. Turner F, Merle A, Gotteland D (2020) Enhancing consumer value of the co-design experience in mass customization. J Bus Res 117:473–483
47. Haase J, Hoff EV, Hanel PHP, Innes-Ker Å (2018) A meta-analysis of the relation between creative self-efficacy and different creativity measurements. Creat Res J 30(1):1–16. https://doi.org/10.1080/10400419.2018.1411436
48. Richter AW, Hirst G, Van Knippenberg D, Baer M (2012) Creative self-efficacy and individual creativity in team contexts: Cross-level interactions with team informational resources. J Appl Psychol 97(6):1282
49. Nohutlu ZD, Englis BG, Groen AJ, Constantinides E (2023) Innovating with the customer: Co-creation motives in online communities. Int J Electron Commer 27(4):523–557
50. Tansley AG (1935) The use and abuse of vegetational concepts and terms. Ecology 16(3):284–307
51. Prahalad CK, Hamel G (1999) The core competence of the corporation. Harv Bus Rev 68(3):79–91
52. Mars MM, Bronstein JL, Lusch RF (2012) The value of a metaphor: Organizations and ecosystems. Organ Dyn 41(4):271–280
53. Blažek, P., Kolb, M., Streichsbier, C., Honetz, S.: The evolutionary process of product configurators. In J. Bellemare, S. Carrier, K. Nielsen, & F. T. Piller (Eds.), Managing Complexity, pp. 161–172, Springer (2017).
54. Grosso C, Forza C, Trentin A (2017) Supporting the social dimension of shopping for personalized products through online sales configurators. Journal of Intelligent Information Systems 49(1):9–35

55. Suzić N, Stevanov B, Ćosić I, Anišić Z, Sremčev N (2012) Customizing products through application of group technology: A case study of furniture manufacturing. Journal of Mechanical Engineering 58(12):724–731
56. Dellaert BGC, Dabholkar PA (2009) Increasing the attractiveness of mass customization: The role of complementary online services and range of options. Int J Electron Commer 13(3):43–70
57. Franke N, Keinz P, Schreier M (2008) Complementing mass customization toolkits with user communities: How peer input improves customer self-design. J Prod Innov Manag 25(6):546–559
58. McDonald, C., Golub, A.: Personalized digital last (A women's example): The tool required to enable mass customization. IEEE Industry Connections, 1–16 (2018).
59. Salvador F, de Holan PM, Piller FT (2009) Cracking the code of mass customization. MIT Sloan Manag Rev 50(3):71–78
60. von Hippel E (2001) Perspective: User toolkits for innovation. J Prod Innov Manag 18(4):247–257
61. Piller, F. T., Ihl, C., Vossen, A.: Customer co-creation: Open innovation with customers. In New Frontiers in Open Innovation, pp. 29–49 (2012).
62. Franke N, Piller FT (2009) Cracking the code of mass customization. MIT Sloan Manag Rev 50(3):71–78
63. Zellweger T, Zenger T (2023) Entrepreneurs as scientists: A pragmatist approach to producing value out of uncertainty. Acad Manag Rev 48(3):379–408
64. Zhang W, Sjoerds Z, Hommel B (2020) Metacontrol of human creativity: The neurocognitive mechanisms of convergent and divergent thinking. Neuroimage 210:116572

Algorithmic and Knowledge-Based Modeling of Bus- and Sectional-Modular Designs: A Seating Layout Case Study

Paul Christoph Gembarski, Alejandro Coronas Firchow, and Shakiran Sandran

Abstract Knowledge-based CAD systems provide innovative solutions for automating design processes and enabling efficient, user-centric product customization. This paper presents a CAD configuration system for designing subway seat layouts, where users can input parameters such as train length, number of seats, and comfort settings. The system outputs a complete seat layout along with the maximum and standard seating capacities. This approach facilitates rapid visualization and evaluation of seat configurations, offering a valuable tool for generating and refining design concepts. The study outlines the development and implementation of this knowledge-based approach, focusing on the integration of parameter control into the CAD environment using Microsoft Excel and Autodesk iLogic. By externalizing parameter control, the system simplifies the process of visualizing and testing layout ideas. The results demonstrate the system's ability to handle complex interdependencies between design variables while providing intuitive customization options while keeping efforts for adaptation and maintenance of the system low.

Keywords Knowledge-based engineering · CAD configuration · Subway seat layout · Design automation · Parameter-driven design

1 Introduction

In recent years, design automation in mechanical engineering has seen remarkable progress. By embedding knowledge into parametric models, designers can quickly create and alter products while adhering to specified design restrictions [1, 2]. In such knowledge-based computer aided design (CAD) modeling, parameter control is facilitated by the use mathematical and logical constraints. Parameter trees and parameter plans serve for defining and managing the dependencies between parameters [3, 4]. In addition to this, the integration of algorithmic design methods, such as script- or macro-based approaches, allows for even generating product variants

P. C. Gembarski (✉) · A. C. Firchow · S. Sandran
Institute of Product Development, Leibniz University Hanover, Garbsen, Germany
e-mail: gembarski@ipeg.uni-hannover.de

M. Manns (ed.), *New Paradigms for Anticipated Uncertainty*, Lecture Notes in Mechanical Engineering, https://doi.org/10.1007/978-3-032-16889-4_16

by combining modules based on their interfaces and established design rules. This is particularly advantageous for bus- or sectional-modular architectures with standard interfaces [5]. As a result, the system can assemble the product without a pre-existing CAD master model [6, 7].

However, building such comprehensive models usually requires software development skills. Since many CAD systems provide basic scripting environments and spreadsheet integration, the question arises *how a combination of such widely accessible tools can be orchestrated to integrate knowledge-based and algorithmic CAD modelling, balancing flexibility and ease of use.* To answer this question, this study explores a method for automating such designs, exemplified by metro seating layouts. The corresponding model is built in Autodesk Inventor using its iLogic language and spreadsheet software. The remainder of the paper is organized as follows: Section two introduces the theoretical background of knowledge-based engineering and algorithmic design before section three clarifies the research objectives of this study. Section four then describes the case study and its implementation. Afterwards, Sect. 5 contains a discussion and critical reflection before section six summarizes and concludes the paper.

2 Theoretical Background

Knowledge-Based CAD refers to an approach where a model's geometry is derived through reasoning capabilities [8]. For a CAD model to be considered knowledge-based, it must be parametric, meaning that its geometry can be altered based on established parameters. Therefore, designers must plan the conceptual configuration of parameters within the model. These parameters can be controlled by logical or mathematical constraints, e.g. using an equation editor [9, 10]. Such parameter control can also be extended using tools like spreadsheets, which offer additional mathematical and statistical capabilities beyond the CAD system. For instance, integrating lookup tables allows efficient selection of standard parts based on geometric or load information, incorporating basic reasoning. Another method to incorporate reasoning into CAD systems is through script languages, macros, or external knowledge-based systems to determine parameters [11].

Unlike simple sequential macros, algorithmic modeling emphasizes the abstraction and modeling of the design process itself to create intelligent, interactive model generators [6]. There are several ways to develop an algorithm-based product model, e.g. using an application programming interface (API), design language, or expert system shell to drive a CAD kernel or to generate polygonal or spline-based geometric models [12]. Additionally, algorithms can be applied to evaluate and optimize the geometry during generation [13]. Algorithmic modeling aims to generate a unique product for each set of customer requirements, which is beneficial for complex geometries and pure configuration designs but often limits subsequent parametric editing of the variant. To create an updated version, requirements and design rules must be altered [6].

3 Research Objectives

This study follows a learning-by-building approach and aims to explore the boundaries of spreadsheets for externalizing parameter control for highly flexible designs that follow the principle of bus-modular or sectional-modular architectures. While these architectures offer a large solution space since modules may be aggregated with maximum freedom of choice, the spreadsheet-driven design approach necessitates a certain degree of predetermination. To solve this contradiction while being applicable for designers without deep programming knowledge, a corresponding method should minimize the reliance on coding. In particular, the system to be built needs to handle flexible module assignments and their respective parametrization with respect to size and positioning constraints.

4 Case Study: Train Layout Design

Designing seating arrangements for urban transit (Fig. 1) is a crucial aspect of railway vehicle engineering, influencing both passenger comfort and overall efficiency [14]. Traditional design methods often rely on iterative prototyping and physical mock-ups, which can be costly and time-consuming [15]. Exploring automation approaches within CAD environments may offer significant efficiency gains.

4.1 Bus-Modular Architecture

This study centers on optimizing the seating layout by focusing on two key parameters: train length and passenger capacity. Accordingly, the seating layout is abstracted into a bus-modular architecture. A framework of standard modules is developed (Fig. 2), which are then algorithmically arranged and placed to meet specific user requirements.

For this study, five standard modules shall be implemented, two sitting modules, a door module, a bicycle stand and an articulation module. The following basic design constraints shall be exemplarily implemented:

1. A minimum of 4 doors in two pairs has to be present in the assembly to facilitate the free flow of people in and out of the train. There is no upper limit.

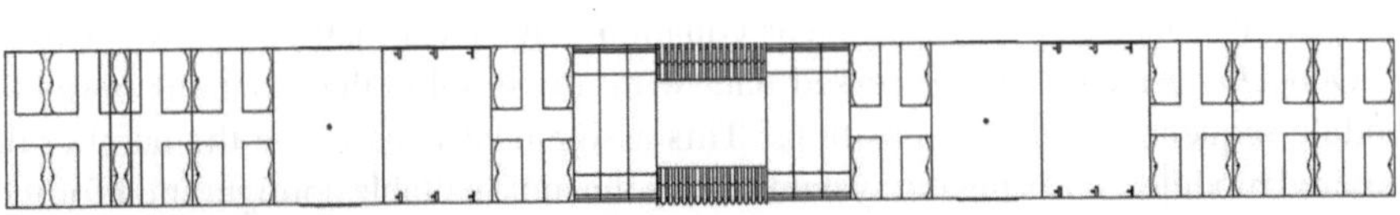

Fig. 1 Example 2D seating layout for an articulated light rail

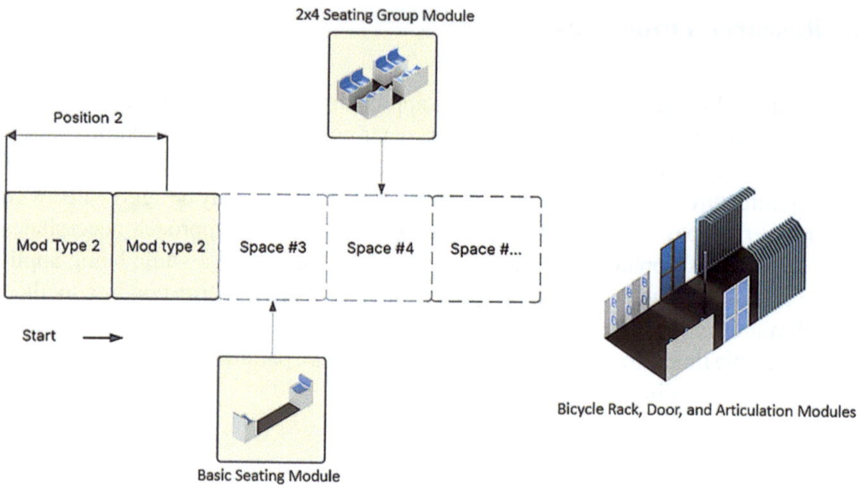

Fig. 2 Bus-modular architecture and basic modules for creating seating layouts

2. If a model reaches a specified length, an articulation has to divide the wagon in a minimum of two parts to enable the movement along curves on the track.
3. For passenger flow efficiency it's important that bicycle rack modules are placed next to a door. This rule can be changed by the end user to adhere to design standards and regulations.

4.2 External Parameter Control and Algorithmic Module Placement

For externalizing the parameter control of the parametric CAD model, Microsoft Excel was chosen for the following characteristics:

- Spreadsheet software is a standard tool used by most engineers in the field.
- It offers different techniques to easily create and enhance the user experience, e.g., conditional formatting facilitating the communication of design boundaries.
- The native automatic parameter import and updating of many commercial CAD systems facilitates the communication between the model and the parameter control without the need of APIs or additional software development skills

The system calculates the required number of modules by running the length and seating capacity through a linear equation solved via matrices. Next, it uses VLOOKUP to match the processed data with a case table that contains possible module sequences and their positions. This assignment is based on the number of required modules, ensuring the system selects the most suitable configuration for the given input parameters.

To overcome the case list of predefined module groupings, Lab Script by Microsoft was implemented into the system to allow JavaScript code to change and read data from the spreadsheet directly into excel, while maintaining legacy systems cross compatibility. JavaScript was decided on based on its good implementation on Excel and its simplicity and user friendliness. The code was organized into three logical parts (Fig. 3). This structured approach facilitates future modifications, simplifies debugging, and improves overall system comprehension.

The initial step involves creating a series that specifies the type of module and the sequence in which they should be arranged. This function draws data from the spreadsheet and returns it to the sheet as a series of cells, each containing a number that corresponds to the type of module to be placed. The second part uses this as input to determine the module quantity and the appropriate location for special modules, e.g. doors, according to the implemented design rules. After positioning them, the output of the first part is overwritten. The third part reads this series and generates a parallel output that defines the position of each module based on their length and the starting plane's position. Currently, this is only implemented for the z-axis, but it could be easily extended, e.g. to populate a second deck for a regional train.

The native spreadsheet integration in Inventor makes it mandatory to follow a strict scheme in the first worksheet (parameter name, value, unit, and comment) to input spreadsheet data from Excel into Inventor parameters. An iLogic loop selects the file, depending on the module type and the predefined file directory, defines a dynamic name for it and constrains it in an assembly file. The constraints are defined in all 3 axis. X and Y axis are set to be flush with the global X and Y plane, this could be changed to allow for more customization by adding a second or third series of positions in the JavaScript code. The Z axis is dependent on the predefined position imported from Excel, this is added as an offset from the global Z plane, therefore placing the modules in order next to each other. To ensure proper alignment and prevent inconsistencies, an assembly update command must be executed after each constraint is applied. This guarantees that the model accurately adjusts to its defined position throughout the assembly process. The loop runs through all modules from 1 to i which is defined as the number of modules numbered in excel, this value is calculated in Excel with a COUNTIF function. Figure 4 shows two example configurations of the system.

4.3 Adaptability and Customization

An advantage of this approach is its adaptability, allowing for rapid customization based on operational requirements of the end user which includes the definition of new modules. The necessary adjustments to the system are outlined below:

- Module Creation: The new CAD model of a module is created in Autodesk Inventor and saved on the project folder.

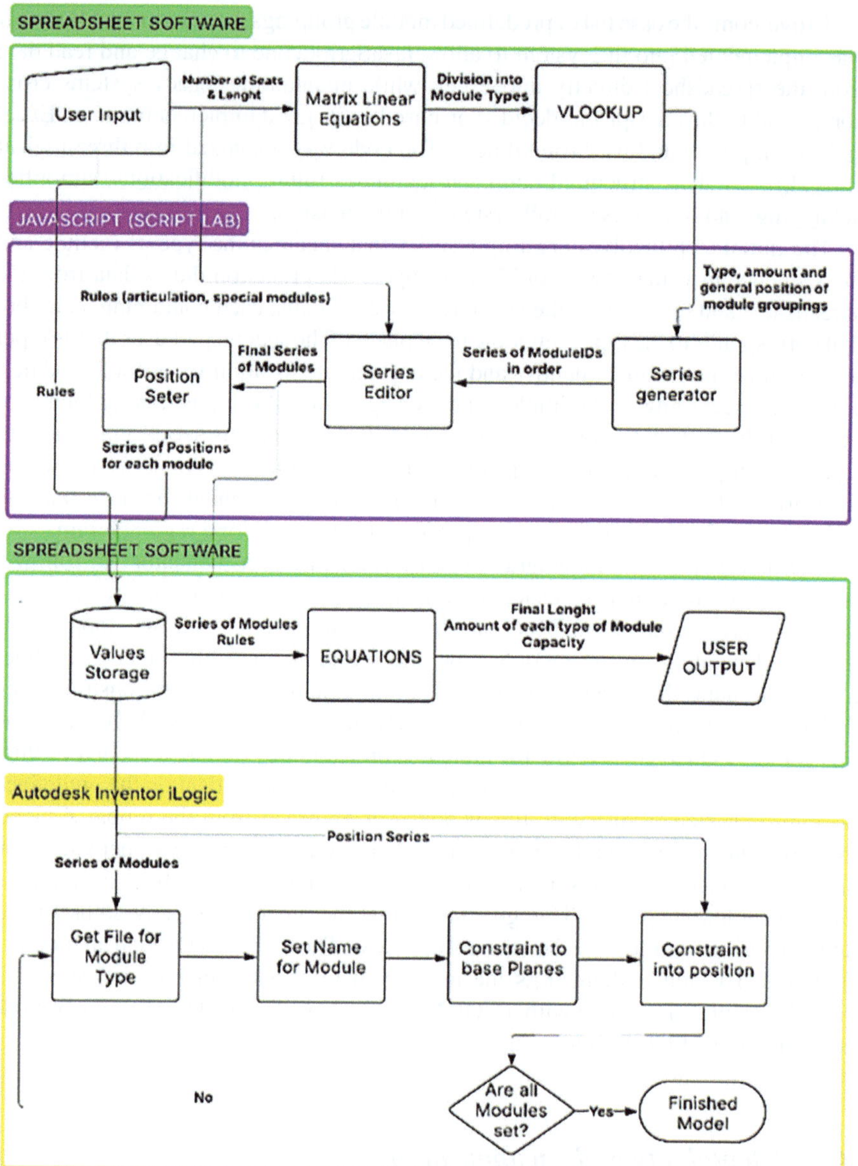

Fig. 3 Flowchart for external parameter control and algorithmic module placement

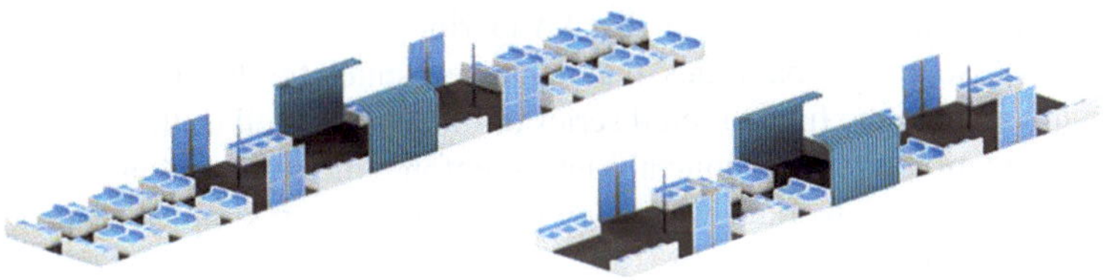

Fig. 4 Solution variants

- Module Definition: The new module is defined within the spreadsheet, including its parameters (dimensions, seating capacity) and functional constraints. A module ID has to be set to call the module on all the different systems, this is used both in JavaScript and iLogic.
- Logical Integration: The series creation logic has to be extended to recognize the new module, ensuring it is placed dynamically based on the design constraints.
- Automated Placement: The positioning rules in the JavaScript-based logic are updated to ensure correct integration into the layout.
- iLogic Extension: It is necessary to add a line in the iLogic code to correlate the module ID with the path of the CAD model of the module.

To evaluate the feasibility, two students from an advanced CAD-course, who were not involved in system creation, were asked to implement a multi-purpose module with foldable seating. The module could be implemented following the above mentioned steps. Nonetheless, a serious feedback was that updates had to be organized across three separate source files. A comparison with a more API-based approach is planned.

5 Assessment and Discussion

The primary objective was to create a logical system completely in Microsoft Excel. To achieve this, it was necessary to simplify the task into multiple subtasks and solve them through matrices. While this approach was effective for simple tasks, it failed to maintain symmetry and produce ordered series. To address these issues, a table of cases was paired with a VLOOKUP function to assign specific commands to each case, but executing these commands necessitated a more advanced logical system.

While Microsoft Excel´s dynamic array manipulation functions can be used for dynamic sequence generation and complex rule implementation, the need for multiple nested functions directly affects the maintainability and adaptability of this model for future changes. Furthermore, the use of array manipulation hinders cross-compatibility with legacy systems (Microsoft Excel 2019 and older). To overcome these constraints, JavaScript was implemented directly into the spreadsheet software in the form of the plugin *Script Lab*. This facilitates the connection of data between the code and the spreadsheet. Additionally, JavaScript provides comprehensive debugging and error isolation capabilities, making it well-suited for this application. It was tasked with creating the final ordered series of modules as well as their position and the implementation of simple logical rules to enforce the summoning of additional special modules. This worked well allowing for rapid and simple expansion and optimization of the code.

As Inventor works with predefined parameters, it is imperative to define a finite number of parameters in the spreadsheet for them to be read by Inventor and used for processing in the iLogic module. In this case, 60 parameters were defined for module type and 60 for positioning data. This limitation does not stem from Excel itself but rather from the constraints of iLogic and the Inventor internal scripting, which does not allow creating variables at runtime. As a result, all necessary parameters for placing the seat modules must be predetermined at build-time, requiring a fixed number of possible module definitions. This restricts the flexibility of the system and limits the number of representable modules without manually extending the predefined parameter set. Nonetheless, for the given use case this is tolerable if the defined variable space is large enough.

6 Summary and Conclusion

Spreadsheet-driven design can integrate parameter control, rule-based logic, and data-driven automation into knowledge-based CAD models. This study examined the potential of using Microsoft Excel as a front-end interface and logical system for configuring metro seating layouts, integrating it with JavaScript for enhanced logic processing and Autodesk Inventor's iLogic for automated 3D model generation.

Excel's historical limitations in logical automation, which necessitated scripting for tasks like module sequencing, have been largely overcome by modern dynamic arrays. However, scripting remains a strategic design choice, offering distinct advantages in readability, debugging, and backward compatibility not always achievable in a pure formula-based approach. On the CAD side, iLogic supports efficient parametric modeling despite dynamic variable limitations. The case combines widely available engineering tools to create flexible and user-friendly design automation solutions. Incorporating advanced scripting environments like Python could further improve automation while remaining accessible to engineers with limited programming skills. The seating layout serves as a placeholder for bus-modular and sectional-modular designs, including aircraft cabin layouts, office space planning, and various transport infrastructure projects.

Furthermore, closed-loop data-driven configuration models, where real-time design decisions are influenced by operational data would be an interesting avenue. In the case of metro seating, this could mean making dynamic adjustments based on passenger flow analysis or adaptive space utilization combining passenger and e.g. parcel transportation within data-driven business ecosystems [16].

References

1. Ortner-Pichler A, Landschützer C (2022) Integration of of parametric modeling in web-based knowledge-based engineering applications. Adv Eng Inform 51:101492. https://doi.org/10.1016/j.aei.2021.101492
2. Aranburu A, Cotillas J, Justel D, Contero M, Camba JD (2022) How does the modeling strategy influence design optimization and the automatic generation of parametric geometry variations? Comput Aided Des 151:103364. https://doi.org/10.1016/j.cad.2022.103364
3. Tang Z, Zou Q, Gao S (2023) A decision-support method for multi-parameter editing of parametric CAD models. Adv Eng Inform 56:101997. https://doi.org/10.1016/j.aei.2023.101997
4. Marchenko M, Behrens BA, Wrobel G, Scheffler R, Pleßow M (2011) A new method of visualization and documentation of parametric information of 3D CAD models. Comput-Aided Des Appl 8(3):435–448. https://doi.org/10.3722/cadaps.2011.435-448
5. Ulrich K (1994) Fundamentals of product modularity. In: Dasu S, Eastman C (eds) Management of design. Springer, Dordrecht. https://doi.org/10.1007/978-94-011-1390-8_12
6. Fuchs D, Bartz R, Kuschmitz S, Vietor T (2022) Necessary advances in computer-aided design to leverage on additive manufacturing design freedom. Int J Interact Des Manuf 16(4):1633–1651. https://doi.org/10.1007/s12008-022-00888-z
7. Tedeschi A, Lombardi D (2014) AAD—algorithms-aided design. Le Penseur, Brienza
8. Raffaeli R, Germani M (2010) Knowledge-based approach to flexible part design. J Eng Des 21(1):7–29. https://doi.org/10.1080/09544820802086996
9. Li H, Gembarski PC, Lachmayer R (2018) Template-based design for design co-creation. In: Proceedings of the 5th international conference on design creativity (ICDC 2018), University of Bath, Bath, UK, pp 387–394
10. Skarka W (2007) Application of MOKA methodology in generative model creation using CATIA. Eng Appl Artif Intell 20(5):677–690. https://doi.org/10.1016/j.engappai.2006.11.019
11. Gembarski PC, Li H, Lachmayer R (2017) KBE-modeling techniques in standard CAD-systems: case study—Autodesk inventor professional. In: Bellemare J, Carrier S, Nielsen K, Piller FT (eds) Managing complexity. Springer, Heidelberg, pp 215–233. https://doi.org/10.1007/978-3-319-29058-4_17
12. Frank G, Entner D, Prante T, Khachatouri V, Schwarz M (2014) Towards a generic framework of engineering design automation for creating complex CAD models. Int J Adv Syst Measurements 7(1):179–192
13. Boretti V, Sardone L, Bohórquez Graterón LA, Masera D, Marano GC, Domaneschi M (2023) Algorithm-aided design for composite bridges. Buildings 13(4):865. https://doi.org/10.3390/buildings13040865
14. Yang J, Shiwakoti N, Tay R (2024) Train passengers' perceptions and preferences for different platform and carriage design features. J Public Transp 26:100085. https://doi.org/10.1016/j.jpubtr.2024.100085
15. Thoreau R, Holloway C, Bansal G, Gharatya K, Roan TR, Tyler N (2016) Train design features affecting boarding and alighting of passengers. J Adv Transp 50(8):2077–2088. https://doi.org/10.1002/atr.1446
16. Gembarski PC, Kammler F (2022) Mass customizing for circular and sharing economies: a resource-based view on outside of the box scenarios. In: Andersen AL et al (eds) Towards sustainable customization: bridging smart products and manufacturing systems. Springer, Cham, pp 1039–1046. https://doi.org/10.1007/978-3-030-90700-6_119

Conceptualizing Compound Features: On Resource- and Capability Adaption in Sustainable Ecosystems

Paul Christoph Gembarski and Friedemann Kammler

Abstract Smart product-service systems (sPSS) revolutionize mass customization by adapting solutions to changing customer requirements, yet demand substantial engineering to ensure valid future configurations. A "maximum configuration" approach installs all potential features regardless of use. While enhancing scalability, this raises concerns about sustainability. Traditionally, sPSS address defined customer problems within optimized supply networks, but especially business ecosystems require broader perspectives regarding complementarity. This paper conceptualizes compound features, derived from systems engineering, as emergent properties of interactions between system components. By treating compound features as resources and capabilities and employing a resource-based configuration strategy, the paper highlights how this approach helps to find capability gaps and align requirements across entities. Furthermore, it discusses compound features within different hierarchical levels of business ecosystems.

Keywords Business ecosystems · Systems engineering · Compound features · Network search · Resource and capability configuration

1 Introduction

Complex offerings such as smart product-service systems (sPSS) enhance mass customization by adapting customer solutions to temporal changes of requirements [1, 2]. Their ability to reconfigure, recombine, and repurpose resources demands significant engineering to ensure future configurations remain valid [3]. One method deploys a full feature set and then digitally limits performance or disables unused functions—such as automotive applications like pay-per-use seat heating or electric

P. C. Gembarski (✉)
Institute of Product Development, Leibniz Universität Hannover, Garbsen, Germany
e-mail: gembarski@ipeg.uni-hannover.de

F. Kammler
Osnabrück University, Osnabrück, Germany
e-mail: fkammler@uni-osnabrueck.de

M. Manns (ed.), *New Paradigms for Anticipated Uncertainty*, Lecture Notes in Mechanical Engineering, https://doi.org/10.1007/978-3-032-16889-4_17

vehicle range extenders [4]. However, the inclusion of latent technical capabilities regardless of usage raises questions about sustainability and engineering efficiency. Moreover, in business ecosystems, the design and operation of sPSSs often span multiple stakeholders, creating a distributed environment that challenges traditional governance, interoperability, and lifecycle management [5, 6].

As in every system, component interactions lead to emergent characteristics and behaviors which can enhance customer perceived value and optimize the effectivity of the sPSS. Systems-of-systems engineering methods tame this complexity by leveraging complementarities, but classical tools—axiomatic design, modular function deployment and the design interface matrix—presume a single, centrally coordinated process [7, 8]. In ecosystems where partners contribute complementary capabilities, specialized frameworks are required to identify, align, and coordinate each actor's contributions.

This paper contributes the concept of compound features as a first stage toward the design, implementation and use design science research pattern [9] and follows the hypothesis that complementarities can be expressed by resources and capabilities, representing how resources are used. In a first step, adapting the resource-based configuration approach [10], the paper demonstrates how this concept aids in identifying capability gaps and establishing cross-entity requirements. Additionally, it outlines the role of compound features across various structural levels of business ecosystems.

2 Theoretical Background

Modularity and the capability of configuring modules according to known rules and restrictions are key characteristics of mass customization [11]. Today, model-based configuration approaches, e.g. based on concepts like the constraint-satisfaction-problem, dominate the state of the art [12]. Although belonging to this category, the resource-based configuration approach reframes the configuration problem as one of matching and balancing resources. Following this approach, every component in a technical system is characterized by the resources it provides (energy, information, installation space, technical interfaces, etc.) and the resources it requires. The configuration task then has the aim of ensuring that—in the assembled system—the supply of these resources from the individual components meets the overall demand imposed by the system's intended features and that resource provision and demand are balanced [10]. Components and their resource profiles are organized within a hierarchical catalogue that spans the solution space, allowing for solutions to be assembled by humans or algorithms to satisfy the requirements. The approach ensures compatibility across generations, provided that updated components offer additional resources compared to previous ones. Any other restrictions such as incompatibilities between components are added as additional constraints. The resource-based configuration approach was implemented as configurator shell, which, besides configuration, supports the exploration of design alternatives and the optimization of resource usage [13, 14].

While the concept recognizes that resources can be supplied by combining components, it does not fully consider the emergent properties that arise from interactions between components. Additionally, the economic aspect is incorporated by linking components to commercial factors such as production costs and potential pricing. When components are integrated, value attribution becomes multifaceted, encompassing aspects like time savings or process optimization, which must also be treated as resources.

3 Retrieval of Compound Features

Taking these two deficiencies as a basis, a concept is needed to define resource provision and demand for emergent properties, which result from the interaction of aggregated components at a higher structural level.

Products create value by addressing specific customer problems through features, i.e. a specific functionality or attribute, designed for this purpose. These features can either directly provide core value (solitaires) or enhance value by complementing other products' features. In case of complementarity, cross-entity requirements exist between the corresponding features. Taking a simple example of a hack saw, the blade has a solitaire feature to cut material while the frame (and its studs) has the solitaire features to hold blade and handle and to ensure cutting depth and clearance. When blade and frame are aggregated on the next higher structural level (in the right way), a complementary feature is the exchangeability of the blade and a cross-entity requirement resulting from the interface between both, which leads either to standardized blade lengths or a feature of adaptability in the frame. In this example, solitaire and complementary features and possible alternatives are the result of an active design process.

In complementary use, a third type of feature emerges organically among products, which we call *compound features* (Fig. 1). They result from (end-user) integration to address unforeseen customer problems, making them difficult to design beforehand and include them in the scope of product specification. Consequently, identifying compound features presents a new methodological challenge.

A second characteristic for describing the interaction of aggregated components is the structural level on which they interact. We suggest an abstraction into three layers, where complementarity can be recognized:

- *Micro*: Complementarity of features,
- *Meso*: Complementarity of entire products and services,
- *Macro*: Complementarity of product domains, i.e. all or a group of related products or services offered by an organization.

From an implementation point of view, we suggest refining the modeling and configuration process from resource-based configuration. All abstraction layers have in

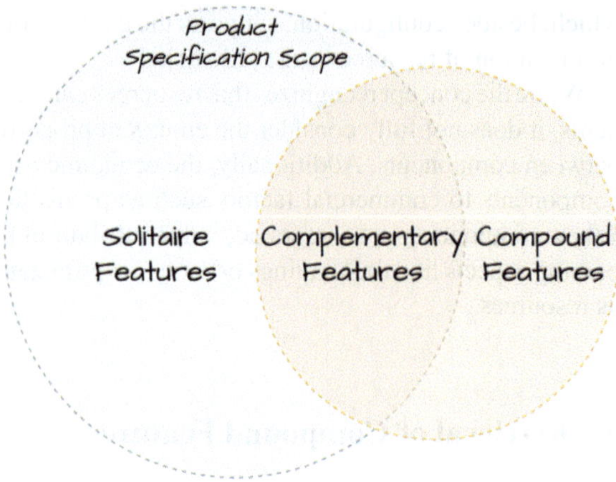

Fig. 1 Product decomposed into three distinct feature sets

common that they can be modeled as a network graph, where nodes represent resources and capabilities. While the first follows the concept of Heinrich and Jüngst and integrates demand and provision functions in the nodes themselves, the latter describe how resources are optimally used, which bridges functional requirements and components. Edges in the graph either represent constraints—such as restrictions on resource flows or compatibility issues—or complementarities that "unlock" additional resources or properties within the connected nodes.

Inference on the graph can be done in two ways: The problem-centered approach introduces a new customer problem and directs the search towards a solving component set expressed by its resources with the goal of finding an optimal, balanced solution. In contrast, the node-centered approach searches for new applications of a component before the occurrence of the particular practical problem to identify and introduce new complementarity constraints.

4 Micro-level Application: Compound Guide Offset Correction

We demonstrate the application of the approach with a scenario from the power tools domain. The scenario includes three customary tools, namely a circular saw, a routing machine, and a guide rail, all of which contribute different solitaire features (Table 1).

The end-user problem originates from the following production schedule: A wood panel should be sewed on a distinct length and then undergo additional processing, such as finishing the cut surface with a router, creating a finger joint, or adding a fillet on the downside. For the initial task, the guide rail is positioned and fixed to facilitate the cut using a circular saw, after which the guide rail is removed. For the second task, a spacer must be attached to the router to establish the offset in relation to the cut

Table 1 Solitaire and complementary features of the power tool system (excerpt)

No.	Feature	Allocation	Type
1	Arbor and blade	Circular saw	Solitaire feature
2	Blade guard	Circular saw	Solitaire feature
3	Height and bevel adjustments	Circular saw	Solitaire feature
4	Guide rail attachment	Circular saw, guide rail	Complementary feature
5	Anti-slip pad	Guide rail	Solitaire feature
6	Splinter guard	Circular saw, guide rail	Complementary feature
7	Multi-tool collet	Router	Solitaire feature
8	Power engine	Router	Solitaire feature
9	Elevator	Router	Solitaire feature
10	Exchangeable baseplate	Router	Solitaire feature

surface, serving as a primary guide. In terms of precision, it is important to note that the circular saw's guide rail can absorb transverse forces in both directions, whereas the router spacer can only counteract forces in one direction, making it susceptible to drift.

In the end, both tools need guiding but use different features and resulting resources for this. Relating the router to the guiding rail of the circular saw and additionally unlocking of the greater stability for the routing operation creates a compound feature between guiding rail and the router. From a requirement point of view, the distance between the guide rail's key and the tool axis of the router or the radial cutting depth respectively must be overcome and the guide rail itself needs to be sufficiently stiff to securely plunge the router from outside without the support of the panel. One solution would be to design a router baseplate that is connectable to the guide rail.

Figure 2 shows an excerpt of the constraint graph. The variables of the features and their resources are hierarchically broken down and added with their domains (not depicted). The constrained resource between saw and guiderail is the slot-key pair with their respective properties (dotted line). The inference strategy involves three levels:

1. Problem-centered: This involves the question whether the guiding function of the router can be related to the guiding function of the rail. Here both are incompatible as they don't have compatible resources since the spacer usually works with a stop bar and the rail has a key. So to speak, both are positives.
2. Node-centered on constraint level: This is looking for needed or provided resources and their properties in other features of the router. In this case, guide rail and baseplate share an interface resource to the work piece, the contact surface. Since no size constraints limit the dimensions of the plate, so that a bigger plate can take a slot according to the guide rail, a new resource constraint is constituted

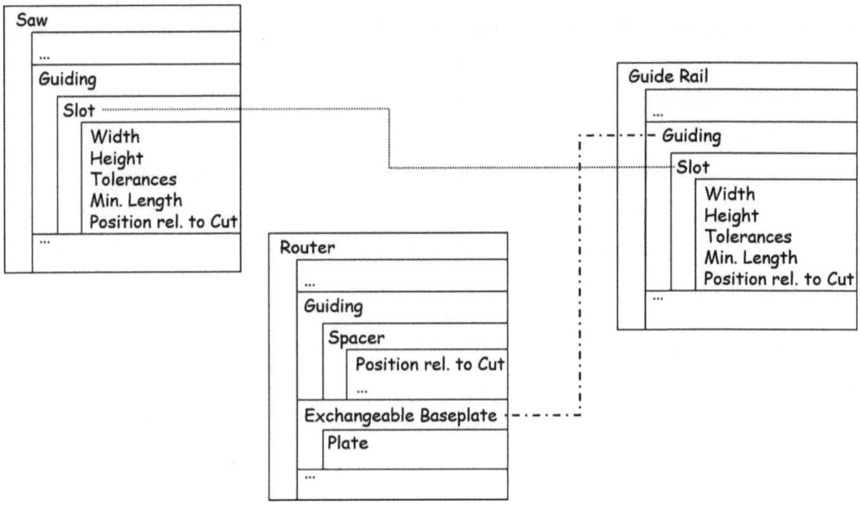

Fig. 2 Constraint graph of the resource-based configuration problem (excerpt)

between both (dash-dotted line) and the specification of the baseplate is modified accordingly to add a slot. Constraints between the features of the router, e.g. between baseplate and elevator, are re-evaluated then.

3. Node-centered including feature or component generation: In case that the ground plate has restrictions that counteract an integration of the slot here and no other feature can add this, there are two solutions: First, a new feature or second, a new component needs to be created. E.g., the configuration system could propose an adapter as new component as a mediator between the guide rail and the router. Such solutions, also adjustable to different tool diameters, are available on the market.

5 Discussion

The example demonstrates the concept of identifying and reasoning about compound features. From a systems engineering perspective, the modification of the system boundary of the guide rail to include also the router has further consequences. When a clearance between rail edge and milling tool is necessary, in case of milling a fillet on the upward surface, this results also in an offset between saw blade and rail edge. A consequence is then that the splinter guard of the rail cannot fulfill its function. A solution then would be to integrate the splinter guard not in the rail but in the baseplate of the saw itself, which is technically difficult since the blade teeth introduce friction and wear on ever the same position of the lip.

The contribution of this study is the justification of compound features as a design objective. The exemplary application given above provides first insights on the applicability of the concept [9]. From a modeling point of view, further research is needed to design and implement a design tool and to integrate existing methods and tools known from systems engineering. However, we see an inference strategy as a promising starting point. If a feature can provide sufficient resources to achieve an additional function or modified accordingly, many context variables need to be considered. Modeling them explicitly seems at least feasible in the example. But thinking this further to a configuration system for business ecosystem offerings in the sense of a computer-aided design tool would result in exhaustive constraining. Taking the described resource-based approach as a basis, an inclusion of semantic web principles, especially for the node-centered inference parts, as well as modern, generative AI-based approaches for the formulation of context variables could be interesting for reducing modeling efforts.

The concept raises the question of whether *compound features* constitute genuinely new attributes or merely represent an integration of existing features utilized in sequential value creation activities. A way to argue this is customer perceived value. This refers to the notion that four types of value—functional, monetary, social, and psychological—contribute to the overall tangible benefits. When divided by the costs incurred, this calculation yields the overall perceived value. In the example above, reducing set-up efforts results in shorter processing times (in the end a monetary benefit), increased robustness since the guiding has to be set-up only once (functional and perhaps psychological benefit) and increased commonality (functional benefit). Therefore, we argue that compound features are valid if they enhance tangible benefits and, consequently, the perceived value.

Furthermore, it must be scrutinized whether there are not already neighboring concepts. By nature, there are similarities with single 9R-strategies, in particular for the repurposing of products. Repurpose strategies involve reinterpreting the features of products, and aligning the requirements they meet with new values [15]. In our view, compound features can contribute here as a basic concept that makes repurposing opportunities systematically shootable.

6 Conclusion and Outlook

The starting point of this study was the observation that physical products often integrate features that are used only in particular variants, creating an inefficient use of resources. While this type of application certainly is valid for physically inseparable features such as with the pay-per-use seat heating system, other complex value proposals promise to allow for combined feature sets that originate from different products, in the design sovereignty of multiple stakeholders. This mechanism has already been introduced by further concepts such as strategic alliances and platforms (cf. [16], pp. 2260), where compatibility of complements is a product of engineering efforts, mostly modularization and standardization. In this paper, we ask beyond this observation, whether the conceptual space also holds a third option, where ex-ante value propositions are fulfilled through inferred feature configuration. We introduced this concept as *compound features* and derive an application of resource-based configuration in order to demonstrate its real-world utility.

The concept should be tested for generalizability in other domains apart from physical products and for different structural levels. Addressing sPSS as offerings from business ecosystems, two testing scenarios from literature could deliver further insight: [17] describes an urban logistics use case, where delivery and pick-up services are integrated by a complementary (in this case perhaps even compound) feature: a vehicle's loading space which can change its attribution from parcel to waste. Compared to the example case of this study, this addresses the meso-level, i.e. two services and an adaptive product as boundary resource and allows for further investigation of the features provided and its interface functionality. In [18], the case of a digital mapping service is analyzed that has an open application programming interface and a provided software development kit. In this way, partner businesses can integrate and tailor the service, e.g. either for location display, route planning or real-time tracking. By examining the integration of digital services, a platform and data as resources in further detail, we see option to learn about macro-level compound feature design.

References

1. Machchhar RJ, Toller CNK, Bertoni A, Bertoni M (2022) Data-driven value creation in smart product-service system design: state-of-the-art and research directions. Comput Ind 137:103606. https://doi.org/10.1016/j.compind.2022.103606
2. Gembarski PC, Lachmayer R (2018) Product-service-systems—what and why developers can learn from mass customization. Enterp Model Inf Syst Arch (EMISAJ) 13:16–11. https://doi.org/10.18417/emisa.13.16
3. Zheng P, Chen CH, Shang S (2019) Towards an automatic engineering change management in smart product-service systems—a DSM-based learning approach. Adv Eng Inform 39:203–213. https://doi.org/10.1016/j.aei.2019.01.002
4. Dhebar A (2023) Preinstalled functionality as a service. Bus Horiz 66(5):643–653. https://doi.org/10.1016/j.bushor.2022.12.004
5. Roundy PT, Bradshaw M, Brockman BK (2018) The emergence of entrepreneurial ecosystems: a complex adaptive systems approach. J Bus Res 86:1–10. https://doi.org/10.1016/j.jbusres.2018.01.032
6. Schoormann T, Kammler F, Gembarski PC, Hagen S et al (2024) Sustainable ecosystems: findings from the NaWerSys workshop series. In: Informatik 2024. Gesellschaft für Informatik eV, Bonn, pp 1779–1797. https://doi.org/10.18420/inf2024_155
7. Gorod A, Sauser B, Boardman J (2008) System-of-systems engineering management: a review of modern history and a path forward. IEEE Syst J 2(4):484–499. https://doi.org/10.1109/JSYST.2008.2007163
8. Nielsen CB, Larsen PG, Fitzgerald J, Woodcock J, Peleska J (2015) Systems of systems engineering: basic concepts, model-based techniques, and research directions. ACM Comput Surv 48(2):1–41. https://doi.org/10.1145/2794381
9. Sonnenberg C, Vom Brocke J (2012) Evaluations in the science of the artificial—reconsidering the build-evaluate pattern in design science research. In: Peffers K, Rothenberger M, Kuechler

B (eds) Design science research in information systems. Advances in theory and practice. Springer, Heidelberg, pp 381–397. https://doi.org/10.1007/978-3-642-29863-9_28
10. Heinrich M, Jüngst EW (1991) A resource-based paradigm for the configuring of technical systems from modular components. In: Proceedings of the seventh IEEE conference on artificial intelligence applications 1991. AAAI Publications, Washington DC, pp 257–264
11. Suzic N, Forza C (2023) Development of mass customization implementation guidelines for small and medium enterprises (SMEs). Prod Plan Control 34(6):543–571. https://doi.org/10.1080/09537287.2021.1940345
12. Felfernig A, Hotz L, Bagley C, Tiihonen J (2014) Knowledge-based configuration: from research to business cases. Morgan Kaufmann, Waltham
13. Jüngst WE, Heinrich M (1998) Using resource balancing to configure modular systems. IEEE Intell Syst Appl 13(4):50–58. https://doi.org/10.1109/5254.708433
14. Sabin D, Freuder E (1999) Optimization methods for constraint resource problems. In: Proceedings of the AAAI workshop on configuration 1999. AAAI Publications, Washington DC, pp 83–89
15. Muñoz S, Hosseini MR, Crawford RH (2024) Towards a holistic assessment of circular economy strategies: the 9R circularity index. Sustainable Prod Consumption 47:400–412. https://doi.org/10.1016/j.spc.2024.04.015
16. Jacobides MG, Cennamo C, Gawer A (2018) Towards a theory of ecosystems. Strateg Manag J 39(8):2255–2276. https://doi.org/10.1002/smj.2904
17. Brettmo A, Browne M (2024) Comparative study on urban freight transport sustainability initiatives: two cases from Sweden. Res Transp Bus Manag 57:101238. https://doi.org/10.1016/j.rtbm.2024.101238
18. Rong K, Hu G, Lin Y, Shi Y, Guo L (2015) Understanding business ecosystem using a 6C framework in internet-of-things-based sectors. Int J Prod Econ 159:41–55. https://doi.org/10.1016/j.ijpe.2014.09.003

Concept for the Sustainability-Oriented Variant Management of Technical Products Supported by Machine Learning

Nikolai Kelbel, Alexander Keuper, and Günther Schuh

Abstract In recent years, product portfolio management (PPM) has faced increasing pressure from various stakeholders due to the sustainability transformation. This resulted in a higher complexity of decision-making and necessitates the systematic integration of sustainability into PPM to achieve corporate sustainability targets. However, incorporating this new dimension introduces target conflicts as well as a significant data demand. This paper presents a concept for the sustainability-oriented variant management of technical products. The concept enables portfolio managers to define, streamline and operationalize the sustainability controlling in accordance with the company's target system. Furthermore, it fosters the scalability of sustainability assessments of products with machine learning approaches. This addresses the industry's need for faster and more practical assessment alternatives with sufficient preciseness to provide transparency for decision-making in multi-variant product portfolios. Finally, the concept supports users in linking the insights with measures of variant management, thus promotes the sustainable development of the product portfolio.

Keywords Product portfolio management · Controlling · Sustainability · Variant management · Machine learning

1 Introduction

As concerns regarding climate change, resource depletion, and environmental degradation intensify, the need for industries to implement sustainable practices becomes increasingly urgent [1]. The manufacturing sector plays a crucial role in influencing

N. Kelbel (✉) · A. Keuper · G. Schuh
Laboratory for Machine Tools and Production Engineering WZL, RWTH Aachen University, Aachen, Germany
e-mail: n.kelbel@wzl.rwth-aachen.de

G. Schuh
Fraunhofer Institute for Production Technology IPT, Aachen, Germany

M. Manns (ed.), *New Paradigms for Anticipated Uncertainty*, Lecture Notes in Mechanical Engineering, https://doi.org/10.1007/978-3-032-16889-4_18

environmental, social, and economic outcomes [2]. Spurred on by the EU's sustainability ambitions and the resulting regulations, such as the Corporate Sustainability Reporting Directive (CSRD), but also by the intrinsic motivation of customers, more and more companies are focusing on transforming their value creation processes. First companies have even set themselves specific targets with the Science Based Target Initiative [3] or are aiming to meet requirements for receiving sustainability awards such as EcoVadis [4]. Sustainability has long been seen as a competitive advantage. However, emerging conflicts of interest with a company's short-term economic goals require precise management and consideration of compromises [5]. This complicates decision-making processes, for example regarding product changes or eliminations, making it essential to establish an efficient controlling process to serve as a basis for decisions [6]. Especially companies with complex multi-variant product portfolios face challenges while trying to create transparency. Due to its interface function between the market, production and product development, the dimension of sustainability now also falls within the remit of product portfolio management (PPM).

Historically, PPM is primarily concerned with the management of variants and meeting market demand [7]. However, the introduction of the sustainability perspective poses challenges for the product and portfolio management due to potential target conflicts and additional requirements. Many companies are uncertain how sustainability considerations can be incorporated into decision-making for portfolio management. Furthermore, obtaining the data basis for these decisions in the context of sustainability is challenging as well [8]. A key component in this effort is the sustainability assessment of products, which is necessary to establish a baseline for the portfolio as well as guide and control future steering initiatives. However, quantitative sustainability assessment methods are time-consuming and thus impractical for multi-variant product portfolios [9]. Alternative options for this are more of a qualitative or estimative nature and therefore imprecise.

Due to a lack of solutions in literature and practice regarding the quantitative integration of sustainability in the decision-making process of variant management this paper aims to address these challenges. By introducing a concept of a methodology for this the following research question is answered: *How can data-based management of the environmental sustainability of technical products be designed in multi-variant product portfolios?* For this the research is built on deductive and inductive research methodologies and utilizes systematic literature reviews, machine learning approaches and adapts established tools of variant management. In Sect. 2 the fundamentals are explained and related work is elaborated. Subsequently, Sect. 3 describes the concept of the methodology. Chapter 4 sums up the findings and gives an outlook.

2 Fundamentals and Related Research

In this chapter, the fundamentals of PPM and sustainability assessment are explained. Subsequently, related work is presented from which the research gap is derived.

2.1 Product Portfolio Management (PPM)

PPM is a component of strategic management. It ensures the long-term viability of a company by aligning its product offerings with strategic targets and goals [10]. Strategic management involves making deliberate choices to position the company favorably in its competitive environment, with PPM playing a key role in this process. A product portfolio is the total of all products actively offered by a company on the market at a given point in time [11]. Portfolios can be structured along two dimensions: horizontal and vertical. The horizontal dimension lists the product lines, representing the portfolio's breadth. The vertical dimension lists the product variants within each product line, defining the portfolio's depth. A product line includes all product variants with similar functional principles, technologies, and application areas [12].

PPM involves strategic choices, resource allocation, project selection, and balancing to ensure successful product innovation [7]. It includes defining new product development projects, making revisions and updates, and discontinuing products to ensure alignment with business objectives [13].

With the growing emphasis on sustainability, PPM must optimize the portfolio to incorporate sustainability [14]. This includes financial, technical, human, and environmental values [15]. Effective PPM requires sustainability assessments within the portfolio to ensure transparency [16]. Additionally, an efficient sustainability controlling must be established to steer the portfolio according to sustainability requirements.

2.2 Sustainability Assessment and Machine Learning

A sustainability assessment is crucial for PPM to pinpoint products that fall short of the company's sustainability objectives. A commonly utilized technique for this purpose is life cycle assessment (LCA), which systematically evaluates the environmental impacts of products throughout their entire lifecycle [17]. The LCA process is standardized via ISO 14040 [18] and ISO 14044 [19]. When LCA results are validated by a third party, the resulting documentation is referred to as an Environmental Product Declaration (EPD), with principles and procedures outlined in ISO 14025 [20].

However, the utilization of LCA and similar methodologies suffer from three major challenges. One is the complexity of precise modelling a product's life cycle and the associated data demand, the second is data quality and uncertainty and the third is their sensitivity to human error [9, 21–25]. Other approaches for assessing the sustainability of products, as simplified LCA described by Kiemel et al. [9], try to reduce the complexity and effort of assessments but trade-in the accuracy of results in return. Consequently, their effective application in managing multi-variant portfolios face significant barriers [26].

Machine learning (ML), a subfield of artificial intelligence, focuses on the development of computer algorithms that improve their performance through experience [27]. The application of ML in sustainable product portfolio management (SPPM) has gained traction to enhance decision-making processes and achieve sustainability goals. ML approaches have high potential in supporting sustainability assessments in large quantities due to their scalability and represents an alternative approach for addressing the complexity of the task [28]. Machine learning is broadly categorized into supervised learning (learning with labeled data), unsupervised learning (pattern recognition with unlabeled data), and reinforcement learning (trial and error learning) [29]. A broadly used approach for developing machine learning models is the Cross-industry standard for data mining (CRISP-DM) [30].

2.3 Related Work

The integration of sustainability into PPM is explored by Seifert et al., who presented a dynamic PPM framework that considers lifecycle stages. Their work highlights the importance of incorporating lifecycle data to inform portfolio decisions, ensuring that products not only meet market needs but also adhere to sustainability targets throughout their lifecycle [31]. In this context, Medini and Boucher focus on the possible target conflicts and discuss an approach for steering product and service offerings towards economic and environmental sustainability. They stress the importance of aligning corporate sustainability targets with product development strategies, providing a pathway for companies to achieve long-term sustainability objectives. Their approach is based on multi-criteria decision-making theory to reduce the dimensionality of sustainability indicators to a manageable amount [32]. Furthermore, multiple other approaches, e.g. from Pinheiro et al. [33] or Villamil et al. [6, 26], exist. However, analyzing the literature revealed their predominantly qualitative and strategic nature.

Sousa et al. introduce an approximate LCA method leveraging learning systems to evaluate the environmental impact of product concepts. This approach provides a preliminary framework for understanding the lifecycle impacts at the early design stages, thus enabling more sustainable decision-making. However, this method lacks the ability to incorporate real-time data for dynamic assessments and focuses only on product development [34]. Song et al. extend this by developing a rapid life-cycle impact screening for chemical products using artificial neural networks (ANNs).

Their method significantly reduced the time required for LCA by employing ANNs to predict lifecycle impacts, allowing for quicker iterations in the product design process. This was a crucial advancement as it enabled real-time adjustments and more agile responses to sustainability challenges [35]. Hannila et al. also examine the challenges and preconditions for data-driven, fact-based PPM at the product level. They identify the critical role of accurate data in enabling financial product assessments and informed decision-making. However, they note the difficulty in obtaining high-quality, comprehensive data across all product lifecycle stages [36].

The reviewed literature indicates a research deficit regarding practical methodologies for integrating sustainability into PPM. The existing approaches lack in addressing the challenges regarding data quality and availability, as well as systematizing decision-making while incorporating the sustainability perspective.

3 Methodology

The presented concept follows a four-step approach (see Fig. 1). In step one a methodology is developed for companies to derive a specific target system in the context of sustainability based on a generic approach. Step two focuses on the linkage of this target system with a practical set of key performance indicators (KPIs) for enabling effective target controlling within the product portfolio. Step three supports this by providing a scalable solution for creating transparency regarding sustainability KPIs in multi-variant portfolios using machine learning approaches. Step four links the sustainability controlling with variant management and enables the derivation of measures for reaching the set sustainability targets.

3.1 Description of the Target Dimensions of Sustainability-Oriented PPM

At first a systematic literature review enhanced by an empirical analysis of practical target dimensions derived from company reports results in a long list of sustainability targets. In accordance with the approach of Vom Brocke duplicates are eliminated, and the object filter of ecological sustainability is applied due to the set focus of this work [37]. The subsequent consolidation includes the hierarchization of company targets depending on their strategic depth and detail level. Subsequently, the content of the targets is checked for relevance in the context of PPM. For this, the term of SPPM (see 2.2) is defined in accordance with existing literature. Target dimensions that do not align with the goals or responsibilities of SPPM are eliminated. The remaining targets are checked for interdependencies and influences onto each other. This is done with a qualitative approach and results in target dimension clusters and hierarchies. In dependence to the origin of the target dimension a categorization is

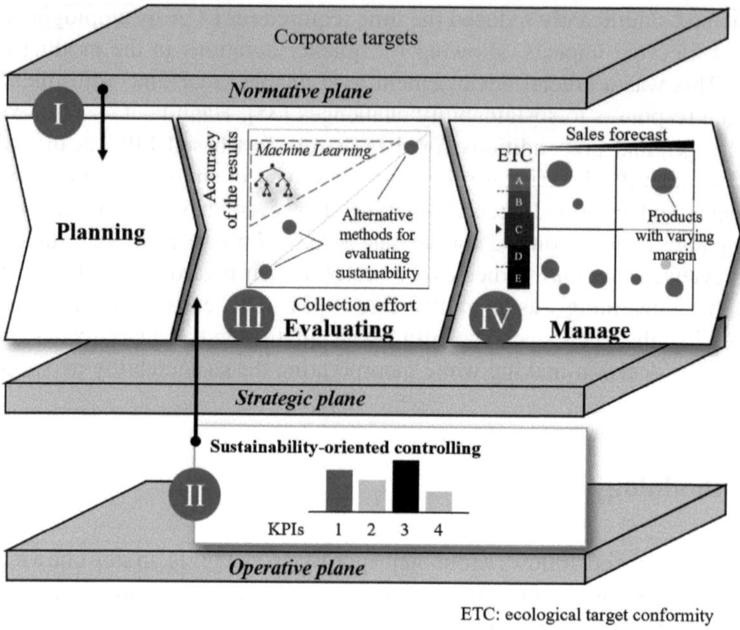

Fig. 1 Target picture of the introduced methodology for integrating sustainability into variant management

done which includes an estimation of the criticality of the target to determine its priority. This generic approach is the basis to enable the derivation of a company-specific target system. This aims to support companies in setting sustainability targets and prioritize them in the case of target conflicts.

3.2 *Identification and Description of Target-Oriented and Sustainability-Relevant KPIs*

By analyzing existing literature as well as sustainability reporting in practice a long list of sustainability KPIs (such as various KPIs regarding air emissions, raw material consumption, energy efficiency or hazardous materials) is derived. Similar to the previous step, the long list is shortened by various filter and consolidation steps. The consolidation includes the assessment of the formal and content wise quality of the KPIs. The assessment criteria are based on existing KPI assessments from the literature, as the criteria by Medini and Boucher [32]. Subsequently, the remaining KPIs are linked to each other as well as to the sustainability targets before. This is done by breaking the targets down to an operational level using the interactive goal panel of Lassenius et al. [38]. In combination with the formal requirements of target controlling, like measurement reference or frequency, this results in descriptions of the KPIs that display their value for the sustainability-oriented controlling. This enables the generation of a standard set of recommended sustainability KPIs for the creation of a controlling system.

3.3 Development of a Method for Predicting a Product's Sustainability KPIs

The research approach of this step is oriented on the CRISP-DM [30]. Firstly, to be scalable the input data for the model needs to be easily available and optimally already consolidated. The identification of this data is part of this step and is done by scanning the literature and interviewing industry representatives. In this context, a developed check list determines if a product information can be used as an input variable in the model. The list contains criteria such as its availability or its distinctiveness. Additionally, LCAs are analyzed to identify relevant data inputs. The data is part of the meta data of the products. Furthermore, as a prerequisite a small set of already conducted LCAs or other ecology assessments is needed for the model to be able to extrapolate the links to the sustainability KPIs. As the performance of a ML-model is improved by minimizing the dimensionality of its input vector the variables are tested for their logical and statistical link to the desired output. Using Pearson' Test the input variables are tested for correlation. Subsequently, a principal component analysis is used to reduce the dimensionality further and focus on variables with significant influence on the output. The output vector is created by adapting the standard set of KPIs that have been identified in 3.2. By analyzing the training data of the model regarding the number of data points in general or the covered intervals of certain input variables an uncertainty scoring is developed that supports the user in estimating the uncertainty of the model's results. While a company is running LCAs this scoring is constantly improved by enlarging the data base. In addition, measures for improving the scoring also indicate which LCA to perform next to improve the prediction capabilities of the model. Finally, modelling involves the development of the model's architecture. For this the XGBoost algorithm is selected due to its ability to handle big data sets efficiently. Furthermore, Bayesian optimization is used for hyperparameter optimization [39]. Training the model involves iteratively adjusting the model to minimize a predefined loss function. For this the Mean Absolute Percentage Error loss function is used. This results in a ML-model that predicts the sustainability of products based on their meta data.

3.4 Development of a Method for Controlling the Sustainability of Products

To operationalize the sustainability controlling the principles of the multi-criteria-decision-making theory are leveraged and an analytic hierarchy process is used to calculate the "ecological target conformity" (ETC). This overarching KPI describes the distance of a product from the company specific sustainability target area. The ETC is calculated by weighting the different target dimensions in accordance with their prioritization and multiplying them with the respective KPI-set. The reduction to one overarching KPI enables a one-dimensional visualization of sustainability within the product portfolio (see left side of Fig. 2). This is done by introducing a scale consisting of five categories (A to E), whereas A is the best rating indicating the best ecological target conformity of a product within the portfolio and E is the worst rating indicating a large gap to a company's sustainability targets. The scale is oriented on the average ETC of the portfolio enabling a relative controlling. The borders of the categories can be set individually by the management influencing the scales sensitivity. Products that show a low ETC are shifting to the center of attention for steering activities within the portfolio. Those products may also be a significant lever for improving the target conformity of the portfolio.

Subsequently, classical PPM tools are enhanced with the ETC. A visualization matrix is developed that is inspired by the BCG-matrix [40]. On the y-axis the ETC is mapped, on the x-axis the sales forecast per product variant is displayed. The size of the dots represents the margin of each variant. The combination of the ecologic and the economic dimensions fosters the resolution of target conflicts. This results in a 2 × 2-matrix that visualizes the biggest levers for improvements in the sustainability context (see right side of Fig. 2). To continuously improve the product portfolio SPPM aims to move all variants to the "Polaris"-quadrant (high sales, high ETC). Depending on the variant's current quadrant this comes with improving one or more of the three described dimensions (ETC, margin, sales forecast). To achieve this SPPM has a toolset with measures like product elimination, modification or new

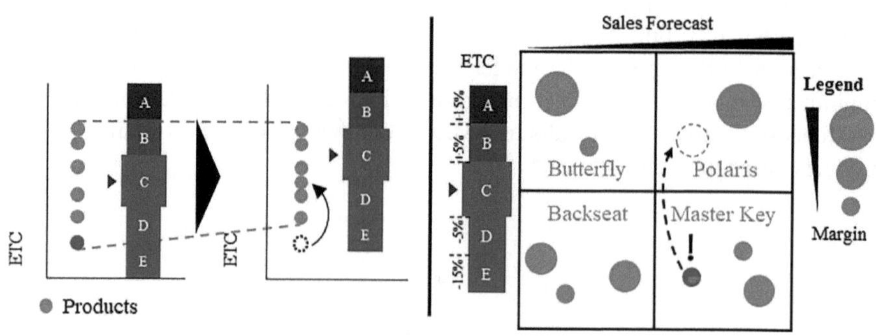

Fig. 2 Controlling portfolio sustainability by means of relative visualization of the variant-specific ecological target conformity (ETC)

product development. Measures that aim to influence the dimensions of sales forecast and margin has already been investigated in previous research [41]. For decision-making regarding the measures SPPM needs to estimate the resulting impact on the sustainability of the portfolio. This is done by using the predicted KPIs of the variants and the estimated sales shifts. When it comes to improving the sustainability of certain products SPPM benefits from a starting point to guide the following processes. For this, firstly, the KPI-set of the variant is analyzed for sustainability weaknesses. Secondly, the insight from this analysis is used to formulate guiding questions for the Research and Development department (R&D) that help to focus their work on certain features of the product. While working on solutions R&D utilizes the developed ML-model to test the impact of their ideas which increases the transparency in early development. Finally, this results in a priority list for SPPM. The list contains variants that need management attention, linked measures of variant management on how to improve the product or the portfolio performance and estimated potentials fueled by informed suggestions of R&D. By using this list SPPM can support their decision-making with data which ultimately fosters efficient sustainability target achievement. Ultimately, the proposed methodology tackles the identified research deficit of lacking sustainability transparency and integration of the dimension into decision-making.

4 Conclusion

Integrating sustainability in decision-making in PPM is often challenging due to potential target conflicts and a lack of transparency. This paper introduces a concept for a data-based management of multi-variant technical product portfolios in the context of ecological sustainability. This is done by systematically selecting KPIs for controlling sustainability targets and substantiating them with estimated data using machine learning. The results are then linked to measures of variant management by creating a framework for multi-target decision-making. In conclusion, this supports PPM and associated R&D activities to become more efficient and effective. The authors are continuing to refine the methodology and compare it with other approaches. The current limitations include the lack of validation which will be subject of future research. For this the methodology is going to be demonstrated with a company example containing company targets, resulting in a specific KPI-set which is predicted via machine learning using real data. In the end the results are going to be linked to concrete measures of variant management.

References

1. Lee H et al (2023) IPCC, 2023: climate change 2023: synthesis report. Contribution of working groups I, II and III to the sixth assessment report of the intergovernmental panel on climate change (Core Writing Team, Lee H, Romero J (eds)). IPCC, Geneva, Switzerland
2. Stock T, Seliger G (2016) Opportunities of sustainable manufacturing in industry 4.0. Procedia CIRP 40:536–541
3. Science Based Targets Initiative (2025) Ambitious corporate climate action. Available at https://web.archive.org/web/20250404083044/https://sciencebasedtargets.org/
4. EcoVadis (2025) Die führende Plattform für Nachhaltigkeits-informationen für globale Lieferketten. Available at https://web.archive.org/web/20250404083029/https://ecovadis.com/de/
5. Liedong TA, Taticchi P, Rajwani T, Pisani N (2022) Gracious growth. Organ Dyn 51:100895
6. Villamil C, Schulte J, Hallstedt S (2022) Sustainability risk and portfolio management—a strategic scenario method for sustainable product development. Bus Strat Env 31:1042–1057
7. Cooper R (1999) New product portfolio management: practices and performance. J of Product Innov Manag 16:333–351
8. Complexity management academy: managing sustainable innovations. Ergebnisbericht der Benchmarking Studie (2024)
9. Kiemel S, Rietdorf C, Schutzbach M, Miehe R (2022) How to simplify life cycle assessment for industrial applications—a comprehensive review. Sustainability 14:15704
10. Riesener M, Dölle C, Schuh G, Lauf H, Jank M-H (2019) Performance-driven and company goal-orientated design of product portfolios: a methodological framework. Procedia CIRP 84:725–730
11. Jank M-H (2021) Produktportfoliosteuerung mittels präskriptiver Datenanalyseverfahren, 1st edn. Apprimus Wissenschaftsverlag
12. Krause D, Gebhardt N (2018) Methodische Entwicklung modularer Produktfamilien. Hohe Produktvielfalt beherrschbar entwickeln, 1st edn. Springer, Berlin Heidelberg
13. Jugend D, Da Silva SL (2014) Product-portfolio management: a framework based on methods, organization, and strategy. Concurr Eng 22:17–28
14. Mohammad J, Pan Y-C (2022) Sustainability, the fourth pillar of project portfolio management—a holistic approach. J Mod Proj Manag 9:198–215
15. Vandaele N, Decouttere C (2011) Sustainable R&D Portfolio assessment. SSRN J. https://doi.org/10.2139/ssrn.1974873
16. Fagarasan C, Cristea C, Cristea M, Popa O, Pisla A (2023) Integrating sustainability metrics into project and portfolio performance assessment in agile software development: a data-driven scoring model. Sustainability 15:13139
17. Sphera (2024) What is life cycle assessment (LCA)? Available at https://web.archive.org/web/20250404083543/https://sphera.com/resources/glossary/what-is-a-life-cycle-assessment-lca/
18. ISO 14040:2006 (2006) Environmental management—life cycle assessment—principles and framework
19. ISO 14044:2006 (2006) Environmental management—life cycle assessment—requirements and guidelines
20. ISO 14025:2006 (2006) Environmental labels and declarations—Type III environmental declarations—principles and procedures
21. Scrucca F et al (2020) Uncertainty in LCA: an estimation of practitioner-related effects. J Clean Prod 268:122304
22. Bamber N et al (2020) Comparing sources and analysis of uncertainty in consequential and attributional life cycle assessment: review of current practice and recommendations. Int J Life Cycle Assess 25:168–180
23. Bicalho T, Sauer I, Rambaud A, Altukhova Y (2017) LCA data quality: a management science perspective. J Clean Prod 156:888–898
24. Lesage P, Mutel C, Schenker U, Margni M (2018) Uncertainty analysis in LCA using precalculated aggregated datasets. Int J Life Cycle Assess 23:2248–2265

25. Hauschild MZ, Rosenbaum RK, Olsen SI (2018) Life cycle assessment. Springer, Berlin
26. Villamil C, Hallstedt SI (2018) Sustain Prod Portfolio Rev EJSD, 7
27. Mitchell TM (1997) Machine learning. McGraw-Hill
28. Saad M, Zhang Y, Jia J, Tian J (2024) Decision tree-based approach to extrapolate life cycle inventory data of manufacturing processes. J Environ Manag 360:121152
29. Han J, Pei J, Tong HD, mining. (2023) Concepts and techniques. Morgan Kaufmann
30. R. Wirth, Jochen Hipp: CRISP-DM: towards a standard process model for data mining (2000)
31. Seifert RW, Tancrez J-S, Biçer I (2016) Dynamic product portfolio management with life cycle considerations. Int J Prod Econ 171:71–83
32. Medini K, Boucher X (2016) An approach for steering products and services offering variety towards economic and environmental sustainability. CIRP J Manuf Sci Technol 15:65–73
33. Paula Pinheiro MA, Jugend D, Demattê Filho LC, Armellini F (2018) Framework proposal for ecodesign integration on product portfolio management. J Clean Prod 185:176–186
34. Sousa I, Eisenhard JL, Wallace D (2001) Approximate life-cycle assessment of product concepts using learning systems. J Ind Ecol
35. Song R, Keller AA, Suh S (2017) Rapid life-cycle impact screening using artificial neural networks. Environ Sci Technol 51:10777–10785
36. Hannila H, Koskinen J, Harkonen J, Haapasalo H (2020) Prod-Lev Profitab JEIM 33:214–237
37. vom Brocke J et al (2015) Standing on the shoulders of giants: challenges and recommendations of literature search in information systems research. CAIS 37
38. Lassenius G, Nissinen M, Rautiainen K, Sulonen R (1998) IEMC'98 proceedings. International conference on engineering and technology management. Pioneering new technologies: management issues and challenges in the third millennium (Cat. No. 98CH36266). IEEE, USA, pp 142–147
39. Koehrsen W (2018) A conceptual explanation of Bayesian hyperparameter optimization for machine learning. Towards Data Science
40. BCG Global (2024) What is the growth share matrix? Available at https://web.archive.org/web/20250404083809/https://www.bcg.com/about/overview/our-history/growth-share-matrix
41. Schuh G, Riesener M Produktkomplexität managen. Strategien—Methoden—Tools, 3rd edn. (Hanser 2018)

Bringing Innovation Systems to Life: Collaborative Upskilling and Customized Connectivity

Paul Blazek and Marton Liszka

Abstract Innovation Systems are dynamic ecosystems that integrate individuals, organizations, policies and processes to drive innovation, economic growth and employment. This work explores collaborative upskilling and customized connectivity as key enablers of innovation. The concept of Focus Parcours is introduced to facilitate specialized, adaptable learning pathways tailored to regional, industry-specific or thematic needs. These pathways strengthen Local, Regional and National Innovation Systems by enhancing targeted education and skills development. Connectivity emerges as a cornerstone of innovation ecosystems, blending physical infrastructures with digital layers to foster seamless integration. Human-centricity remains at the core of this approach, with an envisioned digital overlay connecting the proposed innovation infrastructure and enabling cross-border knowledge exchange within a Global Innovation System. This paper highlights customizable elements and the flexibility of the proposed strategy, enabling a holistic approach to infrastructure usage, tailored upskilling efforts and connectivity solutions to specific goals—whether enhancing regional innovation capacities, fostering industry-specific collaboration or addressing global challenges. Aligning education, infrastructure and collaboration fosters a transnational innovation mindset and sustainable growth.

Keywords Innovation systems · Collaborative upskilling · Customized connectivity · Tailored learning pathways · Focus Parcours

P. Blazek (✉) · M. Liszka
Co-Innovation Cosmos, Vienna, Austria
e-mail: paul@coinnovationfactory.com

M. Liszka
e-mail: marton@coinnovationfactory.com

M. Manns (ed.), *New Paradigms for Anticipated Uncertainty*, Lecture Notes in Mechanical Engineering, https://doi.org/10.1007/978-3-032-16889-4_19

1 Introduction

Innovation systems drive economic growth, employment and technological progress. They have evolved beyond static institutional frameworks toward adaptive, interconnected networks that leverage both physical and digital infrastructures to enhance knowledge transfer and skill development [1]. By integrating human-centric learning, connectivity and interdisciplinary collaboration, innovation ecosystems can maximize their transformative potential.

The Co-Innovation Cosmos framework, introduced in previous research, highlights the importance of collaborative upskilling and customized connectivity as keyenablers of systemic innovation [2]. This work introduces Focus Parcours, a model for modular and customizable learning pathways that are tailored to specific regional, industry-specific or thematic needs. Designed to support cross-border collaboration and lifelong upskilling, Focus Parcours offer a structured yet adaptable format for embedding education within innovation ecosystems. By aligning learning content with real-world challenges and strategic priorities, it addresses critical gaps in current education-to-innovation pipelines. This paper outlines how such customizable education models, combined with strategic infrastructure use and a digital overlay for global knowledge exchange, can drive long-term innovation capacity at local, regional and global levels.

2 Transforming Innovation Systems

National Innovation Systems (NIS) frameworks explain how countries foster innovation through institutions, policies and knowledge flows [3]. Since their emergence in the late 1980s, NIS concepts have evolved to include innovation ecosystems, cross-sector collaboration and adaptive learning mechanisms that drive technological advancement and economic growth [4].

An effective NIS relies on collaboration across government, industry, academia and civil society. The Triple Helix model—emphasizing academia, industry and government—has evolved into Quadruple and Quintuple Helix frameworks, integrating sustainability and societal engagement as key innovation drivers [5].

An effective NIS relies on strong linkages between R&D infrastructure, industry networks, policy frameworks, knowledge diffusion mechanisms and an entrepreneurship culture, fostering innovation through investment in human capital, advanced technologies and institutional adaptability [6]. Increasingly, complex adaptive systems theory has reframed NIS as dynamic environments where small systemic changes can trigger transformative effects [1].

The Co-Innovation Cosmos (CoIC) framework, introduced in previous research, builds upon this understanding by fostering structured, collaborative innovation [2, 7]. Addressing the limitations of traditional, siloed approaches, it emphasizes integrated upskilling, physical and digital connectivity and cross-sector knowledge exchange. Moreover, collaboration is fundamental to innovation systems at local, regional, national, and global levels. The ability to coordinate efforts, facilitate interdisciplinary teamwork and align incentives determines the overall strength of an innovation system.

3 The Four Cs of Workplace Evolution: Change, Customization, Collaboration and Co-Innovation

Workplace transformation is driven by four principles: Change, Customization, Collaboration and Co-Innovation [8, 9]. These elements form a continuous cycle of interactive value creation, where individualization, participation and innovation reinforce one another, ultimately shaping the future of work environments [8].

3.1 Change: Embracing Workplace Transformation

The workplace has undergone a fundamental shift, driven by technological advancements, evolving work habits and new organizational structures [10]. Traditional office settings are being replaced by remote and hybrid work models, fixed schedules are giving way to autonomous work structures and digital tools have replaced analog workflows. This transformation redefines work by emphasizing flexibility, autonomy and adaptability. Embracing this change is the first step toward redefining workspaces that align with the evolving needs of individuals and organizations [9].

3.2 Customization: Individualization as a Key Driver

As organizations adapt to workplace transformation, the demand for customization becomes critical. Employees expect personalized work experiences, including modular workspaces, tailored digital tools and adaptive processes that cater to their specific needs. The shift toward human-centered workplace design ensures that employees feel engaged, productive and valued. However, customization goes beyond the physical workspace; it also includes organizational flexibility—allowing employees to shape their workflows, tools and collaborative processes to optimize efficiency and satisfaction.

Customization fosters ownership, encouraging employees to shape their work environment. When organizations prioritize customization, they not only enhance employee well-being but also lay the foundation for collaboration and innovation [9].

3.3 Collaboration: Unlocking Collective Potential

Collaboration integrates diverse perspectives to co-create solutions. Unlike structured team-based approaches, collaborative work environments encourage spontaneous knowledge exchange and interdisciplinary engagement [9].

The role of physical and digital spaces in enabling collaboration is critical. Well-designed workspaces act as catalysts for interaction, fostering informal communication, creativity and collective problem-solving. The more adaptable and open these spaces are, the more effectively they support collaboration by allowing employees to engage dynamically with colleagues, share knowledge and develop ideas collectively. Collaboration also enhances participation—when employees feel they can contribute meaningfully; they are more likely to engage in innovation processes. By fostering an inclusive, interactive and open culture, organizations can strengthen collaborative mindsets and unlock the full potential of their workforce.

3.4 Co-Innovation: From Collaboration to Systemic Innovation

When collaboration is combined with a culture of openness and participation, it leads to co-innovation—a jointly initiated, incremental innovation process that integrates employees, customers and external partners. Unlike traditional R&D-driven innovation models, co-innovation relies on continuous feedback, experiential engagement and the collective generation of new ideas [9].

Co-innovation thrives in environments that actively support experience-driven engagement. Employees and external stakeholders must be provided with the right tools, spaces and platforms to experiment, iterate and contribute to innovation. This requires a reconfiguration of physical and digital workspaces to enable seamless collaboration, rapid prototyping and interactive knowledge transfer.

A key challenge in co-innovation is bridging the gap between digital efficiency and knowledge-intensive innovation processes. Companies must develop strategies to extract valuable insights from customers and employees without overwhelming them, ensuring that participation in innovation remains accessible and engaging.

3.5 The Interactive Value Creation Cycle (IVC)

The Four Cs—Change, Customization, Collaboration and Co-Innovation—form an iterative cycle, where each component reinforces the next. As work environments evolve (Change), the demand for personalized, flexible workplaces (Customization) grows. This in turn enhances engagement and participation, driving greater interaction and knowledge exchange (Collaboration). Finally, when collaboration is systematically supported, new ideas emerge collectively (Co-Innovation), feeding back into workplace transformation and continuous improvement [9].

For organizations to fully embrace this cycle, they must rethink how they design, structure and utilize workspaces. Modular, technology-integrated and purpose-driven environments can transform offices into active innovation laboratories, fostering creativity, adaptability and sustained business success.

By embedding the Four Cs into workplace strategies, organizations can create a resilient, employee-driven and innovation-focused culture, ensuring that workplaces remain agile, future-ready and capable of sustaining long-term growth in a rapidly changing business landscape.

4 A Holistic Strategy for Systemic Innovation

The Co-Innovation Cosmos (CoIC) model addresses the limitations of isolated innovation hubs by integrating research, industry and entrepreneurship into a scalable, cross-sectoral collaboration network. It connects physical and digital spaces, ensuring broad access to knowledge and resources [2].

At its core, the Co-Innovation Factory (CoIF) acts as a central hub for interdisciplinary collaboration, knowledge transfer and commercialization [8]. Multiple CoIFs across regions link to form a global innovation network, supporting startups, corporations and researchers in refining innovative solutions. The model also includes Focus Parcours, structured programs guiding participants through diverse innovation processes.

Beyond these central hubs, Co-Innovation Labs (CoILs) function as sector-specific R&D units, focusing on fields such as biotechnology, digital health and renewable energy. By directly connecting to CoIFs, CoILs benefit from specialized expertise, funding access and streamlined commercialization pathways [2].

On a local level, Innovation Stations (IS) provide small-scale collaboration spaces, serving as entry points into the broader innovation ecosystem. Unlike larger coworking spaces, IS units facilitate intimate, hands-on collaboration for entrepreneurs, students and researchers while ensuring access to national and global networks.

By integrating these interconnected hubs, the CoIC fosters knowledge exchange, accelerates technological advancements and promotes a participatory innovation culture. Its flexibility and scalability make it a dynamic framework for fostering sustained economic growth and global competitiveness [2].

Traditional innovation frameworks often focus on institutional coordination, funding schemes or siloed R&D efforts without adequately addressing the need for agile, human-centered learning mechanisms. While models like the Triple Helix and national innovation platforms enable structural collaboration, they tend to underemphasize the individual learner's journey and the integration of physical and digital infrastructures.

In contrast, the Co-Innovation Cosmos approach combines spatial, educational and technological components in a cohesive structure. The inclusion of Focus Parcours provides a missing link between system-level coordination and individual capacity-building. By embedding learning directly into innovation spaces and aligning them with thematic or regional goals, the proposed model advances beyond static policy tools toward a participatory, scalable ecosystem model.

5 Collaborative Upskilling

5.1 Need for Continuous Education in Innovation Systems

To keep the Interactive Value Creation Cycle turning, a new approach to knowledge dissemination and consumption is needed. Continuous learning fuels innovation, enabling individuals and organizations to adapt. As industries evolve and technology cycles shorten, lifelong education becomes essential for sustaining innovation-driven growth.

Lifelong learning underpins knowledge-based economies, driving competitiveness. In Local, Regional and National Innovation Systems, education is an ongoing process embedded within professional and industrial networks. Effective innovation ecosystems foster a learning culture that enhances both individual and organizational transformation [11].

Aligning education with industry needs is a key challenge [8]. Traditional models often fail to equip individuals with interdisciplinary and problem-solving skills. Additionally, academia-industry disconnects create mismatches between workforce capabilities and labor market demands. Addressing these gaps requires a shift from static, standardized education toward dynamic, customizable upskilling strategies. A strong example of this shift is 'HEC Paris's 2050NOW Learning Experience', an immersive program combining multimedia learning, industry collaboration and real-world problem-solving [12]. This transdisciplinary approach equips students with skills in sustainability, leadership and technological adaptation while remaining relevant to regional innovation landscapes. Universities worldwide are adopting similar models. Cranfield School of Management's 'Exploring Sustainable Futures' Game uses role-playing to teach sustainability challenges, while UNSW Sydney Business School's 'Diverse Horizons' VR Project immerses learners in diverse lived experiences. These examples reflect a growing trend toward real-world engagement, digital learning tools and problem-solving exercises in continuous education.

The Focus Parcours model, introduced in this paper, builds on this trend by providing tailored, modular learning pathways aligned with regional, industry-specific and thematic needs. By fostering collaborative upskilling, this approach ensures that professionals, entrepreneurs and innovation leaders remain agile in acquiring and applying knowledge.

5.2 *Focus Parcours: A Tailored Learning Approach*

To operationalize continuous education within innovation ecosystems, the Focus Parcours model offers a structured yet flexible learning model that adapts to the diverse needs of innovation stakeholders. The core idea behind Focus Parcours is to create specialized and adaptable learning pathways that are (i) industry-specific (e.g., Digital Transformation, AI and Data Science, Biotech Innovation), (ii) thematic (e.g., Leadership, Intrapreneurship, Sustainable Development) and (iii) regionally oriented (e.g., Innovation for Smart Cities, Local Manufacturing Networks).

Unlike conventional education programs, Focus Parcours emphasizes hands-on, applied learning, integrating real-world industry challenges, mentorship and interdisciplinary collaboration. These modular pathways allow for short-term, skill-targeted learning.

Focus Parcours are characterized by the following key components:

(1) Integration with LIS, RIS and NIS Infrastructure
(2) Customized Learning Tracks
(3) Collaborative Learning Environments
(4) Application-Oriented Methodology.

By embedding these pathways within the broader Co-Innovation Cosmos, Focus Parcours foster a human-centered, lifelong learning ecosystem that drives continuous professional development and strengthens innovation capacity at all levels [7].

5.3 *Leveraging Innovation System Infrastructure for Upskilling*

Effective upskilling relies on strategic use of local, regional and national infrastructure. The Focus Parcours model integrates seamlessly with these systems, ensuring that learning pathways are contextually relevant and readily accessible.

At the community level, innovation hubs and centers serve as catalysts for grassroots educational initiatives. 'Ruta N' in Medellín, Colombia, exemplifies LIS and community-centric upskilling, showcasing how a municipally-owned innovation agency can transform a city's economic landscape [13]. Leading Medellín's transition to a knowledge-based economy, it fosters collaboration among universities, businesses and government entities while providing state-of-the-art facilities and training programs that support continuous learning and entrepreneurial growth.

Regional innovation districts centralize research institutions, startups and enterprises. Boston's Innovation District is an example of a successful RIS that integrates customized learning programs with industry needs [14]. Through its co-location of academia, corporates and incubators, the district fosters a culture of collaboration and skill development, ensuring that upskilling aligns with industry growth.

At the national level, cohesive policies are essential to scaling up local and regional upskilling initiatives. NIS provide a framework for aligning educational programs with national economic goals and technological advancements. Countries that implement coordinated national strategies—such as innovation-focused funding and competency-based education—demonstrate higher workforce adaptability and economic resilience [15]. Silicon Valley exemplifies this approach, where strong ties between universities, government initiatives and private sector investments have fostered a dynamic ecosystem that continuously generates cutting-edge innovations and high-growth industries [11]. Similarly, Saudi Arabia aims to develop its own National Innovation System through Vision 2030, leveraging strategic investments in education, research and technology hubs to diversify its economy and tries to position itself as a global leader in innovation-driven sectors [2].

By incorporating Focus Parcours within this multi-tiered infrastructure, regions can ensure that upskilling is not an isolated effort but a systemic, sustainable initiative.

5.4 Customization and Strategic Integration

The adaptability of Focus Parcours makes them a powerful tool for addressing specific educational and economic objectives. They offer a modular structure, allowing participants to select and combine learning modules that align with their career goals, current skill levels and industry trends. This flexibility ensures that the learning experience is personalized and relevant.

By aligning with current educational policies and workforce development programs, Focus Parcours enhance existing structures rather than replacing them. This alignment facilitates smoother implementation and greater acceptance among stakeholders. For instance, in regions where governments have initiated digital literacy campaigns, Focus Parcours can offer specialized modules that build upon these foundations, advancing participants' skills to more complex areas such as data analytics or cybersecurity.

An interdisciplinary engagement between different stakeholders supports the success of Focus Parcours. Corporations can contribute by identifying skill gaps, providing real-world challenges for learners and even co-developing curriculum content. Educational institutions, on the other hand, bring academic rigor and pedagogical expertise, ensuring that the learning pathways are both challenging and accessible. This collaboration results in a dynamic learning environment that is responsive to industry needs and technological advancements.

To ensure the effectiveness of the Focus Parcours model, it is essential to establish clear metrics for success. These may include participant satisfaction rates, employment outcomes, career progression and contributions to innovation within their respective fields. Regular assessment and feedback mechanisms allow for continuous improvement of the learning pathways, ensuring they remain relevant and impactful.

6 Customized Connectivity

6.1 The Dual Nature of Connectivity: Physical and Digital Layers

In the contemporary innovation landscape, connectivity manifests through both physical and digital infrastructures, each playing a pivotal role in fostering collaboration and knowledge exchange.

Physical spaces such as coworking areas, incubators and innovation hubs serve as tangible environments where individuals and organizations converge to collaborate. For instance, the establishment of data centers, like NextDC's facility in Adelaide, underscores the significance of physical infrastructure in supporting sectors such as government, health, mining, defense and space [16]. This facility not only provides essential IT space but also integrates advanced technologies like artificial intelligence and quantum computing, thereby driving digital innovation and economic growth in the region.

The expansion of digital infrastructures facilitates global knowledge exchange and remote collaboration [15]. Digital infrastructure encompasses the physical hardware and software-based technologies that enable digital services, including IT systems and networks that allow organizations to operate and communicate effectively.

Fostering innovation requires synergy between physical and digital infrastructures, where physical spaces enable face-to-face interactions and digital platforms support continuous collaboration beyond geographical constraints [3]. This integrated approach ensures that innovation networks are both robust and flexible, adapting to the evolving needs of participants.

We envision the Co-Innovation Sphere (CoISphere) as a global digital overlay, enhancing innovation systems by facilitating seamless collaboration across borders [7]. It should act as a unifying digital platform that integrates various innovation ecosystems, enabling stakeholders worldwide to collaborate effectively. By providing a centralized virtual space, it fosters the exchange of ideas, resources, and expertise, thereby accelerating innovation processes and cultivating a global innovation mindset.

6.2 Human-Centricity as the Core of Connectivity

Placing humans at the center of innovation networks ensures that technological advancements align with societal needs and values [17]. These networks prioritize the experiences, well-being and active participation of individuals. By focusing on human needs, such networks foster solutions that are both innovative and socially relevant. For example, human-centered innovation emphasizes understanding and addressing the real needs of people, leading to more effective and meaningful solutions.

Trust and social capital are foundational to effective collaboration [18]. Regular interactions in both physical and digital spaces build relationships and facilitate the exchange of ideas. In human-centered service co-innovation, especially in public services like elderly care, trust and continuous interaction among stakeholders lead to more effective and user-friendly solutions.

Accessible innovation networks enhance creativity and problem-solving. Inclusive models that adapt to the varying needs of participants lead to more comprehensive and resilient solutions [19]. For instance, living labs adopt an open and citizen-centric approach to innovation, involving users in the development process to ensure outcomes that truly meet their needs.

6.3 Customizing Connectivity Solutions for Different Needs

Customizing connectivity strategies enhances innovation network impact. Understanding the unique economic and social objectives of a region allows for the development of targeted connectivity solutions. For example, rural areas may prioritize infrastructure development to support digital access, while urban centers might focus on enhancing existing networks to support high-tech industries.

Different industries have distinct collaboration needs. For instance, the healthcare sector may require secure data-sharing platforms to comply with privacy regulations, whereas the creative industries might benefit from open, collaborative design tools. Customizing connectivity solutions to these specific requirements ensures that innovation processes are both efficient and compliant with industry standards.

Complex global issues, such as climate change and pandemics, necessitate collaborative efforts that transcend national borders. By establishing transnational knowledge networks, stakeholders can pool resources and expertise to develop comprehensive solutions. For example, international consortia working on renewable energy technologies can share research and development efforts, accelerating the deployment of sustainable solutions worldwide.

7 Conclusion

As innovation ecosystems continue to evolve, collaborative upskilling and customized connectivity have become central pillars of competitive, knowledge-based economies. The Focus Parcours model, introduced in this paper, offers a flexible and modular approach to lifelong learning, aligning educational pathways with industry demands and regional goals. At the same time, a transnational digital infrastructure—the envisioned Co-Innovation Sphere—has the potential to enhance knowledge sharing and interdisciplinary collaboration across borders.

To realize the full potential of this framework, coordinated efforts across policy, education and industry will be essential. Future work will focus on piloting Focus Parcours in diverse local and sectoral contexts, refining its modular structure in response to evolving skill demands. Particular attention will be given to how these pathways support workforce adaptability in the face of accelerating digital transformation, automation and changing career models—key dimensions of the future of work.

Further development of the Co-Innovation Sphere will explore how human-centric digital platforms can integrate with physical infrastructure to support hybrid learning and innovation spaces. Research will also examine governance models and metrics that ensure inclusive access, continuous feedback, and measurable outcomes across innovation systems.

By embedding collaborative upskilling and customized connectivity into long-term innovation strategies, this approach supports not only economic competitiveness but also the agility and resilience required for navigating the complexities of future work environments.

References

1. Organization for Economic Cooperation and Development (OECD) (1999) Managing national innovation systems. OECD Publishing, Paris. Available at: https://doi.org/10.1787/9789264189416-en. Accessed 26 Feb 2025
2. Blazek P, Liszka M (2024) Understanding and enhancing national innovation systems: dynamics, challenges and strategies. In: MCP 2024 conference proceedings, Novi Sad, Serbia pp 43–50
3. Organization for Economic Cooperation and Development (OECD) (ed) (1997) Oslo manual: proposed guidelines for collecting and interpreting technological innovation data, 2nd edn. Paris
4. Edquist C, Hommen L (2008) Comparing national systems of innovation in Asia and Europe: theory and comparative framework. In: Edquist C, Hommen L (eds) Small country innovation systems. Globalization, change and policy in Asia and Europe, Cheltenham, Northampton, pp 1–28
5. Leung SKC (2005) Measuring innovation—the experience of Hong Kong, China. In: Conference on innovation systems in Hong Kong and the Chinese Mainland. The Hong Kong University of Science and Technology, Hong Kong
6. Organization for Economic Cooperation and Development (OECD) (ed) (2005) Oslo manual: guidelines for collecting and interpreting innovation data, 3rd edn. Paris
7. Co-Innovation Cosmos (2025) https://www.coinnovationcosmos.com. Accessed 26 Feb 2025
8. Blazek P, Aschenbrenner V (2021) Creating customizable co-innovation spaces. In: Towards sustainable customization: bridging smart products and manufacturing systems. Proceedings of the 8th changeable, agile, reconfigurable and virtual production conference (CARV 2021) and the 10th world mass customization & personalization conference (MCPC 2021), Aalborg, Denmark

9. Blazek P, Aschenbrenner V (2023) Human centered innovation workplaces. In: Galizia FG, Bortolini M (eds) Production processes and product evolution in the age of disruption. CARV 2023. Lecture notes in mechanical engineering. Springer, Cham
10. Innovation and Technology Commission (ITC) (2004) New strategy of innovation and technology development, consultation paper, Hong Kong
11. Blazek P (2012) Hong Kongs digitale Positionierung: die Bedeutung von Informations- und Kommunikationstechnologie-Innovationen für die nachhaltige wirtschaftsräumliche Entwicklung, Göttinger Geographische Abhandlungen, Number 119. Göttingen
12. Financial Times (2025) Teaching award: rising to the challenge. Financial Times. Available at: https://www.ft.com/content/bad75b74-a260-4306-8b9f-9281bc0ca85a. Accessed 26 Feb 2025
13. Ruta N Medellin (2009) Available at: https://rutanmedellin.org/. Accessed 26 Feb 2025
14. Katz B, Wagner J (2014) The rise of innovation districts: a new geography of innovation in America. Brookings: Metropolitan Policy Program, Washington
15. Lallana EC, Uy MN (2003) The information age, UNDP-APDIP, e-ASEAN task force, e-primer, Kuala Lumpur
16. NextDC (2024) NextDC opens $100M Adelaide data centre. Available at: https://www.nextdc.com/news/-unveils-first-tier-iv-data-centre-in-adelaide. Accessed 26 Feb 2025
17. Bergvall-Kåreborn B, Ståhlbröst A (2009) Living lab: an open and citizen-centric approach for innovation. Int J Innov Reg Dev 1:356–370. Available at: https://doi.org/10.1504/IJIRD.2009.022727. Accessed 26 Feb 2025
18. Savchuk K (2019) The five best policies to promote innovation—and one policy to avoid. Stanford Business, 7 Oct 2019, pp X–Y
19. Drucker JM, Kayanan CM, Renski HC (2019) Innovation districts as a strategy for urban economic development: a comparison of four cases. SSRN. Available at: https://ssrn.com/abstract=3498319. Accessed 26 Feb 2025

Model-Based Configuration as a Tool for Discovering Design Knowledge: A Skate Park Case-Study

Paul Christoph Gembarski, Leo Duchan, Julian Schröder, and Julius Vahle

Abstract Knowledge-Based Engineering (KBE) combines engineering expertise with parametric computer aided design models to automate repetitive tasks, enhance design flexibility, and ensure more efficient and accurate results. Inputs may be requirements or distinct parameters. Nonetheless, for comprehensive designs, the complexity, i.e., amount and intertwinement of such inputs may overwhelm users. A way to overcome this is reducing user inputs by augmenting them with surrogate models, e.g., using physics-based calculations. This paper examines KBE and its application in modeling a skate park. The design incorporates essential physical calculations, such as speed dynamics, and user-specific requirements to reduce inputs to basic attributes, such as skill level or body weight, while ensuring both safety and customization. The resulting model of the design solution space then can be used to discover engineering knowledge by analyzing the sensitivity of input changes to the resulting configurations.

Keywords Model-based configuration · Knowledge-based engineering · Design automation · Design knowledge elicitation

1 Introduction

Skate parks have become a crucial part of urban leisure and sports facilities. Designing them involves considering different usage requirements, safety standards, and local conditions. This comprises many routine tasks until designing the single obstacles and aggregating them into a course [1].

In this context, knowledge-based engineering (KBE) provides numerous advantages. By integrating knowledge models with automated design routines, it becomes possible to generate product variants tailored to user needs within a defined solution space [2]. This is valuable not only for individualization but also for the discovery of design knowledge, as it allows for the evaluation of input parameters' influence

P. C. Gembarski (✉) · L. Duchan · J. Schröder · J. Vahle
Institute of Product Development, Leibniz University Hanover, Garbsen, Germany
e-mail: gembarski@ipeg.uni-hannover.de

M. Manns (ed.), *New Paradigms for Anticipated Uncertainty*, Lecture Notes in Mechanical Engineering, https://doi.org/10.1007/978-3-032-16889-4_20

on the design outcome, e.g., through sensitivity analysis [3]. In such solution space models, inputs play a critical role, particularly for comprehensive designs, regarding complexity, such as the number and intertwinement of parameters, which can overwhelm users. One way to address this is by reducing user inputs through surrogate models, such as physics-based calculations or embedded simulation tools [4].

This paper explores the reduction of user inputs and the utilization of the solution space model to uncover design knowledge, using a skate park as an example. The dimensions and placement of obstacles within the course are determined by high-level user inputs, which include body data, obstacle difficulty, obstacle sequence, and a feedback loop based on velocity profile calculations.

In the remainder of this paper, Sect. 2 first explains the theoretical background of KBE and parametric design before Sect. 3 presents the research aim. Section 4 then describes the case study of parametric modeling a skate park. Finally, Sect. 5 discusses the insights gained and concludes the paper.

2 Theoretical Background

The concepts of knowledge-based engineering (KBE) and design automation signify a substantial transformation in computer-aided product modeling. The fundamental principle of KBE is to capture design-relevant knowledge, explicitly map it, and transfer it into automated routines. Rather than concentrating on designing a single product variant, the goal is to create a solution space from which a variant can be easily derived based on specific requirements [5, 6]. E.g., design rules and constraints can be stored in parametric computer-aided design (CAD) models so that adjustments to components or assemblies are carried out by the system itself rather than manually [7]. The parameter control can be further structured and externalized by integrating spreadsheets or algorithmic approaches, e.g. using scripting languages [8]. In combination with other simulation and analysis tools, product variants can be evaluated automatically, e.g., in terms of material usage, stability, or costs [9].

A different approach to knowledge-based engineering is algorithmic modeling, where the focus is on abstracting and modeling the design process itself to create intelligent, interactive model generators [10]. Approaches include using an application programming interface (API), design language, or expert system shell to drive CAD kernels or generate polygonal geometric models, and evaluating and optimizing the geometry during generation by suitable analysis tools [11, 12]. Algorithmic modelling is particularly advantageous for complex geometries and pure configuration designs but often restricts subsequent parametric editing of the variant [13].

In industrial practice, KBE solutions are particularly valuable when customer-specific products have to be realized with a short development time. In addition, KBE promotes better traceability of development decisions, as the stored expert knowledge is documented transparently [14]. Knowledge-Based Engineering (KBE) has been applied across various industries to automate and optimize design processes.

In automotive engineering, KBE systems are used to streamline the development of vehicle components, such as chassis and aerodynamics, by integrating expert knowledge into parametric design frameworks [15]. Similarly, in aerospace, KBE enables the efficient modelling of complex structures like wings and fuselages, facilitating rapid iterations and adherence to strict safety standards [16]. In architecture, parametric design tools driven by KBE principles support the creation of adaptive building structures that respond to environmental inputs, such as light and temperature [17].

3 Research Aim

A crucial challenge in KBE systems is managing the complexity of user input [18]. Instead of requiring users to manually define numerous parameters such as single dimensions and structural details, an optimized approach consolidates inputs into higher-level categories from which model parameters can be reasoned [19]. Rule-bases and constraint networks are still typical implementations here that follow a sequential approach to reason from input to output [20].

This study employs a learning-by-building approach and seeks to investigate an iterative approach. By integrating surrogate models and optimization techniques in the sense of algorithmic modelling, a model-in-the-loop process will be implemented that takes a first set of input parameters to calculate initial outputs and iteratively refines both. This approach aims to reduce cognitive load for users while simultaneously allowing the analysis of dependencies between different inputs and outputs, thereby facilitating the explication of design knowledge that can be fed back into the Knowledge-Based Engineering (KBE) system.

4 Case Study: Knowledge-Based Design of a Linear Skate Park

As case study, a linear skate park was chosen. The aim was to reduce user inputs only to body data of one or more skaters, obstacle difficulty, and obstacle sequence. The obstacles include a quarter pipe as starting ramp, spine, A-frame, and bump (Fig. 1).

Especially, the obstacle difficulty has a far-reaching effect on the primary obstacle dimensions, i.e., height and slope, and the course setting, i.e., the distance between the single obstacles. As a surrogate model, the velocity profile was chosen since it is both applicable to express another aspect of difficulty, which is top speed, acceleration, and pause, as well as to calculate the heights of each obstacle.

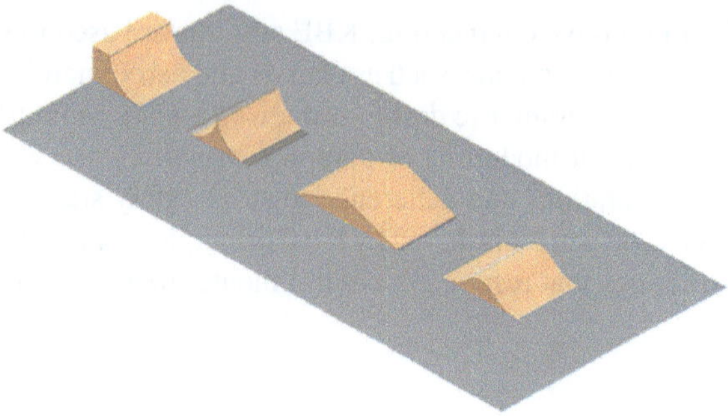

Fig. 1 Skate park with implemented types of ramps

4.1 Knowledge Model and Basic Reasoning

The knowledge model consists of two parts. The first contains basic design rules and constraints that determine the geometry of the obstacles. Besides the strength-based dimensioning of the substructure, this also includes basic safety rules and restrictions from standards, e.g., different minimal widths depending on the height of a ramp.

The second part is driven by the velocity model of the course. Based on the chosen obstacles, the difficulty setting, and the weight of the skater, a first draft of the course is created and the starting speed after the quarter pipe is calculated following the principle of conservation of energy, added by basic loss functions, e.g., for basic friction and drag coefficients. In this version of the model, advanced starting procedures such as kick-start or additional pushing are neglected, but could be implemented as additional energy sources.

After leaving the ramp, the distance to the next obstacle is configured again based on the difficulty level of the course. Essentially, the higher the difficulty, the shorter the distance. However, for safety reasons, a minimum time frame is set to guarantee sufficient response time to approach the obstacle so that the initial configuration may be altered already at this point. Next, the inlet velocity for each next obstacle is used to reason about its slope and height. The premise for spine and A-frame is to transform all kinetic into potential energy so that the speed at the top of these obstacles is zero. The bump is different from this, here a smooth transition from deceleration to acceleration is defined. At this point, the configuration of the skate park is updated with the height and slope parameters from the velocity model and checked for consistency with the geometric design constraints.

Afterwards, the real height profile of the course is fed back into the velocity model to get the speed for each position of the course, which is mapped into the velocity profile shown in Fig. 2. Here, the gradients are analyzed and related to the time to skate the according obstacle. If the gradient is out of a predefined range, the slope of the obstacle is adjusted, the geometry updated, a new velocity profile is calculated, and the analysis is repeated. The process iterates until no more changes to the geometry occur.

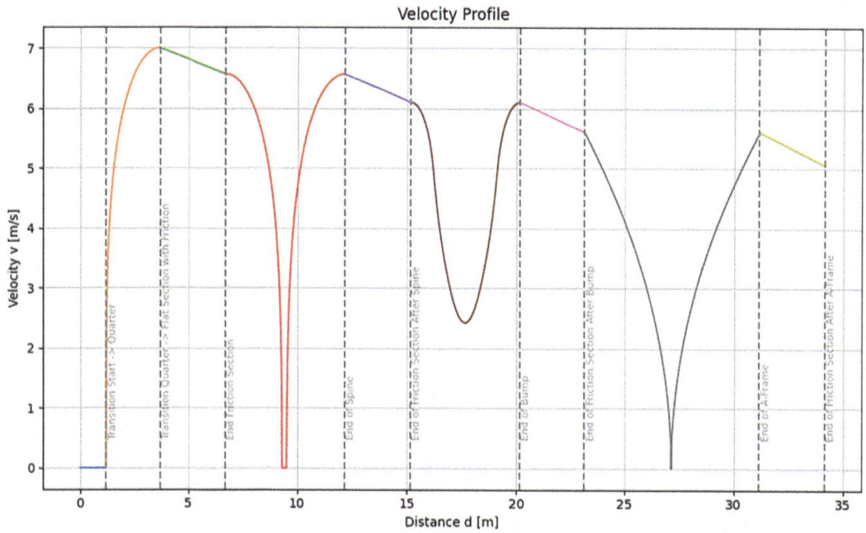

Fig. 2 Velocity profile for a configuration consisting of quarter, spine, bump, and A-frame

4.2 Sensitivity Analysis and Advanced Reasoning

Up to this point, the skate park's configuration has been tailored for either a single skater or a group of skaters with similar weights, similar to a configuration with limit values. To enhance the design's significance, it is essential to explore the impact of the skater's weight and possibly other body data on the course settings and obstacle geometries. This exploration is based on the velocity model and involves comparing two different scenarios and their respective speed profiles and to draw conclusions about necessary changes to the course.

The scenario of an upper limit configuration needs to take lower weight values into account while keeping the geometry the same. This results in obstacles that cannot be skated since the amount of kinetic energy is too low. This enables the calculation of the additional speed required for a lighter skater and the determination of whether this can be achieved, e.g., through additional pushing between the respective obstacles. This calculation also depends on the distance between obstacles and the skaters' anthropometric data, such as leg length, as it's necessary to ensure that the skater can make at least one pushing step. If the course layout needs to be adapted, these inputs can be fed back into the configurator as a new initial configuration, and the dimensioning process is repeated.

When selecting a lower limit configuration scenario, skaters with higher weights are used for comparison within the given layout. This can result in overshooting an obstacle, especially the spine. The surplus kinetic energy can be assessed and linked to the distance between obstacles. Since the velocity model assumes a direct linear ride between obstacles, it is possible to check if a diagonal or zigzag approach to the

Fig. 3 CAD model of the spine case without cover

next obstacle can sufficiently reduce the kinetic energy within the current width or if the obstacle needs to be enlarged or moved. Once again, all changes to the geometry are processed into a new initial configuration, and the configuration iterates.

In such a way, the knowledge-based approach for modelling the skate park can be used to find a good compromise for different skaters.

4.3 CAD Integration and Outputs

Each obstacle was modelled with full details so that it consists of structural beams, retaining walls and deck (Fig. 3). Machine elements such as nails and screws were neglected. The models are parametric, and for better structured parameter management, each obstacle assembly is controlled by a spreadsheet that integrates basic calculations. The course is pre-assembled as a digital master where unused obstacles are suppressed while the active ones are positioned on the course using geometric constraints.

For users interested in building the skate park as a DIY project, an additional feature is provided for cost estimation. The bill of materials for each obstacle is used to determine the total amount of required raw materials (without accounting for cuts). A Python-based web crawler then retrieves current prices from hardware store websites and calculates the cost for each obstacle accordingly (Fig. 4).

5 Discussion and Conclusion

The basic goal of integrating additional algorithmic elements into the sequential reasoning of KBE models and reducing the complexity of user inputs was achieved in the case study. After configuring an initial course, design rules were applied to detail the geometry of the obstacles before the velocity model iterated the design to sufficiently fulfill the requirements of either a narrow or wider group of skaters.

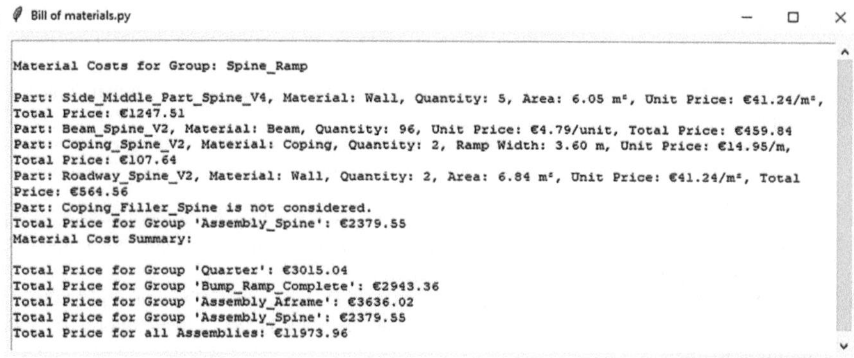

Fig. 4 Cost calculation sheet

However, the case study is not without limitations. One of the main issues lies in the oversimplification of certain physical relations and the skaters' influence on driving dynamics. For instance, neglecting friction on the ramps themselves and restricting the consideration of energy gains from user movements, such as pushing or crouching, means the model only approximates real-world scenarios. The current model also lacks the ability to account for external factors like weather conditions and material variability.

The advanced reasoning concept and there the reference velocity profile encompasses different simplifications to have a single reference, such as the assumption of a linear ride between the obstacles. Another premise relates to the bump in particular where an overshooting is not wanted. In such a way, jumps and tricks are at least hindered.

There are various ways to expand the functionality of the KBE system. Besides defining additional obstacles, a major enhancement would be greater flexibility in course definition. This includes both greater degrees of freedom in obstacle placement and the integration of non-linear, looped layouts. Since the skate park is, by nature, a sectional-modular design, such an implementation would require configuring and sequencing the velocity model according to the chosen obstacles, which is algorithmically feasible and an obvious next step. Another avenue for extension is the previously mentioned integration of an abstraction of energy-gaining user movements into the velocity model. This would also be valuable for addressing safety concerns by simulating dangerous user behavior.

Besides the skate park example, a model-in-the-loop approach as integral part of the reasoning mechanism is interesting for products that contain dynamics and the necessity for (multi-criteria) optimization. By integrating dynamic models into the decision-making process, designers and engineers can better simulate real-world scenarios, predict outcomes, and make informed choices that enhance the overall functionality and reliability of their products.

References

1. Wixon B (2009) Skateboarding: instruction, programming, and park design. Human Kinetics, Champaign
2. Ortner-Pichler A, Landschützer C (2022) Integration of of parametric modeling in web-based knowledge-based engineering applications. Adv Eng Inform 51:101492. https://doi.org/10.1016/j.aei.2021.101492
3. Gembarski PC, Melching J, Plappert S (2023) A knowledge-based engineering system for the planning of networked rainwater harvesting and distribution systems. Sustainability 15(11):8636. https://doi.org/10.3390/su15118636
4. Boretti V, Sardone L, Bohórquez Graterón LA, Masera D, Marano GC, Domaneschi M (2023) Algorithm-aided design for composite bridges. Buildings 13(4):865. https://doi.org/10.3390/buildings13040865
5. Gembarski PC, Li H, Lachmayer R (2017) Template-based modelling of structural components. Int J Mech Eng Robot Res 6(5):336–342. https://doi.org/10.18178/ijmerr.6.5.336-342
6. Zhang LL (2014) Product configuration: a review of the state-of-the-art and future research. Int J Prod Res 52(21):6381–6398. https://doi.org/10.1080/00207543.2014.942012
7. Verhagen WJ, Bermell-Garcia P, Van Dijk RE, Curran R (2012) A critical review of knowledge-based engineering: an identification of research challenges. Adv Eng Inform 26(1):5–15. https://doi.org/10.1016/j.aei.2011.06.004
8. Gembarski PC, Li H, Lachmayer R (2017) KBE-modeling techniques in standard CAD-systems: case study—Autodesk inventor professional. In: Bellemare J, Carrier S, Nielsen K, Piller FT (eds) Managing complexity. Springer, Heidelberg, pp 215–233. https://doi.org/10.1007/978-3-319-29058-4_17
9. Poot LP, Wehlin C, Tarkian M, Ölvander J (2020) Integrating sales and design: applying CAD configurators in the product development process. In: Proceedings of the design society: DESIGN conference, vol 1, pp 345–354. https://doi.org/10.1017/dsd.2020.129
10. Fuchs D, Bartz R, Kuschmitz S, Vietor T (2022) Necessary advances in computer-aided design to leverage on additive manufacturing design freedom. Int J Interact Des Manuf 16(4):1633–1651. https://doi.org/10.1007/s12008-022-00888-z
11. Frank G, Entner D, Prante T, Khachatouri V, Schwarz M (2014) Towards a generic framework of engineering design automation for creating complex CAD models. Int J Adv Syst Meas 7(1):179–192
12. Tedeschi A, Lombardi D (2014) AAD—algorithms-aided design. Le Penseur, Brienza
13. Müller P, Gembarski PC, Lachmayer R (2022) Parametric topology synthesis of a short-shaft hip endoprosthesis based on patient-specific osteology. In: Andersen AL et al (eds) Towards sustainable customization: bridging smart products and manufacturing systems. Springer, Cham, pp 669–676. https://doi.org/10.1007/978-3-030-90700-6_76
14. Amadori K, Tarkian M, Ölvander J, Krus P (2012) Flexible and robust CAD models for design automation. Adv Eng Inform 26(2):180–195. https://doi.org/10.1016/j.aei.2012.01.004
15. De Almeida R (2021) Knowledge-based engineering CAD templates applied in vehicle advanced concepts design. SAE technical paper, 2021-01-5049. https://doi.org/10.4271/2021-01-5049
16. La Rocca G, Van Tooren M (2010) Knowledge-based engineering to support aircraft multidisciplinary design and optimization. J Aerosp Eng 224:1041–1055. https://doi.org/10.1243/09544100JAERO592
17. Cao H (2024) Innovative applications of parametric design and digital tools in architecture: exploring the integration of generative design and BIM technology. Appl Comput Eng 98:79–84. https://doi.org/10.54254/2755-2721/98/20241127
18. Esanakula J, Sridhar NV, Rangadu V (2015) Knowledge-based engineering: notion, approaches and future trends. Am J Intell Syst 5(1):1–17. https://doi.org/10.5923/j.ajis.20150501.01
19. Tang Z, Zou Q, Gao S (2023) A decision-support method for multi-parameter editing of parametric CAD models. Adv Eng Inform 56:101997. https://doi.org/10.1016/j.aei.2023.101997

20. Felfernig A, Hotz L, Bagley C, Tiihonen J (2014) Knowledge-based configuration: from research to business cases. Morgan Kaufmann, Waltham

AI-Driven Transformation of Online Product Configurators: Enhancing User Interaction in Mass Customization

Paul Blazek and Clarissa Streichsbier

Abstract In observing the evolution of product configurators since 2007 in our Configurator Database Research Project we have been conducting analyses on the development of configurators based on various criteria as well as exploring the appearance of their user interfaces. Being a crucial tool in the concept of mass customization the criteria for successful product configurators have changed and adapted to the evolving needs of customers and the technological possibilities. The rise of Artificial Intelligence (AI) is and will massively influence the way how software interaction is taking place. In our paper we examine if AI will augment and enhance a product configurator experience and what features and effects might change in how configurator interaction will take place in the future.

Keywords Mass customization · Product configurators · Artificial intelligence · User experience · Digital innovation

1 Introduction

1.1 Evolution of Product Configurators

Since 2007, the Configurator Database Research Project has systematically analyzed the evolution of product configurators, focusing on various criteria and the design of their user interfaces [1]. As a core component of the mass customization paradigm, the success factors of product configurators have continually adapted to meet changing customer needs and technological advancements. The recent rise of artificial intelligence (AI) is poised to significantly transform software interactions, including the configurator experience. This paper examines how AI may augment and enhance

P. Blazek (✉) · C. Streichsbier
CYLEDGE Media GmbH, Wien, Austria
e-mail: p.blazek@cyledge.com

C. Streichsbier
e-mail: c.streichsbier@cyledge.com

M. Manns (ed.), *New Paradigms for Anticipated Uncertainty*, Lecture Notes in Mechanical Engineering, https://doi.org/10.1007/978-3-032-16889-4_21

product configurators, exploring potential new features, changes in interaction patterns, and the implications for future customer engagement. Our findings aim to provide insights into the evolving role of AI in shaping the next generation of configurator tools.

1.2 Product Configurator Research

Product configurators have become essential tools for enabling mass customization and providing customers with tailored solutions that align with their preferences and needs. Traditionally, these systems rely on predefined rules, static options and manual user selection, requiring customers to navigate complex choices. While such configurators allow a degree of flexibility, they often fail to deliver an intuitive, efficient and truly personalized experience.

In the digital era, web-based product configurators have emerged as transformative tools that bridge the gap between customer aspirations and industrial production capabilities. These interactive software applications enable direct, online communication between companies and consumers, allowing users to design, create or assemble their desired products by selecting specialized components tailored to their needs. This process not only enhances customer engagement but also drives the evolution of mass customization in various industries [2].

We started the Configurator Database Project in 2007 with the objective of aggregating a comprehensive collection of web-based configurators available online [1]. By doing so, the project aimed to monitor the development and diversification of configurator applications across multiple categories, while also providing a detailed snapshot of the status quo and identifying emerging trends in the field. Initially, the database catalogued only a few hundred entries, but it has since expanded significantly to include more than 1500 configurators, as documented on the project's website [1]. This growth reflects not only the widespread adoption of configurator technology but also its increasing relevance across different sectors and geographic regions.

The development and evolution of web-based product configurators underscore a broader transformation in consumer behavior and industrial strategy. As companies increasingly seek to offer personalized experiences, the integration of advanced technologies – such as augmented reality and artificial intelligence – into configurator systems is becoming more prevalent. This ongoing innovation not only enriches the user experience but also positions configurators as critical tools for competitive differentiation in a rapidly evolving digital marketplace.

1.3 Information Delivery and Usability in Online Product Configuration Systems

Online product configuration relies on clear and comprehensive information delivery to reduce uncertainties during the purchasing process. Essential details – such as product options, pricing, delivery terms, after-sales conditions and company capabilities – must be communicated effectively. Tailoring product descriptions to match customers' cognitive expectations, whether by highlighting performance, functions or physical components, minimizes complexity and the overall information load.

Usability is equally crucial for the success of these systems. Intuitive navigation, fast loading times, transparent structure and robust support are key aspects that enable users to interact efficiently with the configurator. Additionally, factors like visual realism, creative freedom and overall enjoyment enhance the user experience, ultimately encouraging customers to make confident purchasing decisions.

2 The Evolution of Artificial Intelligence

2.1 The Role of Artificial Intelligence

The rapid advancement of artificial intelligence is fundamentally transforming e-commerce, revolutionizing how businesses interact with consumers, personalize experiences and streamline operations. Over the past decade, AI has evolved from simple rule-based automation to sophisticated deep learning models capable of generating text, images and personalized recommendations in real time. This shift has been driven by breakthroughs in natural language processing (NLP), machine learning and neural networks, leading to the development of first AI-driven configurators that enable mass customization at scale.

In the early stages of AI adoption, e-commerce relied on basic recommendation algorithms and automated customer support tools. However, the development of transformer-based models such as OpenAI's GPT series and generative adversarial networks (GANs) has expanded AI's capabilities, allowing it to analyze vast datasets, predict consumer preferences and create hyper-personalized shopping experiences. These advancements have directly impacted key areas of digital commerce, particularly in conversational commerce, interactive design and in the future on configurable products.

The convergence of generative AI with visual computing technologies such as augmented reality (AR) and 3D rendering is enhancing online shopping. Customers can now visualize and interact with personalized products before purchase, increasing confidence and reducing return rates. The incorporation of AI-powered conversational agents and voice interfaces is also starting to make customization more accessible, allowing users to configure products through natural dialogue rather than complex menus.

Looking ahead, AI is expected to drive predictive customization, where AI not only responds to user input but anticipates preferences based on behavioral patterns, environmental factors and emerging trends. As AI continues to improve, the future of e-commerce will be defined by seamless, intuitive and deeply personalized shopping experiences. Businesses that effectively integrate AI-driven configurators, real-time customization and conversational commerce will gain a competitive edge in the evolving digital marketplace.

2.2 *Transforming Digital Customer Interaction – The New Role of Conversational Commerce*

The rapid advancement of AI and digital communication technologies has significantly transformed customer engagement in online commerce. One of the most profound developments in this field is "conversational commerce", which integrates real-time communication tools such as chatbots, voice assistants and messaging applications into the decision and purchasing process. This approach facilitates direct and interactive engagement between businesses and consumers, allowing for personalized product recommendations, seamless transactions and instant customer support. Conversational commerce enhances the customer journey by leveraging NLP, AI-driven automation and machine learning algorithms to understand and respond to customer inquiries in real time or near real time. By doing so, it bridges the gap between traditional e-commerce and personalized, human-like interactions.

2.3 *Theoretical Foundations and Industry Implementation of Conversational Commerce*

Conversational commerce is grounded in theories of human–computer interaction (HCI), digital consumer behavior and artificial intelligence-driven automation. Unlike traditional e-commerce, where interactions are static and linear, conversational commerce creates a "dynamic, interactive environment" where customers engage with digital agents in real time.

AI-Driven Personalization and Engagement. A central aspect of conversational commerce is its ability to provide context-aware, personalized recommendations through AI and machine learning algorithms. Studies have shown that personalization enhances customer satisfaction and increases conversion rates by tailoring content, promotions and product suggestions to individual user preferences [3]. AI chatbots and virtual assistants analyze user history, behavior and preferences, thereby optimizing the relevance of recommendations and reducing decision fatigue [4].

Automation and Efficiency in Customer Service. From a business operations perspective, conversational commerce introduces significant efficiency gains through automation. By deploying AI-powered chatbots and automated customer service interfaces, companies can reduce operational costs, improve response times and ensure 24/7 customer support [5]. Unlike human agents, AI-driven conversational systems are scalable, capable of handling multiple inquiries simultaneously while learning from past interactions to improve accuracy and relevance over time.

Practical Applications and Industry Implementation. Conversational commerce is being adopted across various industries to improve customer experience, operational efficiency and digital engagement strategies. Especially in the Retail and E-Commerce conversational commerce can enhance the digital shopping experience by offering personalized styling advice, automated customer service and seamless checkout processes. AI-powered chatbots can guide customers through product selection based on their individual preferences and past purchases, improving customer satisfaction and driving sales [6].

3 AI and Product Configurators

3.1 How AI Will Augment and Enhance a Product Configurator Experience

The integration of AI has the potential to transform the configurator landscape. AI-powered configurators can enhance decision-making through machine learning (ML), NLP, generative AI and computer vision, significantly improving user engagement, efficiency and overall configurability. As AI becomes more advanced, its role in product customization will continue to expand, reshaping digital commerce and the way customers interact with brands.

According to Pine and Gilmore [7], mass customization aims to combine the flexibility of tailored products with the efficiency of mass production. AI takes this concept further by learning from users, predicting preferences and enabling intelligent recommendations in real-time. As AI-driven configurators evolve, they can redefine the balance between consumer choice and manufacturing feasibility.

It therefore appears that AI enables companies to improve the user-friendliness and reduce the potential complexity of online product configurators. The opportunities and benefits of using AI in configuration processes are described below.

AI-Driven Personalization and Recommendation Systems. One of the primary benefits of AI in product configurators is its ability to personalize the user experience dynamically. Traditional configurators typically require users to select from predefined options, which can lead to decision fatigue and a suboptimal experience. AI-driven personalization addresses this issue by analyzing:

- User behavior and preferences from past configurations, browsing history and interaction data.
- Real-time inputs, including text, voice commands and visual elements that indicate customer intent.
- Demographic, contextual and industry-specific data to recommend features and configurations suited to particular market segments.

As Zipf's principle of least effort suggests, users tend to favor solutions that minimize cognitive load [8]. AI-driven configurators simplify decision-making by offering contextually relevant recommendations and reducing the number of steps required to arrive at an optimal configuration. Through collaborative filtering and content-based recommendation algorithms, AI-powered configurators aim to analyze customer profiles and make real-time suggestions, ensuring that users receive options that align with their preferences. This not only accelerates the decision-making process but also improves user satisfaction and conversion rates. So they directly influence user satisfaction and conversion rates by making the process efficient and reducing frustration.

Conversational AI and NLP in Configurators. Conversational AI and NLP can further enhance configurator usability by allowing users to engage in dynamic, intuitive interactions rather than relying on static menus and forms. NLP-powered interfaces improve accessibility and engagement by enabling:

- Conversational chatbots and voice assistants that guide users through configuration in natural language.
- Text and speech-based queries, where customers can describe their needs in a conversational manner (e.g., "I need a minimalist sofa that fits in a small apartment").
- Automated responses that adapt configurations based on user intent and sentiment analysis.

Hirschberg and Manning [9] emphasize that NLP systems enhance human–computer interaction by enabling more natural and efficient communication. AI-driven product configurators powered by NLP can interpret complex user requests and translate them into actionable configurations, significantly reducing friction in the decision-making process. By integrating speech-to-text and text-to-speech capabilities, these configurators become even more accessible, accommodating users with different interaction preferences and improving customer engagement and conversion rates.

Real-Time Optimization and Feasibility Analysis. Product configuration often involves balancing multiple constraints, such as cost, materials, technical feasibility and production timelines. AI optimizes this process by providing real-time validation and constraint-solving algorithms that dynamically assess:

- Compatibility between selected features to prevent invalid configurations.
- Performance-based recommendations that consider functional requirements and customer use cases.

- Predictive analytics to estimate production lead times, cost implications and inventory levels.

According to Kumar et al., AI-powered constraint-solving algorithms improve configurator efficiency by reducing the likelihood of user errors and non-feasible selections. Real-time feasibility checks ensure that customers receive configurations that are not only desirable but also manufacturable and cost-efficient. Generative design approaches further enhance optimization by using AI to generate product configurations based on predefined parameters, offering users the best possible balance of performance, cost and aesthetics.

AI-Powered Visual Configurations and Augmented Reality. The integration of computer vision and AR in product configurators represents another major advancement in user experience. AI-powered visualization enhances configuration by allowing users to:

- See real-time 3D renderings of their customized product.
- Use AI-enhanced image recognition to personalize configurations based on reference images.
- Employ deep learning models to generate hyper-realistic textures, colors, and material previews.

Azuma [10] defines augmented reality as an interactive system that overlays virtual objects onto the real world, enhancing user perception and decision-making. In the context of product configurators, AI-enhanced AR enables customers to visualize products in their real environment, making it easier to assess dimensions, aesthetics and usability before making a purchase decision. This immersive experience increases purchase confidence and significantly reduces product returns, as customers gain a more accurate understanding of what they are ordering.

Automated Learning and Continuous Improvement. AI-driven configurators do not remain static, they continuously evolve based on user interactions, feedback and industry trends. Through reinforcement learning and adaptive AI models, configurators can:

- Identify emerging trends and adjust recommendations accordingly.
- Analyze aggregated user behavior to refine product offerings and UX design.
- Optimize upselling and cross-selling strategies by understanding customer purchasing patterns.

Russell and Norvig [11] emphasize that AI systems become more effective over time as they learn from new data, enabling configurators to offer increasingly accurate and engaging experiences. By leveraging AI for automated insights and predictive analytics, companies can ensure that their product configurators remain competitive, relevant and highly optimized for user engagement.

3.2 Inclusion of People and Expansion of Target Groups Through AI

Regular configurators often use selection options such as sliders, dropdowns or collapsible fields that are a hurdle for certain people or are difficult for them to use. Control via keyboard and screen reader, which people with visual impairments rely on, can also make configuration processes complicated or impossible. Pure voice control and large visualizations may replace complex or multi-stage configuration processes in which many options have to be selected manually.

So in addition to fast and simplified interaction and visual possibilities, the integration of AI in online product configuration processes offers an expansion of the target groups. In specific, the use of AI in the configuration process can facilitate accessibility for people with physical impairments as well as older people. In addition, the Web Content Accessibility Guidelines [12] are becoming increasingly important for companies that offer online services, as the Web Accessibility Guidelines will be enshrined in law from June 2025 [13].

By using AI, companies can create more inclusive, user-friendly and accessible experiences for a wide range of users. The following features are some of the key ways in which AI can help optimize inclusion and accessibility in configuration processes, according to ChatGPT [14].

Voice Interaction and Speech Recognition. AI-supported speech recognition enables users to interact with a configurator using voice commands—hands-free by speaking naturally. This is particularly beneficial for people with mobility impairments or those who have difficulty using a conventional keyboard or mouse.

Text-to-Speech and Screen Readers. By activating screen readers or text-to-speech functions, AI makes the configuration process more accessible for blind or visually impaired users so that they can use the interface without visual input.

Personalized Content and Customization for Diverse Needs. Users with cognitive impairments, learning difficulties or even language barriers benefit from simplified language, clearer choices or content that matches their preferences. AI can also highlight or prioritize features for users with specific needs.

4 Conclusion

AI-powered product configurators have the potential to bridge the gap between user expectations and manufacturable solutions, reshaping the way customers engage with personalized products in the digital age. The integration of AI into product configurators represents a fundamental shift in digital customization. By employing predictive analytics, NLP-driven interfaces, real-time optimization, AI-powered visualization and continuous machine learning, AI enables configurators with the potential to deliver higher levels of personalization, efficiency and usability. As AI technology continues to evolve, future developments should focus on enhancing multimodal interactions, integrating generative AI for real-time customization and addressing ethical considerations in AI-driven decision-making. However, when considering the transformative power of AI, potential constraints – such as computational cost, data privacy concerns, and practical limitations – must not be overlooked and should be discussed.

References

1. The configurator database project. https://www.configurator-database.com. Accessed 19 Feb 2025
2. Blazek P, Partl M, Streichsbier C (2020) Configurator database report 2017/18. Lulu Press Inc., Raleigh
3. Haenlein M, Kaplan A (2021) Artificial intelligence and robotics: shaking up the business world and society at large. J Bus Res 124(C):405–407
4. Huang MH, Rust RT (2018) Artificial intelligence in service. J Serv Res 21(2):155–172
5. Luo X, Tong S, Fang Z, Qu Z (2019) Frontiers: machines vs. humans: the impact of artificial intelligence chatbot disclosure on customer purchases. Mark Sci 38(6):937–947
6. Daugherty P, Wilson HJ, Bianzino N (2021) Human + machine: reimagining work in the age of AI. Harv Bus Rev Press
7. Pine BJ, Gilmore JH (1999) The experience economy: work is theatre & every business a stage. Harvard Business Review Press
8. Zipf GK (1949) Human behavior and the principle of least effort. Addison-Wesley
9. Hirschberg J, Manning CD (2015) Advances in natural language processing. Science 349(6245):261–266
10. Azuma RT (1997) A survey of augmented reality. Presence Teleoper Virtual Environ 6(4):355–385
11. Russell S, Norvig P (2021) Artificial intelligence: a modern approach. Pearson
12. World Wide Web Consortium (W3C). https://www.w3.org/TR/WCAG22/. Accessed 19 Feb 2025
13. Das Portal der Wirtschaftskammern Österreich. https://www.wko.at/ce-kennzeichnung-normen/informationen-zum-barrierefreiheitsgesetz. Accessed 19 Feb 2025
14. OpenAI - ChatGPT. https://chatgpt.com. Accessed 19 Feb 2025

continues to evolve, future developments should focus on enhancing multimodal [illegible] integrating generative AI for real-time customization and addressing ethical considerations of AI-driven decision making. However, when considering [illegible] potential constraints—such as computational [illegible] must not be overlooked [illegible]

References

[illegible]

Changeable, Reconfigurable and Flexible Manufacturing

Classification of Contact Points in Car Body Production to Enable a Mechanization Assessment

Julian Schützenberger, Claus-Dieter Reiniger, and Martin Manns

Abstract The car body manufacturing process is an essential step in automotive production. It not only defines the structure and exterior of the car, but it must also meet all requirements of downstream processes. Therefore, one elementary goal is to ensure the geometric quality of the car body. The assembly of diverse free-formed metal components necessitates the utilization of a wide range of joining racks. Each of them is tailored to the distinctive form of the incoming components, the joining technologies employed and the prescribed joining sequence. Due to numerous influencing factors during the manufacturing process, the joining racks are designed to allow geometry adjustments at every contact point. Despite the extensive level of automation found in car body production, the process of achieving the desired geometry relies heavily on manual tasks. Since individual contact points within the joining rack are adjusted in iterative loops, the production line must be stopped every time. In an effort to improve the flexibility and automation level of the geometry processes within car body manufacturing, the mechanization of the adjustable contact points is imperative. To achieve this objective, it is necessary to first assess each contact point individually. This paper proposes a classification for contact points within joining racks to facilitate their assessment. After introducing five categories, its use is illustrated for a selected mechanization strategy. Further research is needed to determine the applicability of the classification to a full-size automotive series production line.

Keywords Manufacturing systems planning · Flexibility · Car body · Mechanization · Joining rack

---

J. Schützenberger (✉) · M. Manns
PROTECH-Institut für Produktionstechnik, Universität Siegen, Siegen, Germany
e-mail: julian.schuetzenberger@uni-siegen.de

C.-D. Reiniger
Verfahrensentwicklung Rohbautechnologien, Mercedes-Benz AG, Sindelfingen, Germany

M. Manns (ed.), *New Paradigms for Anticipated Uncertainty*, Lecture Notes in Mechanical Engineering, https://doi.org/10.1007/978-3-032-16889-4_22

1 Introduction

In car body manufacturing the geometry of the final assembly is crucial. It must provide structural integrity for the car while simultaneously meeting stringent quality targets to accommodate the exterior and align with downstream processes. The geometry of the assemblies during car body manufacturing is highly influenced by deviations from individual components as well as variations resulting from successive assembly steps. To counter these variations, every contact point in the joining racks of each body assembly is designed to be adjustable. Currently, a manual, experience-based process executed in iterative loops is employed to achieve the desired geometry targets.

Automating this process requires digitally operated actuators for geometry manipulation, positioned appropriately within each joining rack [19]. Prior to addressing each specific tolerance issue, a comprehensive survey of all existing contact points within the assembly must be compiled [14]. Since no distinctive method could be found in the literature, this paper proposes a classification of joining racks based on their general construction conditions. The classification supports the assessment of mechanization strategies regarding their technical feasibility and their implementation expense.

First, an overview of the car body manufacturing process is provided in the state of the art, including an examination of joining racks in car body manufacturing and their operating principle. Next, the classification method for these joining racks is introduced and the application for an assessment of mechanization strategies is shown. The paper ends with a discussion of the findings, a conclusion and an outlook on future research.

2 Objective

This paper introduces a classification method for contact points in joining racks used in car body manufacturing, aiming to facilitate the assessment of mechanization strategies. The joining racks are utilized to define the geometry of assemblies within the car body. Their mechanization is aimed to manipulate the assembly with a controlled movement.

3 State of the Art

3.1 Car Body Manufacturing

The automotive manufacturing process begins with the production of individual components, followed by the car body manufacturing process. Once the body is finished, it is painted and proceeds to the final assembly stage [12]. The car body itself consists of hundreds of individual parts [1, 10], which predominantly feature free-form geometries [2]. The car body manufacturing process is divided into three main production lines [11, 15]. First, the underbody is assembled by joining the rear and front main assembly to the middle section of the passenger cabin. Afterward the side panels are attached and by adding the roof, the base structure is set. Meanwhile doors, hoods and tailgates are produced in an independent process. In the final production line all closures are added to complete the car body [2, 21].

The quality assurance of the entire vehicle, defining tolerances and designing the car body manufacturing process requires a consistent and reproducible positioning and referencing methodology throughout the whole production. For this reason, a global vehicle coordinate system is employed. The origin of the coordinate system is defined at the center of the vehicle's front axle. The X-axis runs in the longitudinal direction of the vehicle, opposite to the driving direction. The Y-axis extends in the transverse direction, while the Z-axis represents the vertical direction [3, 4].

The 3–2–1 principle is used along with the associated reference system to define the position of assemblies according to the coordinate system. Six alignment features are necessary to specify all six degrees of freedom of a rigid object in space. The initial three points define the primary axis. The next two points form the secondary axis, which stands perpendicular to the primary axis. The final point defines the tertiary plane, which is perpendicular to both the primary and secondary axes [3]. In joining racks, the 3–2–1 principle is implemented utilizing a combination of centering pins and planar component supports. One pin aligns with a circular hole and another pin with a slotted hole on the component to prevent rotation. Both have tight tolerances applied [9]. At last, planar supports are used to define the remaining degrees of freedom.

For free-formed sheet metal components, additional clamping points are necessary. These serve two main purposes: they define the geometry despite the low intrinsic stiffness of the components and they enable targeted adjustments to geometric properties of the assembly [3, 17]. This is crucial because tolerances from incoming components and other influencing factors in the manufacturing process would otherwise lead to deviations from the targeted geometry.

3.2 Car Body Joining Rack

Unlike handling or logistic structures, joining racks have high accuracy and repeatability requirements because they are employed to join individual components into one assembly. A more detailed distinction can be made between joining racks that define the geometry of assemblies in car body construction and those in which an already defined assembly is supplemented with additional joining operations to achieve the final mechanical properties [1].

A joining rack consist of a base plate, the centering pins to position the components for the joining process and several clamping points mounted on diversely designed consoles to define the geometry of the finished assembly. To clamp the individual sheet metal components in the joining rack down, pneumatically operated toggle clamps are employed. Each contact point between the rack and the components can be adjusted using distance plates, which allow for fine-tuning in the sub-millimeter range. A dedicated stack of distance plates is provided for each spatial direction requiring adjustment.

4 Classification of Contact Points in Joining Racks

Due to the extensive variety of joining tasks throughout the car body manufacturing process, the variability of individual joining racks is substantial. The proposed classification enables an assessment of contact points by providing a structured survey building upon the existing global vehicle coordinate system. This classification is focused on joining racks which define the geometry of individual assemblies, as these exhibit the most clamping points.

The classification method distinguishes five categories, with each successive category offering a more granular analysis of the contact points, see Fig. 1.

In the *first category* of the classification, the number of adjustable directions is examined. This refers to the direction in which the clamping point can influence the geometry of the assembly according to the vehicle coordinate system. Usually

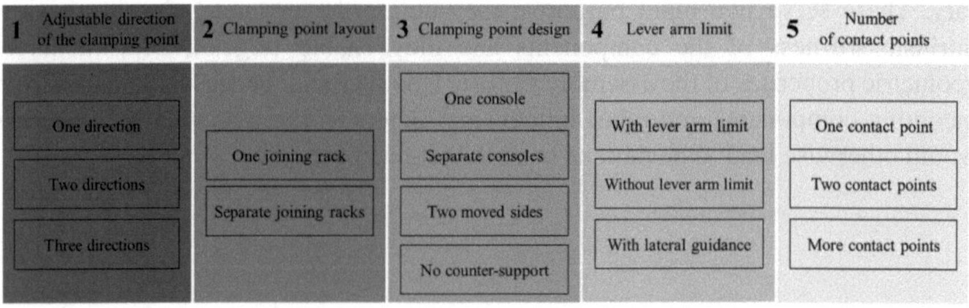

Fig. 1 Classification of contact points within joining racks in car body manufacturing

only one adjustable direction is implemented, but more complex solutions with up to three directions are also possible. These cases occur when spatially rotated or inclined surfaces of components need to be clamped down and adjusted accordingly.

For the *second category*, the overall layout of the clamping point is analyzed. It needs to be determined whether it is entirely contained within a single joining rack or whether two racks form one clamping point on the assembly. It is possible for this to happen if the joining rack has at least two distinct sections. For instance, one can be fixed to the ground while the other can be moved and connected to the first one.

In *category three*, the design of the counter support of the contact point is classified. A clamping point typically comprises two sides, at least one of which can be moved to clamp individual components between them. Due to varying requirements, designs can differ significantly. If sufficient space is available, a fixed counter-support is utilized. The support can be attached to the same console as the movable clamping element, controlled by a toggle clamp or to a separate console. If necessary, both sides of the clamping point are mounted on movable actuators making it easier to insert and remove components. Besides these three scenarios there are instances where the component is clamped down directly without utilizing a counter-support.

To allow additional adjustment options, limit blocks for the movement of the toggle clamps end positions can be integrated. Their design is investigated in *category four*. To ensure consistent positioning while dealing with leveraging forces, lateral guidance can be added. This is particularly important if high leverage forces occur.

Finally, in *category five*, the number of contact points is determined. These are the points at which contour pieces are in direct contact with the component.

A demonstration of the application of the classification is shown in Fig. 2.

The illustrated clamping point utilizes a single adjustable direction in the global vehicle coordinate system and is mounted on a single joining rack, with both the movable and fixed counter-support mounted on the same console. Furthermore, it features a limit block that defines the end position of the toggle clamp lever. Only a single contact point is employed. Starting with category one, which involves observing only the adjustable direction and progressing to category five this method can be applied to other clamping points as well.

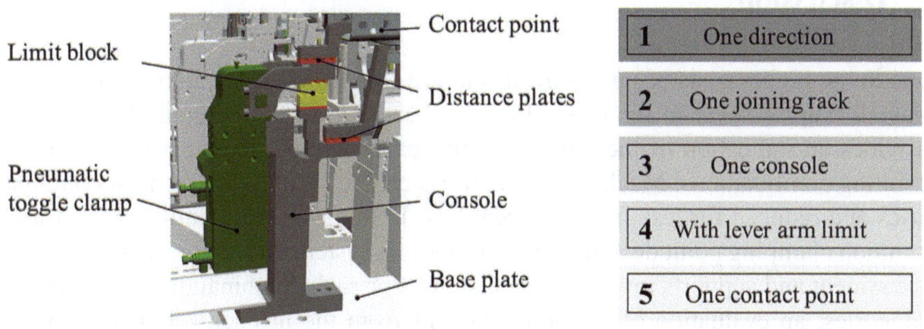

Fig. 2 Exemplary clamping point and its respective categories based on the classification

5 Mechanization of Contact Points

Each combination of the described categories may require a specific mechanization strategy. When implementing a mechanized movement, it is important to ensure that the contact point on the assembly can be moved in its entirety. This work does not assess all individual mechanization strategies. Instead, it exemplifies and evaluates a widely used approach based on the classification. Different mechanization strategies may be evaluated following a similar approach.

One common technical solution is the mechanization of contact points using numerically controlled axes, also known as NC-axes. This approach is being used for over 30 years [20]. Besides electrically driven variants, there are alternative drive options, such as proposals for hydraulic drives [16] or combinations of both [6]. Pneumatic methods are presented as well [8]. Before their implementation, it must be verified whether the accuracy and mechanical load capacity are sufficient for the application. Regardless of the type of drive used, one NC-axis is required for each adjustable direction from the first category. When the clamping point according to the second category is divided into two separate joining racks, it is also necessary to split the driven axes. The clamping point design in the third category further influences the technical implementation. If the counter-support is mounted on the same console as the other side of the clamping point, one drive alone is sufficient. However, dividing it onto different consoles necessitates an adaptation of the utilized axes. Here it is important to bear the available installation space in mind. The implementation is influenced even further by the lever arm limit from the fourth category. Finally, the number of contact points in the fifth category has a significant impact on the feasibility of NC-axes to automate a clamping point. In general, each contact point must be capable of an individual adjustment to its geometric position. That is why typically an independent solution must be found for each contact point. If multiple points are combined with a single mechanized movement the feasibility must be verified.

6 Discussion

Due to the higher costs of mechanized solutions compared to standard designs, it is crucial to apply them in the most effective locations. The structured classification provides an indication of the number of use cases within the categories and their resulting combinations. Mechanization efforts can be focused on the most common cases and reduced for less frequent scenarios. In this way the identification of less common clamping point designs in the production system allows for a differentiated assessment and supports investment decisions. For each combination of the various categories, an evaluation of the added benefit from the mechanization in relation to the necessary expense can be conducted. In Sect. 4, a technical solution that is able to be integrated into existing car body manufacturing systems is discussed.

The separation of clamping points into categories allows for a targeted examination of individual technical solutions and identifies main obstacles. For instance, using NC-axes for three adjustable directions or several individually adjustable contact points combined in one clamping scenario can be very extensive [7]. In contrast, the arrangement shown in Fig. 2 can be effectively mechanized utilizing a single NC-axis to move the entire toggle clamp together with the limit block and the counter-support.

While the classification method provides an overview for common joining racks encountered in car body manufacturing, regardless of the type of joining technology used, it does not cover every specialized configuration. There must be separate considerations made for these cases. Also, there are additional factors apart from the defined categories to consider when selecting an appropriate mechanization. Individual installation space for additional mechanization equipment is one important factor to keep in mind. The available space is highly reliant on the component geometry and the associated collisions especially during the insertion and removal procedures of the assembly. Mechanization can also lead to a reduced stiffness of the clamping point [5, 16]. The impact depends on the technical details of the design. The effort of compensation measures dealing with the reduced stiffness needs to be evaluated with the resulting potential for automation in mind.

There exist mechanization approaches that interfere with the fundamental setup of joining racks or alter the logic of the joining sequence by employing additional positioning features within the individual components [13, 18]. These cannot be assessed using the classification method.

7 Conclusion and Outlook

This work introduces a classification method for categorizing clamping points in the car body manufacturing process. It allows companies to gain a structured understanding of their initial baseline by classifying existing or previous production systems and to evaluate mechanization possibilities. It supports the assessment of potential short-term improvements and facilitates the first steps toward long-term automation solutions.

Future studies can be conducted to assess the applicability of the classification to a full-size production line in car body manufacturing. An analysis could provide a better understanding of the frequency distribution of individual categories and their resulting combinations. With this information, standardized mechanization solutions can be defined. Subsequently, the potential of implementing these solutions can be evaluated for the identified use cases according to the classification. A comparison can then be made between the impact of a low-cost approach, which only mechanizes simple categories and a more extensive scenario.

References

1. Abulawi J (2012) Ansatz zur Beherrschung der Komplexität von vernetzten 3D-CAD-Modellen. 9-12 dissertation, Helmut-Schmidt-Universität Hamburg
2. Birkert A, Haage S, Straub M (2013) Umformtechnische Herstellung komplexer Karosserieteile. Springer, Berlin, Heidelberg
3. Bohn M, Hetsch K (2013) Toleranzmanagement im Automobilbau. Carl Hanser Verlag, München
4. Fang X (2023) Karosserieentwicklung und -Leichtbau. Springer, Berlin, Heidelberg
5. Fritzsche R (2022) Beitrag zur Flexibilisierung automatisierter und hochratenfähiger Anlagentechnik im Karosseriebau. Dissertation, Technische Universität Dresden
6. Fritzsche R, Richter A, Putz M (2017) Automatic adjustment of car body fixtures using artificial intelligence. Procedia CIRP 62:600–605
7. Fritzsche R et al (2018) Kinematikmodule zur flexiblen Positionierung von Vorrichtungskomponenten. DE 10 2018 102 717 A1
8. Ganesh S, Gurunathan SK (2017) Evolutionary algorithms for programming pneumatic sequential circuit controllers. Procedia Manuf 11:1726–1734
9. Kästle C (2016) Simulationsmethode zur Beurteilung der Maßhaltigkeit von rollgefalzten Karosseriebaugruppen im Zusammenbau. Dissertation, Universität Stuttgart
10. Kiefer J (2007) Mechatronikorientierte Planung automatisierter Fertigungszellen im Bereich Karosserierohbau. Dissertation, Universität des Saarlandes
11. Klug F (2010) Logistikmanagement in der Automobilindustrie. Springer, Berlin, Heidelberg
12. Kropik M (2021) Produktionsleitsysteme für die Automobilindustrie. Springer, Berlin, Heidelberg
13. Marvel JA, Bostelman R, Falco J (2018) Multi-robot assembly strategies and metrics. ACM Comput Surv 51(1)
14. Peschke F, Eckardt C (2019) Flexible Produktion durch Digitalisierung: Entwicklung von Use Cases. Carl Hanser Verlag, München
15. Pischinger S, Seiffert U (2021) Vieweg Handbuch Kraftfahrzeugtechnik. Springer Fachmedien, Wiesbaden
16. Putz M et al (2014) Efficient flexibility in car body construction. In: Drossel W-G, Putz M, Landgrebe D (eds) Proceedings of the 7th Chemnitz car body colloquium, Chemnitz
17. Rezaei Aderiani A et al (2020) Optimal design of fixture layouts for compliant sheet metal assemblies. Int J Adv Manuf Technol 110(7–8):2181–2201
18. Schuh G et al (2021) Effects of digital twin simulation modelling on a flexible and fixtureless production concept in automotive body shops. Procedia CIRP 104:768–773
19. Schützenberger J, Reiniger C-D, Manns M (2024) Applicability of control systems for predicting adjustments of body shop fixtures. Procedia CIRP 126:1023–1028
20. Shimokawa K, Jürgens U, Fujimoto T (1997) Transforming automobile assembly. Springer, Berlin, Heidelberg
21. Wemhöner N (2005) Flexibilitätsoptimierung zur Auslastungssteigerung im Automobilrohbau. Dissertation, Rheinisch-Westfälische Technische Hochschule Aachen

Design Principles for Front-End Radical Manufacturing Innovation: Insights from Pharmaceutical Manufacturing

Rasmus Andersen, Casper Schou, Ole Madsen, Astrid H. Lassen, Martin B. Jensen, Tristan Schwörer, Bjørn C. Dueholm, and Oline V. Olesen

Abstract In response to market shifts, manufacturing companies must innovate their production systems to sustain competitiveness. Incremental changes to production systems fail to address these challenges, necessitating radical innovations in manufacturing. This is particularly evident in the pharmaceutical manufacturing industry, where product shortages and an expanding product variety challenge the traditional mass production paradigm. However, there is a noticeable lack of methods and tools to support the initial stages of manufacturing innovation activities in the pharmaceutical and general manufacturing industries. This paper aims to fill this gap by drawing on insights from a radical manufacturing innovation project in a pharmaceutical manufacturing company. Based on the case study findings, six design principles for front-end radical manufacturing innovation processes and potential supporting tools and activities are proposed.

Keywords Manufacturing innovation · Radical innovation · Front-end · Design principles · Case study

1 Introduction

Manufacturing companies must innovate their manufacturing systems to stay competitive. This is especially so in the pharma industry, where drug product shortages [1] and an expected increase in new drug products [2] pose a challenge for conventional pharma manufacturing systems based on a mass production paradigm [3]. Research on manufacturing systems development methods to integrate changeability in such systems to accommodate market uncertainties is well-established

R. Andersen (✉) · C. Schou · O. Madsen · A. H. Lassen · M. B. Jensen · T. Schwörer · B. C. Dueholm
Department of Materials and Production, Aalborg University, Aalborg East, Denmark
e-mail: rasmus@mp.aau.dk

O. V. Olesen
Novo Nordisk Biotech and Rare Disease, Gentofte, Denmark

M. Manns (ed.), *New Paradigms for Anticipated Uncertainty*, Lecture Notes in Mechanical Engineering, https://doi.org/10.1007/978-3-032-16889-4_23

[4–6]. However, such problem-solving methods focus on manufacturing development [7], assuming that requirements and solution principles can be identified and specified upfront. This reliance on fixed requirements specifications and forecasted developments poses a challenge in situations where radical changes are necessary to accommodate an uncertain future. This is referred to as Manufacturing Innovation (MI) [8]. The front-end of MI is concerned with generating ideas for complex problems [9] and is generally considered an essential phase in MI [10]. Despite multiple studies arguing for the importance of gaining insights into this phase of MI [10], in particular, exploring tools and methods to support decision-making [7], research remains limited for this phase of MI [7, 9, 10]. This is demonstrated by, for example, searching Elsevier's Scopus database for "manufacturing innovation" and "front-end," which, at the time of writing, provided just one result, while derivative search strings return less than ten results, indicating a minimal body of knowledge on this phase of MI.

Historically, innovation has focused on product innovations, assuming that manufacturing would later accommodate these requirements [11]. For the early phase of MI, this implies that (i) there is a lack of research to facilitate the scientific and practical maturation of general MI methods, and (ii) the lack of research presents a challenge for practitioners in terms of getting inspiration and insights into specific approaches for early front-end MI activities. There is, therefore, a need to investigate this topic to provide researchers and practitioners with new insights into the workings of the front-end of MI. To address these issues, this paper therefore poses the following research questions:

1. Which design principles can support the front-end of radical manufacturing innovation projects?
2. How can these design principles be implemented in the front-end of radical manufacturing innovation projects?

The remainder of the paper is structured to support answering these questions. Section 2 reviews research related to the front-end phase of radical MI and identifies typical characteristics of this phase. Section 3 presents findings from a case study in a pharmaceutical manufacturing company where activities and tools supporting manufacturing innovation have been developed and experimented with. Section 4 discusses the findings and proposes design principles to facilitate front-end manufacturing innovation after which Sect. 5 concludes on this study.

2 Related Research

Manufacturing innovation has been defined in multiple ways, as described by Larsen [7], ranging from rather generic definitions [12] to more specific definitions highlighting the extent of the innovation [13] or the value resulting from said innovation [14]. However, regardless of which definition is adopted, all focus on the change process. However, innovation does not come automatically; it must be actively fostered [15].

2.1 Radical and Incremental Manufacturing Innovation

The novelty of MI is a key factor in determining the approach for innovation projects within manufacturing organizations. Generally, a distinction is made between incremental and radical innovation, where the former is associated with minor changes and more certainty in process and outcome, often relying on existing commercial solutions [7, 15] or modifications of these. On the other hand, radical innovation is characterized by significant changes in the manufacturing system and supporting or related processes [10] and may involve innovations at both manufacturer and equipment supplier [9]. The scope of changes in radical innovation projects leads to high uncertainty and complexity in such innovation projects [7, 9]. Indeed, these circumstances of radical innovation projects are often linked to the problems for which solutions are sought, as these are typically characterized as complex problems, which are ill-defined and challenging to solve due to dynamic requirements, limited knowledge, and complex system relations [16]. These challenges of radical MI are mainly present in the front-end phase of the innovation process.

2.2 Manufacturing Innovation Process

Achieving the potential benefits expected from system changes necessitates progressing through an MI process. Generally, this involves progressing through three phases [13] and their associated activities:

1. **Front-end**: identification of problems [17] and idea generation [12].
2. **Development**: identification of existing solutions [17] or development of new solutions [13, 17].
3. **Implementation**: implementing the solution or innovation in the manufacturing system [13, 17]

Although considered generic phases of a MI process, variations in the number of phases (see, e.g. [7, 14, 18, 19]) and their suggested activities (see, e.g. [13, 18]) differ. Indeed, there is no "one-size-fits-all" approach to MI [7, 12, 15, 17]. Nevertheless, a common theme in most studies on MI processes is their emphasis on the iterative nature of such projects (see, e.g., [19]). The importance of iterations in MI processes is linked closely to the degree of novelty, or radicality of the change, to the system in scope [16].

2.3 Processes Supporting Front-End Manufacturing Innovation

The three-phased MI process can be broken down into sub-phases related to each innovation project phase. While several studies suggest headlines setting the scope of the front-end phase, such as Dooley and O'Sullivan [11] arguing that MI starts with idea generation, few studies focus on the structured activities, or processes, that comprise the front-end in MI.

Kurkkio et al. [10] propose a four-phase process for the front-end of MI, including (i) informal start-up, (ii) formal idea-study, (iii) formal pre-study, and (iv) formal pre-project. Additionally, several activities associated with each of the four phases are suggested. In a study on the front-end of process innovation, Frishammar et al. [20] likewise suggest a four-phased approach to this part of the innovation process, covering (i) strategic assessment, (ii) solution exploration, (iii) risk assessment, and (iv) process change realization. However, while Kurkkio et al. [10] highlight the iterative nature of the process, this aspect is not emphasized by Frishammar et al. [20]. While not explicitly focusing on elaborating on the sub-phases comprising the front-end phase of an innovation project, Braun [17] argues for the importance of familiarity with the problem as an initial "zeroth phase" of the MI process. Fitting the four-stage MI process by Braun [17] into the overall three-phased process would imply that the front-end consists of two sub-phases: (i) context familiarity and (ii) problem identification.

2.4 The Front-End of Radical Manufacturing Innovation

The front-end phase of radical MI projects, often characterized as "fuzzy," underscores the absence of clear-cut solutions and defined processes [9, 10]. It is inherently iterative, involving diverse knowledge-gathering activities [10]. While problems may originate within the manufacturing company, potential solutions typically arise externally [9]. Recognized as essential to the innovation project's success, the front-end phase demands careful attention [10].

Although all projects are unique, the review of related research in previous sections suggests several characteristics typical of the front-end phase of radical MI can be identified. These characteristics include:

- Ill-defined process and scope [16]
- Dynamic solution requirements [16]
- Limited knowledge [16]
- Inherently iterative process [10]
- Involves diverse knowledge-gathering activities [10]
- Solutions typically arise externally [9]

Some of these characteristics carry forward to later stages in the MI process. For example, the iterative nature of the innovation process is consistent throughout [19]. Another example is solution requirements, which may be adjusted as knowledge increases during the solution concept and design phases.

3 Case Study on Accommodating Characteristics of the Front-End of Radical Manufacturing Innovation

This section presents findings from the front-end phase of an MI project by a large European pharmaceutical manufacturing company. Like many other pharmaceutical manufacturing companies, the company is challenged by a trend towards higher quality requirements and more product variants in smaller production volumes. This has highlighted the need for production systems capable of efficiently accommodating these needs while maintaining the strict quality standards characteristic of the pharmaceutical industry. This represents a divergence from the typical mode of operation within the case company and most of the industry, which prompted the case company to initiate a radical MI project to provide a suitable solution. The project was initiated with a university, and the core project team consists mainly of manufacturing engineering researchers, with project management and administration handled mostly by case company representatives.

The case highlights how the project team has developed and experimented with different activities and tools to accommodate the challenges posed by this phase's characteristics and their effect on the innovation process.

3.1 Ill-Defined Processes and Scope

The MI project is based on an agile approach, following the case company's guidelines for project management, although tailored to the project's specifics. The agile approach allowed the project to be planned in smaller increments as knowledge of the domain, requirements, and potential solutions grew. Furthermore, the ability to plan project activities in smaller increments allowed the project team to invest additional resources in investigating specific aspects, such as further clarifying technical requirements. The agile approach allowed the project team to be more dynamic towards uncovering the relevance or potential of technologies discovered during meetings with subject-matter experts (SMEs). However, devoting resources to these smaller deep dives on technologies or company-specific aspects meant that resources were diverted from earlier planned activities. Thus, obtaining a good balance between exploration and exploitation sometimes proved challenging.

3.2 Dynamic Solution Requirements

The project relied on two major approaches to periodic reviews: (1) internal project requirement reviews among the core project team members and (2) so-called "Partnership Days," which were held every six months and involved broader representation from stakeholders in the case company. The focus of the internal reviews was to ensure alignment among core project members concerning the scope and target of the solution. In contrast, the purpose of the Partnership Days focused more on ensuring that the concepts and potential technology solutions presented corresponded with the challenges experienced by the case company, as represented by the multidisciplinary team of SMEs and stakeholders attending these events. Simultaneous access to multiple experts facilitated the application of Delphi-style processes for achieving consensus on important aspects, such as the importance of specific system requirements.

The internal requirement review meetings were not formalized to the same degree as the Partnership days and were mainly used ad hoc to reiterate the project scope and objectives. However, the uncertainty of solution requirements early in the project meant that early concepts deemed relevant would be revised as requirements were changed and clarified.

Partnership days proved useful to "pressure test" recommendations or conclusions made by the core project team as the prolonged engagement with stakeholders allowed for, e.g., uncovering missing relevant aspects in the proposed concepts or discovering technologies potentially applicable to the solution. Nevertheless, it was experienced that due to the somewhat abstract representations of proposed concepts and technology implementations early in the project, some stakeholders found it difficult to visualize the intended solutions in the case company's context.

3.3 Limited Knowledge

A key aspect of the project was establishing an advisory board composed of SMEs from the case company. The purpose of the advisory board was to provide detailed insights into various relevant aspects, including, among others, quality control, environmental monitoring, product portfolio management, production line operation, and aseptic technologies. The advisory board was intended as a dynamic board where members' composition would change according to the project phase and areas investigated so that the most relevant SMEs would always be represented.

As most of the core project team comprised researchers from a national university, access to the SMEs on the advisory board proved very valuable in speeding up the knowledge-gathering process and allowing for rapid feedback on proposed solution aspects. Through the advisory board, it was possible to quickly locate relevant employees whenever specific questions or data needs were identified.

3.4 Inherently Iterative Process

Besides using Partnership Days to review and further develop solution concepts, several workshops involving the core project team were held. These workshops were intended to facilitate radical innovation and "thinking outside the box" of current pharmaceutical production. Initial workshops focused on generating "system archetypes," which were designed to emphasize specific features of production systems, such as system flexibility, modularity, or mobility. Later workshops focused on continually improving and adjusting system concepts based on the discovery of new technologies or case-specific insights. The inputs from engaging with company SMEs and meetings with technology suppliers resulted in some system archetypes being disregarded or hybrid system concepts comprised of several archetypes being proposed.

A potential key technology was identified early in the project. The university partner had lab facilities where this technology was available for experimentation. To promote the exploration of the key technology's capabilities, several student projects revolving around the use of this technology were established early in the project. These experiments and associated demonstrators proved valuable in evaluating the technology's application in the case company. Showcasing the technology demonstrators at the Partnership Days proved helpful in gaining further insights into their potential use and related challenges in the case company and allowed SMEs to visualize the abstract system concepts more easily.

3.5 Involves Diverse Knowledge-Gathering Activities

The university researchers in the core project team had little to none prior experience with the pharmaceutical manufacturing industry. Consequently, there was a need for rapid knowledge acquisition within this domain. Case company SMEs facilitated this through several onboarding classes spanning multiple disciplines, such as drug manufacturing processes, quality requirements, and regulatory aspects. Besides providing a knowledge foundation, the onboarding classes also introduced the SMEs to the project team, facilitating future collaborations such as data collection or solution concept reviews.

Subject-matter experts were used extensively throughout the early phases of the project. The relatively close interaction with the SMEs and the possibility to rapidly arrange meetings or workshops meant that expert insights could be integrated into analyses or solution maturation much quicker than otherwise, thereby assisting in faster progression through the iterations. Furthermore, SMEs provided, for example, technology or case insights that would either be difficult or even impossible to obtain through traditional literature reviews.

Part of familiarizing with the pharmaceutical manufacturing domain, potentially relevant solutions, and technologies involved addressing the existing research literature and searching commercial systems and technology suppliers for state-of-the-art solutions. These searches proved valuable sources of inspiration as several interesting technologies and systems design concepts were identified, forming the input to developing the system archetypes described in Sect. 3.4.

3.6 Solutions Typically Arise Externally

Manufacturing innovation in the pharmaceutical sector also happens at technology and system suppliers, where novel technologies and system architectures are being developed for commercial use. With significant experience and insight into the industry, these suppliers represented sources of inspiration but also raised several questions concerning choices of solution designs or potential challenges in regulatory compliance. As such, the ability to visit and meet with suppliers at trade exhibitions and on-site visits allowed the project team to gain insights into the suppliers' perspectives on specific technologies or solution principles. In some instances, it was found that suppliers were working on principles similar to system concepts proposed by the project team. This provided additional belief in these concepts and allowed the project team members to get the suppliers' reflections on specific issues or potential uses of a particular technology or system architecture. In some cases, these insights sparked additional investigations into the potential for integrating a particular technology into existing system designs, which were then reviewed with SMEs to evaluate the feasibility of the case company.

Reviews of research and commercial literature were, as described in Sect. 3.5, also utilized to gain inspiration on state-of-the-art technologies and solutions.

4 Discussion

From the case study findings in Sect. 3, several design principles can be deduced directly addressing the characteristics of the front-end of radical manufacturing innovation. In agreement with Larsen et al. [16], Sect. 3.1 demonstrates that a *flexible innovation process* can accommodate the ill-defined nature of the front-end phase in MI by adapting the process to project dynamics, such as changing requirements or needs for acquiring new knowledge. Changing solution requirements can more effectively be accommodated by adopting *periodic requirement reviews* with SMEs and stakeholders from relevant disciplines—thereby supporting the findings of Simms et al. [21]—such as product development, production management, or equipment operators, as exemplified in Sect. 3.2. The limited knowledge often associated with the early phases of radical innovation projects can more effectively be accommodated by *engaging with stakeholders from multiple disciplines*, as described in Sect. 3.3, such as internal or external SMEs. Section 3.4 demonstrates that the frequent iterations typical of the early phases of radical MI projects can be augmented by adopting *early and frequent solution concepting and prototyping*, providing valuable insights that feedback into earlier stages of the innovation process, thereby providing the potential for higher-quality solutions. Simms et al. [21] furthermore found that demonstration of such artifacts was favored by project participants. The uncertainty and limited knowledge typically also result in a broad range of knowledge-gathering activities to gradually increase knowledge and simultaneously reduce the uncertainty of the MI project. Among the *diverse knowledge-gathering activities* described in Sect. 3.5, some can be relevant for identifying potential solutions, such as literature reviews or SME interviews [22]. Section 3.6 demonstrates that, since solutions for radical MIs are typically found outside the company seeking radical innovation in their manufacturing, *searching external sources* such as technology suppliers [9] or research organizations may provide suitable solutions. The proposed design principles described in this section are summarized in Table 1 and presented in relation to their related characteristics and supporting activities as applied in the case study.

While several of these activities are identical to those suggested in the literature, see Table 1, others have been developed and experimented with as part of this study. However, the activities and their effect on the innovation process reported in Sect. 3 represent only this case. Consequently, these activities' suitability or composition should not be considered a blueprint for other MI projects but rather serve as inspiration. Furthermore, this study does not claim completeness regarding the activities described in Sect. 3 and reported in Table 1. Additional relevant supporting activities are expected to be developed or identified, forming an interesting aspect for further

Table 1 Proposed design principles and their related front-end MI characteristics

Front-end characteristics	Design principles	Supporting activities
Ill-defined process and scope	Ensure flexibility in the innovation process	Agile process [15, 16, 23]
Dynamic solution requirements	Adopt periodic requirements reviews	Multidisciplinary workshops [21, 22]; Delphi studies [22]
Limited knowledge	Engage a multidisciplinary team of stakeholders	Engage internal [22] and external [21] SMEs; Engage innovation users [7]
Inherently iterative process	Perform frequent and early concerting and prototyping	Laboratory tests [10, 16]; Technology demonstrators [21]
Involves diverse knowledge-gathering activities	Apply multi-modal knowledge-gathering	Literature reviews [22]; Engage internal [22] SMEs; Open/closed supplier search [9]
Solutions typically arise externally	Scout external sources for inspiration	Open innovation collaboration [23]; Open/closed supplier search [9]; Literature reviews [22]

Activities supporting the implementation of each design principle are likewise proposed research. In extension to this, establishing a "toolbox" of activities to support the front-end of radical MI processes is highly interesting for future research.

5 Conclusion

This paper has focused on manufacturing innovation's critical yet underexplored front-end phase. A review and analysis of existing research identified six characteristics of the front-end of radical manufacturing innovation processes.

A case study from a large European pharmaceutical manufacturing company was reported on. The case study highlighted the importance of agile processes, periodic requirement reviews, and diverse stakeholder engagement for navigating dynamic requirements and evolving contexts. Engaging with a multidisciplinary team expedited knowledge gathering, while interactions with technology suppliers and review of literature improved solution ideation through external insights. Frequent prototyping and iterations allowed the project team to respond to emerging findings, leading to solution concepts better suited to the needs of the pharmaceutical industry. These findings led to a proposal of six corresponding design principles to aid in designing front-end radical manufacturing innovation processes. Several examples of applied activities to implement these design principles were also presented.

In summary, this paper contributes to the current body of knowledge by proposing several design principles to support the early stages of manufacturing innovation projects.

References

1. Hohmann L, Löbnitz L, Menke C, Santhirakumaran B, Stier P, Stenger F, Dufour F, Wiese G, Zur Horst-Meyer SZ, Kusserow B, Zang W, Nirschl H, Kockmann N (2018) Continuous downstream processing of amino acids in a modular miniplant. Chem Eng Technol 41:1152–1164. https://doi.org/10.1002/ceat.201700657
2. Diel B, Manzke C, Peuker T (2013) Flexible biomanufacturing processes that address the needs of the future. In: Eibl D, Eibl R (eds) Disposable bioreactors II. Springer Berlin Heidelberg, Berlin, Heidelberg, pp 207–237. https://doi.org/10.1007/10_2013_220
3. Siiskonen M, Malmqvist J, Folestad S (2020) Integrated product and manufacturing system platforms supporting the design of personalized medicines. J Manuf Syst 56:281–295
4. Andersen A-L, Brunoe TD, Nielsen K, Rösiö C (2017) Towards a generic design method for reconfigurable manufacturing systems. J Manuf Syst 42:179–195
5. Andersen A-L, Andersen R, Napoleone A, Brunø TD, Kjeldgaard S, Nielsen K, Sørensen DGH, Raza M, Bilberg A, Rösiö C, Boldt S, Skärin F (2023) Paving the way for changeable and reconfigurable production. REKON Press
6. Bortolini M, Galizia FG, Mora C (2018) Reconfigurable manufacturing systems: literature review and research trend. J Manuf Syst 49:93–106. https://doi.org/10.1016/j.jmsy.2018.09.005
7. Larsen MSS (2023) Manufacturing innovation: engineering an innovation approach to industry 4.0. https://doi.org/10.54337/aau621655437

8. Lassen AH, Larsen MSS (2024) Manufacturing innovation for industry 4.0: an innovation capability perspective. JMTM
9. Chaoji P, Martinsuo M (2022) Managers search practices at the front end of radical manufacturing technology innovations. Creat Innov Manage 31:636–650. https://doi.org/10.1111/caim.12524
10. Kurkkio M, Frishammar J, Lichtenthaler U (2011) Where process development begins: a multiple case study of front end activities in process firms. Technovation 31:490–504. https://doi.org/10.1016/j.technovation.2011.05.004
11. Utterback JM, Abernathy WJ (1975) A dynamic model of process and product innovation. Omega 3
12. Dooley L, O'Sullivan D (2000) Systems innovation: managing manufacturing systems redesign. Int J Comput Integr Manuf 13:410–421. https://doi.org/10.1080/09511920050117900
13. Yamamoto Y, Bellgran M (2013) Four types of manufacturing process innovation and their managerial concerns. Procedia CIRP 7:479–484. https://doi.org/10.1016/j.procir.2013.06.019
14. Larsson L (2020) A conceptual framework for production innovation
15. Bessant J, Lamming R, Noke H, Phillips W (2005) Managing innovation beyond the steady state. Technovation 25:1366–1376. https://doi.org/10.1016/j.technovation.2005.04.007
16. Larsen MSS, Laursen ES, Lassen AH (2023) Managing manufacturing innovation: four types of problems and matching innovation processes. In: Galizia FG, Bortolini M (eds) Production processes and product evolution in the age of disruption. Springer International Publishing, Cham, pp 691–699. https://doi.org/10.1007/978-3-031-34821-1_75
17. Braun E (1981) Constellations for manufacturing innovation. Omega 9:247–253. https://doi.org/10.1016/0305-0483(81)90031-1
18. Hofer A, Brandl F, Bauer H, Haghi S, Reinhart G (2020) A framework for managing innovation cycles in manufacturing systems. Procedia CIRP 93:771–776
19. Larsen MSS, Lassen AH, Schou C (2023) Manufacturing innovation: a heuristic model of innovation processes for industry 4.0. ASI 6:98. https://doi.org/10.3390/asi6060098
20. Frishammar J, Lichtenthaler U, Richtnér A (2013) Managing process development: key issues and dimensions in the front end. R&D Manage 43:213–226. https://doi.org/10.1111/radm.12011
21. Simms C, Frishammar J, Ford N (2021) The front end in radical process innovation projects: sources of knowledge problems and coping mechanisms. Technovation 105:102214. https://doi.org/10.1016/j.technovation.2020.102214
22. Gärtner Q, Hofer A, Reinhart G (2021) Identification and systematization of strategic technology demands in manufacturing. Procedia CIRP 104:32–37. https://doi.org/10.1016/j.procir.2021.11.004
23. Larsen MSS, Lassen AH (2020) Design parameters for smart manufacturing innovation processes. Procedia CIRP 93:365–370. https://doi.org/10.1016/j.procir.2020.04.068

Systematic Literature Review to Derive Basic Elements for Model-Based Product Development

Gereon C. Bönsch, Alexander Keuper, and Günther Schuh

Abstract Model-based systems engineering is becoming increasingly widespread in the manufacturing industry. The reason for this lies in the potential for effectiveness, efficiency and complexity control offered by MBSE. In addition to the question of suitable tools, however, the design of appropriate product development processes for sustainably implementing MBSE poses particular challenges for companies. In order to make the necessary adjustments to existing processes, companies need a uniform understanding of the elements to be adapted. This paper presents the findings from a systematic literature analysis of scientific work regarding product development and MBSE. As a result, activities, roles and models were identified as basic elements of the development processes to be adapted. A consolidated list of the identified elements provides an overview of the roles, activities and models for MBSE described in the literature. The results provide users with orientation for process analysis and support the standardized description of processes in model-based product development. They provide the basis for further organizational design.

Keywords Product development · MBSE · Process modeling · Literature review

1 Introduction

Model-based approaches are becoming increasingly significant in modern product development, as they promise improvements in efficiency, effectiveness, and the ability to manage complexity [1, 2]. Despite their theoretical advantages, many companies struggle to successfully implement Model-Based Systems Engineering (MBSE) in practice [3]. A key challenge lies in the inadequacy of existing processes and organizational structures which often do not align with the requirements of model-based approaches. Empirical studies and industry feedback consistently highlight the adaptation and improvement of organizational structures as one of the most

G. C. Bönsch (✉) · A. Keuper · G. Schuh
Laboratory for Machine Tools and Production Engineering (WZL) of RWTH Aachen University, Aachen, Germany
e-mail: g.boensch@wzl.rwth-aachen.de

M. Manns (ed.), *New Paradigms for Anticipated Uncertainty*, Lecture Notes in Mechanical Engineering, https://doi.org/10.1007/978-3-032-16889-4_24

critical success factors for MBSE implementation [4]. Companies frequently face difficulties in identifying the necessary adjustments to their processes and structures to enable a seamless transition to model-based practices. This paper aims to address this challenge by identifying and characterizing the basic elements to be adapted, thereby offering a foundation for a structured approach to MBSE. After an introductory motivation, relevant fundamentals regarding product development, MBSE and process description are discussed. Subsequently, the methodology for conducting the literature review is presented. Findings, including an excerpt of the shortlist of basic elements, are then discussed, followed by a summary and outlook.

2 Fundamentals

This section explains the basic terminology necessary for further understanding, covering product development, MBSE and basic elements.

Product development is a structured process that transforms initial ideas into market-ready products. It typically involves stages such as planning, design, prototyping, testing, and commercialization. Effective product development requires cross-functional collaboration to ensure technical feasibility and market success. Various established methodologies support this process, including the product development approach by PAHL/BEITZ and process models outlined in standards such as VDI 2206 [5–7].

Model-Based Systems Engineering MBSE is a model-driven approach that supports key product development activities, from requirements definition through design to validation [8–10]. At its core, MBSE relies on an overarching system model that represents both the architecture and behavior of the system. This central model acts as the primary information repository, linking domain-specific models such as CAD data or simulation results to ensure consistency and traceability [11].

Basic elements in (product development) processes are commonly described using activities and roles, with activities being defined as a set of tasks or work steps and roles being the personal resources responsible for carrying out certain activities [12]. In the context of this work, we also consider models as basic elements with models comprising system descriptions for a specific purpose [8]. Figure 1 illustrates the relationship between the basic elements. An activity is performed by a role using a model.

3 Methods

The literature review in this paper follows a structured five-step process inspired by XIAO AND WATSON as illustrated in Fig. 2 and described in detail below [13].

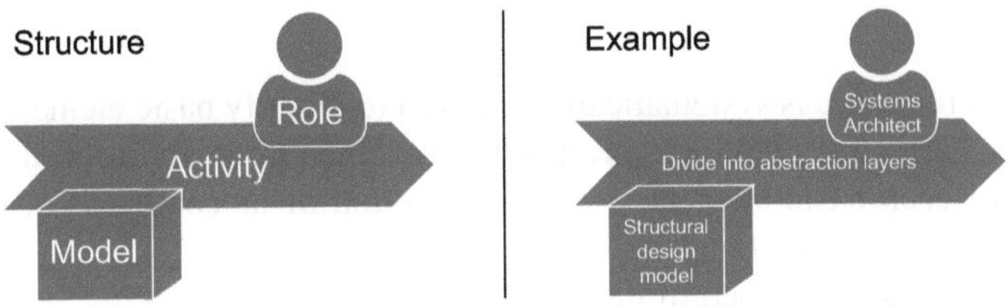

Fig. 1 Structure and example of the interaction of the basic elements role, activity and model

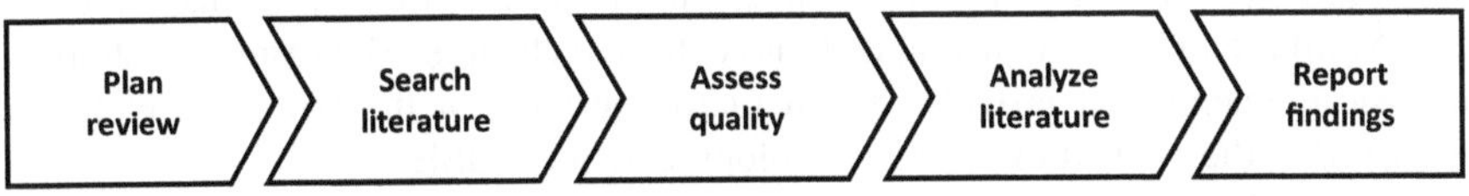

Fig. 2 Approach of the literature review based on [13]

Plan Review

The first step involves establishing a clear focus for the literature review. This includes selecting appropriate academic databases and identifying relevant search terms. Keywords that represent the research topic are determined, along with synonyms and alternative expressions. Boolean operators are then used to create combinations of these terms, forming comprehensive search strings. Search terms were created by combining "fundamental elements", "core constructs", "roles", "activities", and "models" with "product development" and "(model-based) systems engineering". These were then used in Google Scholar and IEEE Xplore to ensure a structured and comprehensive literature review.

Search Literature

Using the search strings developed in the previous step, targeted searches are conducted in the chosen databases. This process is iterative, allowing for the review and refinement of results by incorporating additional keywords or synonyms as needed. This ensures that the search comprehensively covers the research area. The identified sources are evaluated for accessibility, and their bibliographic details are recorded. Initially, over 500 sources, including research papers, edited volumes, and textbooks, were identified.

Assess Quality

The preliminary literature pool is screened for relevance to the research topic. Exclusion criteria may include factors such as publication date, use of specific terminology, or frequency of keyword occurrences. Initially, titles and abstracts are reviewed to eliminate irrelevant sources. A more detailed examination of the remaining sources, including full-text analysis, is conducted to finalize the literature pool.

A total of 19 contributions form the final literature pool for this work [9, 14–31].

Analyze Literature

The selected literature is systematically analyzed to identify basic elements relevant to the study. Particular attention is given to basic elements containing product-related insights. For each basic element, a brief description of its context is documented, creating a basic elements longlist.

Over 200 entries were part of the initial longlist.

Report Findings

The identified basic elements are organized and grouped based on thematic similarities. Similar entries are consolidated into cohesive clusters. The result is a shortlist of basic elements that provides a structured overview of the key basic elements discussed in the context of product development and MBSE.

In total, 131 basic elements were identified.

4 Results

This section presents the outcome of the literature review. Initially, it discusses the overarching findings, followed by a summary of the basic elements shortlist and an analysis of the content clusters within that list.

Table 1 presents an excerpt from the shortlist of basic elements. The first column features a unique serial number, while the second column lists the name of each basic element. The third column categorizes each element—whether it is a role, activity or model—and indicates its relevance within the development process. The fourth column provides a brief description of each element, helping to characterize and distinguish it from the others in the list.

Roles

A total of 37 roles have been identified. This points to the many perspectives that must be captured in product development. Since there is seldom a uniform understanding of roles across different contexts, each role is associated with distinct responsibilities and contributions. For instance, the System Architect plays a central role by creating the high-level systems architecture. They are often very involved in orchestrating the MBSE process. In contrast, the VV&A expert ensures that the system meets specified requirements. Focussing on the models themselves, they rather provide guidance than coordinating with others.

Models

A total of 32 models have been identified. Some directly support a certain phase of product development while overarching models integrate various aspects of the development process. Requirements Diagrams visually represent system requirements and their relationships. They help teams understand what needs to be achieved while

Table 1 Basic elements in product development based on [9, 14–31]

	Basic elements	Category	Context
R1	Methodology researcher	Roles	Develops and refines MBSE methodologies to improve processes
R2	Systems architect	Roles	Defines systems architecture, translates requirements into models, and ensures consistency across abstraction levels
R3	Internal project manager	Roles	Coordinates MBSE processes, aligns stakeholders, and manages resources
R4	Workflow designer	Roles	Structures and optimizes MBSE development processes, ensuring seamless tool integration and collaboration
R5	Simulation expert	Roles	Performs simulations to analyze system behavior, optimize designs, and validate functional performance
…			
R37	VV&A experts (Verification, Validation and Accreditation)	Roles	Ensures that models meet quality standards, comply with requirements, and are properly validated for decision-making
M1	Requirements diagrams	Models	Capture and organize system requirements, ensuring traceability and consistency throughout the development process
M2	Use case diagrams	Models	Capture functional requirements by modeling interactions between users and the system, ensuring a user-centered development approach
M3	Activity diagrams	Models	Model workflows and dynamic behavior, supporting the analysis and optimization of system processes
M4	Sequence diagrams	Models	Capture the temporal flow of interactions over time, helping identify bottlenecks and improving communication in MBSE-driven development
M5	Structural design models (SDM)	Models	Define the system architecture, ensuring consistency and integration across all other models
…			
M32	Simulation models	Models	Enable virtual testing and performance analysis, reducing development risks and supporting design validation
A1	Problem definition and analysis	Activities	Identify and analyze problems, requirements, and constraints to establish a solid foundation for system development
A2	Division into different abstraction layers	Activities	Structure system information across hierarchical levels to manage complexity and enhance traceability

(continued)

Table 1 (continued)

	Basic elements	Category	Context
A3	Generation of system views	Activities	Create visual representations of the system to support analysis, communication, and decision-making in MBSE
A4	Safety assessment	Activities	Evaluate potential hazards and risks to ensure compliance with safety standards
A5	Verification and implementation	Activities	Confirm that the system meets specified requirements through testing and analysis
…			
A62	Configuration management	Activities	Control system changes to maintain consistency, traceability, and integrity throughout development

ensuring that all requirements are considered during development. Often, Structural design models, showcasing the composition of system components and their hierarchical relationships via block definition diagrams, are at the core of MBSE.

Activities

A total of 62 activities have been identified which serve as overarching categories rather than exhaustive lists. The detailed listing of activities can vary significantly depending on the specific project which is why broader categories have been established to facilitate organization and management. For instance, Problem Definition and Analysis involves identifying core issues that need to be addressed and analyzing them to understand their implications on the project, setting a solid foundation for subsequent development efforts. Another key activity is Verification and Implementation where teams ensure that developed solutions meet specified requirements through rigorous testing processes before full-scale implementation. The generation of system views is often central to linking MBSE activities to the overall development procedures in companies.

The interplay between roles, activities, and models is fundamental to successful product development. Each role contributes specific expertise to various activities; for example, a Systems Engineer may lead requirements gathering while collaborating with Product Managers to ensure alignment with business goals. Similarly, activities often rely on models to inform decisions; design analysis requires accurate systems models to evaluate potential solutions effectively. Understanding these connections allows organizations to optimize their processes by aligning roles with relevant activities, while leveraging models as essential tools in the development lifecycle. Companies (re-)structuring their MBSE activities may refer to this work as a reference. The classification of models may aid in choosing appropriate tool support. By comparing their organizational structure to the described roles, missing responsibilities may be identified.

5 Conclusion

Organizations engaged in model-based product development often encounter a complex array of processes during their transition to MBSE. This paper provides a comprehensive overview of the basic elements critical for successful MBSE implementation. We have identified 131 basic elements categorized into roles, activities, and models.

By clearly defining these basic elements, companies can effectively map their current and future processes within an integrated framework. This structured approach does not only aid in comparing various methodologies, but also lays the groundwork for designing targeted processes that promote efficient and effective MBSE practices.

However, it is important to acknowledge certain limitations inherent in this study. The quality of our literature review largely depends on the breadth and rigor of the sources consulted; variations in terminology or inadequate search strategies may lead to significant literature being overlooked. Given the rapid advancements in this field, this paper represents only a snapshot of current knowledge. Current literature also focuses mostly on larger companies with more advanced MBSE approaches, thus applicability for small businesses' first steps into MBSE may be limited.

Future research endeavors could build upon or refine the identified basic elements. Validation within practical business environments will be crucial to ascertain the applicability and effectiveness of this framework. Further investigations should examine how these basic elements can serve as foundational inputs for developing tailored processes that meet specific industry requirements. To this end, it should be investigated how sets of roles, activities and models that fit the company-specific context and requirements can be derived.

References

1. Riesener et al (2021) Understand and control complexity
2. Pinzone et al (2020) A framework for operative and social sustainability functionalities in human-centric cyber-physical production systems. Comput Ind Eng 139
3. Majumder (2023) A domain-driven model generation framework for CPS
4. Di Carlo et al (2021) Retrofitting a process plant in an industry 4.0 perspective for improving safety and maintenance performance. Sustainability 13(2)
5. Feldhusen et al. (eds) (2013) Pahl/Beitz Konstruktionslehre. Springer Vieweg, Berlin, Heidelberg
6. Schuh (2020) Sustainable innovation. [S.l.]: Morgan Kaufmann
7. VDI/VDE 2206 (2021) Entwicklung mechatronischer und cyber-physischer Systeme—Inhaltsverzeichnis: Entwicklung mechatronischer und cyber-physischer Systeme, VDI 2206, Verein Deutscher Ingenieure VDI, Berlin
8. Weilkiens (2019) Systems engineering mit SysML/UML. Heidelberg, dpunkt.verlag
9. International council on systems engineering (2020) Systems Engineering Vision 2020
10. Bernhard et al (2023) Sustainability in the internet of production
11. Zhang et al (2021) Towards holistic system models including domain-specific simulation models based on SysML. Systems 9(4)

12. Laue et al (2021) Prozessmanagement und process-mining. De Gruyter, Berlin, Boston
13. Xiao et al (2019) Guidance on conducting a systematic literature review. J Plann Educ Res 39(1)
14. Kantelberg (2018) Gestaltung agiler Entwicklungsprozesse technischer Produkte. Dissertation, Rheinisch-Westfälische Technische Hochschule Aachen and Apprimus Verlag
15. Keuper (2024) Kompetenzbasierter Ressourceneinsatz in Multientwicklungsprojektumgebungen. Dissertation/Dissertation, Apprimus Verlag and Rheinisch-Westfälische Technische Hochschule Aachen
16. Xu et al (2006) A framework for product lifecycle management system
17. Berschik et al (2023) MBSE within the engineering design community—an exploratory study. Proceedings of the design society, vol 3
18. Bretz et al (2016) A concept for managing information in early stages of product engineering by integrating MBSE and workflow management systems
19. D'Ambrogio et al (eds) (2014) Modeling and simulation-based systems engineering handbook. CRC Press, Boca Raton, Florida
20. Eigner et al (2014) Modellbasierte virtuelle produktentwicklung. Springer Vieweg, Berlin, Germany
21. Gadzo E. (2021) Modellbasierte Umsetzung des Anforderungsmanagements zur Unterstützung der Entwicklung eines Experimentalsatelliten
22. Kößler et al (2018) Integration of MBSE into existing development processes—expectations and challenges
23. Krause et al (2022) Design methodology for future products. Springer International Publishing, Cham
24. Mandel et al (2021) A model-based systems engineering approach to support continuous validation in PGE—product generation engineering
25. Micouin M (2013) Model based systems engineering. Wiley-ISTE, London
26. Mital et al (2014) Product development. Elsevier, Amsterdam, Boston
27. Morgan et al (2019) The Toyota product development system. CRC Press, Taylor & Francis Group, Boca Raton, FL
28. Schmidtner et al (2020) Towards an adaptive reference model for agile and hybrid frameworks in automotive development
29. Sheard (1996) Twelve systems engineering roles. INCOSE Int Symp 6(1)
30. Tschirner et al (2015) Tailoring model-based systems engineering concepts for industrial application
31. Wu et al (2007) A framework of generating collaborative and dynamic new product development process

Scalable Cell for the Production of Sheet Metal Car Body Components

Michael Schiller, Peter Frohn-Sörensen, Michael Jonek, Walid Elleuch, Marco Fries, Ireneus Wior, Izel Gediz, Anja Elser, Christopher Heftrich, Tony Joost, Meik Ebbinghaus, Marvin Engel, Dario Winterberg, Michael Keckeisen, Martin Manns, and Bernd Engel

Abstract Ensuring independence through technological openness and accelerated digitalization will be a key challenge for the automotive industry in the future, especially for automotive suppliers. Many body parts that cannot always be manufactured profitably with current production technologies are among the components that OEMs usually outsource to suppliers. This is due to the large number of variants that require smaller batch sizes. Instead of cost-intensive and uneconomical high-volume production, flexible geometric customization with variable variants and capacities could make such production batches more attractive and profitable. Scaling variant-rich body parts for small series production requires the use of new

M. Schiller (✉) · M. Jonek · W. Elleuch · M. Manns
Chair of Production Automation and Assembly, University of Siegen, Siegen, Germany
e-mail: Michael.Schiller@uni-siegen.de

P. Frohn-Sörensen · B. Engel
Chair of Forming Technology, University of Siegen, Siegen, Germany

M. Fries
Educational Technologies for the Digital Transformation, State Distance-Learning University, Hagen, Germany

I. Wior · I. Gediz · M. Keckeisen
TWT GmbH, Stuttgart, Germany

A. Elser
ISG Industrielle Steuerungstechnik GmbH, Stuttgart, Germany

C. Heftrich
MPA Technology GmbH, Burbach, Germany

T. Joost
Voestalpine Automotive Components Dettingen GmbH and Co. KG, Dettingen, Germany

M. Ebbinghaus
HMT Heldener Metalltechnik GmbH and Co. KG, Attendorn, Germany

M. Engel
Franz Hof GmbH, Haiger, Germany

D. Winterberg
VIA Consult GmbH and Co. KG, Olpe, Germany

M. Manns (ed.), *New Paradigms for Anticipated Uncertainty*, Lecture Notes in Mechanical Engineering, https://doi.org/10.1007/978-3-032-16889-4_25

production methods. For this reason, SkaLaB is developing and testing a highly adaptable production cell for car body components that can be expanded in every dimension. This enables the production process to be individually adapted to the respective component. The aim is to reduce manufacturing costs for new, geometrically different component variants. A specially developed planning tool creates the corresponding production concept. By digitally mapping the production steps, a cost-, efficiency- and resource-optimized production strategy can be selected depending on the quantity. Flexible, disruptive processes are integrated into existing, proven workflows within the production cell.

Keywords Scalable · Flexible · Production · Resilient · Forming technology

1 Introduction

The reduction of mass alone is no longer sufficient to promote the sustainable use of future-oriented technologies. Increasing automation, component complexity, and function integration are critical factors. Companies must embrace innovative approaches, even at the risk of replacing long-established technologies. Systematic technology assessments help evaluate the implementation of new materials and manufacturing methods [1]. Digitization and Artificial Intelligence enhance adaptability to changing conditions, such as material property fluctuations, and enable production adaptability, as shown in Fig. 1.

A survey by the University of Siegen highlights that most sheet metal suppliers are SMEs relying on progressive forming but struggle with economic production below 100,000 units annually due to setup and tooling costs [2].

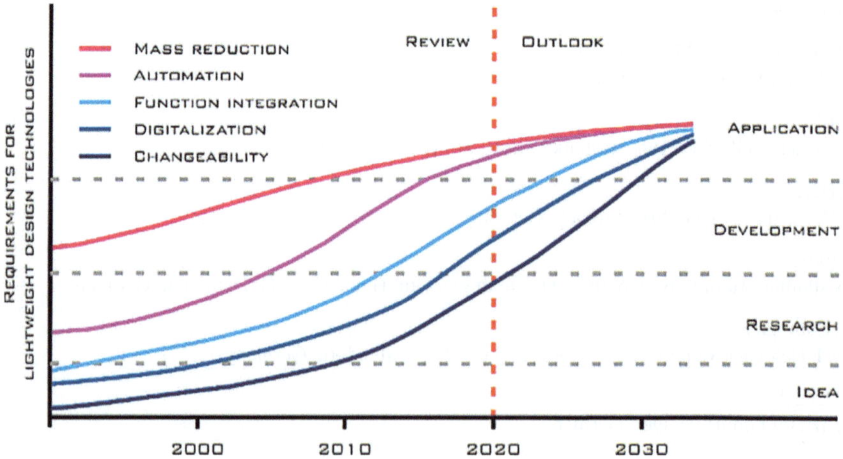

Fig. 1 Lightweight manufacturing 2030+, following [1]

Mass production of sheet metal parts using progressive forming offers precision and efficiency through simultaneous operations. However, the process requires significant investment in high-capacity presses, especially in sectors like the automotive. Flexible technologies such as swivel bending enable product variation in material, angle, and thickness, providing viable alternatives to conventional high-volume methods [3, 4].

Mechanical and plant engineering increasingly uses simulation models, summarized under the digital twin concept, to support the entire production lifecycle [5, 6]. Despite its growing adoption, virtual commissioning with digital twins, faces limitations due to incomplete virtual system representation and insufficient model depth. This results in extended commissioning times, which can threaten competitiveness when adapting to new product variants [7].

2 State of the Art

To support the detected gap of the critical annual batch size for the production of automotive components below 100,000 units per year, flexible forming technologies are being developed, tested and optimized. For this purpose, scalable manufacturing processes such as swivel bending and die bending are considered. There are significant process extensions such as 3D-swivel-bending [2, 8]. The design of a process is supported by an associated, scientifically designed process window. A desired component geometry can be classified and examined for feasibility within this window.

These new bending techniques, which are representative of flexible manufacturing processes, have the potential to replace the conventionally tool-intensive production of body parts in progressive dies, which is strongly linked to the geometry. Not only the low-tool production, but also the overall smaller dimensioning of the plant technology to be used is illustrated by the example of the substitution of press technology by bending techniques. Accordingly, the concept presented makes an important contribution to the efficient, sustainable use of resources and energy input, both in terms of the tools used in the processes and the plant technology required. In most cases, the equipment and tools of flexible manufacturing processes can be used across a wide range of product variants. The development of a scalable production concept requires an equally scalable approach to production and method planning for the entire workflow.

Recognition of Geometry Features

Despite advances in 3D modeling, 2D technical drawings remain indispensable in manufacturing [9]. While 3D models hold extensive information, they are often not shared or only available in rudimentary formats (e.g., STL) loosing valuable information, and requiring high computational resources for processing [10–12]. In contrast, 2D drawings are easier to access, more portable, and don't require specialized hardware or software, making them more practical for a variety of manufacturing tasks

[10–12]. Recent advancements in object recognition, using deep neural networks like YOLO (You Only Look Once) [13] or Faster Region-based Convolutional Neural Network (R-CNN) [14] have expanded their application to extract information from technical drawings [15]. Dillenhöfer et al. [16] achieve 89% accuracy in detecting geometric features but focus solely on gearboxes, highlighting the need for more generalized solutions in flexible manufacturing. Furthermore, dimensions and tolerances are essential for complete processing. Haar et al. and [17]. demonstrate effective text extraction via optical character recognition (OCR). The combination of object recognition with OCR offers a promising foundation for extracting both geometric and textual features, supporting automated manufacturing processes.

Assignment of Geometry Features to Manufacturing Processes

Recent approaches to assign geometry features to manufacturing processes typically rely on rule-based systems supported by design feature libraries. CAD-integrated feature recognition is used to match design elements with predefined process steps. While early-stage selection often depends on empirical rules or heuristics, adaptability to novel geometries remains limited. AI-based tools and digital twin methods are increasingly explored to improve flexibility and reduce reliance on manual planning [18].

Digital Twin

Recent advances in production planning emphasize the integration of digital twins, originally developed in the aerospace sector and now gaining traction in manufacturing. These virtual counterparts of physical systems enable real-time monitoring, predictive maintenance, and dynamic reconfiguration through synchronization with sensor data. Semantic data models further support simulation-driven decision-making across various stages of production. Despite promising developments, research remains in early stages, with frameworks and industrial applications still evolving [19].

Digital twins for the simulation of manufacturing cells can be used to evaluate cell layouts regarding different key values such as manufacturing time or cost. However, the automatic generation of such simulations according to layouts or manufacturing steps requires the adaption of the material flow and the simulation logic automatically, which is usually very demanding.

Manufacturing

A central Programmable Logic Controller (PLC) in the cells controls the process of production modules, gauges and robots. Driven by the urgent requirement to enhance both, efficiency and adaptability, mobile robots need to be incorporated into flexible manufacturing processes. This achievement marks a significant step forward in the context of modern industry. These systems leverage mobile robots for tasks such as material handling, transportation, and even direct manufacturing operations, enhancing the flexibility and productivity of production lines. In the design and implementation, several key factors must be considered. Autonomous mobile manipulators, such as OMNIVIL [20], are engineered with high maneuverability and robust sensor configurations that enable them to handle both static and dynamic uncertainties. This design not only improves operational performance but also guarantees safe collaboration between humans and robots. Similarly, holonomic mobile robots, like DOMINO [21] are developed with omnidirectional capabilities, allowing for efficient space utilization and effective obstacle avoidance in complex manufacturing environments. Despite these technological advancements, several challenges remain in deploying mobile robots within FMS. The complexity of these robotic systems, coupled with safety concerns inherent in human–robot interactions, requires advanced monitoring and preventive strategies [20]. Additionally, real-time control is crucial; dynamic assignment and control of robots are essential to maximize production efficiency and adapt swiftly to disruptions [22]. Recent models that employ techniques such as Double Deep Q-Networks have shown promise in addressing these real-time control challenges [23].

In conclusion, mobile robots are crucial in improving the flexibility and efficiency of manufacturing systems. Through the integration of advanced scheduling algorithms and innovative design, these robots have the potential to significantly improve production processes. However, ongoing research is imperative to overcome the remaining challenges of complexity, safety, and real-time control, ensuring that the full potential of mobile robotics in flexible manufacturing environments can be realized.

3 Development of a Scalable Production Cell

In SkaLaB—Scalable Centre for Flexible Manufacturing of Customizable Sheet Metal Components, it will be possible for the first time to adapt the manufacturing process for specific components within series production. By splitting up the individual production steps and processes, a considerable increase in flexibility can be achieved, which ultimately leads to a reduction in production costs for different body variants. The geometries created using modern production technologies are displayed in process windows specific to the forming process, which makes handling easier for both the user and the adaptive design assistant. Manufacturing does not focus on the complete production of components, but on the production of individual geometric features. These features can be produced flexibly using different manufacturing processes and are not dependent on a single forming process [24]. Furthermore, both the hardware and the control programming of the SkaLaB cell must be designed in such a way that they can be easily reconfigured. To this end, the tools of the forming machines are developed in such a way that they can be produced quickly using rapid tooling and can be exchanged easily using mechanical interfaces [24]. In order to ensure an efficient connection between upstream processes and downstream processes, intelligent networking of the individual production steps through the use of a process generator is necessary.

SkaLaB is based on a fully digital and transparent manufacturing concept. A process generator automatically plans a flexible process chain based on the product's

CAD data using artificial intelligence. The result is a flexible process chain that can be optimized in terms of factors such as production costs, production time, loading capacity or CO_2 footprint. This concept can also be transferred to forming technology such as additive manufacturing. A machine-independent process generator that only uses the geometry and material data of the component as well as information about the production resources as input values and independently determines the production plan and control for the SkaLaB cell helps to reduce the barriers to the introduction of new technologies. The process generator, equipped with integrated technology and knowledge memory, enables the immediate use of the technology with minimal expertise, so that companies do not need their own equipment or extensive know-how on site. In this way, companies in the automotive supply industry can avoid expensive IT investments.

Importantly, the knowledge repository is interoperable with other systems via standardized interfaces, making the SkaLaB cell a key component for adaptable supply chains. A simplified representation of the SkaLaB process generator is shown in Fig. 2. The schematic representation of the process generator follows four central aspects.

Aspect 1: The product design is evaluated through feature recognition on a CAD file using a trained AI model.

Aspect 2: The development process of the process generator begins with the mapping of manufacturing processes, in which a rule-based knowledge database is created and shall be continuously updated by a learning unit. This process generator shall be transferred from a realized manufacturing construction kit to an interactive, knowledge-based engineering system in the design department. The system remains open across all manufacturing stages, learns continuously through its assistance functions and collects implicit knowledge from traditional manufacturing processes. The manufacturing toolbox thus supports design and planning by providing early insights through the activation of the process generator.

Aspect 3: In SkaLaB, production planning is carried out by the process generator, which is divided into two steps: rough planning and detailed planning. Rough planning is a rule-based approach and incorporates expert knowledge expressed as rules. It enables an initial assessment of manufacturability as well as a rough planning of the production chain for manufacturing the geometric properties of the component, which narrows down the possible solution space in terms of production methods, process sequence and process parameters. Thus, the multi-dimensional solution space is reduced in its dimensions and the remaining dimensions are narrowed down in their feasible value range. The expert system PyKE (Python Knowledge Engine) is used for reasoning, for which the rule knowledge base must be available as PyKE rules. The user is supported by a rule specification tool when translating the user's knowledge into PyKE rules (an example of basic knowledge that can be presented as a rule: "Bending process 1 can bend metal sheets with a thickness of up to 3 mm."). In rough planning the process capability of the various process chains based on the existing knowledge base are checked. In this way, process chains that cannot produce the component in accordance with the requirements can be sorted out in advance.

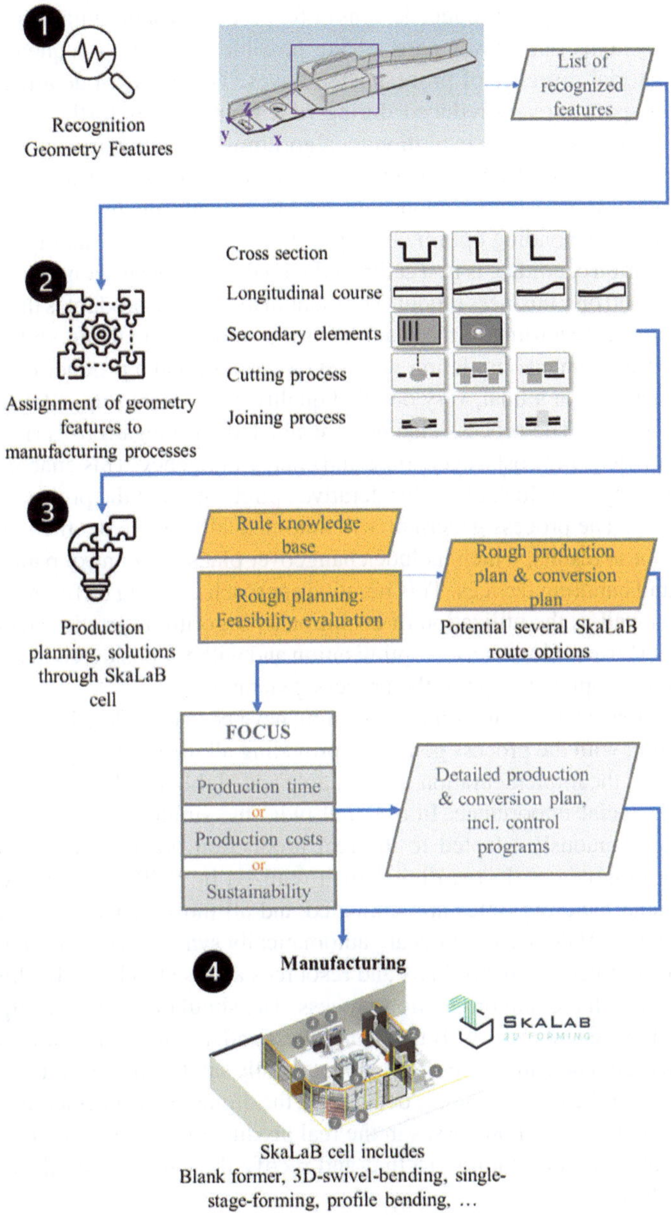

Fig. 2 Schematic of the SkaLaB process generator, following [24]

Especially for complex product designs or for product designs with many features, rule-based systems only cover parts of the possible production sequences due to the combinatorial diversity of process alternatives. Therefore, in detailed planning the solution space already reduced during rough planning is further searched for promising solutions utilizing evolutionary algorithms. Evolutionary algorithms are a class of heuristic search methods that mimic biological evolution. The process generator uses both discrete and continuous representations of the manufacturing processes to formulate corresponding optimization problems. The solution quality of a manufacturing solution candidate is first estimated with the help of an economic feasibility evaluation based on a business model. The manufacture of the product is then geometrically simulated according to the manufacturing solution candidate with the aid of a production equipment simulation with ISG's virtuose and a product is thus virtually manufactured. In a final, very detailed quality assessment step, a finite element simulation is used. The process chains are evaluated with regard to various criteria such as manufacturing costs, cycle time and energy efficiency. This enables iterative virtual quality cycles which allow for iterative optimization of the process chain and its parameters. The process generator will automatically derive detailed production plans from these results, which include changeover plans and control parameters for machines and handling devices. This iterative approach utilizing different evaluation strategies as well as the utilization of evolutionary algorithms provide many opportunities to perform multi-criterion optimization and will be considered in more detail in the later development stage of the process generator.

Aspect 4: Once the simulation and planning have been completed, actual production can begin, with the process generator providing all the necessary planning data. In this context, the implementation of a scalable and highly flexible SkaLaB forming center is of crucial importance. In a digital twin, the simulation of the production process is continuously adapted to the real production process carried out in the SkaLaB forming center. In the digital twin realized in NVIDIA's Omniverse, the production and handling times are simulated and all movements of the automated guided vehicles (AGVs) and robots are automatically synchronized. To this end, the signal traces left behind by products and resources are used to keep the digital twin synchronized with the real production process. The simulation with the digital twin runs parallel to the real process. If the real process and the simulation deviate significantly from each other, an unexpected and potentially safety-relevant situation arises. As soon as a significant deviation is detected in the digital twin, an emergency stop is triggered, which stops all processes in the real production cell. Manual intervention by the operator is then required to find and rectify the cause of the deviation and resume production.

The SkaLaB manufacturing cell has manufacturing technologies such as additive manufacturing (laser metal deposition, 3D printing), subtractive manufacturing (laser cutting), deformation (bending and pressing), and joining (spot welding) that help to manufacture a product prototype. In the production cell, there is also a gravity rack as an interface for the employees outside the cell to load and unload the unprocessed and processed sheets. Material flow and handling are performed by robotic arms equipped with various instruments such as finger grippers, vacuum grippers, and screwdrivers. In addition to product logistics, these instruments can also reconfigure production machines. The manufacturing process sequences will be sent to the two mobile manipulators, which enable material flow between different workstations or cells while avoiding obstacles and optimizing their paths in real-time, cf. Figure 2.

The core flexibility of the SkaLaB production concept lies in the sequence generation provided by the process generator (ProGen). There is no predefined order of manufacturing steps, as each new production has different features that require different machinery, tooling, and manufacturing sequences. To summarize, SkaLaB is a flexible concept with economic advantages that can adapt to any product and production cell layout, as long as the required machines and processes meet SkaLaB's standards, including geometric dimensions, tolerance, and functionality.

A reconfigurable production cell requires that manufacturing modules, quality inspection systems and handling technology can be used individually. Necessary for this is an independent and automatic exchange of information between individual modules. In the SkaLaB cell, the process control at each resource is decentralized on local PLCs. At the material flow level, a central Master-Controller controls production. Communication between the Master-Controller and the individual PLCs mainly takes place via OPC UA. In the automation pyramid, the control of the modules from the cell moves from the field level to the operations management level:

- Company level—ProGen: planning of the setup and production process; definition of machine, tool, setup, assembly, manufacturing, testing and other required parameters
- Operations management level—Master-Controller: control, monitoring and automation of the setup and the material flow of the production process
- Field level—Cell-Modules: Setup and production in digital twin and in lab cell

Economical Feasibility Evaluation in a Business Model:

The previous elaborations and findings were then transferred into a business model with integrated budgeting, which is described in more detail in the following chapter. The development of a potential business model was based on Osterwalder's Business Model Canvas [25]. The approach selected for SkaLaB's practical market entry was "Industry as a Service" [26, 27], operating as an independent company with its own SkaLaB machinery housed in rented facilities. SME customers, who come from various sectors such as the automotive supply, aerospace and mechanical engineering industries, can quickly obtain a feasibility, cost and delivery time quote for geometrically demanding sheet metal components by simply uploading 2D/3D-CAD files. The upload should be web-based and possible 24/7, 365 days a year. What is new about SkaLaB compared to existing solutions is the flexible selection of optimization parameters (production time, production costs, sustainability) for a sheet metal component as well as the reconfigurability of production processes and automated creation of component-specific tools. Although previous solutions already offer the processing of CAD files, they are tied to rigid production processes with non-interlinked or reconfigurable systems and focus primarily on simple sheet metal geometries with higher personnel deployment. In the initial phase, staffing

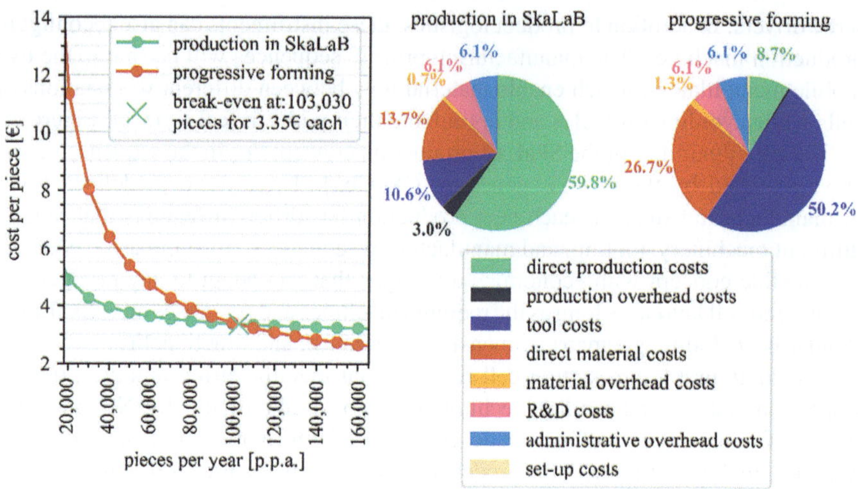

Fig. 3 Economic efficiency analysis in the benchmark of SkaLaB production with conventional progressive forming

for the SkaLaB cell will be limited to a single individual in a multifunctional role, responsible for overseeing all organizational areas, including purchasing, operations, delivery, and financial matters. Support will be provided by a working student position. Income will be generated per order produced and delivered. It is assumed that the SkaLaB cell in a developed phase is going to be able to operate in a daily three-shift operation and enable production six days a week with a yearly output of around 250,000 sheet metal components. An associated integrated planning calculation has shown that the SkaLaB cell is able to achieve a positive result from the fourth financial year after its foundation, if all machines are purchased directly. A comparison of the SkaLaB cell with the classic progressive forming shows an economic advantage of the SkaLaB cell for component volumes of less than 100,000 units per year, as illustrated in Fig. 3.

4 Conclusions

In the traditional progressive forming of sheet metal components, forming stages that are firmly connected to each other are used to produce the components, whereby an optimization in one stage has an effect on all the others. Several components are then joined together to form assemblies. With the SkaLaB cell, individual adaptation to the requirements of individual components is possible for the first time. By splitting up the individual steps and production processes, a considerable increase in flexibility is achieved. This should help to reduce the manufacturing costs for new geometry variants. However, in order to maintain manufacturability and productivity, these newly designed steps must be intelligently controlled and coordinated via the process generator.

The automotive product portfolio often includes car body components. Due to the high number of variants and small batch sizes, these components may not always be produced cost-effectively by SMEs due to the limitations of current production technologies. A viable solution for SMEs lies in systemic support and the implementation of advanced, flexible production technologies. Similar to additive manufacturing, where production planning is based on design data in a process generator, this method can be transferred to the production of sheet metal components. A process-based, machine-independent process generator can help reducing the barriers to implement innovative manufacturing techniques. The SkaLaB process generator uses only the geometry and material data of the component to be manufactured as input information and uses this to independently derive the production planning and control for the SkaLaB cell. This enables the technology to be used immediately with minimal specialist knowledge, as the process generator is equipped with an integrated technology and knowledge memory. This eliminates the need for special equipment on site or extensive internal know-how.

In addition, the process generator can be used for remote production [28], enabling centralized control of production networks. This helps companies in the automotive supply chain to avoid significant IT investments. The prerequisite for this is that the knowledge repository is compatible with the systems of all OEMs and suppliers. SkaLaB thus enables the creation of adaptive supply chains via standardized interfaces.

Acknowledgements The authors acknowledge the financial support by the Federal Ministry for Economic Affairs and Climate Action in the project SkaLaB (project number 13IK025).

Funded by
the European Union
NextGenerationEU

Supported by:
Federal Ministry
for Economic Affairs
and Climate Action

on the basis of a decision
by the German Bundestag

References

1. Göddecke J, Meschut G, Gude M, Lieberwirth H, Tekkaya E, Zaeh M, Stegelmann M, Müller M, Böhme K, Krampitz T, Zöllner M, Hahn M, Schmitz F, Hofer A, Grohmann S (2020) FOREL-Wegweiser: Handlungsempfehlungen für den ressourceneffizienten Leichtbau. https://ris.uni-paderborn.de/record/20318. Accessed 25 Apr 2022
2. Schiller M, Frohn-Sörensen P, Engel B (2022) 3D-Swivel-bending—a flexible and scalable forming technology. In: Inal K, Levesque J, Worswick M, Butcher C (eds) NUMISHEET

2022. Springer International Publishing, Cham, pp 125–139. https://doi.org/10.1007/978-3-031-06212-4_12
3. Klocke F (2017) Fertigungsverfahren 4: Umformen, Springer, Berlin, Heidelberg. https://doi.org/10.1007/978-3-662-54714-4
4. Chatti S (2006) Production of profiles for lightweight structures, BoD–Books on Demand
5. VDI 4499 Blatt 1—Digitale Fabrik—Grundlagen (2008). https://www.vdi.de/richtlinien/details/vdi-4499-blatt-1-digitale-fabrik-grundlagen. Accessed 23 Jan 2023
6. Umsetzungsstrategie Industrie 4.0: Ergebnisbericht der plattform industrie 4.0 (April 2015) (n.d.). https://www.plattform-i40.de/IP/Redaktion/DE/Downloads/Publikation/umsetzungsstrategie-2015.html. Accessed 23 Jan 2023
7. Hale KS, Stanney KM (2014) Handbook of virtual environments: design, implementation, and applications, CRC Press
8. Schiller M, Engel B (2022) 3D-swivel-bending (SB^3). Procedia CIRP 112:406–411. https://doi.org/10.1016/j.procir.2022.09.028
9. Haar C, Kim H, Koberg L (2023) AI-Based engineering and production drawing information extraction. In: Kim K-Y, Monplaisir L, Rickli J (eds) Flexible automation and intelligent manufacturing: the human-data-technology nexus. Springer International Publishing, Cham, pp 374–382. https://doi.org/10.1007/978-3-031-18326-3_36
10. Lin Y-H, Ting Y-H, Huang Y-C, Cheng K-L, Jong W-R (2023) Integration of deep learning for automatic recognition of 2D engineering drawings. Machines 11:802. https://doi.org/10.3390/machines11080802
11. Content first (n.d.) Proceedings of the 6th ACM multimedia systems conference. https://dl.acm.org/doi/10.1145/2713168.2713169. Accessed 21 Mar 2025)
12. Gattullo M, Scurati GW, Fiorentino M, Uva AE, Ferrise F, Bordegoni M (2019) Towards augmented reality manuals for industry 4.0: a methodology. Robot Comput-Integr Manuf 56:276–286. https://doi.org/10.1016/j.rcim.2018.10.001
13. Redmon J, Divvala S, Girshick R, Farhadi A (2016) You only look once: unified, real-time object detection, pp 779–788. https://www.cv-foundation.org/openaccess/content_cvpr_2016/html/Redmon_You_Only_Look_CVPR_2016_paper.html. Accessed 21 Mar 2025
14. Girshick R (2015) Fast R-CNN, pp 1440–1448. https://openaccess.thecvf.com/content_iccv_2015/html/Girshick_Fast_R-CNN_ICCV_2015_paper.html. Accessed 21 Mar 2025
15. Bhanbhro H, Hooi YK, Hassan Z (2022) Modern approaches towards object detection of complex engineering drawings. In: 2022 international conference on digital transformation and intelligence (ICDI), pp 01–06. https://doi.org/10.1109/ICDI57181.2022.10007400
16. Dillenhöfer F, Doksanbir A, Kraske JL, Künne B (2023) Automatic generation of training data for ai object detection in terms of technical drawings in engineering. In: Auer ME, Langmann R, Tsiatsos T (eds) Open science in engineering. Springer Nature Switzerland, Cham, pp 643–651. https://doi.org/10.1007/978-3-031-42467-0_60
17. Villena Toro J, Wiberg A, Tarkian M (2023) Optical character recognition on engineering drawings to achieve automation in production quality control. Front Manuf Technol 3. https://doi.org/10.3389/fmtec.2023.1154132
18. Zhang H, Qi Q, Tao F (2022) A multi-scale modeling method for digital twin shop-floor. J Manuf Syst 62:417–428. https://doi.org/10.1016/j.jmsy.2021.12.011
19. Negri E, Fumagalli L, Macchi M (2017) A review of the roles of digital twin in CPS-based production systems. Procedia Manuf 11:939–948. https://doi.org/10.1016/j.promfg.2017.07.198
20. OMNIVIL—an autonomous mobile manipulator for flexible production (n.d.) https://www.mdpi.com/1424-8220/20/24/7249. Accessed 21 Mar 2025
21. Đokić L, Jokić A, Petrović M, Miljković Z (2021) Design and development of a holonomic mobile robot for material handling and transportation tasks. In: Karabegović I (ed) New technologies, development and application IV. Springer International Publishing, Cham, pp 709–716. https://doi.org/10.1007/978-3-030-75275-0_78
22. Bhatta K, Chang Q (2024) Enhancing production in robot-enabled manufacturing systems: a dynamic model and moving horizon control strategy for mobile robot assignment. J Manuf Sci Eng 147. https://doi.org/10.1115/1.4066978

23. Bhatta K, Huang J, Chang Q (2022) Dynamic robot assignment for flexible serial production systems. IEEE Robot Autom Lett 7:7303–7310. https://doi.org/10.1109/LRA.2022.3182822
24. Schiller M, Frohn-Sörensen P, Schreiber F, Morez D, Manns M, Engel B (2024) Smart design and additive manufacturing of bending tools to improve production flexibility. Manuf Lett 41:91–102. https://doi.org/10.1016/j.mfglet.2024.09.013
25. Osterwalder A, Pigneur Y (2010) Business model generation: a handbook for visionaries, game changers, and challengers. Wiley
26. Vandermerwe S, Rada J (1988) Servitization of business: adding value by adding services. Eur Manag J 6:314–324. https://doi.org/10.1016/0263-2373(88)90033-3
27. Schöllhammer O, Schmitt JL, Nebauer S, Bauernhansl T (2023) Everything-as-a-service Geschäftsmodelle für die industrie. In: Bauernhansl T (ed) Handbuch industrie 4.0: band 1: produktion. Springer, Berlin, Heidelberg, pp 277–293. https://doi.org/10.1007/978-3-662-58532-0_95
28. Schiller M, Heftrich C, Engel B (2021) Remote production. Procedia CIRP 99:242–247. https://doi.org/10.1016/j.procir.2021.03.099

The Learning Factory Approach: From Knowledge Transfer to Knowledge Creation

Lasse Christiansen, Jonas Frendrup, Maria S. S. Larsen, and Astrid H. Lassen

Abstract The learning factory approach has proven valuable in education, lifelong learning and research over the past decade. Learning factories have well-defined goals and enhance participants' knowledge and experience with specific technologies. However, existing research often lacks systematic conclusions on how the design of scenarios affects learning outcomes, thus hindering theory generation. The CIRP Collaborative Working Group's morphology is an initial attempt to characterise learning factories, but further study is needed to understand the scenario design's impact. This paper uses AI-supported bibliometric analysis to identify trends in learning modes, target groups, topics and didactics. A review of 421 studies found that most research has focused on transferring closed-case learning outcomes to students and professionals, with a growing trend of knowledge creation in open-ended, interdisciplinary learning factories. These insights conceptualise learning factories as a multimodal concept, requiring distinct design principles based on the learning target and supporting strategies for cost reduction, knowledge extension and innovation.

Keywords Learning factories · Bibliometric analysis · Upskilling · Education

1 Introduction

Learning factories are valuable tools for competence development in manufacturing [1, 2]. Defined as an environment for competence development and/or innovation, they simulate authentic technical and organisational processes in a changeable setting resembling a real value chain manufacturing a physical product, facilitated by

L. Christiansen (✉) · J. Frendrup
University College Northern Denmark, Aalborg SV, Denmark
e-mail: lch@ucn.dk; lc@mp.aau.dk

L. Christiansen · A. H. Lassen
Department of Materials and Production, Aalborg University, Aalborg E, Denmark

M. S. S. Larsen
Bluestar PLM, Aalborg, Denmark

M. Manns (ed.), *New Paradigms for Anticipated Uncertainty*, Lecture Notes in Mechanical Engineering, https://doi.org/10.1007/978-3-032-16889-4_26

formal, informal and non-formal learning [3]. Learning factories serve both research validation and education/training purposes [2], and they enable researchers to test hypotheses and research results in a closed environment. Validating research in a learning factory is less complicated than using a manufacturing system in industry [1, 4] because they allow continuous testing without interfering with ongoing operations [5].

Learning factories used for education and training have well-defined learning goals with setups and designs that do not change significantly between courses [1, 2, 6, 7]. Hence, there is a constructive alignment between the learning activities and the learning outcome [8]. Learning factories emphasise the enhancement of the participants' (e.g. students, engineers, managers or operators) knowledge of and experience with installing and using technologies [9, 10]. Learning factories are often built to demonstrate particular technologies, such as reconfigurability and changeability [11], process improvements [12], resource and energy efficiency [13], and Industry 4.0 [9].

Research interest in learning factories is high, particularly since 2024. A search for "learning factory" on Google Scholar results in more than 8000 hits, of which approximately 15% have been published since 2024. However, much of the research lacks maturity for journal publication (only 142 out of 8000) and often focuses on explorative empirical cases without synthesising results to support theory generation. Hence, structured understanding and theory development for concepts such as how design aspects affect learning outcomes is lacking. Enke [7] found that publications on learning factories tended to focus on the technical aspects of the learning factory rather than didactics, despite learning being the main objective. Larsen et al. [2] found that most research focused on the competence development of shop-floor employees, such as setting up and operating Industry 4.0 solutions, and research on management's competence development was limited, despite Industry 4.0 often needing top-down initiation.

The learning factory morphology by the CIRP Collaborative Working Group [1, 14] is considered an initial work characterising learning factories. The morphology combines research into a framework that describes a learning factory according to seven dimensions: operational model, targets and purpose, process, setting, product, didactics, and metrics [3]. To fully understand how the design of a learning factory scenario affects learning outcomes, the relationships between dimensions need to be addressed.

From the initial temporal distribution pattern identified at the introductory stage of this research, it seems likely that the focus and application of learning factories have changed over time. Relationships between learning factory designs and the aims of learning outcomes may be detectable by studying the documented differences emerging in the literature over time. This paper study existing research on learning factories and answer the following question: How have the scenarios *of learning factories developed over the years, and how has this influenced the learning outcomes produced?*

2 Methods

To explore this question, we conducted a systematic bibliometric analysis using data from the Web of Science, December 2024. Our search strings included "Learning factory" OR "Learning Factories" OR "Teaching factory". The design principles of the learning factories scenarios were operationalised as learning structure (open vs. closed). Learning outcomes were characterised as knowledge transfer or knowledge creation.

Closed learning structures with low learner agency corresponded to all learners following a specific proven path to a defined outcome. This could be, for example, a step-by-step operation worksheet in which one correct procedure produces an expected result. Open learning structures with high learner agency encourage exploration, problem-solving and learner choice. This can be achieved through a project-based learning task in which learners decide on the design, the construction method and how to test it. There are many solutions and outcomes.

The classification of knowledge creation and knowledge transfer, as different mechanisms for providing knowledge, is central in education [15]. In this research, the analysis adopted the following definitions:

- *Knowledge creation* is the process of generating, developing and sharing new knowledge through interactions, problem-solving and experimentation. It often involves collaborative learning, discovery and innovation, in which individuals actively construct new insights rather than simply receiving existing information. Examples include project-based learning, research-driven education and experiential learning, in which students actively develop new ideas, frameworks and applications.
- *Knowledge transfer* involves disseminating existing knowledge from one source (e.g., a teacher, textbook or expert) to another (e.g., students, employees or trainees). This can be one-way (passive learning) or interactive (applied learning), but it fundamentally revolves around conveying pre-existing knowledge rather than creating new insights. Examples include traditional lectures, structured training programmes and instructional content in which learners receive predefined information.

We limited the search to conference and journal papers but applied no additional exclusion criteria regarding language, publication year or document type. This search yielded 472 papers, all of which were included for further analysis. The retrieved documents were downloaded as PDFs for automated content extraction and analysis.

We employed an LLM-based agent (ChatGPT 4o) to extract key metadata and systematically categorise the papers. Each paper was processed with a structured prompt instructing the AI to generate a table with the following details [16]:

- Paper title, year of publication, and authors
- Learners (subjects in the learning factory)
- Topic (learning objectives)
- Learning structure (open-ended or closed)

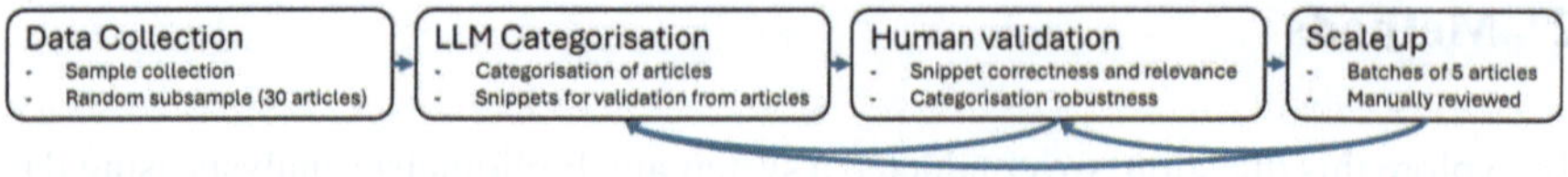

Fig. 1 Flowchart of the data extraction method

- Knowledge type (knowledge creation or knowledge transfer)
- Tags (justification for knowledge classification)
- Text snippets (at least three unedited excerpts from the paper)

To ensure accuracy, a manual validation process was conducted. The authors reviewed the extracted text snippets and examined whether the AI's classification was justified based on the evidence provided. If necessary, classifications were corrected, or the paper was excluded from classification if the text snippets did not sufficiently support either category. The process was handled by the authors and was never automated, meaning that each file was individually uploaded and carefully checked for prompt adherence and hallucinations. A flowchart of this is seen in Fig. 1.

Additionally, we conducted a check of 30 randomly selected papers, verifying that the retrieved text snippets were accurately extracted from the original publications. This step ensured the integrity of the AI-driven extraction and classification process.

3 Results

After exclusions, the literature sample contained 421 research papers from journals and conference proceedings published between 1996 and 2024. Of these, 219 were labelled as focusing on knowledge transfer, while 202 focused on knowledge creation. Before 2014, the research volume was low, with only five papers a year included. The early years focused on knowledge transfer, with knowledge creation building slowly as a learning outcome. Figure 2 shows the fractional distribution between knowledge creation and knowledge transfer, the total number of papers per year, and their target group.

It is noteworthy that knowledge creation surpassed knowledge transfer as the focus in learning factory publications in 2023 after the two had been on par since 2019. The dip in knowledge transfer coincides with the change in the outlet of the learning factory conference (CLF 2020–2023), which was not indexed in Scopus, as well as a slight production dip in the rest of the field due to COVID-19. As the learning factory conference was re-incorporated in Scopus in 2024, the dip filled again. In addition to this trend, it was found that the knowledge transfer and knowledge creation tracks were present in the early years of the learning factory field. The first included study, from 1996, focused on knowledge creation among shop-floor workers [17]. Many descriptions of early learning factories focused on transferring tactile knowledge through hands-on experiences and theoretical knowledge [18]. This remains a central

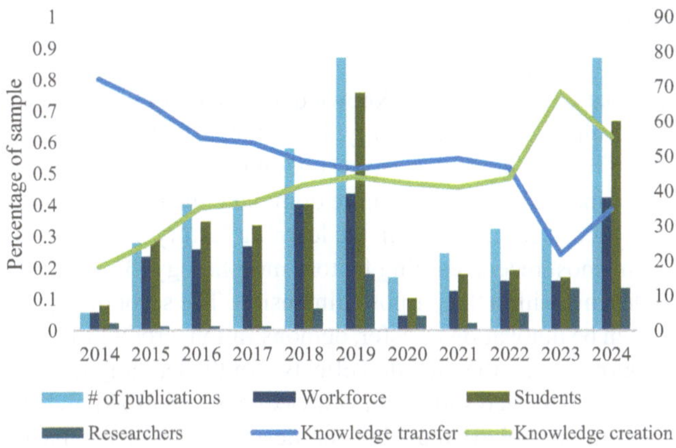

Fig. 2 Fractional distribution between knowledge creation and knowledge transfer focus in the literature sample

point of research on learning factories, whose main purpose is the transfer of study and well-established concepts [19].

Expanding the agenda beyond knowledge transfer within knowledge creation was the most distinct trend. Furthermore, with growing computational power, simulation and risk-free training for risky tasks have gained traction. These learning factories can involve a modifiable physical environment with processes and products similar to a factory, as well as virtual environments [20]. The ability to make significant changes in an open-ended setting was frequently mentioned in research on knowledge creation learning factories, where all papers were labelled as open-ended. In 177 of 219 results, knowledge transfer publications were labelled as closed cases. This shows that knowledge creation and open-ended scenarios are closely linked, while knowledge transfer can be applied in more closed settings, which might also be the simplest to design. As the transfer is most effective when the setting is authentic [21], this can make sense, as adaptability and authenticity can be a trade-off [22]. Table 1 shows these relationships. As Enke et al. pointed out, the didactic dimensions of learning factories are often underdeveloped; hence, these discussions could be of assistance for the learning factory community moving forward [7].

Table 1 Number of closed and open-ended learning factories and target groups for knowledge transfer and creation within the literature sample

	Closed	Open-ended	Workforce	Students	Researchers
Knowledge transfer	177	42	118	136	26
Knowledge creation	0	202	98	159	30

4 Discussion

The results show that a transition from knowledge transfer to knowledge creation is occurring in the learning factory community. This change could be a natural extension of the development of learning factories, as skill and competency research in manufacturing and other industries becomes more dominant due to factors such as labour shortages. Hence, theory development in the learning factory field must also encompass this trend. The most recent learning factory morphology [3], presented in 2024, somewhat includes this within the purpose dimension. The secondary purpose of the learning factory can be innovation transfer, demonstration, production or innovation. However, these purposes are the intended effects, not the learning mechanisms.

The reason for the identified shift from knowledge transfer to knowledge creation could originate from several sources, but the development of technical didactics, along with a higher focus on project-, problem- and action-based learning, may be a contributing factor. Furthermore, the focus on contextual skills or twenty-first century competencies [23] in training and education may also have contributed to more open-ended learning factories. Multifaceted learning goals have been incorporated into educational development along with core technical skills for given professions [24]. The didactical dimension of the learning factory morphology addresses the learning scenario strategy, which can be open or closed, along with instruction and demonstration. However, this is not addressed when the goal is to create knowledge within the learner group.

Considering this, further expansion of the learning factory morphology could be argued for. It could also be argued that, with the current definitions, the change could indicate a subtle link between the didactics and the purpose of a given learning factory. After all, knowledge creation is a collaborative process that might take place due to the designed learning factory scenario rather than as a purpose.

It should also be noted that a knowledge creation activity (e.g. an upskilling activity) might not be the desired outcome of a learning factory. For example, it could be a viable approach to transferring the understanding that reclaiming reusable parts within a hardware appliance is complicated to a group of managers [25]. Hence, future development in the learning factory field must acknowledge that knowledge creation is a valuable opportunity and can be applied through proper design.

5 Conclusion

This paper demonstrates how the focus of the learning factory field has moved from mainly knowledge transfer in the early years towards knowledge creation. As this transition occurs, theory building needs to follow, and learning factory users should be reminded that functioning knowledge transfer is also a vital goal.

Our study revealed a connection between open-ended learning environments and knowledge creation in learning factories. Specifically, many sessions classified as knowledge creation were conducted in open-ended formats, while closed-ended sessions exclusively aligned with knowledge transfer. This raises important questions for future research, particularly regarding whether knowledge creation inherently depends on open-ended structures or whether various degrees of openness can facilitate knowledge creation in different ways. Further investigation is necessary to explore whether closed-ended learning settings can also support knowledge creation by integrating elements of open-endedness or by establishing intermediate degrees of openness that foster innovation and active learning. Additionally, we observed a noticeable transition from a predominant focus on knowledge transfer to an increasing emphasis on knowledge creation in learning factors. This shift prompts an enquiry into the underlying reasons: Has the transition occurred due to knowledge transfer becoming well established and, therefore, less of a research novelty? Or have educational and industrial needs evolved, necessitating more explorative, innovative and collaborative learning frameworks? Further investigation into these dynamics could provide valuable insights into learning factories' long-term development and research priorities.

Acknowledgements The EU Regional Development Fund funded this study.

References

1. Abele E, Metternich J, Tisch M (2019) Learning factories. Concepts, guidelines, best-practice examples
2. Larsen MSS, Lassen AH, Nielsen K (2019) Process innovation in learning factories: towards a reference model, presented at the advances in production management systems. Production management for the factory of the future: IFIP WG 5.7 international conference, APMS 2019, Austin, TX, USA, September 1–5, 2019, Proceedings, Part I. Springer, pp 658–665
3. Abele E, Metternich J, Tisch M, Kreß A (2024) The variety of learning factory concepts. In: Learning factories: featuring new concepts, guidelines, worldwide best-practice examples. Springer, pp 97–127
4. Küsters D, Praß N, Gloy Y-S (2017) Textile learning factory 4.0–preparing germany's textile industry for the digital future. Procedia Manuf 9:214–221
5. Pisano GP (1997) The development factory: unlocking the potential of process innovation. Harvard Business Press
6. Kreimeier D, Morlock F, Prinz C, Krückhans B, Bakir DC, Meier H (2014) Holistic learning factories—a concept to train lean management, resource efficiency as well as management and organization improvement skills. Procedia Cirp 17:184–188
7. Enke J, Glass R, Metternich J (2017) Introducing a maturity model for learning factories. Procedia Manufacturing 9:1–8. https://doi.org/10.1016/j.promfg.2017.04.010
8. Biggs J (1999) What the student does: teaching for enhanced learning. High Educ Res Dev 18(1):57–75
9. Faller C, Feldmüller D (2015) Industry 4.0 learning factory for regional SMEs. Procedia Cirp 32:88–91
10. Karre H, Hammer M, Kleindienst M, Ramsauer C (2017) Transition towards an industry 4.0 state of the LeanLab at Graz university of technology. Procedia Manuf 9:206–213
11. Brunoe TD, Mortensen ST, Andersen A-L, Nielsen K (2019) Learning factory with product configurator for teaching product family modelling and systems integration. Procedia Manuf 28:70–75

12. Rybski C, Jochem R (2016) Benefits of a learning factory in the context of lean management for the pharmaceutical industry. Procedia CIRP 54:31–34
13. Hvidsten T, Vangen FJ, Laursen ES, Christiansen L (2022) Sustainability issues across educational disciplines in learning factories. Social Science Research Network, Rochester, NY. Accessed 05 Apr 2022. [Online]. Available: https://papers.ssrn.com/abstract=4075268
14. Kreß A et al (2023) Revision of the learning factory morphology, presented at the proceedings of the 13th conference on learning factories (CLF 2023), 9–11 May 2023, Reutlingen, Elsevier, pp 1–7
15. Scott D (2017) Knowledge creation transfer and commodification. In: Peters MA (ed) Encyclopedia of educational philosophy and theory. Springer, Singapore, pp 1–5. https://doi.org/10.1007/978-981-287-532-7_632-1
16. Christiansen L, Frendrup J, Lassen AH (2025) Prompt for learning factory paper characterisation. Accessed 06 Jun, 2025. [Online]. Available: https://www.ucviden.dk/da/publications/prompt-for-learning-factory-paper-characterisation
17. Alasoini T (1996) A learning factory: experimenting with adaptable production in Finnish engineering workshops. Int J Human Factors Manuf 6(1):3–19
18. Ssemakula ME, Aguwa CC, Ellis D, Kim K-Y, Liao G, Sawilowsky SS (2011) Manufacturing integrated learning lab (MILL): a curriculum model for hands-on manufacturing education, presented at the 2011 ASEE annual conference and exposition, pp 22–1039
19. Ahmad MH, Moshood TD, Nawanir G (2024) The influence of industrial revolution 4.0 in the implementation of the learning factory at the University of Malaysia Pahang. J Appl Res Higher Educ 16(1):89–103
20. Vázquez-Hurtado C et al (2024) Virtual twin for the smart factory as a tool to enable robotics skills acquisition, presented at the conference on learning factories. Springer, pp 113–120
21. Aarkrog V (2011) A taxonomy for teaching transfer skills in the Danish VET system. Nordic J Vocat Educ Training 1(1):1–13
22. Christiansen L, Nellemann C, Frendrup J, Zhang Y, Bøgevald Bjørnsten T, Heidemann Lassen A (2023) Didactical considerations for further education learning factories: a three case study. In: 13th conference on learning factories (CLF 2023). Reutlingen. Accessed 05 Jun 2023. [Online]. Available: https://papers.ssrn.com/abstract=4469259
23. Van Laar E, Van Deursen AJ, Van Dijk JA, De Haan J (2017) The relation between 21st-century skills and digital skills: a systematic literature review. Comput Hum Behav 72:577–588
24. Christiansen L et al (2022) A framework for developing educational industry 4.0 activities and study materials. Educ Sci 12(10):659
25. Petrusch N, Briese C, Lange A, Oertwig N, Kohl H (2023) Learning factory for circular economy and the corporate environment. Rochester, NY, p 4456416. https://doi.org/10.2139/ssrn.4456416

A Concept for Reconfigurable Gripping Systems for Flexible Disassembly Systems

Marc-André Weismüller, Oliver Petrovic, and Christian Brecher

Abstract The circular economy offers a potential solution to a multitude of contemporary challenges, including those pertaining to resource consumption, import dependency, and economic growth. This is accomplished by facilitating the circular reuse of products through the implementation of various R-strategies. Of these, remanufacturing is a particularly promising approach, as it aims to preserve as much of the value-creating activities that were previously invested in the product as possible. This is founded upon the disassembly of end-of-life (EoL) products, which may exhibit considerable variation in geometric dimensions, arrangement of parts, and materials. The high variance of EoL products, especially evident in WEEE (Waste Electrical and Electronic Equipment) products significantly complicates automation and imposes extremely high demands on the flexibility of gripper systems. This challenge currently hinders the economic operation of automated disassembly systems, as complex and highly adaptable gripper systems are required, which are often customized for single-product applications. Accordingly, this paper proposes a conceptual framework for a flexible and reconfigurable gripping system for robotic disassembly. This system is capable of automatically reconfiguring to process a variety of products. To develop this system, a functional analysis is conducted, in which various gripping systems are analyzed to identify requirements and functions of adaptable handling systems for the disassembly of WEEE components. This is an important step toward developing more agile, flexible, and cost-effective disassembly systems.

Keywords Disassembly · Gripping system · Reconfiguration

M.-A. Weismüller (✉) · O. Petrovic · C. Brecher
Laboratory for Machine Tools and Production Engineering (WZL), RWTH Aachen University, Aachen, Germany
e-mail: m.weismueller@wzl.rwth-aachen.de

O. Petrovic
e-mail: o.petrovic@wzl.rwth-aachen.de

C. Brecher
e-mail: c.brecher@wzl.rwth-aachen.de

M. Manns (ed.), *New Paradigms for Anticipated Uncertainty*, Lecture Notes in Mechanical Engineering, https://doi.org/10.1007/978-3-032-16889-4_27

1 Introduction

The concept of the circular economy (CE) has gained increasing attention as a potential solution to some of the most pressing environmental and economic challenges of our time. By extending the lifecycle of products and materials, CE aims to minimize waste, reduce the need for resource extraction, and lessen the environmental impact of industrial manufacturing. A central element of the CE is the implementation of various R-strategies (e.g., reuse, repair, refurbish, remanufacture, and recycle) that aim to keep products, components, and materials in use for as long as possible [1]. Among these, remanufacturing stands out as the strategy with the greatest potential for value retention, offering significant environmental and economic benefits while enhancing resilience by reducing dependency on imports [2, 3]. Remanufacturing involves disassembling end-of-life (EoL) products, restoring components to like-new condition, and reassembling them into functional products.

Despite its advantages, remanufacturing faces challenges like technical disassembly complexities and variability in quality, types, and amounts of returned products. This is especially a problem for the disassembly and makes it very complex and non-deterministic. Therefore the disassembly is often performed manually, which leads to high costs and makes it economically unattractive for many products. To overcome this economic obstacle, the throughput of disassembly systems must be increased. Given the unpredictability of the type and quantity of EoL products exiting the market, the system must be flexible and capable of processing diverse products. This can be achieved by automated platform-based disassembly systems [4]. The emergence of those systems has placed substantial demands on the handling systems, and fixtures as they must adapt to a vast range of product types and conditions. Therefore the flexibility of the disassembly system is dependent on the capabilities of the gripper, which plays an important role in enabling cost-effective disassembly operation [5]. This is especially relevant for the disassembly of Waste Electrical and Electronic Equipment (WEEE). Manual disassembly of WEEE is cost- and labor-intensive, and due to a shortage of skilled workers, it is not economically feasible on an industrial scale. An adaptive gripping system can significantly address these challenges by enabling handling of diverse devices [6].

In this paper we address this problem by developing a novel concept for a modular gripper with a high adaptability that can be used in the handling of a very wide product spectrum and therefore is able to decrease disassembly cycle time and investment cost for automated disassembly of WEEE. The gripper can automatically adapt to specific product and be reconfigured to be used in different disassembly applications of WEEE. The gripper intended application is the full automated handling of WEEE products and sub-components from collection groups 1 and 3 as defined by EU Directive 2012/19/EU. We achieve this by deriving key requirements from a requirements analysis grounded in current regulations, the dismantling process of the selected WEEE complemented by a functional analysis based on VDI 2221 and VDI 2206 regarding the state-of-the-art in reconfigurable and disassembly grippers.

The main contribution of this paper is a novel concept for a modular and reconfigurable adaptive gripper system that can cope with the specific requirements of WEEE dismantling of WEEE product groups 1 and 3.

2 State of the Art of Reconfigurable Gripping Systems

The majority of systems, presented in research, are constituted of numerous sub-components, including, but not limited to, two-jaw grippers, suction cups and clamping mechanisms and gripping fingers with multiple degree of freedom (DoF). Further they often include one or more actuators for moving and relocating these sub-components [7]. The DoF of these actuators is dependent on the respective application. Due to the high product variety, many grippers are based on actuated multi joint gripper fingers [8]. According to Yeung [9] this is a prerequisite and a reconfigurable gripper must feature multiple fingers with reconfigurable positions to effectively handle objects of varying shapes. Each finger should include movable joints to adapt to different part shapes and sizes [10]. The design of a reconfigurable gripper should prioritize simplicity, lightweight design, and position correction. Designs that are minimalist in nature, incorporating a reduced number of components and straightforward mechanisms, are conducive to cost reduction and enhanced reliability. Moreover, the incorporation of intrinsic position correction obviates the requirement for precise object positioning [9]. To achieve this, many systems use tendon-driven finger actuators [8, 11]. New research efforts are being invested in shape memory alloy (SMA) tendons to create high-performance, multi-joint, reconfigurable gripping actuators. This SMA-actuators are compact yet powerful electric actuator-sensor systems that can perform a wide variety of 3-dimensional motion sequences [12]. Besides tendon-driven actuators, SMA are embedded in flexible materials or used as a solid-state joint to generate flexible grasping motions [13]. Further simplification of reconfigurable gripping systems as well as weight and installation space reduction of the gripping systems is achieved by exploiting the sensory capabilities of the SMA materials inherent self-recognition functions and eliminating the need for sensors [14].

In addition to the named features, many reconfigurable gripping systems in research incorporate attachment or detachment mechanisms that facilitate the replacement of the entire end effector or sub-components of the gripper. This enables such grippers to change the relative orientation of the gripping fingers to one another, thereby enabling them to perform a variety of grasping types [15, 16]. In this instance as well, SMA actuators are employed for miniaturised, multidirectional coupling mechanisms [17]. These make it possible to exchange gripping elements of the gripping system and thus adapt it to the specific gripping and disassembly requirements of the workpieces.

In addition to scientific research, there are simple industrial reconfigurable gripping solutions available on the market. The TÜNKERS modular gripper system [18] is a notable example of this, utilising fasteners for the reconfiguration of the gripper. This necessitates manual intervention and produces reconfiguration downtimes.

Current research highlights a gap in automated reconfigurable gripping systems for WEEE, as existing solutions focus on specific models within product groups rather than versatile handling across different groups, as also stated by Foo et al. [19].

3 Requirement Analysis

The following requirements for a reconfigurable gripper system are derived from an examination of the requirements arising from the specified product range and the components to be disassembled. Subsequently, a functional analysis of state-of-the-art systems is conducted to identify additional requirements for the gripping system.

3.1 Analysis of WEEE Products

The EU Directive WEEE 2012/19/EU, the RoHS Directive 2011/65/EU, and the ElektroG3 are key regulations in the EU and Germany that guide the dismantling of WEEE. These directives specify which components must be removed from WEEE products, thereby establishing requirements for the gripper system.

WEEE products vary significantly in dimensions, shapes, and materials and the individual product differ in shape an appearance due to their EoL characteristics. An analysis of WEEE products from collection categories 1 and 3, as outlined in the EU Directive WEEE 2012/19/EU, reveals that most components are either prismatic or flat rectangular in shape, with overall dimensions ranging from 200 to 2000 mm. Additionally, these components often feature appendices and, being designed for manual assembly, lack interfacing surfaces or predispositions that would simplify their grasping.

Material-related challenges arise from the structural characteristics of these components. Some have compliant plastic walls or size irregularities due to the molding process, necessitating the use of varying assembly and disassembly forces to address potential interferences. Moreover, the variety of materials used in these components often requires employing different gripping technologies, as the center of mass may be offset from the geometrical center and material properties influence the choice of handling methods. Additional complexities stem from the presence of fragile objects or surfaces, movable parts of the product, and areas not intended for load application due to mechanical instability. Moreover, for white industry electro-mechanical components, many variants exist in size and functional features.

Table 1 Common functions of reconfigurable grippers from literature review

Nr	Function	Source
1	Reposition gripper sub-components	[8, 16, 20–25]
2	Exchange gripper sub-components	[15]
3	Intrinsic position correction to environment	[25–27]
4	Determine part presence in gripper	[23]

3.2 Functional Analysis of State-Of-The-Art Reconfigurable Grippers

Table 1 presents an excerpt of functions derived from a function analysis of grippers, based on VDI 2221, as seen in the current state of the art. The analysis highlights that a key function is the repositioning of gripper sub-components, allowing adaptation to products with diverse shapes and geometries by adjusting elements like gripping fingers, clamping brackets, or suction cups. Furthermore, the exchange of components is identified as a significant function, as it facilitates alternative gripping strategies and broadens the range of processable products.

3.3 Requirements for a Reconfigurable Gripping System

The analyses in Sects. 3.1 and 3.2 identify two essential capabilities required for a reconfigurable gripping system to dismantle the selected WEEE.

The first is the ability to add and remove sub-components, and the second is the capacity to change the position and orientation of those sub-components on the gripper. These abilities enable the adaption to the diverse product return conditions, encompassing alterations in shape, size, weight, appearance, tear and wear, as well as the presence or absence of components. The reconfiguration must be automated to be efficiently included in the ongoing disassembly process performed by the robot, as certain information about the condition of the product can only be obtained after the first dismantling steps. For example, knowledge about the components inside a housing can only be determined after it has been opened.

Additionally, it is of great importance that products are grasped firmly and the precise position inside the gripper is known to prevent unintended object movement and mispositioning within the gripping system and to facilitate the integration into subsequent automated processing procedures. The remaining function are listed in Table 2.

Table 2 Requirements for a reconfigurable gripper for WEEE disassembly

Nr	Requirement
1	Automated exchange of sub-components and gripping technologies
2	Automated relocating and orientating gripper sub-components
3	Relocation of gripper component in the range of 200 to 2000 mm
4	Modular design with stiff connections

4 Concept Description

To achieve the objective of flexible and automated exchange of gripper sub-components, as outlined in requirement 1, our concept incorporates an SMA-based coupling mechanism. Figure 1 presents a CAD concept of this mechanism. The SMA coupling mechanism is used to attach or detach certain sub-components to the gripper automatically. By utilizing SMA wires the mechanism can be design more compact, lightweight and yet be powerful enough to ensure a stiff connection of the sub-components. The mechanism has a diameter of 25 mm and a height of 20 mm, and consists out of two main parts. The first is an outer bushing which is driven by an SMA wire to release the mechanism for the coupling process. The second is the inner shaft which, together with balls, forms the female part of the coupling mechanism. A biasing spring is used to keep the system locked in an inactive state (see Fig. 1). The SMA wire is activated by electrical heating through Joule heating and is insulated from both thermal and electrical effects by a PTFE tube. When activated, the SMA wire contracts and compresses the biasing spring. This moves the outer bushing down and releases the locking mechanism by releasing the balls. To improve energy efficiency, the mechanism is passive-active, meaning that the SMA wire only needs to be heated during the coupling process.

To meet requirement 2 of Table 2, our concept differentiates between adaptation to product shape and geometry due to the diverse product mix, and adaptation to geometrical deviations arising from varying return conditions of the WEEE. To adapt to large product shape and geometry differences in size and shape, our concept comprises a modular system of mechanical building blocks. These mechanical building blocks are beams, blocks or plates of a similar structure like the Euro-Gripper-Tooling system [18] or other industry standard construction profiles.

This ensures a modular adoption of the gripper frame to enables large geometric adjustments. This can be done by manual brackets or by using the described SMA coupling mechanism. On this adjustable gripper frame, different gripper sub-components can be positioned and exchanged via the SMA coupling mechanism. This enables the large adjustments of the position of gripper sub-components and adaption of those to the respective WEEE.

In addition to its capacity for adaptation to product size through the repositioning of the building elements and SMA coupling mechanism, the gripper is equipped with one or more gripping fingers that exhibit multiple DoF, enabling it to adapt to geometrical deviations resulting from varying return conditions of the WEEE.

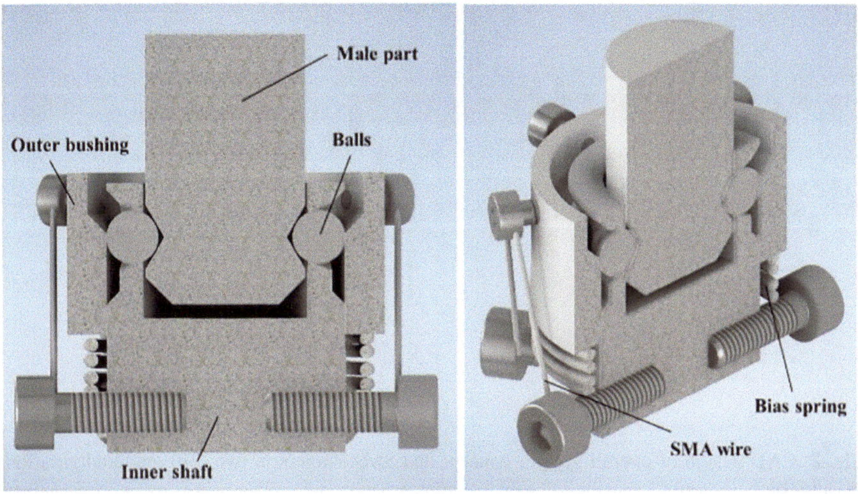

Fig. 1 CAD design of the proposed SMA coupling mechanism

This ability compensates any deformed, broken or missing gripping elements of the product. The concept of such a finger is shown in Fig. 2 and consist of multiple phalanges actuated by SMA wires. This kinematic structure enables the fingertip to move in all three Cartesian dimensions. This movement can be utilized to reorient gripping elements attached to the fingertip, such as suction cups or magnetic grippers, for holding objects within the gripper. Additionally, the bare fingertip itself can be used to clamp and secure the product within the gripper. This overall concept is shown in two configurations in Fig. 2. To power the SMA actuators in the overall systems, the individual actuators are connected to the robot's power supply via the robot flange.

5 Conclusion and Outlook

This work presents the first step towards a modular and flexible implementation of a reconfigurable gripper system. By combining automated reconfiguration capabilities with precise adjustment over position and orientation of gripper subcomponents, this system contributes to the advancement of platform-based disassembly and remanufacturing processes, laying the groundwork for more sustainable and resource-efficient manufacturing practices.

The concept consists of two different adjustment mechanisms, which together enable automatic adaptation to the specified product sizes. The first mechanism involves the utilisation of actuated multi-DoF gripping fingers, which are employed to compensate for geometrical deviations resulting from varying product return conditions. The second mechanism facilitates broader, discrete adjustment through the

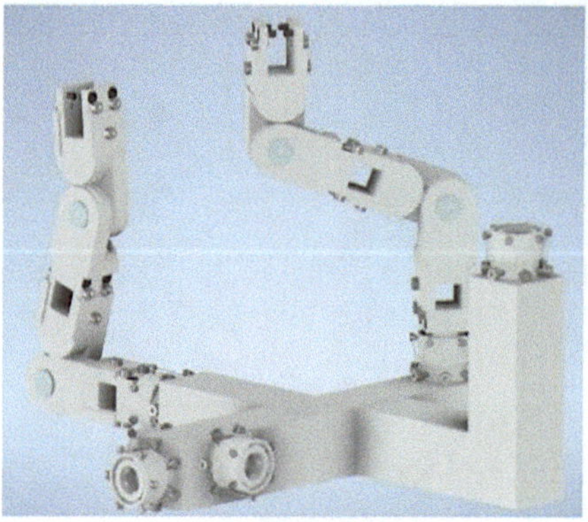
(a)

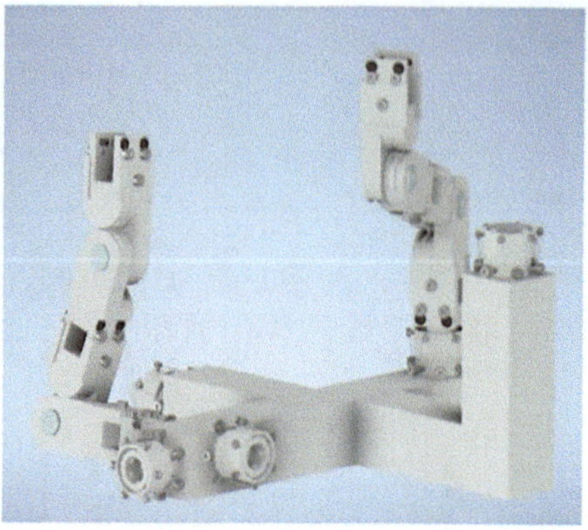
(b)

Fig. 2 CAD design of overall gripper concept and SMA fingers in two different configurations (**a**) and (**b**)

exchange of building blocks, thereby ensuring adaptability to diverse product sizes and shapes. Further sub-components can be exchanged, therefore enabling adoption of the gripping technology. This way, the system can be adapted to the demands of the disassembly of WEEE products and enables handling of a wide range of product types.

As a future research objective, the system concept needs to be evaluated. This will involve implementing prototypes of both the SMA coupling mechanism and the SMA finger actuator. These components will first be tested individually before being integrated into a complete system comprising coupling mechanisms, finger actuators, and mechanical building blocks. The fully assembled gripping system will then be validated on a laboratory scale within an automated disassembly robot cell for different WEEE products to assess its performance.

References

1. Potting J et al (2017) Circular economy: measuring innovation in the product chain. [Online]. Available: https://www.pbl.nl/uploads/default/downloads/pbl-2016-circular-economy-measuring-innovation-in-product-chains-2544.pdf
2. Butzer S, Schötz S (2016) Map of remanufacturing processes landscape. [Online]. Available: https://www.remanufacturing.eu/assets/pdfs/ERN_DeliverableReport_WP3_Processes_final_for_upload-1.pdf
3. Sharma V et al (2016) Identification of major drivers and roadblocks for remanufacturing in India. https://doi.org/10.1016/j.jclepro.2014.11.082
4. Poschmann H et al (2020) Disassembly 4.0: a review on using robotics in disassembly tasks as a way of automation. https://doi.org/10.1002/cite.201900107
5. Friedmann M, Fleischer J (2022) Automated configuration of modular gripper fingers. https://doi.org/10.1016/j.procir.2022.02.157

6. Brazier JP, Prasetyo J (2023) Robotic solution for the automation of e-waste recycling. https://doi.org/10.59097/jasae.v1i1.9
7. Schmitt J et al (2011) Disassembly automation for lithium-ion battery systems using a flexible gripper. [Online]. Available: https://doi.org/10.1109/ICAR.2011.6088599
8. Reddy CE et al (2021) Development and analysis of reconfigurable robotic end-effector for machining and part handling. https://doi.org/10.7166/32-1-2293
9. Yeung BHB (2002) Development of a 6 DOF reconfigurable gripper for flexible fixtureless assembly. https://doi.org/10.1109/ROBOT.2002.1013469
10. Han S et al (2022) Screwdriving gripper that mimics human two-handed assembly tasks. https://doi.org/10.3390/robotics11010018
11. Park S-J, Jeon H-J, Meng J-H, Song J-B (2024) Gripper with adaptability to environmental constraints. https://doi.org/10.1109/LRA.2024.3402627
12. Motzki P, Rizzello G (2023) Smart shape memory alloy actuator systems and applications. https://doi.org/10.5772/intechopen.1002632
13. Jani MJ et al (2014) A review of shape memory alloy research, applications and opportunities. https://doi.org/10.1016/j.matdes.2013.11.084
14. Motzki P, Seelecke S (2022) Industrial applications for shape memory alloys. https://doi.org/10.1016/B978-0-12-803581-8.11723-0
15. Nurpeissova A et al (2022) An open-source reconfigurable robotic gripper with detachable fingers. https://doi.org/10.1109/MESA55290.2022.10004392
16. Ahmed AK et al (2022) Design of a new modular reconfigurable gripper. https://doi.org/10.22541/au.166186989.90702457/v1
17. Saab W et al (2019) A review of coupling mechanism designs for modular reconfigurable robots. https://doi.org/10.1017/S0263574718001066
18. TÜNKERS Maschinenbau GmbH. Guideline for modular gripping system TGT series. [Online]. Available: https://www.cava.com.tr/upload/files/dosyalar/2020/tgt-gripper-manual-43914-6105300463.pdf. Accessed 17 Sept 2024
19. Foo G, Kara S, Pagnucco M (2022) Challenges of robotic disassembly in practice. https://doi.org/10.1016/j.procir.2022.02.085
20. Li G et al (2015) A reconfigurable three-finger robotic gripper. https://doi.org/10.1109/ICInfA.2015.7279534
21. Yumbla F et al (2025) An open-source 3D printed three-fingered robotic gripper for adaptable and effective grasping. https://doi.org/10.3390/biomimetics10010026
22. Nefzi M et al (2006) Development and design of a multi-fingered gripper for dexterous manipulation. https://doi.org/10.3182/20060912-3-DE-2911.00026
23. Lu Z et al (2022) GTac-gripper: a reconfigurable under-actuated four-fingered robotic gripper with tactile sensing. https://doi.org/10.1109/LRA.2022.3181370
24. Spiliotopoulos J et al (2018) A reconfigurable gripper for dexterous manipulation in flexible assembly. https://doi.org/10.3390/inventions3010004
25. Motzki P (2018) Advanced design and control concepts for actuators based on shape memory alloy wires. [Online]. Available: https://publikationen.sulb.uni-saarland.de/bitstream/20.500.11880/27177/1/Dissertation_pm__final.pdf
26. Cong Q, Fan W, Zhang D (2024) TacFR-gripper: a reconfigurable fin-ray-based gripper with tactile skin for in-hand manipulation. https://doi.org/10.3390/act13120521
27. Godaba H et al (2021) A two-fingered robot gripper with variable stiffness flexure hinges based on shape morphing. https://doi.org/10.1109/IROS45743.2020.9341554

Economic Justification of Reconfigurable Manufacturing Systems: Insights from the Swedish Automotive Industry

Gary Linnéusson, Filip Skärin, Itxaso Andoaga Ayestaran, and Carin Rösiö

Abstract Reconfigurable Manufacturing Systems (RMS) offer a compelling solution to anticipating uncertainty in volatile markets. Despite their potential value, their adoption is often constrained by industrial investment models that narrowly focus on short-term business cases, overlooking lifecycle considerations and the economic value of flexibility. This oversight can be traced to inadequate systematic risk management of flexibility dimensions and insufficient planning for extended equipment use in production system design. This paper explores how companies can better anticipate market uncertainty in their economic investment models, a prerequisite for valuing reconfigurability effectively. This study identifies critical barriers to adopting RMS through a literature review and interviews with representatives from four manufacturing companies. Findings reveal a gap in including and quantifying the economic and sustainability advantages, particularly regarding lifecycle and flexibility considerations caused by volume, product, and product mix uncertainties. Instead, current investment models include short-term cost metrics, leaving long-term adaptability undervalued.

Keywords Reconfigurable manufacturing systems · Economic justification · Anticipated uncertainty · Investment models · Sustainability

G. Linnéusson (✉)
School of Engineering Science, University of Skövde, Skövde, Sweden
e-mail: gary.linneusson@his.se

F. Skärin · C. Rösiö
School of Engineering, Jönköping University, Jönköping, Sweden

I. A. Ayestaran
Faculty of Engineering, Mondragon University, Mondragon, Spain

M. Manns (ed.), *New Paradigms for Anticipated Uncertainty*, Lecture Notes in Mechanical Engineering, https://doi.org/10.1007/978-3-032-16889-4_28

1 Introduction

Reconfigurable Manufacturing Systems (RMS) provides a profound solution to managing uncertainty in uncertain markets. RMS makes it possible to adapt to changes such as volume fluctuations and new product introductions [1, 2], driven by sustainability efforts and regulations. In addition, RMS provides the possibility of incrementally adapting the production system, thus enabling efficient coping mechanisms to better anticipate uncertainties [1, 3]. Despite the potential value of RMS, wide industry-spread adoption is constrained by current industrial investment models and processes that often overlook RMS's lifecycle benefits. Instead, these models and processes focus narrowly on current business cases and product lifecycles [4]. This oversight can be traced to inadequate systematic risk management of flexibility dimensions and insufficient planning for extended equipment use during the planning and design of production systems. Consequently, these constraints lead to difficulties in providing an equally valued RMS, compared to more flexible and dedicated solutions. Consequently, this results in the procurement of a production system incapable of efficiently managing uncertainties. Investments occurring in the industry are commonly based on the Net Present Value (NPV) method to economically evaluate an investment utilizing discounted cash flows in a single lifetime scenario [5]. A scenario grounded in certainty of knowing, e.g., the demand and new product introductions throughout the entire lifetime of the production system, which manufacturing companies in need of RMS seldom face [6]. Various financial models and equations, especially emphasizing the inclusion of the benefits of RMS exist in the literature. For instance, previous literature includes proposing a simplistic model for investment evaluation wherein RMS is justified by considering volume and product flexibility and associated costs [6], or a fuzzy multi-objective mixed integer optimization model used to calculate the economic benefits of RMS, emphasizing multiple product demand [7]. Although these examples provide a partial solution, insights are still lacking in this complex matter. A broader picture of the uncertainties must be coordinated and mapped to cope with the issue and surpass economic calculations and optimization modeling. This paper, thus, explores the complexity and key elements required to support adequate investments in RMS, laying the groundwork for positioning RMS as the core principle in production development to address uncertainty and achieve sustainability through adaptability.

2 Method

This research follows the approach of scoping reviews [8], which includes evidence synthesis based on pre-determined inclusion criteria, in our case, to identify and map key elements to justify RMS investments. In the literature review, we applied multiple search strings to scope in relevant literature, considering reconfigurability with a lifecycle perspective, addressing economic motivations. The review helped map key elements to justify RMS investments and to define the empirical studies. The Scopus citation database was used to conduct the literature searches [9]. Eight search strings were applied, including variants of the terms reconfigurability, changeability, modularity, life cycle, manufacturing, machining, financial, investment, cost, automation, indicator, and KPI. The review resulted in 385 hits of abstracts that were reviewed and narrowed down to 80 potentially relevant papers for a careful read throughout.

To identify the state of practice, a multiple-case study was carried out, including four manufacturing companies in the automotive industry. Relevant roles were identified and interviewed, including in total ten people, ranging from high management level responsible for strategic development and machine investments, responsible for the conceptual phase in the production development process, roles working with business control and with the economic follow-ups at the financial department, and preparation engineers. In total, six interviews were recorded, including both onsite and digital meetings. All but one interview was conducted by two of the researchers, except for one, where only one of the researchers participated. The applied interview guide included open-ended questions with a generic part, with questions to map the companies' self-perception of working with RMS, if they evaluate uncertainty, future change needs, reuse of machines, how new investments are evaluated, complementary engineering tools, the process for machine selection, etc. For each company, there was also a specific part in the interview guide based on a preceding document study of their investment models. Regarding investment models: in Case A they were in the process of redefining their model and we had only access to the obsolete one; in Case B we got access to their model and a exemplary case; in Case C we accessed a cleaned model; in Case D they did not want to share their model so we carried out a presential observation.

3 Conceptual Framework for RMS Investments

Attempts to systematically address the economic justification of RMS are scarce in the literature [10]. Hence, we in this paper collect various contributions from literature to allow for a more complete and sustainable economic evaluation of RMS, synthesized into a conceptual framework depicted in Fig. 1. At the center is the investment model, considered as a hub for collecting information impacting the design of the production system. The included elements are seen as areas adding value to a systematic analysis to appropriately value RMS designs, these are: Anticipated uncertainty; Changeability measurement; Economic factors; and Performance factors.

Anticipated uncertainty: To compile a proper risk assessment and decision basis in the investment model, it is crucial to consider the required flexibility planning [11]. It is also vital to consider the entire lifecycle of the investment, as the purpose of RMS is both to prolong its useful phase and ease upgrades to new product features [12]. This includes selecting an apt expected lifecycle and assessing the anticipated

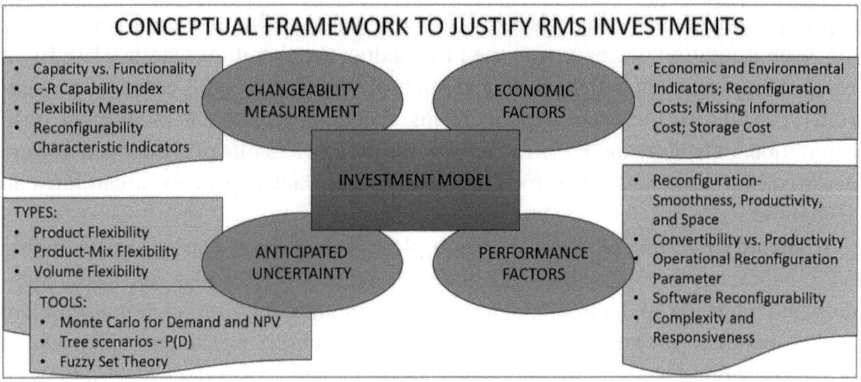

Fig. 1 A conceptual framework for a systematic approach to investment decisions

change drivers: how many new product introductions (*product-mix flexibility*); the need to handle several product families (*product flexibility*); and estimating demand uncertainty (*volume flexibility*). Various tools have been proposed to consider uncertainty in RMS literature. Mainly, *Monte Carlo* simulation has been used to consider demand uncertainty [13–15] and to explore various design scenarios assessed through formulas for NPV and a customized Changeability–Reconfigurability (C-R) Capability Index [10]. Uncertainties have also been explored by applying *Decision Tree* models, such as product demand, customer and product changes, new materials, and legislations, see, e.g., [16, 17]. *Fuzzy Set Theory* enables implementing ranges of uncertainty and has been applied to investigate various interest rates, NPV cash flows [18], and market uncertainties [7].

Changeability measurement: An appropriate level of changeability is beneficial to mitigate uncertainties [13], yet a lack of consensus among researchers on universal reconfigurability indicators causes application challenges for companies [19]. Being able to vary capacity and functionality is fundamental to RMS [15]. Costs associated with expanding *Capacity vs. Functionality*, or both, need to be considered as investments and the operational consequences of various costs, such as potential integration downtime, output reductions, and investments in additional capacity [13]. A *C-R Capability Index,* including time for integration, resource availability, servicing time, and more, see [10], to economically consider changeability throughout a production system lifecycle has been proposed. Considering *Flexibility Measurement* is relevant to minimize flexibility costs during a production system lifecycle [20]. Considering only *volume flexibility*, a method to assess flexibility lists configuration cost factors, such as area cost, energy, labor, etc. [16]. In another proposal, flexibility measurements considering only product flexibility were introduced to reduce costs and enable quick responses to changes [21]. Lastly, scholars argue there is a need to relate changeability measurements to the *Reconfigurability Characteristic Indicators*, see [19, 22], i.e., convertibility, scalability, customization, modularity, integrability, and diagnosability [23].

Economic factors: Diverse proposals on including economic factors exist. These include: *Economic and environmental indicators*, considering configuration, maintenance, and energy costs [24]; *Reconfiguration costs,* for installing new modules or moving equipment [17], adding or removing machines or modules together with integration and testing costs [25], and 'switching costs' [16], including investments to expand functionality and capacity as reconfiguration costs; and *Missing Information Cost* [20], calculating costs for lacking reconfiguration ability to highlight an appropriate analysis in the conceptual phase. Complementary to these costs is the *Storage Cost* for machines or modules stored away while not part of any RMS [17].

Performance factors: To evaluate the reconfigurability of an RMS design in the planning phase various proposals on indicators have been proposed. Assessing factors such as *Reconfiguration Smoothness* (how easy it is to add machines or modules), *Reconfiguration Productivity* (defining demand fulfillment and capacity scalability), and *Space efficiency* have been proposed [26]. Discrete Event Simulation has been argued in support of distinguishing between performance indicators and indicators for system changeability, helpful to identify an *Operational Reconfiguration Parameter* termed 'Resource use rate' [19], through operational rules for adaptations to changed volume demands. Also fuzzy multi-criteria decision model has been proposed [18], as most literature ignores reconfiguration time, where they underline the importance of considering set-up time, production time and switching time to accurately evaluate RMS. In [10], an investment model is proposed, including the performance measure of *Software Reconfigurability* to consider programming changes in RMS. To make an appropriate tradeoff between *Complexity and Responsiveness*, multiple objectives, i.e., maximize NPV and RMS responsiveness, and minimize structural complexity of RMS, have also been explored [7].

4 Empirical Results

This chapter presents the empirical findings from the four case studies. The aim is to provide a descriptive overview of how RMS-related investment considerations are currently handled in practice. Each case outlines key features of the company's investment logic, including how flexibility, reuse, lifecycle aspects, and cost factors are considered. The findings are presented without direct theoretical interpretation, which follows in the subsequent chapter where the cases are discussed in relation to the conceptual framework developed from the literature.

4.1 Case Study A

In this company, the production systems are characterized by RMS both at a system and a machine level. Reconfiguration implies, e.g., splitting up the production lines to reuse the machinery for new products, refurbishing machines to extend the length of life, and changing fixtures and tooling in the machine to fit new products.

During the study, this case company was developing a new investment model that aimed for higher accuracy in the calculations to optimize the time spent on calculations. The new model will be implemented in commercial software connected to the ERP system for automatic collection of data. In their current investment model, they consider the product price, the cost of investment in the machinery, and the number of operators, while in the new model, factors like power consumption of the machines, detailed tool consumption, and maintenance cost will be considered.

Currently, they do not carefully analyze change drivers (*e.i, product-mix flexibility, product flexibility, volume flexibility*), as they strive to have machines that are as flexible as possible. In investment projects, there is no formalized process to follow, they rely on current engineering skills. Before making new investments, they strive to reuse existing equipment, based on parameters such as machine condition, lifetime of the machine, and cost to repair if a refurbishment is needed. In investments, they prioritize machine systems with lower complexity and decoupled design to enable reconfiguration and reuse, despite knowing that this flexibility might imply an initial extra cost.

4.2 Case Study B

This company is mostly equipped with very specialized machinery and, thus, a low level of reconfigurability. In investments, they primarily try to reuse the machines they already have prior to the calculations in the investment model. If investments are required, decisions are based on the specific next product to be produced, not having a long-term perspective, which they identify as a systematic problem.

When evaluating the profitability of an investment, the pay-off time and the interest rate are considered. The process of selecting a machine starts by analyzing the market and available technologies, then establishing relations with potential suppliers, and ends by calculating costs and capacities. Future volumes and potential product variations are considered in this process; however, with a limited range. The calculations focused only on the upcoming product, considering factors like energy consumption, waste, material, scrap, operator, and maintenance. Set-ups are not considered, meaning that the machines were selected to produce only one product of the current product portfolio. In their investment model, they calculate the residual value of the lifespan of the machine in case its lifetime exceeds the project timeframe, but they don't include this value in any part of the model to highlight the economic benefit this may provide (*e.i, comparing two solutions with varying lifecycle*). NPV was used to analyze the investment costs.

4.3 Case C

The production systems in this company include RMS solutions combined with dedicated equipment (DMS). In the past 5–10 years, a tendency to opt for reconfigurable solutions was prominent, to achieve product variations and volume variations. Even if they initially identify the need for change when new products are introduced and consider the product's critical dimensions and volumes to map these characteristics to the production machine requirements, the process they follow favors DMS. The company plans to buy 50% fewer machines in the following 5 years, therefore, an initial consideration is whether they should invest in a new machine or refurbish an existing one in their facilities. To evaluate the profitability of an investment in a new machine/system, they follow a standardized process where direct and indirect costs are considered simultaneously. They include two parts: a profitability calculation and, in a later stage of a project, a Life Cycle Cost (LCC) calculation, where the data from the Profitability calculation is inserted into the LCC calculation file. The LCC is carried out cross-functionally, while the Profitability analysis is conducted by the business controller. In the LCC calculation file, both reconfiguration costs and set-up costs are ignored. The technique used to analyze the investment costs was NPV.

Future demands and uncertainties are analyzed and predicted twice a year in a cross-functional team related to the development of the products, planned and extant capacities, and investment plans. They prefer to invest in long-term solutions that could encompass future change needs and have tools to calculate the residual value of the investment, and if it surpasses the lifespan required for the project, but this is not considered in the formal calculations and analysis that bases the decisions.

4.4 Case D

The production systems are currently not characterized by RMS. To evaluate the need for change, they follow a structural procedure, carried out by a cross-functional team, comprising activities of understanding the product changes and similarities, mapping these characteristics to the machines, and analyzing whether the current production needs any changes. Investments in machines are made to produce a single product, and the ability to handle long-term changes is not explicitly considered in investment calculations, but is regarded as adding value to an option. Also, reusing existing machines is prioritized. In comparing options, the technical requirements are compared with available existing options, considering retooling costs, new fixture costs, and the residual value of the machine (the depreciation value). Due to this mindset, their lines are a mix of new and reused machines.

An LCC model is, when considered necessary, used to evaluate the profitability of an investment, considering costs of investment, installation, energy consumption, downtime, environmental costs, and maintenance costs. The lifespan of the investment options is not quantified in the analysis, but they claim to be able to compare machines concerning different lifetimes. They argue that trying to prolong the lifetime of the equipment is something inherent in the mindset of the experienced engineers who compose the cross-functional team that discusses investment decisions.

The process of selecting a machine includes discussing delivery time, investment, and strategic alliances with potential suppliers. They tend to opt for the cheapest option after applying this strategy. To predict future demand, they rely on the customers' predictions and usually adapt to the capacity request each project calls for. Changes in product variations are considered at a strategic level, but it is not integrated with the investment model. Volume variation, on the other hand, is to some extent considered. Currently, they use simulation to evaluate set-up times to decide whether they should invest in dedicated machines or flexibility. They make a difference between generic and specific costs, which signals a certain level of awareness of reconfiguration costs. The technique used to analyze investment costs was the Internal Rate of Return.

5 Discussion

This section discusses how the empirical findings relate to prior literature. Companies acknowledge the importance of flexibility, lifecycle thinking, reuse, and cost awareness, yet these aspects are typically addressed in fragmented, experience-based ways. A structured investment logic that integrates these dimensions is largely missing, mainly due to underdeveloped organizational capabilities to implement a systematic RMS investment approach. A proposed conceptual framework emphasizes building four key organizational capabilities: anticipating uncertainty, assessing flexibility, defining economic value, and using performance indicators to support RMS. The case studies show that while companies consider reuse and lifecycle characteristics, they lack formal tools to quantify these benefits or compare alternatives with different expected lifespans. Strategic intent may exist, but supporting structures and routines are absent, and the investment models remain conventional and short-term oriented.

A gap within the literature is discovered, as many proposed models are fragmented, scenario-specific (e.g., product or volume flexibility), or lack lifecycle integration. Few offer comprehensive methods to assess RMS value across investment horizons. This dual gap—between theory and practice, and within theory—limits practical applicability and theoretical development. Addressing this requires hybrid investment models combining financial evaluation (e.g., NPV, LCC) with engineering-based assessments of changeability, uncertainty, and sustainability. These models must remain practical for industry, with routines that embed them into decision-making, involving cross-functional collaboration and shifting the mindset from evaluating machines to evaluating adaptability. Hence, for practice, RMS investment readiness depends not only on technical feasibility but also on building the above-mentioned organizational abilities, where scenario-based tools can support more balanced trade-offs. Moreover, for research, the focus should shift toward making theoretical models actionable and embedding them in real workflows. Future studies should explore how companies internalize sustainable investment learning and create conditions for sustained inclusion of changeability. Finally, this study reframes RMS adoption as a question of organizational capability, where investment models are central to how firms evaluate, implement, and sustain adaptability under uncertainty.

6 Conclusions

This study explored how companies can better anticipate uncertainty in their investment models to enable the adoption of RMS. A conceptual framework was developed, identifying four key capabilities: anticipating uncertainty, measuring relevant types of flexibility, defining long-term economic value, and applying performance indicators.

Empirical findings show that although companies value flexibility, reuse, and lifecycle thinking, these are rarely integrated into formal investment processes. Instead, short-term economic metrics dominate, supported by limited tools or routines to assess long-term adaptability. This confirms the presence of critical barriers—both organizational and conceptual. A key capability gap limits the ability to evaluate RMS systematically, especially due to the lack of methods to quantify flexibility and sustainability benefits over time. This may sustain the reliance on short-term economic metrics and the risk of favouring dedicated solutions. A second gap is identified in the literature: existing RMS investment models are fragmented or too abstract, and rarely account for lifecycle trade-offs. This dual gap restricts both practical and theoretical progress.

The findings underscore that investment models are not merely tools for economic calculation, but core mechanisms through which organizational capability is enacted. Addressing the capability gap requires hybrid models that combine financial evaluation with engineering-based assessments, alongside routines that embed these approaches into cross-functional practice. This research contributes by offering a framework to guide capability-building around RMS investments and by positioning investment logic as a central enabler of sustainable, reconfigurable production strategies under uncertainty.

References

1. Koren Y, Shpitalni M (2010) Design of reconfigurable manufacturing systems. J Manuf Syst 29:130–141. https://doi.org/10.1016/j.jmsy.2011.01.001
2. Koren Y, Gu X, Badurdeen F, Jawahir IS (2018) Sustainable living factories for next generation manufacturing. Procedia Manuf 21:26–36. https://doi.org/10.1016/j.promfg.2018.02.091
3. Koren Y, Heisel U, Jovane F, Moriwaki T, Pritschow G, Ulsoy G, Van Brussel H (1999) Reconfigurable manufacturing systems. CIRP Ann Manuf Technol 48:527–540

4. Brunoe TD, Napoleone A, Andersen AL, Nielsen K (2021) Impact of different financial evaluation parameters for reconfigurable manufacturing system investments. IFIP Adv Inf Commun Technol 631 IFIP:479–487. https://doi.org/10.1007/978-3-030-85902-2_51
5. Siziba S, Hall JH (2021) The evolution of the application of capital budgeting techniques in enterprises. Glob Finance J 47. https://doi.org/10.1016/j.gfj.2019.100504
6. Olsson F, Werthén A, Andersen A-L (2021) A real options approach for NPV investment evaluation of changeable manufacturing. In: Andersen A-L et al (eds) CARV and WMCP 2021. Springer, pp 130–137
7. Kuzgunkaya O, ElMaraghy HA (2008) Economic and strategic perspectives on investing in RMS and FMS. Int J Flex Manuf Syst 19:217–246. https://doi.org/10.1007/s10696-008-9038-8
8. Peters MDJ, Marnie C, Colquhoun H, Garritty CM, Hempel S, Horsley T et al (2021) Scoping reviews: reinforcing and advancing the methodology and application. Syst Rev 10
9. Mongeon P, Paul-Hus A (2016) The journal coverage of Web of Science and Scopus: a comparative analysis. Scientometrics 106:213–228. https://doi.org/10.1007/s11192-015-1765-5
10. Brad S, Murar M, Brad E (2018) Design of smart connected manufacturing resources to enable changeability, reconfigurability and total-cost-of-ownership models in the factory-of-the-future. Int J Prod Res 56:2269–2291. https://doi.org/10.1080/00207543.2017.1400705
11. Du J, Jiao Y, Jiao JA (2006) Real-option approach to flexibility planning in reconfigurable manufacturing systems. Int J Adv Manuf Technol 28:1202–1210. https://doi.org/10.1007/s00170-004-2418-7
12. Boldt S, Rösiö C, Linnéusson G (2023) Mapping production capabilities: proposing support towards changeable production. In: Galizia FG, Bortolini M (eds) Production processes and product evolution in the age of disruption. CARV 2023. Lecture notes in mechanical engineering. Springer, Cham, pp 330–337. https://doi.org/10.1007/978-3-031-34821-1_36
13. Andersen AL, Brunoe TD, Nielsen K, Bejlegaard M (2018) Evaluating the investment feasibility and industrial implementation of changeable and reconfigurable manufacturing concepts. J Manuf Technol Manag 29:449–477. https://doi.org/10.1108/JMTM-03-2017-0039
14. Skärin F, Abdelmageed ME, Linnéusson G, Rösiö C (2022) Supporting manufacturing investment decisions in new product introductions through line balancing techniques. In Proceedings of the advances in transdisciplinary engineering. IOS Press BV, vol 21, pp 89–100
15. Andersen AL, Brunoe TD, Bockholt MT, Napoleone A, Hemdrup Kristensen J, Colli M, Vejrum Wæhrens B, Nielsen K (2023) Changeable closed-loop manufacturing systems: challenges in product take-back and evaluation of reconfigurable solutions. Int J Prod Res 61:839–858. https://doi.org/10.1080/00207543.2021.2017504
16. Alexopoulos K, Mourtzis D, Papakostas N, Chryssolouris G (2007) DESYMA: assessing flexibility for the lifecycle of manufacturing systems. Int J Prod Res 45:1683–1694. https://doi.org/10.1080/00207540600733501
17. Manzini M, Urgo M (2019) An approach for the robust design of a reconfigurable assembly cell with 7-axis robot. Int J Comput Integr Manuf 32:777–797. https://doi.org/10.1080/0951192X.2019.1636408
18. Abdi MR (2009) Fuzzy multi-criteria decision model for evaluating reconfigurable machines. Int J Prod Econ 117:1–15. https://doi.org/10.1016/j.ijpe.2008.06.009
19. Beauville dit Eynaud A, Klement N, Roucoules L, Gibaru O, Durville L (2022) Framework for the design and evaluation of a reconfigurable production system based on movable robot integration. Int J Adv Manuf Technol 118:2373–2389. https://doi.org/10.1007/s00170-021-08030-1
20. Tolio T, Urgo M (2013) Design of flexible transfer lines: a case-based reconfiguration cost assessment. J Manuf Syst 32:325–334. https://doi.org/10.1016/j.jmsy.2012.12.003
21. Anthony C, Zakariae M, Armand B (2023) Measuring and identifying appropriate levels of flexibility: a novel approach and use case in the automotive industry. In: 3rd international conference on innovative research in applied science, engineering and technology, IRASET 2023

22. Rösiö C, Aslam T, Srikanth KB, Shetty S (2019) Towards an assessment criterion of reconfigurable manufacturing systems within the automotive industry. Procedia Manuf 28:76–82
23. Dolgui A, Hashemi-Petroodi SE, Kovalev S, Kovalyov MY (2021) Profitability of a multi-model manufacturing line versus multiple dedicated lines. Int J Prod Econ 236. https://doi.org/10.1016/j.ijpe.2021.108113
24. Singh PP, Madan J, Singh H (2021) Economically sustainable configuration selection in reconfigurable manufacturing system. In: Singh H, Singh Cheema PP, Garg P (eds) Sustainable development through engineering innovations 2021. Springer, Singapore, pp 457–466
25. Abbas M, Elmaraghy H (2018) Functional synthesis of manufacturing systems using co-platforming to minimize cost of machines and system changes. J Mech Des Trans ASME 140. https://doi.org/10.1115/1.4038069
26. Yang J, Son YH, Lee D, Do Noh S (2022) Digital twin-based integrated assessment of flexible and reconfigurable automotive part production lines. Machines 10. https://doi.org/10.3390/machines10020075

Design of a Flexible Tool System for the Scalable Production of Secondary Forming Elements

Michael Schiller, Peter Frohn-Sörensen, Martin Freitag, Martin Manns, and Bernd Engel

Abstract The starting point of this study is the increasing number of variants with simultaneously decreasing batch sizes of body sheet metal components. The suppliers' current production and plant technologies are increasingly reaching their economic limits. The SkaLaB joint project aims to provide economically viable solutions through scalable and flexible production processes and strategies. Car body sheet metal components have small-scale, locally distinctive yet characteristic geometries that realize the functional properties of the component. In production, secondary forming elements represent critical areas whose manufacture poses particular challenges and limits the flexibility and universality of the tools. A flexible and scalable tool system for the universal forming of sheet metal components is being developed for the separate production of secondary forming elements, independently of the main forming process. Within the scope of this work, individual groups of secondary forming elements are classified, and an order of these elements is created. In addition, a tool set for two particularly relevant secondary forming elements is designed and tested experimentally. Furthermore, a spring-loaded clamping system for rubber pad forming is developed, which enables a localized application of this process within a larger sheet metal blank.

Keywords Secondary forming elements · Localized rubber pad forming · Restrainer system

M. Schiller (✉) · M. Manns
Chair of Production Automation and Assembly, University of Siegen, Siegen, Germany
e-mail: Michael.Schiller@uni-siegen.de

P. Frohn-Sörensen · M. Freitag · B. Engel
Chair of Forming Technology, University of Siegen, Siegen, Germany

M. Manns (ed.), *New Paradigms for Anticipated Uncertainty*, Lecture Notes in Mechanical Engineering, https://doi.org/10.1007/978-3-032-16889-4_29

1 Introduction

New car registrations in the automotive industry have been stagnating since 1990 [1], while the variety of models and car bodies is increasing sharply—a trend that is being further reinforced by electromobility [2–4]. This leads to decreasing batch sizes, which makes the production of sheet metal components economically challenging, as these have traditionally predominantly been manufactured using conventional processes such as progressive forming. These processes are tool-bound, require high investments, large systems and long set-up times [5]. Small and medium-sized companies in particular are coming under pressure, because many component derivatives can no longer be produced in a cost-covering way.

This paper aims to develop and validate a tool system for the flexible and scalable production of secondary forming elements (SFE) independently of the main forming process.

The paper is structured as follows: After outlining the challenges of current production technologies in the automotive sector, secondary forming elements are classified, and benchmark geometries are identified. The applied rubber pad forming process and the tool design are then described in detail, followed by experimental validation. The findings are discussed with reference to future optimization strategies.

A survey conducted by the Chair of Forming Technology at the University of Siegen revealed that suppliers are already frequently falling short of the optimum quantities of over 100,000 units per year—call-offs are approaching critical values of around 50,000 units per year [6]. To meet these challenges, the joint project "Scalable Center for the Production of Sheet Metal Body Components" (SkaLaB) was launched. The aim is to develop flexible production centres that enable cost-effective small series production. A central strategy of these centres is to break down complex components into successively manufacturable sub-geometries. One group of these sub-geometries are the so-called secondary forming elements (SFE), such as beads or collars. SFEs are technically demanding forming zones that sometimes require additional production stages [7, 8] which makes tools and processes that include SFE complex and expensive. They also specify the tool to one or a few products that can be produced.

In conventional progressive stamping, secondary forming elements are components of the main tool. If only the position or size of one of these elements changes between two body component variants, a new or modified tool is required, which is also associated with set-up times. These inflexibilities and restrictions in production are to be avoided in SkaLaB by producing the SFE independently of the main forming and outside the main tool. This takes place in the secondary forming station within the SkaLaB production cell. A tool system suitable for the secondary forming station is developed and designed as part of this work.

2 Methodology

In order to define the requirements for the mold system of the secondary forming station, it is first necessary to appropriately define, differentiate and classify secondary forming elements in the SkaLaB project. This is followed by a brief overview of the rubber pad forming process used, as well as its possibilities and limitations. The sheet metal used for the tests is characterized and the systematic structure of the tool system is explained.

2.1 Order of Secondary Forming Elements

In SkaLaB, ten car body sheet metal components were selected as a reference for the requirements from the current product repertoire of the alliance partners. Additionally, two desired components were designed. These contain production steps such as holes, beads, nubs or collars, which are commonly referred to as secondary forming elements in forming technology. There is currently no generally valid and universally usable classification and definition for SFE. Over the last 60 years, there have been various approaches and attempts, but no uniform standard has emerged [9, 10]. Rather, limited and manageable classification systems have been created for specific purposes and applications [7, 8]. Despite the diversity of these systems, a common thread can be recognized in the definitions: main forming elements (MFE) are shape-dominating geometric elements of the workpiece shape. In contrast, there are the local and limited SFEs, which can be regarded as the main carriers of the workpiece functions [7]. As a consensus in the workpiece consideration, holes are counted among the secondary forming elements. As there is no standardised definition of SFE, an element order was created for the sheet metal components within the project. This can be seen in Fig. 1.

To determine the requirements for the tool system, all reference and desired components were examined regarding the occurrence and characteristics of SFE. Particularly characteristic is a non-perpendicular truncated cone nub arranged in the curved surface, with largest volume and a 32 mm long semicircular bead with a small volume and sharp radii. These elements mark the range of geometric requirements. They were intended to serve as benchmark elements.

2.2 Rubber Pad Forming

Rubber pad forming (RPF) is a forming process in which an elastic rubber pad is used as the flexible tool half. It enables the production of components with high surface quality at low tool costs. Further advantages are the simple tool design and handling. On the other hand, there is a high force requirement and increased wear of the rubber

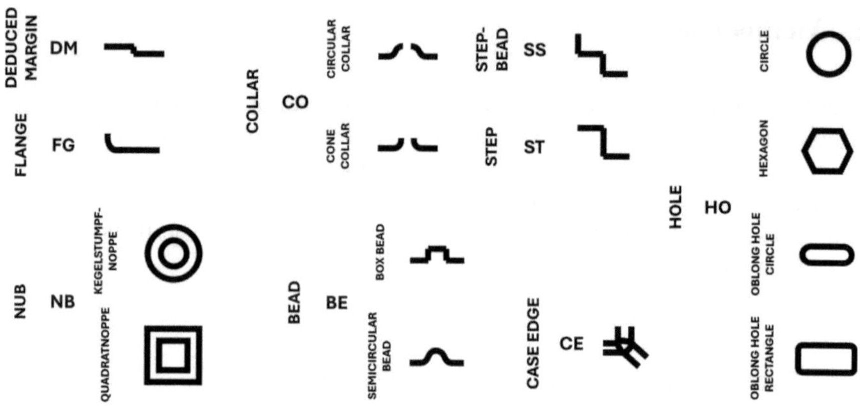

Fig. 1 Order of secondary forming elements

pad [11, 12]. For this reason, there exist special RPF presses, which can deliver high forming forces. RPF is particularly used in the automotive and aerospace industries.

One important process is the Guerin process, developed in California at the end of the 1930s. During this process, the sheet metal is formed between a rigid base with moulding blocks and a rubber cushion in a fixed box. The process can be considered male (sheet metal is moulded over a solid) or female (sheet metal is moulded into a solid). Rubber pads consist of either one piece or several layers, some with different degrees of hardness [13].

The Marform process supplements the Guerin process with a blank holder to prevent wrinkling and enable higher drawing ratios. The Maslennikov process even allows punch-less deep drawing with high drawing ratios [13]. Studies show that the use of a rubber cushion without a restrainer is also possible under certain conditions [14].

When penetrated by another body, the elastomer exhibits fluid-like behaviour, it builds up hydrostatic pressure on all sides. This pressure can get very high and can be used for forming or calibrating sheet metal parts, provided that the rubber cushion cannot move out of its way. To withstand these forces, a housing is required around the rubber pad. This housing is called a restrainer. The necessity for this to exceed the blank size in prior RPF process variants represents a limitation that can be overcome.

2.3 Sheet Metal Material

Two grades of low-alloyed deep-drawing steel were used for the initial design of the process. The first is a 1 mm thick DC04, the second a 2 mm thick DC01. Mechanical properties are shown in Table 1.

Table 1 Mechanical properties of the materials used

	Unit	DC04 (1.0338) [15]	DC01 (1.0330) [16]
Blank thickness	mm	1	2
Modulus of elasticity	GPa	210	194 (*)
Yield strength	MPa	210	170–220 (*)
Tensile strength	MPa	270–350	375–385 (*)
Min. elongation at fracture	%	38	28

Values marked with (*) were determined by own tensile tests

2.4 Tool System Development

Overall concept. As mentioned above, the cone nub and the semicircular bead of a SkaLaB component emerged as benchmark elements. To form these elements in a flat blank, the corresponding surface contours were milled as negatives in a mould made of tool steel. The moulds can be seen in Figs. 4 and 5. These moulds are screwed onto a reconfiguration-enabling threaded plate using four-cylinder head screws. It contains two cylindrical pins for precise positioning of the moulds. The matrices in turn are equipped with a fitting hole and fitting slot on their underside. This makes positioning fast and easily repeatable. The idea follows that of the zero-point clamping system.

The restrainer assembly is aligned and secured against the upper adapter plate using the same principle. Before starting work on the hydraulic press, only the adapter plates need to be aligned once. The moulds can then be easily exchanged without any alignment. The moulds can therefore also be positioned fully automated. To ensure that the air can escape from the cavities completely, vent holes are drilled near the lowest point of the cavity. Due to the flexibility of the rubber pad and the fact that there are no interlocking machine components, it is not necessary to guide the moulds towards each other during the stroke. The sheet metal blank is positioned in the correct position relative to the mould by an additively manufactured sheet metal guide. The sheet metal guide can be used for several SFEs, if the blank shapes are the same, see Fig. 3. The guide can be produced quickly and resource-efficiently in-house. The lightweight sheet metal guide can be (re)positioned by robots, regardless of the mould used. The design of the tool system was based on simulation and analytical calculation.

From a design engineering perspective, separating the production of SFE from the main forming introduces specific challenges regarding tolerance control. While the isolated forming allows for geometric freedom and late-stage customization, it also requires additional alignment mechanisms to ensure position and orientation tolerances remain within $\pm$ 0.2 mm. These tolerances are currently achievable through precise locating pins and robot-guided blank positioning. However, cumulative positioning deviations remain a critical factor that needs to be addressed through improved referencing strategies.

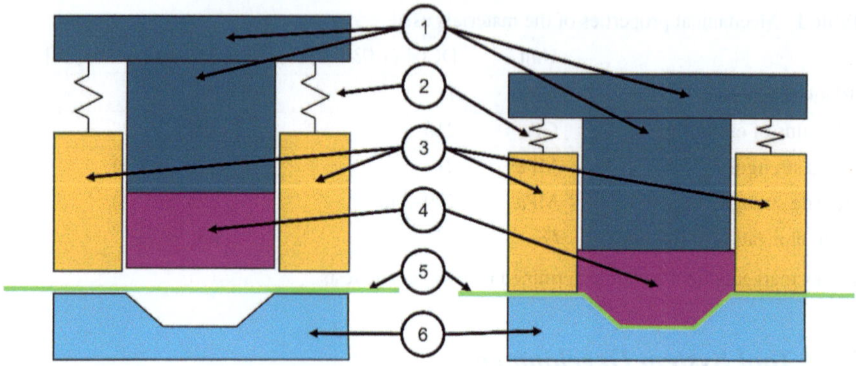

Fig. 2 Restraining system for local RPF. (1) Restrainer bottom, (2) Compression springs, (3) Restrainer side, (4) rubber pad, (5) blank, (6) forming die

Restraining system for local RPF. As the rubber pad must be larger than the sheet metal in conventional processes, this represents a significant limitation for the application within the SkaLaB cell. To overcome this limitation, a new concept that allows the RPF to used locally in the sheet metal was developed. The base of this solution are movable restrainers. The new concept is shown in Fig. 2. Simultaneously to the conventional RPF, the sheet metal is pressed into a mould cavity of the die by an elastic cushion and the forming pressure is generated via a punch-like base part. However, now the cushion is held laterally by movable retainers, which move with the stroke of the base part and thereby fulfil a defined hold-down function. The restrainer itself is placed above the component and holds the cushion in position through friction. The restrainer returns to its original position when releasing the forming pressure, due to its spring suspension.

The central element of the assembly is the restrainer-bottom, which transmits the pressing force into the forming zone and guides the blank holder device. The assembly has an axially symmetrical design. Eight helical compression springs with external diameters of Ø 12.5 mm and Ø 20 mm are fitted to generate the blank holder force. These springs are available in four different stiffness classes. Depending on the configuration, four or eight springs can be installed, enabling a total of 24 different hold-down characteristics. Forces from 840 to 7833 N can be achieved at maximum spring travel.

Rubber pad. Polyurethane (PU) is chosen as the material for the rubber pad as it shows a high chemical and mechanical resistance [17] and offers a wide range of adjustable hardness levels [18]. In this study a hardness variation between 20 and 70 Shore A were tested, in order to investigate their influence. Studies show that softer pads enable better mould filling, but are more susceptible to wear and wrinkling [19]. A circular mould with a diameter of 68 mm was chosen for the rubber pad. This geometry covers both the nub and the semicircular bead, which means that only one restrainer assembly is required.

3 Results and Discussion

The rubber pads were cast especially with a hardness of 38 Shore A. Sheet metal blanks measuring 130 mm × 130 mm were used. The forming tests were carried out on a single-acting hydraulic press with a maximum press force of up to 600 kN. It provides a digital display of the current force with an accuracy of 1 kN.

First, the sheet is inserted. Then the piston movement is started, and the blank holder is moved onto the sheet at a speed of 5 mm/s, compressing the compression springs. The retainer-side section slides along the sides of the rubber pad and the outer surface of the restrainer-bottom section. After the blank holder has been placed on the sheet metal, it takes approx. one second for the elastic rubber pad to touch the sheet metal. The rubber pad is then pressed into the cavity. The pressure in the rubber pad and the forces on the components of the restrainer increase, as well as the frictional force between the sheet metal and the rubber pad, or the restrainer and the rubber pad (Fig. 3).

When a critical force is exceeded, the restrainer side piece is pressed upwards so that it lifts off the sheet. The deformation force drops abruptly to a fraction of the previous maximum force. The rubber pad is pushed past the sharp inner edge of the restrainer and through the resulting gap between the sheet and the blank holder. It undergoes considerable wear, resulting in deep cracks, coarse abrasion and, in some cases, breakage of entire areas of the rubber pad. There is no further deformation of the blank. Due to this unintentional lifting of the blank holder, only deformation forces of up to 323 kN could be achieved in the tests.

There is neither a chamfer nor a radius on the inner edge of the blank holder towards the rubber pad, which means that any resulting static pressure cannot explain the lifting of the blank holder. There are two possible explanations for this undesired effect:

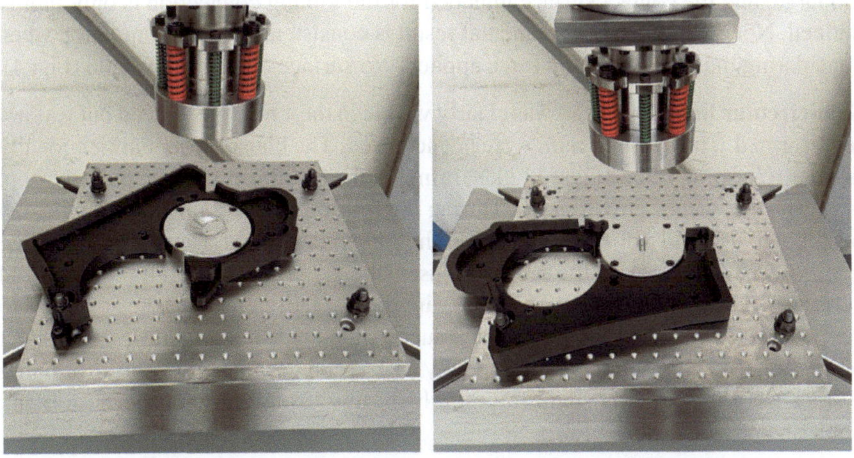

Fig. 3 Finished tool system, in configuration for nub (left) and bead (right)

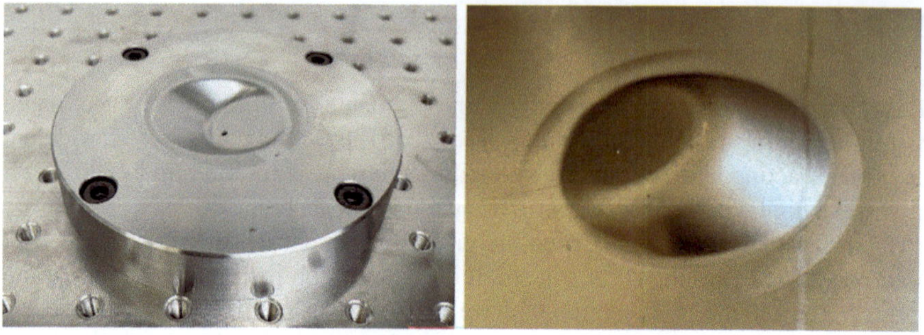

Fig. 4 Formed nub (right) and its forming die (left)

- The sheet metal pushes the blank holder upwards sufficiently during forming, allowing the rubber pad to apply pressure to the front face of the blank holder.
- Friction between the flowing rubber pad and the inner wall of the blank holder is responsible for pushing it upwards.

Cone nub. The nub was successfully formed in the 1 mm sheet metal. Even the slight waviness in the flange area of the nub is clearly defined. The surface structure of the die is visible along the radii, and the contour of the vent hole is distinctly imprinted in the bottom area of the nub. However, the target radii have not yet been fully formed. No wrinkling or other undesired deformations were observed. Apart from the imprinting of the tool surface and the vent hole contour, no surface defects were discovered. The overall result shows strong similarity to the previously simulated outcome. However, the simulation assumed complete forming under maximum press force, which was not achieved in this case. During the forming of the nub in the 2 mm sheet metal, only 126 kN of the intended 600 kN press force could be applied. The nub is clearly visible yet its formation remains incomplete. The resulting radii are too large, and the contact surface intended for the joining connection is only faintly defined. Nonetheless, the experimental result is consistent with the simulation when considering the actual forming force applied. No surface defects are visible (Fig. 4).

Semicircular bead. The bead was clearly visible in the 1 mm sheet metal but was not fully formed. The required radii were not achieved. The blank holder unintentionally lifted at a forming force of just 112 kN, bringing the forming process to an early stop. No surface defects were observed. During the forming of the 2 mm sheet metal, the bead contour could not be replicated in the material, even at a maximum forming force of 123 kN. As with the previous cases, the experimental results for the bead correlate well with the predictions of the simulation. Differences can be attributed to simplifications in the simulation model, such as idealized material flow and neglect of frictional heating effects. Additionally, the spring dynamics of the restraining system were modeled as static forces, which may explain discrepancies in blank holder behavior (Fig. 5).

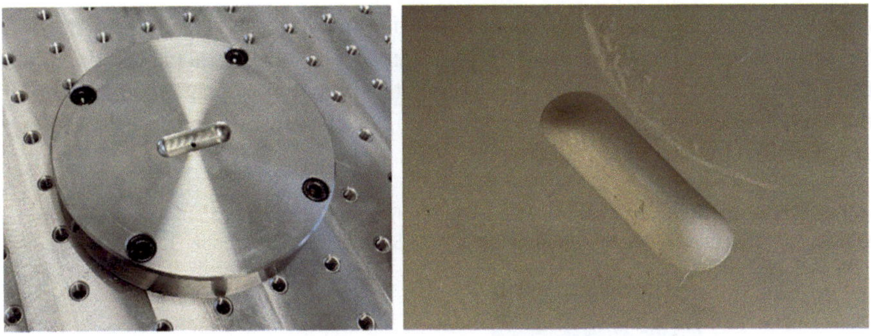

Fig. 5 Formed bead (right) and its forming die (left)

Further Experiments. The trials conducted so far demonstrate that the process works in principle but yields insufficient results. The primary reason is that the maximum press force cannot be reached due to the premature lifting of the blank holder. When softer rubber pads (20–30 Shore A) were used for the forming operations, an even earlier unintended lifting and extrusion of the rubber pad were observed. Consequently, only lower forming forces could be applied, and the geometries were not formed to a satisfactory degree. The rubber pads used in these cases exhibited significantly increased wear. Conversely, when harder rubber pads (60–70 Shore A) were used, the effects were reversed. The blank holder lifted from the sheet later in the process, allowing higher press forces to be achieved. The resulting geometries were formed with significantly better definition, and the rubber pads showed reduced wear during operation.

An increase in the stiffness of the blank holder spring assembly also resulted in delayed blank holder lift-off. To address the problem of unintended blank holder lift, several design modifications are under evaluation. These include: (1) implementing chamfered restrainer edges to reduce stress concentrations, (2) using higher-stiffness spring configurations, and (3) adding mechanical interlocks to delay blank holder movement. This enabled the application of higher forming forces for the same forming tasks.

4 Conclusion

As part of the SkaLaB project, an innovative approach was developed to improve the cost-efficient production of small batch sizes of automotive body sheet metal components. The core concept of this work is the outsourcing of complex secondary forming elements to a separate forming station using rubber pad forming. A movable blank holder enables the local forming of individual geometry features within a larger sheet. Initial simulations and trials confirm the viability of this process extension but also highlight the need for further investigation regarding the rubber pad and blank holder systems. The tooling system has shown promising results to date, but needs to be further adapted to meet the specific requirements of sheet metal part production in the automotive industry. Further improvements will focus on design modifications to the blank holder to prevent unintentional lifting, such as integrating anti-lift contours or active locking mechanisms.

Acknowledgements The authors acknowledge the financial support by the Federal Ministry for Economic Affairs and Climate Action in the project SkaLaB (project number 13IK025).

Funded by the European Union
NextGenerationEU

Supported by:
Federal Ministry for Economic Affairs and Climate Action
on the basis of a decision by the German Bundestag

References

1. Verband der Automobilindustrie (VDA) (2024) Anzahl der Pkw-Neuzulassungen in Deutschland von 1955 bis 2023 und Prognose für 2024 (in Millionen). https://de.statista.com/statistik/daten/studie/74433/umfrage/neuzulassungen-von-pkw-in-deutschland/. Accessed 21 Dec 2024
2. Transport & Environment (2019) Anzahl angebotener Elektroautomodelle in Europa nach Herstellern von 2012 bis 2025. https://de.statista.com/statistik/daten/studie/1032064/umfrage/anzahl-angebotener-elektroautomodelle-in-europa-nach-herstellern/. Accessed 21 Dec 2024
3. Horváth & Partners (2019) Anzahl angebotener Modelle von Personenkraftwagen mit Elektroantrieb in Deutschland von 2013 bis 2022. https://de.statista.com/statistik/daten/studie/1076875/umfrage/anzahl-der-modelle-an-elektrofahrzeugen-in-deutschland/. Accessed 21 Dec 2024
4. Mandat M (2015) Das Dilemma mit der Vielfalt. Produktportfolios der fünf großen deutschen Autohersteller in Deutschland: 1990 bis 2014, München
5. Engel B, Frohn P, Hillebrecht M, Knappe A (2017) Inkrementelles Schwenkbiegen zur kostengünstigen Herstellung von Trägerstrukturen in Elektrofahrzeugen. In: Piskun A (ed) Karosseriebautage Hamburg 2017. Proceedings. Springer Fachmedien Wiesbaden, Wiesbaden, pp 219–234
6. Schiller M, Frohn-Sörensen P, Engel B (2022) 3D-Swivel-Bending—a flexible and scalable forming technology. In: Inal K, Levesque J, Worswick M, Butcher C (eds) NUMISHEET 2022. Proceedings of the 12th international conference and workshop on numerical simulation of 3D sheet metal forming processes. The minerals, metals & materials series, 1st edn. Springer, Imprint Springer, Cham, pp 125–139
7. Breitsprecher T (2019) Entwicklung eines selbstlernenden Assistenzsystems zur automatischen Akquisition von konstruktionsrelevantem Fertigungswissen. VDI Verlag
8. Werner M (2022) Automatisierte Extraktion fertigungsrelevanter Merkmale aus Geometriedaten umformtechnischer Bauteile und deren Analyse zur feature-basierten Fertigungsprozessgestaltung. Dissertation, Technische Universität Chemnitz; Verlag Wissenschaftliche Scripten

9. Zimmermann D (1967) Eine allgemeine Formenordnung für Werkstücke. Gestaltung, Handhabung und Rationalisierungserfolg. Dissertation, Technische Universität Stuttgart
10. Opitz H (1966) Werkstückbeschreibendes Klassifizierungssystem. Dissertation, RWTH
11. Geueke M, Frohn-Sörensen P, Reuter J, Padavu N, Reinicke T, Engel B (2021) Structural optimization of additively manufactured polymer tools for flexible sheet metal forming. Procedia CIRP. https://doi.org/10.1016/j.procir.2021.11.226
12. Thiruvarudchelvan S (1993) Elastomers in metal forming: a review. J Mater Process Technol. https://doi.org/10.1016/0924-0136(93)90008-T
13. Ramezani M, Ripin ZM (2012) Rubber-pad forming processes. Technology and applications. Woodhead Publishing in mechanical engineering. WP Woodhead Publ, Oxford
14. Niknejad A, Rezaee N, Asl FJ (2015) Experimental investigation of Teflon-pad forming on circular metal blanks using a concave die. J Manuf Process. https://doi.org/10.1016/j.jmapro.2015.07.001
15. Salzgitter Flachstahl GmbH (2022) DC04. Weiche Stähle zum Kaltumformen. https://www.salzgitter-flachstahl.de/fileadmin/footage/MEDIA/gesellschaften/szfg/informationsmaterial/produktinformationen/kaltgewalztes_feinblech/deu/dc04.pdf. Accessed 14 Apr 2025
16. Salzgitter Flachstahl GmbH (2022) DC01 (CR1). Weiche Stähle zum Kaltumformen. https://www.salzgitter-flachstahl.de/fileadmin/footage/MEDIA/gesellschaften/szfg/informationsmaterial/produktinformationen/kaltgewalztes_feinblech/deu/dc01.pdf. Accessed 14 Apr 2025
17. Thiruvarudchelvan S (2002) The potential role of flexible tools in metal forming. J Mater Process Technol. https://doi.org/10.1016/S0924-0136(02)00077-8
18. Wolff Kunststoffe (2024) Polyurethan (PU/PUR)—Wolff Kunststoffe. https://wolff-kunststoffe.de/qualitaetsuebersicht/polyurethan-pupur/. Accessed 9 Jan 2025
19. Frohn-Sörensen P, Geueke M, Tuli TB, Kuhnhen C, Manns M, Engel B (2021) 3D printed prototyping tools for flexible sheet metal drawing. Int J Adv Manuf Technol. https://doi.org/10.1007/s00170-021-07312-y

9. Zimmermann D (1969) Eine allgemeine Formenordnung für Werkstücke. Grundlage, Handhabung und Rationalisierungserfolg. Dissertation, Technische Universität Stuttgart
10. Opitz H (1966) Verschlüsselungsrichtlinien ... Dissertation, [illegible]
11. Gencke M, Hecker-Spoerer B, Zonka A, [illegible] N, [illegible] T, [illegible] B (2021) [illegible] tool [illegible] for flexible sheet metal forming. [illegible]
12. [illegible]
13. [illegible] M, [illegible] (2012) [illegible]
14. [illegible]
15. [illegible]

Open Access This chapter is licensed under the terms of the Creative Commons Attribution 4.0 International License [illegible] which permits use, sharing, adaptation, distribution and reproduction in any medium or format, as long as you give appropriate credit to the original author(s) and the source, provide a link to the Creative Commons license and indicate if changes were made.

The images or other third party material in this chapter are included in the chapter's Creative Commons license, unless indicated otherwise in a credit line to the material. If material is not included in the chapter's Creative Commons license and your intended use is not permitted by statutory regulation or exceeds the permitted use, you will need to obtain permission directly from the copyright holder.

Design and Experimental Validation of a Soft, Shape-Programmable Multi-fold Gripper

Florian Schreiber and Martin Manns

Abstract The increasing need for flexibility in industrial automation has prompted the development of adaptable grippers capable of handling objects with varying shapes and sizes. This paper introduces a novel pneumatically actuated multi-fold gripper with a programmable deformation mechanism. The core of the gripper is a flexible ring actuator with an internal air channel. By inserting star-shaped forms during the first actuation, folds formation in the ring's inner wall can be controlled, allowing for the definition of discrete folds. Experimental tests using inserts with 4, 5, and 8 tips demonstrated successful and repeatable programming of the ring's deformation pattern, which remained stable over multiple actuations and after a one-week rest period. Attempts to generate only two folds were unsuccessful, likely due to uncontrolled stress distribution and material limitations. The multi-fold gripper is manufactured additively using a combination of SLS and PolyJet technologies, enabling rapid customization. This method provides a simple, low-cost approach to pre-shaping soft grippers, reducing the energy losses typically associated with deformation. The concept of using localized structural memory introduces a new design pathway for soft robotic actuators. Future work should explore automatic deformation programming through embedded structural features created during the additive manufacturing process.

Keywords Soft robotics · Multi-fold gripper · Ring gripper · Additive manufacturing · Programmable deformation

1 Introduction

In industrial automation, robotic grippers are commonly used to handle objects. These grippers are typically made of rigid materials and are designed for specific tasks, establishing contact with the object through force closure, form closure, or

F. Schreiber (✉) · M. Manns
PROTECH—Institut für Produktionstechnik, Universität Siegen, Siegen, Germany
e-mail: florian.schreiber@uni-siegen.de

M. Manns (ed.), *New Paradigms for Anticipated Uncertainty*, Lecture Notes in Mechanical Engineering, https://doi.org/10.1007/978-3-032-16889-4_30

a combination of both principles. However, increasing product diversity and shortening product life cycles demand a higher degree of flexibility in handling [1]. As a result, grippers must frequently be adapted or redesigned, which entails significant investment.

A promising approach to increasing flexibility is the use of soft grippers [2]. These grippers are made of compliant materials such as silicone and can adapt to objects with varying geometries due to their material flexibility. Pneumatically actuated soft grippers, in particular, offer advantages in terms of simple control. Despite these benefits, soft grippers are rarely used in industrial environments. This is primarily due to limited durability and insufficient gripping forces [3], which can result from energy losses caused by material deformation and a lack of structural stiffness.

A key challenge is the development of a soft gripper whose initial shape closely matches the geometry of the target object, allowing for a maximum portion of the applied force to be converted into effective gripping force. At the same time, the gripper must remain sufficiently adaptable to reliably handle objects with unknown or varying geometries. This trade-off between shape conformity and adaptability forms the basis of the present work.

2 State of the Art

Soft robotic grippers have gained increasing attention as a promising alternative to conventional rigid grippers in flexible automation scenarios. Among them, pneumatically actuated grippers—commonly referred to as pneunets (pneumatic network)—are widely studied due to their simple actuation mechanism, safe interaction with objects, and the ability to conform to various shapes [4]. These systems typically rely on elastomer-based chambers that expand upon pressurization, enabling bending or grasping movements. While such designs are advantageous in terms of adaptability and compliance [5], they often suffer from limited gripping force and insufficient shape control, especially in repetitive or industrial-scale applications [6].

Recent advances in additive manufacturing have opened new avenues for customizing soft grippers, allowing the integration of internal channels and compliant structures directly during fabrication [7]. This has led to the development of modular or reconfigurable gripper systems, where individual fingers or components can be replaced or repositioned depending on the application [8]. However, these approaches still require mechanical interfaces or external control systems, which can increase complexity and cost.

To address limitations in deformation control, several studies have explored the concept of shape programmability in soft actuators. Methods include the use of thermo-mechanical processes [9], shape memory polymers [10], or localized material property variations to induce preferred bending patterns [11]. Additionally, kirigami- or origami-based structures have been proposed to guide deformation through predefined cuts or folds [12]. These methods, while innovative, often rely on complex fabrication processes or require multi-material printing.

The concept of imprinting deformation patterns via mechanical constraints during initial actuation—such as inserting rigid forms—remains underexplored. Such a method could enable simple and repeatable programming of gripping behavior without the need for embedded sensors or active control.

3 Objectives

This study presents an innovative, adaptable multi-fold gripper. The gripper's ring-shaped actuator can be geometrically configured to match the shape of the target object, functioning as a centrically closing multi-fold gripper. The number of folds can be predefined prior to initial deployment. This work introduces the gripper's design and investigates its adaptability through experimental analysis.

4 Methodology

4.1 Design and Operating Principle of the Multi-fold Gripper

The design of the multi-fold gripper is based on the design of the previously published ring gripper [13]. The fundamental structure of the multi-fold gripper is presented in Fig. 1. The core component is a flexible ring with an integrated air channel. Compressed air is supplied via an inlet on the left side, while a cleaning outlet on the opposite side is sealed after manufacturing.

During operation, the object to be handled is placed centrally within the ring. Increasing the internal air pressure causes the ring to expand. The surrounding housing prevents outward expansion, directing the motion inward toward the center of the gripper. This results in a clamping effect, enabling the secure handling of the object. In addition to supporting the flexible ring, the housing also serves as the interface for robotic mounting.

As internal pressure rises, the inner walls of the ring are pushed toward the center. While the top and bottom surfaces expand, the inner surface undergoes compression. If the pressure increases further, this compression can lead to localized wrinkling, especially at points with minor manufacturing or material imperfections. This effect is intentionally leveraged to "program" the deformation behavior of the gripper by influencing the locations where folds form. For this purpose, predefined mold structures are placed within the gripper prior to its initial actuation, thereby guiding the fold formation as the internal pressure within the ring is increased.

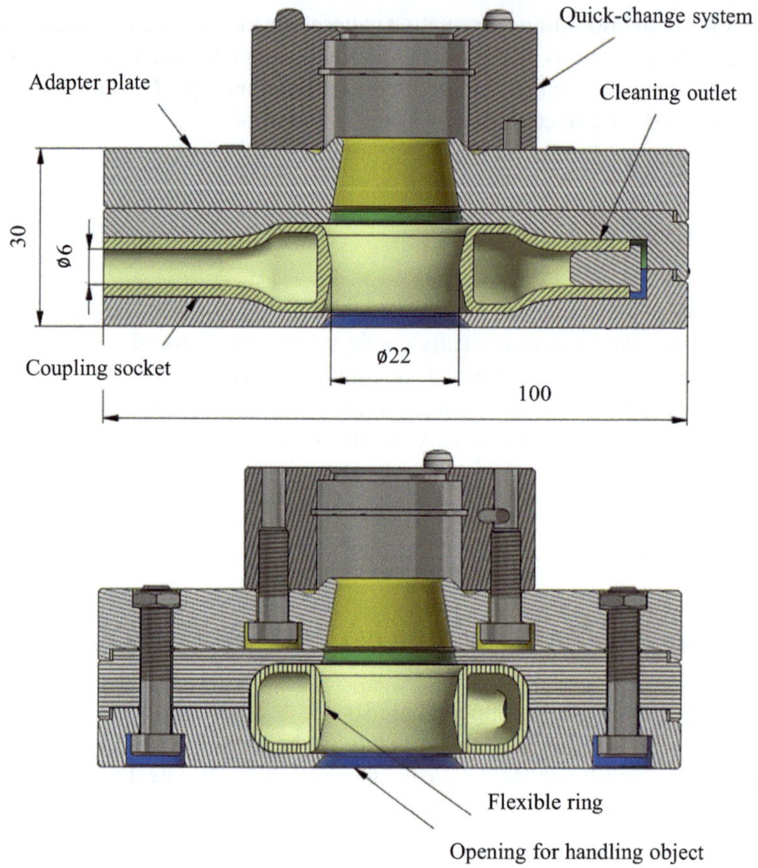

Fig. 1 Cross-section of the multi-fold gripper

4.2 Experimental Setup for Shape Programming

To evaluate programmable deformation, the gripper is vertically aligned and monitored via a camera capturing the ring's interior. Prior to initial actuation, geometrically shaped inserts—featuring 2, 4, 5, or 8 protruding tips—are positioned within the ring opening. These star-shaped inserts inhibit expansion at defined regions while leaving gaps that facilitate targeted folds formation.

After placement, the internal pressure is increased to 500 kPa and maintained for 60 s. The pressure is then released, the insert is removed, and the gripper is re-actuated without the insert. The objective is to determine whether the pre-programmed fold pattern is preserved. Each geometry is tested with three new grippers that have not been previously actuated.

Table 1 Mechanical parameters of the TPU Luvosint material

Mechanical parameter	TPU Luvosint
Compressive elastic modulus	20 MPa
Tensile elastic modulus	27 MPa
Ultimate tensile elongation	5.2
Ultimate tensile stress	20 MPa
Poisson ratio	0.45
Density	1200 kg/m^3

4.3 Manufacturing of the Multi-fold Gripper

The entire multi-fold gripper, except for screws, is additively manufactured. The flexible ring is fabricated using a selective laser sintering (SLS) process on an EOS P-series printer, utilizing the TPU powder Luvosint. The corresponding material properties are summarized in Table 1. Due to the self-supporting nature of the SLS process, no support structures are required for the fabrication of the ring cavities. To remove residual powder from the air channel, the ring is cleaned post-fabrication using compressed air and ultrasonic vibration.

The housing is produced on a Keyence Agilista 3200 3D printer using the PolyJet process and consists of the transparent, urethane-based material AR-M2, allowing internal inspection during operation. The star-shaped inserts are printed from PLA using a Prusa MK3S+.

5 Results

Programming of the ring using inserts with 4, 5, and 8 tips is successfully demonstrated. Figure 2 presents representative results for the five-folds configuration. The figure illustrates how the five-tip insert is positioned within the ring prior to the initial actuation. During pressurization, the ring expands but is only able to deform in the spaces between the tips of the insert. After maintaining a pressure of 500 kPa for 60 s and subsequently removing the insert, a repeated actuation without the form reveals that the ring deforms according to the previously "programmed" shape.

Figure 3 presents the results for inserts with four and eight tips. Similarly, the characteristics of the folds were controllable through the application of the predefined insert structures.

The attempt to generate only two folds using a two-tip insert was unsuccessful. Instead, four folds formed—two in each of the unrestrained regions where the expansion was not constrained by the tips of the insert.

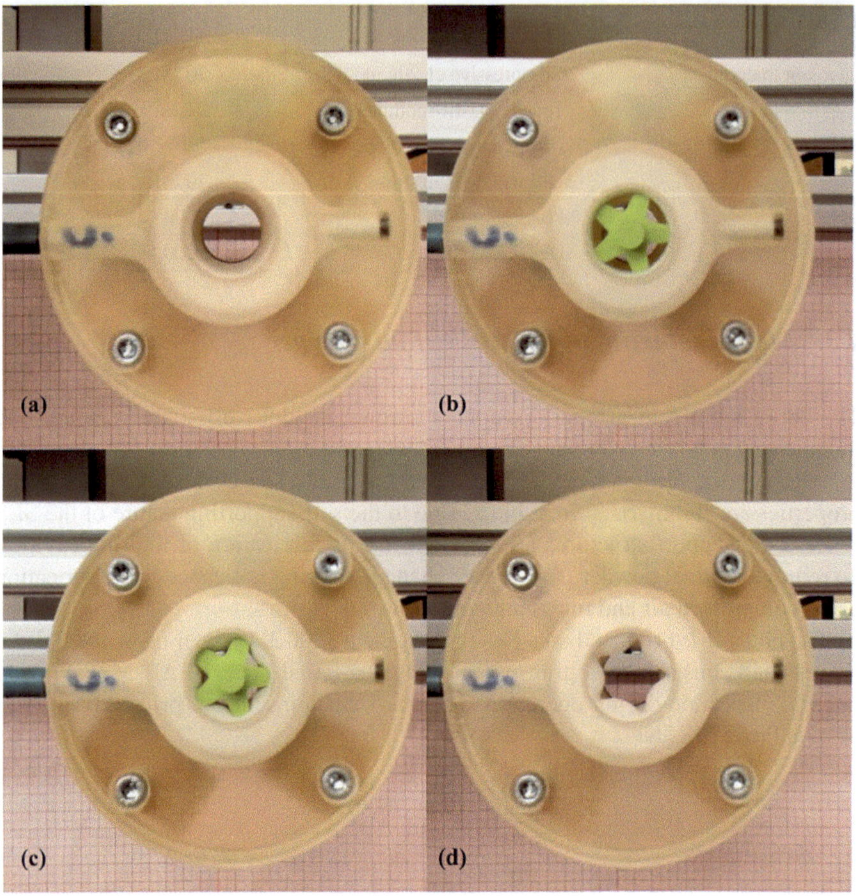

Fig. 2 Programming of the multi-fold gripper: **a** initial shape at 0 kPa, **b** inserted form with 5 tips, **c** deformation of the ring with inserted form, **d** identical ring expansion without insert

6 Discussion

The experimental results demonstrate that the deformation of the ring can be effectively controlled by using inserts with 4, 5, or 8 tips. In all cases, the number of folds formed in the ring corresponded precisely to the number of unconstrained segments, resulting in 4, 5, and 8 folds, respectively. Based on these findings, it can be reasonably assumed that configurations with 6 or 7 folds could also be achieved using appropriately designed inserts. This provides a promising method to roughly tailor the gripper's internal shape to match the geometry of a target object, potentially enhancing gripping efficiency and contact stability. Potential applications of this prototype include the handling of delicate items (e.g., pralines of varying shapes) or flexible products (e.g., cables).

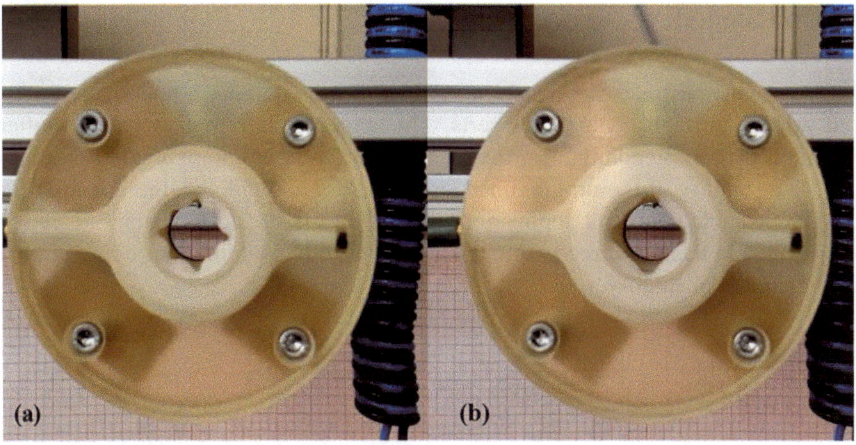

Fig. 3 Multi-fold gripper with 8 folds (**a**) and 4 folds (**b**)

A particularly noteworthy observation is the persistence of the programmed deformation over time. A total of 50 actuation cycles were performed, each involving a pressure ramp from 0 to 500 kPa and back to 0 kPa, with each pressure level maintained for 10 s. Notably, the fold configurations remained unchanged even after one week, indicating long-term stability of the programmed deformation, at least under the tested conditions. In future studies, it could also be investigated whether reprogramming a ring to achieve a different number of folds is feasible.

The unsuccessful attempt to program a two-fold configuration is likely due to the high compressive forces generated within one half of the ring, which led to the spontaneous formation of additional folds. Alternatively, insufficient stiffness in the material may have allowed unwanted deformations in regions not constrained by the insert. Future investigations could explore the use of alternative materials or optimized wall geometries to enable stable two-fold configurations.

While a lower limit for the number of folds is determined, a maximum of eight folds was successfully tested. However, it is reasonable to assume that the number of folds is subject to an upper limit, as the width of each fold decreases with increasing fold count. Due to the material's stiffness and incompressibility, it is likely that a material-specific minimum fold width exists, which would limit the number of achievable folds based on the ring's circumference. This aspect should be investigated in future work.

While the use of form inserts represents a straightforward and rapid programming method, it remains a manual procedure. A promising alternative approach could involve the integration of artificial structural weaknesses during additive manufacturing. These predefined features would locally favor folds formation during pressurization, enabling tool-free, automatic programming of the ring geometry. Exploring such embedded design strategies holds significant potential and warrants further investigation.

7 Conclusion

This work demonstrates the feasibility of shape-programming a pneumatic multi-fold gripper through geometric inserts. The achieved deformation patterns offer a reliable way to preconfigure gripping behavior without complex control strategies. The stability of the programmed shape over time highlights the potential for use in repeatable industrial applications. Future efforts should focus on integrating programmable features directly into the design, enabling autonomous configuration and broader applicability.

References

1. European Commission, Directorate-General for Research, Innovation (2011) Europe 2020 flagship initiative Innovation Union: SEC(2010) 1161, communication from the Commission to the European Parliament, the Council, the European Economic and Social Committee and the Committee of the Regions. https://doi.org/10.2777/27497
2. Onal CD, Rus D (2012) A modular approach to soft robots. In: 2012 4th IEEE RAS & EMBS international conference on biomedical robotics and biomechatronics (BioRob). IEEE, pp 1038–1045
3. Tennakoon CL, Kulasekera AL, Gopura RARC et al (2024) PneuNet based hybrid soft gripper for multi-shape object handling. IEEE Access 12:190158–190168. https://doi.org/10.1109/ACCESS.2024.3515265
4. Ilievski F, Mazzeo AD, Shepherd RF et al (2011) Soft robotics for chemists. Angew Chem 123:1930–1935. https://doi.org/10.1002/ange.201006464
5. Su H, Hou X, Zhang X et al (2022) Pneumatic soft robots: challenges and benefits. Actuators 11:92. https://doi.org/10.3390/act11030092
6. Shintake J, Cacucciolo V, Floreano D et al (2018) Soft robotic grippers. Adv Mater e1707035. https://doi.org/10.1002/adma.201707035
7. Truby RL, Lewis JA (2016) Printing soft matter in three dimensions. Nature 540:371–378. https://doi.org/10.1038/nature21003
8. Low JH, Khin PM, Han QQ et al (2022) Sensorized reconfigurable soft robotic gripper system for automated food handling. IEEE/ASME Trans Mechatron 27:3232–3243. https://doi.org/10.1109/TMECH.2021.3110277
9. Wang W, Li C, Cho M et al (2018) Soft tendril-inspired grippers: shape morphing of programmable polymer-paper bilayer composites. ACS Appl Mater Interfaces 10:10419–10427. https://doi.org/10.1021/acsami.7b18079
10. Wang W, Tang Y, Li C (2021) Controlling bending deformation of a shape memory alloy-based soft planar gripper to grip deformable objects. Int J Mech Sci 193:106181. https://doi.org/10.1016/j.ijmecsci.2020.106181
11. Schönfeld D, Chalissery D, Wenz F et al (2021) Actuating shape memory polymer for thermoresponsive soft robotic gripper and programmable materials. Molecules 26. https://doi.org/10.3390/molecules26030522
12. Li S, Stampfli JJ, Xu HJ et al (2019) A vacuum-driven origami "magic-ball" soft gripper. In: 2019 international conference on robotics and automation (ICRA). IEEE, pp 7401–7408
13. Schreiber F, Manns M, Morales J (2019) Design of an additively manufactured soft ring-gripper. Procedia Manuf 28:142–147. https://doi.org/10.1016/j.promfg.2018.12.023

Development of a Software Tool for the Company-Individual and Change-Specific Support of Processes in Manufacturing Change Management

Jan-Philipp Rammo, Clément Rouvelle Romegoux, Moritz Goeldner, and Michael F. Zaeh

Abstract Manufacturing companies face a dynamic environment that demands frequent adaptations of production processes and infrastructure. These so-called Manufacturing Changes (MCs) differ in scope, complexity, and cost, making standardized approaches insufficient for addressing specific company needs. Effective decision making requires tailored solutions that consider both company-individual and change-specific requirements. This paper presents a software tool that systematically supports MCs by analyzing upcoming changes and applying a correlation model to derive necessary actions. The tool integrates change characterization with decision-support functionalities to enhance the effectiveness and the efficiency of MC handling. Developed as a web-based application using HTML, CSS, JavaScript, and Python, the tool provides actionable insights to optimize decision making. Initial tests demonstrated its potential to support adaptive manufacturing environments, enabling companies to respond proactively to production changes. This solution represents a significant step forward in managing MCs by combining a scientific methodology with a practical application to enhance operational flexibility.

Keywords Manufacturing changes · Change management · Software tool · Decision support · Flexibility

1 Introduction

In today's production landscape, manufacturing companies need to be able to react flexibly to a range of external and internal influences [1]. For example, changing customer requirements or new regulations can force companies to adapt their

J.-P. Rammo (✉) · C. R. Romegoux · M. Goeldner · M. F. Zaeh
Institute for Machine Tools and Industrial Management, Technical University of Munich, Garching Near Munich, Germany
e-mail: jan-philipp.rammo@iwb.tum.de

M. Manns (ed.), *New Paradigms for Anticipated Uncertainty*, Lecture Notes in Mechanical Engineering, https://doi.org/10.1007/978-3-032-16889-4_31

production by implementing new machines, processes, or tools. These changes to a factory are referred to as Manufacturing Changes (MCs) [2]. The process of dealing with these MCs is referred to as Manufacturing Change Management (MCM) and considers the organization and control of the process of making any alteration to a factory [3]. An efficient MCM is becoming increasingly important in the ever-more volatile environment in which manufacturing companies operate [4].

Therefore, Rammo and Zaeh [5] proposed a methodology for the efficient and effective handling of MCs that streamlines the change process of manufacturing companies and suggests suitable methods and digital tools (M&DT) for a better implementation of the change.

The methodology (Fig. 1) is initialized by incorporating company-specific information. This information includes, in particular, the company-individual change process [6], the change attributes used to evaluate the changes [7], and the M&DT available or applicable in the company [8]. In the *Operational phase*, a pending change is then evaluated with the previously selected change attributes, and based on a correlation model, the necessary process steps and the suitable M&DT are calculated.

The requirements for this methodology [9] focus particularly on the applicability and usability in an industrial context. For this reason, a software tool was designed and developed as part of this contribution to guide the user through the individual steps of the methodology.

In theory, software development involves designing, testing, and maintaining applications using structured methodologies [10]. A key distinction is made between front-end development, which handles the user interaction, and the back-end development, which manages the server-side logic and databases [10]. This distinction is particularly relevant for this contribution, as both aspects play a crucial role in the software's architecture. This contribution details the concept and development of the software, compares it with existing approaches from the literature, and provides a first indication of the software's applicability in practice.

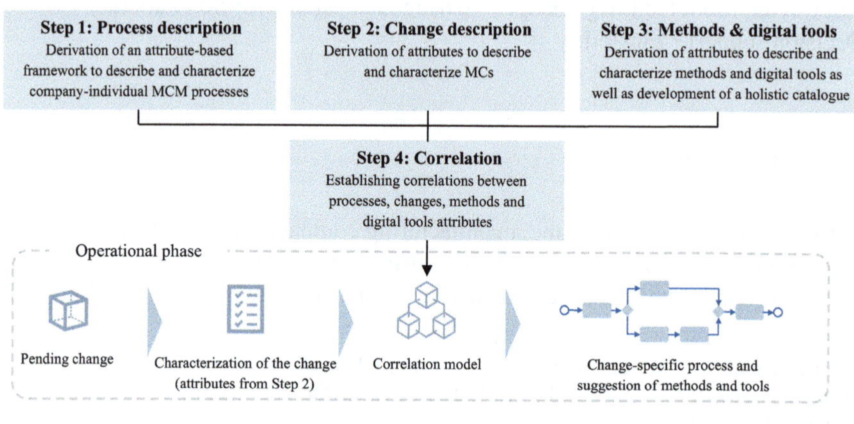

Fig. 1 Structure of the overall methodology [6]

The remainder of this contribution is structured as follows. Section 2 presents the state of the art and the derived research gap. Section 3 depicts the methodological approach to developing the software. Section 4 details the results regarding the technical framework and the software architecture. Section 5 discusses the results and the limitations, gives an outlook on future research, and summarizes the contribution.

2 State of the Art

This section summarizes the state of the art by focusing on publications that directly address software tools or algorithms designed for managing Manufacturing Changes (MCs). The identified publications showed that some approaches have already addressed the software- and algorithm-based support of changes in manufacturing. As an overview, Ahmad et al. [11] compared different information models used for product changes emphasizing traceability, dependency mapping, and propagation prediction as key software features in change management. Several studies explored methods and developed tools to predict and manage the impacts of changes. The Change Propagation Method tool proposed by Ou-Yang and Cheng [12] uses graph-based algorithms to model dependencies and evaluate change impacts. Similarly, the Design Change Prediction Matrix tool proposed by Siddart and Sakar [13] assesses ripple effects in design and production through dependency matrices. Hindmarch et al. [14] proposed a support tool based on a Dependency Structure Matrix to analyze change impacts while preserving historical design knowledge.

Several studies also highlighted tools supporting modular and scalable approaches. One such tool is the modular framework proposed by Rafaelli et al. [15], which models product dependencies to manage changes in product families. Ibbs et al. [16] developed a tool to integrate the recognition, the evaluation, and the implementation of changes to enhance decision-making in complex projects. The software tool developed by Malatesta et al. [17] supports the change process of products in the consumer goods sector. They highlighted significant time and cost savings achieved through the software tools.

Decision-support tools and visualization solutions play a crucial role in several publications. Glegg et al. [18] developed a visual management tool which provides intuitive interfaces and real-time feedback for change project planning and evaluation. The agent-based system developed by Habhouba et al. [19] offers decision-making support by automating change evaluation. Also, the impact of agile development on change management was explored with tools and frameworks that support iterative workflows [20]. Macke et al. [21] discussed advancements in smart MCM, focusing on software tools that enhance automation and responsiveness to dynamic environments. These approaches and tools can also be used in software development itself [22].

It can, therefore, be concluded that various software programs have already been developed in the area of change management in production. Most of them focus on predicting change propagation and dependencies in product changes. Only Malatesta et al. [17] and Macke et al. [21] consider the change process, which is why the software tool developed in this contribution incorporates the fundamental concepts from the process-oriented approaches, such as change process modeling and change description. However, a tool that deals with the flexibilization of the process or the selection of suitable M&DT does not yet exist and is seen as a research gap for this contribution.

3 Methodology

The development of the software tool followed an iterative methodology, enabling continuous adjustments based on feedback from the industry partner. The process adhered to Adaptive Software Development (ASD) principles [23], which emphasize flexibility and close collaboration with the industrial partner. Initially, the requirements were gathered through a joint workshop with the industrial partner, where specific needs were defined. The tool was then iteratively developed and regularly presented to the partner for review and feedback. In the final phase, the software was applied to a real-world use case provided by the industrial partner. It was tested in an experimental setup within a learning factory, simulating a manufacturing change scenario. Additionally, the tool was demonstrated to experts from other companies to gather further insights and validate its applicability in various contexts. This approach allowed for the continuous refinement of the tool, ensuring that it met the specific requirements of the industrial partner while also addressing broader needs within the manufacturing sector.

4 Results

This section describes the results by discussing the technical framework (Sect. 4.1) and the software architecture (Sect. 4.2).

4.1 Technical Framework

The technical aspects of the tool can be divided into two main parts: the front end and the back end. The front end is built using *Hypertext Markup Language* (HTML), *Cascading Style Sheets* (CSS) and *JavaScript*. It includes the user interface (UI) and the input registration. The back end is used for data processing and computation and is written in *Python*. The framework of the application is provided by the *Flask* library, which allows communication between the front end and the back end of the application. Array calculations are facilitated by the *NumPy* library, and most of the data is stored in *Comma-Separated Values* (CSV). The *pandas* and *csv* libraries are used to access and manipulate these files. The *os* module simplifies file path manipulation, and the visualizations are handled by the package *pygraphviz*.

4.2 Software Architecture

The following section presents the architecture of the software using a hybrid diagram that combines elements of both a navigation and a sequence diagram (Fig. 2). The front end, at the center of the diagram, provides users with access to different pages, which guide them through the *Initialization* and *Operational* phases. Additionally, the diagram provides an overview of the user inputs on each page (displayed on the right) and the corresponding back-end processing of these inputs (shown on the left). In the first step of the *Initialization* phase, users upload or manually input a company-individual change process, either in Business Process Management Notation (BPMN) format or as a matrix. The system stores this data in CSV format and generates a BPMN model. On *Page 1.2*, users input process attributes, which are also saved as matrices in CSV format. On *Page 1.3*, users specify the relevant change attributes for the organization, and these are stored as a vector in CSV format. Finally, during the *Initialization* phase, users select suitable M&DT from a predefined catalog and can add custom methods and tools through an input form. These selections are saved as a CSV matrix in the back end. Once the *Initialization* phase is complete, users move to the *Operational* phase. Here, the user assesses an upcoming change by entering relevant data via an input form, which is stored as a vector in the back end. Using the stored matrices and an embedded correlation model, the software identifies the relevant process steps and M&DT required to handle the specific change. The results are presented on *Page 2.2*, where an updated BPMN diagram is displayed, showing the assigned M&DT for each process step.

5 Industrial Application

The software tool was evaluated in collaboration with an automotive supplier. The company's MCM process was modeled, and past changes were then analyzed to assess how the process adapted and which M&DTs were suggested. The evaluation showed that the tool's ability to change-specifically adapt the MCM process and to suggest appropriate methods and digital tools improved process efficiency, decision-making and documentation. In addition to the industrial case study, a simulation game in a controlled environment further tested the tool's effectiveness. The results demonstrated a reduction in change execution times, especially for smaller changes, and a decrease in documentation effort due to the tool's automated storage of change process data. The feedback highlighted several strengths, including the tool's ability to adapt the MCM process to different changes, improve decision-making, and reduce documentation efforts through automation. However, the initial setup

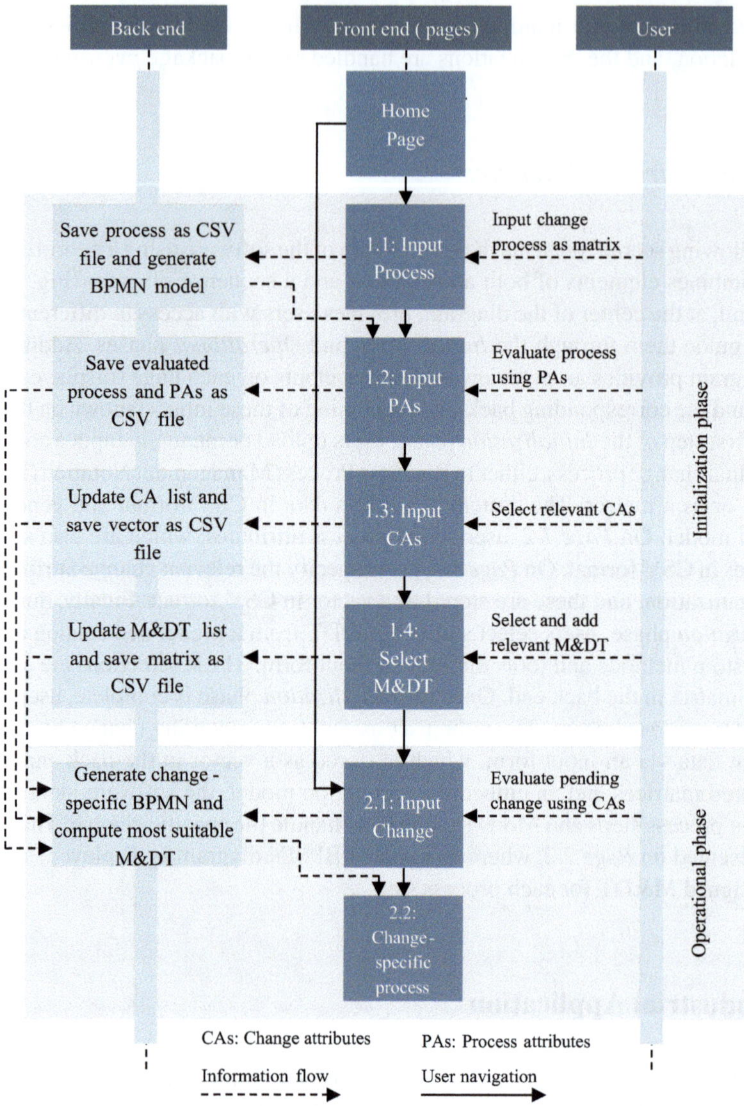

Fig. 2 Overview of the software architecture

was time-consuming and could benefit from better onboarding and support. Users also suggested making customization more intuitive and enhancing guidance during the Initialization phase. Additionally, the tool's ability to handle complex changes with many process steps and M&DTs could be improved for broader applicability. Overall, the industrial application validated the tool's potential to enhance MCM by streamlining processes and improving transparency, despite the initial setup effort required.

6 Discussion

The developed software tool significantly enhances the applicability and usability of the proposed methodology for MCM. Key contributions include the tool's adaptability, allowing users to customize company-specific processes, integrate custom M&DT, and adjust the correlation model without needing to modify the code. This flexibility makes the tool suitable for a wide range of manufacturing environments.

Improvements in MCM were demonstrated through the tool's ability to tailor the process to different types of changes, improve decision-making, and reduce documentation efforts. Testing with an industry partner confirmed the tool's practical utility in supporting these objectives. However, some limitations were identified, particularly the time-consuming initial setup phase, which could be streamlined with better onboarding and support. Additionally, the tool performed well for smaller changes but needs further optimization to handle complex changes with multiple M&DT and process steps. Furthermore, the core logic, such as the correlation model, was not addressed in this paper but will be explored in future research focused on the algorithmic aspects of the methodology. Future research will also focus on integrating responsibilities and stakeholder management into the software to improve collaboration and decision-making across teams. Additionally, enhancements will be made to streamline the setup process, integrate a workflow feature for better guidance through the change process, and improve the tool's ability to handle more complex change scenarios.

7 Summary

Manufacturing companies operate in an increasingly dynamic environment, requiring frequent production adaptations. Standardized approaches often fail to address company-specific and change-specific needs, necessitating tailored decision-making tools. This paper introduces a software tool that supports MCs by integrating company-individual processes and change-specific decision-support functionalities. Developed as a web-based application using HTML, CSS, JavaScript, and Python, the tool suggests tailored process steps and suitable M&DT. The testing with an industry partner validated its utility in optimizing decision-making and change processes. The feedback highlighted the need for a workflow feature to guide users through the complete change management. This tool advances MCM by providing a software which combines a scientific methodology with practical adaptability.

References

1. Bauer H, Haase P, Sippl F, Ramakrishnan R, Schilp J, Reinhart G (2020) Modular change impact analysis in factory systems. Prod Eng Res Dev 14(4):445–456
2. Rammo J-P, Mujcinovic S, Zaeh M (2024) Manufacturing change management – a case study-based analysis of change processes in industrial practice. In: Conference on production systems and logistics
3. Koch J (2017) Manufacturing change management—a process-based approach for the management of manufacturing changes. Dissertation, Technische Hochschule München, München
4. Rammo J-P, Graf J (2023) Manufacturing change management – a survey on current challenges. The state of digitalization and the application of change impact analysis in industrial practice. In: 5th conference on production systems and logistics
5. Rammo J-P, Zaeh M (2024) Towards a change-specific and company-individual manufacturing change management. In: 57th CIRP conference on manufacturing systems 2024 (CMS 2024)
6. Rammo J-P, Arifagic A, Agyekum E, Zaeh MF (2024) Development of a holistic framework for the description and characterization of company-individual Manufacturing Change Management processes. Prod Eng Res Dev. https://doi.org/10.1007/s11740-024-01307-w
7. Rammo J-P, Custovic N, Zaeh MF (2024) Development of a comprehensive description model for manufacturing changes. In: 2024 IEEE international conference on industrial engineering and engineering management (IEEM), Bangkok, Thailand, pp 984–988
8. Rammo J-P, Agyekum E, Perez O, Zaeh MF (2025) Development of a catalog and a description model for methods and digital tools in Manufacturing Change Management. Prod Eng Res Dev
9. Rammo J-P, Agyekum E, Stang J, Zaeh M (2024) Requirements elicitation for a flexible, digital, and methodological manufacturing change management based on a literature review and expert interviews. In: 57th CIRP conference on manufacturing systems (CMS 2024)
10. Stober T, Hansmann U (2010) Agile software development: Best practices for large software development projects. Springer, Berlin, Heidelberg
11. Ahmad N, Wynn D, Clarkson J (2011) Information models used to manage engineering change: a review of the literature 2005–2010. In: International conference on engineering design
12. Ou-Yang C, Cheng MC (2003) Developing a PDM/MRP integration framework to evaluate the influence. Int J Adv Manuf Technol
13. Siddharth L, Sarkar P (2017) A methodology for predicting the effect of engineering design changes. Procedia CIRP 60:452–457
14. Hindmarch H, Gale AW, Harrison RE (2010) A support tool for assessing the impact of design changes during built environment projects. In: 2010 IEEE international conference on industrial engineering and engineering management, Macao, China, pp 1583–1587
15. Raffaeli R, Malatesta M, Marilungo E, Germani M (2013) An approach for managing engineering changes in product families. In: Proceedings of the ASME 2013 international design engineering technical conferences and computers and information in engineering conference
16. Ibbs W, Wong C, Kwak YH (2001) Project change management system. J Manag Eng
17. Malatesta M, Raffaeli R, Mengoni M, Germani M (2013) Supporting the modification process of products through a change management tool. International conference on engineering design
18. Glegg SMN, Ryce A, Brownlee K (2019) A visual management tool for program planning, project management and evaluation in paediatric health care. Eval Program Plann 72:16–23
19. Habhouba D, Desrochers A, Cherkaoui S (2006) Engineering change management and decision-making assistance using software agent. In: 2006 Canadian conference on electrical and computer engineering, Ottawa, ON, Canada, pp 1694–1697
20. Cvitak LD, Car Z (2010) Impact of agile development implementation on configuration and change management in telecom domain. In: MIPRO 2010
21. Macke N, Rulhoff S, Stjepandic J (2016) Advances in smart manufacturing change management. In: Advances in transdisciplinary engineering, transdisciplinary engineering: crossing boundaries. IOS Press

22. Perez M, Navas A, Parada HA, Duenas JC (2015) A tool for the analysis of change management processes in software development cycles. In: Proceedings of the 17th international conference on enterprise information systems (ICEIS-2015)
23. Qureshi RJ, Hussain SA (2008) An adaptive software development process model. Adv Eng Softw 39(8):654–658

Supply Chains and Production Strategies

Balancing Between Efficiency Value and Service Value in Delivering Customized Solutions

Junsong He and Miia Martinsuo

Abstract Companies that deliver customized solutions through projects are interested both in the efficient resource use in their manufacturing process and complementing a core product with services to fulfill customers' specific needs, that is, efficiency value and service value. Firms face tensions, trade-offs, and synergies in aligning efficiency value and service value. This study focuses on the development and delivery of customized solutions in business-to-business (B2B) settings, aiming to uncover project actors' perceptions of efficiency value and service value and the mechanisms used to balance them. A qualitative exploratory study conducted in two software companies and two shipyards reveals that service value and efficiency value are perceived through three aspects: benefits and sacrifices (from both short-term and long-term perspectives), related lifecycle phases and activities, and various enablers. Ten balancing mechanisms harmonize the tensions and trade-offs between these two values. Internal balancing mechanisms include standardizing, modularizing, configuring, reusing, and scaling. External balancing mechanisms comprise co-creating, negotiating, adapting, bargaining, and segmenting. These findings enhance our understanding of value perceptions from a dual-lens perspective and illuminate balancing mechanisms in project business and solution delivery. The examination of software firms and shipyards broadens mass customization research by providing empirical evidence from unconventional contexts.

Keywords Customized solutions · Projects · Service value · Efficiency value · Value balancing

1 Introduction

Businesses are increasingly concerned with what kind of value they can offer to their customers, while also bringing commercial value to themselves. Value in business-to-business (B2B) settings refers to a compromise of benefits and sacrifices [1], and

J. He (✉) · M. Martinsuo
Department of Mechanical and Materials Engineering, University of Turku, Turku, Finland
e-mail: junsong.he@utu.fi

M. Manns (ed.), *New Paradigms for Anticipated Uncertainty*, Lecture Notes in Mechanical Engineering, https://doi.org/10.1007/978-3-032-16889-4_32

it is often considered from the perspective of customers and other stakeholders [2]. Value is multi-dimensional—it can cover economic, ecological, social, and aesthetic dimensions—and can be viewed from multiple perspectives [3]. In project business, identifying the most critical values and delivering them through implementing a project contribute to both project success and overall business performance [4]. This study deals with B2B firms' pursuit of multiple aspects of value simultaneously.

Customized solutions such as turnkey machinery complemented with installation and training; or IT systems supported with consulting are integrated offerings combining products and services to resolve complex customer needs [5]. Delivering customized solutions involves making tradeoffs, for example, between accommodating specific customers' needs and maximizing productivity [5].

Service value and efficiency value are central to this tradeoff. Service value implies an emphasis on maintaining the customer relationship [6, 7]. Solution providers deliver this value by responding to customer needs, enhancing customer loyalty, extending lifecycles for new business, and ultimately creating differentiation that contributes to project success [6, 7]. Efficiency value emphasizes maximizing productivity. Solution providers deliver this value by utilizing digitalization, streamlining processes, enhancing collaboration and communication among stakeholders, and thereby converting inputs into outputs more effectively [8, 9].

Previous research has clarified what constitutes value and how to create it [1–4], and what are the tensions and trade-offs of value-related elements [10–12] in project business and related domains. Further research is needed to explore the perception of two critical values and the mechanisms that support their balance and synergy in project delivery. The purpose of this study is to deepen the understanding of tensions, trade-offs, and synergies between efficiency and service value in customized solutions. The goal is to reveal the mechanisms of balancing these values in different project settings. Two research questions will guide the study:

- How do project actors perceive service value and efficiency value in projects that deliver customized solutions?
- How (through what kinds of mechanisms) do project actors balance service value and efficiency value in these projects?

This study focuses on business-to-business (B2B) solution delivery projects, particularly those in industries with customization complexity, such as software development and shipbuilding. Project actors here refer to the key stakeholders involved in delivering customized solutions. The research excludes projects that solely concentrate on services, mass-market solutions, and external factors such as cultural differences.

2 Theoretical Background

Prior research has extensively examined the paradoxes and tensions that exist within project business and related domains such as customization, services, and innovation management. Studies have analyzed the interplay between customization and standardization in maintaining customer loyalty [10], the exploration–exploitation trade-offs in driving innovation [11], and the tension of flexibility and stability in achieving organizational harmony [12]. These dual pursuits are closely related to values, and there is a need for further knowledge on value-related performances in different contexts, as selecting and implementing the right values is critical to project success and overall business performance [2].

Firms need to apply some balancing mechanisms to create harmony and synergy in delivering customized solutions. Previous research offers a solid foundation on what constitutes value and how to create it [1–4] and clarifies specific values such as "added value" [13] and "strategic value" [14]. However, there remains a notable gap in balancing two critical values, efficiency and service value. The search of balance will require examining enablers [15] at the macro level and exploring lifecycle stage relationships [16] at the micro level. Moreover, adopting a "both-and" perspective rather than an "either-or" compromise [17] further helps achieve synergy between these dual values.

This research adopts customized solution projects as the investigation context, selects service value and efficiency value as critical values, and analyzes the tensions between these two values and the mechanisms for balancing them.

3 Research Method

We implemented a qualitative exploratory interview-based study, to search responses to the how-type questions [18]. We selected four firms operating in the shipbuilding and software industries as representatives of different contexts for delivering customized solutions. Shipyards deliver highly customized vessels where each project is unique and operations are precisely planned and controlled, offering an ideal context to explore how service value and efficiency value are harmonized in settings that demand precision and tailored solutions. In contrast, software companies adapt standardized modules to meet customer needs through flexible processes, providing a complementary perspective on balancing these values in rapidly evolving, scalable environments. All target firms operate internationally, their main operations are in a European country (different for each firm), they are medium to large, and they are considered successful in their business. They all offer unique and/or customized solutions to their customers, but the extent of customization differs.

Data were gathered primarily through individual, in-depth interviews [18] with employees representing various phases in the solution design and delivery process, such as customer interface, customer needs translation and solution design, and implementation and execution. In total, we conducted 12 interviews with such key informants that have first-hand knowledge of customized solution design and delivery. The interviewees included top managers (business development manager, managing director, head of product lifecycle management), middle managers (product managers, sales managers), and experts (chief data officer, consultant). Additionally, some secondary data were gathered from the firms' internal documents and public sources, to offer background information and verify some interview findings.

A semi-structured interview outline was developed for the interviews. The questions covered the nature of the customized solutions, the process of solution delivery, perceptions of service value and efficiency value, and internal and external balancing mechanisms of these values. We encouraged informants to answer questions based on their previous project delivery experiences. The interviews lasted 45–60 min, they were held either face to face or online, and they were audio-recorded and transcribed.

To analyze the data, we compared and interpreted the interview transcripts, researcher's notes, and relevant secondary data. Guided by our pre-understanding of theory and an initial open-ended reading of the transcripts, the analysis proceeded in the following steps. First, each insight was assigned a code [19, 20] if it appeared in at least one primary source; data to service value and efficiency value were coded separately from those referring to internal and external balancing mechanisms. Second, we identified 3 key aspects in the perception of service and efficiency values—benefits and sacrifices (short-term and long-term), related lifecycle phases, and enablers—and developed 10 themes (Fig. 1) related to the balancing mechanisms. Third, we categorized these ten themes into internal and external balancing mechanisms and synthesized the findings to construct a comprehensive understanding of how project actors perceive and balance service and efficiency values. Additionally, we conducted a comparative analysis across industries (shipbuilding versus software) and roles (managing, customer-facing, and operational/engineering) to reveal nuanced differences in these perceptions and balancing strategies. Our theoretical pre-understanding initially guided the coding process, and we revisited the theory, data, and coding structures iteratively until stable, coherent findings emerged [19, 20].

4 Findings

4.1 Project Actors' Perception of Service Value and Efficiency Value

The empirical evidence (in Fig. 2) illustrates how project actors perceive service value and efficiency value. Their perception of these two values mainly focuses on three aspects.

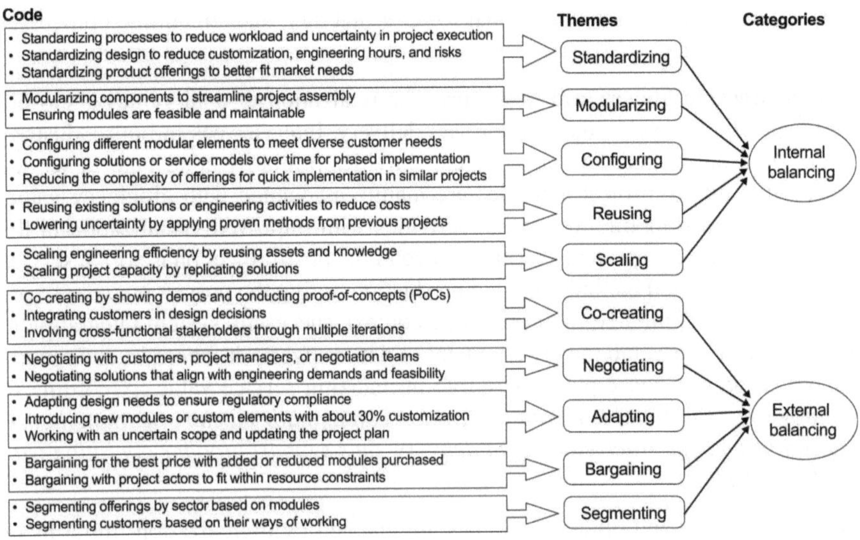

Fig. 1 Coding for internal and external values balancing mechanisms

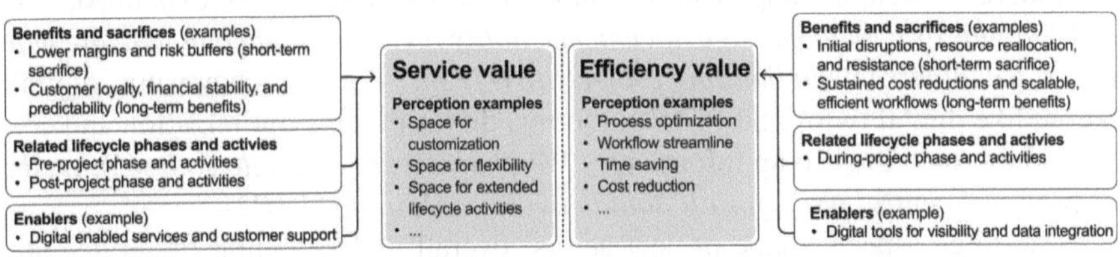

Fig. 2 Project actors' perception of service value and efficiency value

Benefits and sacrifices (short-term and long-term). Both values refer to a compromise of benefits and sacrifices and could be considered and achieved through a single project (short-term) or a sequence of multiple projects (long-term). Project actors in different project contexts understand these differently. In shipbuilding, managing and customer-facing actors perceive short-term sacrifices—such as accepting lower margins and investing in risk buffers—as necessary investments that secure long-term customer loyalty, financial stability, and predictability in highly customized projects. According to an interviewee: *"Create a long-term relationship with customers is a long win. If the customer is not happy, then next time they will seek a different owner [SB2P2]."* While operational and engineering teams focus on achieving immediate efficiency gains by streamlining production processes and optimizing labor allocation, which helps reduce the complexity and high costs inherent in customized projects. As an interviewee stated, *"Each ship requires around 700,000 to 1,000,000 man-hours, so even a 10% efficiency gain leaves a large portion of costs tied to labor [SB2P2]."* Conversely, in the software industry, project actors view the short-term costs of accommodating customer needs (for instance, adapting standardized modules) as strategic investments that foster scalability, and streamlined processes over the long term.

Related lifecycle phases and activities. The realization of dual values is closely tied to specific lifecycle phases in project delivery. Interviewees in both industries indicate that pre-project and post-project activities primarily contribute to service value, while during-project activities are mainly associated with efficiency value. For example, in shipbuilding, design activities with improved visualization in the pre-project phase and proactive maintenance in the post-project phase drive service value, and digitalized engineering and production during the project phase enhance efficiency value. Similarly, in the software context, successful system integration in the pre-project phase and the development of data insights, dashboards, and reports in the post-project phase play a key role in realizing service value. According to an interview, *"PLM and IoT data integration can monitor machine working hours, material usage, and track equipment delivery, thus support broader operation with insights [SW1P2]."*

Enablers. Service and efficiency values are enabled by digitalization. In terms of service value, digital tools such as cloud platforms and document management systems facilitate rapid software updates, prompt notifications, and efficient revisions, thereby strengthening customer support. As one interviewee explained, *"We use PLM platform to do manage documents, notify revisions, and communicate with multi stakeholders in shipbuilding projects [SW1P1]."* For efficiency value, digitalization, like digital twins and PLM platforms, drives process optimization and cost reductions. According to an informant: *"We aim to build a 'single source of truth,' meaning one central place for all data to avoid confusion [SB1P2]."*

These three aspects shape project actors' overall understanding of service and efficiency values. For service value, project actors' perception examples include the space for customization, flexibility, and extended after-project activities. For efficiency value, perception examples encompass process optimization, streamlined workflows, time savings, and cost reductions. Moreover, these companies pursue both values simultaneously throughout the project process, necessitating a balance between them.

4.2 Mechanisms to Balance Service Value and Efficiency Value

The empirical evidence (in Fig. 1) shows the internal and external mechanisms that project actors use to balance service value and efficiency value and create harmony potentially driving project success.

Internal Balancing. Mechanisms such as standardizing, modularizing, configuring, reusing, and scaling help harmonize the dual demands of satisfying customer needs and maintaining efficient operations. For example, a modular design enables different product combinations to be formed quickly, supporting both operational efficiency and customized outcomes: *"We have a technology platform, and it consists different modular elements. They are like Lego bricks, and you can just combine them into combination based on customer needs [SW2P1]."* Design reuse increased serviceability but also efficiency: *"To improve efficiency, the shipyard reuses designs, such as propulsion systems, allowing earlier documentation submission to suppliers, which can speed up production [SB1P1]."* These actions work by aligning different internal actors—those focused on sales, product development, implementation, and delivery—to work with flexibility and efficiency. For instance, standardizing and configuring processes ensure that even customized projects follow common templates, while modularizing and reusing designs enable faster turnaround and reduce re-engineering efforts.

External Balancing. Actions such as co-creating, negotiating, adapting, bargaining, and segmenting facilitate alignment between the customer side and the solution provider side. These external mechanisms are crucial for ensuring that client expectations are met while also maximizing productivity. Through co-creation and negotiation, project actors engage with customers to tailor solutions that reflect their unique needs. As one informant explained: *"We try to provide a good service with our customers, and they are happy, and they come back, and we try to be flexible [SB2P2]." At* the same time, adapting and bargaining help accommodate differing priorities between external stakeholders, ensuring that service enhancements do not compromise efficiency.

5 Conclusions

These findings advance our understanding of value creation in project business, customization, and innovation research in three ways. Firstly, service value and efficiency value, as two more specific values, are identified and clarified, thereby adding to prior value emphasis like "added value" [13] and "strategic value" [14]. Secondly, the study introduces a dual-value lens for understanding the trade-offs between two values by incorporating insights from dual pursuits [10–12], which fills the gap regarding these critical values in value creation. Thirdly, by responding to the need for a more nuanced understanding of values [15, 16] from a dual-value perspective, the study adds both macro-level aspects, such as enablers, and micro-level aspects, such as related lifecycle phases and activities, to the overall understanding of values. By integrating these findings with previous knowledge [1–4], value can be understood more holistically in terms of what constitutes value, how to create it, and how to perceive it through a dual-value lens.

The resulting thematic framework offers a new comprehensive approach to creating harmony and synergy between values [17] and resolving tensions and trade-offs that are critical for project success [10–12]. Furthermore, evidence from software firms and shipyards contributes to mass customization research in unconventional contexts [21]. Unlike traditional mass customization industries, software firms provide digital applications while shipyards deliver highly tailored vessels, representing extreme cases of service-driven and engineering-intensive customization.

The findings have implications for practitioners in B2B solution delivery. Identification of values and related trade-offs helps practitioners communicate and negotiate priorities. The balancing mechanisms help them structure actions to create synergy and harmony, to consequently achieve project success.

The focus on software firms and shipyards, the selection of informants, and the number of informants and firms constrain the broader applicability of the findings. The use of qualitative analysis limits generalizability. In the future, researchers could employ quantitative methods, such as surveys, to test the findings and explore individual-level views and other industry contexts. Additionally, this study adopts a relatively static perspective, failing to consider potential shifts and changes within the project lifecycle. Therefore, we recommend that future research examine how tensions and balancing methods evolve throughout different project phases and assess the impact of project shifts on these mechanisms. Lastly, researchers could explore the implications of project-level value balancing at the firm, supply chain, and network levels.

Acknowledgements This research was conducted as part of the EU Horizon project SEUS (Smart European Shipbuilding, Grant agreement nr. 101096224), funded by European Union. We gratefully acknowledge the support of this funding, and the partner organizations involved in this research.

References

1. Ahola T, Laitinen E, Kujala J, Wikström K (2008) Purchasing strategies and value creation in industrial turnkey projects. Int J Project Manage 26(1):87–94
2. Minerbo C, Brito LAL (2021) An integrated perspective of value creation and capture: a systematic literature review. J Busi Ind Mark 37(4):768–789
3. Martinsuo M (2020) The management of values in project business: adjusting beliefs to transform project practices and outcomes. Proj Manag J 51(4):389–399
4. Artto K, Ahola T, Vartiainen V (2016) From the front end of projects to the back end of operations: managing projects for value creation throughout the system lifecycle. Int J Project Manage 34(2):258–270
5. Brady T, Davies A, Gann DM (2005) Creating value by delivering integrated solutions. Int J Project Manage 23(5):360–365
6. Lusch RF, Nambisan S (2015) Service innovation. MIS Q 39(1):155–176
7. Witell L, Snyder H, Gustafsson A, Fombelle P, Kristensson P (2016) Defining service innovation: a review and synthesis. J Bus Res 69(8):2863–2872
8. Serrador P, Turner R (2015) The relationship between project success and project efficiency. Proj Manag J 46(1):30–39
9. Sun X, Zhu F, Sun M, Müller R, Yu M (2020) Facilitating efficiency and flexibility ambidexterity in project-based organizations: an exploratory study of organizational antecedents. Proj Manag J 51(5):556–572

10. Kasiri LA, Cheng KTG, Sambasivan M, Sidin SM (2017) Integration of standardization and customization: impact on service quality, customer satisfaction, and loyalty. J Retail Consum Serv 35:91–97
11. Clauss T, Kraus S, Kallinger FL, Bican PM, Brem A, Kailer N (2021) Organizational ambidexterity and competitive advantage: the role of strategic agility in the exploration-exploitation paradox. J Innov Knowl 6(4):203–213
12. Grote G, Kolbe M, Waller MJ (2018) The dual nature of adaptive coordination in teams: balancing demands for flexibility and stability. Organ Psychol Rev 8(2–3):125–148
13. De Chernatony L, Harris F, Riley FDO (2000) Added value: its nature, roles and sustainability. Eur J Mark 34(1/2):39–56
14. Martinsuo M (2019) Strategic value at the front end of a radical innovation program. Proj Manag J 50(4):431–446
15. Lenka S, Parida V, Wincent J (2017) Digitalization capabilities as enablers of value co-creation in servitizing firms. Psychol Mark 34(1):92–100
16. Bilge P, Emec S, Seliger G, Jawahir IS (2017) Mapping and integrating value creation factors with life-cycle stages for sustainable manufacturing. Procedia CIRP 61:28–33
17. Smith W, Lewis M (2022) Both/and thinking: embracing creative tensions to solve your toughest problems. Harvard Business Press, Boston
18. Yin RK (2009) Case study research: design and methods, vol 5. Sage, Thousand Oaks, CA
19. Gioia DA, Corley KG, Hamilton AL (2013) Seeking qualitative rigor in inductive research: notes on the Gioia methodology. Organ Res Methods 16(1):15–31
20. Locke K, Feldman M, Golden-Biddle K (2022) Coding practices and iterativity: beyond templates for analyzing qualitative data. Organ Res Methods 25(2):262–284
21. Semini M, Haartveit DEG, Alfnes E, Arica E, Brett PO, Strandhagen JO (2014) Strategies for customized shipbuilding with different customer order decoupling points. Proc Inst Mech Eng Part M: J Eng Marit Environ 228(4):362–372

Data-Driven Order Management for Built-to-Stock Products in Multi-variant Series Production

Jonas Strecker, Simon Duerr, and Stephan Daurer

Abstract In multi-variant series production, companies face the challenge of ensuring factors like flexibility and efficiency in planning and ordering despite external and internal turbulences, while simultaneously guaranteeing production as well as procurement stability. To address this challenge, the concept of planned orders has the potential to holistically optimize over the entire planning and ordering process from generating fully specified product configurations, followed by the scheduling of these to valid virtual production programs, until the assignment of incoming customer and stock orders. In doing so, precise material requirements forecasting and simulation-based analyses of future production scenarios are directly integrated in the order-to-delivery (OTD) process, facilitating the early identification and avoidance of potential bottlenecks and risks. This paper introduces a data-driven approach that optimizes the distribution of build-to-stock (BTS) orders across the global sales network, enabling the recommendation of demand-oriented product configurations to dealers. Therefore, several data sources such as historical orders, product structure and market as well as dealer information are taken into account to train the developed algorithm. Coupled with the consideration of existing constraints and given uncertainties, the method offers the ability to allocate pre-planned orders across the dealer network, with the objective to align future customer demand optimally. The presented approach is validated by a real-world use case of the Dr. Ing. h.c. F. Porsche AG (PAG) to demonstrate the potential of significantly improving the allocation of orders to the dealers in terms of fulfilling upcoming customer requirements, while guaranteeing stable production and procurement processes.

Keywords Planned orders · Mass customization · Data-driven technologies · Order fulfillment · Uncertainty

J. Strecker (✉) · S. Duerr
Dr. Ing. h.c. F. Porsche AG, Stuttgart, Germany
e-mail: jonas.strecker@porsche.de

S. Daurer
DHBW Ravensburg, Ravensburg, Germany

M. Manns (ed.), *New Paradigms for Anticipated Uncertainty*, Lecture Notes in Mechanical Engineering, https://doi.org/10.1007/978-3-032-16889-4_33

1 Introduction

In recent years, multi-variant series production has been increasingly influenced by a volatile, uncertain, complex, and dynamic environment [1]. In order to remain competitive in an era of globalization, cost pressures, and rapidly changing customer demands, the need of optimizing internal as well as external processes throughout the entire supply chain is omnipresent. The goal is to achieve cost reductions, shorter lead times, and a product and service quality tailored to customer requirements [2]. Mass Customization (MC) emerged as a successful method to handle the challenges of a dynamic environment. As described in [3], the goal of MC is to reach a high customer satisfaction by providing each customer its individualised product while maintaining a low-cost and efficient production without extensions in lead time. A distinction in order fulfillment strategies is made between build-to-order (BTO) and BTS [4]. While in BTO-systems the production is initiated by receiving customer specific orders, BTS refers to production for stock based on forecasted customer demand [5, 6]. BTO configurations remain mostly stable, while BTS orders have to be aligned at an early stage with future customer demand. This difference makes it much more complex to handle BTS products. Therefore, the approach proposed in this paper describes a data-driven method for an intelligent recommendation of BTS products across a global sales network by creating so called planned orders targeting to achieve two primary objectives: the fulfillment of customer requirements and the optimization of production stability. The main contributions are briefly as follows: (1) Identification of the research gap with regard to uncertainty in order fulfillment focusing on the concepts of virtual-build-to-order (VBTO) and planned orders. (2) Data-driven allocation of BTS products across a global sales network. (3) Identification of further research potential to enhance the proposed approach.

The paper is structured as follows: In Sect. 2, the literature review is introduced. In Sect. 3, an approach for managing BTS products is presented, which is used in Sect. 4 to be applied on a real-world problem of the PAG. Section 5 provides a conclusion and an outlook for future work.

2 Literature Review

2.1 Order Fulfillment Under Uncertainties

The primary objective of the entire order fulfillment process in multi-variant series production is to realize short delivery times for the products specified by the customer. A reliable OTD process enhances planning reliability and minimizes overall costs across the entire supply chain, from procurement to manufacturing and distribution [7]. In addition to BTS and BTO, VBTO is another innovative order fulfillment strategy characterized by a planning pipeline consisting of product configurations [5]. This pipeline is divided into a virtual and a physical part. The physical part consists of products that cannot be altered, as they are either in production or located in the factory or dealer warehouses. The virtual part contains all products until the start of production which can be configured according to the customer's requirements. To fill the virtual part, fully specified virtual orders are generated that ideally correspond to actual customer demand. VBTO enables the optimal allocation of a product to a customer along the entire pipeline [5].

2.2 *Planned Orders*

The concept of planned orders serves as the fundamental for VBTO. The description of the planned order object has to be identical to a customer order object, except for the assignment to the customer. The concept enables the anticipation of customer requirements by generating fully specified planned objects and subsequently allocating real customer orders to the planned orders. This allows the integration of sales, production and material requirements planning in the medium and short-term planning horizon. The result is a more efficient utilization of resources to satisfy customer requirements. This approach fosters stability and efficiency within the multi-variant production and its supply chains, even in the absence of customer orders within the medium-term planning horizon [8].

The concept of the planned order is divided into three components. First, a finite number of valid, buildable product configurations must be generated, each defined by multiple options and governed by product model-specific buildability rules. To create a valid production plan, the generated configurations are scheduled into the production, while considering all relevant limitations, particularly factory restrictions and supplier capacities. Finally, incoming customer orders are allocated to suitable, available planned orders with the objective of minimizing the costs of adjustments and delivery date postponements [9].

2.3 *Optimization Problems in Multi-variant Production*

The field of Operations Research (OR) deals with the analysis of practical problems that arise in the context of planning processes with the aim of making well-founded decisions through the use of mathematical methods. The main task is to represent a real problem in the form of a mathematical optimization or simulation model and its solution [10]. Within the domain of production planning OR-models seek to optimize specific aspects of resource allocation and production control. Especially in the area of multi-variant production, mathematical optimization approaches have been introduced in the literature in various forms [1]. While in [8, 9] approaches are presented using Integer Programming (IP) and Linear Programming (LP), Constrained Programming (CP), metaheuristics or hybrid variants are often used to handle complex scheduling problems [11].

3 Concept Elaboration

3.1 *Research Gap and Problem Definition*

As described in Sect. 2 the concept of VBTO and planned order offer the potential to improve flexibility and stability in production planning. An important functionality of the order fulfillment strategy is the optimal allocation of incoming orders to the already planned elements. However, it is not defined how planned orders that cannot be allocated to real customer orders can be optimally distributed into the sales network. Data-driven technologies in the area of OR emphasize the potential to handle challenges in multi-variant production.

Dealers often lack solid decision-making tools for selecting product configurations optimized for future market-specific demand. A tailored set of BTS recommendations for dealers H within the global sales network is required. For this purpose, a set of planned orders P has to be scheduled in a virtual way at an early planning stage. Each planned order $p \in P$ is derived from a set of vehicle configurations C and scheduled as a virtual data object in the production pipeline. Each configuration $c \in C$ consists of a set of options O and is considered buildable, meaning that all constraints and exclusions among options $o_i, o_j \in O$, where $i \neq j$, are properly taken into account. Furthermore, sales planning defines the planned sales volume v_h for each dealer $h \in H$. The total planned sales volume equals the number of scheduled planned orders P. Due to the numerous possibilities, several data sources need to be considered to anticipate take rates for each option $o \in O$. Each take rate e_{oh} is assigned to a dealer h, an option o and defines the percentage of orders, which are equipped with the option o. The methodology for accurately approximating future take rates is beyond the scope of this paper and is assumed to be given. This area, particularly concerning the automotive industry, has been explored in previous studies, see [12].

3.2 *Data Utilization*

To develop a concept applicable to the defined problem, the necessary data basis is first specified. For the allocation of unassigned planned orders, a set of dealers $h \in H$ is identified. A set of generated configuration proposals $c \in C$ is defined, from which the planned orders $p \in P$ will be selected in the subsequent process. Each c and p represent a feasible configuration, consisting of a specific number of options $o \in O$. Each $p \in P$ is confirmed to align with current production and supplier limits and is already scheduled in the production. Additionally, the sales plan V is required, specifying the planned sales volume for each dealer $h \in H$. For this specific problem, only the open planned sales volume V, not yet covered by existing orders, is needed. Products in multi-variant series production should always be optimally adapted to customer requirements. To reflect customer needs, the forecasted take rates $e \in E$

of options at the dealer level are required. In general, the database related to the presented approach can easily be extended with additional input data, such as further constraints or objectives.

3.3 Data-Driven Order Management for Built-to-Stock Products

This section introduces a concept for solving the challenge of allocating pre-planned orders across the dealer network by translating the defined problem into a mathematical optimization model that includes the necessary constraints and an objective function. The concept is separated into two kinds of methods, see Fig. 1: Algorithm I introduces an approach to recommend BTS products to the sales network keeping in mind, that the amount of planned orders is equal to the capacity limit of the volume sales plan. In contrast, Algorithm II optimizes the recommendation of BTS products by taking into account all possible configuration suggestions, which are not pre-validated against capacity constraints.

While Algorithm I offers the advantage of applying production stability, Algorithm II can contribute to provide a greater flexibility to fulfill the objective due to the larger search space of possible recommendations. The details are described in the following:

Algorithm I: Optimized Allocation of Planned Orders
Initially, the three essential constraints are defined. (1) defines that each decision variable x_{hp} assumes only values 0 or 1, ensuring that each planned order p is either assigned to a dealer h or not. The equality constraint (2) ensures that the number of planned orders assigned to each dealer h corresponds exactly to the dealer's available sales volume v_h. This results in a complete allocation, preventing slots from remaining unused or dealers from being assigned too many planned orders. The equality constraint (3) must be specified to prevent any planned order from being assigned to multiple dealers.

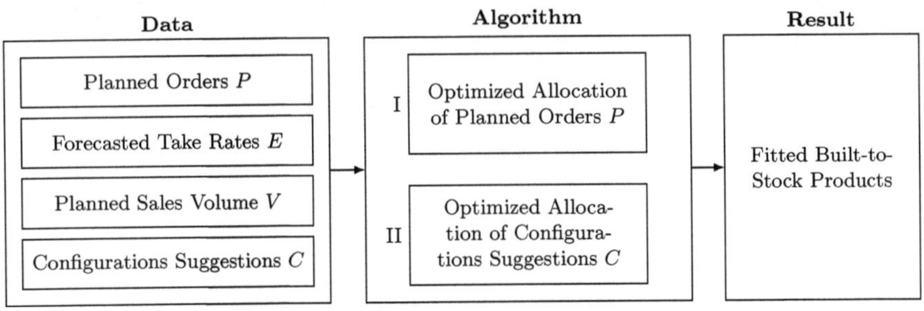

Fig. 1 Process model: data module, algorithm module, and output module

$$x_{hp} \in \{0, 1\}, \forall h \in H,\ p \in P \tag{1}$$

$$\sum_{p \in P} x_{hp} = v_h, \forall h \in H \tag{2}$$

$$\sum_{h \in H} x_{hp} = 1, \forall p \in P \tag{3}$$

The overarching goal of optimization is to increase the probability of a sale by aligning the distributed configurations with future customer needs. To represent customer needs as accurately as possible, the optimization should approximate the predicted take rates E. Therefore, it is essential that the take rates closely reflect future customer demand. The objective function minimizes the deviation between the forecasted take rates and the average take rates realized by the allocated planned orders for each option and dealer, thereby equally penalizing both over- and underestimation of the forecast values:

$$\min\ f(x) = \sum_{o \in O} \sum_{h \in H} \left| \mathrm{e}_{oh} - \frac{1}{\sum_{p \in P} x_{hp}} \sum_{p \in P} x_{hp} \cdot b(o, p) \right| \tag{4}$$

- $f(x)$: value of the objective function that is to be minimized.
- e_{oh}: forecasted take rate of option o for dealer h.
- $b(o, p)$: returns 1 if the option o is part of planned order p, else 0.
- $\sum_{p \in P} x_{hp}$: sum of the assigned plan orders.
- $\frac{1}{\sum_{p \in P} x_{hp}} \sum_{p \in P} x_{hp} \cdot b(o, p)$: average achieved take rate of option o for dealer h over all assigned planned orders.

Algorithm II: Optimized Allocation of Configuration Suggestions

The Algorithm II is similar to the Algorithm I described above. Instead of the set of planned orders P, the configuration suggestions C are distributed. Accordingly, P must be replaced by C. As these are not already scheduled planned orders, the configurations C can be allocated multiple times to a dealer h and not every configuration has to be allocated. This results in the following constraints ensuring that the planned sales volume v_h is covered by selecting a subset of C:

$$x_{hc} \in \mathbb{N}_0 \quad \forall h \in H, c \in C \tag{5}$$

$$\sum_{c \in C} x_{hc} = v_h \quad \forall h \in H \tag{6}$$

4 Real-World Application

4.1 Use Case Specification

To evaluate the approach, real-world data was used to generate 10 000 vehicle configurations of the model type "Taycan (Basic)" of the PAG by using the approach presented in [13]. Each configuration $c \in C$ is a combination of 235 options. In order to validate the robustness of the approach for the process of allocating BTS products, a simulation environment with 20 dealers–to depict a realistic dealer network–and 100 independent runs was constructed. To ensure reproducibility, a fixed seed of 42 was initialized. In each run, forecasted take rates E for all options O and sales volume V are sampled at the dealer level. To obtain forecasted take rates E per dealer h, uniform random values were generated for each option o, respecting all feasibility rules. For each dealer $h \in H$, a sales volume v_h is drawn from a discrete uniform distribution over [0, 200]. In the planning phase, configurations are randomly scheduled as planned orders until the target volume is reached; afterwards a subset of size U_h, drawn from a discrete uniform distribution over $[0, v_h]$, is randomly selected as customer orders. Next, the open sales volume v_h for each dealer $h \in H$ is allocated using the following methods: random allocation of already scheduled planned orders P as baseline, optimized allocation of planned orders P (see Algorithm I) and optimized allocation of configuration suggestions C (see Algorithm II). The optimization models were formulated via Google OR-Tools and solved by the CP-SAT solver, using a 200-second time limit per instance. Finally, the deviation from the forecasted take rates was evaluated using the Mean-Absolute-Error (MAE), which provides an interpretable average deviation and penalizes exceeding and undershooting the take rates equally.

4.2 Evaluation Results

Figure 2 validates the concept by demonstrating that both proposed approaches achieve a substantially lower MAE than a random selection. A lower MAE indi-

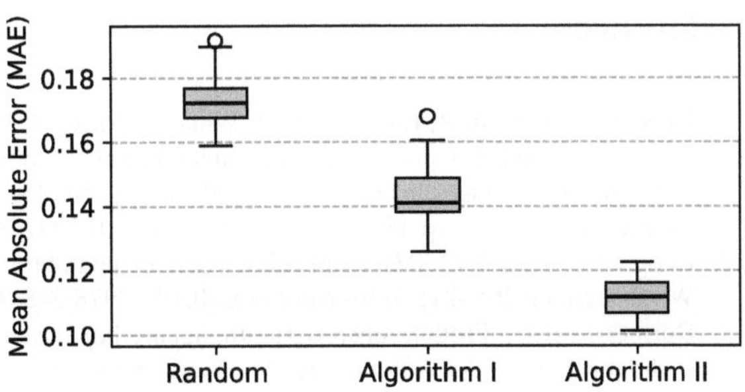

Fig. 2 Comparison of the methods based on the deviation from forecasted take rates

cates a closer alignment with the forecasted take rates, thereby ensuring a better fulfillment of customer demand. The spread of results in all three methods can be explained by the simulation of randomized variable assignments. However, data-driven allocation methods provide a significantly better approximation of take rates. In particular, the configuration suggestion allocation (see Algorithm II) shows a significantly lower MAE than the planned order allocation (see Algorithm I). This outcome aligns with the expectation that a broader range of configurations–despite the increased complexity–yields more accurate solutions. In conclusion, these results must be interpreted in the context of the specific use case: While Algorithm II achieves the best possible take rates, Algorithm I maintains production stability by relying solely on already planned orders.

5 Conclusion and Future Work

In this paper, an approach for data-driven order management for BTS products fitted to customer demand was introduced. The concept can be used to directly allocate already planned orders or configuration suggestions to dealers. The evaluation on a real-world use case demonstrated its potential for maintaining production stability and fulfilling customer demand in multi-variant series production. The work provides high potential for future research. The relationship between meeting forecasted take rates and sales probability to explore the economic impact of an optimized BTS product recommendations for the dealers is worth to be investigated. For that purpose, a comparison between the recommended dealer stock and the real sales probability including the evaluation of purchasing relevant options can prove the benefits. Additionally, the transferability of the concept to other industries offers a promising way to test its efficiency beyond the automotive sector. Furthermore, the development and optimization of methods to predict take rates at a more precise and granular level could improve BTS allocations, since the take rates serve as the basis for the optimization problem and directly determine the effectiveness of the allocation. By focusing on these areas, the concept could be refined and adapted to meet broader and more diverse challenges in data-driven order management.

References

1. Duerr S, Lamprecht R, Kauffmann M, Winter J, Alexy H, Huber M (2021)Development of an integrated data-driven process to handle uncertainties in multi-variant production and logistics: a survey. In: Advances in automotive production technology–theory and application. Springer Vieweg, Wiesbaden, pp 486–494. https://doi.org/10.1007/978-3-662-62962-8_56
2. Chopra S, Meindl P (2007) Supply chain management. Strategy, planning & operation. Gabler, Wiesbaden, pp 265–275. https://doi.org/10.1007/978-3-8349-9320-5_22
3. Dörmer J (2013) Produktionsprogrammplanung bei variantenreicher Fließproduktion: untersucht am Beispiel der Automobilendmontage. Springer

4. Klug F (2018) Logistikmanagement in der Automobilindustrie. Springer, Berlin, Heidelberg. https://doi.org/10.1007/978-3-662-55873-7
5. Brabazon PG, MacCarthy B (2006) Fundamental behaviour of virtual-build-to-order systems. Int J Prod Econ 104(2):514–524. https://doi.org/10.1016/j.ijpe.2006.02.007
6. Meyr H (2004) Supply chain planning in the German automotive industry. OR Spectr 26(4). https://doi.org/10.1007/s00291-004-0168-4
7. Lochmahr A, Wildemann H (2007) Die Einführung logistischer Konzepte in Theorie und Praxis–Fallbeispiele Kapazitätsmanagement. Springer, Berlin, pp 509–524. https://doi.org/10.1007/978-3-540-72556-5_14
8. Duerr S (2023) Entwicklung eines datenbasierten Prozesses zur Steuerung der Auftrags- und Planungssysteme in der variantenreichen Serienproduktion unter Unsicherheit. Stuttgarter Beiträge zur Produktionsforschung. Fraunhofer IPA, Stuttgart. https://doi.org/10.18419/opus-13980
9. Buergin J, Beisecker J, Fischer S, Geier B, Tutsch H, Mercamp S, Lanza G (2017) Generation of planned orders and their matching with customer orders in multi-variant series production. Procedia CIRP 61:499–504. https://doi.org/10.1016/j.procir.2016.11.159
10. Domschke W, Drexl A, Klein R, Scholl A (2015) Einführung in operations research. Springer, Berlin, Heidelberg, pp 1–16. https://doi.org/10.1007/978-3-662-48216-2_1
11. Hashimoto H, Kubo M (2016) Supply chain optimization: a survey: models, algorithms, decision support systems, and applications. Innov Supply Chain Manag 10:9–18. https://doi.org/10.14327/iscm.10.9
12. Gebhardt J, Detmer H, Madsen AL (2003) Predicting parts demand in the automotive industry–an application of probabilistic graphical models. In: Proceedings of the international joint conference on uncertainty in artificial intelligence (UAI'03, Acapulco, Mexico), Bayesian modelling applications workshop
13. Duerr S, Lamprecht R, Kauffmann M, Huber M (2021) Reinforcement learning based optimization of Bayesian networks for generating feasible vehicle configuration suggestions. In: 2021 IEEE 17th international conference on automation science and engineering (CASE), pp 16–22. https://doi.org/10.1109/CASE49439.2021.9551428

Emotional Durability Through Mass Customization in the Fashion and Apparel Industry

Daniella Besse, Jocelyn Bellemare, Claudia Déméné, and Marie-Eve Faust

Abstract Extending the lifespan of clothing by strengthening the user-object bond is becoming a growing subject of research to reduce the environmental impact of the fashion industry. Meanwhile, mass customization, based on the use of flexible manufacturing systems and digital configurators, enables the creation of garments which are tailored to individual preferences while maintaining production efficiency, which seems to be an interesting avenue to explore a long lasting the user-object relationship. This study assesses the economic, and organizational implications of integrating mass customization into current value chains to reinforce emotional durability. It also critically examines the operational challenges, such as the logistical complexity and carbon footprint of customized production. This empirical analysis is based on a survey conducted in March 2025 with ten Quebec-based fashion companies involved in or exploring mass customization. Results indicated a strong perception of sustainable potential, especially regarding product durability and inventory efficiency. Moreover, there is limited consumer demand for personalization despite interest. These findings highlights the need for improved communication and operational streamlining. By focusing on the evolution of business models, this paper aims to inform both researchers and practitioners about the transformative potential of these approaches and proposes concrete pathways to support their adoption in the fashion industry.

Keywords Personalized design · Emotional durability · Sustainability in fashion · Eco-design · Circular design strategies

D. Besse · C. Déméné
Laval University, Québec, QC, Canada
e-mail: daniella.besse.1@ulaval.ca

J. Bellemare (✉) · M.-E. Faust
University of Quebec in Montreal (UQAM), Montreal, QC, Canada
e-mail: bellemare.jocelyn@uqam.ca

M. Manns (ed.), *New Paradigms for Anticipated Uncertainty*, Lecture Notes in Mechanical Engineering, https://doi.org/10.1007/978-3-032-16889-4_34

1 Introduction

The fashion industry is widely recognized as one of the most environmentally damaging sectors after oil and gas. Currently, it is standing at a critical juncture during intensifying ecological concerns. It is responsible for up to 10% of global carbon emissions. A figure that, according to some sources, rivals the annual emissions of major economic blocs like European Union [1], showing how disproportionately this sector impacts on the planet.

A significant drainage of water sources happens due to its operations through resource-heavy dyeing and finishing processes. This alone contributes to over a third of its environmental footprint [2]. At the same time, the laundering of synthetic garments releases half a million tons of microfibers into the oceans each year. It is equivalent to 50 billion plastic bottles [3], which greatly increases marine pollution. UNFCCC projections suggest that emissions from textile manufacturing could rise by 60% by 2030 if no action is taken [4]. Thus, it is highly urgent to adopt sustainable business models. The *Valuing our Clothes: the Evidence Base. Technical Report* analyzed that extending the lifespan of a garment by just three months of use would result in a 5–10% reduction in each of its carbon, water, and waste footprints, and extending it by nine months would result in a 22–33% reduction [5]. As consumers are becoming more aware of the environmental and social impact of their choices, they also aspire for unique as well as meaningful products. This dual imperative to meet consumer expectations and at the same time minimizing environmental impact makes innovation a much sough-after industry strategy. In this context, mass customization could be considered as a promising solution to reinforced emotional durability and reduced the environmental impact of the industry. Emotional durability is an approach which is centered on the affective experience as well as perceived value of products [6]. It aims to strengthen consumer attachment to their clothing. This consequently extends their lifespan. At the same time, mass customization relates to combining the principles of industrial production with product individualization, through the coordinated integration of practices such as modular design, process flexibility, and digital technologies [7]. It provides the possibility of designing garments perfectly tailored to consumer preferences. Thus, mass customization appears to be an interesting strategy to support emotional sustainability. What could be the mechanisms which are necessary for the integration of mass customization into current business models? How to analyze the emotional impact of mass customization on customers?

This work aims to explore these questions through multidimensional analysis. On one hand, we will examine the tools and methods that form the foundation for mass customization, such as digital configuration systems and modular production approaches. Their role in adapting to individual consumer needs will be emphasized. On the other hand, we will analyze how emotional durability is understood, integrated and when possible, measured within current practices. While these two concepts are distinct, we consider mass customization as one potential strategy to foster emotional durability by enabling a deeper user-product connection. Moreover, this introduction contextualizes the structural constraints and strategic possibilities associated with the use of these technologies in the fashion sector. In doing so, it initiates a deep reflection on the future of the sector which is striving to balance innovation, sustainability, and consumer satisfaction. Through this analysis, we aim to identify concrete pathways for integrating these concepts into viable business models which relate to the global sustainability goals.

2 Literature

The fashion industry has long been overlooked in sustainability-related research. Recently, it has started to receive more attention. This shift is largely due to its environmental footprint, social impacts, and lack of transparency. These three dimensions were identified by Gonçalves and Silva in their review of sustainability scoring in the apparel sector [3]. One emerging area of interest is emotional durability and its application to fashion. This concept explores how strengthening the emotional bond between people and garments might help extend product lifespans. It is particularly relevant, as it could offer a potential response to the ecological crisis.

This section reviews both pioneering and recent studies related to mass customization and emotional durability. It also highlights how these two approaches intersect and how they might contribute to more sustainable fashion practices.

The concept of emotional durability originated in Jonathan Chapman's work *Emotionally Durable Design: Objects, Experiences and Empathy* [6]. Chapman states that overconsumption is the result of an emotional void which is left by ephemeral objects designed to be quickly replaced. According to him, meaningful objects which are capable of promoting lasting emotional attachment can counteract this trend. This concept was initially explored in the field of industrial design. Further, this idea was brilliantly transposed to fashion by Kate Fletcher in her seminal book *Sustainable Fashion and Textiles: Design Journeys* [8]. Fletcher emphasizes that clothing, as carriers of personal stories and meanings, can go beyond their utilitarian function and could become catalysts of emotional engagement [9]. Ezio Manzini expands this reflection by linking social innovation and sustainability [10]. He demonstrates the way in which co-creation (a design approach that empowers consumers to participate in the creative process) can generate products with high emotional value while at the same time addressing specific needs. These works were fundamental for understanding of emotional durability, integrating social, psychological, and aesthetic dimensions.

At the same time, mass customization is becoming a fertile ground for combining sustainability and innovation. This approach, detailed in the work of Pine and Gilmore (*The Experience Economy*, 1999), is based on the idea that consumers want more than just products—they crave experiences [11]. In fashion, this aspiration means a growing demand for personalized items that show unique identity of every individual. Although mass customization is not in itself a strategy of emotional durability, it can act as a powerful lever to support it. Particularly when personalization strengthens the user's emotional attachment to a garment, increasing its perceived value and lifespan. Studies such as those by Maldini et al. confirm that mass customization can extend the lifespan of a garment. If specific preferences are integrated into the design, these garments become less likely to be perceived as obsolete [12]. Additional research by Fletcher and Grose further supports this idea [9]. Their work shows that involving users in the design process (whether through co-design or customization) fosters emotional engagement, which in turn can encourage consumers to keep their garments longer.

Moreover, these studies highlight the importance of flexible production models and digital systems that make customization more accessible and relevant to sustainability goals. By reducing overproduction, optimizing inventories, and creating garments that better match user needs, mass customization contributes to more efficient and less wasteful value chains. In a world marked by overabundance, Carl Honoré provides a strong critique of speed and instant gratification as cultural norms [13]. His advocacy for slowing down resonates particularly in the fashion industry, where relentless production and consumption cycles are the cause behind rapid obsolescence. Tim Cooper, in *Longer Lasting Products: Alternatives to the Throwaway Society* [14], supports this further and argues that product longevity, combined with meaningful design, could transform our relationship with consumption. Finally, Stone and Farnan (*The Dynamics of Fashion*, 2018) situate these debates within a more pervasive context [15]. He analyzes the tensions between technological innovation, consumer expectations, and sustainability. This shows that there is importance of understanding how traditional business models can evolve toward practices that integrate social and environmental values.

2.1 Gaps Between Theory and Practice

The existing literature converges to show that mass customization could be an interesting tool to explore to answer emotional durability in the fashion industry. Together, they provide answers to the complex challenges faced by the fashion industry. These work by reducing overproduction, extending product lifespans, and promoting mindful consumption. However, several questions are still there: Is the concept of emotional durability understood by design practitioners and how is it assessed within companies? Is it evaluated using frameworks comparable to those applied in eco-design? Furthermore, how do companies that employ personalization for marketing purposes differ from those that adopt it as a strategy to foster sustainability? And finally, how can the ecological impacts of customization processes, particularly in terms of logistics and technology, be mitigated? The objective of this paper is twofold: first, to understand the underlying mechanisms involved in integrating mass customization and emotional durability within the fashion industry, and second, to assess the implications of these strategies on business models.

3 Methodology

This study employed a mixed exploratory approach. It combined qualitative analysis of literature with an empirical evaluation of related industrial case studies. Data collection was carried out through an online survey conducted in March 2025. The study population consisted of fashion manufacturers, brand owners, and other industry stakeholders. The final sample included a total of ten participants who met the selection criteria. The inclusion criteria consisted of being based in Quebec, having a minimum of 5 years of experience in clothing design and manufacturing, and being actively involved in mass customization. Among the respondents, six are micro-businesses (1–10 employees), two are small businesses (11–50), and two are medium-sized enterprises (51–250).

4 Results

The fashion industry is changing rapidly. There is an increasing emphasis on integrating more sustainable practices and responding to individual needs of the customers. Mass customization relates to this trend. It enables brands to offer unique garments and at the same time strive to reduce waste. In this study, 10 fashion-related companies based in Quebec responded to the survey. The results showed that interest in customization is high. Eight of the respondents already offer some form of customization, and five of them have integrated it into their core business model. However, these companies do not clearly recognize mass customization as a means of fostering emotional durability, likely because they do not fully understand the concept. This is notable, given that all of the respondents claim to offer sustainable fashion. Despite these efforts, implementation remains inconsistent and faces significant challenges.

4.1 Misunderstanding of Emotional Durability?

In response to the multiple-choice question, "What sustainability strategies are currently implemented within your company?," nine out of ten companies selected "emotional durability." However, when asked, "Do you measure consumers' emotional attachment to your products?" four admitted being unfamiliar with the subject, and four lacked the necessary skills or resources to develop such initiatives. Four companies indicated that they assess emotional durability through customer loyalty and repeated purchases, which is proof of brand attachment, but not necessarily product attachment. This is not the only instance in which companies claiming to have adopted emotional durability (ED) provided responses suggesting a lack of understanding of the concept or a different interpretation than that found in the literature. For instance, one company referred to "offering repair services" as a way to measure ED, and another cited using sustainable materials as an ED strategy, suggesting confusion between physical and emotional durability.

4.2 *Challenges of Integrating Emotional Durability into Business Models*

Interviews with companies regarding the barriers to integrating emotional durability into their offerings revealed a strong consensus around economic constraints. Forty percent of respondents said that their current business models do not accommodate this approach well. Half of the companies mentioned a lack of the necessary knowledge, expertise, and internal resources to develop such initiatives. Furthermore, two companies reported that the cost of producing emotionally durable products would be too high.

4.3 *Positive Perception of Emotional Durability Through Mass Customization*

The results revealed that mass customization is perceived as a significant factor in more sustainable fashion practices, particularly with regard to emotional durability, as nine respondents associated it with garments having a longer lifespan. This concept has several concrete advantages, as shown in Fig. 1. Respondents also associated mass customization with on-demand production, which helps prevent overstocking. Seven respondents claimed that mass customization contributes to reducing material waste. Half of the companies emphasized its potential to optimize inventory management.

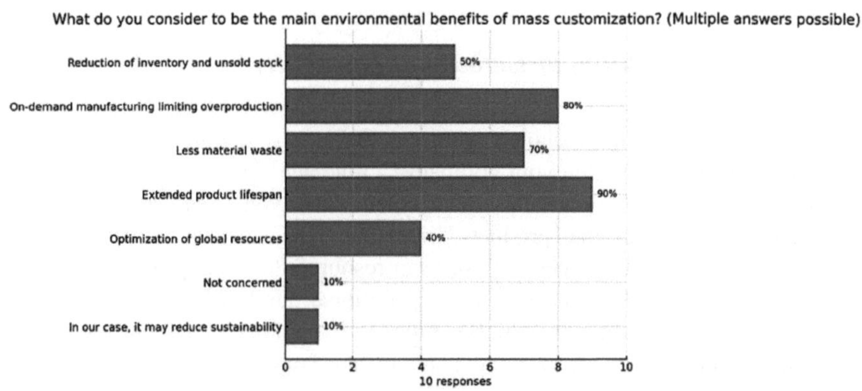

Fig. 1 Perceived environmental benefits of mass customization by Québec fashion firms

4.4 A Lack of Measurement

However, apart from this overwhelmingly positive perception, none of the companies offering customized products have implemented specific metrics to measure consumer attachment. Although personalization is often described as a tool for emotional sustainability, it remains largely unexplored in practice. This reveals a gap between belief and evidence, and highlights the need for concrete, quantifiable indicators. Very few have assessed its environmental impact. While half of the companies expressed interest in evaluating the carbon footprint of customization, only one has done so. Thus, although the link between customization and sustainability is widely acknowledged in theory, its real impact on reducing the fashion industry's environmental footprint remains largely unproven. The concept of demand-driven production aligns with sustainability goals, but there is still little data to support its effectiveness. Without measurable results, it is difficult to claim that customization significantly reduces the industry's carbon emissions.

4.5 Gradual but Incomplete Adoption

The results show that approximately 8 out of 10 surveyed companies offer some form of customization. This is implemented in various ways, including color and pattern selection (7 companies), morphological adjustments (4), material choices (5), and unique product features such as embroidery (4). However, the degree of integration varies depending on the size of the company, the type of business model, operational capabilities, and market positioning. Among the companies not engaged in customization, half cited a focus on other forms of consumer adaptation. Additionally, four companies pointed to operational and logistical complexity as key barriers, while one mentioned a lack of appropriate technologies. These reasons are illustrated in Fig. 2. Companies positioning themselves as sustainable fashion businesses were generally more advanced in adopting customization. They perceived it as a way to better align production with demand and to reduce overstock. In contrast, companies targeting mass markets through standardized offerings reported higher barriers due to the cost and infrastructure requirements associated with personalized production.

Despite these challenges, results found that 80% of companies are willing to participate in a pilot project which combines mass customization and emotional durability. This figure shows that there is a strong interest in the approach. Although there are still uncertainties which exist related to its implementation and long-term viability.

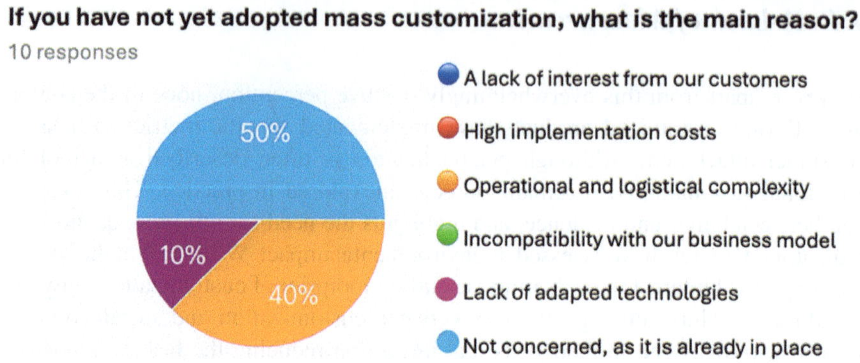

Fig. 2 Main barriers to mass customization as perceived by Quebec fashion firms

5 Discussion

Although it is being viewed as a promising future trend, pervasive adoption will depend on overcoming structural, economic, as well as consumer-related barriers. A central issue is the actual environmental impact. Furthermore, customization may increase energy use or complexity of logistics. This could counterbalance its ecological advantages, especially if the actual impact on users and the promise of longer-lasting garments are not measured. Clear performance indicators are needed to evaluate life-cycle impact and carbon footprint of mass customization. Without this data, it will be difficult to justify large-scale changes in production models. The cost remains one of the main barriers to customization. Standardized production benefits from economies of scale, while customization incurs higher costs due to stock management, specialized materials, and logistics. One strategy is to automate certain processes using technologies such as 3D printing, artificial intelligence, and digitally assisted manufacturing. Another is to develop hybrid models where customization is partial, combined with standardized elements and targeted adjustments. Reducing production and delivery times also remains a major challenge. For customers to fully adopt customization, brands must accelerate the process without compromising quality. However, its widespread adoption still requires adjustments. The strong interest shown by companies in a pilot project indicates a willingness to explore this approach further.

The next steps should involve testing new economic models and measuring the emotional impact of customization. Yet, although emotional durability is increasingly theorized, its application in professional fashion design remains limited. As shown by Kam and Yoo [16], only one out of eleven South Korean designers interviewed identified emotional durability as a relevant approach. This suggests a weak diffusion of the idea within the industry. In contrast, European studies such as Fletcher (2012–2018) [8, 9] in the UK and Bjerregaard [17] in Sweden highlight a growing interest in implementing emotional durability in practice. Together, these works emphasize the need to create accessible frameworks to support its integration into mass customization strategies.

5.1 Gap Between Supply and Consumer Demand

While most of the companies acknowledged that there is strategic promise of mass customization, the demand from consumers is nothing more than hesitancy and indifference. Only three brands have noted a modest increase in sales linked to personalized offerings. None of the brands reported any significant breakthrough. Several factors could be responsible for this disconnect. First, the perceived cost of personalized garments is often a major deterrent. Conditioned by the low prices and immediate gratification offered by fast fashion, many consumers are reluctant to pay more for clothing that reflects their individuality. Second, longer production times are also a big challenge in an era where expectations of immediacy are dominant. The made-to-order nature of customization inevitably causes delays and this clashes with the rhythm of 24–48-h deliveries that e-commerce has normalized. Finally, there is a lack of consumer awareness. The benefits of customization are uniqueness, higher quality, and environmental value. These have remained under communicated and largely misunderstood by the public. To reduce this gap, companies must not only make the customization process smoother as well as faster, but also invest in storytelling, educating consumers about what truly sets personalized products different from others. Only then can customization hope to move from niche to norm.

As previously noted, none of the participating brands reported significant breakthroughs. However, it is possible to speculate this may be due to emotional attachment to the most recently purchased product, which could reduce the need to make another purchase. Research on emotional durability remains incomplete without examining the perspective of consumers. This highlights the need to develop analytical tools aimed at capturing emotional durability from the user's point of view. The future of fashion may well depend on a balance between innovation, responsible production, and adapting to consumer expectations.

Ten companies represent a relatively limited sample. This does not undermine the empirical nature of the study, but it does restrict the generalizability of the findings. It is therefore essential to consider this research as exploratory and primarily indicative, especially if the objective is to identify trends, perceptions, or user sentiments.

6 Conclusion

Mass customization offers an alternative to traditional mass production by aligning supply with actual consumer demand. This study reveals that although most surveyed companies have already integrated some level of customization, full-scale implementation remains constrained by logistical, economic, and technological challenges. However, despite these perceived benefits, concrete data on its emotional and environmental impact is still lacking. Without clear metrics to quantify product attachment, reductions in carbon footprint, material waste, and energy use, it is difficult to determine whether customization truly supports sustainability or merely reflects a shift in production methods. Moreover, a gap exists between supply and consumer demand. While brands increasingly offer customization options, consumers remain hesitant, often due to higher costs, longer lead times, and limited awareness of its benefits. To reduce this gap, brands must better communicate the value of customization by highlighting advantages in quality, exclusivity, and environmental impact. Despite these challenges, most respondents expressed interest in participating in a pilot project to explore the integration of mass customization and emotional durability. This strong interest suggests that with the right strategies, such as lowering costs, streamlining production, and evaluating consumer attachment, customization could become a more attractive and impactful practice.

References

1. Goldsworthy K, Harris J, Hassan MM, Solomon L (2022) Catalyst report on circular fashion and textiles. University of the Arts London
2. Quantis International (2018) Measuring fashion: environmental impact of the global apparel and footwear industries study. Quantis, Boston
3. Gonçalves A, Silva C (2021) Looking for sustainability scoring in apparel: a review on environmental footprint, social impacts and transparency. Energies 14(11):3032
4. WRAP (2013) Valuing our clothes: the evidence base. Technical report
5. United Nations Framework Convention on Climate Change (UNFCCC) (2018) Fashion industry charter for climate action. UNFCCC, Bonn
6. Chapman J (2005) Emotionally durable design: objects, experiences and empathy, 2nd edn. Routledge, London
7. Trentin A, Sandrin E, Suzic S, Grosso C, Forza C (2021) Reconciling product flexibility with cost, delivery, and quality: the importance of bundling mass customization practices. J Intell Manuf 32:2025–2043
8. Fletcher K (2008) Sustainable fashion and textiles: design journeys, 2nd edn. Earthscan, London
9. Fletcher K, Grose L (2012) Fashion and sustainability: design for change. Laurence King Publishing, London
10. Manzini E (2015) Design, when everybody designs: an introduction to design for social innovation. MIT Press, Cambridge
11. Pine BJ, Gilmore JH (1999) The experience economy: work is theatre and every business a stage. Harvard Business Review Press, Boston
12. Maldini I, Duncker L, Sihvonen J (2019) Assessing the impact of design strategies on clothing lifetimes, usage, and volumes: the case of product personalization. Sustain Prod Consumption 19:132–142

13. Honoré C (2004) In praise of slow: challenging the cult of speed. HarperOne, New York
14. Cooper T (2010) Longer lasting products: alternatives to the throwaway society. Gower Publishing, Farnham
15. Stone E, Farnan S (2018) The dynamics of fashion, 4th edn. Bloomsbury, New York
16. Kam S, Yoo Y (2022) Practice of sustainable fashion design considering customer emotions and personal tastes. Front Psychol 13, Article 976471. https://doi.org/10.3389/fpsyg.2022.976471
17. Bjerregaard Jensen M (2021) Emotional durability strategies and future integration. Master's thesis, University of Borås, Sweden

Exploring the Flexibility and Impact of Hybrid Production Models in Mass Customization Within the Fashion and Apparel Industry

Anne Léger and Jocelyn Bellemare

Abstract Mass customization in the fashion and apparel industry falls between two extremes: on one end, highly personalized designs tailored to individual needs, and on the other, standardized production aimed at efficiency and profitability. Through a qualitative survey, this article explores how hybrid customization production models can bridge this gap by catering to specific consumer demands while maintaining operational flexibility. The findings suggest that these models enable a two-way interaction between brands and consumers, allowing for continuous adjustments to better meet individual preferences. However, they also place pressure on value chains, particularly in areas such as human resource management, material procurement, and production planning. A conceptual framework specific to the fashion industry is introduced to highlight the key challenges to consider, as well as the opportunities from operational, consumer, and sustainability perspectives. The study concludes with recommendations for implementing hybrid models in fashion, such as combining standardized base production with strategic customization options. These approaches have the potential to enhance customer satisfaction, improve profitability, and reduce environmental impact.

Keywords Operational flexibility · Value chain management · Consumer expectations · Sustainability · Supply chain challenges

A. Léger · J. Bellemare (✉)
School of Business and Management (ESG), University of Quebec in Montreal (UQAM), Montreal, QC, Canada
e-mail: bellemare.jocelyn@uqam.ca

A. Léger
e-mail: leger.anne@uqam.ca

M. Manns (ed.), *New Paradigms for Anticipated Uncertainty*, Lecture Notes in Mechanical Engineering, https://doi.org/10.1007/978-3-032-16889-4_35

1 Introduction

The fashion industry has developed in the context of accelerated globalization. This has resulted in an ultra-competitive system where speed of execution and innovation have become significant. Although this dynamic has caused unprecedented growth, it is now leading to economic, social, and environmental imbalances. The acceleration of 'fast fashion' engendered by the growing demand for prompt collection renewal has further entrenched pre-existing patterns of overproduction and hyper-consumption. This situation is causing an accumulation of unsold goods and the intensive exploitation of natural resources [21].

This dominant business model, which is based on enhanced production cycles and an oversupply of products and trends, is the cause of increasing pressure on global supply chains. These kinds of exploitative practices put workers as well as natural ecosystems at risk [5]. The linear fashion model based on design, production, consumption, and disposal is quite unsustainable and needs further scrutiny from experts [6].

Faced with rapidly evolving consumer expectations, brands are now compelling to adjust their strategies. While some fast-fashion giants are leveraging hyper-digital personalization to accelerate production [2], others strive to balance efficiency with social and environmental responsibility. This shift reflects changing buying patterns, as consumers increasingly value both sustainability and personalization [18]. Despite increasing awareness for responsible practices, "producing less but better" is still a complex challenge to implement in the fashion and apparel industry. Its very infrastructure is built on mass production and economies of scale, leading to significant overproduction and waste [9]. This model, known for its opacity, promotes excessive turnover of trends, while demand remains subject to economic, cultural, and social influences [5, 9]. On the other hand, full customization, though, fulfills customer expectations, but it is costly and difficult to scale, apart from being resource intensive [1].

In many other sectors, hybrid customization production models are becoming popular because of its simultaneous efficiency and personalization with the help of optimizing industrial processes [4, 22]. It helps brands to meet consumer expectations without compromising production efficiency, reducing excess inventory, and optimizing resources [12]. Hybrid customization production can be seen as a derivative or transitional form of mass customization. While mass customization aims to provide strategic limitation of choices (e.g., color, fit, modular elements) through scalable and efficient production systems. Hybrid customization, by contrast, adopts a more flexible and selective approach. It allows partial personalization at specific stages of the value chain. This model enables companies to experiment with customization while maintaining operational efficiency. It is especially useful when full-scale mass customization is too costly or complex to implement.

Though successful in industries like automotive and electronics, its potential in fashion is still underexplored. Only a few studies in industry have examined how hybrid models could reconcile sustainability and supply chain optimization. The fashion industry provides a compelling context to explore how companies balance uncertainty and predictability. Brands must continuously navigate fluctuating and highly specific demands, such as garment fit, while managing resources to minimize overproduction and waste. Traditional mass customization can be difficult to implement in this context, as more advanced forms of personalization, like made-to-measure, which represent a market of interest and need, require significant resources and operational flexibility. Hybrid customization, by contrast, offers a more adaptable model, allowing companies to move between lower degree of customization (e.g., color or monogramming) and higher degrees of customization (e.g., tailored fit), each carrying distinct operational implications. This flexibility could better reflect the industry's fragmented and unpredictable market, enabling brands to differentiate themselves from the saturated, standardized offer while maintaining their identity and amplifying that of the consumer. Therefore, it is essential to explore whether these models can support more responsible production by reconciling predictability to improve resource efficiency with the need for flexibility and profitability. In order to better meet consumer needs and enhance accessibility in the fashion and apparel industry.

2 Literature Review

Hybrid models are on the rise in many industries [7]. Key instances could be BMW and Tesla which have implemented modular models that offer customization on specific product aspects without disrupting entire manufacturing process (Hui et al. 2024). In furniture retail, IKEA follows a similar logic, allowing consumers to adapt their furniture to needs while maintaining optimized production system [11].

If the fashion industry draws inspiration from these practices, it must, however, contend with unique constraints, particularly the ephemeral nature of trends and the uncertainty of stock management (Casciani 2024). The fast-paced evolution of fashion makes things complicated for the implementation of hybrid models, which require a delicate balance between anticipation and responsiveness. New technologies including digital manufacturing, artificial intelligence, and on-demand production offer innovative solutions for integrating customization without compromising profitability [13]. Some initiatives, such as Unspun, MiAdidas, and Nike By You, already demonstrate how hybrid customization can be successfully implemented by leveraging these digital tools [10].

Despite these advancements, the widespread implementation of hybrid customization still has several obstacles. Logistics and material traceability remain major challenges, especially in the context of responsible production (Cuc 2023). Moreover, influencers and social media have an increasing impact on consumer practices and could serve as a strategic lever for normalizing sustainable customization (Jacobson and Harrison 2021). However, one of the key challenges remains the gap between consumer intentions and actual purchasing behaviors. Several studies show that while consumers express support for customization and more responsible fashion, their shopping habits are still heavily influenced by price and delivery speed [14, 20]. Moreover, although hybrid models can strengthen customer loyalty by offering products that align with their expectations [15], this loyalty can only be sustained if companies invest in awareness and education strategies to promote more thoughtful consumption [16].

This underscores a persistent tension between growing sustainability awareness and the rising demand for personalized products in reaction to the uniformity and saturation of mass-market offerings. To remain relevant, the industry must navigate this duality commercially, by meeting consumer desires and standing out in a crowded market; socially, by valuing producer communities and cultural contexts; and environmentally, by contributing to the broader ecological transition.

Although some pioneering initiatives (such as those by Adidas, Nike, and Unspun) have demonstrated the feasibility of this model, there is still much to explore to maximize its benefits. Given these findings, it is necessary to structure a conceptual framework that integrates three essential dimensions for the successful implementation of hybrid customization: technological and operational optimization, sustainable product and practices, and consumer education and engagement. This confirms that while hybrid customization is progressively emerging as a solution to reconcile flexibility and sustainability in the fashion industry, its large-scale deployment depends on the ability of a company to address key challenges: reducing operational costs and supply chain pressure and meeting customers need while encouraging more responsible purchasing behaviors.

Here, several main questions arise: (1) how can these hybrid models be optimized to balance profitability, flexibility, and reduction in environmental impact?, (2) how can a supply chain be structured to meet personalized demand without generating excessive costs or logistical inefficiencies?, and finally (3) what strategies should be adopted to guide consumers toward more thoughtful and sustainable purchases, without encouraging unnecessary consumption under the pretext of personalization?

This literature review highlights the urgent need to continue research on hybrid customization as a means of accessing best practices in operational efficiency and sustainability. Nevertheless, this current study is based on the central question: *How do fashion companies perceive hybrid customization production in terms of efficiency, profitability, and sustainability?*

3 Methodology

Given the limited academic research on the hybrid customization production in fashion and apparel industry, an exploratory qualitative approach was adopted to gather rich and grounded, practice-based perspectives that may not be otherwise documented, to establish a foundation for further research. This study employed a qualitative approach to study the links between product customization and socio-environmental responsibility in the fashion industry. It aims to explore the perceptions of fashion manufacturers about the impact of sustainability practices on mass customization and to assess whether this approach can help to reduce textile waste by promoting responsible production which is more closely aligned with actual consumer needs.

The province of Québec provides a relevant context for this study given its strategic position in North America and its concentration of diverse actors across the fashion value chain [3]. Experience-based insights from field experts were favored to capture the perceptions of informed and stakeholders [17]. Participants were then intentionally selected based on specific criteria: being located in Québec, having at least five years of experience in garment manufacturing, and being actively involved in mass customization. Data was collected through an online survey conducted in winter 2025 and was administered to fashion manufacturers, brand owners, and other industry stakeholders. A semi-structured approach was used, combining open-ended questions with multiple-choice options that included an "other" category. This design aimed to guide responses in line with the research objectives while allowing participants to freely express their perceptions. It also enabled the emergence of rich, sometimes unexpected, qualitative insights grounded in their lived experience [8]. Of the 78 invited participants, 7 responded.

The survey provided straightforward responses while allowing for nuanced perspectives specific to each respondent's experience. Open-ended answers helped clarify intentions and added context to the closed-ended data. The survey was structured in seven sections: an introduction with theoretical background and definitions; respondent profile; familiarity with customization; and four sections assessing perceptions of hybrid customization in terms of operations, socio-environmental impacts, customer experience, and future market and technological prospects. It concluded with three open-ended questions to further explore respondents' views.

Open-ended responses were examined qualitatively to identify recurring ideas, key concerns, and contrasting viewpoints. Responses were grouped based on relevance to the main themes of the study of operational practices, environmental considerations, and consumer engagement. This interpretive process supported the exploratory nature of the research and helped to highlight shared patterns as well as divergent perspectives. Although this study is limited due to its small sample size and its geographic focus on Quebec, it provides valuable information about the opportunities and constraints related to integration of customization into responsible production models.

4 Results

The findings revealed a wide range of professional backgrounds among respondents, including brand or product managers, fashion designers, and business owners. The results of the survey provided a clearer understanding of the perspectives of participants and their role for the adoption of hybrid customization. Most participants reported long-standing experience in the fashion industry have over 10 years of experience in the industry.

The results also confirmed a strong interest among fashion manufacturers in integrating more sustainable business models, while also highlighting the challenges associated with hybrid customization. Customization is perceived as a strategy to better align supply with demand and reduce overproduction, but it also raises economic and logistical concerns.

From an environmental perspective, opinions remained divided. While many acknowledged that targeted production can helps reduce textile waste. Others feared it may encourage overconsumption by making a more flexible offering too attractive. The survey results also showed that most companies which adopted intensive customization recognize the importance of raising consumer awareness about responsible consumption. Therefore, it becomes essential to promote product sustainability by emphasizing quality and longevity rather than just the ability to customize.

The results of the survey suggest that several companies already offer advanced forms of customization (tailored adjustments, choice of cuts, etc.). However, some still provide basic customization (color selection, monogramming, minor adjustments). This highlights that a change to more personalized models is underway but still in development for many companies.

Many respondents indicated having experimented with mass customization to varying degrees. Some mentioned a more intensive adoption, while others reported limited or exploratory implementations on a smaller scale. This variation reflects a significant interest in this approach, though operational and financial barriers are still there. One of the main advantages of hybrid customization lies in its potential to strengthen the interaction between brands and consumers.

Several respondents reported that hybrid customization had a positive effect on customer loyalty. Those who had implemented such models generally perceived an improvement in client engagement, often estimating the impact to be around 4 out of 5 on a subjective scale. However, its implementation presents major operational challenges, particularly regarding customer expectations and logistical complexity. The need for individualized order tracking and clear communication on lead times and production constraints are central concerns which were raised by respondents.

In hybrid customization, many participants emphasized the need to restructure production processes and invest in specialized workforce skills. Unlike mass production, it requires continuous training and infrastructure adaptation for flexible management of customized orders. Moreover, the companies surveyed indicated that a lack of automation in certain process stages results in higher costs and increased inefficiencies. Logistical complexity and delivery time management emerged as major challenges. To address these issues, some participants reported reducing the number of customization options to optimize profitability. While others focused on improving demand forecasting tools to better align production with demand. These strategies were generally viewed as necessary to balance operational feasibility with economic viability. Another major challenge frequently mentioned by participants was related to supply chain management. Several respondents claimed that there is limited access to sustainable materials. This complicates the implementation of truly responsible practices. Maintaining a balance between having enough inventory for customization and avoiding overproduction was also described as a persistent issue. To address this issue, some companies have adopted on-demand production or pre-order strategies. This strategy helps align supply with actual demand in a better synchronize supply with actual consumer demand, while minimizing excess inventory and reducing waste.

Many participants considered artificial intelligence and 3D modeling tools as the main enablers to optimize customization while minimizing its environmental impact. Other technological advancements, such as production automation and logistics flow optimization, were also indicated as priorities for making these models more viable. Several participants shared information about the future of hybrid customization. One respondent emphasized that "*artificial intelligence and data analysis will play an important le for predicting trends and optimizing production while engaging customers in a loyalty loop.*" Another respondent highlighted the importance of having a balance between automation and human intervention, stating that "*even with smart tools, maintaining human oversight in the process is quite important to ensure quality and customer satisfaction.*"

Hybrid customization represents an opportunity with high potential for a more sustainable fashion industry, but its success depends on its strategic integration and the ability of companies to overcome economic and logistical challenges.

5 Discussion

The results of this study underscore the importance of hybrid models in the transition between mass production and full customization. These models offer a flexible as well as a scalable alternative to traditional manufacturing models. These approaches allow companies to adapt their supply to consumer expectations and meanwhile limit their overproduction. However, it was found that challenges associated with logistics, human resource management, and demand forecasting are looming. Some respondents identified logistics and lead time management as major challenges. Customization does not automatically guarantee more responsible consumption.

Although it can help in reducing waste as it aligns production with actual needs [19], it may also encourage excessive purchasing behaviors if perceived as too accessible and easy to obtain. Several participants shared the view that targeted production helps to reduce textile waste, noting that it allows for better alignment between supply and demand. However, others expressed concern that it encourages overconsumption. This divergence of perspectives shows that there is a need for a well-structured approach which combines responsible practices with consumer awareness.

5.1 Balancing Personalization and Efficiency in Fashion

The evaluation of hybrid models in the fashion industry reveals that there are tensions between customization and efficiency. In contrast to industries such as automotive or electronics where standardization and modularity are already well-established (Hui et al. 2024). The fashion industry faces unique challenges, especially due to rapid trend cycles and volatile demand (Casciani 2024). These constraints make it difficult to implement hybrid strategies that require balancing flexibility with efficiency.

This makes things complicated for the adoption of standardized strategies and forces brands to rethink their approach to ensure economic as well as environmental viability. Several respondents mentioned that they prioritize reducing the number of customization options as a way to balance personalization with operational efficiency and profitability. The integration of digital tools, such as artificial intelligence and predictive analytics, could improve production flow management. However, their adoption remains limited by cost and implementation complexity. A number of respondents considered AI and 3D modeling tools as main enablers for optimizing customization and at the same time minimizing its environmental impact. While hybrid models have shown great potential, their effectiveness depends on the ability of a company to balance innovation, resource management, and trend anticipation.

5.2 A Structured Framework for Responsible Hybrid Customization

Some of the key takeaways from this study are the need for structured conceptual framework to integrate hybrid customization responsibly. This framework is based on three pillars:

- Sustainable design and processes: ensuring high-quality materials, extended product lifecycles, environmentally friendly and low-waste production.
- Consumer engagement and education: raising awareness to prevent disguised overconsumption under the guise of customization. Promoting mindful purchasing behavior through transparency, education, and storytelling (Jacobson and Harrison 2021).
- Operational optimization: several companies already offer advanced customization, while others provide basic personalization.

In this context, brands need to invest in optimized logistics strategies, such as pre-orders and on-demand production, to minimize excess inventory and improve model profitability. Many respondents also expressed struggling to balance maintaining sufficient stock for customization with the need to avoid overproduction. Figure 1 presents a customization intensity model, illustrating the two extremes: high-intensity pure customization versus an easily standardized baseline level.

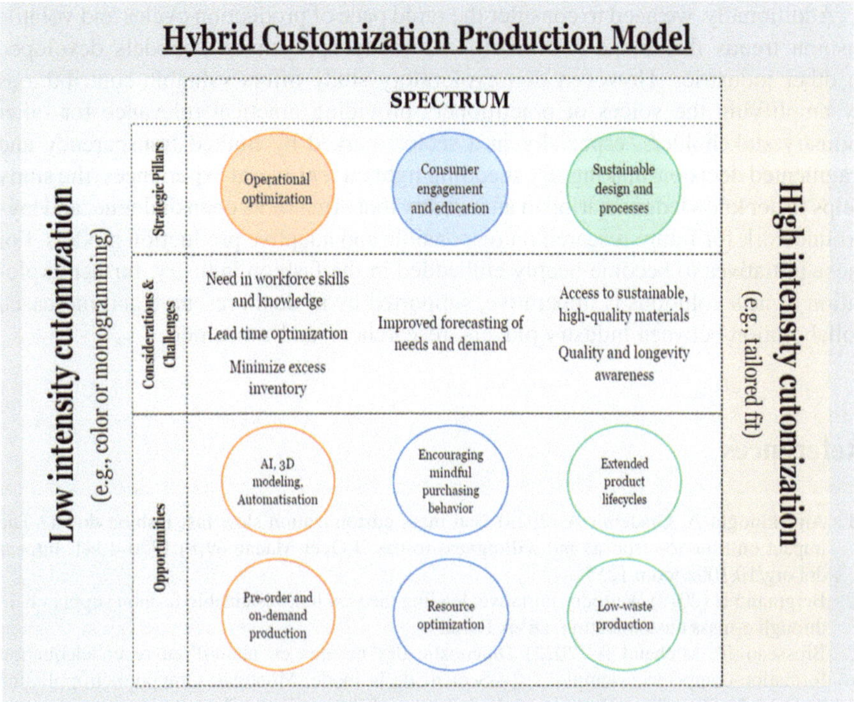

Fig. 1 Hybrid customization production model for fashion and apparel industry, illustrating challenges and opportunities of a hybrid production approaches between low and high intensity customization from an operational, consumer and sustainability perspective

6 Conclusion

The results of this study reveal that there is a diversity of approaches related to the way fashion companies apply customization and how it enables sustainability principles. This makes it challenging to assess the actual impact of customization on responsible production and consumption. If customization is to play a constructive role in the transition to a more sustainable fashion industry, it must not be seen only as a marketing tool. Instead, it should be considered a means of designing high-quality, durable products that truly align with the needs of the consumers. Thus, there requires a paradigm shift related to perception of customization. It means that rather than a passing trend, it should be embedded in a structural transformation. It should aim to extend the lifespan of garments and reduce textile waste. The study also highlights the potential of hybrid customization production models to reconcile personalization and sustainability.

We acknowledge the potential bias introduced by the intentional selection of participants, as well as the limitations of a small sample size, which restrict the generalizability of the findings to the broader fashion industry.

Additionally, we need to consider the rapid pace of production cycles and volatile fashion trends that make it difficult to directly apply hybrid models developed in other industries. However, this exploratory study offers valuable contributions by amplifying the voices of practitioners providing practical relevance for other industry stakeholders, especially in a sector marked by limited transparency and fragmented decision-making. By shedding light on real-world experiences, the study helps foster knowledge-sharing in an industry that often lacks open dialogue, and lays groundwork for future research on responsible and adaptive production models. For these initiatives to become deeply embedded in the fashion industry, further exploration of new solutions is imperative, supported by in-depth research and increased collaboration between industry players, researchers, and consumers.

References

1. Alptekinoğlu A, Örsdemir A (2023) Can mass customization slow fast fashion down? The impact on time-to-disposal and willingness-to-pay. J Oper Manag 69(8):1320–1341. https://doi.org/10.1002/joom.1255
2. Bergmann H (2012) Wettodry initiative: leading the way to a sustainable fashion supply chain through e-mass customization. DiVA Portal
3. Brosseau JP, Michaud A (2022) Diagnostic des besoins en main-d'œuvre et adéquation formation-compétences-emploi 2.0—Secteur de la mode. Montréal: Grappe métropolitaine de la mode, Conseil emploi métropole & Raymond Chabot Grant Thornton
4. Casciani D, Chkanikova O (2023) Fashion craftsmanship 4.0: learning experience about Industry 4.0 technologies for hybrid digital fashion-tech products, processes, and business model design. In: BW conference proceedings 2023. OAPEN Library, pp 213–225
5. Dickson MA, Eckman MJ, Loker S (2009) Social responsibility in the global apparel industry. Fairchild Books, New York
6. Ellen MacArthur Foundation (2017) A new textiles economy: redesigning fashion's future. Available at: https://emf.thirdlight.com/file/24/IwnEDbfI5JTFoAIw_2QI2Yg-6y/A-New-Textiles-Economy_Summary-of-Findings_Updated_1-12-17.pdf
7. Fani S et al (2022) Design product-service systems by using a hybrid approach: the fashion renting business model. Sustainability 14(9):5207. https://doi.org/10.3390/su14095207
8. Fenneteau H (2015) L'enquête: entretien et questionnaire, 3rd edn. Dunod, Paris
9. Fletcher K (2014) Sustainable fashion and textiles: design journeys, 2nd edn. Earthscan from Routledge/Taylor & Francis Group, London
10. Ghui G, Al Mamun A, Masukujjaman M, Makhbul ZKM (2024) The relationship between mass customization and sustainable performance: the role of firm size and global E-commerce. Heliyon 10(6):11. https://doi.org/10.1016/j.heliyon.2024.e17185
11. Hong Y, Wang X, Xue Y, Zhang J (2024) A sustainable supply chain design for personalized customization in industry 5.0 era. IEEE Xplore. https://doi.org/10.1109/ICIM53746.2024.10480134
12. Lee YK (2021) Transformation of the innovative and sustainable supply chain with upcoming real-time fashion systems. Sustainability 13(3):1081. https://doi.org/10.3390/su13031081
13. Longo F, Padovano A (2021) Towards a mass customization in the fashion industry: an evolutionary decision aid model for apparel product platform design and optimization. Comput Ind Eng 161:107675. https://doi.org/10.1016/j.cie.2021.107675
14. Mehta S et al (2022) "Sustainable attitude"—a modest notion creating a tremendous difference in the glamorous fast fashion world: Investigating moderating effects. Soc Bus Rev 18(2). https://doi.org/10.1108/sbr-10-2021-0205

15. Nogueira L et al (2023) The impact of customer-centric sustainability on brand relationships. Sustainability 15(16):12212. https://doi.org/10.3390/su151612212
16. Park H, Kim Y (2023) An empirical test of the triple bottom line of customer-centric sustainability: the case of fast fashion. Fashion Text 3:25. https://doi.org/10.1186/s40691-016-0077-6
17. Savoie-Zajc L (2007) Comment peut-on construire un échantillonnage scientifiquement valide? Rech Qual 5:99–111
18. Shaver J, Yan RN (2022) Examining sustainable consumption behaviors through the mass customization context: emotional product attachment and environmental attitude. J Sustain Res 4(2):205–225. https://doi.org/10.20900/jsr20220022
19. Steiner F (2014) A contribution toward a research agenda: identifying impact factors of mass customization on environmental sustainability. Int J Ind Eng Manag 5(4):169–178
20. Stringer C et al (2020) Do ethical concerns and personal values influence the purchase intention of fast-fashion clothing? J Fash Mark Manag 24(1):99–120. https://doi.org/10.1108/jfmm-01-2019-0011
21. Todeschini BV, Cortimiglia MN (2017) Innovative and sustainable business models in the fashion industry: entrepreneurial drivers, opportunities, and challenges. Bus Horiz 60(6):759–770. https://doi.org/10.1016/j.bushor.2017.07.003
22. Yan WJ, Chiou SC (2020) Dimensions of customer value for the development of digital customization in the clothing industry. Sustainability 12(11):4639. https://doi.org/10.3390/su12114639

Spare Parts Supply Capabilities and Equipment Coverage: A Systematic, Data-Driven Portfolio Assessment Approach

Simon Klarskov Didriksen, Kristoffer Wernblad Sigsgaard, Christian Brunbjerg Jespersen, and Niels Henrik Mortensen

Abstract Recent disruptions to the global supply chain have increased maintenance organizations' focus on spare parts supply resilience and securing the continuous supply of critical spare parts for their production plants. Studies have identified several characteristics of spare parts that challenge planning for spare parts availability. Studies also indicate an industry need to reduce the technical expertise required to implement theoretical models for the management and alignment of spare parts supply capabilities and equipment coverage. To address these challenges, this study proposes a systematic, data-driven portfolio assessment approach to conducting risk-based assessments of spare parts supply capabilities and spare parts coverage of the maintainable asset. The approach is tested in a case study on a scope of 15,427 spare parts in a major offshore oil and gas company. The case study demonstrates enhanced decision-making due to the improved, risk-based overview. The approach also requires low modelling competencies and can be modelled in business intelligence software utilizing existing computerized maintenance management system data. Practitioners can use existing data to assess current spare parts coverage and strategically adjust supply chain capabilities to reach the desired portfolio coverage.

Keywords Spare part management · Supply chain resilience · Maintenance · Data-driven approach · Portfolio assessment

1 Introduction

Maintenance in equipment-intensive industries is imperative in ensuring continuous operation and stable production output. In most cases, maintenance cannot be conducted if spare parts are unavailable, which is why stock-keeping and spare parts

S. K. Didriksen (✉) · K. W. Sigsgaard · C. B. Jespersen · N. H. Mortensen
Department of Civil and Mechanical Engineering, Technical University of Denmark, Kongens Lyngby, Denmark
e-mail: simdi@dtu.dk

M. Manns (ed.), *New Paradigms for Anticipated Uncertainty*, Lecture Notes in Mechanical Engineering, https://doi.org/10.1007/978-3-032-16889-4_36

suppliers exist to support maintenance [1]. To avoid equipment down-time, stockpiling remains the strategy of choice for organizations, but this may not increase security due to, e.g., high annual spare part obsolescence rates of up to 23% [2].

In recent years, disruptive events in the global supply chain have resulted in a greater focus by maintenance companies on spare parts supply chain resilience and the risks associated with spare parts coverage of critical equipment. Intermittent and irregular spare parts demands, custom designs with strict quality control from original equipment manufacturers (OEMs) [3], a large volume of spare parts in the portfolio [2, 4], and the need for fast responses to equipment failures [5] all limit supply chain resilience. Additionally, the supply of spare parts tends to include long lead times as a result of specialization and custom designs [6]. Computerized maintenance management systems (CMMSs) are used to capture, store, and manage information for amongst others, spare parts management (SPM), equipment, and maintenance. However, Teixeira et al. [4] identify difficulties in data collection for decision-making due to large spare parts portfolios and excessive volumes of information to consider.

Many forecasting and classification methodologies have been introduced in the body of literature, but it remains a continuous challenge for organizations to forecast demand and adjust their spare parts portfolios to cover future maintenance demands [7–9]. Bacchetti et al. [10] report that inadequate stock management policies and low-quality demand forecasts are frequently found in companies. Martínez et al. [11] describe a general lack of simple applications for maintenance management to identify patterns of deficiencies, and Bacchetti et al. [10] conclude that practitioners prioritize applications that are easy to implement. According to Ayu Nariswari et al. [12], the holistic thinking of decision-makers is currently not embraced by mathematic models, as these models only consider quantitative criteria. This indicates both that the applicability of complex mathematical approaches is low in practice and that there is a need for practical approaches enabling holistic decision-making.

This study presents a systematic and data-driven approach for risk-based spare parts portfolio assessment of spare parts supply capabilities and spare parts coverage of the maintainable asset. To guide this study and develop the approach, the following research questions were posed:

> How can systematic risk assessment in SPM be facilitated by the evaluation of supply chain capabilities and asset coverage from the spare parts portfolio using CMMS data?

The structure of this paper includes Sect. 2 presenting the research methodology, while Sect. 3 reviews relevant literature. Section 4 describes the developed approach, Sect. 5 presents the use of the approach in a case company, and Sect. 6 discusses and concludes the study.

2 Research Methodology

The study presented in this paper closely adheres to design research methodology [13]. The research process focused on the development and initial validation of a systematic, data-driven spare parts portfolio assessment approach through iterative development in collaboration with a case company. Several cycles were conducted validating both the identified criteria used in the approach as well as the insights gained from using the approach. The approach was developed based on knowledge gained from literature on SPM and from practices in the case company. A lack of an overview in the SPM domain was observed in the case company. This concerned the ability to differentiate between vital and nonvital spare parts relative to the maintainable assets, as well as conducting fact-based and asset-oriented argumentation for stock management policy change. The systematic assessment approach was developed as a solution to solve these problems. To validate its applicability, testing was conducted on CMMS data from a major offshore oil and gas production company operating 50 installations in the North Sea. A data model was built on historical data for maintenance, spare parts, and equipment extracted from the case company's CMMS with the time scope of 2010 to 2024.

3 Theoretical Background and Related Work

According to Scarf et al. [1], spare parts stock exists to facilitate the maintenance of production equipment. This equipment are in many cases installed with a bill of materials (BOM) of spare parts [14]. However, a BOM may not be applicable as a spare parts list due to e.g. high obsolescence rates presented by Cohen et al. [2]. Spare parts stock changes over time, which can be explained by high difficulty in forecasting the demand, as the demand can be irregular, sporadic, and intermittent [1, 15]. As noted by Teixeira et al. [4], the volumes of spare part portfolios are often large and challenging to manage over time and the volume of information to consider in the decision-making is excessive.

For systematic SPM, Cavalieri et al. [16] introduce a decision-making framework that gathers and concretizes existing SPM methodologies into an approach that practitioners can apply. The framework includes the following steps: (1) part coding, (2) part classification, (3) part forecasting, (4) stock management policy, and (5) policy test and validation. These steps are widely acknowledged in improving spare parts portfolios.

Most SPM studies discuss how demand forecasting and classification can be used for determining stock management policies [4]. However, continuous assessment of these policies in relation to both the spare parts portfolio and the equipment portfolio remains underprioritized. Cavalieri et al. [16] suggest that the final step, policy test and validation, should be conducted prior to implementation of a policy and that simulations could be the methodology of choice. However, simulation efforts are often time-consuming and mathematically complex.

Many modelling approaches and methods exist for improving the spare parts portfolio, but they remain either maintenance-oriented or inventory-oriented [1]. Van Horenbeek et al. [17] report that unlimited and immediate availability of spare parts are often assumptions found in maintenance models. This contradicts real-world spare parts availability. Scarf et al. [1] discuss an increasing gap between methodologies introduced in theory and the application of them in practice: the mathematical complexity of theoretical models is increasing, but industry needs simple models.

4 Proposed Assessment Approach

This section presents the systematic data-driven portfolio assessment approach shown in Fig. 1 for continuous review of current spare parts supply chain capabilities and equipment coverage. The approach is intended for TO-BE goal setting and adjustment for future alignment of the spare parts portfolio with the maintainable asset. Each step is detailed in the following subsections.

4.1 Linking the Maintainable Asset Portfolio and the Spare Parts Portfolio

The first step concerns linking the maintainable asset portfolio with the spare parts portfolio in a format that allows data models to be developed. If a data model is developed in the first step, a continuous assessment cycle should be enabled. The CMMS data required for linking the portfolios are historical maintenance demand, asset hierarchy, equipment criticality, all spare parts registered in the CMMS and coding for each spare part in current stock management policy, lead time, and supplier information. Linkage of the spare parts portfolio and the maintainable asset portfolio is done at the lowest level of the asset hierarchy. A decomposition of maintainable assets may be asset, system, subsystem, equipment, and BOM components from top level to bottom level, respectively. The two portfolios are linked through spare parts

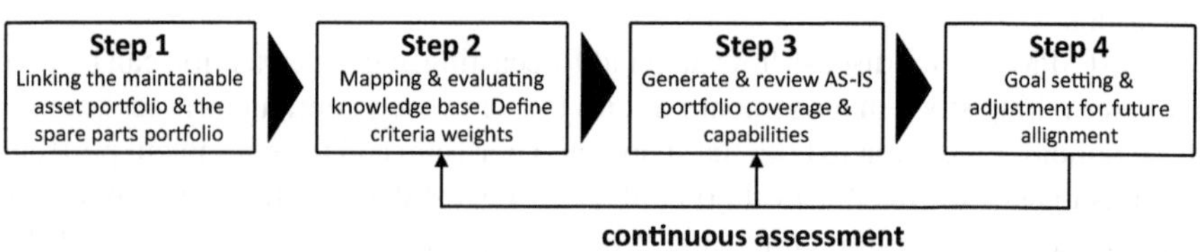

Fig. 1 The systematic, four-step approach

mentioned in the BOM and in historical maintenance as a collective list of spare parts per equipment.

4.2 Mapping and Evaluation of the Knowledge Base and Criteria Weighting

The second step focuses on evaluation of the knowledge base and on the definition of criterion weights resulting in a quantitative assessment basis. First, the knowledge base is mapped through sharing matrices. For the spare parts portfolio, criteria are mapped per spare part. For the maintainable asset, the criteria are mapped per equipment. Between the portfolios, each spare part is mapped to each piece of equipment. The availability and variance of criteria can be assessed to verify the decision basis.

4.3 AS-IS Review of Equipment Coverage and Supply Chain Capabilities

The third step concerns an AS-IS overview of the current spare parts portfolio, the capability of the spare parts supply chain, and the coverage of the maintainable asset portfolio. The overview is to be used as a performance check of the current spare parts portfolio. As presented in Fig. 2, in the first view, all application areas and criticalities are aggregated to each spare part of the spare parts portfolio. The aggregation is done from the BOM and the historical maintenance found in the maintainable asset hierarchy as well as from the supply chain characteristics.

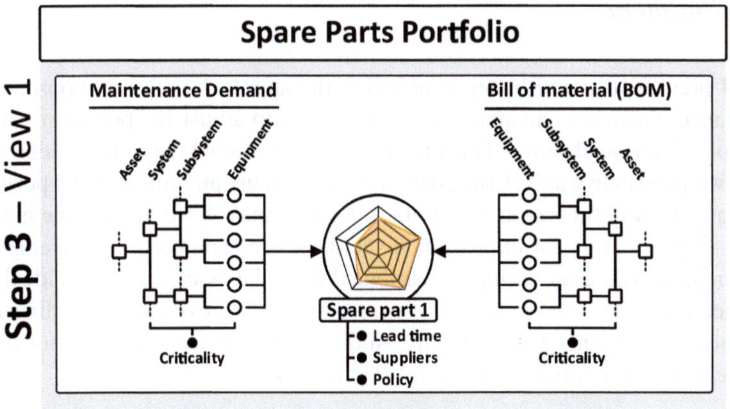

Fig. 2 AS-IS supply capability and equipment coverage overview for spare parts in the portfolio

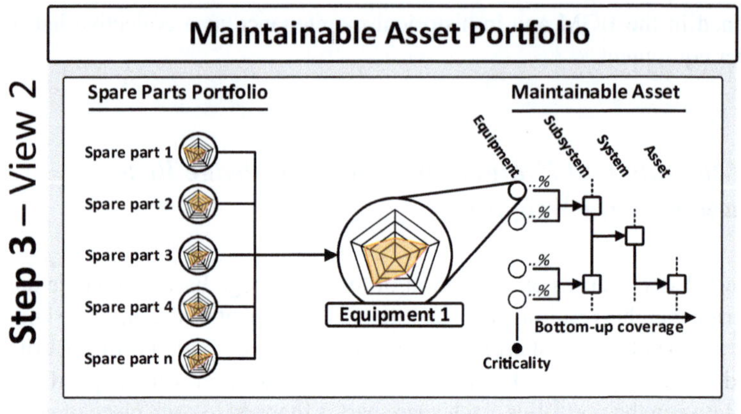

Fig. 3 AS-IS overview of the accumulated supply chain capabilities and equipment coverage from the spare parts for individual equipment bottom-up to the entire asset

All spare parts related to the individual equipment and each spare part assessment are combined as presented in Fig. 3. This combination, presented in the second view, serves as a summarized coverage indicator for the individual equipment. Depending on the criticality of the equipment and the amount of equipment in a subsystem, the percentage weight of the equipment can be multiplied with the supply chain criteria coverage. Through a bottom-up approach, the spare part coverage assessment can be reviewed on various levels of the maintainable asset hierarchy as well as when comparing equipment classes, such as all pumps or turbines.

4.4 *TO-BE Goal Setting and Adjustment for Future Alignment*

Figure 4 presents the fourth step concluding the approach with an opportunity for maintenance managers and spare parts controllers to adjust the portfolio coverage and supply chain capabilities. This may be done by defining tangible strategies from a top-down perspective as TO-BE goal settings for future alignment of the portfolios.

The qualitative criteria variants found in step 2 may serve as tangible elements to select in goal setting for improvements to the spare parts portfolio coverage. In a top-down approach, goal setting is distributed from the asset to the individual spare part. Therefore, the importance of a spare part can be reviewed according to the application area variance and criticality. Additionally, part over-/under-prioritization can be reviewed in relation to the supply chain capability coverage.

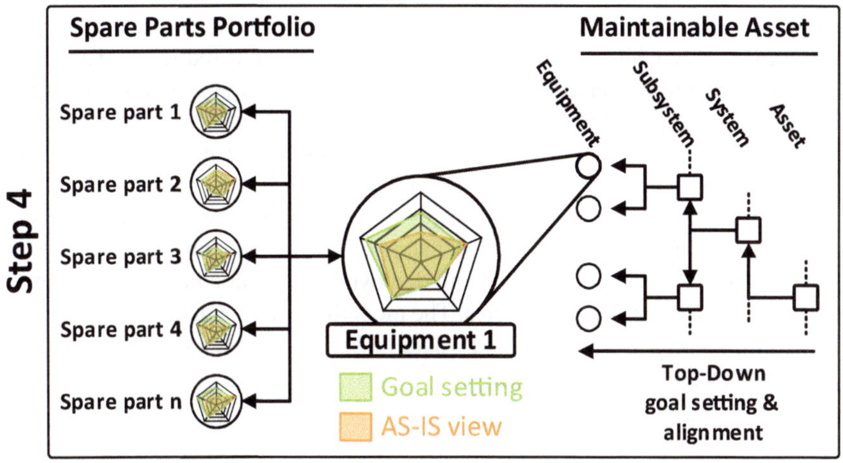

Fig. 4 Step 4 includes TO-BE goal setting and top-down adjustment for future alignment of spare parts supply chain capabilities and equipment coverage of spare parts

5 Approach Applicability in the Case Company

The approach was tested in a case company on a scope of 15,427 spare parts, which were to be updated with new stock management policies. The policies were decided by internal maintenance planners, viewed as equipment experts. The maintenance manager was concerned about signing the update due to the lack of a risk-based overview. The concerns were related to the risk of discarding spare parts vital for operations, stockpiling, and pressure from management to reduce the high stock level. The maintenance manager signed the update 32 weeks after it was ready. The approval was a result of several iterations of risk presentation generated from the utility of the approach.

Criticality was available for 95% of the spare parts due to the application of both BOM and demand linkage to the equipment. Policies existed for all spare parts. Overall, 69% of the scope was relatable to suppliers. Lead time was available for 16% of the scope due to a substantial portion being slow-moving or dead. Lead time was weakly trusted by decision-makers. Therefore, logistics and procurement removed all lead times and updated them with trusted once for all purchased and transported spare parts over the last 3 years. A goal was set to update the lead time every quarter. In addition, efforts could be further invested in acquiring lead times from known suppliers while identifying suppliers for more spare parts in the same process.

Only 35% of the spare parts used in maintenance over the previous 3 years were consumed on the registered BOM; the remaining 65% had been used at other locations. This indicated the validity of combining the demand and BOM spare parts into a combined overview. The scope was perceived as spare parts vital for operations. However, after the assessment was conducted, it was concluded that 8% of the scope was in fact nonvital and did not require purchase. The overall supply capability score increased by 15% while resulting in a 10% scope stock value reduction. The scoring focused on policy conservatism where critical spare parts unit investment yielded higher score. The percentage of parts with just-in-time policies was reduced by 1%, while the proportions of parts with a single- or a multi-unit policy were increased by 1% and 8%, respectively. These findings were translated to the company as to which parts of the maintainable asset were affected by stock coverage removal, stock coverage lowering, and stock coverage increase. The approach was considered easy to implement, as it can be modelled and deployed in common business intelligence (BI) software and spreadsheet queries. The modelling was conducted in a few weeks by two people with front-end CMMS data table spreadsheet extractions that were commonly used in the case company.

6 Discussion and Conclusion

If spare parts are unavailable, this can have severe consequences for a plant's production, such as downtime becoming lead-time dependent. Therefore, continuous review and reallocation of stock management policies should be conducted. Literature shows a need from industry to reduce the complexity of theoretical models. This study presents a systematic, data-driven approach to conducting risk-based portfolio assessment of spare parts supply capabilities and the spare parts portfolio coverage of the maintainable asset. The approach is a low complexity solution that can be applied using BI software and existing CMMS data. Maintenance organizations can apply the approach without mathematic modelling skills. A case study was conducted with a condensed scope of 15,427 spare parts. Therefore, this study only serves as an indication of the approach's applicability. Future work should evaluate the approach with full portfolios. The case company needed risk-based insights into stock management policy reallocation. Therefore, future work should focus on end-to-end testing of the approach in goal-setting for future alignment of multiple supply chain capability criteria and equipment coverage requirements.

References

1. Scarf P, Syntetos A, Teunter R (2023) Joint maintenance and spare-parts inventory models: a review and discussion of practical stock-keeping rules. IMA J Manag Math 35:83–109. https://doi.org/10.1093/imaman/dpad020
2. Cohen MA, Agrawal N, Agrawal V (2006) Winning in the aftermarket. Harv Bus Rev 84(5):129–138
3. Bhattacharyya SS, Kulkarni O, Mishra A (2022) Study of emerging avenues in supply chain resilience; the case of integration of additive manufacturing with spare parts procurement. Benchmarking Int J 30(10):4100–4118. https://doi.org/10.1108/BIJ-03-2022-0163

4. Teixeira C, Lopes I, Figueiredo M (2024) Spare parts stock management: classification and policy assignment. FME Trans 52(2):257–270. https://doi.org/10.5937/fme2402257T
5. Murthy DNP, Solem O, Roren T (2004) Product warranty logistics: issues and challenges. Eur J Oper Res 156(1):110–126. https://doi.org/10.1016/S0377-2217(02)00912-8
6. Liu P, Huang SH, Mokasdar A, Zhou H, Hou L (2014) The impact of additive manufacturing in the aircraft spare parts supply chain: supply chain operation reference (scor) model based analysis. Prod Plann Control 25(13–14):1169–1181. https://doi.org/10.1080/09537287.2013.808835
7. Scala NM, Rajgopal J, Needy KL (2014) Managing nuclear spare parts inventories: a data driven methodology. IEEE Trans Eng Manag 61(1):28–37. https://doi.org/10.1109/TEM.2013.2283170
8. Balugani E, Lolli F, Gamberini R, Rimini B, Regattieri A (2018) Clustering for inventory control systems. IFAC-PapersOnLine 51(11):1174–1179. https://doi.org/10.1016/J.IFACOL.2018.08.431
9. Vazquez Hernandez J, Elizondo Rojas MD (2024) Improving spare parts (MRO) inventory management policies after COVID-19 pandemic: a Lean Six Sigma 4.0 project. TQM J 36(6):1627–1650. https://doi.org/10.1108/TQM-08-2023-0245
10. Bacchetti A, Plebani F, Saccani N, Syntetos AA (2013) Empirically-driven hierarchical classification of stock keeping units. Int J Prod Econ 143(2):263–274. https://doi.org/10.1016/J.IJPE.2012.06.010
11. Martínez LB, Márquez AC, Gunckel PV, Andreani AA (2013) The graphical analysis for maintenance management method: a quantitative graphical analysis to support maintenance management decision making. Qual Reliab Eng Int 29(1):77–87. https://doi.org/10.1002/QRE.1296
12. Ayu Nariswari NP, Bamford D, Dehe B (2019) Testing an AHP model for aircraft spare parts. Prod Plann Control 30(4):329–344. https://doi.org/10.1080/09537287.2018.1555341
13. Blessing LTM, Chakrabarti A (2009) DRM, a design research methodology. In: DRM, a design research methodology, p 397. https://doi.org/10.1007/978-1-84882-587-1
14. Stip J, Van Houtum GJ (2020) On a method to improve your service BOMs within spare parts management. Int J Prod Econ 221:107466–107466. https://doi.org/10.1016/J.IJPE.2019.08.001
15. Pinçe Ç, Turrini L, Meissner J (2021) Intermittent demand forecasting for spare parts: a critical review. Omega 105:102513. https://doi.org/10.1016/j.omega.2021.102513
16. Cavalieri S, Garetti M, MacChi M, Pinto R (2008) A decision-making framework for managing maintenance spare parts. Prod Plann Control 19(4):379–396. https://doi.org/10.1080/09537280802034471
17. Van Horenbeek A, Buré J, Cattrysse D, Pintelon L, Vansteenwegen P (2013) Joint maintenance and inventory optimization systems: a review. Int J Prod Econ 143(2):499–508. https://doi.org/10.1016/j.ijpe.2012.04.001

Bridging the Gap: A Conceptual Framework for Aligning Internal Capabilities and Strategic Goals for Technology Integration in Manufacturing Companies

Louise Paaske Jørgensen, Mikkel Østergaard, Søren Odgaard, Zsigmond Csaba Dósa, Vita Jutinskaitė, and Torben Tambo

Abstract This study proposes a conceptual framework to support strategic decision-making in manufacturing firms integrating new technologies. Addressing a gap in the literature on effective technology adoption, the framework aligns organizational goals with operational capabilities to sustain competitive advantage. Adopting a top-down perspective, it offers a structured methodology for evaluating and implementing technological innovations. The framework is grounded in a comprehensive review of relevant theories and delineates key decision-making components critical to manufacturing contexts. To evaluate its practical relevance, the framework was applied to two industrial case studies, enabling empirical analysis in distinct organizational settings. The findings demonstrate the framework's utility in guiding technology integration strategies and ensuring alignment with corporate objectives. This research contributes to the field of strategic operations management by offering a systematic, theory-informed model that supports coherent and goal-oriented technology adoption in manufacturing environments.

Keywords Technology integration · Technology intelligence · Project prioritization · Operational efficiency · Technology risk assessment

1 Introduction

In an era of rapid technological advancements, manufacturing companies must integrate emerging technologies to remain competitive. However, successful integration requires strategic alignment between internal capabilities and business objectives, bridging the gap between technological potential and practical implementation [1].

L. P. Jørgensen · M. Østergaard · S. Odgaard · Z. C. Dósa · V. Jutinskaitė · T. Tambo (✉)
Department of Business Development and Technology, Aarhus University, Herning, Denmark
e-mail: torbento@btech.au.dk

M. Manns (ed.), *New Paradigms for Anticipated Uncertainty*, Lecture Notes in Mechanical Engineering, https://doi.org/10.1007/978-3-032-16889-4_37

Manufacturing firms employ structured methodologies like gap analysis and impact-effort matrices to assess readiness and prioritize investments [1, 2]. Additionally, Total Quality Management (TQM) principles support technology adoption through leadership commitment, continuous improvement, and stakeholder engagement [3].

This paper is founded in an observed gap in research in integrating new technologies in existing manufacturing systems. The complexity of digital transformation highlights the need for governance structures and project management frameworks to ensure success [4, 5]. Technology foresight methodologies, such as the Cisco Technology Radar, enable organizations to anticipate industry trends and disruptions [6]. Zero-defect manufacturing (ZDM) principles and data-driven decision-making enhance quality control and operational efficiency [7, 8]. As sustainability gains prominence, structured technology assessment methodologies align technological adoption with environmental and economic considerations [9]. However, the innovation management literature lacks concrete tools for aligning foresight with operational readiness. As Boe-Lillegraven and Monterde [6] note, "the radar enables strategic discussion, but translation into execution is often left unaddressed." Similarly, Fang et al. [10] emphasize that "firms struggle to convert technological intelligence into actionable investment decisions." From a manufacturing perspective, Powell et al. [7] observe that "digital manufacturing initiatives often lack a coherent framework that connects project-level efforts to enterprise-level goals." Too and Weaver [4] further argue that "effective project governance requires alignment between strategic intent and operational decision-making, which remains underdeveloped in current practice." These gaps underscore the need for an integrative framework that bridges foresight, operational capability, and strategic alignment.

This paper presents a conceptual framework for aligning internal capabilities with strategic goals to facilitate technology integration in manufacturing. As a research question, this is expressed as: How can internal capabilities be aligned with strategic goals to facilitate effective technology integration in manufacturing? By leveraging structured decision-making and governance mechanisms, firms can navigate digital transformation and achieve a sustainable competitive advantage.

The study uses a design science research (DSR) approach within a pragmatic paradigm, incorporating design thinking elements. Primary qualitative data was gathered through semi-structured and unstructured industry interviews, secondary data was sourced from databases of Aarhus University Library.

This paper is organized in the following way. First, the theoretical background is presented drawing on innovation and production engineering, then the framework development is presented. After this, there is the framework is tested and the application scope is unfolded. Finally, there is a discussion and a conclusion.

2 Theoretical Background

Technology integration in manufacturing companies necessitates a strategic alignment between internal capabilities and overarching business objectives. This alignment ensures that technological advancements contribute directly to competitive advantage and operational efficiency. Effective integration involves structured decision-making processes, capability assessment, and strategic goal-setting.

A structured approach to technology integration must first evaluate existing capabilities through gap analysis, a method that identifies discrepancies between current performance and strategic objectives [1]. This process is complemented by the impact-effort matrix, which prioritizes initiatives based on their potential impact and the effort required for implementation [2]. These tools enable organizations to allocate resources efficiently and maximize return on investment.

Total Quality Management (TQM) principles play a pivotal role in integrating new technologies. Critical success factors identified in TQM, such as leadership commitment, continuous improvement, and stakeholder engagement, provide a foundation for technology adoption [3]. Similarly, design thinking, as a user-centered approach, fosters innovation by ensuring that technology integration aligns with both operational needs and sustainability objectives [11].

Digital transformation projects in manufacturing often face challenges in achieving successful outcomes due to a lack of coherent strategy and governance [4]. The balanced scorecard methodology provides a structured framework to bridge this gap by linking strategic objectives to performance metrics [1]. Additionally, technology foresight techniques, such as the Cisco Technology Radar, help in anticipating technological trends and preparing adaptive strategies [6].

Zero-defect manufacturing (ZDM) exemplifies the importance of advanced technology adoption in enhancing product quality. This concept integrates real-time monitoring, predictive maintenance, and automation to minimize defects and improve efficiency [7]. For successful implementation, organizations must establish robust requirements engineering processes, which ensure that technology solutions align with business needs [12]. Illustrated in Fig. 1.

Project governance frameworks are crucial for managing technology integration efforts. An effective governance structure facilitates resource allocation, risk management, and stakeholder engagement, thereby enhancing project outcomes [5]. Moreover, technology intelligence and digitalization strategies enable companies to remain competitive by leveraging data-driven decision-making processes [8].

The evolution of technology assessment methodologies underscores the necessity for systematic evaluation of new technologies. These assessments consider economic, social, and environmental implications, ensuring that technology integration aligns with broader sustainability goals [9]. Additionally, structured methodologies such as IDEF0 and DSM aid in modeling information management processes, thereby improving workflow efficiency [13].

In summary, aligning internal capabilities with strategic goals in technology integration requires a multi-faceted approach encompassing gap analysis, TQM,

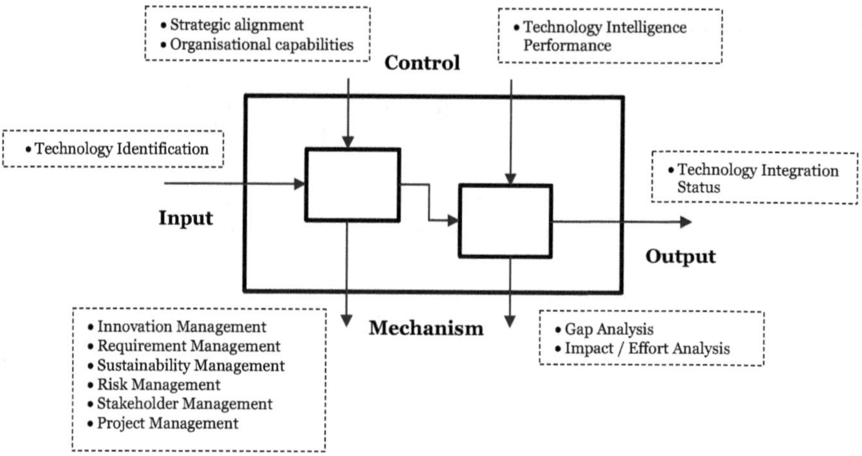

Fig. 1 IDEF-0 inspired framework development

design thinking, governance frameworks, and advanced manufacturing strategies. By leveraging structured methodologies and data-driven insights, manufacturing firms can achieve sustainable technological advancements and long-term competitive advantage.

3 Framework Development

By reviving a series of empirical insights, a conceptual framework has been modelled to allow navigation within technology integration (T.I.) decision-making. The main objective is to support decision-making for T.I. by investigating the controlling parameters: strategic alignment, organizational capabilities, and T.I. performance. These are resolved from the mechanism represented by six management areas and a modified gap and impact/effort assessment.

The six management areas concern critical technology integration perspectives, suggesting relevance for further gap measuring within the operational capabilities and strategic alignment. Assessing these areas further guides the integration of technologies and supports the decision on the chosen technology by indicating match alignment. The approach enables the match score to be held against the impact of the technology, communicated through the performance objectives.

By measuring the gap between strategy and capabilities as a match score and holding it against the performance score, decision-making is supported. This indicates whether the company should run the technology integration, hold it on standby, stop the integration, add resources, and run it, or outsource and run it. The T.I. status assumption is based on match and performance scores. The framework integrates theoretical and practical considerations, forming a holistic approach while providing specific guidance for organizations in decision-making on technology integration. See Tables 1 and 2. Figure 2 illustrates the suggested decision-making flow and principles.

Figure 2 outlines the logic for evaluating alignment between strategic objectives and operational capabilities across the six key management areas of Table 2. Each area is assessed using a four-level scale (from low to high), generating separate scores for both strategic intent and current capability as illustrated below.

Table 1 Theoretical basis - conceptual framework

Reference	Perspective	Transformation	Subject	Application
Unsal and Centidamar [14] Fang et al. [8]	Introduction		Motivational factors	Identified drivers
	Model formulation	• Creative design thinking	(diagnosis/guiding frame/coherent actions) (input/output/control/mechanism)	Guiding frame structure
Malmström et al. [15] Murrell [16] Stanford [17]				Decision tool: 1
			MOT-OS-MATRIX	
Slack and Lewis [18] D'Antone and Santos [19] Tidd and Bessant [20] Fang et al. [8] Pohl and Rupp [21] Dörr et al. [12] Buhl et al. [11] Xia et al. [22] Hallstedt [10] Aquilani et al. ([3]) Grunwald [9] Unsal and Centindamar [14] Patanakul [23]	Strategic alignment capabilities: Innovation management Requirements management Sustainability management Risk management Stakeholder management Project Management			
Silva et al. [24] Too and Weaver [5] Powell et al. [7] Boe-Lillegraven and Monterde [6]	Gap assessment		Gap analysis	Decision tool: 2
Balanced Scorecard Institute [1]				
SixSigma.us [2] Patanakul [23]	Strategic impact assessment		Impact / effort analysis	Decision tool: 3
	Decision insight		T.I. Status	Output insight

Table 2 Conceptual framework overview

MOT-OS-MATRIX	Strategic alignment	Organizational capabilities	Gap	Strategic impact assessment	T.I. status
Innovation management Requirements management Sustainability management Risk management Stakeholder management Project management	Where should the level of I.M. Be to satisfy our S.A. At what level does the RQ.M. Support the alignment with S.A At what level does the S.M. support the S.A. At what level should the R.M. Be for securing S.A. At what level should the S.M. be for securing S.A. At what level should the P.M. be for securing S.A.	At what level will S.A. be to align with O.C. At what level will S.A. be to align with O.C. At what level will S.A. be to align with O.C. At what level will S.A. be to align with O.C. At what level will S.A. be to align with O.C. At what level will S.A. be to align with O.C.	Gap parameter = Score parameter →	Impact score (0–3). + Match score (0–3). →	Hold Kill Run Add resources & run Outsource & run
Diagnosis			*Guiding frame*		*Coherent action*

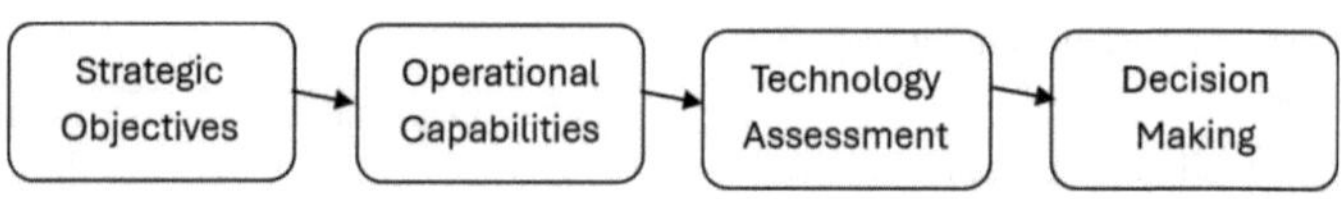

Fig. 2 Guiding principles

The gap between these is quantified for each area, with higher gaps indicating misalignment. These gap scores are then aggregated to produce an overall "match score," reflecting the degree of strategic fit. Separately, the estimated operational impact of the proposed technology integration is scored on a 1–3 scale. Combining the match score with the impact score positions the initiative within a decision matrix that suggests appropriate actions, such as "run," "hold," or "outsource and run."

4 Testing and Application of Framework

The framework integrates established models such as the Impact-Effort Matrix [2], Balanced Scorecard logic [1], and maturity assessment frameworks [18, 21], but extends them by addressing the specific challenge of technology integration in manufacturing. It introduces a dual-axis evaluation logic—strategic alignment (match score) and operational impact (performance score)—to support structured, adaptable decision-making. This design builds on and complements portfolio models like Patanakul's, embedding capability-strategy fit and simplifying cross-project prioritization.

The case companies were selected based on their contrasting yet complementary profiles, offering diverse industry perspectives and levels of digital maturity. One operates globally with a strategic focus on maintaining domestic production, while the other assessed 42 digital initiatives across its manufacturing portfolio. This deliberate contrast allowed the framework's applicability to be tested across both broad strategic planning and large-scale digital transformation contexts. Selection criteria emphasized relevance to technology integration, organizational capability alignment, and portfolio-based decision-making. These cases were thus well-suited to test the framework's versatility in real-world manufacturing scenarios, highlighting its ability to support structured decision-making across varied organizational settings. The case companies are kept anonymous but are manufacturing companies in the Nordic countries.

Tables 3 and 4 illustrate how qualitative assessments of strategic objectives and organizational capabilities are translated into quantitative scores across six management areas. Each area is rated on a four-level scale, representing maturity from low to high. These levels were intentionally selected to balance clarity and analytical depth, making them effective for workshop settings and compatible with standard maturity models. The resulting match and impact scores guide placement within the decision matrix, helping firms determine whether to proceed, adjust, outsource, or discontinue initiatives. This method enhances transparency and offers a robust basis for practical technology integration decisions in complex manufacturing environments.

Testing included (1) expert interviews, (2) experts filling out Table 4 and aligning across the companies, (3) scenario "play" with the experts. Through testing of the conceptualized framework in the case companies, the relevance and potential of offering a structured approach to evaluate technology projects were assessed. The case companies identified the matrix component as applicable for a visual and analytical element to spark a conversation and guide the company in making the right decisions. Figure 3 visualizes the overall identified potential and future opportunities for the framework.

The decision outcomes guide the technology management team using a graphical representation, see Fig. 3. Projects are categorized as "run," "add resources and run," "hold," "outsource and run," or "kill." A "run" decision indicates strong alignment with strategy, capabilities, and impact.

Table 3 Strategic objective levels

I#	Level 1	Level 2	Level 3	Level 4
Statement	We lack any formal structure or guidelines for managing <insert>, and it is not considered a strategic priority	We have developed basic guidelines based on the outcomes and learnings from previous projects, beginning to acknowledge the importance of <insert> management	We consistently follow established <insert> processes identified as helpful, adopting and applying relevant industry standards across the organization	We adhere to well-recognized industry standards and prioritize research and development within <insert> management, continuously improving our processes and strong culture of creativity
Innovation management	<innovation>	<innovation>	<innovation>	<innovation>
Match alignment		X		
Requirements management	<requirements>	<requirements>	<requirements>	<requirements>
Match alignment			X	
Sustainability management	<sustainability>	<sustainability>	<sustainability>	<sustainability>
Match alignment		X		
Risk management	<risk>	<risk>	<risk>	<risk>
Match alignment			X	
Stakeholder management	<stakeholders>	<stakeholders>	<stakeholders>	<stakeholders>
Match alignment		X		
Project management	<projects>	<projects>	<projects>	<projects>
Match alignment		X		

If the score suggests "add resources and run," additional capabilities should be developed internally to strengthen alignment while maintaining strategic control. A "hold" decision indicates limited strategic alignment and low impact, suggesting the need for further assessment or capability enhancement. The "outsource and run" category implies high strategic alignment but limited internal capacity or impact, making external partnerships a practical option. Finally, a "kill" score reflects both

Table 4 Organisational capabilities levels

2#	Level 1	Level 2	Level 3	Level 4
Innovation management	We have basic operational skills focused on established technologies, relying on existing technical competencies with minimal training required	We are able to embrace to proven technologies, training programs for new but established tools, reflecting moderate technical flexibility	We possess skills in researching and integrating emerging technologies, demonstrating high adaptability and a continuous learning mindset with technical foresight capabilities	We have expertise in pioneering new technologies with strong R&D capabilities, leading in technological innovation and driving advancements across the industry
MatchAlignment		**X**		
Requirements management	We rely on our current skill sets with limited capacity for new learning, supported by a basic training infrastructure	We enhance existing skills and access training that builds upon current competencies, showing openness to incremental learning	We have robust learning and development programs, acquiring new competencies and supporting cross-functional skill expansion within the organization	We implement advanced training initiatives with the capability to master cutting-edge skills, fostering a culture that encourages redefining industry roles through new competencies
MatchAlignment		**X**		
Sustainability management	We have an understanding of basic compliance standards with minimal sustainability training and limited environmental awareness	We develop skills in efficiency improvement and provide training in waste reduction techniques, exhibiting moderate environmental consciousness	We have expertise in integrating sustainability into our strategies, practicing sustainable methods and reporting with high environmental and social responsibility	We lead in sustainability innovation with advanced skills to create significant environmental and social impact, influencing industry sustainability standards proactively
MatchAlignment		**X**		
Risk management	We possess basic awareness of risk factors with minimal risk assessment skills and limited mitigation planning	We have fundamental risk management skills, able to react to issues as they arise with basic problem-solving capabilities	We are proficient in proactive risk assessment, developing and implementing mitigation strategies with regular monitoring abilities	We exhibit advanced expertise in risk analysis, capable in predictive analytics with strong implementation skills, leading best practices

(continued)

Table 4 (continued)

2#	Level 1	Level 2	Level 3	Level 4
MatchAlignment			X	
Stakeholder management	We have basic communication skills with minimal engagement with stakeholders and limited feedback mechanisms	We provide updates and solicit feedback periodically, practicing basic relationship management with moderate communication abilities	We possess strong collaborative skills, actively engaging stakeholders and proficient in relationship building and maintenance	We have expertise in partnership development with advanced communication and engagement skills, capable of involving stakeholders strongly in decision-making
MatchAlignment		X		
Project management	We focus on maintaining existing processes with minimal project management practices in place. Projects are handled informally, and there is limited support for structured project planning and execution	We have basic project management capabilities, enabling small-scale projects with some degree of planning and control. Moderate openness to adopting standardized tools and methodologies	We possess skills to effectively plan, execute, and monitor projects using established project management processes. Resources are allocated for project management training, and we support the use of project management software and tools	We demonstrate leadership in project management with advanced skills in managing complex projects and programs. A culture is embedded that values project management best practices, positioning us as industry influencers in project excellence
MatchAlignment			X	

low impact and weak alignment, signaling discontinuation as the most viable course of action.

Classification in 5 action categories is grounded in portfolio management principles and strategic decision-making literature, aligning with approaches such as Rumelt's strategic filters and Patanakul's project prioritization models. The two-dimensional evaluation logic builds on gap analysis and impact-effort matrices, combining strategic relevance (match score) with expected performance (impact score) to guide action.

Testing was conducted through semi-structured and unstructured interviews in two industrial cases. One company applied the framework retrospectively to global technology projects, while the other used it prospectively across 42 digital manufacturing initiatives (2022–2024). Case companies valued the matrix for its clarity and

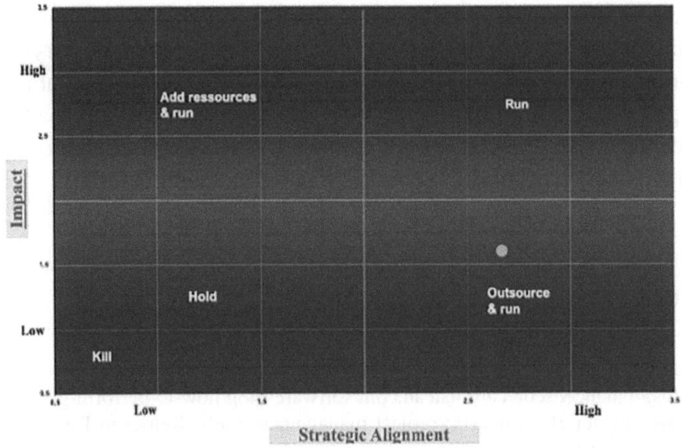

Fig. 3 Matrix impact/strategic alignment

decision-support function. The structured scoring and visual output enabled transparent discussions and evidence-based prioritization, confirming the framework's practical relevance and methodological coherence.

5 Discussion and Conclusion

This study proposes a conceptual framework to support manufacturing firms in navigating the complex decision-making processes involved in technology integration. By aligning organizational capabilities with strategic objectives and assessing technological impact, the framework offers a structured, theory-informed methodology that bridges the gap between technological potential and practical implementation.

The results enable a structured categorization of technology initiatives—"run," "add resources," "hold," or "outsource"—based on quantified alignment and estimated impact. This approach contributes scientifically by operationalizing strategic alignment [18] and translating gap analysis [1] into a visual, decision-oriented framework. It advances the literature on technology integration readiness by merging concepts from technological intelligence [8] and project portfolio management [5] into a coherent, scalable tool. The contribution lies in bridging theory and practice through a quantifiable, replicable decision-support model.

The framework was empirically tested in two case companies, generating industry-based insights into its practical relevance. It enabled structured evaluations of strategic alignment, operational capabilities, and performance outcomes, supporting evidence-based decisions on whether to proceed, revise, or terminate technology projects.

While demonstrating utility, the study identifies avenues for refinement. Case feedback underscores the need for customization to accommodate varying strategic contexts and capability levels. Furthermore, enhancing the matrix's ability to visualize and prioritize multiple projects could increase its value in broader portfolio management. Overall, the framework contributes a scalable decision-support tool tailored to the challenges of digital transformation in manufacturing.

References

1. Balanced Scorecard Institute. How to perform gap analysis. Retrieved December 15, 2024, from https://balancedscorecard.ltslean.com/software-blog/how-to-perform-gap-analysis/
2. 6Sigma.us. Impact effort matrix: project management tools. Retrieved December 15, 2024, from https://www.6sigma.us/project-management/impact-effort-matrix/
3. Aquilani B et al (2017) A systematic literature review on total quality management critical success factors. TQM J. 29(1):184–213
4. Clausen PA, Clausen JBB, Lassen AH, Henriksen B (2024) Why project success in manufacturing digitalization remains elusive. J. Manuf. Technol. Manag.
5. Too EG, Weaver P (2014) The management of project management: a conceptual framework for project governance. Int J Proj Manag 32(8):1382–1394
6. Boe-Lillegraven S, Monterde S (2015) Exploring the cognitive value of technology foresight: the case of the Cisco Technology radar. Technol Forecast Soc Chang 101:62–82
7. Powell D et al (2022) Advancing zero defect manufacturing: a state-of-the-art perspective and future research directions. Comput Ind 136:103596
8. Fang S, Moreno Brenes A, Brusoni S (2023) Technology intelligence and digitalization in the manufacturing industry. Res Technol Manag 66(5):22–33
9. Grunwald A (2009) Technology assessment: concepts and methods. In: Meijers A (ed) Handbook of the philosophy of science. Elsevier, pp 1101–1127
10. Hallstedt SI (2017) Sustainability criteria and sustainability compliance index for decision support in product development. J Clean Prod 140(1):251–266
11. Buhl A et al (2019) Design thinking for sustainability. J Clean Prod 231:1248–1257. https://doi.org/10.1016/j.jclepro.2019.05.259
12. Dörr J et al (2008) Implementing requirements engineering processes: using cooperative self-assessment and improvement. IEEE Softw 25(3):71–77
13. Manenti G et al (2019) Functional modelling and IDEF0 to enhance and support process tailoring in systems engineering. In: In 2019 Intl symposium on systems engineering. IEEE, pp 1–8
14. Unsal E, Cetindamar D (2015) Technology management capability: definition and its measurement. Eur Int J Sci Technol 4(2):181–196
15. Malmström J, Pikosz P, Malmqvist J (1999) Complementary roles of IDEF0 and DSM for the modeling of information management processes. Integr Comput-Aided Eng 6(1):95–103
16. Murrell A. Summary of Richard Rumelt's good strategy, bad strategy. From https://www.alexmurrell.co.uk/summaries/richard-rumelt-good-strategy-bad-strategy/
17. Stanford d.school (2010) An introduction to design thinking: process guide. Stanford University Institute of Design
18. Slack N, Lewis M (2020) Operations strategy, 6th edn. Pearson
19. D'Antone S, Santos JB (2016) When purchasing professional services supports innovation. Ind Mark Manag 58:72–186. https://doi.org/10.1016/j.indmarman.2016.05.024
20. Tidd J, Bessant J (2014) Strategic Innovation Management. Wiley
21. Pohl K, Rupp C (2015) Requirements engineering fundamentals: a study guide for the certified professional, IREB compliant, 2nd edn. Rocky Nook

22. Xia D et al (2017) Sustainable technology selection decision-making model for enterprise in supply chain. J Clean Prod 141:1337–1348
23. Patanakul P (2020) How to achieve effectiveness in project portfolio management. IEEE Trans Eng Manag 69(4):987–1000. https://doi.org/10.1109/TEM.2020.2964316
24. Silva JF et al (2022) Decision making in the process of choosing and deploying industry 4.0 technologies. Gestão & Produção 29:e163

Battery Systems Logistics in Automotive Supply Chains: Success Factors and Best Practices for Efficiency and Sustainability

Henrik Uebach

Abstract With the development of electromobility, car manufacturers have established and expanded the value chain for lithium-ion batteries. Specialised logistics for battery systems can help improve the environmental performance of the supply chain by increasing its efficiency and sustainability, taking into account the requirements for volumes and characteristics of battery systems and cell modules. One way to limit the carbon footprint is to shift the transport of battery products from road to rail. Focusing on the design of the transport system for battery systems and cell modules, our design starts with the special load carrier (SLC), as both goods are class nine dangerous goods according to the Agreement concerning the International Carriage of Dangerous Goods by Road (ADR). Based on the results of a functional analysis, the racks should be designed to fulfil the specific load carrier functions for cell modules and battery systems. The combination of SLC design and cargo space optimisation results in the best possible use of the volume and mass capacity of freight wagons for intercompany transport by rail. Special trailers are used for internal transport with electric trucks. All loading/unloading, handling, storage and transport processes are fully automated to ensure efficient internal material flow. An application example shows that around 14,300 tonnes of CO_{2e} can be saved per year for the supply chain of cell modules and battery systems by rail. This result points the way to further practices and improvements in the sustainability of automotive or other supply chains.

Keywords Battery systems · Sustainable logistics · CO_2 savings · Railway transport · Freight space optimisation

H. Uebach (✉)
U-CI Uebach Consulting Innovations GmbH, Freudenberg, Germany
e-mail: henrik.uebach@u-ci.de

M. Manns (ed.), *New Paradigms for Anticipated Uncertainty*, Lecture Notes in Mechanical Engineering, https://doi.org/10.1007/978-3-032-16889-4_38

1 Scope of the HV Battery Supply Chain

HV batteries are modular in design. Depending on the desired performance of the HV battery, a corresponding number of cell modules are installed. The largest and heaviest components are the housing parts and the cell modules. As the housing parts can be manufactured at the battery site, their logistics and the transport of smaller components such as busbars, cell controllers, connectors, fasteners, castings and cross sections are not considered below.

The supply chain success factors described therefore relate to the transport of HV battery systems from a battery assembly plant to a vehicle assembly plant and the transport of cell modules from the module manufacturers to the battery assembly plant. Both transport goods are subject to the legal transport regulations of the ADR (Agreement concerning the International Carriage of Dangerous Goods by Road), as they are categorised as Class 9 dangerous goods [1]. Such transport requires compliance with special regulations and increased safety standards.

The supply chain can consist of the following elements (Fig. 1).

It all starts with the production of the cell modules in the cell module manufacturer's factory (1). The cell modules are transported by rail (2) to the battery plant's logistics centre (LC, 3). In the LC, the cell modules are automatically unloaded from the wagon, stored and loaded (4) into an e-truck shuttle, which transports the cell modules to the logistics centre (6) of the battery factory (5). There, the cell modules are automatically unloaded, stored (6a) and, like the other components (7), fed into the battery housings for installation (8).

The fully assembled HV batteries are stored in the logistics centre. HV batteries called off for dispatch are automatically removed from storage, loaded and driven to the LC by electric lorry shuttle, alternating with cell modules, where they are unloaded and temporarily stored. The HV batteries are then loaded automatically, alternating with the unloading of the cell modules. The HV batteries are transported by rail to the vehicle assembly plant (9), where the transport processes are also automated. The empties (10) are returned by rail to the cell module manufacturer, where the load carrier cycle is completed.

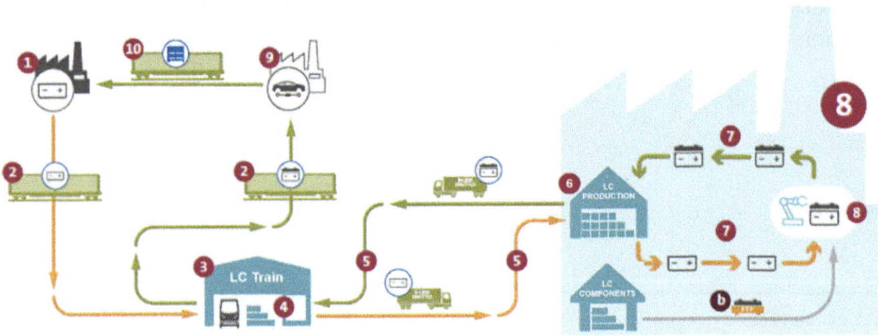

Fig. 1 Example of a supply chain network for cell modules and HV batteries

This paper outlines the key success factors, best practices, and techniques that contribute to the efficiency and sustainability of transportation and logistics in the supply chain for high-voltage (HV) batteries and cell modules.

2 Key Material Flow Elements of the Supply Chain System for HV Batteries and Cell Modules

2.1 Special Load Carrier

The core element of the supply chain for automotive batteries and cell modules for their production is the special load carrier for safe and efficient transport. With ReBox-Pool, Kampker et al. describe an innovative logistic concept based on a modular loading carrier concept [2] and a business model for the operation of pools of modular special load carriers [3].

Thanks to its modular design, the special load carrier (SLC) can accommodate both the suppliers' cell modules and the fully assembled HV batteries for this special supply chain and fulfils the necessary storage and protection functions with load securing elements. The load securing must be verified with a drop test in order to fulfil the regulations of the German Dangerous Goods Transport Act (GGBefG).

The SLC fulfils the transport functions by optimally utilising the volume and weight capacities of the rail wagon and truck means of transport. To fulfil the handling functions, the SLC is designed for automatic and manual conveying, storage and handling (stacking/destacking). For loading and external transport, the SLC are stacked on base pallets. To fulfil the information functions, the SLC is uniquely labelled with a barcode; transponders are recommended.

2.2 Modified Railway Wagons and Truck Shuttle

The freight wagons and trucks are to be determined in terms of load volume and permissible load weight in conjunction with the loaded special load carrier by means of freight space optimisation. An example selection can be a four-axle sliding wall wagon of the Habbii(II)ns type. The H17 can carry 98 SLCs, 98 HV batteries, or the necessary cell modules. The railway wagons are equipped with simple load securing devices that prevent the load from shifting in the event of impact. Conventional loading for emergency operation and universal use is possible.

The truck shuttles can be used for shuttle transport of cell modules and HV batteries between a logistics centre and an assembly plant or for in-house transport. Loading and load securing can be fully automated to maximise efficiency and minimise the number of shuttles. The use of e-trucks can further optimize CO_2 emissions.

2.3 *Fully Automated Internal Material Flow and External Transport*

Designed for the logistics functions of conveying, storing, handling, stacking/unstacking and loading/unloading, the system enables a fully automated internal material flow. For this purpose, internal transport units can be formed by stacking the load carriers with HV batteries or cell modules several times, depending on the possibilities of the conveyor and storage technology. Individual SLC or stacks of up to 7 SLC can be conveyed and stored, as the loading units can consist of stacks of seven.

The SLC are stacked on base pallets for loading/unloading and external transport. The planning basis should be that single-variety transport units are formed in order to maximise storage capacities and material flow performance.

For external material flow and transport by rail and lorry, loading and transport units are formed that meet the available freight space and legal requirements for transport.

2.4 *Formation of Loading Units for Rail Transport*

The formation of loading units is subject to legal regulations, the UCI (International Union Of Railways) Loading Guidelines: Code of practice for the loading and securing of goods on railway wagons [4].

Loading Guidelines Sect. 3.3 states: The load should be distributed as evenly as possible on the wagon, without exceeding the maximum mass per axle. The load distribution should be such that the following ratios are not exceeded: For 2-axle wagons: ratio of 2:1 between the masses per axle. For bogie wagons: ratio of 3:1 between the masses per bogies, ratio of masses per wheel: 1.25:1″. Calculation of the ratio of masses per axle for bogie wagons:

An example calculation is to be carried out for load distribution with a Habbiins 17 (H17) from Transwaggon [5]. Two battery types, for example midsize and maxsize, are to be loaded and transported in a stack according to type. A load unit consists of a stack with one basic pallet (GP) and seven special load carriers, each with one HV battery of the same type. To ensure the required form fit, the stack heights must be the same. An H17 wagon has 14 wagon spaces for the 14 load units. The criteria for wagon load distribution are

- Ratio of masses per axle for bogie wagons and
- Compliance with a total mass of 90 tonnes, a bogie must not be loaded above 45 tonnes and an axle must not be loaded above 22.5 tonnes.

Checking shifting loading patterns for the two types of battery stacking units midsize and maxsize shows that not just any combination of stack types may be loaded (Fig. 2).

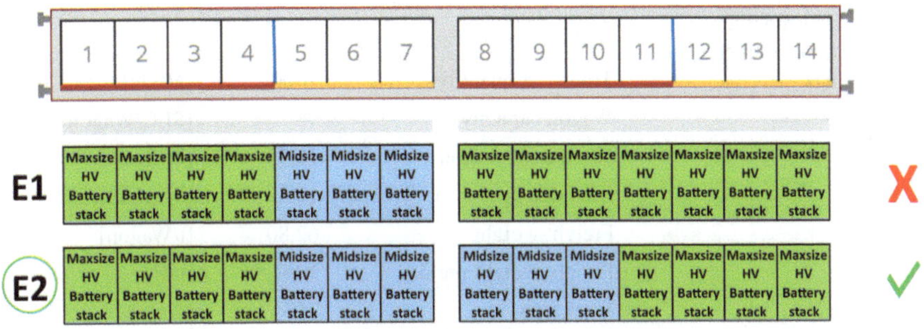

Fig. 2 One not permitted (E1) and one permitted loading (E2) example load combination

Table 1 Decision tool for checking the criteria for loading and securing of goods on railway wagons

	Distance	Width	Stack type	Mass	Weight	Lever arm1 (related to centre) (E2)	torque (E2)	Comment
Centre of stack 01	850 mm	750 mm	2	65,00 kg	0,638 KN	19,024 m	12,137 KNm	/
Centre of stack 02	2.350 mm	1.500 mm	2	4480,00 kg	43,949 KN	17,524 m	770,162 KNm	/
Centre of stack 03	3.850 mm	1.500 mm	2	4480,00 kg	43,949 KN	16,024 m	704,239 KNm	/
Centre of stack 04	5.350 mm	1.500 mm	2	4480,00 kg	43,949 KN	14,524 m	638,315 KNm	/
Centre of stack 05	6.850 mm	1.500 mm	2	4480,00 kg	43,949 KN	13,024 m	572,392 KNm	/
Centre of stack 06	8.350 mm	1.500 mm	2	4480,00 kg	43,949 KN	11,524 m	506,468 KNm	/
Centre of stack 07	9.850 mm	1.500 mm	2	4480,00 kg	43,949 KN	10,024 m	440,545 KNm	/
Centre of stack 08	12.114 mm	1.500 mm	2	4480,00 kg	43,949 KN	7,760 m	341,044 KNm	/
Centre of stack 09	13.614 mm	1.500 mm	2	4480,00 kg	43,949 KN	6,260 m	275,121 KNm	/
Centre of stack 10	15.114 mm	1.500 mm	2	4480,00 kg	43,949 KN	4,760 m	209,197 KNm	/
Centre of stack 11	16.614 mm	1.500 mm	2	4480,00 kg	43,949 KN	3,260 m	143,274 KNm	/
Centre of stack 12	18.114 mm	1.500 mm	2	4480,00 kg	43,949 KN	1,760 m	77,350 KNm	/
Centre of stack 13	19.614 mm	1.500 mm	2	4480,00 kg	43,949 KN	0,260 m	11,427 KNm	/
Centre of stack 14	21.114 mm	750 mm	2	4480,00 kg	43,949 KN	-1,240 m	-54,497 KNm	/

Load on bogie	B1	394,632 KN	40,227 t	1	0,9
			< 45 t	:	:
Load on bogie	B2	442,213 KN	45,078 t	1,12	1
Load on bogies	Summe E	836,845 KN	85,305 t	< 1 : 3	
			< 90 t		

A decision table for solving the combinatorial problem can be constructed as follows and shown for example E1 (Table 1).

It can be seen that due to the overall high stacking loads, the bogie load ratio of max. 3:1 is not effective, but the maximum permissible bogie load is the decisive criterion.

3 Sustainability of Transporting HV Batteries and Cell Modules by Rail

In order to examine the sustainability of rail transport compared to road transport, an approximate calculation of the energy consumption of both modes of transport for the following supply chain example is to be made: 1600 HV battery systems per day, 250 days a year, are to be transported from a battery assembly plant to a vehicle assembly plant, and the required cell modules from a cell module supplier to the

Table 2 Cell-module transportation parameter and CO_{2e} savings

Rail transport HV batteries		
Transport quantity:	1600	[Modules-SLC/d]
Wagon capacity	98	[SLC/Wagon]
Number of wagons	16,33	[Wagon/d]
Rounded up	17	[Wagon/d]
Freight weight	62,80	[t/Wagon]
Freight weight per day	1067,6	[t/d]
Workdays	250	[d/a]
Freight weight per year	266900,0	[t/a]
GHG emissions as CO_{2e} (WTW)	TS truck	TS train
Truck	44	0
Train	0	11
CO_2 Savings	33 t/d	
	8250 t/a	

battery assembly plant. The empty special load carriers are returned by rail from the vehicle assembly plant to the cell module supplier.

The calculations are carried out using the EcoTransIT tool [6]. The approximate calculation bases and results for the three transport routes are listed below. The main parameters and results for the transport of the cell modules from the cell module supplier to the battery assembly plant are as follows (Table 2).

A wagon load for the transportation of HV Batteries should consist of 6 stacks of midsize and 8 stacks of maxsize batteries. The main parameters and results for transporting HV batteries from the battery assembly plant to vehicle assembly plant are as follows (Table 3).

The main parameters and results for transporting of empty SLC from vehicle assembly plant to the cell module supplier are as follows (Table 4).

The total savings of the three routes for the transport of cell modules, HV batteries and empty SLCs is 14,322.50 tonnes, rounded to 14,300 tonnes of CO_{2e} per year, compared to truck transport.

4 Summary, Critical Evaluation and Outlook

4.1 Summary

The paper outlines the key elements for successful rail transport of cell modules and batteries in production networks for larger energy storage systems. The optimal design of the load carriers for both cell modules and HV batteries, in coordination with the choice of wagons, enables the optimal use of the freight space and the

Table 3 HV-battery transportation parameter and CO_{2e} savings

Rail transport HV batteries		
Transport quantity:	1600	[HV batteries/d]
Wagon capacity	98	[SLC/Wagon]
Number of wagons	16,33	[Wagon/d]
Rounded up	17	[Wagon/d]
Freight weight	61,70	[t/Wagon]
Freight weight per day	1048,9	[t/d]
Workdays	250	[d/a]
Freight weight per year	262225,0	[t/a]
GHG emissions as CO_{2e} (WTW)	TS truck	TS train
Truck	23,65	0
Train	0	4,67
CO_2 Savings	18,98 t/d	4745 t/a

Table 4 Empty SLC transportation parameter and CO_2 savings

Rail transport empty SLC		
Transport quantity:	1600	[Empty-SLC/d]
Wagon capacity	98	[SLC/Wagon]
Number of wagons	16,33	[Wagon/d]
Rounded up	17	[Wagon/d]
Freight weight	15,70	[t/Wagon]
Freight weight per day	266,9	[t/d]
Workdays	250	[d/a]
Freight weight per year	66725,0	[t/a]
GHG emissions as CO_{2e} (WTW)	TS truck	TS train
Truck	7,53	0
Train	0	2,22
CO_2 Savings	5,31 t/d	
	1327,5 t/a	

creation of a round trip of the special load carriers. The environmental benefits of rail transport compared to road transport are demonstrated by the calculated CO_2 savings of 14,300 tonnes.

4.2 *Critical Evaluation*

The calculations of transport volumes and daily requirements are examples and not fully calculated. The transport quantities of cell modules required are based on the

HV battery types to be produced. As maxsize HV batteries contain more modules than midsize HV batteries, there will be variations in the number of SLC with cell modules required. This, together with the possibility of more cell modules being transported in an SLC than are required for the largest HV battery, results in an imbalance between the demand for SLC for cell modules and HV batteries. It does not take into account the quantities and transport required to redress the balance. The economics need to be assessed on a case by case basis.

The lack of prospects for maintaining and expanding the rail infrastructure could slow down growth in the future. In order to organise transport on the scale described, wagons need to be procured and transport needs to be purchased. The economic viability of the investment required for rail transport, including investment in stations, handling and storage technology and much more, is not considered in this paper.

Rail transport requires good coordination in the supply chain. Short-term changes in production schedules or unforeseen problems with rail transport may require truck transport.

4.3 Outlook

The benefits can also be demonstrated in applications in other industries. In addition to CO_2 savings, significant reductions in carbon dioxide, nitrogen oxides, non-methane hydrocarbons, sulphur dioxide and particulates can be achieved.

If the necessary transport volumes are available, if load carriers and rail wagons can be coordinated, and if the necessary operating resources and rail transport can be organised, the economic and environmental advantages of rail transport will lead to more transport being shifted from road to rail.

References

1. https://unece.org/transport/road-transport/about-adr. LAst acccesed 26 Feb 2025
2. Kampker A, Franzkoch B, Wesch-Potente C, Brökelmann I (2011) ReBox-Pool–Innovative logistic concept based on a modular loading carrier concept. In: Thoben K-D, Stich V, Imtiaz A (eds) 17th International Conference on Concurrent Enterprising (ICE). 20–22 June 2011, Aachen, Germany. Forschungsinstitut für Rationalisierung e.V. der RWTH Aachen; International Conference on Concurrent Enterprising; ICE. Piscataway, NJ: IEEE
3. Kampker A, Burggräf P, Swist M, Ivanescu (2012) Geschäftsmodell für den Betrieb von Pools modularer Sonderladungsträger. Zeitschrift für wirtschaftlichen Fabrikbetrieb 2012(12):932–936
4. Loading Guidelines. Code of practice for the loading and securing of goods on railway wagons, Volume 1 Principles, Version 01/04/2024, 8th edition, ISBN 978-2-7461-3365-5
5. https://www.transwaggon.group/images/pdf/wagons-specs/HB17_Habbiins.pdf. Last accessed 26 Feb 2025
6. https://www.ecotransit.org/en/emissioncalculator/ of 2024. Last accessed 28 Feb 2025

Towards Intelligent Supply Chain Reconfiguration: A Framework Integrating Dynamic Knowledge Graph and AI-Driven Optimization

Chaouki Saidi, Nadia Hamani, Mounir Benaissa, Benjamin Rolf, Tobias Reggelin, and Sebastian Lang

Abstract In a dynamic and uncertain global environment, supply chain reconfiguration has emerged as a solution for managing disruptions and enhancing operational resilience. This paper presents an intelligent reconfiguration framework developed following an in-depth analysis of the existing literature on reconfiguration, the application of knowledge graph and the integration of Artificial Intelligence in supply chain reconfiguration. This comprehensive literature review highlights gaps in current methodologies, particularly in achieving efficient reconfiguration. In response, we propose a novel framework that leverages real-world data, Digital Twin modeling, Dynamic Knowledge Graph and Artificial Intelligence-driven Decision Making to enable intelligent supply chain reconfiguration. This framework facilitates the continuous evaluation and optimization of supply chain configurations, while incorporating key performance indicators such as sustainability. By using the Artificial Intelligence to enhance predictive capabilities and decision-making, the framework ensures that supply chains can seamlessly adapt to disruptions and evolving requirements. This work is driven by the need for sustainable and resilient supply chain that maintain a competitive edge in dynamic environments, a goal that can be achieved through efficient dynamic reconfiguration.

Keywords Reconfigurable supply chain · Dynamic knowledge graph · Artificial intelligence decision-making · Dynamic optimization · Digital Twin

---

C. Saidi · N. Hamani
MIS Laboratory—UR UPJV 4290, University of Picardie Jules Verne, Amiens, France

C. Saidi (✉) · M. Benaissa
OLID-OASIS Laboratories, Higher Institute of Industrial Management of Sfax, Sfax, Tunisia
e-mail: chaouki.saidi@u-picardie.fr

B. Rolf · T. Reggelin · S. Lang
Institute for Engineering of Products and Systems, Otto-von-Guericke-University Magdeburg, Magdeburg, Germany

M. Manns (ed.), *New Paradigms for Anticipated Uncertainty*, Lecture Notes in Mechanical Engineering, https://doi.org/10.1007/978-3-032-16889-4_39

1 Introduction

Supply chain management has evolved significantly (digitalization, pandemic-related adjustments) in recent years due to the increasing challenges of uncertainty and Covid-19. In this context, Reconfigurable Supply Chain (RSC) has emerged as a key strategic solution, enabling companies to quickly adapt their configurations in response to unforeseen disruptions. In addition, the concept of Supply Chain 5.0 emphasizes the integration of advanced technologies such as Artificial Intelligence (AI), digital twin (DT) and knowledge graph (KG) to create more sustainable, resilient, intelligent and human centric SC. However, despite technology progress, several challenges (uncertainty, disruptions) remain. First, reconfiguration knowledge management remains a major issue due to the complexity of data and the diversity of relations within supply chains [1]. Existing approaches also face limitations in assessing key performance indicators, often focusing primarily on reconfigurability characteristics, while overlooking critical issues such as sustainability and resilience. Furthermore, existing RSC approaches often lack flexibility and the ability to adapt to changing conditions [2].

To address these challenges, this paper proposes the use of a dynamic knowledge graph as the basis for reconfiguration knowledge management. That allows the structuring and optimization of all SC knowledge, while enabling the integration of AI for prediction and decision making. The objective is to develop an intelligent reconfiguration framework that combines the benefits of KG, DT and AI to create an adaptive methodology that meets current changes, particularly in terms of sustainability, resilience and human centricity.

This paper is organized as follows: Sect. 2 provides a detailed literature review on supply chain reconfiguration, the use of knowledge graph and the integration of AI, optimization and digital twin in this context. Section 3 presents the proposed framework for intelligent supply chain reconfiguration. Finally, Sect. 4 concludes the paper and provides perspectives for future research.

2 Background

This section will explore the state of the art in RSC, the KG application, and the AI role in enhancing decision-making and optimization within RSC.

2.1 Reconfigurable Supply Chain

Supply chain reconfiguration has been introduced as a solution to address disruptions. It involves adjusting the supply chain entities and their relations to better manage unexpected events. The adjustments considered may include adding, removing, or reorganizing [2].

Kelepouris et al. [3] introduced the concept of RSC, drawing inspiration from Reconfigurable Manufacturing Systems (RMS). Nepal et al. [4] proposed evaluation criteria for reconfiguration based on total cost and compatibility. Biswas [5] developed an approach based on five reconfigurability characteristics, using the "Total Interpretive Structural Modeling" (TISM) method to analyze the relations between variables. Later, Biswas et al. [6] defined a reconfigurability index using these characteristics. Dolgui et al. [1] expanded the field of supply chain reconfiguration by proposing a framework that integrates concepts such as resilience and digital twin. They introduced the ideas of dynamic meta-structures and autonomous services to adapt to changes. Nie et al. [7] proposed a mixed-integer programming model aimed at minimizing total cost while addressing disruptions. Zidi et al. [8] defined a reconfigurability index based on six reconfigurability characteristics (modularity, integrability, convertibility, diagnosability, scalability, and customization). Sardesai and Klingebiel [9] introduced a method for analyzing resilience using a composite indicator that includes total cost, service level error, and configuration capacity. Alexopoulos et al. [10] proposed a methodology for reconfiguring production systems to enhance their supply chain resilience. Recently, Saidi et al. [11] have proposed the reconfigurable supply chain modeling on KG as a solid foundation for reconfiguration knowledge management and [12] for human centricity in RSC. Their work is based on their study of ontological work in supply chain management [13]. Other works, such as Tu et al. [14] have explored supply chain reconfiguration as suppliers selection problem.

Based on the analysis of supply chain reconfiguration studies, the next section will focus on the KG application to identify future research directions.

2.2 *Reconfigurable Supply Chain Knowledge Graph*

The KG technology was developed as a model designed to effectively structure complex knowledge and use reasoning techniques to infer new knowledge. It is described in the literature as a multi-relational graph that connects entities (nodes) through edges (relations), allowing for better structuring and knowledge understanding [15].

Advances in KG offer promising perspectives, particularly for accurate modeling of SC entities [11]. Aziz et al. [16] focused on predicting hidden relations within the SC-KG. Kosasih and Brintrup [17] also investigated predicting missing links to improve SC visibility. Their approach treats visibility as a link prediction problem to facilitate supplier evaluation. Using an uncertain KG, Brockmann et al. [18] focused on risk management while predicting hidden SC relations. In addition, Kosasih et al. [19] developed a risk management framework by integrating missing link prediction, contributing to improved resilience. Given the growing interest in KG and AI, Karam et al. [20] combined Bayesian network with KG to improve resilience. Liu et al. [21] introduced a methodology for predicting hidden relations using different reasoning techniques based on supplier interactions. Other work, such as Deng et al. [22], explored the extraction and automated KG construction to make supply chain management (SCM) more autonomous. Düggelin and Laurenzi [23] emphasized the value of using KG to improve resilience, while Zhou et al. [24] proposed a forecasting framework to anticipate market trends and strengthen resilience. Yang et al. [25] proposed a framework based on the KG for risk management, which helps to improve its resilience.

Despite the growing interest in KG in various domains, their application in SC remains limited [11]. KGs provide enhanced visibility and predictive capabilities, and represent a promising research direction [26]. However, their practical use in supply chain reconfiguration is still lacking [11].

2.3 AI and Optimization in the Supply Chain Reconfiguration

Artificial intelligence has been widely adopted in SCM [27], but its application to RSC remains largely unexplored. While there is no existing work directly addressing AI-driven optimization in RSC [11], significant advances in AI for SCM provide a promising foundation for future developments [27]. The ability of AI to process and analyze large amounts of data has proven to be instrumental in improving decision-making process, particularly in areas such as forecasting, supplier selection and risk management. For example, Machine Learning algorithms have demonstrated their effectiveness in uncovering hidden patterns and improving the predictive SC capabilities [28].

The KG-DT integration has shown significant potential for enhancing system intelligence and adaptability, particularly in areas such as resource selection [29] and RMS [30]. KG provides a structured representation of knowledge, enabling better reasoning and decision making, while DT creates dynamic virtual replicas of physical systems, enabling predictive analysis and simulation [31]. In the literature, this combination has been used to improve operational efficiency, optimize resource allocation and support dynamic decision making. However, its application in RSC remains unused, highlighting an opportunity to explore its positive impact in this area.

Optimization can play a key role in RSC, ensuring that supply chain configurations are both efficient. AI also can contribute to this by enabling the evaluation of multiple configurations [11] and identifying the best one based on various performance criteria [2]. Recent studies in SCM and RMSs highlight the AI success in solving complex optimization problems such as minimizing costs and reducing risks [30]. Despite these advances, the integration of AI for dynamic optimization in intelligent RSC is still an underexplored area. The lack of research in AI-driven RSC optimization highlights the need for innovative methodologies. By combining AI techniques with dynamic KG, future work can improve both the transparency and effectiveness of reconfiguration strategies. The hybrid AI-Optimization can help RSC by enabling the selection of the most sustainable and resilient configuration in rapidly changing environments [31].

The comprehensive review of the RSC, the KG use, and the integration of AI, optimization, and DT reveals significant progress, but also notable gaps. While research has explored the use of KG for structuring and analyzing SC knowledge, AI-driven techniques for predictive analysis and optimization, these efforts remain largely independent. Existing work lacks a framework that dynamically combines these technologies to address real-world disruptions and ensure continuous RSC optimization. In addition, challenges such as incomplete integration of data from multiple sources, limited data extraction and collection, and lack of a feedback mechanism for iterative improvement have been identified.

To address these gaps, we propose a conceptual framework that integrates KG, AI, optimization and DT to enable an intelligent approach to RSC. This framework aims to overcome existing limitations by providing an adaptive and holistic solution, paving the way for smarter and more responsive supply chain.

3 Framework for Intelligent Reconfigurable Supply Chain

Based on the reviewed literature, we propose an RSC framework that integrates KG, AI, optimization techniques and DT. This framework aims to contribute to improving supply chain responsiveness and reconfigurability by combining comprehensive knowledge structuring, digital modeling, AI, and continuous optimization.

The proposed intelligent reconfiguration framework is structured into key interdependent stages, each of which is expected to contribute to enhancing RSC efficiency and resilience. The process begins with data acquisition and integration, where real-world knowledge from multiple sources is collected and structured into a dynamic KG. This KG serves as a central repository of interconnected knowledge, supporting advanced reasoning, relations discovery, and context-aware decision making. The DT modeling provides a virtual representation of the RSC, which can be used to simulate and analyze different reconfiguration scenarios in real time. This step allows actors to explore potential outcomes and assess the impact of disruptions before implementing adjustments in the physical system. The AI-driven optimization engine is employed to evaluate and suggest optimal configurations. These techniques can leverage advanced algorithms to support reconfiguration decisions aligned with configuration selection criteria such as sustainability, cost minimization, and resilience. Once the best configuration is identified, the next stage focuses on the implementation and continuous monitoring of reconfiguration solutions.

By incorporating real-time feedback and ongoing evaluation, the system may help ensure that SC operations remain agile and capable of dynamically adapting to changing requirements and unforeseen disruptions. This integrated approach offers a potential contribution to RSC methodologies by addressing some of the limitations identified in the existing literature and providing a flexible path towards intelligent RSC. The proposed reconfiguration Framework is illustrated in Fig. 1.

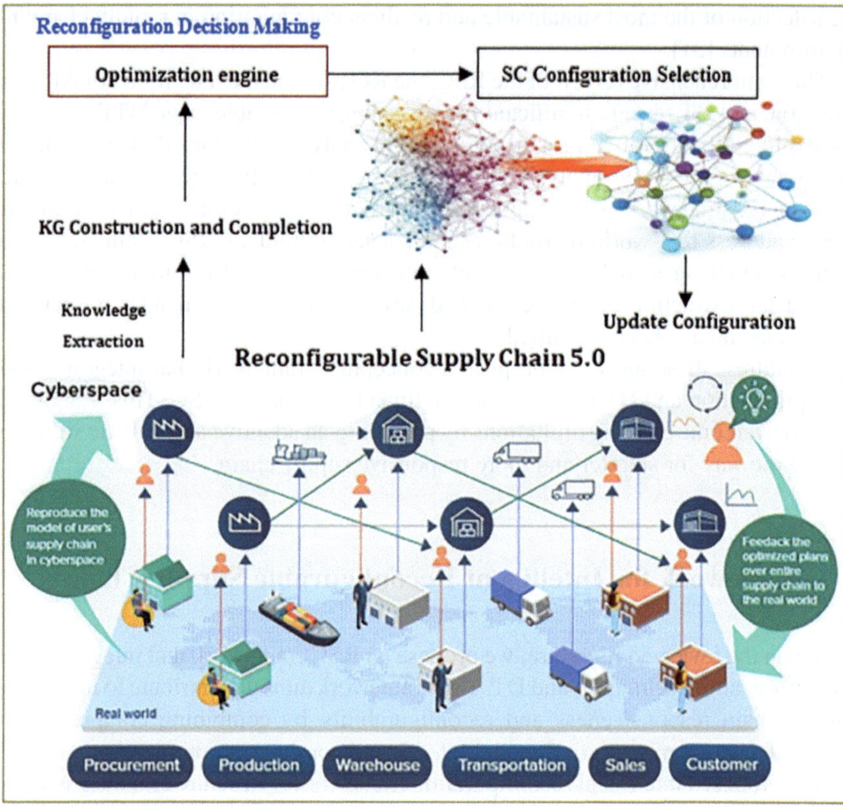

Fig. 1 Framework for intelligent reconfigurable supply chain

The framework is designed as an iterative process that integrates real-time data collection, advanced modeling and optimization to dynamically address disruptions. Each stage works in synergy to ensure that the RSC remains resilient and adaptive. Below we outline the key stages that define the framework and demonstrate their interconnected functionality.

- ***Data collection***

The process begins with real-time data collection from multiple sources, such as IoT sensors and other digital tools. This data includes information such as material flows, inventory levels, production capacity and transportation schedules. Once collected, this data will be used to build a DT model that virtually represents the RSC. Comprehensive and accurate data collection provides a solid foundation for creating a DT model.

- ***Knowledge Graph Construction***

Based on the semantic model, a KG is created to organize the knowledge. Entities and their relations are formalized as nodes and edges within the KG. AI plays a key role in this stage by enriching the KG with missing data through knowledge completion, enrichment or prediction of future events such as demand fluctuations. This transforms the KG into a centralized foundation for knowledge analysis and management, and can improve decision-making processes.

- ***Optimization and Configuration Selection***

This step allows to generate and evaluate multiple RSC configurations, taking into account specific constraints such as available resources. The objective is to select the most effective configuration that achieves an optimal balance between performance and resilience. Advanced AI techniques, including metaheuristics and prediction can be used to address these complex challenges. These techniques allow different configurations to be compared against predefined criteria, and facilitate the configuration selection that offers the best trade-off for effectively dealing with disruptions.

- ***Implementation and validation***

The selected configuration is not immediately implemented in the real world. It is first tested in the DT environment to simulate its operation. This validation stage verifies the feasibility and effectiveness of the configuration in a controlled environment before it's deployed in the real world. Naturally, this stage acts as a safe testing ground, reducing the risk before actual deployment. After validation, the best configuration is deployed in the real RSC. The KG is updated to reflect the new parameters and relations, which ensures that is ready for future analysis. The link between validation and implementation is based on the successful results of the DT.

By carrying out these steps, we guarantee the continuous RSC configurations evaluation to respond to changes.

4 Conclusion

The proposed intelligent Reconfigurable Supply Chain framework leverages the integration of Knowledge Graph, AI-Optimization techniques, and Digital Twin to address some of the dynamic challenges faced by modern supply chains. By enabling a structured representation of Reconfigurable Supply Chain knowledge, advanced modeling capabilities, and AI-driven decision support, this framework has the potential to contribute to enhancing sustainability and resilience in the face of disruptions. Additionally, the continuous evaluation and optimization processes embedded in the framework may support Reconfigurable Supply Chain in remaining competitive within an ever-changing environment. While the framework presents promising capabilities, it also opens perspectives for further research and development. One important direction is its application to real-world supply chain, incorporating real-time data streams to potentially improve responsiveness. Moreover, the integration of sustainability indicators into the optimization engine could help align the framework with economic, environmental, and social objectives. Finally, exploring the use of hybrid models that combine simulation and AI techniques may provide deeper insights into complex reconfiguration scenarios. These perspectives suggest that the proposed framework could serve as a useful foundation for future advancements in supply chain management.

References

1. Dolgui A, Ivanov D, Sokolov B (2020) Reconfigurable supply chain: the X-network. Int J Prod Res 58(13):4138–4163
2. Zidi S, Kermad L, Hamani N, Zidi H (2023) Reconfigurable supply chain selection: literature review, research roadmap and new trends. Appl Sci 13(7):4561
3. Kelepouris T, Wong CY, Farid AM, Parlikad AK, McFarlane DC (2006) Towards a reconfigurable supply network model. In: Intelligent production machines and systems. Elsevier Science Ltd., pp 481–486
4. Nepal B, Monplaisir L, Famuyiwa O (2011) A multi-objective supply chain configuration model for new products. Int J Prod Res 49(23)
5. Biswas P (2017) Modeling reconfigurability in supply chains using total interpretive structural modeling. J Adv Manag Res
6. Biswas P, Kumar S, Jain V, Chandra C (2019) Measuring supply chain reconfigurability using integrated and deterministic assessment models. J Manuf Syst 52:172–183
7. Nie D, Li H, Qu T, Liu Y, Li C (2020) Optimizing supply chain configuration with low carbon emission. J Clean Prod 271:122539
8. Zidi S, Hamani N, Kermad L (2021) New metrics for measuring supply chain reconfigurability. J Intell Manuf 33(8):2371–2392
9. Sardesai S, Klingebiel K (2023) Maintaining viability by rapid supply chain adaptation using a process capability index. Omega 115:102778
10. Alexopoulos K, Weber M, Trautner T, Manns M, Nikolakis N, Weigold M, Engel B (2023) An industrial data-spaces framework for resilient manufacturing value chains. Procedia CIRP 116:299–304
11. Saidi C, Hamani N, Benaissa M, Rolf B, Reggelin T, Lang S (2025a) Modeling reconfigurable supply chains using knowledge graphs: towards supply chain 5.0. Prod Eng:1–24
12. Saidi C, Yaddaden A, Hamani N, Benaissa M (2025b) GraphRAG for human centricity in supply chain reconfiguration. In: International conference on intelligent and fuzzy systems. Springer Nature Switzerland, Cham, pp 624–631
13. Saidi C, Rolf B, Hamani N, Benaissa M, Reggelin T, Lang S (2025c) A semantic framework: ontology-driven knowledge management for reconfigurable supply chains in the supply chain 5.0 era. IFAC-PapersOnLine 59(10):166–171
14. Tu Y, Li W, Song X, Gong K, Liu L, Qin Y, Liu M (2024) Using graph neural network to conduct supplier recommendation based on large-scale supply chain. Int J Prod Res:1–14
15. Wang Q, Mao Z, Wang B, Guo L (2017) Knowledge graph embedding: a survey of approaches and applications. IEEE Trans Knowl Data Eng 29(12):2724–2743
16. Aziz A, Kosasih EE, Griffiths RR, Brintrup A (2021) Data considerations in graph representation learning for ssupply chain networks. arXiv preprint https://arxiv.org/abs/2107.10609
17. Kosasih EE, Brintrup A (2022) A machine learning approach for predicting hidden links in supply chain with graph neural networks. Int J Prod Res 60(17):5380–5393
18. Brockmann N, Elson Kosasih E, Brintrup A (2022) Supply chain link prediction on uncertain knowledge graph. ACM SIGKDD Explorations Newsletter 24(2):124–130

19. Kosasih EE, Margaroli F, Gelli S, Aziz A, Wildgoose N, Brintrup A (2022) Towards knowledge graph reasoning for supply chain risk management using graph neural networks. Int J Prod Res:1–17
20. Karam N, Matini S., Laas R, Hoppe T. A Hybrid knowledge graph and Bayesian network approach for analyzing supply chain resilience
21. Liu Y, He B, Hildebrandt M, Buchner M, Inzko D, Wernert R, Tresp V (2023) A knowledge graph perspective on supply chain resilience. arXiv preprint arXiv:2305
22. Deng J, Chen C, Huang X, Chen W, Cheng L (2023) Research on the construction of event logic knowledge graph of supply chain management. Adv Eng Inform 56:101921
23. Düggelin W, Laurenzi E (2024) A knowledge graph-based decision support system for resilient supply chain networks. In: International conference on research challenges in information science. Cham, Springer Nature Switzerland, pp 66–81
24. Zhou S, Liu Y, Liu Y (2024) A market convergence prediction framework based on a supply chain knowledge graph. Sustainability 16(4):1696
25. Yang Y, Peng C, Cao EZ, Zou W (2024) Building resilience in supply chains: a knowledge graph-based risk management framework. IEEE Trans Comput Soc Syst
26. Kosasih E, Papadakis E, Baryannis G, Brintrup A (2023) Explainable artificial intelligence in supply chain management: a systematic review of Neurosymbolic approaches. Int J Prod Res
27. Kaushik M, Sathe M, Shyni S, Mary C, Singh B, Tiwari M, Ravikiran P, Kakade P (2023) A review of artificial intelligence applications in supply chain management
28. Zhou L, Zhang C, Liu F, Qiu Z, He Y (2019) Application of deep learning in food: a review. Compr Rev Food Sci Food Saf 18(6):1793–1811
29. Zhao Z, Zhang M, Chen J, Qu T, Huang GQ (2022) Digital twin-enabled dynamic spatial-temporal knowledge graph for production logistics resource allocation. Comput Ind Eng 171:108454
30. Mo F, Rehman HU, Monetti FM, Chaplin JC, Sanderson D, Popov A et al (2023) A framework for manufacturing system reconfiguration and optimisation utilising digital twins and modular artificial intelligence. Robot Comput Integr Manuf 82:102524
31. Listl, F G, Dittler D, Hildebrandt G, Stegmaier V, Jazdi N, Weyrich M (2024) Knowledge graphs in the digital twin: a systematic literature review about the combination of semantic technologies and simulation in industrial automation. arXiv preprint https://arxiv.org/abs/2406.09042

The Effect of Learning Factories on Manufacturing Innovation in SMEs

Lasse Christiansen, Jonas Frendrup, and Astrid H. Lassen

Abstract Small- and medium-sized enterprises (SMEs) are vital to the economy across various business domains, including manufacturing. However, as manufacturing principles and technologies evolve, SMEs often struggle to fully benefit from emerging knowledge and innovations. Their limited capacity to absorb and apply expertise in physical and data-driven processes hinders manufacturing innovation. One way to build this capacity is through professionals' engagement in learning factory activities. Learning factory approaches have been proposed as a means to support the development of knowledge and competencies essential for manufacturing innovation. To advance the understanding of how knowledge and skills gained through learning factory approaches influence manufacturing innovation, this study explores the question: "What role do diverse learning factory approaches play in facilitating SMEs' transition to new manufacturing technologies and operational strategies?"

Keywords Manufacturing innovation · Learning factory · Small- and medium-sized enterprises

1 Introduction

Manufacturing innovation has become a critical driver of competitiveness and efficiency for small- and medium-sized enterprises (SMEs) [1]. As technological advancements and digital transformation continue to reshape the manufacturing landscape, SMEs must adapt to remain viable [2]. However, these enterprises often face challenges in fully leveraging new technologies due to limited resources, knowledge

L. Christiansen (✉) · J. Frendrup
University College Northern Denmark, Aalborg, SV, Denmark
e-mail: lch@ucn.dk; lc@mp.aau.dk

L. Christiansen · A. H. Lassen
Department of Materials and Production, Aalborg University, Asalborg E, Denmark

M. Manns (ed.), *New Paradigms for Anticipated Uncertainty*, Lecture Notes in Mechanical Engineering, https://doi.org/10.1007/978-3-032-16889-4_40

gaps and organisational inertia [3]. The ability to assimilate and implement innovative manufacturing practices is essential for SMEs to sustain growth and enhance productivity [4].

When SMEs pursue digital transformation, managerial competence and employee skills are critical [5]. Employees must develop the capabilities to operate new technologies, while managers need strategic insight to guide digital transformation [6]. Many SMEs struggle because decision-makers often are unable to assess the impact of new technologies, align investments with business goals and drive organisational change. Without active engagement at all levels, digital tools risk being poorly integrated, underutilised or misaligned with company needs. Overreliance on external consultants for digitalisation can hinder long-term innovation by limiting internal knowledge development and reducing the organisation's capacity for agile decision-making [7].

SME digital transformation is most effective when companies collaborate with external partners while building internal capabilities [8]. Many SMEs struggle with adopting new technologies due to isolated efforts or overreliance on consultants [9]. Cross-industry collaborations—with universities, technology providers and peer companies—provide diverse expertise and accelerate learning. However, these partnerships must be balanced with internal development to avoid dependency and foster independent innovation [9].

Learning factories offer a promising solution to the knowledge and skill gaps by simulating real manufacturing environments for hands-on training [10]. Abele et al. define learning factories as environments that support competence development and innovation through authentic technical and organisational processes, mimicking real-world value chains [11]. Unlike traditional courses, they combine technical training with problem-solving, strengthening expertise and innovation capacity [12]. Structured around purpose, didactics and operational models, learning factories ensure experiential learning that is both relevant and adaptable [13]. This supports SMEs in adopting advanced manufacturing technologies and building domain-specific knowledge [14].

Research on learning factories has led to significant advancements in higher education, vocational training and research settings [15]. An enterprise's actual factory can also function as a learning factory if it is sufficiently adaptable and guided by a didactic framework that enables knowledge creation and transfer among workers, managers and experts [11]. Since SMEs frequently struggle with transferring knowledge related to advanced manufacturing technologies, this approach shows promise for improved outcomes [16]. Knowledge transfer is effective when the gap is clearly defined and can be addressed by existing knowledge [17]. However, in many cases, the gap must first be identified and bridged by creating new, interconnected knowledge [18]. This involves unique combinations of processes, products and IT systems—especially in complex or legacy environments where knowledge brokers may lack relevant experience.

These limitations highlight the need for a deeper understanding of how different learning factory designs impact SMEs in integrating new manufacturing technologies and methods into their operations. To address this, this article examines how

SMEs engage with learning factory methodologies, develop knowledge and competencies and enhance their innovation capacity. This study investigates the relationship between improvement strategies and learning factories, assessing how SMEs leverage existing knowledge to integrate new insights gained from learning factory experiences.

This research seeks to answer the following question: "What role do diverse learning factory approaches play in facilitating SMEs' transition to new manufacturing technologies and operational strategies?"

2 Methods

This study applies a qualitative research approach to a sample of 29 companies. This purposive sample of learning factory cases was selected to provide information-rich insights. The selected companies are participants in a European Digital Innovation Hub focused on digitalising small- and medium-sized manufacturing enterprises. Enrolment in the Hub requires formal SME status, according to the EU definition [19]. All sampled companies had participated in the project for more than six months, allowing knowledge brokers to meaningfully assess their progress.

Data collection was conducted through quarterly interviews with the knowledge brokers involved in the project, with 5 guiding questions in an open discussion. These brokers collaborated with personnel within the enterprises responsible for technology integration. While the primary collaboration was with owners or managers, technicians and blue-collar workers also participated in some activities. The data collection was embedded within the engagement phase, enabling a comprehensive capture of multiple dimensions of SME innovation efforts:

- The nature of digitalisation initiatives undertaken within the SMEs
- The enterprise's current innovation efforts, including ongoing projects and strategic focus
- The competency mix, identifying the skills and expertise present within the organisation
- The process and product structure, examining how production systems and workflows are organised.
- The business model, assessing how mechanisms for value creation, delivery and capture align with innovation efforts

In doing so, the study distinguishes between knowledge that is provided, knowledge that is transferred and subsequently applied and knowledge that is created—supporting SMEs in bridging the gap between technological advancements and practical application. To evaluate the effects on SMEs, a knowledge broker systematically assessed whether the enterprise had been provided with the requested knowledge, had transferred and implemented that knowledge or had created new knowledge beyond the original request.

Based on these evaluations, enterprises were categorised into three groups:

1. Knowledge transfer with limited effects
2. Knowledge transfer with expected implementation effect
3. Knowledge creation with additional implementation effects

Following this categorisation, a thematic analysis was conducted to identify common patterns and trends in SMEs' digitalisation efforts.

2.1 The Learning Factory Approach to Technology Implementation in SMEs

This study employs a learning factory approach to industrial collaboration, in which six dimensions govern how learners engage with industrial learning. These dimensions are purpose, didactics, operational model, product, processes and setting [11].

The learners in this context are industrial practitioners, prompted to seek digital knowledge to support manufacturing innovation through a European Digital Innovation Hub. The study applies the following understanding of learning factory principles:

First, while the overarching purpose of the activities was digitalisation, this was operationalised through a gemba-walk with the knowledge broker to conduct an in situ assessment of the enterprise [20]. Following this, the enterprise and the knowledge broker collaboratively developed a digital roadmap, outlining potential technologies, processes and digital routines.

The enterprise then began working within the framework of this roadmap, with the knowledge broker providing follow-ups and clarifying relevant tasks. This constituted the operational model of the learning factory activity, where the product, process and setting were all embedded within the enterprise itself.

Although "didactics" is typically associated with classroom education, in this context, digital upskilling required a didactic approach that emphasised linking the enterprise's digitalisation goals to its existing knowledge at the product, process and setting levels.

3 Results

The 29 SMEs surveyed in this study applied various digital manufacturing technologies, including carbon dioxide assessment through production data, product configuration, automated manufacturing and production simulation. They did not possess this technology prior, or only to a minor extent. The enterprises were divided into the three categories described in Sect. 2. Five enterprises were categorised as knowledge transfer with limited effect, 17 as knowledge transfer with expected implementation effect and 7 as knowledge creation with additional implementation effects. Furthermore, upon investigation, common factors were identified within each category. An overview is provided in Table 1.

Table 1 Overview of the identified mechanics

Purpose	Category	Description	Common Characteristics	Challenges Identified
Digital transformation	Knowledge transfer with limited effect	SMEs received knowledge but failed to implement it effectively	• Lack of prior knowledge required for digitalisation • Absence of skilled personnel to internalise knowledge • No structured digitalisation plan	• Skills gap preventing meaningful application • Limited strategic commitment to digital transformation
	Knowledge transfer with expected implementation effect	SMEs successfully adopted transferred knowledge and implemented digital solutions	• Identified and integrated off-the-shelf digital solutions • Engaged in structured projects for digital adoption (e.g. ERP and automation tools) • Tasked specific employees with digitalisation responsibilities	• Digitalisation was often incremental rather than transformative • Limited ability to experiment with custom solutions
	Knowledge creation with additional implementation effects	SMEs went beyond knowledge transfer, generating new digital capabilities and innovations	• Iterative digital improvements within existing infrastructure • Strengthened data utilisation, e.g. ESG reporting or automation • Integrated digital tools into production workflows creatively	• Required ongoing engagement with knowledge brokers • Necessitated a culture of experimentation and iterative learning

3.1 Knowledge Transfer with Limited Effect

In enterprises where knowledge transfer had a limited impact, it was evident that they lacked the necessary prior knowledge to internalise the insights provided by the knowledge broker and apply them for commercial purposes. This challenge was often linked to a shortage of qualified personnel, creating a skills barrier that must be addressed. To overcome this, enterprises would need to acquire, outsource or internally develop the required competencies before advancing with digitalisation in the given domain.

3.2 Knowledge Transfer with Expected Implementation Effect

Enterprises that achieved the expected results from knowledge transfer exhibited two key traits in their digitalisation efforts. First, some identified off-the-shelf solutions to their challenges after engaging in the project and refining their understanding of digitalisation. These companies sought out IT providers and integrated additional services compatible with their existing infrastructure. The main barriers before this stage were a lack of awareness about specific solutions, failure to recognise them as viable options or an unclear problem definition within the learning factory setting. While these off-the-shelf solutions may not qualify as manufacturing innovations, they proved effective and satisfactory for the enterprises involved.

The second major trend among enterprises that succeeded with the planned transfer of digital knowledge was implementation through structured projects. A set of employees was tasked with investigating the feasibility of digital tools, such as ERP systems or new manufacturing technologies. While this structured approach effectively facilitated the expected technology transfer, it also had limitations. Due to a restricted capacity for small-scale testing and incremental improvements, these projects typically achieved their intended results but did not generate additional value beyond initial expectations.

3.3 Knowledge Creation with Additional Implementation Effects

Lastly, some enterprises experienced internal knowledge creation that went beyond their original purpose for engaging in the project. A common factor among these enterprises was their approach to strengthening digital integration through incremental iterations and the continuous improvement of digital processes within their existing infrastructure. For example, this approach enabled enhancements such as expanding the data pool used for ESG reporting and sustainability efforts, or automating customer-specific product generation—all within the company's current systems.

4 Discussion and Conclusion

The findings of this study suggest that learning factory approaches can serve as an effective means for enterprises to gain the insights needed to drive manufacturing innovation. These approaches achieve this by upskilling individuals or teams within the organisation and fostering internal competency development, rather than relying on external hiring or costly consultancy services. While not the only viable approach, this aligns with prior research, emphasising that building internal competencies is a prerequisite for successful digital transformation [21]. The incremental steps that support knowledge creation, combined with knowledge-backed supervision during small-scale integration, help activate and embed content knowledge within the enterprise. Simultaneously, the upskilling needs of specific employees can also be addressed within the learning factory framework.

These effects will only emerge if the enterprise has allocated strategic resources to pursue digitalisation. If an enterprise expects a quick fix or relies on an off-the-shelf solution for a problem that requires a more tailored approach, it may struggle due to a lack of meta-cognitive knowledge. The complexity of implementation may exceed expectations and lack sufficient support. Therefore, it is essential to move quickly through small, meaningful and value-adding implementation steps, rather than pursuing extensive, time-consuming or costly projects.

Instead of viewing digital transformation as a one-time technology upgrade, management must establish mechanisms for continuous learning, where insights from collaborations are systematically integrated into company processes [22]. Encouraging internal teams to engage with external experts, apply new knowledge and experiment with digital solutions ensures that innovation becomes an ongoing, internally driven process—rather than a sporadic or outsourced initiative [23, 24].

Hence, it is also vital, when preparing learning factory activities, to target SMEs without overcomplicating the process. As the setting within manufacturing enterprises is often complicated, the demonstration of new digital technologies should be approached with care. Learning curves can accumulate, and when additional challenges—such as digitising a heritage production setup—are introduced, the full potential of a new software system may be overshadowed by excessive complexity.

Connecting digital knowledge with both silent and tactile expertise in the enterprise is a challenging task, and this research illuminates how this can be done for digital knowledge that the enterprise itself can take in. Other means must be implemented if the knowledge is outside the enterprise's grasp. If an enterprise, for instance, seeks to buy competence by hiring a new employee, the social integration required to be able to link to silent knowledge is a significant barrier. Hence, a learning factory approach is most applicable when the digitalisation process remains within the enterprise's capability to manage and integrate internally.

Future research should explore how a learning factory approach to digitalisation in SMEs facilitates internal industrial innovation. Specifically, studies should investigate how knowledge-informed incremental implementations and experiments contribute to developing product- and process-specific digitalisation knowledge. Furthermore, research should examine how this digital knowledge enables radical innovation in manufacturing, including its impact on throughput, margins, sustainability documentation and customisation. This could be in both industrial case studies and broader surveys.

Acknowledgements This research was carried out in the European Digital Innovation Hub Add-Smart.

References

1. Del Giudice M, Scuotto V, Papa A, Tarba SY, Bresciani S, Warkentin M (2021) A self-tuning model for smart manufacturing SMEs: effects on digital innovation. J Prod Innov Manage 38(1):68–89
2. Ghobakhloo M, Iranmanesh M, Vilkas M, Grybauskas A, Amran A (2022) Drivers and barriers of industry 4.0 technology adoption among manufacturing SMEs: a systematic review and transformation roadmap. J Manuf Technol Manage 33(6):1029–1058
3. Christiansen L, Borregaard T, Antonsen MG, Laursen ES, Brunoe TD (2022) Implementing virtual prototyping for the production of customized products: An SME study. In: Towards sustainable customization: bridging smart products and manufacturing systems, Andersen AL, Andersen R, Brunoe TD, Larsen MSS, Nielsen K, Napoleone A, Kjeldgaard S, (eds), in Lecture Notes in Mechanical Engineering. Springer International Publishing, Cham, pp 762–769. https://doi.org/10.1007/978-3-030-90700-6_87
4. Larsen MSS, Lassen AH, Schou C (2023) Manufacturing innovation: a heuristic model of innovation processes for industry 4.0. Appl Syst Innov 6(6):98
5. Li F (2020) Leading digital transformation: three emerging approaches for managing the transition. Int J Oper Prod Manage 40(6):809–817
6. Müller JM, Buliga O, Voigt K-I (2018) Fortune favors the prepared: how SMEs approach business model innovations in industry 4.0. Technol Forecast Soc Chang 132:2–17
7. Chesbrough H (2004) Managing open innovation. Res Technol Manage 47(1):23–26
8. Lassen AH, Laugen BT (2017) Open innovation: on the influence of internal and external collaboration on degree of newness. Bus Process Manage J 23(6):1129–1143
9. Sagala GH, Őri D (2024) Toward SMEs digital transformation success: a systematic literature review. IseB 22(4):667–719
10. Petrusch N, Schumacher B, Schwiethal A, Kohl H (2021) Evaluation of a level-based industry 4.0 competency model by the example of an internationally distributed learning factory, Rochester NY 3864147. https://doi.org/10.2139/ssrn.3864147
11. Abele E, Metternich J, Tisch M, Kreß A (2024) The variety of learning factory concepts. In: Learning factories: featuring new concepts. Worldwide Best-Practice Examples, Springer, Guidelines, pp 97–127
12. Bitsch G, Senjic P (2021) Learning factories on demand Rochester. In: Proceedings of the conference on learning factories (CLF). NY, p 3864330. https://doi.org/10.2139/ssrn.3864330
13. Enke J, Glass R, Metternich J (2017) Introducing a maturity model for learning factories. Proced Manuf 9:1–8. https://doi.org/10.1016/j.promfg.2017.04.010

14. Christiansen L, Nellemann C, Frendrup J, Zhang Y, Bøgevald Bjørnsten T, Heidemann Lassen A (2023) Didactical considerations for further education learning factories: a three case study. In: 13th conference on learning factories (CLF 2023), Reutlingen. Accessed Jun. 05, 2023. [Online]. Available: https://papers.ssrn.com/abstract=4469259
15. Daniele F, Confalonieri M, Pedrazzoli P, Graf A, Ferrario A, Foletti M (2021) Implementation of a learning factory for research, education and training: the SUPSI mini-factory. Rochester, NY, 3858404. https://doi.org/10.2139/ssrn.3858404
16. Corral de Zubielqui G, Lindsay N, Lindsay W, Jones J (2019) Knowledge quality, innovation and firm performance: a study of knowledge transfer in SMEs. Small Bus Econ 53(1):145–164. https://doi.org/10.1007/s11187-018-0046-0
17. Aarkrog V (2011) A taxonomy for teaching transfer skills in the Danish VET system. Nordic J Vocational Educ Train 1(1):1–13
18. Gilchrist AL (2015) How should we measure chunks? A continuing issue in chunking research and a way forward. Front Psychol 6. https://doi.org/10.3389/fpsyg.2015.01456
19. SME definition—European Commission (2025) Accessed Mar 09, 2025. [Online]. Available: https://single-market-economy.ec.europa.eu/smes/sme-fundamentals/sme-definition_en
20. Powell D, Coughlan P (2020) Corporate lean programs: practical insights and implications for learning and continuous improvement. Proced CIRP 93:820–825
21. Hansen AK, Christiansen L, Lassen AH (2024) Technology isn't enough for industry 4.0: on SMEs and hindrances to digital transformation. Int J Prod Res 0(0):1–21. https://doi.org/10.1080/00207543.2024.2305800
22. Awad JAR, Martín-Rojas R (2024) Digital transformation influence on organisational resilience through organisational learning and innovation. J Innov Entrep 13(1):69. https://doi.org/10.1186/s13731-024-00405-4
23. Couñago-Blanco E, Depino-Besada NI, Ferrer-Serrano M, López-Manuel L (2024) Collaboration as an enabler for digital transformation: the helix paradigm. In: Machado C, Davim JP (eds) Management for digital transformation. Springer International Publishing, Cham, pp 161–182. https://doi.org/10.1007/978-3-031-42060-3_8
24. Meissner D, Kotsemir M (2016) Conceptualizing the innovation process towards the 'active innovation paradigm'—trends and outlook. J Innov Entrep 5(1):14. https://doi.org/10.1186/s13731-016-0042-z

Process Design for Digitally Assisted Location Selection for Sustainable Factories

Günther Schuh, Esben Schukat, Siyuan Wang, Felix Paßlick, and Carsten Engeln

Abstract Available locations for new factories and the suitability of these locations changes, especially due to climate change. Long-term consequences and high costs entailed by factory location selection become more difficult to predict. The decision process for factory location therefore becomes more complex, uncertain and time-consuming. We aim to develop a data-driven digital decision support application for factory location selection. This paper presents the process to be implemented in that system. We research and refine established processes to enable data-driven, objective and sustainable factory location selection. Finally, we give an outlook on the decision support system to be developed.

Keywords Factory planning · Location selection · Site selection · Decision support · Sustainability

1 Introduction and Motivation

Factory location selection is critical for strategic planning in production and supply chain management. It identifies sites for new production facilities by reducing many locations to the most suitable alternative. [1] The resulting strategic decision for a new location has long-term impact on operational efficiency and business success [2].

The **higher-level process** encompassing factory location selection may be production network design or supply chain design [1, 3], because many firms operate within a network of geographically dispersed locations [2]. Alternatively, factory location selection may be seen as part of factory planning [2]. The German standard for factory location selection defines it as search, evaluation and selection of a geographical location for a factory and a fundamental task in factory planning [4]. Factory location decisions significantly impact both production network and factory planning [3].

G. Schuh · E. Schukat · S. Wang · F. Paßlick · C. Engeln (✉)
Laboratory for Machine Tools and Production Engineering (WZL), RWTH Aachen University, Aachen, Germany
e-mail: c.engeln@wzl.rwth-aachen.de

M. Manns (ed.), *New Paradigms for Anticipated Uncertainty*, Lecture Notes in Mechanical Engineering, https://doi.org/10.1007/978-3-032-16889-4_41

General challenges in location selection present optimization potential. The risk in factory location selection is high since the decision is costly and difficult to reverse, often requiring significant investment and shaping a firm's long term costs [2]. Mistakes can result in severe financial loss. The process often takes several years from initial assessment to final decision. Schönsleben [1] It is often revisited every few years due to changing external conditions and evolving strategies [5]. The inherent uncertainty is increased by dynamic markets [3]. Infrastructure, logistics, labor and many further requirements are assessed. Gathering reliable data therefore can be both difficult and time-consuming. MacCarthy and Atthirawang [6] Careful planning and evaluation are essential [3].

The **research agenda** this work follows was proposed by RIESENER ET AL. and aims for objective and sustainable data-driven decision support for factory location selection in a digital application [7]. This work proposes a location selection process that is well defined for implementation into the application and addresses unresolved challenges. Based on these considerations, the guiding research question of this paper is: How can the factory location selection process be structured to enable data-driven, objective, and sustainable decision-making within a digital support system? The following sections address the main challenges of factory location selection, analyze existing procedural models, and derive a structured process design. The paper concludes with a discussion of the results and an outlook on future research.

2 Process Challenges

Four challenges need to be addressed to develop a robust factory location selection process to be implemented in the data-driven decision-making support system.

Challenge one is the holistic consideration of factors across global, regional and local selection to achieve the highest decision quality [1, 3]. An iterative process progressively eliminating locations from global to local level helps to manage complexity. But initially excluded global locations may outperform others when evaluated at the local level [8]. Location selection must therefore look for the global rather than a local optimum. However, the effort in the decision-making process must not increase.

Challenge two is the availability and quality of data. Location selection often heavily depends on available data [6]. The difficulty in obtaining comprehensive and reliable data often limits the number of considered factors, affecting the decisions quality. Many potential data sources include industrial institutions such as chambers of commerce, industrial publications such as trade journals and firm directories, statistical institutions such as government agencies and bulletins, and private institutions and publications such as business contacts and magazines [9]. Often insufficient data description prevents automated data extraction resulting in high effort in data search. Data availability is often limited because of paywalls, or it does not exist. The suitability for representing location factors and thus relevance of datasets varies considerably. The precision of data may be limited, e.g. for relative measures, indices or rankings instead of absolute values. The trustworthiness may be limited, if data origins or modelling approach are not disclaimed. [10] This limits the rationality and objectivity of location selection [11]. Rationality may be increased by validation of all sources and data sets. This runs counter to the objective of minimizing the effort and is often beyond the capacity of firms. Due to poor quality and limited availability of data, assumptions are often made based on top-level factors. High effort data usage often leads to decreased accuracy of results [12].

Challenge three is the complexity of the factors to be considered. Strategic decisions often consider an insufficient number of factors [13]. Assessing interrelated location factors is increasingly important [14] and enables robust decisions [7]. E.g., the requirement 'low land prices' may depend on 'low groundwater table'. If such interdependencies are not considered, factors may be overemphasized, potentially impairing decisions. Statistical methods can help to capture interdependencies, but they require often difficult to identify mathematically formulations. Warnecke et al. [9] Interdependencies are often estimated rather than empirically derived. The relevance of location factors also depends on the processes carried out at the selected location. E.g., labor costs are more important for high degrees of manual labor, while availability of highly skilled labor may be more critical for high automation [15]. Reducing uncertainty while considering many factors and research, selection and preparation of factors is challenging [7]. Identifying and evaluating interrelated factors must be both low effort and accurate.

Challenge four is the consideration of sustainability. The range of factors increases as sustainability gains importance. Sustainability has become a key element of corporate strategy, driven by legislation, societal expectations and investor demands. International agreements like the Paris Agreement, the EU Green Deal, and industry regulations increase the importance of sustainable decisions. Gutglück et al. [16] Increasing sustainability regulations like CO_2 taxes and tariffs must be considered [17]. Consumer preferences increase the demand for sustainability. Investors increasingly prioritize firms that adhere to ESG principles and targets towards 1.5 °C goal. Gutglück et al. [16] Natural factors such as sunshine hours, wind conditions, temperature or water availability influence a locations sustainability potential. These factors vary between locations and are important in location selection [17, 18]. Green Factory KPIs must be considered [7]. Comprehensive assessments integrate environmental, social and economic dimensions [19].

3 Process Analysis

In the following section, the process challenges are systematically linked to the phases of existing location selection processes. This analysis forms the basis for the structured process design presented later in the paper. All four challenges are explicitly addressed to improve the robustness and applicability of the decision-making process. **Systematic literature research** was conducted following the STARLITE [20] method to identify process models. The sampling strategy included relevant publications on related topics. The type of studies was limited to fully reported processes. The approach was to search in databases followed by forward–backward-search. The range of years was limited to 2010 to 2025 except for the forward–backward-search. The limits were set to English scientific papers. The inclusion of topic focused on engineering and operations research. The terms used resulted in following search string: ("factory" OR "production" OR "manufacturing") AND ("site selection" OR "location selection" OR "site assessment" OR "location assessment") AND ("Sustanab*" OR "Data"). The sources included IEEE Xplore, Pascal & Francis, Scopus, Web of Science and Springer Link. The filtering followed the PRSIMA method [21]. 8951 publications were identified, leading to 6276 publications without duplicates. 3566 were excluded due to off-topic journal, 2418 due to off-topic title and 225 due to off-topic abstract. 53 full texts remained. The **general process** for factory location selection follows typical similarities. Strategies, especially corporate strategy [3] or production strategy [5], are inputs. The assessment of requirements fulfillment depends on the chosen strategy [1]. **Requirements** define which factors need to meet which requirements [3]. Knock out requirements exclude locations not fulfilling them. They include hard requirements, which may be numerical or not and have no benefit if exceeded, and minimum requirements, which must meet a numerical minimum and have a benefit if exceeded [22]. **Factors and criteria** are sometimes used synonymously [2], while other times factors are assessed if they meet the requirements defined in criteria [9]. In this work a factor is understood as the summary of many criteria so that for any factor or criterium a requirement can be compared to a value depending on the location to be assessed. The **scope** is successively refined through global, then regional and then local location selection [22]. The process may be started with any of these three scopes depending on the strategic input regarding what alternatives to consider [3]. The **methods** for selection are mostly qualitative [1]. 90% of firms use benefit analysis and the other most used methods are competitor analysis and empirical benchmark [23]. Quantitative methods include net present value method, cost calculation and transport cost minimization [24]. The **selection of processes** to be analyzed meets three requirements. First, the selected processes focus on holistically describing process steps for practitioners to enable management decisions via comprehensible reasoning. Literature was disregarded if it mainly focused on describing singular steps or methods, e.g. in this context popular approaches for multi criterial decision making like analytical hierarchy process and approaches for consideration of uncertainties like fuzzy logic. Second, the selected processes aim for general applicability instead of focusing on singular use cases. Third, the processes were cited latest 15 years ago and therefore reflect current knowledge.

Five processes for factory location selection were analyzed. The first by Eversheim [9] is cited by Burggräf et al. [25] and Grundig [24]. The second by Meyer [26] is cited by Schönsleben [1]. The third by VDI [27] is not cited but used by the German standard VDI 5200 PART 3:2016- - 05 [3] and cited by Verhaelen et al. [28]. The fourth was developed by Vierschilling [8]. The fifth was developed by Pawellek [2].

The **analysis approach** structured numbered steps and challenges into derived phases. The definition of required data was assumed to precede the description via acquired data. The following figure Fig. 1 shows the analyzed processes.

The **results of the analysis** are three main insights. First insight: Factory location selection mainly consists of phases one, two, and three, regarding requirements, description and assessment. Phases zero and four regarding initiative and decision, are not always described. Second insight: The iterative refinement is not defined regarding when to define global, regional and local requirements, when to describe locations via data and when to use rough or fine selection. Third insight: No process addresses the four challenges. Challenge 1, holistic consideration of factors, arises during phase 3, assessment of locations. It arises if global or regional locations are excluded that would be preferred if they were considered in further steps. Challenge 2, data availability and quality, arises during phase 3, description of locations. It arises if selected factors, criteria and measurable indicators cannot be represented by available data in the desired quality. This leads to either expensive data acquisition or disregard of desired assessments. Challenge 3, complexity of factors, arises during phase 3, description of locations. It arises if factors, criteria, indicators and data sets are selected disregarding their interdependencies leading to distorted assessment. Challenge 4, consideration of sustainability, arises during phase 1, definition of requirements. It arises if requirements for sustainability are not known or not linkable to established requirements e.g. monetary objectives, or regulatory compliance.

4 Process Design

The **objective** of the process design was to address the described challenges. Steps were mostly derived from Eversheim [9] and VDI 2012 [2], with novel additions. Steps may be skipped, as is usual [3]. The following figure Fig. 2 shows the developed factory location selection process with 5 phases, 9 steps, 28 sub steps and outputs per sub step.

Phase 0, initiative, consists of two steps. Step 0.1, input examination at the corporate level, may result in a mandate to revise the location strategy if corporate objectives and strategy require it. Step 0.2, location strategy development on network level, results in a location strategy with a target production network structure that defines role, processes and products per location and thereby possibly including new locations to be selected in the subsequently described process. Both steps are taken from the german standard VDI 5200 Part 3:2016- - 05 [3] and its basis by VDI 2012 [27].

Phases 1, global, 2, regional, and 3, local, each consist of the two steps rough selection and fine selection. Rough selection assesses knock out requirements that are selected and assessed if they are either met or not. Knock out requirements may be either quantitative or qualitative and either case-independent or case-dependent. Case-independent requirements relate to just locations themselves and may be assessed for each location individually, e.g. a location´s GDP. Case-dependent

Original process Recent citation	Eversheim 1999 Burggräf 2022, Grundig 2021		Meyer 2006 Schönsleben 2024, Wiendahl 2014	VDI 2012 VDI 2016, Verhaelen 2021		Vierschilling 2022	Pawellek 2014		
Phase 0 **Initiative**	0. Initiative via decisions in corporate objective planning			1. Input examination on corporate level 1.1 Examination of corporate objectives 1.2 Examination of corporate strategy 1.3 Determination of need for action 2. Location strategy development on network level 2.1 Analysis of product, process and network structure 2.2 Development of network alternatives 2.3 Decision for location strategy					
Phase 1 **Requirements**	1. 1.1 Location requirements from corporate objectives	Iterations 1. Global 2. Regional 3. Local		3. Location level 3.1 Creation of requirements profiles	Iterations 1. Global 2. Regional 3. Local	1. Requirements definition based on predefined suggestions 2. Weighting of requirements	1. Objective planning 1.1 Objective of the location regarding products, market, and quantity 1.2 Search area regarding countries, free trade areas, accessible sales areas etc.		Challenge 4 **Consideration of sustainability**
Phase 2 **Description**	1. … 1.2 Location description			3. … 3.2 Search for potential locations		3. Data identification 4. Data acquisition 5. Data standardization 6. Data documentation	2. Rough location selection 2.1 Pre-selected macro locations 2.2 Factor selection and weighting	3. Fine location planning 3.1 Pre-selected micro locations, factor selection and weighting	Challenge 2 **Data availability and quality** Challenge 3 **Complexity of factors**
Phase 3 **Assessment**	2. Rough selection based on knock out requirements 3. Fine selection 4. Ranking		1. Global pre-selection of countries 2. Selection of locations and functional range on country level 3. Local pre-selection of around 10 to 30 locations in a long list 4. Local pre-selection 5. Local location selection	3. … 3.3 Assessment …		7. Location ranking 8. Optional refinement	2. … 2.3 Evaluated macro locations 2.4 Selected macro locations	3. … 3.2 Final selection of micro locations with benefit analysis, factors and weightings 2.3 Ranking of micro locations with benefit analysis	Challenge 1 **Holistic consideration of factors**
Phase 4 **Decision**	5. Optimal location		6. Decision	3. … 3.3 … and selection			4. Decision e.g. after location visits, conversations with administration and business development agency and rough investment calculation		

Fig. 1 Five analyzed location selection processes

Phase	Step	Output	Challenges
0 Initiative	0.1 Input examination on corporate level		
	0.1.1 Review of corporate objectives	→ Strategic corporate objectives related to location selection	
	0.1.2 Comparison with corporate strategy	→ Strategic corporate objectives to be met by location selection	
	0.1.3 Determination of need for action	→ Possibly mandate for revision of the location strategy	
	0.2 Location strategy development on network level		
	0.2.1 Analysis of product, process and network structure	→ Ideal production network structure allocating future processes and products	
	0.2.2 Development of network alternatives	→ Real production network alternatives considering locations and constraints	
	0.2.3 Decision for location strategy	→ Location strategy with network structure possibly including new locations	
1 Global	1.1 Global rough location selection		
	1.1.1 Global knock out requirements	→ Global knock out requirements	4
	1.1.2 Global location description	→ Global location data for selected requirements and locations	2
	1.1.3 Global location limitation	→ Global locations for further consideration	
	1.2 Global fine location selection		
	1.2.1 Global detailed requirements	→ Global weighted requirements with assessed interdependencies	3
	1.2.2 Global location description	→ Global location data and knowledge for selected requirements	
	1.2.3 Global location ranking	→ Global location ranking for available data	
	1.2.4 Global location limitation	→ Global locations for further consideration	
2 Regional	2.1 Regional rough location selection		
	2.1.1 Regional knock out requirements	→ Regional knock out requirements	4
	2.1.2 Regional location description	→ Regional location data for selected requirements	2
	2.1.3 Regional to global consistency check	→ Further possibly suitable global locations according with regional assessment	1
	2.1.4 Regional location limitation	→ Regional locations for further consideration	
	2.2 Regional fine location selection		
	2.2.1 Regional detailed requirements	→ Regional weighted requirements with assessed interdependencies	3
	2.2.2 Regional location description	→ Regional location data and knowledge for selected requirements	
	2.2.3 Regional location ranking	→ Regional location ranking for available data	
	2.2.4 Regional location limitation	→ Regional locations for further consideration	
3 Local	3.1 Local rough location selection		
	3.1.1 Local knock out requirements	→ Local knock out requirements	4
	3.1.2 Local location description	→ Local location data for selected requirements	2
	3.1.3 Local to regional consistency check	→ Further possibly suitable regional locations according with local assessment	1
	3.1.4 Local location limitation	→ Local locations for further consideration	
	3.2 Local fine location selection		
	3.2.1 Local detailed requirements	→ Local detailed requirements with weighting and interdependencies	3
	3.2.2 Local location description	→ Local location data and knowledge for selected requirements	
	3.2.3 Local location ranking	→ Local location ranking for available data	
4 Decision	4 Final location selection	→ Selected site	

Fig. 2 Developed factory location selection process

requirements relate to more than locations themselves and may only be assessed considering further case-specifics, e.g. any locations distance to another case-specific location. Rough selection ends with selecting locations for further consideration. Fine selection assesses detailed requirements that are selected, analyzed in terms of their interdependencies, possibly more closely selected regarding their interdependencies, weighted and then evaluated in terms of their degree of fulfillment.

The degrees of fulfillment and weightings result in a performance level for each location creating a ranking of locations. Fine selection ends with selecting locations for further consideration. These steps in phases 1 to 3 are taken from the Eversheim [9] except for the following changes. Sub steps 1.1.1, global, 2.1.1, regional, and 3.1.1, local knock out requirements address challenge 4, consideration of sustainability. They should therefore use a holistic requirements catalog that considers sustainability. Such a catalog is outlined to various extend by Eversheim [9], Grundig [24], Meyer [26], Schönsleben [1], VDI 2012 [27], Verhaelen et al. [28] and Pawellek [2] but not provided by any of the sources of the analyzed processes. Sub steps 1.1.2, global, 2.1.2, regional, and 3.1.2, local, address challenge 2, data availability and quality. They should therefore use a holistic data catalog that specifies availability and quality. Such a catalog is outlined by Eversheim [9], Grundig [24], VDI 2012 [27], but not provided by any of the sources of the analyzed processes. Sub steps 1.2.1, global, 2.2.1, regional, and 3.2.1, local detailed requirements, address challenge 3, complexity of factors. There the mentioned holistic requirements catalog also transparently explains the requirements interdependencies. The dependencies would otherwise need to be analyzed during fine selection [9]. Sub steps 2.1.3, regional to global, and 3.1.3, local to regional consistency check, are new additions to address challenge 1, holistic consideration of factors. A review of previous global or regional limitations enables including previously excluded locations if regional or local knock out requirements are applied to the knock out limitation of the previous phase.

Phase 4, decision, concludes the location selection process by selecting a site.

5 Discussion, Conclusion and Outlook

The **result of this work** is a process for factory location selection. Compared to other research in this field the innovation of the process is that it is clearly defined from phase zero to four including outputs per step. Additionally, it addresses the outlined challenges of holistic consideration of factors, data availability and quality, complexity of factors and consideration of sustainability. The benefit is that it is suitable for integration in the data-driven decision support system for factory location selection.

The **approach** was to analyze, combine and extend five established holistic processes into a process design, which addresses the challenges. Works by EVERSHEIM [9] and VDI [27] were the main inspiration. A limitation remains in the refinement of the process through expert opinions and validation in various use cases.

As an **outlook** the authors follow the outlined research agenda. They aim to integrate the process into the developed prototypical application. Upcoming research and development include automation potentials for data acquisition and data analysis for trend identification and forecast via artificial intelligence as well as visualization in the user interface.

Acknowledgements Funded by the Deutsche Forschungsgemeinschaft (DFG, German Research Foundation) under Germany's Excellence Strategy—EXC-2023 Internet of Production—390621612.

References

1. Schönsleben P (2020) Integrales logistikmanagement, 8th edn. Springer, Berlin Heidelberg, Berlin, Heidelberg
2. Pawellek G (2014) Strategieplanung. In: Pawellek G (ed) Ganzheitliche fabrikplanung. Springer, Berlin Heidelberg, Berlin, Heidelberg, pp 83–147
3. Verband Deutscher Ingenieure (2016) Fabrikplanung, phasenmodell zur gestaltung globaler produktionsnetzwerke, VDI 5200 Blatt 3, 2016–05
4. Verband Deutscher Ingenieure (2011) Fabrikplanung, planungsvorgehen, VDI 5200 Blatt 1, 2011–02
5. Wiendahl H, Richardt J, Nyhuis P (2014) Handbuch fabrikplanung, 2nd edn. Hanser, München, Wien
6. MacCarthy BL, Atthirawang W (2003) Factors affecting location decisions in international operations—a Delphi study. Int J Oper Prod Manag 7:794–818
7. Riesener M, Adlon T, Schukat E, Salzwedel J, Engeln C, Passlick F (2023) From complexity to clarity in sustainable factory planning: a conceptual approach for data-driven integration of green factory KPIs in manufacturing site selection. In: Herberger D, Hübner M (eds) Conference on production systems and logistics (2023–2). Publish-Ing., Hannover, pp 874–881
8. Vierschilling S (2022) Automatisierung der standortauswahl in der fabrikplanung, 1st edn. Apprimus Verlag, Aachen
9. Warnecke J, Wiendahl H, Brankamp K, Eversheim W, Henn G, Kühnle H (1999) Fabrikplanung. In: Eversheim W, Günther S (eds) Produktion und management 3. Springer Berlin Heidelberg, Berlin, Heidelberg, , pp 1–117
10. Riesener M, Schukat E, Wang S, Engeln C (2024) Data assessment for effective regional factory location selection. In: 18th CIRP conference on intelligent computation in manufacturing engineering, Elsevier, Amsterdam, pp 1–5
11. Maier G, Tödtling F (2006) Regional-und Stadtökonomie 1, 4th edn. Springer Wien New York, Wien, New York
12. Benfer M, Ziegler M, Gützlaff A, Fränken B, Cremer S, Prote J, Schuh G (2019) Determination of the abstraction level in production network models. In: 52nd CIRP conference on manufacturing systems. Elsevier, Amsterdam, pp 1–6
13. Afra S, Aders C (2000) Shareholder value konzepte: eine untersuchung der DAX 100-unternehmen, 2nd edn. KPMG Consulting GmbH, Frankfurt am Main
14. Nyhuis P, Großhenning P, Mühlenbruch H (2006) Das Globale varianten produktionssystem. Zeitschrift für wirtschaftlichen Fabrikbetrieb 101(5):282–286
15. Witthohn C (2019) Konfiguration globaler produktionsnetzwerke im Spannungsfeld von Kosten-und Risikominimierung, 1st edn. Apprimus Verlag, Aachen
16. Gutglück C, Schäfer T, Essers L, Götz M (2023) Sustainability @ automotive, 1st edn. Deloitte, München

17. Welsing M (2023) Bewertung der ökologischen Nachhaltigkeit von produktionsnetzwerkkonfigurationen, 1st edn. Apprimus Verlag, Aachen
18. Lanza G, Ferdows K, Kara S, Mourtzis D, Schuh G, Váncza J, Wang L, Wiendahl H (2019) Global production networks: design and operation. In: CIRP annals—manufacturing technology, vol 68(2). Elsevier, Amsterdam, pp 823–841
19. Chen L, Olhager J, Tang O (2014) Manufacturing facility location and sustainability: a literature review and research agenda. Int J Prod Econ 149:154–163
20. Booth A (2006) "Brimful of STARLITE": toward standards for reporting literature searches. J Med Libr Assoc JMLA 94(4):421–429
21. Moher D, Liberati A, Tetzlaff J, Altman D (2009) Preferred reporting items for systematic reviews and meta-analyses: the PRISMA Statement. Open Med 3(2):123–130
22. Eversheim W, Robens M, Witte K (1977) Entwicklung einer systematik zur verlagerungsplanung, 1st edn. Westdeutscher Verlag GmbH, Opladen
23. Schuh G, Gützlaff A, Adlon T, Schupp S, Endrikat M, Schlosser T (2022) Datenbasierte standortauswahl. Zeitschrift für wirtschaftlichen Fabrikbetrieb 117:258–263
24. Grundig C (2021) Fabrikplanung, 7th edn. Hanser eLibrary, München
25. Burggräf P, Schuh G, Dannapfel M, Esfahani M, Hoeschen A, Ivanescu S (2021) Standortplanung. In: Burggräf P, Schuh G (eds) Fabrikplanung. Springer Berlin Heidelberg, Berlin, Heidelberg, pp 63–121
26. Meyer T (2006) Globale standortauswahl—einflussfaktoren. In: Abele E (eds) Handbuch globale produktion. Hanser, München, pp 36–100
27. Verband Deutscher Ingenieure Fachausschuss Fabrikplanung (2012) Strategien und nachhaltige wirtschaftlichkeit in der Fabrikplanung. In: Standorte gezielt auswählen, investitionen sicher planen, 1st edn. Beuth, Berlin
28. Verhaelen B, Peukert S, Lanza G (2021) Site selection processes in global production networks. In: Friedli T, Lanza G, Remling D (eds) Global manufacturing management. Springer International Publishing, Cham, pp 101–115

Empirical Study on Asset Fingerprinting with Natural Markers to Improve Supply Chain Traceability Using Digital Product Passports

Marco Buecheler, Maximilian Hentsch, Frithjof Dorka, Grant Richards, and Daniel Palm

Abstract The use of natural markers can improve the consistent and forgery-proof traceability of assets in sustainable supply chains. When using fingerprint technology, the inherent surface structure of an asset is captured by an imaging system and then compressed into a digital asset fingerprint which is stored in a database. To identify an asset, a newly generated fingerprint is compared against those in the database. This approach is particularly relevant for enhancing consistent asset traceability to meet regulatory requirements and support initiatives such as the digital product passport. To consistently locate and align an asset's fingerprint region of interest, this work investigates the use of bounding symbols. Five bounding symbol shapes are empirically evaluated for their recognition performance. Additionally, four open source algorithms (ORB, BRISK, SIFT and pHash DCT) to create asset fingerprints are compared and evaluated in terms of identification confidence, processing speed, and memory requirements to assess their feasibility for tracing assets throughout life cycles. The experimental setup examines aluminum raw castings in the form of medallions. Results reveal differences in the identification confidence and resource usage among the tested algorithms. Notably, pHash DCT is more than three orders of magnitude faster than the algorithms used for feature matching, requires the least storage space, and still provides sufficient identification confidence.

Keywords Digital product passport · Identification technology · Fingerprint technology · Perceptual hashing · Feature matching

M. Buecheler (✉) · M. Hentsch · F. Dorka · G. Richards · D. Palm
Reutlingen University, Reutlingen, Germany

Purdue University, West Lafayette, IN, USA
e-mail: marco.buecheler@student.reutlingen-university.de

M. Manns (ed.), *New Paradigms for Anticipated Uncertainty*, Lecture Notes in Mechanical Engineering, https://doi.org/10.1007/978-3-032-16889-4_42

1 Introduction

Asset traceability is crucial for sustainable value creation, particularly in the transition from a linear to a circular economy where assets can have multiple life cycles [1]. Traceability refers to the systematic ability to understand the historical path and movement of an asset throughout its life cycle(s). This enables transparency and informed decision-making across supply chain stakeholders [2]. Legislative initiatives like the digital product passport (DPP) under the European Green Deal [3] or the United States Uyghur Forced Labor Prevention Act [4] aim to enhance the traceability of assets. DPPs will store asset related data generated along the value chain and product life cycle [5]. A prerequisite is the linkage of the physical asset with the corresponding digital data. Therefore, consistent asset identification (ID) systems and technologies are essential.

2 State-of-The-Art

Asset ID technologies like QR codes or RFID tags have limitations in the consistency, durability and security as they are artificially applied to assets [1, 5, 6]. A promising approach for consistent asset identification is the use of inherent asset surface features as natural markers [8]. These natural markers increase the consistency and trustworthiness of the ID system [1, 5]. One ID technology using natural markers is referred to as fingerprint technology. When creating a fingerprint of an asset, the unique asset surface features are captured with an imaging system in a predefined region of interest (ROI) [6, 8]. Bounding symbols (BSs) aid the localization and alignment process of the fingerprint ROI [10]. An image of the ROI is captured and compressed into a digital representation using a suitable algorithm. The digital fingerprint is then either stored in a database (when newly registering an asset) or compared to the ones already stored in the database (when identifying an asset). As described by [11], open source algorithms used for feature matching like ORB [12], BRISK [13], or SIFT [14] can be used to create digital asset fingerprints. These three algorithms detect and describe keypoints in an image. In the identification process, the keypoints are used to match two genuine fingerprints, hence one image pair. Another possibility to create asset fingerprints is called perceptual hashing [8]. Perceptual hash algorithms (e.g., pHash DCT) create similar hash values for images with comparable visual structures. Both algorithm types do not create completely different fingerprints if surface alterations (e.g., scratches) occur [6, 8], enabling secure asset life cycle identification if the alterations remain within a defined threshold. Using the algorithms for DPP applications is lacking scientific investigation.

3 Requirements for Consistent Asset Traceability Using DPPs and Study Objectives

Requirements An asset traceability system in a circular economy must meet certain requirements: The asset marker needs to be consistent (unambiguous and available throughout an asset's life cycle) [1], easily locatable [3], and linked to a globally unique (single-occurrence) identifier [6]. The identifier must be accessible for all parties in the supply chain [1]. A suitable sensor (camera) has to encode the identifier information from the marker [1, 5, 14] in the ROI. In a last step, the DPP is provided from a data management solution.
Study Objectives As the ROI must be locatable, five *BS shapes* with an orientation attribute are investigated. Moreover, four *algorithms to create fingerprints* are explored to determine their feasibility in DPP use case scenarios.

4 Experimental Setup, Procedure and Statistics

The paper conducts two experiments (A and B). Both experiments use a Manta G-2040B camera with an optical lens (Schneider-Kreuznach APO-Componon 4/60) and a LED light ring (LUMIMAX CRC CD 50FL-ALD-W). Cast aluminum medallions with five BSs on each medallion (MED) were designed and simple randomly sampled for experiment A (n_{MED}=40) and B (n_{MED}=80). The objectives, procedures, and statistical analysis are explained below.
Experiment A This experiment employs a randomized complete block design and applies a one-way analysis of variance with blocking to assess the effect of BS type on ROI recognition performance. Each of the 40 medallions is set as a block to isolate the production variances of the individual medallions. The dependent variable is the similarity score based on Hu Moments [16]. Figure 1 illustrates the BS detection process to calculate the score. The independent variable is the BS type, consisting of five levels as Fig. 2 shows. The null hypothesis states that there is no significant difference in the mean similarity scores between the reference group and a specific non-reference group at a significance level α=0.050.
Experiment B Based on experiment A, the BS shape with the highest ROI recognition performance is selected for experiment B. Four algorithms to create fingerprints are compared in the second experiment: ORB, BRISK, SIFT, and pHash DCT. The primary performance metric is the matching score (determining the identification confidence), calculated as $n_{inliers}/N_{total}$ for ORB, BRISK and SIFT [11] and *1-Hamming distance* for pHash as no keypoints are matched [7]. Other metrics under investigation are processing speed and storage requirements, all compared descriptively. Figure 3 shows the fingerprinting process: The first step (a) is to rotate the image. Then the ROI (red 500 × 500 pixel area) is positioned in the BS center (b). Features are extracted from the ROI using one of the four algorithms. Two sets of 80

Fig. 1 Bounding symbol detection process. Note: (1) Original image. (2) Blurred Image. (3) Thresholded image. (4) All detected contours. (5) Detected reference contour. (6) Closest contour match

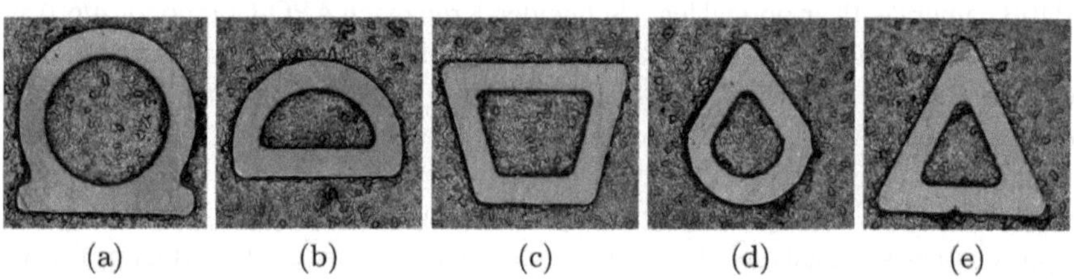

Fig. 2 Detected bounding symbol shapes. **a** Circle. **b** Half circle. **c** Trapezoid. **d** Cone. **e** Triangle

images are later matched (c) and a score calculated for each image pair. Green (inliers) and red (outliers) lines show matched keypoints after RANSAC [17] verification.

5 Results and Discussion of Empirical Study

Results and Discussion for Experiment A Table 1 presents the analysis of variance results. The randomized complete block design indicates a significant effect of the factor *Shape* on the similarity score ($F(4, 156) = 6.95$, $p < 0.001$, $\alpha = 0.05$). This provides strong evidence to reject the null hypothesis, suggesting that not all BSs perform equally well in terms of similarity to the individual reference shapes. The blocking factor *Medallion* does not significantly affect the response ($p = 0.830$), indicating that inter-medallion variability does not confound recognition performance.

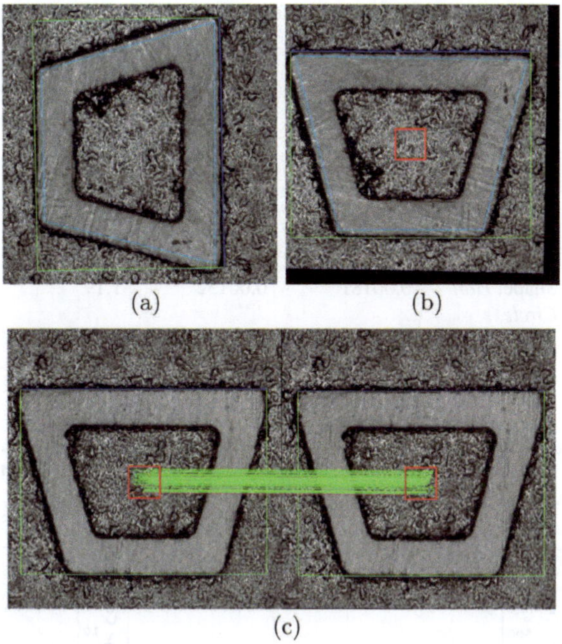

Fig. 3 Fingerprint creation process. **a** Original Image. **b** Rotated image with positioned ROI (Red Square). **c** Two matched image pairs with extracted and matched ORB keypoints

Table 1 Analysis of variance results for shape and medallion factors

Source	Degrees of Freedom	Adjusted Sum of Squares	Adjusted Mean Square	f-Value	p-Value
Shape	4	0.001290	0.000323	6.95	0.000
Medallion	39	0.001393	0.000036	0.77	0.830
Error	156	0.007240	0.000046		
Total	199	0.009923			

The coefficients (coef) in Table 2 represent the estimated effect size of each shape relative to the reference constant *Trapezoid* (lowest mean of the sample group). Based on the results of the regression analysis, the shapes *Cone* ($p = 0.010$) and *Triangle* ($p < 0.001$) are significantly different from the *Trapezoid* at the $\alpha = 0.050$ significance level. The shape *Circle* is marginally significant ($p = 0.050$). Additional data is recommended to confirm this effect. *Half Circle* does not show any statistically significant difference from *Trapezoid* ($p = 0.235$).

Table 2 Coefficient estimates for shapes

Term	Coef	SE Coef	t-Value	p-Value	95%-Confid Intervall
Constant: *Trapezoid*	0.01770	0.00320	5.54	0.000	[0.01136, 0.02404]
Shape: *Circle*	0.00300	0.00152	1.97	0.050	[0.00001, 0.00599]
Shape: *Cone*	0.00398	0.00152	2.61	0.010	[0.00099, 0.00697]
Shape: *Half Circle*	0.00181	0.00152	1.19	0.235	[−0.00118, 0.00480]
Shape: *Triangle*	0.00761	0.00152	5.00	0.000	[0.00462, 0.01060]

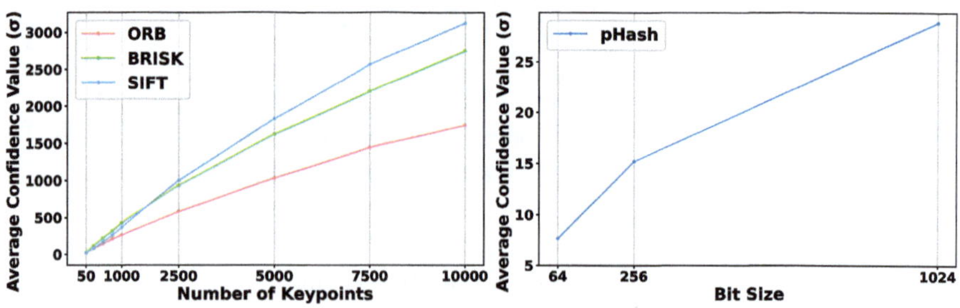

(a) Algorithms used for Feature Matching. (b) Perceptual Hash DCT.

Fig. 4 Identification confidence versus number of keypoints/bit size

5.1 Results and Discussion for Experiment B

Identification Confidence and Matching Scores All four algorithms under investigation successfully identified the 80 medallions of the sample group. As seen in Fig. 4, the confidence curves flatten (logarithmic tendency) for an increased number of extracted surface features (casting surface).

The identification confidence values differ between algorithm groups and in-between algorithm levels (keypoint number and hash length dependent). To get a better impression on how far the genuine and imposter matches are away from each other, Fig. 5 shows the inter (imposter matches represented in yellow) and intra (genuine matches represented in blue) scores as histogram plots. The gap between the farthest right part of the inter distance distribution and the farthest left part of the intra distance distribution is notably bigger for the algorithms used for feature matching (e.g., ORB) compared to the perceptual hash algorithm (showing the difference in confidence levels). However, pHash still has a sufficient identification confidence to reliably identify all medallions of the sample group.

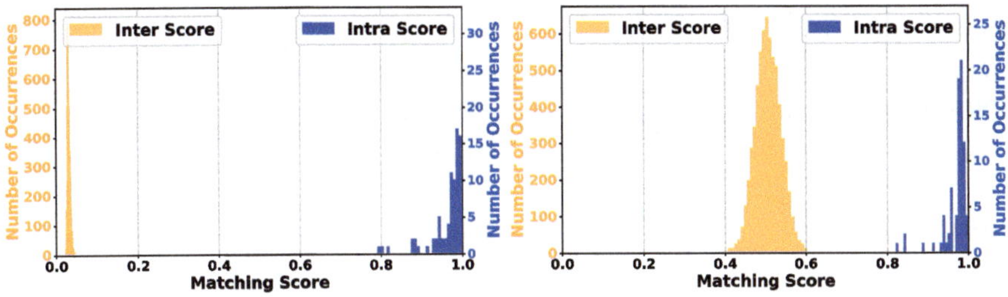

(a) 1,000 keypoints ORB. (b) 256 bit pHash.

Fig. 5 Inter and intra scores

Table 3 Algorithm performance comparison

Algorithm	Total fingerprint Creation time [s]	Total fingerprint Matching time [s]	Total Time [s]
ORB (1,000 keypoints)	2.0861	402.1175	404.2036
BRISK (1,000 keypoints)	60.0778	405.6453	465.7231
SIFT (1,000 keypoints)	24.6664	426.0428	450.7092
pHash DCT (256 bit)	0.3289	0.0732	0.4021

Identification Speed Table 3 shows a compact comparison of fingerprint creation, feature matching, and total (creation and matching) time across all four algorithms under investigation for the two image sets of the sample group. For pHash DCT, the total time is less than half a second to successfully identify all 80 medallions of the dataset. All three algorithms used for feature matching needed more than 6.5 min (at a level of 1,000 keypoints extracted). The pHash algorithm (256 bit) therefore is around three orders of magnitude faster compared to each of the three algorithms used for feature matching (when extracting 1,000 keypoints). One reason for the rapid comparison of the perceptual hash compared to the three other algorithms is the small size of the created fingerprints (see page 453 for the comparison of storage characteristics). Due to its fast processing times, pHash DCT is well suited for mass production where many parts must be registered in a short time period.

Storage Characteristics The four algorithms under investigation show a linear increase in storage size when increasing the keypoint number or bit size respectively. Since ORB and BRISK use binary descriptors, their storage requirements are significantly lower than SIFT (see Table 4), which relies on floating-point numbers. A single pHash fingerprint (256 bit) requires the same storage space as a single ORB keypoint, making pHash the most storage-efficient algorithm among the four.

Table 4 Algorithm storage comparison

Algorithm	Fingerprint Size [KB]
ORB (1,000 keypoints)	31.2500
BRISK (1,000 keypoints)	62.5000
SIFT (1,000 keypoints)	500.0000
pHash DCT (256 bit)	0.03125

6 Summary, Conclusions, and Recommendations

There are three aspects to consider when choosing a suitable algorithm for the given use case scenario to consistently trace assets throughout life cycles using fingerprint technology and DPPs. First, the required confidence for part identification must be determined. Second, the available computational resources must be assessed. Third, the storage capacities must be quantified.

Algorithms used for feature matching like ORB, BRISK and SIFT offer high identification confidence. However, with an increased number of keypoints extracted, the high confidence comes at the cost of speed and storage. In contrast, pHash DCT offers a reliable, fast, and low storage alternative. This makes the perceptual hash algorithm well suited for given DPP use case scenarios, where low storage requirements and fast processing times are beneficial.

Therefore, for small batch sizes (e.g., luxury products), algorithms used for feature matching or perceptual hashing are suitable. However, for mass produced parts (e.g., bolts), perceptual hash algorithms are preferable.

For future research, it is recommended to explore the feasibility of using deep learning-based algorithms for asset fingerprinting tasks as part of DPP application scenarios. Additionally, scientific investigations on the suitability of fingerprint technology for surfaces that experience wear over an asset's life cycle need to be conducted.

References

1. Künster N, Dorka F, Bücheler M, Palm D (2024) Traceability as a key element of circular value creation - Identification of objects for circular supply chains. wt Werkstattstechnik online 114:150–155. https://doi.org/10.37544/1436-4980-2024-04-42
2. Olsen P, Borit M (2013) How to define traceability. Trends in Food Sci Technol 29(2):142–150. https://doi.org/10.1016/j.tifs.2012.10.003
3. European Commission (2024) Ecodesign for sustainable products regulation. https://eur-lex.europa.eu/eli/reg/2024/1781/oj/eng. [Accessed: 09-February-2025]
4. US Customs and Border Protection, Uyghur Forced Labor Prevention Act. https://www.govinfo.gov/app/details/PLAW-117publ78. [Accessed: 09-February-2025]
5. Walden J, Steinbrecher A, Marinkovic M (2021) Digital product passports as enabler of the circular economy. Chem Ing Tec 93(11):1717–1727. https://doi.org/10.1002/cite.202100121

6. Palm D, Künster N, Dorka F, Bücheler M (2024) Architectures and systems for identifying assets in circular supply chains using digital product passports and the asset administration shell. In: Herberger D, Hübner M (eds) Proceedings of the CPSL 2024. Hannover, pp 898–905. https://doi.org/10.15488/17776
7. Wigger B (2020) Inhärente Oberflächenstrukturen als Identifikationsmerkmal zur markierungsfreien Einzelteilverfolgung. [Doctoral Thesis, Universität Stuttgart]. https://doi.org/10.18419/opus-11408
8. Schmid-Schirling T, Koerber T, Förste A (2022) Trace components by fingerprint every second. IST Int Surface Technol 15(2):44–47. https://doi.org/10.1007/s35724-022-0545-z
9. Wigger B, Koinzer I, Meissner T, Barth M, Zimmermann A (2019) Robust and fast part traceability in a production chain exploiting inherent, individual surface patterns. Robot Comput-Integrated Manufact 63:101925. https://doi.org/10.1016/j.rcim.2019.101925
10. Hippenmeyer H, Moosmann T (2016) Automatische Identifikation für Industrie 4.0, 1st edn. Springer Vieweg, Berlin, Heidelberg. https://doi.org/10.1007/978-3-662-52701-6
11. Takahashi T, Ishiyama R (2014) FIBAR: fingerprint imaging by binary angular reflection for individual identification of metal parts. In: 2014 fifth international conference on emerging security technologies, pp 46–51. https://doi.org/10.1109/EST.2014.25
12. Rublee E, Rabaud V, Konolige K, Bradski G (2011) ORB: an efficient alternative to SIFT or SURF. In: 2011 international conference on computer vision, pp 2564–2571. https://doi.org/10.1109/ICCV.2011.6126544
13. Leutenegger S, Chli M, Siegwart R (2011) BRISK: binary robust invariant scalable keypoints. In: 2011 IEEE international conference on computer vision (ICCV). IEEE, Piscataway, NJ, pp 2548–2555. https://doi.org/10.3929/ethz-a-010025751
14. Lowe DG (2004) Distinctive image features from scale-invariant keypoints. Int J Comput Vis 60(2):91–110. https://doi.org/10.1023/B:VISI.0000029664.99615.94
15. McFarlane D, Sheffi Y (2003) The impact of automatic identification on supply chain operations. Int J Logist Manag 14(1):1–17. https://doi.org/10.1108/09574090310806503
16. Hu M-K (1962) Visual pattern recognition by moment invariants. IRE Trans Inf Theory 8(2):179–187. https://doi.org/10.1109/TIT.1962.1057692
17. Fischler MA, Bolles RC (1987) Random sample consensus: a paradigm for model fitting with applications to image analysis and automated cartography. In: Fischler MA, Firschein O (eds) Readings in computer vision. Morgan Kaufmann, pp 726–740. https://doi.org/10.1016/B978-0-08-051581-6.50070-2

Production Planning, Operation and Control

Optimizing Offer Management: Reuse Strategies for Efficient Production System Planning

Max Eichenwald and Rainer Müller

Abstract The special machinery sector is distinguished by a high level of complexity, driven by the consistently growing demand for customer-specific special solutions. In today's volatile market and highly competitive landscape, suppliers must be able to respond swiftly with tailored offers for specialized machinery. It is therefore essential to adopt a success-oriented approach to offer management, ensuring the detailed preparation of offers. One potential strategy for addressing these challenges in the design process is to prioritize the reuse of individual units (artifacts) from previous projects in new customer projects. Information to support the design process and the calculative preparation of an offer is relevant here. This article provides an overview of reuse strategies for the design of production systems. The approaches and mechanisms of reuse are described in more detail. We provide an overview of current research gaps in this field and a derivation of possible further developments for the development of a supplementary methodology for use in offer management in mechanical engineering.

Keywords Offer management · Artifact reuse · Manufacturing system design · Special machinery sector · Methodology

1 Introduction

A trend topic in recent years in the field of (special) mechanical engineering is the constant demand for customized products. According to a VDI survey, 43% of respondents indicated a noticeable increase in individuality towards customized

M. Eichenwald (✉) · R. Müller
Centre of Mechatronics and Automation Technology gGmbH, Saarbrücken, Germany
e-mail: max.eichenwald@zema.de

R. Müller
e-mail: rainer.mueller@zema.de

R. Müller
Chair of Assembly Systems, Saarland University, Saarbrücken, Germany

M. Manns (ed.), *New Paradigms for Anticipated Uncertainty*, Lecture Notes in Mechanical Engineering, https://doi.org/10.1007/978-3-032-16889-4_43

special solutions [1]. This gives rise to external (market complexity) as well as internal complexity drivers (e.g. product and production program complexity, technology complexity or planning complexity) [2–4]. Early design phases in particular, specifically offer management (OM), must adapt to these new circumstances. This shows that, analogous to the required complexity of the system (here: manufacturing system), the design of offers is also becoming more demanding. In relation to the general market, a shift can be seen in the form of customization towards projects in the context of *develop-to-order* or *engineer-to-order* [5]. These special solutions can also be described as *products or solutions that require explanation* [6] and require the creation of offers based on project planning as part of OM. With regard to success-optimized OM, different objectives can therefore be derived, which are closely related within the company to the resolution of complexity in order to reduce offer times, increase transparency or improve efficiency [7]. The aim of this paper is to highlight different approaches to complexity reduction and to identify current research gaps. The insights gained will flow into the development of a methodology.

2 State of the Art

2.1 Offer Management

Order processing in the plant business can be broken down into the phases of *offer processing* (or *OM*) and *order processing* [8]. Processes for the organization and evaluation of inquiries and the control of offer processing can be summarized in the OM phase [7]. A number of works can be found in the literature that explain the necessary processes within this phase in more detail [7, 8]. All procedures are divided into three parts, so that a general distinction can be made between the phases of *offer preparation*, *offer creation* and offer *post-processing*. The documentation of the solution (i.e. the overall conceptual design) and, derived from this, the price to be estimated are essential components of the offer to be prepared [5]. Consequently, there is potential in the optimization of system design processes, as well as in the estimation of costs, for example through the inspection and reuse of already known components. The reuse of knowledge from old projects could lead to an acceleration of the process flow as well as a reduction of uncertainties in this early phase [9, 10].

Schuh sees standardization and subsequent modularization as the foundation of reuse. As an effect of modularization, with reference to reuse, he derives a *"reduction in development effort by using modules that have already been developed"* [11].

In addition, Ehrlenspiel for example, shows that the use of reusable units has a strong influence on the accuracy of the cost calculation. In an exemplary calculation, a degree of reuse of 60% can reduce the imprecision by 16% in an early planning phase [9].

For this reason, the next section will take a closer look at the topic of reuse.

2.2 *Reusability*

According to VDI Guideline 2343 [12], the term *reuse* is defined as the *"use of a device or its components for the same purpose for which it was developed"*. VDI Guideline 3695 [13] goes on to state, that the aim of reuse is "*to be able to use an artifact* [= reusable unit—material or immaterial] *in as many customer projects as possible*". In addition, VDI Guideline 3695 defines target states (A-E) that describe the character of reuse in more detail. Analogous categorizations can be found in Maga [14] and Griss [15]. Maga distinguishes here between ***Ad-hoc reuse (AR)***—Employees reuse on the basis of experience -, ***Opportunistic reuse (OR)***—controlled reuse within a project -, ***Systematic reuse (SR)***—Reuse is organized within the organization—and ***Product-line-based reuse (PLR)***—Models from which reusable units can be derived (Griss [15] names this category ***Designed Reuse***). Concrete reuse mechanisms for the use and adaptation of artifacts can be identified. Schröck [16] and Mahler [17] name the mechanisms *copy&paste*, *libraries*, *type-instance concepts* and *templates*. A number of methodologies can be identified in the literature that expand on the topic of reuse and demonstrate its application in different product life phases (shown in Table 1). The PRISMA systematic approach was used to identify relevant sources. Scientific databases (*Google Scholar*, *Springer Link & Professional* and *Scopus*) were searched. Specifically, the terms *Wiederverwendung*; *reuse*; *Artefakte*; *engineering*; *manufacturing* were filtered, narrowed down to a period from 2010 onwards. 258 hits were found in Google Scholar. Many of the sources refer to the reuse of software systems or other industries (N = 188)—these were neglected After reviewing the abstracts, the number of relevant papers was reduced to 24. Following a detailed review, nine relevant papers were identified. (Kiefer [18] was subsequently included by reviewing the sources). A basic distinction can be made according to application sector—automotive, mechanical engineering. Furthermore, these works implement different reuse mechanisms. It can be noted that the mechanisms of *libraries* and *type-instance concepts* are widespread, so that these approaches generally follow a systematic reuse approach. The *copy&paste* mechanism is only considered in the margin in current publications. *copy&paste* is a type of ad hoc reuse and thus an unsystematic reuse. Blunt copying and pasting is difficult to reproduce retrospectively in a standardized manner and is limited by the lack of availability of central information.

3 Research Gap

Table 2 provides an overview of identified gaps that has been mentioned in the identified reuse concepts. For this purpose, the methodologies described in the reuse concepts were analyzed and identified gaps were extracted. For each article, we systematically analyzed the sections on discussion, methodology and implementation to find explicit or implicit indications of research gaps. We focused on the categories

Table 1 Overview of identified reuse concepts

Source	Sector	Tech. focus	Categorization of reuse	Life-cycle-phase
Herzog [19]	Automotive	High	SR	Engineering/ planning of new systems
Hell [10]	Automotive	High	SR	Design phase
Röpke [20]	Automotive	Low	SR	Development phase of systems
Schröck [16]	Process industry	High	SR	Acquisition and planning
Maga [14]	ME*	Mid	SR, PLR	Engineering
Meixner et al. [21]	ME*	Mid	SR	Engineering of production systems
Kiefer [18]	Automotive	High	SR	Concept and detailed planning
Klein [22]	ME*	High	SR	Concept phase
Reuter [23]	ME*	High	SR	Engineering
Kluge [24]	ME*	High	SR	Rough planning phase

*General **m**echanical **e**ngineering sector

of knowledge gap (KG), practical knowledge gap (PKG), empirical gap (EG) and methodological gap (MG) proposed by Miles et al. [25]. It was found that a number of studies address the same research gaps. In general, it can be stated that further research is needed to ensure the standardization and consistency of data and centralized, as well as cross-system accessibility. The current employee-based dependency on knowledge can also be highlighted. The decoupling of knowledge from individual persons and thus the company-internal diversification represents a **relevant optimization case.**

Building on the scope of OM given in Sect. 2.1 and the existing approaches to reuse, logical further developments result from the existing research gaps. For this reason, the requirements that a reuse methodology must fulfill in order to expand existing concepts within the scope of OM are discussed in more detail below. The aim of this is to answer the research question: '*What requirements should be placed on a system optimized for reuse with a focus on OM, and which subsystems can be derived from this?*'.

Table 2 Identified gaps in research

#	Gap (Type)	Identified reuse concept, according to:										Σ
		[19]	[10]	[20]	[16]	[14]	[21]	[18]	[22]	[23]	[24]	
1	Integration of data from third parties not considered (KG)			x				x			x	3
2	Complexity of data maintenance (PKG)		x	x				x	x	x		5
3	Evaluation of reusable units based on experiential knowledge (EG)	x	x	x	x		x				x	6
4	Entire life cycle not sufficiently considered (KG)	x	x	x								3
5	PPR modeling not sufficiently system-based implemented (MG)	x					x			x		3
6	Insufficient inclusion of monetary attributes of the artifacts (KG)	x							x		x	3
7	Insufficient decoupling between process and resource (MG)						x					1
8	Missing link with existing systems (PKG)		x	x	x	x		x	x		x	7
9	Reusable units cannot be adapted to new situations (KG)					x			x			2
10	Insufficient support for searching and identifying reusable units (PKG)		x						x	x		3

4 Conceptual Considerations|Methodology

In general, it can be noted, that the research gaps can be grouped under the topics of *Standardization and Data Accessibility (S)*, *Knowledge Provision (K)*, *Modelling and Design (M)* and *Calculation (C)*. For this reason, the requirements are structured in a similar way. A definition of the requirements for closing these gaps is given in Table 3. The structure of the requirements was based on the explanations given in Schröck [16]. Based on the gaps identified in Table 2 and the requirements derived in Table 3, a general methodology (see Fig. 1) for optimizing OM can now be developed. The basis of the methodology is the standardized description of projects and production system elements (**Module 1**). As stated in the requirements, a link or direct integration into existing internal company software solutions (e.g. ERP) is necessary. As part of OM, the technical feasibility is often checked with the aid of historical data and knowledge of projects (reusable elements) that have already been completed. The identification of these is crucial in terms of reducing processing times, but also for increasing the accuracy of offers by using known knowledge (**Module 2**).

The design of the production system is based on the identification of known reusable units. Basically, the design results from the interdependencies between product, process and resource (PPR). Process planning is based on the product analysis. As part of the performance coordination, the process steps are divided into stations and ultimately the functional system components are assigned to these stations (**Module 3**). Based on the hierarchical assembly system design and the bill of materials (BOM), the costs of individual components or the entire system are estimated (**Module 4**). Individual module components have already been implemented

Table 3 Module creation and requirement specification regarding the categories *S, K, M, C* related to identified gaps

Requirement		Gap
S1	The description of reusable units must be systematic	1, 2
S2	Reusable units must be instantiated and adapted	2, 9
S3	Reusable units must be accessible to stakeholders at all times	3
S4	Reusable units must be accessible across projects	4
S5	Developed approaches must be integrable into enterprise-internal tools	1, 8
S6	The integration of reuse approaches must be possible in the OM of manufacturing companies	1, 4
K1	Reusable units must be identifiable	2, 3, 4, 10
K1	The situation-based suggestion of reusable units must be possible	8, 10
K3	Project planning should be decoupled from the implicit knowledge of individual employees	3, 4, 10
M1	Reusable units and newly developed artifacts must be combinable	2, 3, 4, 10
M1	The modeling of system design must be based on the relationships between product, process, and resource	8, 10
M3	The modeling of the system must be based on the functional alignment between process parameters and reusable units	3, 4, 10
M4	Reusable units and newly developed artifacts must be combinable	5
C1	Reusable units must contain both technical and monetary information	1, 6
C2	Estimated costs for newly developed artifacts must be based on monetary information of reusable units	6

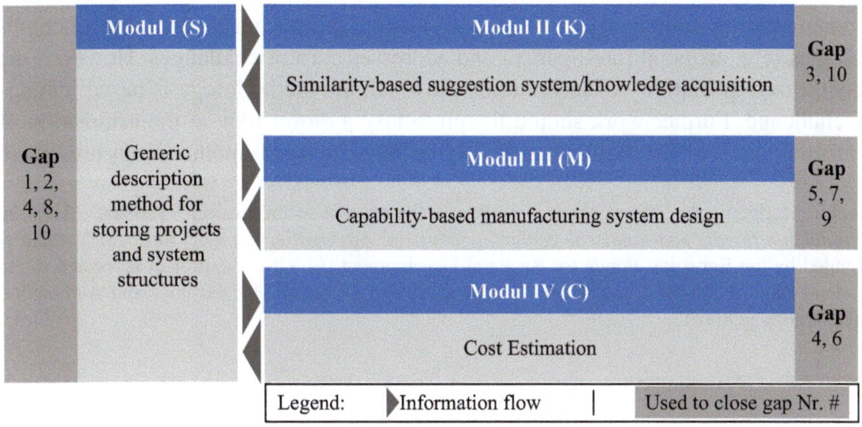

Fig. 1 Overall overview of the methodology for supporting employees in project planning

and validated in more detail in previous work. In [26], the concept of the administration shell was presented as a currently promising approach to standardization. A recommender system was presented in [27] to provide knowledge in the form of relevant reusable units. The purpose of this system is to present similar elements on the basis of the current situation. To optimize the modeling, the PPR engineering approaches presented in [28] can be used. Finally, an AI-based approach for estimating unknown elements was given in [29]. **Module 1** is to be understood as a basic module for centralized data storage (within an ERP system). Subsequent modules access this information (REST interface) so that, with regard to the modularity of the individual modules, they are initially to be understood as independent of each other. The overall methodology given in this publication results from the approaches already published. This paper serves to summarize the individual modules and provides an overarching methodology for optimizing OF in mechanical engineering.

5 Discussion and Outlook

This paper examines strategies of reuse to reduce complexity in mechanical engineering. The focus is on the early phase of OM (see Sect. 2.1) In this work, types of reuse, reuse mechanisms and current methods were discussed in more detail (see Sect. 2.2). From the analysis of the identified approaches, a number of research gaps were identified (see Table 2). This results in requirements for the development of an advanced methodology (see Table 3). These serve as the basis for a methodology presented (see Fig. 1). In the context of this work, reference was made to existing publications by us. These specify the respective sub-modules of the methodology in more detail. Further validation of the overall methodology is now pending. The presented approach aims at a broad and accessible applicability in the OM of manufacturing companies. It combines existing approaches with current methodologies (e.g. artificial intelligence) and addresses current challenges. However, the centralized availability of information on which the methodology is based remains a challenge. Further work should therefore take a closer look at the acquisition of information from historical projects as a starting point for the methodology presented.

Acknowledgements This paper was written in the context of the project VProSaar ("*Verteilte Produktion für die saarländische Automotivindustrie: Nachhaltig, Vernetzt, Resilient*") which is funded by the European Fonds for Regional Development (EFRE). The project is located in the federal state of Saarland and is executed by ZeMA—Centre for Mechatronics and Automation Technology gGmbH.

References

1. Schmidt H (2008) VDI-studie zum Angebotsmanagement. Ergebnisse einer Umfrage bei ca. 300 Industrieunternehmen im B2B-Geschäft. https://www.wiwi.uni-siegen.de/wirtschafts_informatik/publikationen/dateien/pdf/aufsatz_vdi_studie.pdf. Accessed on 2 Jan 2023
2. Schoeneberg K-P (2014) Komplexitätsmanagement in unternehmen. Springer Fachmedien Wiesbaden, Wiesbaden
3. Kirchhof R (2003) Ganzheitliches Komplexitätsmanagement. Grundlagen und methodik des umgangs mit komplexität im unternehmen. Beiträge zur Produktionswirtschaft. Deutscher Universitätsverlag, Wiesbaden, s.l.
4. Bliss C (2000) Management von komplexität. Gabler Verlag, Wiesbaden
5. Wiedemann M (2016) Methodik zur auslastungsorientierten Angebotsterminierung für hochvariante produkte mit kundenindividuellen leistungsanteilen. Dissertation, Techn. Univ München
6. Binckebanck L, Hölter A-K, Tiffert A (2020) Führung von vertriebsorganisationen. Springer Fachmedien Wiesbaden, Wiesbaden
7. VDI Verein Deutscher Ingenieure e.V. (2010) Angebotsdokumente im vertrieb komplexer technischer produkte blatt 2. Beuth Verlag GmbH (4504). https://www.vdi.de/richtlinien/details/vdi-4504-blatt-1-angebotsmanagement-im-industrieguetergeschaeft
8. Kleinaltenkamp M, Plinke W, Geiger I (eds) (2013) Auftrags-und projektmanagement. Springer Fachmedien Wiesbaden, Wiesbaden
9. Ehrlenspiel K, Kiewert A, Lindemann U, Mörtl M (2021) Kostengünstig entwickeln und konstruieren. Kostenmanagement bei der integrierten produktentwicklung, 8th edn. Springer Vieweg Berlin, Heidelberg
10. Hell K (2018) Methoden der projektübergreifenden wiederverwendung im anlagenentwurf—konzeptionierung und realisierung in der automobilindustrie. Dissertation, Otto-von-Guericke-Universität Magdeburg
11. Schuh G (2012) Innovationsmanagement, 2nd edn. VDI-Buch, vol 3. Springer Vieweg, Berlin
12. VDI Verein Deutscher Ingenieure e.V. (2024) Recycling elektrischer und elektronischer Geräte. Grundlagen und Begriffe. Beuth Verlag GmbH, Berlin (2343)
13. VDI Verein Deutscher Ingenieure e.V. (2010) Engineering von anlagen. Evaluieren und optimieren des Engineerings. Beuth Verlag GmbH, Berlin (3695)
14. Maga CR (2013) Adaptierbares wiederverwendungskonzept für die entwicklung von automatisierten Systemen. IAS-Forschungsberichte, vol 2013,4. Shaker, Aachen
15. Griss M (2003) Reuse comes in several flavors. Achieving a better return on software series. https://www.martin.griss.com/pubs/WPGRISS02.pdf. Accessed on 3 Nov 2024

16. Schroeck S (2016) Interdisziplinäre wiederverwendung im engineering automatisierter anlagen, Helmut-Schmidt-Universität/Universität der Bundeswehr Hamburg
17. Mahler C (2014) Automatisierungsmodule für ein funktionsorientiertes automatisierungsengineering. Zugl.: Hamburg, Helmut-Schmidt-Univ., Diss., 2014. Fortschritt-berichte VDI Reihe 20, rechnerunterstützte verfahren, vol 457. VDI-Verl., Düsseldorf
18. Kiefer J (2007) Mechatronikorientierte planung automatisierter fertigungszellen im bereich karosserierohbau. Zugl.: Saarbrücken, Univ., Diss., 2007. Schriftenreihe produktionstechnik, vol 43. LFT Lehrstuhl für Fertigungstechnik/CAM Univ. des Saarlandes, Saarbrücken
19. Herzog J (2023) Entwicklung einer fähigkeitsbasierten planungsmethode zur analyse der wiederverwendbarkeit von anlagen am beispiel von schraubprozessen in der automobilendmontage. Dissertation, Otto-von-Guericke-Universität Magdeburg
20. Röpke H (2019) Entwicklung einer methode zur risikobeurteilung bei der wiederverwendung von entwurfselementen im anlagenengineering. Dissertation, Otto-von-Guericke-Universität Magdeburg
21. Meixner K, Rinker F, Waltersdorfer L, Lüder A, Biffl S (2023) Organizing reuse for production systems engineering with capabilities and skills. at—automatisierungstechnik. https://doi.org/10.1515/auto-2022-0120
22. Klein PW (2014) Methode zum engineering von produktionsanlagen durch wiederverwendung von modulen, universität Siegen; Institut für Produktionstechnik. https://dspace.ub.uni-siegen.de/handle/ubsi/909
23. Reuter A (2013) Definition eines mechatronischen Informationsmodells zur modellierung von automatisierungskomponenten und maschinen. Zugl.: Stuttgart, Univ., Diss., 2013, 1st edn. Robotik und Automation. Dr. Hut, München
24. Kluge S (2011) Methodik zur fähigkeitsbasierten planung modularer montagesysteme. Zugl.: Stuttgart, Univ., Diss., 2011. IPA-IAO Forschung und Praxis, vol 510. Jost-Jetter, Heimsheim
25. Miles DA (2017) A taxonomy of research gaps: identifying and defining the seven research gaps. In: Doctoral student workshop: finding research gaps-research methods and strategies, Dallas, Texas, pp 1–15
26. Eichenwald M, Molter J, Müller R (2024) Digital twin: automatic generation of asset administration shells (AAS). Procedia CIRP[in print]
27. Eichenwald M, Karkowski M, Müller R (2023) AI-based recommender system for optimization of the offer management in special machine engineering. In: Galizia FG, Bortolini M (eds) Production processes and product evolution in the age of disruption. Lecture Notes in Mechanical Engineering. Springer International Publishing, Cham, pp 774–781
28. Eichenwald M, Molter J, Kuhn D, Müller R (2024) Production system ontology for continuous capability-based engineering. Procedia CIRP. https://doi.org/10.1016/j.procir.2024.03.018
29. Eichenwald M, Karkowski M, Müller R (2024) Generalizable AI pipeline for the cost estimation of manufacturing system structures in special machine engineering. Procedia CIRP. https://doi.org/10.1016/j.procir.2024.08.314

Conception of an Automated Graph-Based Assembly System Development Process for a Balancing Scooter

Timo Schuchter, Theresa Breckle, Ralf Stetter, Markus Till, Patricia Derksen, and Stephan Rudolph

Abstract Using a balancing scooter as an example, the development of an automated design process for a hybrid assembly system is illustrated. A framework of graph-based design languages is used for the modeling, starting with an ontology which defines the concepts and their relationships in Unified Modeling Language (UML) syntax. Based on this, a design graph is generated using a rule-based approach. Furthermore, the question of how manual assembly steps can be simulated and analysed as part of an automated design process is addressed. All processes are automatically derived on the basis of the product model and the associated joining information, which form the basis for a 2D- and 3D-layout of an assembly cell. The manual assembly system is evaluated via an Unreal Engine 5 interface using the created layout. The virtual execution of the assembly enables the recording of movement data, which serves as the basis for an economic simulation. The results of this simulation provide the foundation for continuous adaptation of the assembly system, ensuring that the assembly process is designed in accordance with economic principles. The architectural design allows for the validation of hybrid assembly systems in their developmental phase.

T. Schuchter (✉) · T. Breckle · R. Stetter · M. Till
Department of Mechanical Engineering, Ravensburg-Weingarten University (RWU), Weingarten, Germany
e-mail: timo.schuchter@rwu.de

T. Breckle
e-mail: theresa.breckle@rwu.de

R. Stetter
e-mail: ralf.stetter@rwu.de

M. Till
e-mail: markus.till@rwu.de

P. Derksen
Department of Biosystems Engineering, University of Manitoba, Winnipeg, MB, Canada
e-mail: derkse48@myumanitoba.ca

S. Rudolph
Institute for Aircraft Design, University of Stuttgart, Stuttgart, Germany
e-mail: rudolph@ifb.uni-stuttgart.de

M. Manns (ed.), *New Paradigms for Anticipated Uncertainty*, Lecture Notes in Mechanical Engineering, https://doi.org/10.1007/978-3-032-16889-4_44

Keywords Production layout · Assembly system design · Ergonomic manufacturing · Virtual reality · Graph-based design languages

1 Introduction

One major challenge of current product development processes is to master the constantly rising system complexity driven by the interactions between different disciplines [1, 2]. The complexity of hardware and software in current mechatronic and cyber-physical systems forces many engineering fields to follow an organized, systematic and well-structured design process and, for assistance in the multifaceted processes, the engineers are increasingly exploiting modeling languages [3, 4]. In the scope of model-based systems engineering (MBSE) the substitution of traditional statements by models is expected to provide several benefits, for instance enhanced mapping, traceability, system decomposition as well as a fostering of automatic analysis and synthesis techniques [5, 6]. Another major challenge is the compatibility of models and applications which is a necessity for an unrestricted implementation of MBSE and the relevant processes need to be anticipated [7, 8]. In this context, it is extremely important to fulfill the concrete needs of different stakeholders as well as to consider a multitude of boundary conditions and to dissolve possible contradictions [9]. In this context of interconnected models from requirements to detailed geometry and behavior, this word describes an automated design process for a hybrid assembly system. The process employs a framework of graph-based languages and an ontology defining the elements and their relationships is described using the Unified Modeling Language (UML).

Consequently, the central research question can be formulated as: *"How can an automated development process for a product and its assembly system be sensibly realized for an unrestricted model interconnection?"*

Several research groups worldwide are working on solutions for interconnected and consistent product and processes models for the unrestricted implementation of MBSE. The most current results are summarized in the next sentences. Paehler and Matthiesen [7] deliver a structured overview of linkable and non-linkable product models mainly in embodiment design. Grüble et al [13] describe a combined geometric and kinetic data model in MBSE of robotic cells (Grüble 2024). Pietrusewicz [3] addresses the issue of metamodeling of Domain-Specific Languages (DSL) for designing complex mechatronics systems and proposes the so-called Meta-Object Facility (MOF) approach to metamodeling. Salado and Wach [5] propose an approach to construct true model-based requirements in SysML in order to model requirements as input/output transformations. Wyrwich et al [1] propose an extension of the established "Motego" method for modeling structures and, through this, enabling the seamless parametric modeling of function-oriented mechanical system architectures. Mandel et al [9] investigate methodical approaches for assessing and improving the maturity of the system of objectives. The following search terms were used in the search: MBSE, assembly sequence, graph-based design languages, automated design process, Gabow algorithm, Breadth First Search, Petri net, motego, UML, joining, joining techniques and assembly planning. A freely accessible web search engine was used for this purpose.

It can be concluded that many methodical and tool elements for an unrestricted MBSE implementation are already developed or currently under development, but that a consistent development process for the interconnected synthesis of a mechatronic product and its assembly system is not yet realized. In addition, an approach is used in which consistent modeling is implemented that is characterized by end-to-end modeling from the product to the plant [10]; the method described in this paper is consequently characterized by consistent modeling.

2 Background

One of the major challenges in all areas of engineering is the digital consistency of data, the digital continuity of processes and the digital interoperability of tools in the modeling of systems, system components and processes. This is especially true in assembly planning, because models for product components, fasteners, assembly operations, resources, assembly layouts and assembly systems need to generated and sensibly interconnected. A promising approach for this are graph-based languages (GBDLs). In GBDLs all entities are expressed in a system of vocabulary and rules using an analogy to natural languages. The object-oriented representation of the underlying ontologies is based on the unified modeling language (UML) and with the *Design Cockpit 43* [11] by IILS GmbH, Trochtelfingen, a compilation tool is commercially available. This software was chosen because it is characterised by a high degree of flexibility and guarantees complete control over the central data model.

Assembly planning is a crucial phase in manufacturing planning that ensures efficient and cost-effective product assembly. Key tasks include defining assembly sequences to determine the optimal order for component assembly, allocating resources such as tools, robots, and human labor, evaluating manufacturing constraints like geometric limitations and material properties [15]. Recent advancements in digitalization and automation have transformed assembly planning, presenting both opportunities and challenges. A tool integrating Petri nets and graph-based design languages enables automated assembly sequence optimization, reducing manual intervention [16]. Complex and customized products often require flexible planning approaches that can adapt to different designs and configurations, particularly when handling complex product variants [17] and planning workstation layouts for manual work [18]. To tackle complexity in assembly design, a matching algorithm using graph-based models improves automated resource selection [19].

3 Automated Design Process

A balancing scooter is used as a model example. This model was selected due to the fact that it has a sufficient variation of various components to allow an adequate description of the method. In addition, identical components are available in different quantities. Figure 1 shows the architecture. It essentially consists of five steps [10]. In these steps, the model, which is available in the form of a graph, is gradually enriched and transformed into different models in order to achieve rich and consistent modeling.The process of enriching and transforming models is also referred to in the specialist literature as model-to-model (M2M). The architecture includes pre-processing, in which the given model must be prepared, and the extension and enrichment of the prepared model into a product model. The assembly sequence is then derived and an assembly model is generated from this. Finally, production planning is carried out on the basis of this model. In all steps, a domain-specific model, such as the engineering bill of materials (E-BOM), is derived through a model-to-text (M2T) transformation. This architecture and the underlying rich central data model also make it possible to validate the model at any time using VR (Virtual Reality) and UE (Unreal Engine). The architecture is described in more detail in the following chapters.

3.1 *Enriched Product Model*

The first two transformations of the architecture are shown in the lower part of Fig. 1. The first transformation is product preparation (preprocessing). It can take two different forms. The product can only exist as a CAD model that contains the geometry in the form of assemblies (asm) and parts (p). In this case, the geometry must be assigned to an ontology. This means that a UML model must be created and a model instantiated based on it, which then maps all existing components including the geometry [10]. In this case, the generated model only contains the information that is available about the geometry. However, the best case is when the model has

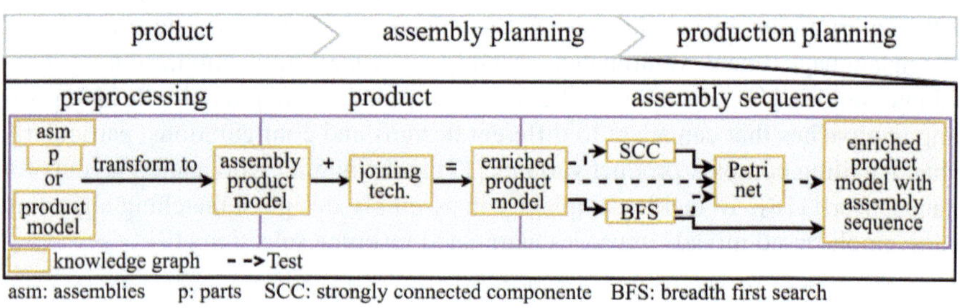

Fig. 1 Model transformations in product and assembly system development

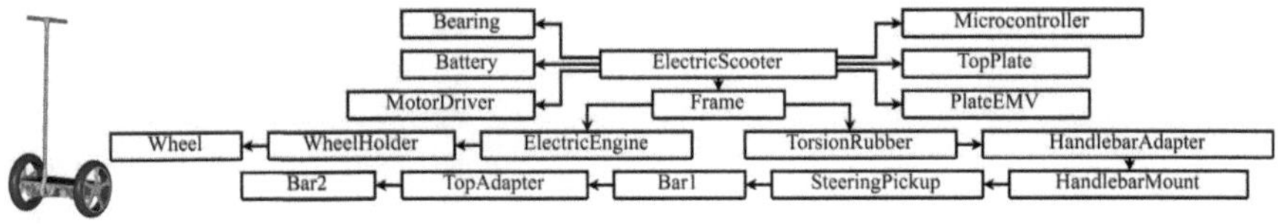

Fig. 2 Product ontology of the balancing scooter

already been created using graph-based design languages and MBSE. In this case, the model is already available as a rich model in the form of a graph. In this case, no new anthology needs to be created and the model only needs to be transformed, as the model is not necessarily available in a structure suitable for assembly planning due to the design methodology. The transformation of the product model into an assembly model is also often necessary in the state of the art. The aim of the transformation is to obtain a structure that arranges assemblies and components in a way that is suitable for assembly. Such a structure is characterised, among other things, by clearly recognizable assemblies, completeness and simple connections. This paper uses a model that was designed using graph-based design languages. For this reason, an ontology exists in the form of a UML class diagram, which is shown in Fig. 2. It serves as a vocabulary and abstractly describes the structure of the scooter. The structure of the associations was selected so that it corresponds to the criteria described above. The association is the connection of components that are directly linked to each other. This results in a liaison graph [22] after instantiation. All other classes not mentioned only occur once in the further specification and have no deeper branches. All classes inherit from the class GeometryObject. This class contains all associations and information that are relevant for the further steps. In the case mentioned above, where there is only a geometry without a domain model, this class could be used to instantiate an explicit model without having to create a product-specific ontology. The Fig. 3 shows the product model after transformation or instantiation. This model is now enriched with the joining information in accordance with DIN 8593 [23] (see Fig. 1). This transformation inserts the joining technology and all necessary information between two joined components. The enriched product model is shown in the Fig. 3.

3.2 *Assembly Planning*

The assembly sequence is now derived on the basis of the enriched product model. There are a number of different options for this [24]. In this paper, the search algorithms Breath-First-Search (BFS) and the Gobow algorithm are used to identify and group partially or fully similar assembly steps and to determine assemblies or components with strong dependencies. Figure 1 shows the possible paths. The dotted paths are optional. A Breath First Search (BFS) [25] is used to search the graph for fully or

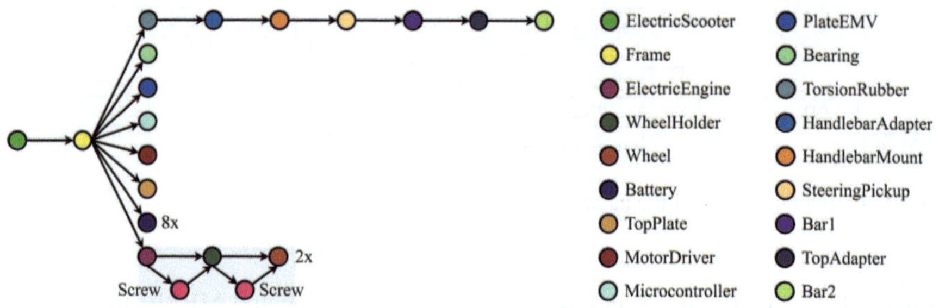

Fig. 3 Product model of the scooter as a graph and example of joining technology. Section of the enriched product model (blue area)

partially similar assemblies. The components are analysed for geometric aspects and joining techniques. If fully or partially similar joining techniques are found, these are summarised in assembly sets. This step is important in order to simplify the complexity and further structure the model. This step is absolutely essential. Next, the Gabow algorithm is used to search for strongly connected components (SCCs) [25]. The aim is to identify further assemblies or constellations that are highly dependent on each other. A Petri net can now be generated based on this result and the assembly sets. As the complexity of the Petri net increases exponentially, two methods were developed. The scooter consists of 26 components. If a Petri net were to be generated on the basis of these components, it would contain more than $4 \cdot 10^{26}$ different sequences. This combinatorial problem can be reduced by searching for SCCs and grouping them into assembly sets. The model is subdivided so that there is only one Petri net at assembly level and a number of smaller nets for each sub-assembly. If no or only a few SCCs are found, only the assembly set information can be used. In the case of the scooter, there are individual assembly sets for all batteries, ElectricEngine, WheelHolder, Wheel and TopPlate. Assembly sets were also found for all other classes, but these are ignored in the following example to simplify understanding. This approach also proves to be suitable for larger or more complex models, as the algorithms used are generally highly efficient and demonstrate satisfactory runtime behaviour. It is also possible to divide the underlying central data model into sub-models. These can then be analysed separately. It is expected that larger or more complex models will result in a significant reduction in complexity, as partially or fully similar assemblies accumulate. This assumption needs to be further investigated. Figure 4b (right) shows the simplified CAD model with only one component per assembly; this simplification is possible because the connection techniques and components are completely similar. This reduces the combinatorial problem. The Petri net shown in Fig. 4b is created on this basis. In cases where the number of potential paths can be classified as small, it is possible to run through the design process for each individual path. An evaluation can then be carried out using the assembly system. It is also possible to define further requirements, such as the use of identical joining techniques for as long as possi-

(a) Scooter CAD model, reduced to selected number of assembly sets

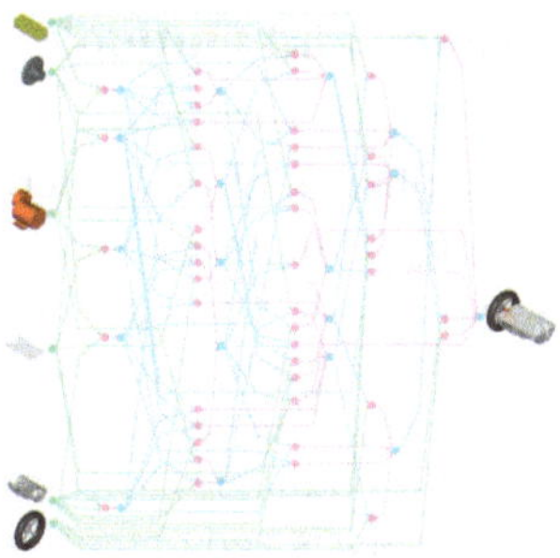

(b) Petri net based on the assembly sets

Fig. 4 Model reduced by assembly set and resulting Petri net

ble or the sequential inadmissibility of certain joining techniques. Furthermore, this approach can also be applied to sub-assemblies, reducing complexity and allowing decisions to be made at macro and micro levels.

4 Results and Discussion

The automated design process for the balancing scooter assembly system has enhanced the integration of product models through a graph-based framework, representing an advancement over existing approaches. Initial CAD models were transformed into enriched product models, including essential joining information. Algorithms like BFS and the Gabow algorithm may reduce assembly sequence complexity by grouping components into assembly sets, resulting in manageable Petri nets for assembly operations. This research highlights the potential of graph-based design languages to improve consistency and interoperability in assembly planning and demonstrates a methodology that enables the integration of the search for an assembly sequence into the development of an assembly plant, including all underlying information. The customizable BFS allows the definition of influencing variables and the consideration of the similarity of assembly steps. Future research will focus on refining ontologies and frameworks to better accommodate complex assembly scenarios and integrate AI optimization techniques.

5 Conclusions and Outlook

The central contribution of this paper is a consistent and interoperable process for assembly planning. This process is based on graph-based design languages and allows the integrated representation of product components, fasteners, assembly operations, resources, assembly layouts and the assembly system. Currently, this system of interconnected models is still under development and further research is planned to expand and consolidate the ontologies that enable this system. Further research will concentrate on further analysis and optimization possibilities including the application of AI.

References

1. Wyrwich C et al (2024) Seamless function-oriented mechanical system architectures and models. Eng 5:301–318
2. Ren Y et al (2024) Model-based reliability systems engineering. Springer
3. Pietrusewicz K (2019) Metamodelling for design of mechatronic and cyber-physical systems. Appl Sci 9:376
4. Buchmann R et al (2024) Large language models: expectations for semantics-driven systems engineering. Data Knowl Eng 102324
5. Salado A, Wach P (2019) Constructing true model-based requirements in SysML. Systems 7:19
6. Binder C (2024) Model-driven basic engineering of flexible production systems with the RAMI toolbox
7. Paehler L, Matthiesen S (2024) Mapping the landscape of product models in embodiment design. Res Eng Des 1–22
8. Li Z, Faheem F, Husung S (2024) Collaborative model-based systems engineering using dataspaces and SysML v2. Systems 12:18
9. Mandel C et al (2023) Improving system of objectives maturity through systematic reuse of knowledge using ontology-based knowledge representations. TU Ilmenau
10. Schuchter T, Till M, Stetter R, Rudolph S (2025) Digital integrated design and assembly planning processes for sports vehicles using the example of a skateboard. Vehicles 7:22
11. IILS mbH (2025) Design Cockpit 43®. See https://www.iils.de for details
12. Elwert M et al (2022) Digital function modeling in graph-based design languages. Appl Sci 12:5301
13. Grüble T et al (2024) Combined geometric and kinetic data model in model-based systems engineering of robotic cells. Procedia CIRP 128:156–161
14. Rudolph S (2002) Übertragung von Ähnlichkeitsbegriffen. Habilitation Thesis, Fakultät Luft- und Raumfahrttechnik, Universität Stuttgart, Stuttgart, Germany
15. Breckle T et al (2021) Multi-criteria evaluation within concept planning phase of assembly system design. Procedia CIRP 97:296–301
16. Schopper D, Tonhäuser C (2021) A user-friendly planning tool for assembly sequence optimization. In: Weißgraeber P, Heieck F, Ackermann C (eds) Advances in automotive production technology–theory and application, pp 28–36
17. Burkhart J et al (2021) A digital modelling approach for design configuration and manufacturing exemplified by a self-balancing scooter. Procedia CIRP 91:214–219
18. Breckle T et al (2022) Digital design and planning process of workstations of an assembly system for a customer-individual last-mile electric vehicle. Procedia CIRP 106:197–202

19. Brovkina D, Riedel O (2021) Matching algorithm for automated resource selection within assembly line design. In: IEEE 3rd Eurasia conference on IOT, communication and engineering (ECICE), pp 627–632
20. Zech A et al (2021) Automated generation of clamping concepts and assembly cells for car body parts for the digitalization of automobile production. In: Weißgraeber P et al (eds) Advances in automotive production technology, pp 293–301
21. Breckle T, Manns M, Kiefer J (2021) Assembly system design using interval-based customer demand. J Manuf Syst 60:239–251
22. Riggs RJ, Hu SJ (2013) Disassembly liaison graphs inspired by word clouds. Procedia CIRP, In: Forty sixth CIRP conference on manufacturing systems 2013, vol 7, pp 521–26. https://doi.org/10.1016/j.procir.2013.06.026
23. Deutsches Institut für Normung. DIN 8593: Fertigungsverfahren Fügen–Begriffe, Einteilung. Beuth Verlag, Berlin
24. Su Y, Mao H, Tang X (2021) Algorithms for solving assembly sequence planning problems. Neural Comput Appl 33(2):525–34. https://doi.org/10.1007/s00521-020-05048-6
25. Kozen DC (1992) The design and analysis of algorithms. Springer, New York, NY. https://doi.org/10.1007/978-1-4612-4400-4

Proposing a Multi-domain and Life-Cycle Based Analysis Process for Service Architecture Design: A Case Study from a Danish Third-Party Logistics Company

Anton Seistrup Hermann, Frederik Holm Nielsen, and Niels Henrik Mortensen

Abstract As companies increasingly adopt service-based business models, the design of coherent and scalable service architectures becomes essential. While literature highlights the need for service analyses as a precursor to architectural design, it lacks concrete methodologies that integrate data from various life-cycle stages and organizational domains. This study addresses this gap by proposing a structured analysis method for identifying hierarchical decompositions across digital, physical, and mental sources–spanning sales, invoicing, and operations. The research question guiding this work is: how can the scope of service architecture analyses be systematically expanded to include more preexisting system decompositions, in order to reduce knowledge-based implementation barriers? The proposed method is tested in a case study at a Danish third-party logistics (3PL) company, revealing structural and semantic misalignments across organizational systems. By increasing transparency and highlighting misaligned service structures, the analysis improves the foundation for conceptualizing and implementing service architectures.

Keywords Logistics services · Service architecture · Service life-cycle · Service system analysis · Variation management

1 Introduction

In recent years, there has been a shift toward service-based business models, with companies across various industries increasingly offering services as products. From car rentals to industrial service equipment, this transformation is driven by growing demand in service-oriented markets [1, 2].

A. S. Hermann · F. H. Nielsen (✉) · N. H. Mortensen
The Technical University of Denmark, Lyngby, Denmark
e-mail: fhol@dtu.dk

A. S. Hermann
e-mail: asehe@dtu.dk

M. Manns (ed.), *New Paradigms for Anticipated Uncertainty*, Lecture Notes in Mechanical Engineering, https://doi.org/10.1007/978-3-032-16889-4_45

Based on product modularization principles and systems theory, scholars have explored how service portfolios might be improved through platform thinking and modularization [3]. However, while much of the literature focuses on reviewing the concepts of service architectures and modules, relatively few studies examine the actual design phase and long-term challenges associated with their implementation [4–6].

Despite this, existing research on mass customization for physical products highlights substantial obstacles when moving from an Engineer-To-Order (ETO) model to a mass-customizable product-portfolio. With services tending to face the same challenges as physical products [7], this study presents a method to improve the collection of structured organization knowledge, as part of the contribution to the research area of the conceptualization and implementation of service architecture.

This article is the result of a research collaboration with a Danish third-party logistics (3PL) warehouse provider, aimed at developing and deploying a *warehouse service architecture*. In this stage, the primary focus is on the initial analysis phase before architecture conceptualization begins. The general proposal is to utilize an multi-domain life-cycle analysis methodology to uncover hierarchical decompositions of greater variety, including those embedded in IT systems, training materials, and operational structures. These decompositions are often comparable to *architectures* and can provide valuable insights into service standardization and future system change propagation [8].

To demonstrate the method's applicability, this study presents a case study within a 3PL warehouse organization, where pre-existing hierarchical structures from multiple sources were systematically analyzed. By expanding the scope of analysis beyond conventional sources, the study aims to increase the implementability of the future service architecture and improve efficiency in the conceptualization phase.

2 Literature Review

The development of modular and platform-based architectures in service settings builds upon decades of work in product modularization. This literature review examines existing theoretical foundations and methodological gaps, with special attention to the role of digital structures and the limitations of current practices in uncovering pre-existing system decompositions.

While the transition from Engineer-to-Order (ETO) product offerings to platform-based or modular portfolios is discussed in academic literature, the primary focus has often been the move from mass production of physical products to mass customization. In this context, Bertram et al. [9] highlights how such transitions in ETO businesses often stall. One of the primary challenges identified is the lack of systematically structured information across organizational systems–manifested in difficulties accessing relevant data, fragmentation across non-interacting systems, and insufficient management of existing knowledge. Additionally, the structure of IT systems were documented as not being ready for proposed portfolio changes, showing how, rather than supporting architecture-driven development, many IT infrastructures are revealed to be constraints [9].

While these observations stem from studies of physical product development, the underlying mechanisms are likely represented in service architectures as well, as they carry many of the same characteristics as their product counterparts [7]. Generally, modularization and architectural thinking are long-standing research areas grounded in general systems theory, with foundational work on hierarchical decomposition laying the conceptual basis for much of the field [10]. As modularization gained traction in industrial applications, key contributions by scholars such as Ulrich and Schilling advanced the understanding of product modularity, particularly within product development and management [11, 12]. In parallel, the concept of service design emerged from marketing and management theory, initially focusing on customer experience and value creation in service delivery contexts [13]. With the rise of servitization and the development of hybrid product-service offerings, researchers began to adopt architectural thinking in service contexts–recognizing services as complex systems that, like products, could exhibit modular and platform-based characteristics [14]. Initially, services were conceptualized primarily as processes, but the field has evolved to emphasize configurability, reuse, and customer-centric value creation [14].

Despite these developments, a recurring challenge in service research remains the lack of standardized definitions for concepts such as "service," "module," "platform," and "architecture" [2, 3, 5]. Just as physical products can be decomposed into hierarchical levels, services too can be analyzed at multiple layers of abstraction [5, 15]. While a detailed discussion on this topic is certainly relevant, it is beyond the scope of this paper. To achieve clarity, the following definitions are going to be utilized in the paper:

- (A *service* is) "the application of specialized competences (skills and knowledge), through deeds, processes, and performances for the benefit of another entity or the entity itself" [16]
- (A service *module* is) "one or several service elements offering one service characteristic" [17]
- (A *service architecture* is) "the way that the functionalities of the service system are decomposed into individual functional elements to provide the overall services delivered by the system [18]"

Given the conceptual parallels between physical and service systems, several tools originally developed for product modularization have also been adapted for use in service settings. Notable among these is the Product Family Master Plan (PFMP) or Product Variant Master (PVM), which are frameworks used to document modular structures through "part of" and "kind of" relationships using different general views, such as *customer*, *engineering*, and *part*-view [15, 19]. PFMPs have been used for both physical products and service portfolios to map system elements across multiple levels of abstraction [20].

However, such tools assume that the analysts are able to uncover the most critical systemic decompositions with little structure. The existing analysis guidelines tend to rely on interviews, sales material, and bill of materials, while offering limited incentive to analyze digital structures such as relational databases [15, 19]. This confined focus can cause a hindrance to the further transition into a platform-based or modular portfolio as argued by Bertram et al. [9]. Additionally, product decompositions might change over time, with Harlou [15] arguing that architectural decompositions can be found across all life-cycle stages where products interact with organizational systems. However, this observation has yet to be translated into specific analytical guidelines or methodologies.

This points towards a methodological research gap: although much has been written about modularization and architectures in both products and services, relatively little guidance exists on how to identify and structure relevant decompositions in the early stages of architecture work–particularly those hidden within IT systems.

This raises a central research question: **how can the scope of service architecture analyses be systematically expanded to include more pre-existing system decompositions, in order to reduce knowledge-based implementation barriers?**

In addressing this question this paper proposes the first iteration of a structured analysis method which utilizes an *analysis matrix* designed to uncover pre-existing hierarchical decompositions across different domains and life-cycle stages. Particular attention is paid to identifying decompositions embedded in the digital domain, to strengthen the early phases of service architecture development, which historically has been guided towards interviews and product structures [19].

3 Multi Domain Life-Cycle Analysis

Building on the challenges identified in the literature, this section introduces an analysis approach that adds structure to the early stages of architecture design. The analysis framework was conceptualized by combining two perspectives from the literature review: first, the recognition that architectural decompositions exist not only at the point of design but throughout the service life-cycle; second, a growing awareness that critical decompositions are often encoded in digital systems. The resulting concept, multi-domain life-cycle analysis, aims to systematically identify and compare hierarchical structures across both traditional artifacts and digitally embedded representations.

3.1 Analysis Concept

The proposed analysis process is designed to guide users in designing and using an *analysis matrix* based on both **life-cycle stages** and **relevant domains**. The analysis matrix has two axis, in which life-cycle stages populate the column headers, and domains populate the row headers.*Domains* refer to the nature of origin of a given hierarchy–namely, the **digital**, **physical**, or **mental** realm.

The method consists of **three structured steps**, each addressing limitations identified in the service architecture literature:

- **Step 1: Identifying relevant life-cycle stages** aims to define the *organizational scope* of the analysis by identifying departments and stages involved in the delivery and monetization of services. This guides the user to include more diverse inputs.
- **Step 2: Structuring and documenting inputs** involves categorizing collected materials according to the three domains.
- **Step 3: Visualizing hierarchical structures** enables comparison across domains, which can reveal semantic and structural misalignments. These visualizations support the identification of latent architectures and inform decisions around standardization and change propagation.

The following sections provide detailed guidance on how to carry out each of these steps.

Step 1: Identifying Relevant Life-Cycle Stages

The first step is to determine which life-cycle stages are relevant (and accessible). The selection of life-cycle stages may be informed either by mapping company departments responsible for different parts of the service flow or by referencing generalized product life-cycle stages (e.g., design, production, use, service, and recycling) [19]. This broader framing ensures that the analysis includes inputs beyond just customer-facing or sales materials and also reflects the internal organizational reality.

A practical starting point is to review an organizational chart with a high-ranking employee, as organizational departments often reflect their products' complexity [21]. Generally, looking for departments involved in either the operation of or the money stream related to the service-product, constitutes a good first step. These departments could be, but aren't limited to, *Marketing, Sales, Operations, IT, Data Warehousing*, and *Finance*, representing general life-cycle stages such as Strategy, Design, Integration, Operation & Maintenance [19]. These life-cycles stages are then used to populate the columns of the analysis matrix as shown in the example in Fig. 1.

Domain \ LCS	Strategy	Design	Integration	Operation & Maintenance
Mental				
Physical				
Digital				

Fig. 1 Example of matrix grid

Domain \ LCS	Strategy	Design	Integration	Operation & Maintenance
Mental	*Input*	*Input*	*Input*	*Input*
Physical	*None*	*None*	*Input*	*Input*
Digital	*Input*	*Input*	*Input*	*Unknown*

Fig. 2 Example of matrix with inputs

Step 2: Structuring and Documenting Inputs

Once the relevant life-cycle stages have been identified, the next step is documenting inputs using the analysis matrix created for this purpose. An example matrix can be seen in Fig. 2, in which *None* represents no available data source, while *Input* covers any available data source. *Unknown* designates that it is not known wether or not a data source can connect the respective *life-cycle stage* and *domain.*

Following the completion of step 1, the relevant life cycle stages are included in the columns of the analysis matrix. The other dimension is represented by the relevant domains: the *Physical Realm* which could be rooms, layouts, models, and physical product structures, the *Digital Realm* (slide decks, database structures, and software systems) as well as the *Mental Realm* (mindset Models, human decision-making processes, explanations, etc.).

An input in this matrix is a source of data which can be visualized into a structural hierarchy. In some cases, inputs may belong to multiple domains, such as *sales material* existing in both digital and physical formats, in which then the input should be assigned to the category with the most perceived relevance. A documented instance of this scenario is presented in the later section of the case study and is illustrated in Fig. 5.

The framework is specifically designed to be flexible and adapt to available data and accessibility rather than enforcing a fixed series of inputs. Given the diversity of services, the primary goal is to generally encourage the search for preexisting hierarchical decomposition in areas that may not have been previously considered, and help systemize the process of gathering inputs, to make it clear if there are any unknowns.

Step 3: Visualizing Hierarchical Structures

Once the inputs have been mapped, they need to be visualized. A common approach is a standard hierarchical system decomposition [18]. As an example, three levels could be utilized: a high-level service overview (0), module-level (1) and service-level (2) as seen in Fig. 3.

Following these three steps, the analysis produces a clearer and more structured overview of the hierarchical structures present within the case company. These domain-specific inputs can then be interpreted through relevant architectural views–such as *customer*, *engineering*, and *part* perspectives–which are commonly applied

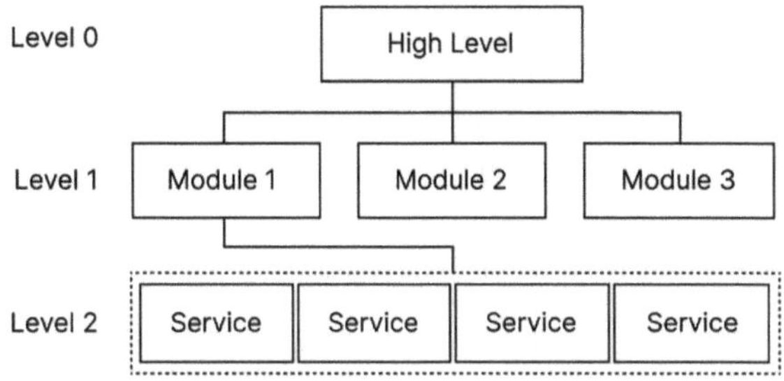

Fig. 3 Example of decomposition

in tools like the PFMP or Product Variant Master PVM [15, 19]. In this way, the analysis matrix serves as a preparatory method that generates structured input for subsequent use in architecture design frameworks such as the PFMP. The following section documents how this approach has been applied within a case company.

3.2 *Case Study—Danish 3PL Company*

To explore the practical use of the proposed methodology, a case study was conducted within a Danish warehouse 3PL company following the initiation of a standardization research project. Specifically, the case company in focus has previously been following ETO-type processes when designing service-offerings, but is now focused on transitioning to a warehouse service architecture as a means to reduce costs. In general, 3PLs, is a service industry performing logistics or supply chain responsibilities for company partners. Warehouse services are a subset of logistics services focusing on getting client goods into storage and then off to another warehouse destination or an end-consumer.

As the logistics market evolves, modern large-scale warehousing has become complex, with the inclusion of diverse and demanding customers [22]. Additionally, it is due to the diverse needs of warehouse customers that many ETO-type processes exists within 3PL companies [22]. In order to control the flow of products, large IT systems–warehouse management systems (WMS)–are integrated into the buildings. Additionally, warehouses might consist of automated solutions as well as unique product configuration setups, such as product assembly or garment steaming. Most modern warehouses also use action-based cost structures for their services, relying on WMS data and time tracking systems, which makes warehouse companies highly dependent on the structural integrity of their IT systems.

Step 1: Identifying the Stages and Domains of the Life-Cycle

A decision was made to map the company departments based on one of the following stages of the life-cycle: Strategy, Design, Integration, Operation & Maintenance.

Life-Cycle Stage	Strategy	Design	Integration	Operation & Maintenance		
Domain \ Department	Sales		Integration & Maintenance		Operations	Finance
Mental						
Physical						
Digital						

Fig. 4 Empty matrix for 3PL company

Life-Cycle Stage	Strategy	Design	Integration	Operation & Maintenance		
Domain \ Department	Sales		Integration & Maintenance		Operations	Finance
Mental	*Interviews*		*Interviews*		*Interviews*	*Unknown*
Physical	*None*		*None*		*Internal Warehouse Floor Layout*	*Unknown*
Digital	*Sales Slides*, Rate Sheets**		*Integration Training Material*		*WMS Data, Time-tracking Data, Invoice Data, SOPs*	*Account Structure*

Fig. 5 Input matrix full

The department inputs were gathered by interviewing key stakeholders who *sell* the product, *make* the product, and *invoice* the product. The rationale behind this selection was to ensure the inclusion of all perspectives that directly influence service structuring. From that, the following matrix was created, as seen in Fig. 4.

Step 2: Data Collection and Analysis

The inputs were gathered and analyzed over six months, including sales material, employee training material, data from invoices, time-tracking systems, warehouse management systems, and Enterprise Resource Planning (ERP) systems. Additionally, a series of informal interviews and workshops were executed with sales, IT, data warehousing, business analytics, and operations personnel. Using the matrix as seen in Fig. 5, the inputs were sorted, creating an overview of what life-cycle stages and domains had been covered.

This step showed, that the current analysis has some unknowns in regards to finance. For the analysis, this was considered acceptable, as the account hierarchy was included and several other stakeholders were in weekly dialogue with finance. This example showcases the main purpose of this method—to make conscious decisions on whether or not enough insights has been gathered, reducing the risk of facing undiscovered barriers further down the line.

Step 3: Visualizing and Identifying Misalignments

Visualizing the inputs was found to be a complex task, as numerous different hierarchies were discovered, with some systems having many unique decompositions. A selection of four relevant examples can be seen in Fig. 6. Example ① and ④

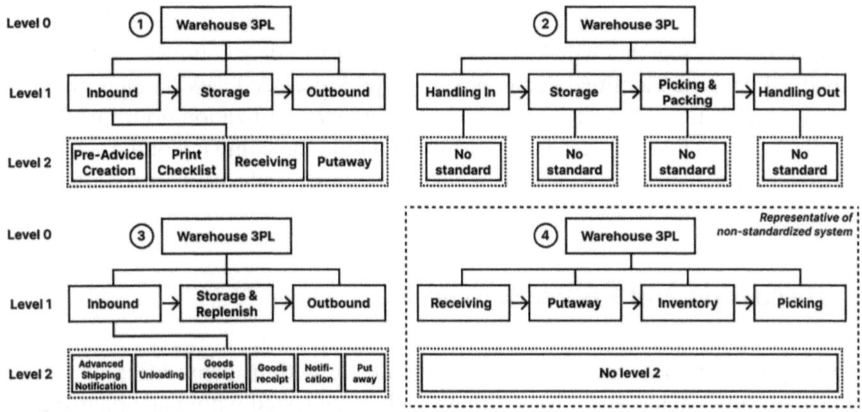

Fig. 6 Different hierarchies visualized

were designed based on IT-systems, with ② and ④ coming from invoices and sales material, all critical elements in the warehouse service life-cycle.

When comparing the structures, both structural and semantic misalignments became evident. For example, the term *Inbound* (① & ③) and the process *Receiving + Putaway* (④) refer to the same warehouse operations, these being terms commonly used to describe the process of accepting new goods into a warehouse. However, the hierarchical levels and naming conventions vary across these systems. In system ①, *Receiving* and *Putaway* are categorized at level 2, whereas in system ④, they are treated as separate first-tier entities, making them semantically different.

Overall, the analysis revealed a variety of different hierarchies with both semantic and structural misalignments, representing mental, physical, and digital decompositions. The digital domain, in particular, provided significant insight into multiple possible decompositions, which were later used as inputs for a service-PFMP. These different hierarchical structures were carried into the architecture conceptualization phase, including critical information on the change propagation implications for each concept.

Case Study Findings

One of the most significant outcomes of this study was the company's first systematic overview of its internal misalignments across different life-cycle stages and domains–an overview that had not previously been visualized. By identifying structural and semantic inconsistencies, the company was able to recognize inefficiencies in its existing service structuring and increase its focus on future standardization, which is expected to support a more structured approach to implementing a warehouse service architecture within the organization.

To further evaluate the findings, the resulting service-PFMP was reviewed in a workshop setting together with key stakeholders from the case company. This session helped validate the identified decompositions, provided feedback on their practical relevance, and supported alignment around future standardization priorities. It also contributed to a more informed understanding within the company of the scope and resource requirements associated with pursuing a large-scale standardization effort.

4 Discussion and Future Research

The proposed analysis method aims to be suitable within many different service industries and systems, but it has currently only been tested within an organization, where IT systems were a critical part of invoicing and operation. In other service domains, the connection between systems might not be as strong, and therefore, this method of analysis may not hold the same value as demonstrated in this case study. However, as noted in the literature review, the absence of a structured way to gather of knowledge and the incompatibility of IT systems with change, have been documented in several other cases involving physical products [9].

Specifically for the case study, the analysis helped document many semantic misalignments between the systems in focus, i.e. *Handling In* and *Inbound*, or *Storage* and *Inventory*. It's not clear what the impact of semantic misalignments is on service systems, but some authors are arguing for the creation of a business glossaries to ease communication of e.g. data [23]. In addition, employees within the case company would often comment that there are discrepancies between the understanding of how services are *named* and what job is associated with said service, highlighting the need for a common service understanding. In light of this observation, the inclusion of semantic alignment should be considered an important aspect of service architecture implementation, and as such, a focus for future research studies.

Additionally, the scope of architecture implementation is unique for each case. For the case company, it was relevant to consider multiple life-cycle phases to be included within the architecture-driven standardization effort. For other projects and organizations, architecture projects, might be solely for development or sales, which changes the degree of relevance of a multi-life-cycle stage analysis tool. That being said, the case company showed appreciation towards the presented work and expressed a desire to create value through multi-dimensional architecture thinking.

Furthermore, the scope of this paper does not cover the evaluating of the architecture concepts that emerge from this analysis. The current method only helps to reveal multiple potential decompositions embedded within systems. Future research should address how to quantify the *best* fit by balancing theoretical architectural performance, such as Fixon's "Multi-dimensional framework" [24], with the cost of architecture implementation.

Finally, while this study focused on the initial analysis phase, future work could explore how the output of this method is best integrated into the subsequent steps of the service architecture design and implementation process.

5 Conclusion

This study addresses a critical gap in service architecture design by proposing a structured analysis method that expands the scope of early-stage analysis to include a broader range of pre-existing system decompositions. While prior literature highlights the importance of preparatory analysis before architecture conceptualization, it often emphasizes conventional sources such as interviews and Bill of Materials–often neglecting the wealth of structural information embedded in IT systems and operational data. This narrow focus can contribute to implementation barriers, including dispersed knowledge, inconsistent terminology, and fragmented IT infrastructures that complicate the shift toward modular or architecture-based portfolios.

In response, this research introduces a method based on multi-domain and life-cycle analyses. The approach guides users through three structured steps: identifying relevant life-cycle stages, organizing inputs across digital, physical, and mental domains, and visualizing the collected hierarchical structures for cross-comparison. The goal is to systematically uncover decompositions that already exist within the organization but have not been aligned or interpreted as architectural elements.

The method's practical application was explored through a case study at a Danish third-party logistics (3PL) warehouse provider. Applying the analysis matrix revealed previously undocumented structural and semantic misalignments across various organizational systems, including conflicting naming conventions and inconsistent hierarchical levels for equivalent warehouse operations.

By broadening the analysis to include operational frameworks and IT-based structures, the method facilitates a more comprehensive and systematic foundation for service architecture development. It directly supports the identification of latent architectures and helps reduce knowledge-related barriers that can impede implementation. In doing so, it addresses a previously overlooked aspect of the design process: the alignment of internal decompositions as a basis for standardized architecture work.

Although the method proved effective in an IT-intensive logistics environment, its applicability to other service industries–particularly those with less integrated systems–remains an open question. Future research should examine the method's transferability, explore the organizational impact of semantic and structural misalignments, and investigate how the resulting decompositions can be evaluated and prioritized based on factors such as implementation cost and system readiness.

Ultimately, this study underscores the importance of conducting broad, structured, and data-informed service analysis as a prerequisite for developing implementable service architectures. Such analysis is essential to supporting standardization efforts and enabling adaptable, scalable service models in increasingly complex business environments.

References

1. Hsuan J, Persson M (2024) Servitization, modularity, and innovation in supply chain management. In: The Palgrave handbook of supply chain management. Springer International Publishing, pp 1441–1466
2. Starr MK (2010) Modular production-a 45-year-old concept. Int J Oper Prod Manag 30:7–19
3. Brax SA, Bask A, Hsuan J, Voss C (2017) Service modularity and architecture–an overview and research agenda. Int J Oper Prod Manag 37(6):686–702
4. Greve E, Krause D (2019) Long-term effects of modular product architectures: an empirical follow-up study. Procedia CIRP 84:731–736. Elsevier B.V
5. Iman N (2016) Modularity matters: a critical review and synthesis of service modularity
6. Broekhuis M, van Offenbeek M, Eissens-van der Laan M (2017) What professionals consider when designing a modular service architecture? Int J Oper Prod Manag 37(6):748–770
7. Strøm EM, Münsberg TM, Hvam L (2023) Identifying potential applications of service configuration systems in a logistics company. Tech. Rep
8. Reddi KR, Moon YB (2009) A framework for managing engineering change propagation. Tech. Rep
9. Bertram CA, Mueller GO, Lokkegaard M, Mortensen NH, Hvam L (2020) Why engineer-to-order portfolio rationalization stalls: challenges in standardization, modularization, platform design, and mass customization. In: Proceedings of the design society: DESIGN conference, vol 1. Cambridge University Press, pp 2265–2273
10. Simon HA (1962) The architecture of complexity. Tech. Rep
11. Ulrich K (1994) Fundamentals of product modularity. Tech. Rep
12. Schilling MA, Berman S, Arnold P, Chesebrough H, Hoffman A, Hibbard J, Paun D, Balasubramanian P (1999) Towards a general modular systems theory and its application to inter-firm product modularity. Tech. Rep
13. Bowen DE, Fisk RP, Bateson JE, Berry LL, Bitner MJ, Brown SW, Chase RB, Edvardsson B, Grönroos C, Parasuraman A, Schneider B, Zeithaml VA (2023) Learning from the pioneering founders of the service research field. J Serv Manag 34:605–630
14. Posselt T (2018) A service-dominant definition of servitization. In: Organizational competence for servitization. Springer Fachmedien Wiesbaden, pp 59–91
15. Harlou U (2006) Developing product families based on architectures—Contribution to a theory of product families. Tech. Rep., Technical University of Denmark
16. Vargo SL, Lusch RF, Smith RH, Hunt S, Laczniak G, Malter A, Morgan F, O'brien M (2004) Evolving to a new dominant logic for marketing. Tech. Rep.
17. Pekkarinen S, Ulkuniemi P (2008) Modularity in developing business services by platform approach. Int J Logist Manag 19:84–103
18. Voss CA, Hsuan J (2009) Service architecture and modularity*. Tech. Rep
19. Hvam L, Mortensen NH, Riis J (2008) Product customization
20. Strøm EM, Münsberg TM, Hvam L (2024) A model to describe logistics service architecture based on product architecture. In: Proceedings of the design society, vol 4. Cambridge University Press, pp 795–804
21. Conway ME (1968) How do committees invent? Tech. Rep
22. Münsberg TM, Hvam L, Tsintzou L (2022) Optimizing the integration process of new customers in a 3PL warehouse
23. Sarkar P (2015) Data as a service: a framework for providing reusable enterprise data services. Tech. Rep
24. Fixson SK (2005) Product architecture assessment: a tool to link product, process, and supply chain design decisions. J Oper Manag 23:345–369

Carbon Footprint Reduction of Queueing Systems in Manufacturing

Sebastian Schötz and Philipp Gölzer

Abstract Manufacturing companies have a strong impact on climate change due to their immense degree of greenhouse gas emissions. Hence, it is mandatory that these companies take measures to reduce their greenhouse gas emissions. A promising measure is the reduction of the carbon footprint caused by single processes within manufacturing process networks. Various approaches to reduce the carbon footprint of single manufacturing processes have already been described in literature. However, there is a scientific need for action when it comes to the systematic reduction of the carbon footprint of queueing systems which are typically an integral element of manufacturing process networks. Therefore, this paper presents a novel approach that aims to assess and reduce the carbon footprint of manufacturing queueing systems systematically. First, a general queueing system is defined and its variables that contribute to the carbon footprint in manufacturing are determined. Subsequently, a mathematical model is designed which enables an abstract description of the interactions of different input variables on the carbon footprint of queueing systems in manufacturing. Afterwards, recommendations to reduce the carbon footprint of queueing systems are developed based on the previous findings. Finally, the approach is validated for a specific manufacturing queueing system with a discrete-event simulation study. The results give foundations for further research on dependencies within manufacturing process networks.

Keywords Manufacturing · Queueing systems · Carbon footprint · Greenhouse gas emissions · Sustainability

1 Introduction

Climate change represents a major challenge of humanity nowadays. Increasing extreme weather conditions (e.g. droughts, storms) and rising sea levels are just some effects of climate change [1]. Greenhouse gas emissions have a significant impact

S. Schötz (✉) · P. Gölzer
Nuremberg Institute of Technology Georg Simon Ohm, Nuremberg, Germany
e-mail: sebastian.schoetz@th-nuernberg.de

M. Manns (ed.), *New Paradigms for Anticipated Uncertainty*, Lecture Notes in Mechanical Engineering, https://doi.org/10.1007/978-3-032-16889-4_46

on global warming and hence on climate change. Especially the emissions of carbon dioxide (CO2) caused by human activities represent a considerable contribution towards global warming [2]. Since various greenhouse gases have a different impact on global warming, an index has been introduced to quantify their impact universally: The so-called global warming potential expresses the effect of greenhouse gases as the mass of CO_2 emissions that correspond to the equivalent effect [3]. The main contributors of global greenhouse gas emissions are electricity and heat, transport, manufacturing and construction [4]. Hence, it is mandatory that manufacturing companies take measures to reduce their greenhouse gas emissions.

A promising measure is the reduction of the carbon footprint emitted by single processes (e.g. machining, material handling) of manufacturing process networks. An integral element that occurs multiple times within manufacturing process networks are queueing systems, realized e.g. as buffer stock combined with a downstream tooling machine. These systems are essential in manufacturing, for example due to varying cycle times within a process sequence.

As shown in Fig. 1 a single server queueing system comprises a waiting line and a server while in manufacturing the server is represented by a single manufacturing process and the customers by semi-finished products [5–8].

The Kendall notation, a common notification for the characterization of queueing systems, consists of different elements, each of which indicates a characteristic of a queueing system [9]. For additional information, see e.g. [5, 7–9].

Queueing systems in manufacturing impact the carbon footprint of manufacturing facilities, e.g. a roller conveyor as buffer stock prior to assembly station requires dedicated floor space, lighning and heating—additional greenhouse gas emissions.

These additional requirements lead to increased greenhouse gas emissions and hence, an increased carbon footprint. Based on a literature review, no reference about the quantitative contribution of manufacturing queueing systems to the global greenhouse gas emissions has been found. This gap can be explained by a lack of systematic approaches to assess the carbon footprint of queueing systems in manufacturing quantitatively. Against this background, it is reasonable to investigate the assessment and reduction of the carbon footprint caused by queueing systems in manufacturing.

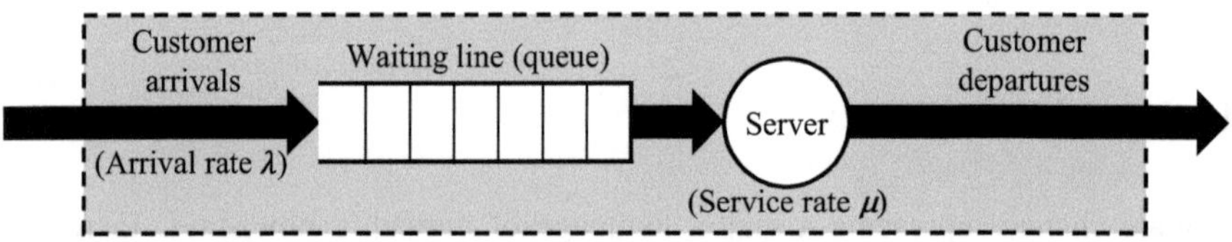

Fig. 1 Single server queueing system structure [5–8]

2 State of the Scientific Knowledge and Need for Action

According to Goa et al. different standards for assessing the carbon footprint exist but vary regarding different characteristics like system boundaries [10]. Life cycle assessment standards like DIN EN ISO 14040 [11] and the DIN EN ISO 14044 [12] are typically applied when focusing on the reduction of the carbon footprint in manufacturing. In addition to these primarily holistically oriented standards, studies which are addressing the carbon footprint of particular elements of manufacturing process networks have already been described. Nilakantan et al. address the reduction of robotic assembly line systems [13]. Elbadawi et al. emphase the carbon footprint of 3D printing focusing on pharmaceutical manufacturing, where the carbon footprint for manufacturing ten sample printlets varies from 11.60 g CO_2-eq up to 595.61 g CO_2-eq [14]. Lin et al. focus on computer numerical controlled machining operations, where the machining of a sample gear with wet cutting causes 736.8 g CO_2-eq and with high-speed dry cutting causes 498.7 g CO_2-eq [15]. Other studies focus on other potential elements of manufacturing process networks, e.g. [16, 17].

In summary, various standards for assessing respectively reducing the carbon footprint holistically exist and studies which are addressing the carbon footprint of single elements of manufacturing process networks are described in literature. However, to the best of our knowledge, there is still a scientific need for action regarding the exploration of the carbon footprint of queueing systems in manufacturing. Therefore, as well as due to the high importance of carbon footprint reduction in manufacturing and the frequent occurrence of queueing systems in manufacturing process networks, this paper describes a novel approach that aims to assess and reduce the carbon footprint of individual queueing systems in manufacturing systematically.

3 Approach

The approach is divided into three consecutive steps. First, requirements which are relevant for the further approach are defined. Subsequently, a mathematical model is developed for an abstract description of the interactions of different input variables on the carbon footprint of queueing systems in manufacturing. Afterwards, recommendations to reduce the carbon footprint of queueing systems are developed based on the previous findings.

3.1 Requirements

In this subsection, the characteristics of the queueing system model are specified by determining a general queueing system that typically occurs in manufacturing.

Afterwards, input variables as well as the output variable of the approach are defined prior to the determination of various scenarios.

In literature, different studies which are focusing on queueing systems in manufacturing base on the Kendall notation M/M/1, e.g. [18]. In addition, Wu et al. mention, that M/M/1 and approximations of G/G/1 are often taken as basis when focusing on the performance of single machines [19]. Therefore, the M/M/1 queueing system has been determined as basis for the further development. This general queueing system can be characterized as follows [5, 8, 9]:

- Exponential distribution of interarrival times of semi-finished products
- Exponential distribution of service times
- Single server
- Infinite maximum number of semi-finished products in the waiting line
- Infinite number of potential semi-finished product
- FIFO (First-in-First-Out) service discipline

The M/M/1 queueing system has already been mathematically described, e.g. [20, 21]. Therefore, existing mathematical descriptions from [20, 21] are taken into consideration within the further development.

The defined input and output variables of the mathematical model are described in Table 1. These variables have been defined with consideration of the approaches' purpose.

The degree of abstraction of the input variables has been selected under the consideration to design the approach comprehensive but not unnecessarily complicated. This ensures a high degree of practicability and applicability of the approach since the input variables may depend on various factors. For instance, the carbon footprint of the waiting line per hour W_{CF} may depend on the carbon footprint for heating the area of the waiting line, the carbon footprint for the proportionate electrical energy for

Table 1 Input and output variables

Variable	Symbol	Unit	Characteristic
Service rate (i.v.)	μ	pcs./h	Controllable
Arrival rate (i.v.)	λ	pcs./h	Non-controllable
Carbon f.p. of the waiting line per hour (i.v.)	W_{CF}	kg CO_2-eq/h	Non-controllable
Carbon f.p. of the idle server per hour (i.v.)	S_{ICF}	kg CO_2-eq/h	Non-controllable
Carbon f.p. of the processing server per hour (i.v.)	S_{PCF}	kg CO_2-eq/h	Non-controllable
Fixed carbon f.p. of the processing server per process (i.v.)	S_{FCF}	kg CO_2-eq	Non-controllable
Average carbon f.p. of the queueing system per semi-finished product (o.v.)	Q_{CF}	kg CO_2-eq	/

i.v. input variable, *o.v.* output variable, *f.p.* footprint the manufacturing facility lighting etc. To determine the carbon footprint-based input variables (W_{CF}, S_{ICF}, S_{PCF} and S_{FCF}) the aforementioned standards for assessing the carbon footprint can be applied.

Furthermore, the following assumptions have been made in order to develop a clear and comprehensive approach:

- The carbon footprint of the waiting line per hour W_{CF} does not depend on the total waiting time per semi-finished product.
- The carbon footprint of the idle server per hour S_{ICF} does not depend on the total idle time per semi-finished product.
- The carbon footprint of the processing server per hour S_{PCF} as well as the fixed carbon footprint of the processing server per process S_{FCF} do not depend on the total processing time per semi-finished product.

The carbon footprint-based input variables are considered as non-controllable. In addition, the arrival rate λ is considered as non-controllable since it typically depends on upstream elements of manufacturing process networks and this approach focuses solely on the isolated optimization of single queueing systems as elements of manufacturing process networks. In contrast, the service rate μ is defined as controllable input variable since it depends on the server as an integral element of the queueing system.

The average carbon footprint of the queueing system per semi-finished product Q_{CF} has been defined as output variable. It represents the average greenhouse gas emissions of a semi-finished product when passing through a queueing system in manufacturing.

In order to give practical relevant recommendations to reduce the carbon footprint in manufacturing, three various scenarios that frequently occur in manufacturing have been defined (see Fig. 2).

<table>
<tr><th>Scenario 1 (S1)</th><th>Scenario 2 (S2)</th><th>Scenario 3 (S3)</th></tr>
<tr><td rowspan="2">Carbon footprint for processing a single semi-finished product is <u>not dependent</u> on the service rate. ($S_{FCF} \geq 0$; $S_{PCF} = 0$; $S_{ICF} > 0$)</td><td colspan="2">Carbon footprint for processing a single semi-finished product is <u>dependent</u> on the service rate. ($S_{FCF} \geq 0$; $S_{PCF} > 0$)</td></tr>
<tr><td>Carbon footprint of the processing server per hour <u>exceeds</u> the carbon footprint of the idle server per hour. ($S_{PCF} > S_{ICF}$)</td><td>Carbon footprint of the processing server per hour <u>subceeds</u> the carbon footprint of the idle server per hour. ($S_{PCF} < S_{ICF}$)</td></tr>
<tr><td>Example: Complete charging of an empty built-in battery via a battery storage device. The duration of the charging process has no impact on the stored energy.</td><td>Example: Fused deposition modeling process with variable layer thickness. Energy for heating is only consumed when processing and furthermore, depends on the process duration.</td><td>Example: Industrial furnace process for heat-holding preheated parts. More energy is consumed to hold heat in idle mode than in processing mode due to no heat supply of input parts in idle mode.</td></tr>
</table>

Fig. 2 Defined scenarios

In scenario 1, the carbon footprint for processing a single semi-finished product equals always only a fixed amount which is independent on the average processing time per semi-finished product. As an example, the duration to completely charge an empty built-in battery via a battery storage device has no impact on the amount of the charged energy which determines the amount of the carbon footprint of the process. Whereas in scenario 2 and scenario 3, the carbon footprint for processing a single semi-finished product depends on the processing duration and therefore on the service rate. The examples which are mentioned for each individual scenario are idealized to provide a proper understandability.

3.2 Mathematical Model

This subsection describes the mathematical model. Firstly, the average carbon footprint of the queueing system per semi-finished product Q_{CF} is calculated as follows, with t_w represents the average waiting time per semi-finished product, n_w the average number of waiting semi-finished products, t_i the average server idle time per semi-finished product and t_p the average processing time per semi-finished product:

$$Q_{CF}(\mu) = W_{CF} \cdot \frac{t_w}{n_w} + S_{ICF} \cdot t_i + S_{PCF} \cdot t_p + S_{FCF} \quad (1)$$

The average waiting time per semi-finished product t_w is calculated with the service rate μ and the arrival rate λ (see equation (2), partly adapted) [20, 21].

$$t_w(\mu) = \frac{\lambda}{\mu \cdot (\mu - \lambda)} \quad (2)$$

The average number of waiting semi-finished products n_w is calculated with the average waiting time per semi-finished product t_w and the arrival rate λ (see equation (3), partly adapted)[20, 21].

$$n_w(\mu) = \frac{\lambda^2}{\mu \cdot (\mu - \lambda)} \quad (3)$$

The average server idle time per semi-finished product t_i is calculated with the probability that the server is in idle mode and accordingly a semi-finished could be processed immediately P_i (see equation (4), partly adapted)[20, 21].

$$t_i(\mu) = P_i \cdot \frac{1}{\lambda} = \left(1 - \frac{\lambda}{\mu}\right) \cdot \frac{1}{\lambda} \quad (4)$$

The calculation of the average processing time per semi-finished product t_p is based on the service rate μ.

$$t_p(\mu) = \frac{1}{\mu} \tag{5}$$

In summary, the average carbon footprint of the queueing system per semi-finished product Q_{CF} can be calculated as follows.

$$Q_{CF}(\mu) = W_{CF} \cdot \frac{1}{\lambda} + S_{ICF} \cdot \left(1 - \frac{\lambda}{\mu}\right) \cdot \frac{1}{\lambda} + S_{PCF} \cdot \frac{1}{\mu} + S_{FCF} \tag{6}$$

3.3 Recommendations

Based on the mathematical model, general graphs have been created for the predefined scenarios. These graphs illustrate the average carbon footprint of the queueing system per semi-finished product Q_{CF} and its summands over the service rate μ. Then, recommendations to reduce the carbon footprint in manufacturing queueing systems have been developed. The general graphs and the recommendations are shown in Fig. 3.

For a swift improvement of the average carbon footprint of the queueing system per semi-finished product Q_{CF}, the appropriate scenario has to be identified and the corresponding recommendation can be implemented directly without focusing on any of the equations of the mathematical model.

In scenario 1 and scenario 3, a decrease of the service rate μ leads to a reduction of the average carbon footprint of the queueing system per semi-finished product Q_{CF}. This can be explained as follows: A reduction of the service rate μ is directly linked to an increase of the average processing time per semi-finished product t_p. This, in turn, leads to a reduction of the average server idle time per semi-finished product t_i and therefore to a decrease of the average carbon footprint per semi-finished product.

With regards to common industrial practice, an increase of the average processing time per semi-finished product t_p may at first seem disadvantageous due to a supposed reduced throughput of the queueing system. However, to avoid that an infinite average number of semi-finished products are in the waiting line n_w, which is crucial in

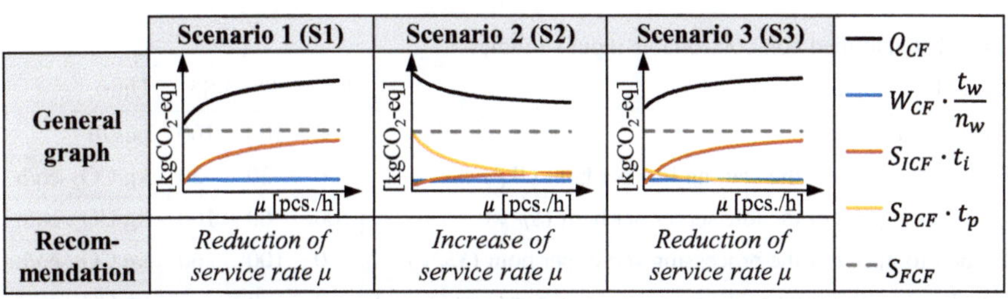

Fig. 3 General graphs and recommendations

manufacturing practice, the service rate μ has to be greater than the arrival rate λ ($\mu > \lambda$) [20, 21]. Hence, the throughput of a queueing system depends on the arrival rate λ which has been defined as a non-controllable variable.

In scenario 2, the processing server causes a higher carbon footprint per hour compared to the idle server ($S_{PCF} > S_{ICF}$) and therefore, an increase of the service rate μ (which is directly linked to a reduction of the average processing time per semi-finished product t_p) leads to a reduction of the average carbon footprint of the queueing system per semi-finished product Q_{CF}.

In addition, it has to be noted that if the carbon footprint of the processing server per hour is equal to the carbon footprint of the idle server per hour ($S_{PCF} = S_{ICF}$), a change of the service rate μ does not have any impact on the average carbon footprint of the queueing system per semi-finished product Q_{CF}. However, this scenario is not covered in the approach since it provides no possibility to reduce the carbon footprint.

These recommendations are non-quantified. For quantified recommendation the mathematical model has to be applied. The mathematical model can also be applied for detailed analyses, like an impact assessment or a numerical comparison of the product carbon footprint prior and after taken actions.

4 Validation

The developed approach has been validated by comparing the results with results of a discrete-event simulation study. The simulation study has been conducted with the software Plant Simulation version 2302 using a simulation period of 50 days and statistic reset after 10 days. For the validation, non-controllable input variables have been quantified for each individual scenario: S1, S2, S3 (see Table 2).

Table 3 gives an overview of the results. To ensure a clear visualization on the one hand and a proper accuracy on the other hand the results are rounded to the third decimal.

It has to be mentioned that the results of the simulation study are generally higher than the results of the developed approach. As the simulation stops after a defined simulation period, the carbon footprint is calculated for all products in the queueing system even for those, who have not yet been processed.

Table 2 Quantified non-controllable input variables

Variable	S1	S2	S3	Unit
Arrival rate (λ)	100	100	100	pcs./h
Carbon footprint of the waiting line per hour (W_{CF})	20	20	20	kg CO_2-eq/h
Carbon footprint of the idle server per hour (S_{ICF})	100	60	100	kg CO_2-eq/h
Carbon footprint of the processing server per hour (S_{PCF})	0	100	60	kg CO_2-eq/h
Fixed carbon footprint of the proc. server per process (S_{FCF})	2	2	2	kg CO_2-eq

S1 = Scenario 1; S2 = Scenario 2; S3 = Scenario 3

Table 3 Average carbon footprint of the queueing system per semi-finished product

μ [pcs./h]	105	115	125	135	145	155	165	175	185	195
Q_{CF} (S1-ap.) [kg CO_2-eq]	2.248	2.330	2.400	2.459	2.510	2.555	2.594	2.629	2.659	2.687
Q_{CF} (S1-sim.) [kg CO_2-eq]	2.252	2.334	2.404	2.463	2.514	2.559	2.598	2.632	2.663	2.691
Q_{CF} (S2-ap.) [kg CO_2-eq]	3.181	3.148	3.120	3.096	3.076	3.058	3.042	3.029	3.016	3.005
Q_{CF} (S2-sim.) [kg CO_2-eq]	3.183	3.150	3.122	3.098	3.078	3.060	3.045	3.031	3.018	3.007
Q_{CF} (S3-ap.) [kg CO_2-eq]	2.819	2.852	2.880	2.904	2.924	2.942	2.958	2.971	2.984	2.995
Q_{CF} (S3-sim.) [kg CO_2-eq]	2.823	2.856	2.884	2.907	2.928	2.946	2.961	2.975	2.987	2.998

S1 = Scenario 1; S2 = Scenario 2; S3 = Scenario 3; ap. = developed approach; sim. = simulation study

Comparing the results of the developed approach as predicted values with the results of the simulation study as observed values the mean absolute percentage error (MAPE) equals 0.116% and the root mean square deviation (RMSD) equals 0.0033 kgCO_2-eq. These scores prove the validity of the mathematical model.

Figure 4 provides a graphical representation of the simulation study results. Here, the progressions of the individual scenarios correspond to the progressions within the general graphs and hence, prove the validity of the determined recommendations.

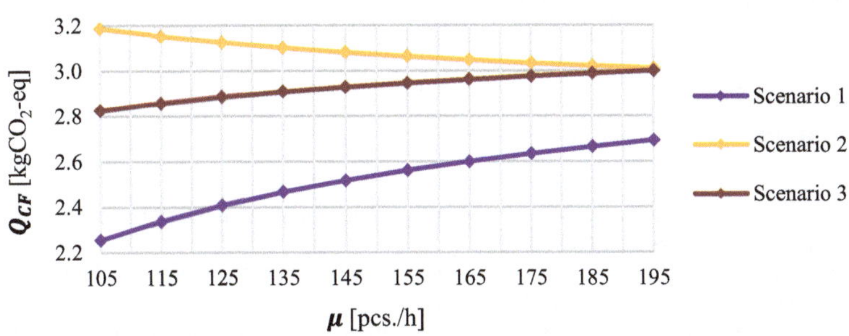

Fig. 4 Graphical representation of the results of the discrete-event simulation study

5 Summary and Outlook

This paper describes a novel approach to reduce the carbon footprint in manufacturing queueing systems. Since different studies which are addressing the carbon footprint of particular elements of manufacturing process networks already exist, this paper focuses on an element that has not been addressed regarding the carbon footprint. Therefore, it provides a supplement towards sustainable manufacturing systems.

Based on existing mathematical descriptions about basic characteristics of queueing systems, a novel carbon footprint-based mathematical model has been developed. This model describes systematically the interactions of different input variables on the carbon footprint of queueing systems in manufacturing for the first time. Additionaly, it enables a quantitative assessment of the carbon footprint of a single semi-finished product when passing through a queueing system in manufacturing. Furthermore, recommendations to reduce the carbon footprint in manufacturing queueing systems are provided which can be implemented directly without focusing on the developed mathematical model. A validation with a discrete-event simulation study approves the validity of the developed approach.

Since this approach gives first insights, a deeper investigation about the assumptions made in this paper is promising to achieve a further detailing of the approach. Also, additional basic queueing systems (e.g. M/G/1) can be investigated regarding the scope of this paper. Furthermore, the developed approach can be validated regarding its feasibility to reduce the carbon footprint of queueing systems in other fields like traffic engineering. Moreover, the developed approach can be applied in further studies to gain knowledge about different effects of queueing systems, e.g. contribution of greenhouse gas emissions on entire manufacturing process networks respectively on the global greenhouse gas emissions.

References

1. United Nations Homepage. https://www.un.org/en/climatechange/science/causes-effects-climate-change. Accessed on 10 Feb 2025
2. European Commission Homepage. https://climate.ec.europa.eu/climate-change/causes-climate-change_en. Accessed on 10 Feb 2025
3. Lashof DA, Ahuja DR (1990) Relative contributions of greenhouse gas emissions to global warming. Nature 344:529–531
4. Ritchie H, Rosado P, Roser M (2025) Breakdown of carbon dioxide, methane and nitrous oxide emissions by sector. In: Our world in data homepage. https://ourworldindata.org/emissions-by-sector. Accessed on 10 Feb 2025
5. Nielsen BF, Nielson F, Pilegaard H, Smith MJA, Yüksel E, Zeng K, Zhang L (2014) Roadmap document of stochastic analysis. MT-LAB
6. He D, Li R, Huang Q, Lei P (2014) Maximum entropy principle based estimation of performance distribution in queueing theory. PLOS ONE 9(9):1–9
7. Hedtstück U (2013) Simulation diskreter prozesse. Methoden und anwendungen. Springer Vieweg, Berlin Heidelberg
8. Herzog A (2021) Simulation mit dem warteschlangensimulator. Mathematische Modellierung und Simulation von Produktions- und Logistikprozessen. Springer Gabler, Wiesbaden

9. Kendall DG (1953) Stochastic processes occurring in the theory of queues and their analysis by the method of the imbedded markov chain. Ann Math Stat 24(3):338–354
10. Gao T, Liu Q, Wang J (2014) A comparative study of carbon footprint and assessment standards. Int J Low-Carbon Technol 9(3):337–243
11. International Organization for Standardization: ISO 14040:2006 (2006) Environmental management—life cycle assessment—principles and framework
12. International Organization for Standardization: ISO 14044:2006 (2006) Environmental management—life cycle assessment—requirements and guidelines
13. Nilakantan JM, Li X, Tang Q, Nielsen P (2017) Multi-objective co-operative co-evolutionary algorithm for minimizing carbon footprint and maximizing line efficiency in robotic assembly line systems. J Clean Prod 156:124–136
14. Elbadawi M, Basit AW, Gaisford S (2023) Energy consumption and carbon footprint of 3D printing in pharmaceutical manufacture. Int J Pharm 639:1–8
15. Li C, Tang Y, Cui L, Li P (2023) A quantitative approach to analyze carbon emissions of CNC-based machining systems. J Intell Manuf 26:911–922
16. Lin W, Yu DY, Zhang C, Liu X, Zhang S, Tian Y, Liu S, Xie Z (2015) A multi-objective teaching−learning-based optimization algorithm to scheduling in turning processes for minimizing makespan and carbon footprint. J Clean Prod 101:337–347
17. Stavropoulos P, Panagiotopoulou VC (2022) Carbon footprint of manufacturing processes: conventional vs Non-Conventional. Processes 10:2–14
18. Baek JW, Moon SK (2014) The M/M/1 queue with a production-inventory system and lost sales. Appl Math Comput 223:534–544
19. Wu K, McGinnis L, Zwart B (2008) Queueing models for single machine manufacturing systems with interruptions. In: Mason SJ, Hill RR, Mönch L, Jefferson T, Fowler JW (eds) Proceedings of the 2008 winter simulation conference. IEEE, pp 2083–2092
20. Brown DA (2011) Decision science and queueing theory. An introduction with call center simulation
21. Taylor III BW (2013) Introduction to management science, 11th edn. Pearson, Upper Saddle River

Data Acquisition Pipeline for the Generation of a Digital Twin for Large Compliant Thin-Walled Structures

Guillaume Tréheux, Dominik Schäfer, Alexander Moriz, Lars Köttner, Tobias Weber, Amon Göppert, and Robert H. Schmitt

Abstract The aerospace industry's drive to improve fuel efficiency and reduce emissions has led to the widespread adoption of lightweight, thin-walled structures in aircraft assemblies. However, these flexible structures are susceptible to deformation under gravitational and external forces, causing assembly deviations and potential misalignment. To counteract the misalignment caused by part deformation in joining processes, shims are specifically designed to fill the gaps between components. Yet, despite advances in metrology, the process of shimming remains largely manual, since each assembly's unique gap profile requires physical measurement, fitting, and adjustment, which hinders full automation. A promising solution is the use of a digital twin to simulate and predict part deformation during the joining process. The predicted deformation resulting from the digitalization of the compliant structures to be joined can be used for designing shims in a virtual assembly environment or as input for adaptive tooling systems specifically designed to compensate for the deformation. This work introduces a concept for a test cell equipped with a large-scale metrology system to capture the data needed for constructing a digital twin of manufactured large, compliant, thin-walled structures. Corresponding data pipelines, designed to meet FAIR data principles, combine a time-series database for continuous measurement data with a non-relational database for storing measurement metadata. Preliminary results demonstrate the test cell's ability to provide the data required for applying inverse form finding methods to approximate the LCTS gravity-free state shape and for training data-driven physics informed models for the prediction of the deformation behavior. These findings lay the groundwork for future digital twin development and offer insights to improve alignment precision in the assembly of large, compliant, thin-walled structures.

G. Tréheux (✉) · D. Schäfer · A. Moriz · A. Göppert · R. H. Schmitt
Chair of Intelligence in Quality Sensing, Aachen, Germany
e-mail: guillaume.treheux@wzl-iqs.rwth-aachen.de

L. Köttner · T. Weber
Boeing Technology Innovation Europe, Munich, Germany

R. H. Schmitt
Fraunhofer Institute for Production Technology (IPT), Aachen, Germany

M. Manns (ed.), *New Paradigms for Anticipated Uncertainty*, Lecture Notes in Mechanical Engineering, https://doi.org/10.1007/978-3-032-16889-4_47

Keywords Virtual assembly · Smart manufacturing · Sensor integration · Digital twin

1 Introduction

Managing gaps to ensure structural integrity presents a challenge in the assembly of aircraft fuselages, particularly those involving large, compliant, thin-walled structures (LCTS) such as skin elements. These lightweight structures, while advantageous for reducing overall aircraft weight and enhancing fuel efficiency, are inherently susceptible to deformation under gravitational and external loads. Such deformations often cause misalignments and deviations during assembly, necessitating precise gap management to maintain structural integrity. When part-to-part gaps in fuselage assembly exceed 0.3 mm, aircraft manufacturers must resort to shimming–a procedure in which custom-shaped spacers, either solid or liquid, are used to fill the gaps between components [1]. The shimming process remains largely manual and time-consuming, involving the placement of parts on an assembly jig in their designated positions, followed by manual gap measurement using gauges. After measuring the gaps, shims are prepared while the components are temporarily removed from their assembly positions. After inserting the shims, the components are repositioned for joining, with repeated measurements required to ensure precise alignment.

1.1 State of the Art

Several approaches have been introduced for the reduction of the required time for shimming in part-to-part gaps: One of the root causes for gaps in assemblies and thus resulting in the need for shimming is shape deviations caused by manufacturing errors. Different methods for assessing the shape deviation of non-rigid parts using optical techniques have been explored by Thiébaut et al. [12] and Radvar-Esfahlan and Tahan [8], eliminating the need for specialized inspection fixtures or jigs. Both studies employ finite element analysis to simulate and predict deformations resulting from gravity and fixturing effects, allowing distinction between intrinsic flexible deformations and geometric deviations due to manufacturing errors. Schmick et al. [9] introduced a novel automated shimming procedure by integrating optical measurements, virtual assembly, and additive manufacturing. Initially, the joining surfaces are optically measured to assess the gap between two parts prior to physical assembly. A mesh is generated from the filtered point clouds of the structures. The meshes are then virtually aligned to simulate the inter-part gap. The predicted gap geometry is extracted from the simulation, and a solid shim is additively manufactured to match it. While this procedure significantly reduces in-line process time and manual work, it is limited to rectangular and circular geometries. In contrast, other works such as Franciosa et al. [3] have integrated physics-based simulation of compliant parts into their approach. The framework proposed in [3] incorporates in-line measurement systems as input for finite element simulations. Additionally, the authors apply a flexible alignment method that accounts for part deformations and enables the prediction and fabrication of custom shims without manual measure–fit–adjust loops. A different approach was presented by Manohar et al. [6], who combined sparse sensor optimization with robust feature extraction, focusing on historical production data from a representative aircraft. The machine learning model was trained on historical production data to identify low-dimensional patterns in shim gap distribution. Furthermore, the model was used to optimize sensor placement, reducing the number of measurements required to predict gap distribution. Instead of using shims, a novel approach could involve applying controlled loads via adaptive tooling, guided by a deformation model of the LCTS. This would keep gaps below the shimming threshold, thereby eliminating the need for shimming in assemblies. Near real-time deformation modeling requires high computational efficiency. Physics-informed machine learning (ML) models can outperform conventional finite element methods (FEM) once trained [4]. Unlike FEM, these models rely on observational data, with data requirements for training, testing, and validation varying by framework. The volume of data needed depends on prior knowledge of boundary conditions and the coefficients in the governing partial differential equations [4].

2 Test Cell Concept and Data Acquisition

Developing and validating physics-informed ML models requires high-quality observational data on the deformation behavior of large, compliant, thin-walled structures (LCTS) under varying boundary conditions. To this end, a dedicated test cell was designed to capture both geometric and physical data relevant to predicting deformation behavior through a digital twin. This section describes the large-scale metrology system integrated into the test cell, focusing on its hardware components and the supporting data infrastructure. The first subsection details the measurement hardware. The second outlines the hierarchical and modular data architecture that enables efficient storage, processing, and retrieval in accordance with FAIR data principles, thereby ensuring data quality [11].

2.1 Hardware of the Test Cell

In this work, the large compliant thin-walled structures considered as test objects for the developed test cell are aircraft panels made of CFRP, which represent one-sixth of the total circumference of a representative aircraft. Figure 3 summarizes the overall workflow of the test cell by distinguishing three subsystems: the scanning system, the coordinate tracking or large-scale metrology system (LSM) and the actuation system. Due to the large size of the panels (2×3 m), a large-scale metrology system

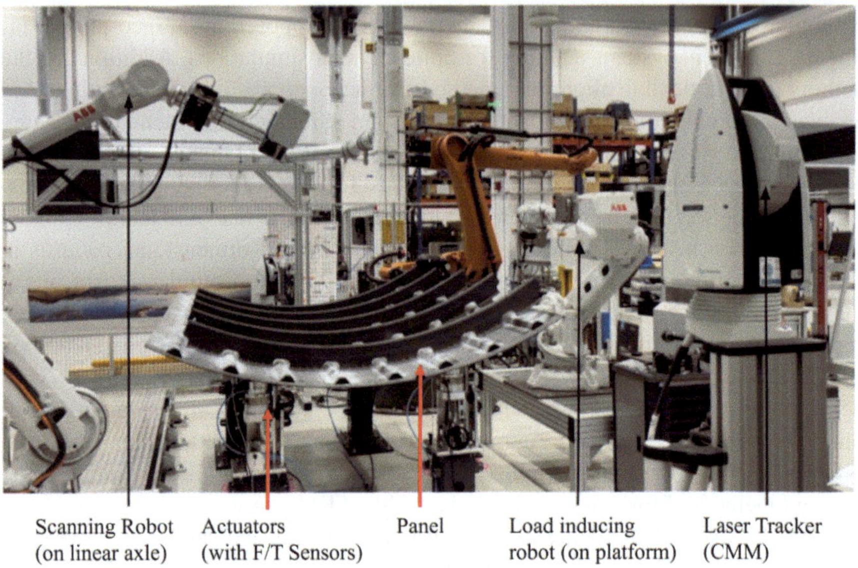

Fig. 1 Test cell set-up for the measurement of the geometry of large compliant thin-walled structures

(LSM), as conceptually introduced in [10], is required. To establish a spatial reference for the acquired sensor data, a global reference system is implemented using a laser tracker. Consequently, all measurement systems are aligned with the laser tracker's base coordinate system. The test cell, which was developed in this work, is shown in Fig. 1. The aircraft panel is placed on a fixture that comprises four linear actuators with vacuum suction cups for clamping the panel during the experiment. Each actuator is equipped with a force-torque sensor (*Schunk FT-AXIA130*) capable of measuring axial forces up to 4000 N in nominal mode, and an inertial measurement unit (*TDK IMU IIM20670*). The force-torque sensor measures the magnitude of the loads exerted by the fixture on the panel, while the inertial measurement unit determines the angle at which the load is applied. Spherically mounted retroreflectors are placed at the base of the actuators to reference them within the global reference system (GRS). The robot on the left in Fig. 1 is mounted on a linear axle, adding an additional degree of freedom for sequentially scanning the panel. The robotic arm's end-effector, which holds the structured light scanner, is equipped with a 6D target to reference both the scanner's position and orientation within the GRS. The complete end-effector of the scanning subsystem can be seen in Fig. 2. Various stereo camera systems have been tested on the test cell, including the *Zivid 2+ M60*, shown in Fig. 2, and the *MicroEpsilon surfaceControl 3D 2500*, shown in Fig. 1. Because the scanner's measurement area is smaller than the full LCTS surface, multiple measurements must be stitched together within the GRS to ensure spatial consistency. To recreate loaded test conditions consistently and repeatably, the test cell offers two methods: manual adjustment of a fixture-mounted linear actuator or automated load-

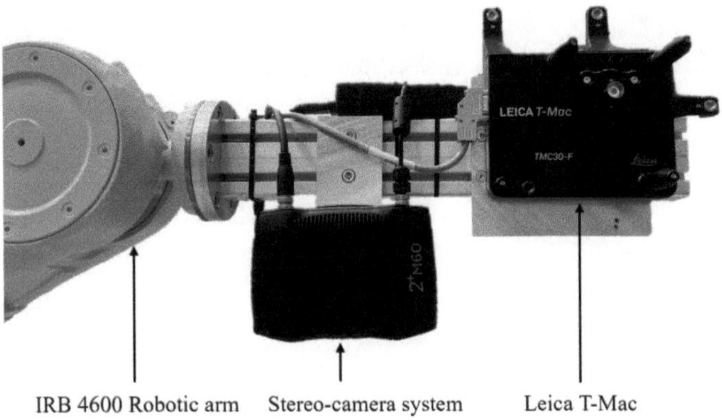

Fig. 2 Scanning robots' end-effector with a Zivid 2+ M60 camera system and a Leica T-Mac for tracking the pose of the end-effector

ing via an industrial robot. The integrated linear actuators can impose displacements, each with a stroke of up to 30 cm. Alternatively, an additional industrial robot–an ABB IRB 4600-2.55–applies pointwise, force-controlled loads. The robot is shown on the right in Fig. 1. This robot is equipped with an FT sensor (*SCHUNK FTN-Omega 160*) capable of measuring forces and torques in six degrees of freedom, with a maximum measurable axial force of 6250 N. The end-effector of this load-inducing robot features a tooling ball for applying pointwise loads. Additionally, pose of the tool center point (TCP) can be tracked using a 6D target in this case a *Leica T-Mac*-system. To ensure seamless integration and coordination between the various hardware components and the digital data processing pipeline, a structured control and communication system is required. This is achieved through a modular programmable logic controller (PLC) system, which acts as the interface between the software modules introduced in Sect. 2.2 and the hardware components introduced in Sect. 2.1. By facilitating real-time control and data synchronization across the physical and virtual domains, the PLC system ensures the accuracy and consistency of sensor measurements. Additionally, a manager PLC module serves as the master clock, synchronizing all sensor data measurements within the LSM.

2.2 *Modular Data Infrastructure*

The data infrastructure for acquiring the characteristics of the LCTS has been designed to support the implementation of FAIR data principles, which define data as findable, accessible, interoperable, and reusable [7, 11]. The data collected from the sensor systems introduced in 2.1 is stored in a time-series database, specifically an *InfluxDB* instance, which is locally hosted within a Docker container. This ensures findability and accessibility, as the data is indexed by timestamps and sen-

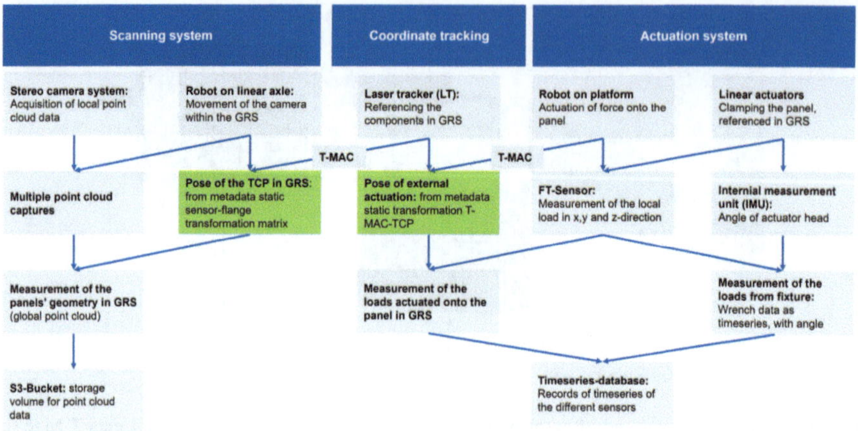

Fig. 3 Overall workflow of the test cells' subsystems and functional description of its components

sor IDs, allowing precise retrieval and analysis over time. Additionally, metadata associated with the measurement–such as the sensor location, calibration details, and experiment ID–is managed in a non-relational database, specifically a locally hosted MongoDB instance. Moreover, static transformation matrices–shown in green in Fig. 3–are stored in the metadatabase. These include, for instance, the sensor-to-flange transformation matrix of the stereo camera system in the GRS and the transformation matrix from the 6D target to the TCP of the load-inducing robot. By structuring metadata in this way, the system ensures interoperability, allowing integration with various analysis tools and external databases. For high-volume datasets such as structured light scanner point clouds, the recorded measurements are stored in a cloud-based storage solution, specifically an S3 bucket. The corresponding metadata entry in MongoDB includes a direct reference to the file's location, ensuring that both structured and unstructured data remain accessible and reusable within the digital twin environment [11]. A middleware layer has been implemented to harmonize the heterogeneous data streams originating from sensors, robots, and optical systems. In the presented test cell, the Robot Operating System 2 (ROS2) serves as this middleware [5]. Its communication architecture facilitates the storage of metadata through ROS2 service calls, enables the collection of sensor data via ROS2 topics, and allows for the triggering of optical systems through ROS2 actions. The use of ROS2 as middleware enables the introduction of a cycle time on the master PLC, which converts incoming UDP sensor signals into ROS2 topics using the built-in *TimeSynchronizer*.

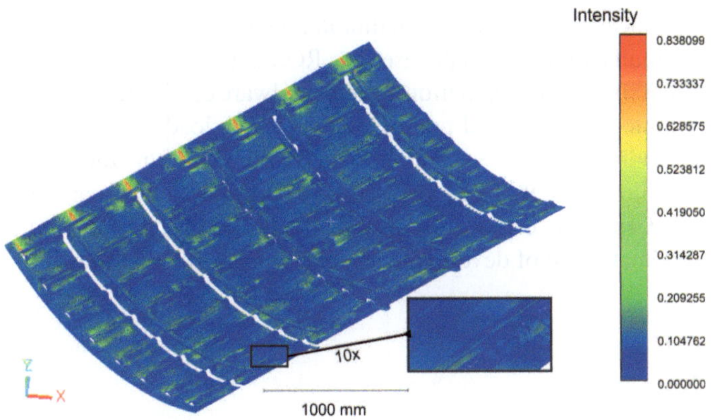

Fig. 4 Intensity of the point cloud of the entire panel scanned with the Zivid 2+ M60 system

3 Preliminary Results

The test cell produces experimental results as a point cloud representing the complete scan of the panel (see Fig. 4), along with records of the applied loads. These load records include force, torque, and the actuation angle applied by the holding fixture on the panel. The experimental records are documented and made accessible through the infrastructure presented in 2.2. The variations in point cloud intensity are caused by the reflective properties of the CFRP material. The green regions on the scan, particularly near the leftmost frame, indicate areas where an antireflective spray coating has been applied, with a thickness ranging from 10 to 50 microns. Given that the Zivid 2+ M60 structured light scanner has a depth resolution typically on the order of 100 microns or greater at the working distances used, the coating thickness falls below the system's spatial resolution limit. As a result, the coating layer does not manifest as a separate surface in the acquired point cloud and does not introduce measurable geometric errors. However, the coating's impact on surface reflectivity can affect intensity values and may introduce minor noise, which can be mitigated through post-processing.

4 Conclusion and Outlook

This work established a test cell for capturing point clouds of deformed LCTS under measurable loads. The implemented LSM demonstrated its capability to generate complete point clouds by fusing multiple captures from a stereo camera system. A key contribution is the FAIR data infrastructure supporting findability, accessibility, interoperability, and reusability of measurement data, ensuring organized and easily retrievable datasets. The modular LSM design enables flexible and scalable adaptation to various measurement needs, facilitating integration with diverse sensors and seamless communication with databases via ROS2. This framework enhances real-time data processing and coordination among hardware components. Future research will focus on integrating optical point cloud data with load measurements to apply inverse form-finding methods, such as those in [2], for predicting the LCTS's gravity-free shape. While Claus et al. [2] targeted isotropic sheet metal parts, this test cell is tailored for CFRP materials with anisotropic properties. These predictions will support the ultimate goal of developing an accurate deformation model of the panel.

References

1. Brieskorn L, Vargas M (2021) Gap measurement for thermoplastic adhesive joining of aircraft frames. J Adhes 98(6):647–655. ISSN: 1545-5823. https://doi.org/10.1080/00218464.2021.2009806
2. Claus F, Hagen H, Hamann B (2021) Calculating the gravity- free shape of sheet metal parts. Int J Adv Manuf Technol 113(11):3401–3417. ISSN: 1433-3015. https://doi.org/10.1007/s00170-021-06702-6
3. Franciosa P et al (2019) A digital twin approach for smart assembly of aircraft skin panels with mechanical fasteners
4. Karniadakis GE et al (2021) Physics-informed machine learning. Nat Rev Phys 3(6):422–440. ISSN: 2522-5820. https://doi.org/10.1038/s42254-021-00314-5
5. Macenski S et al (2022) Robot operating system 2: design, architecture, and uses in the wild. Sci Robot 7(66). ISSN: 2470-9476. https://doi.org/10.1126/scirobotics.abm6074
6. Manohar K et al (2018) Predicting shim gaps in aircraft assembly with machine learning and sparse sensing. J Manuf Syst 48:87–95. ISSN: 0278-6125. https://doi.org/10.1016/j.jmsy.2018.01.011
7. Mitchell SN et al (2022) FAIR data pipeline: provenance-driven data management for traceable scientific workflows. Philos Trans R Soc A Math Phys Eng Sci 380(2233). ISSN: 1471-2962. https://doi.org/10.1098/rsta.2021.0300
8. Radvar-Esfahlan H, Tahan S-A (2012) Nonrigid geometric metrology using generalized numerical inspection fixtures. Precis Eng 36(1):1–9. ISSN: 0141-6359. https://doi.org/10.1016/j.precisioneng.2011.07.002
9. Schmick F, Luders NO, Wollnack J (2016) Automated assembly of large CFRP structures: adaptive filling of joining gaps with additive manufacturing. In: 2016 IEEE international symposium on assembly and manufacturing (ISAM). IEEE, pp 126–132. https://doi.org/10.1109/isam.2016.7750725
10. Schmitt RH et al (2016) Advances in large-scale metrology–Review and future trends. CIRP Ann 65(2):643–665. ISSN: 0007-8506. https://doi.org/10.1016/j.cirp.2016.05.002
11. Schmitt RH et al (2024) Leveraging measurement data quality by adoption of the FAIR guiding principles. tm-Tech Mess 91(9):441–453. ISSN: 2196-7113. https://doi.org/10.1515/teme-2024-0040
12. Thiébaut F et al (2016) Evaluation of the shape deviation of non rigid parts from optical measurements. Int J Adv Manuf Technol 88(5):1937–1944. ISSN: 1433-3015. https://doi.org/10.1007/s00170-016-8899-3

Layout Optimization of Robotized Production Systems Using a Mass-Spring-Damper-Model in a Brownfield Example

Ramez Awad and Katharina Barbu

Abstract A well-designed layout is essential for efficient production, which significantly contributes to a manufacturing company's economic success. In the era of mass customization, production planners must adapt resources in the layout to accommodate various product variants. However, they face constraints in time and resources, limiting their ability to explore the complete solution space for an optimal configuration. Meta-heuristics like genetic algorithms and simulated annealing frequently produce suboptimal solutions and are susceptible to parameter tuning. This paper presents a novel layout optimization approach that converts the layout problem into a mass-spring-damper system, where the equilibrium state represents the ideal layout for resource arrangement. It applies this approach to a brownfield layout problem involving 20 resources, utilizing robots as the handling mechanism. It analyzes the results, assesses the performance of the method, and discusses its limitations. Finally, it outlines potential avenues for future research to address these limitations.

Keywords Layout optimization · Brownfield · Robotized production systems

1 Motivation and Problem Statement

The manufacturing sector accounted for about a fifth of the German Gross Domestic Product (GDP) and a sixth of all employed persons in 2023 [1]. Its significance for the German economy is considerable. Within the manufacturing context, a factory layout pertains to the planned arrangement of physical facilities to improve workflow, reduce handling costs, and increase productivity [2]. This encompasses the placement of machinery, equipment, and workstations in relation to one another and the total available space. Many factories expand organically, placing new production lines and machinery in available spaces rather than optimal locations, resulting in "bad" layouts [3]. An ineffective layout can lead to increased defects, waste, and

R. Awad (✉) · K. Barbu
Fraunhofer IPA, Stuttgart, Germany
e-mail: ramez.awad@ipa.fraunhofer.de

M. Manns (ed.), *New Paradigms for Anticipated Uncertainty*, Lecture Notes in Mechanical Engineering, https://doi.org/10.1007/978-3-032-16889-4_48

higher costs, while a well-designed layout can boost efficiency and lower costs [4]. Voigt defines factory layout planning and optimization as "the optimized decision regarding the type, quantity, and spatial arrangement of production resources" (translated to English from [5]). The challenge is that the solution space for this decision problem is typically continuous and very large, making it difficult to solve [6]. As a result, production engineers often start with a known layout and can only explore "neighboring" solutions, representing just a small portion of the overall solution space [7].

2 State-of-the-Art

Production layouts are manually designed and optimized using 3D modeling and simulation tools. Larger companies, where reconfiguring existing production lines is a common task, establish design rules as "best practices" to follow. The layout optimization problem for automated production systems, which utilize stationary robots to transfer parts from resource A to resource B, is a specific aspect of the broader facility layout problem (FLP) [8]. This has been a research topic since the early 1950s [9]. Solution approaches to the FLP can be categorized into exact methods, heuristics, and meta-heuristics [10].

- **Exact Methods**, attempt to identify the global optimum. Examples of such methods are Cutting Plane Algorithms (CPA), Branch-and Bound Algorithms (B&B), and Dynamic Programming (DP) [11]. They are perfectly suited to solving small instances of the FLP; up to $n = 15$ resources. For larger problem instances, the computation time becomes prohibitive [10].
- **Heuristic** approaches can be categorized into construction methods, which iteratively allocate machines or departments to build a layout [12], and improvement methods, which enhance a given layout [13]. As complexity increases, these heuristics may perform poorly, resulting in suboptimal solutions. Additionally, the final layout often depends on the initial layout chosen, leading to inconsistent results.
- **Meta-heuristics** are "high-level problem-independent algorithmic [frameworks] that [provide] a set of guidelines or strategies to develop heuristic optimization algorithms" [14]. Examples are Genetic Algorithms (GAs), Simulated Annealing (SA), Tabu Search (TS), Ant Colony Optimization (ACO), and Particle Swarm Optimization (PSO) [15–19]. Their drawbacks include sensitivity to parameterization, limited scalability for large problems (although not as severe as exact methods), and the risk of getting trapped in local optima [20].

Hybrid approaches that combine the previously mentioned methods also face similar drawbacks [21]. This highlights the need for a solution approach that harnesses the strengths of existing methods while addressing their limitations.

3 The Mass-Spring-Damper Layout Optimization (MSDLO-) Method

The MSDLO method transforms the layout optimization problem into a mass-spring-damper system, which includes the following components:

- Bodies with Masses: These represent the resources to be organized or other elements within the layout.
- Springs: These connect the mass bodies, illustrating the attraction or repulsion relationships among the resources.
- Dampers: Incorporated in parallel with the springs, dampers serve to reduce oscillations over time.
- Additional Forces and Variables: Elements such as gravity and friction are included to model "financial" interdependencies.

The process begins with an initial placement of the mass bodies, which then oscillate until the system reaches equilibrium, i.e. the sum of forces on each collision body equals zero. The arrangement of the bodies/resources at this equilibrium state is considered the optimal configuration. The foundational theorems and transformation rules are detailed in [22], where the MSDLO method was applied to a greenfield layout problem involving five resources and compared with solutions generated by a genetic algorithm (GA). The findings indicate that while the MSDLO method excels at optimizing handling distances and costs, it falls short of the GA in terms of area efficiency and is significantly influenced by the initial resource placement. Thus, the MSDLO method is not the efficient and effective global optimizer the authors hypothesized it to be. However, its performance in the greenfield example justifies further exploration as a local optimizer. This paper aims to evaluate the performance of the MSDLO method in a larger instance of FLP under brownfield conditions. See Table 1 for key differences.

Table 1 Key differences between the research presented here and the research presented in [22]

Key differences	Reference [22]	This paper
Problem size	n = 5 resources (small)	n = 20 resources (large)
Layout shape	Regular shape (square)	Irregularly shaped layout
Start conditions	Greenfield, i.e. empty layout	Brownfield, i.e. 17 existing resources are already placed
KPIs	Handling cost Area savings	Handling cost Area savings Installation cost, i.e. cost of moving existing resource

4 Validation Example and Results

4.1 Problem Description

To evaluate the MSDLO method under "brownfield" conditions against the performance of a GA, a modified car subassembly layout with 17 resources is used that supports process sequences for variants A and B. The new production program for the upcoming 12 four-week periods (five workdays per week, three shifts per day, eight hours each) includes: 64,800 units of product variant A using process sequence A, 43,200 units of product variant B using process sequence B, and 64,800 units of product variant C using process sequence C. Table 2 merges the list of handling processes across all three process sequences, indicating the number of process iterations for the upcoming 12 × four-week production schedule.

Product variant C differs from variant A in several aspects. A different part is inserted in Fixture 1 at the start of the assembly sequence, which is too wide for Conveyor Elevators 1 and 2, necessitating the addition of Conveyor Elevator 5, matching the dimensions and specifications of Conveyor Elevator 4. Additionally, an extra part must be machined and inserted in Fixture 3 for welding. Due to technical and time constraints, two more resources are required: Material Rack 2 and CNC Machine 2, which should be accessible to Handling Robot 4.

Table 2 List of handling processes across all three process sequence variants and iterations of each process in the new production program

#	Variant	Source resource	Handling resource	Destination resource	# Iterations
01	A/B/C	Conveyor Elevator 1	Handling Robot 1	Fixture 1	172,800
02	A	Conveyor Elevator 2	Handling Robot 1	Fixture 1	64,800
03	B	Conveyor Elevator 3	Handling Robot 1	Fixture 1	43,200
04	C	Conveyor Elevator 5	Handling Robot 1	Fixture 1	64,800
05	A/B/C	Welding Robot 1	Welding Robot 1	Fixture 1	172,800
06	A/B/C	Fixture 1	Handling Robot 2	Fixture 2	172,800
07	B	Material Rack 1	Handling Robot 5	CNC Machine 1	43,200
08	B	CNC Machine 1	Handling Robot 5	Fixture 2	43,200
09	A/B/C	Welding Robot 2	Welding Robot 2	Fixture 2	172,800
10	A/B/C	Fixture 2	Handling Robot 3	Fixture 3	172,800
11	C	Material Rack 2	Handling Robot 4	CNC Machine 2	64,800
12	C	CNC Machine 2	Handling Robot 4	Fixture 3	64,800
13	A/B/C	Welding Robot 3	Welding Robot 3	Fixture 3	172,800
14	A/B/C	Fixture 3	Handling Robot 4	Conveyor Elevator 4	172,800

4.2 Key Performance Indicators (KPIs) for Evaluating the Layouts

Handling Cost (C_H) is calculated by multiplying the distance (d_{ij}) of the route between the source resource i and destination resource j, with the number of iterations (itr_{ijv}) in a cycle and the number of cycles in a planning period (vol_v) and a cost factor related to distance (c). The depreciation cost is calculated at 0.70 € per km. The robot's average energy consumption is assumed to be 2.0 kWh, leading to an energy cost of 0.61 €/km at current energy prices of 0.55 €/kWh and an average speed of 0.5 m/s. Consequently, the total resulting cost factor is – 1.31 €/km.

$$C_H = \sum_v \sum_i \sum_j vol_v * itr_{ijv} * d_{ij} * c \qquad (1)$$

Area Savings (A_S) equal the usable areas (A_a) multiplied by themselves and then again by an area savings factor (c_A). Larger areas are rewarded since they are more flexible in their usage. In 2023, the average rent for industrial areas in greater Stuttgart was 8 €/m^2 per month. A machine and its peripherals require an average space of 12 m^2 (3 m × 4 m). To achieve monthly savings of approximately 96 € for this area, the savings factor is set at 0.67 €/m^4.

$$A_s = \sum_a A_a * A_a * c_A \qquad (2)$$

Installation Cost (C_I): The Installation Cost for existing resources is outlined in Table 3. The Installation Cost for additional resources is assumed to be 0 € since it will not differ much from one layout solution to the other. When moving a non-handling resource along with the associated robot, it is assumed that the effort required to adjust the trajectories will double, which allows for the addition of fixed costs.

4.3 Solutions Obtained by Genetic Algorithm

A GA that discretizes the solution space into 50 mm steps for the x and y directions and 5° rotations around the z-axis is used. Due to the low percentage of valid layout solutions, GA required seeding with 10 valid layouts to generate any valid solutions. A total of 31 experiments were conducted to capture the stochastic nature of the GA. Table 4 summarizes the statistical evaluation of the experiments. The average computation time is 5875 s (01:37:55). The best layout solution is depicted in Fig. 1.

Table 3 Break-down of Installation Cost per existing resource in the brownfield layout

Resource	Fixed Cost	Variable Cost per m distance traveled & ° rotated
Narrow Material Rack 1	60 € + 120 € = 180 €	1.00 €/m and 1.00 €/° Assuming an hourly rate of 60 €/h, a moving speed of 1 m/min, and a rotation speed of 1°/min, respectively
CNC Machine 1	2080 € + 1 × 120 € = 2,200 €	
Fixture 1 & 3	60 € + 3 × 120 € = 420 €	
Fixture 2	60 € + 4 × 120 € = 540 €	
Handling Robot 1	120 € + 4 × 120 € = 600 €	
Handling Robot 2–4	120 € + 2 × 120 € = 360 €	
Handling Robot 5	120 € + 3 × 120 € = 480 €	
Welding Robot 1–3	120 € + 1 × 120 € = 240 €	
Conveyor Elevator 1–4	250,000 €	10,000 €/m

Table 4 Statistical evaluation of the high-resolution GA (n = 31) for the brownfield example

Variable	Handling cost [€]	Installation cost [€]	Area savings [€]	Aggregated KPI [€]
Mean	– 10,028.57	– 1846.59	+ 144,508.56	+ 132,615.41
Median	– 9907.21	– 1837.60	+ 141,117.23	+ 129,383.70
Standard deviation	247.06	554.65	10,554.45	10,178.71
Minimum	– 10.601.10	– 3208.80	+ 128,738.18	+ 117,041.32
Maximum	– 9674.41	– 1142.60	+ 170,601.19	+ 158,148.13
95% Confidence interval for mean	– 10,119.18 to – 9937.96	– 2068.01 to – 1661.17	+ 140,637.68 to + 148,379.45	+ 128,882.32 to + 136,348.49

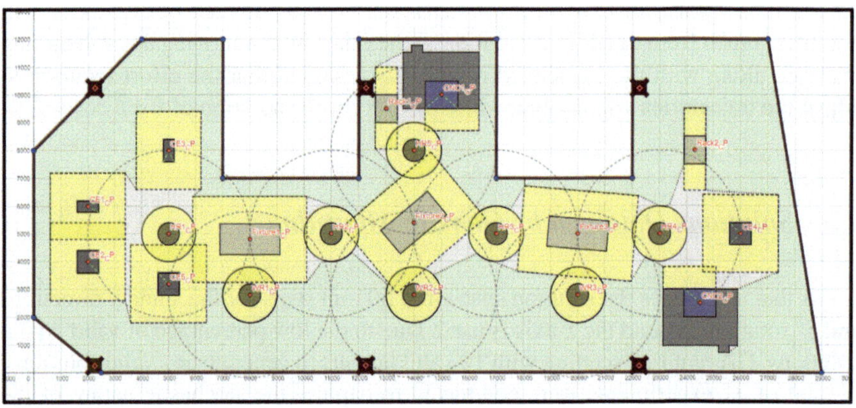

Fig. 1 2D specification of GA's brownfield layout solution for the high-resolution grid

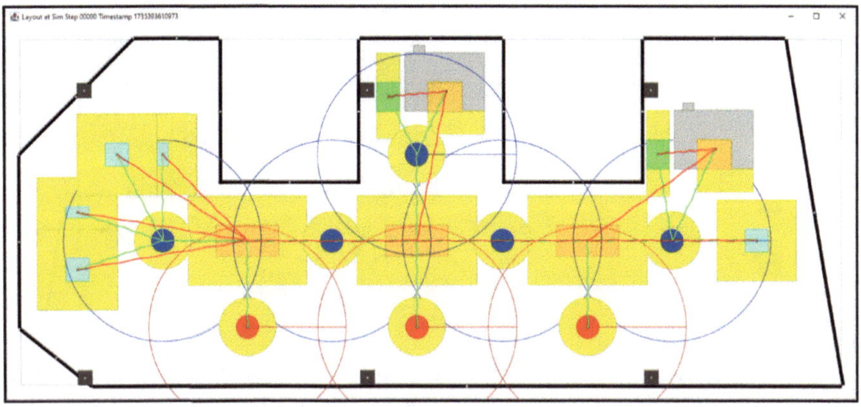

Fig. 2 Screenshot of brownfield layout at the end of step 03 of MSDLO method, where "reachability" SDLs are rendered green, and "closeness" SDLs are rendered red

4.4 Solution Obtained by MSDLO

Step 1: The layout's boundaries, obstacles, and existing resources are represented as collision body groups. The static friction coefficients were calculated by equating the static friction force with the fixed installation costs of the resource. Similarly, the "work" required to move a resource is equated with the variable installation costs.

Step 2: New resources are also modeled as collision body groups and manually positioned in the layout.

Step 3: Relationships between resources are analyzed to create spring-damper-links (SDLs), resulting in 19 direct relationships (reachability SDLs between handling and non-handling resource, i.e. non-handling resource must be within the reach of the associated handling resource) and 11 indirect relationships (closeness SDL between two non-handling serviced after each other by the same handling resource in order to minimize distances for "pick&place" processes), leading to the initial layout in Fig. 2.

Step 4: The system is allowed to oscillate until it stabilizes after about 34 s.

Step 5: The final resource placements are extracted (see Fig. 3), and KPIs are computed (see Table 5).

5 Discussion of Results

The **GA** needed ten valid layout solutions for seeding, which took about three hours to engineer (– 450.00 €), resulting in a total cost of – 450.65 €. GA solutions required moving three to eight resources, leading to higher Installation Costs (– 1846.59 € vs – 0.00 €), slightly lower Handling Costs (– 10,028.57 € vs – 11,122.28 €), and

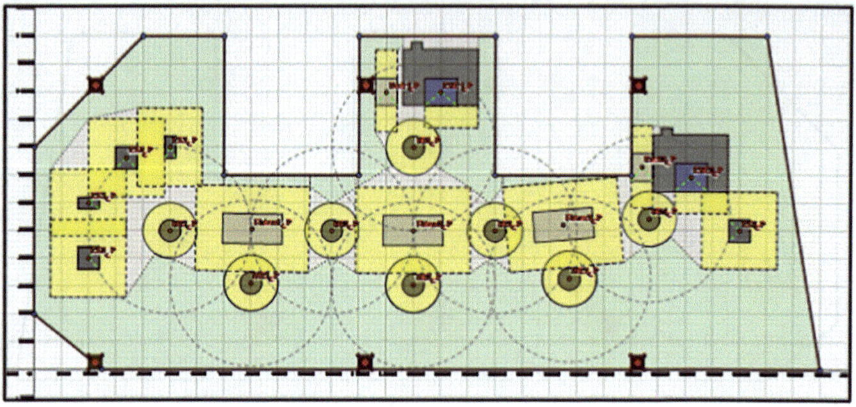

Fig. 3 2D specification of MSDLO's brownfield layout solution for high-resolution grid

Table 5 KPIs of MSDLO brownfield layout solutions for planning horizons 12 months

Variable	Handling cost [€]	Installation cost [€]	Area savings [€]	Aggregated KPI [€]
Value	– 9521.76	– 1936.07	+ 146,579.52	+ 135,121.69

significantly better Area Savings (+ 144,508.56 € vs + 93,910.31 €) compared to the best manual layout.

The **MSDLO** method took about 30 min (– 50.00 €) to establish the initial layout for placing three new resources and computed in 34 s with a cost of – 0.00 €. It achieved the lowestHandling Cost (– 9521.76 €) but had a slightly higher Installation Cost than the GA's average (– 1936.07 € vs – 1846.59 €). However, its Area Savings were slightly better than the GA's average (+ 146,579.52 € vs + 144,508.56 €).

The MSDLO method's computation time was significantly shorter than GA (34 s vs 1 h 37 min 55 s) because it evaluated only one solution, while the GA assessed an average of 232,063 layout solutions. GA's computation time rose from 341 s (greenfield, n = 5 resources [22]) to 5,875 s (brownfield, n = 20 resources), demonstrating exponential growth in requirements. In contrast, the MSDLO method's time increased from 27 to 34 s (+ 25.93%).

6 Conclusion and Outlook

Can the layout optimization problem be solved "better" in terms of effort and quality of the results using the MSDLO method? The short answer is **"It depends!"**.

The MSDLO method performs better with "anchor" resources, making it ideal for brownfield applications. It directly influences handling distances through its spring-damper links, resulting in a significant positive impact on minimizing Handling Cost (C_H). Additionally, by shortening distances, the MSDLO method indirectly reduces the total area occupied by the resources, leading to a substantial positive effect on maximizing Area Savings (S_A). Furthermore, the method computes relatively quickly. However, as demonstrated in [22], the MSDLO method has certain limitations. Its results depend on the initial placement of resources, and if no "anchor" resources are provided, the outcomes become "undirected." Additionally, the results can deteriorate in negative edge cases. Hence, the authors want to evaluate in future work combining GA and MSDLO, i.e. using GA as a coarse global optimizer followed by the MSDLO for local fine-tuning of the layout. Increasing the GA's population size and the number of generations for the GA to yield valid layouts without required seeding (thus saving engineering effort) will also be considered.

Acknowledgements The authors thank Dr. Kai Pfeifer for his valuable discussions and support. This work would not have been possible without Prof. T. Bauernhansl, the institute director. The authors used ChatGPT 4.0 to improve the clarity and conciseness of the paper, and after reviewing and editing the content, they take full responsibility for the publication's content.

References

1. DESTATIS (2025) Erwerbstätige und Arbeitnehmer nach Wirtschaftsbereichen (Inlandskonzept) 1000 Personen [Table]
2. Ramírez A (2024) The key steps for factory layout design
3. Collins III D (19.04.223) Optimizing factory layout for efficiency and cost savings
4. Okpala CC, Chukwumuanya O (2016) Plant layouts' analysis and design. Int J Adv Eng Technol 201–206
5. Voigt K-I (2018) Layoutplanung: definition: was ist "Layoutplanung"?
6. Zhou E, Murray AT (2024) Facility layout and spatial configuration efficiency assessment. Netw Spat Econ
7. Fechter M, Seeber C, Chen S (2018) Integrated process planning and resource allocation for collaborative robot workplace design. Procedia CIRP 72:39–44
8. Kusiak A, Heragu SS (1987) The facility layout problem. Eur J Oper Res 29:229–251
9. Bellman R (1952) On the theory of dynamic programming. Proc Natl Acad Sci 38:716–719
10. Drira A, Pierreval H, Hajri-Gabouj S (2007) Facility layout problems: a survey. Annu Rev Control 31:255–267
11. Eaves BC, Zangwill WI (1971) Generalized cutting plane algorithms. SIAM J Control 9:529–542
12. Seehof JM, Evans WO, Friederichs JW et al. (1966) Automated facilities layout programs. In: Weinberg DF, Mancina WP (eds) Proceedings of the 1966 21st National conference on -. ACM Press, New York, New York, USA, pp 191–199
13. Prasad NH, Rajyalakshmi G, Reddy AS (2014) A typical manufacturing plant layout design using CRAFT algorithm. Procedia Engineering 97:1808–1814
14. Sörensen K, Glover F (2013) Metaheuristics. Encycl Oper Res Manag Sci 62:960–970
15. Gen M, Lin L (2023) Genetic algorithms and their applications. In: Pham H (ed) Springer handbook of engineering statistics. Springer London, London, pp 635–674

16. Matai R (2015) Solving multi objective facility layout problem by modified simulated annealing. Appl Math Comput 261:302–311
17. Prajapati VK, Jain M, Chouhan L (2020) Tabu search algorithm (TSA): a comprehensive survey. In: 2020 3rd International conference on emerging technologies in computer engineering: machine learning and Internet of Things (ICETCE). IEEE, pp 1–8
18. Liu J, Liu J (2019) Applying multi-objective ant colony optimization algorithm for solving the unequal area facility layout problems. Appl Soft Comput 74:167–189
19. Gad AG (2022) Particle swarm optimization algorithm and its applications: a systematic review. Arch Computat Methods Eng 29:2531–2561
20. Adrian AM, Utamima A, Wang K-J (2015) A comparative study of GA, PSO and ACO for solving construction site layout optimization. KSCE J Civ Eng 19:520–527
21. Moslemipour G, Lee TS, Loong YT (2018) Solving stochastic dynamic facility layout problems using proposed hybrid AC-CS-SA meta-heuristic algorithm. IJISE 28:1
22. Awad R, Guizani E (2025) Layout optimization of robotized production systems using a mass-spring-damper model. In: 58th conference on manufacturing systems

Cost Allocation in Modular Product Development: Reviewing Existing Practices and Identifying Opportunities

Morten Nørgaard, Jakob Meinertz Grønvald, and Niels Henrik Mortensen

Abstract Modular product structures are a proven design method for offering a wide variety of customized products to the market while maintaining a low internal product and process complexity and creating several cost benefits for product manufacturers throughout their value chain. However, the economic benefits of modularity-induced effects have proven challenging to quantify using traditional costing methods. This leaves modular product structures at a disadvantage during the product development phase. Since it comes with a high upfront development cost, the economic benefits gained later are challenging to assess accurately. Effecting the overall product design process. This study aims to provide an overview of the cost-allocating methods currently used to evaluate the economic benefits of modularization. It will examine the existing cost accounting methods for the product life cycle's development phase. It will contribute to a broader understanding of current methods' capability and enlighten potential issues and opportunities in existing practices. Finally, opportunities for improving existing resource allocation methods, which can potentially refine cost allocation systems, are discussed. This would improve the accuracy and detailing of cost when evaluating product portfolios with highly complex, customized, or modularized products and function as decision support for design engineers when evaluating existing products in the design process.

Keywords Product family design · Modularity · Life cycle analysis and design · Product/process development · Product platform design · Decision support

1 Introduction

Production technologies constantly need development to adapt to the ever-evolving market demands for new and improved products. The available technologies are limited to the choice of design made by product developers, who are challenged by adhering to industry requirements and expectations while simultaneously trying to

M. Nørgaard (✉) · J. M. Grønvald · N. H. Mortensen
Technical University of Denmark, Lyngby, Denmark
e-mail: mornor@dtu.dk

M. Manns (ed.), *New Paradigms for Anticipated Uncertainty*, Lecture Notes in Mechanical Engineering, https://doi.org/10.1007/978-3-032-16889-4_49

keep down product and development costs [1, 2]. The complexity of products and processes contributes to the growing difficulty of making optimal design decisions [3, 4].

Modular product architecture is a design principle that provides a structured framework for designers to create customizable products while maintaining a low process complexity [4–7]. This provides economic benefits for the product manufacturers throughout the product life cycle. However, quantifying the economic benefits of modularity-induced effects using traditional costing methods, such as Traditional Cost Management (TCM) and Activity-Based Costing (ABC), has proven to be challenging, either resulting in low accuracy or resource-heavy solutions [8–10]. A modular product strategy entails high upfront development costs, making it less favorable, especially when future economic benefits across the product life cycle are difficult to quantify [11]. This might weaken the long-term decision-making in product development. Pushing the product design process away from modular product architectures and adding complexity to the product and process [5, 12, 13].

In this study, we aim to review existing practices by creating an overview of the cost-allocating methods currently used to evaluate the economic benefits of modularization in the development phase. The study revealed that several quantification methods are used in combination to assess the effects. They are measured through four economic targets: time, costs, quality, and flexibility.

Through this study, we can identify potential shortcomings of existing methods and how they might be deficient for an accurate evaluation to support the decision-making of design engineers. Thus, it facilitates a discussion of potential areas of improvement that can be studied in future research.

2 Theoretical Background

This section reviews the literature on Management Accounting and Modular Product Design to explore costing methods for assessing product cost performance and evaluating the economic benefits of modularity.

2.1 Management Accounting

In the 1980s and 1990s, several cost management approaches surged from the literature, which is still the foundation of methods used today. The work of Atkinson et al. [14], Drury [15], and Horngren et al. [16] introduced several cost and management accounting methods to measure the cost of products, services, and customers [14]. Table 1 lists some of the most widely known and commonly used Cost Management Approaches (CMA).

Each CMA suits a purpose dependent on the company's product strategy. For example, Target Costing is a systematic approach to developing products with a

Table 1 Cost management approaches in product development following [17, 18]

Product-oriented CMA	Monetary CMA for internal use
Target costing	Life cycle costing (LCC)
Value engineering	Traditional cost management (TCM)
Quality function deployment (QFD)	Activity-based costing (ABC)
Design for X (DFX)	Time-driven ABC (TDABC)
Kaizen costing	
Total cost of ownership (TCO)	
Strategic costing	

predetermined cost and margin. Cost constraints, therefore, define the functionality and components of new products [19]. Kaizen Costing, which is based on lean principles, is the ongoing pursuit of cost reductions after introducing the product to the market [14]. Cost management approaches are product-oriented and guide how costs are analyzed, allocated, and minimized. Monetary CMA for internal use is implemented in cost management systems to sustain cost management approaches and track, analyze, and manage costs efficiently. They work on a resource or monetary basis for internal evaluation. The most widely used CMAs in costing systems are Traditional Cost Management and Activity-Based Costing, which deliver valuable information about the use of resources related to products along the value chain [8, 16, 20]. This information should arguably be one of the key aspects of making product design decisions [21].

However, as Mertens [8] discusses, these costing systems might not consistently provide the most accurate estimates of cost allocation. This is presumed to lead to errors in decision-making, especially when applying simple TCM systems, which cannot encompass the complexity of the non-unit-based cost measures [8]. ABC promises more accurate and detailed cost information but requires extensive use of resources to implement and maintain [22]. Therefore, ABC's usage in the industry is limited due to its resource-heavy nature and limited generalizability when implementing tools and methods [8, 11, 23].

2.2 *Modular Product Design*

Different researchers in the literature have defined the concept of a modular product structure in various ways. The approaches for modularizing products are thus numerous, but they share a common purpose of reaping the potential benefits of having a modular product architecture. These benefits include reduced part numbers, increased component commonality, fewer tools and production process investments, and increased batch sizes [23–25].

Hackl et al. [25] conducted a study to gather the effects of having a modular product structure found in the literature. The work was collected in the Modularity

Impact Model (IMF), which spans the product life cycle phases. Based on this study, Greve et al. [9] embraced the opportunity to create measures for evaluating each of the effects caused by modularity. These measures were used to build a tool for quantifying the impact of modularity throughout the product life cycle on a product level. The work on the extended IMF is the latest addition within this strand of research to attempt quantification of the modular effects to improve decision support in design [9]. The measures are within the categories of cost, time, flexibility, and quality. They are not all possible to assess with a cost measure.

However, this work still faces a challenge similar to what previous research has pointed out as one of the key issues for implementing modular product architectures: Integrated financial planning. The existing costing systems do not encompass a modular build since they were initially developed for mass-produced products [5, 26, 27]. This leaves an area open for further investigation. In the following sections, we will examine what methods are currently being applied for this purpose and the potential challenges regarding evaluating cost performance in a decision-support setting.

3 Research Method

In this study, we would like to review the current costing methods used to evaluate the effects of modularity based on a literature study. This would provide an opportunity to discuss the limitations of the existing cost systems when making design decisions for a modular product architecture. The literature study takes its offset in the following published materials within the area of cost management accounting, effects of product variety, and modular product architectures [8, 9, 17, 18, 20, 21, 23, 25]. These sources provided a thorough entry into assessing the economic benefits of modularization after employing existing costing methods. The references and citations of these sources led to additional material, which has been included in the study, as well as related material from textbooks and attended conferences. This study cannot be deemed an exhaustive review but comprises some of the most recent materials published within this area of research. Thus, it allows us to discuss future aspects based on our findings.

The analysis focuses on the actual quantification methods used, not limited to the previously introduced cost management approaches. After reviewing the literature, a list of the effects of modularity in the development phase was made. A categorization of the quantification methods used in assessing these effects was conducted. This revealed that TCM and ABC played a significant role in the existing approaches to quantify the cost performance of the modularity effects. These terms, therefore, appear in their own column in the result table. The quantification of the remaining effects seemed to be scattered across a multitude of methods, which can be found under 'Methods' in Table 2.

Thus providing a clear overview of cost methods used in the literature to assess the impact of modularity. The results are elaborated on in the next section.

Table 2 Results from the categorization of costing methods used for evaluating the impact of modularity

Modularity effect	TCM	ABC	Methods	Reference
Improve outsourcing of development task	–	x	TDABC	[9]
Improve parallel development/prototyping	–	x	ABC: Critical path following modularization	[28]
Improve adaptability upgradeability	–	–	No quantification method was found	[6]
Improve reliability of product (Component reuse)	–	–	Percentage: Share of already tested components	[9]
Increase the probability of spreading error	–	–	Change Propagation Analysis	[29]
Decrease in failure rate	–	x	TDABC	[9]
Improve decoupling of risk through module sharing	–	–	No quantification method was found	[6]
Improve/reduce product innovation	–	–	Product Innovation Index Using Linkograph Analysis	[30]
Reduce freedom of design/ product differentiation	–	–	Benefit cost analysis	[31]
Improve variant derivation	–	x	TDABC	[9]
Increase initial development investment	–	–	Sum(Initial development investment)	[9]
Reduce development cost per unit	–	x	TDABC	[9]
Reduce time for part search	–	x	TDABC	[9]
Reduce documentation and coordination effort	–	x	TDABC	[9]
Improve part administration	–	x	TDABC	[9]
Simpler implementation of new processes	–	–	No quantification method was found	[32]
Increase in volume	x	–	TCM	[33]
Increase in weight	x	–	TCM	[33]
Increase in no. of unused features	–	–	Percentage: Number of unused features	[9]

4 Results

Reviewing the literature on cost management revealed several costing methods used for evaluating product portfolios from a cost-performance perspective. They are designed to provide insight into product cost performance. However, studies in Design Research highlight challenges in quantifying modularity effects using existing costing systems. The lack of accuracy might affect the design decision hindering the most optimal design solutions [8, 11, 26].

The different modularity effects are categorized as being quantifiable using TCM, ABC, or other methods. Table 2 presents the modularity effects and categorization of quantitative methods found in the literature.

The categorization of methods used to evaluate the effects of modularity in the development phase provides an overview for further discussion. The results suggest that ABC plays a major role in assessing the impact of modularization. TCM takes on a more minor role in the development phase but will likely have a more significant effect in the later stages of the product life cycle [34]. Several benefits are challenging to convert into quantifiable and comparable measures. These methods are not further categorized. For a few effects, no quantitative evaluation method was found. An important remark in this study is that the modularity effects are not weighted and are considered equally important. To better understand the results, a few examples from the table are described below:

The first effect, "Improve outsourcing of development task," is currently assessed using Time-Driven ABC. The necessary data and a specific cost are computed based on the effects of modularization. Assessing the impact, "Improve reliability of product (component reuse)," will provide a percentage of reused components. However, without additional information, this result does not offer any comparable value with the remaining effects. It is a value out of context when it comes to thoroughly evaluating the products. Contrary to the effect, "Increase weight/volume" will provide a specific, traceable cost measure that can easily be compared between product.

The argument is that the costing methods presented only provide limited insight into the allocation of cost for modular products and services. Since the allocation is either too simple or resource-heavy and product—or project-dependent, the reviewed literature does not provide any proven holistic or generalizable methods for quantifying the cost performance of modular product portfolios. Instead, it relies on a combination of different costing methods developed without considering modular product structures and their indirect effects only on a product level.

5 Discussion and Outlook

The results revealed several findings. Quantifying modularity effects in the development phase heavily relies on ABC, specifically time-driven ABC. A multitude of quantification methods are used in combination to evaluate the effects of modularity. Only a few of them are directly related to a cost measure.

How does this affect the decision-making process of design engineers regarding modular product architectures in the product development phase?

It might weaken the decision-making basis for design engineers in choosing the most optimal design form. The way existing costing systems are structured does not currently provide accurate practices for evaluating the economic effects of modularity. Thus, modularization as a design principle is at a disadvantage compared to non-modular or standardized product development following our current cost management practices [8, 11, 23].

The current quantification methods for evaluating the modularity effects are spread across different types of measurements or economic targets, complicating the comparison of design choices. It is challenging to consider all impacts if they are evaluated on different measures, possibly resulting in sub-optimal design choices. A suggestion would be to evaluate all effects following the same measurement to make product designs comparable. E.g., converting all effects into a cost-measure. This would make it possible to more accurately compare the induced cost of a product through each life cycle phase. Improving the evaluation of product portfolios with complex, customized, and modularized products. Ultimately, strengthen the decision support for design engineers in the development phase.

Future work following this study would be to categorize the usage of costing methods throughout the remaining life cycle phases, such as sales, procurement, production, and service. Providing a complete overview of all costing methods used for assessing the effects of modularity across the product life cycle and laying the foundation for developing a costing framework. To provide a standard measure across all modularity effects related to direct and indirect costs, strengthening the decision support for design engineers when developing new products.

References

1. Kwapisz J, Infante V, Cameron BG (2019) Commonality opportunity search in industrial product portfolios Virginia Infante. Int J Technol Manag 81:258–273
2. ElMaraghy H, Schuh G, Elmaraghy W et al (2013) Product variety management. CIRP Ann Manuf Technol 62(2):629–652. https://doi.org/10.1016/j.cirp.2013.05.007
3. Windheim M, Hackl J, Gebhardt N, Krause D (2016) Assessing impacts of modular product architectures on the firm: a case study. In: International design conference. Dubrovnik-Croatia
4. Windheim M, Gebhardt N, Krause D (2018) Towards a decision-making framework for multi-criteria product modularization in cooperative environments. Procedia CIRP 70:380–385. https://doi.org/10.1016/J.PROCIR.2018.02.027
5. Jiao J, Simpson TW, Siddique Z (2007) Product family design and platform-based product development: a state-of-the-art review. J Intell Manuf 18(1):5–29. https://doi.org/10.1007/s10845-007-0003-2
6. Gershenson JK, Prasad GJ, Zhang Y (2003) Product modularity: definitions and benefits. J Eng Des 14(3):295–313
7. Bonvoisin J, Halstenberg F, Buchert T, Stark R (2016) A systematic literature review on modular product design. J Eng Des 27(7):488–514. https://doi.org/10.1080/09544828.2016.1166482
8. Mertens KG (2020) Measure and manage your product costs right-development and use of an extended axiomatic design for cost modeling
9. Greve E, Fuchs C, Hamraz B et al (2022) Knowledge-based decision support for concept evaluation using the extended impact model of modular product families. Appl Sci (Switzerland) 12(2). https://doi.org/10.3390/app12020547

10. Thyssen J, Israelsen P, Jørgensen B (2006) Activity-based costing as a method for assessing the economics of modularization-a case study and beyond. Int J Prod Econ 103(1):252–270. https://doi.org/10.1016/j.ijpe.2005.07.004
11. Israelsen P, Jørgensen B (2011) Decentralizing decision making in modularization strategies: overcoming barriers from dysfunctional accounting systems. Int J Prod Econ 131(2):453–462. https://doi.org/10.1016/J.IJPE.2010.12.020
12. Bertram CA, Mueller GO, Lokkegaard M, Mortensen NH, Hvam L (2020) Why engineer-to-order portfolio rationalization stalls: challenges in standardization, modularization, platform design and mass customization. In: Proceedings of the design society: DESIGN conference. Cambridge University Press, vol 1, pp 2265–2273
13. Hvam L, Hansen CL, Forza C, Mortensen NH, Haug A (2020) The reduction of product and process complexity based on the quantification of product complexity costs. Int J Prod Res 58(2):350–366. https://doi.org/10.1080/00207543.2019.1587188
14. Atkinson AA, Banker RD, Kaplan RS, Young SM (2001) Management accounting, 3rd edn. Prentice Hall, Upper Saddle River, NJ
15. Drury C (2015) Cost and management accounting—an introduction. Thomson Learning
16. Horngren CT, Datar SM, Rajan MV (2009) Cost accounting a managerial emphasis, 13 edn. Pearson International Edition
17. Wouters M, Morales S (2014) The contemporary art of cost management methods during product development. Advances in Management Accounting 24:259–346. https://doi.org/10.1108/S1474-787120140000024008
18. Nielsen S, Melander P, Jakobsen M (2001) Moderne Økonomistyringsværktøjer i Danmark År 2001
19. Navissi F, Sridharan VG (2017) Determinants of target costing adoption: a research note. J Manag Account Res 29(1):67–77. https://doi.org/10.2308/jmar-51501
20. Siguenza-Guzman L, Van Den Abbeele A, Vandewalle J, Verhaaren H, Cattrysse D (2013) Recent evolutions in costing systems: a literature review of time-driven activity-based costing. Review of business and economic literature. https://lirias.kuleuven.be/retrieve/275810. Accessed 6 Oct 2022
21. Park J, Simpson TW (2008) Toward an activity-based costing system for product families and product platforms in the early stages of development. Int J Prod Res 46(1):99–130. https://doi.org/10.1080/00207540600825240
22. Kaplan RS, Cooper R (1998) Cost & effect: using integrated cost systems to drive profitability and performance. Harvard Business Press
23. Ripperda S, Krause D (2017) Cost effects of modular product family structures: methods and quantification of impacts to support decision making. J Mech Des Trans ASME 139(2). https://doi.org/10.1115/1.4035430
24. Nørgaard M, Grønvald JM, Christensen CKF, Mortensen NH (2024) Challenges in product variant costing—a case study. In: Proceedings of the design society. Cambridge University Press, vol 4, pp 3003–3012
25. Hackl J, Krause D, Otto K et al (2020) Impact of modularity decisions on a firm's economic objectives. J Mech Des Trans ASME 142(4). https://doi.org/10.1115/1.4044914
26. Fixson SK (2005) Product architecture assessment: a tool to link product, process, and supply chain design decisions. J Oper Manag 23(3–4):345–369. https://doi.org/10.1016/j.jom.2004.08.006
27. Simpson TW, Marion T, De Weck O, Höltä-Otto K, Kokkolaras M, Shooter SB (2006) Platform-based design and development: current trends and needs in industry. In: ASME 2006 International design engineering technical conferences & computers and information in engineering conference
28. Levy FK, Thompson GL, Wiest JD (1963) ABCs of the critical path method. Harv Bus Rev 41:98–109
29. Eckert C, Clarkson PJ, Zanker W (2004) Change and customisation in complex engineering domains. Res Eng Des 15:1–21

30. Lingannavar R, Yammiyavar P (2019) Product innovation index using linkograph analysis. In: Chakrabarti A (ed) Research into design for a connected world. Smart innovation, systems and technologies. Springer, Singapore, pp 297–306
31. Zangeneh A, Jadid S, Rahimi-Kian A (2008) Normal boundary intersection and benefit–cost ratio for distributed generation planning. Int. Trans. Electr. Energy Syst 20:97–113
32. Greve E, Krause D (2019) Long-term effects of modular product architectures: an empirical follow-up study. Procedia CIRP 84:731–736
33. VDI 2225 Blatt 1: 1997-11 (1997) Design engineering methodics-engineering design at optimum cost-simplified calculation of costs. Beuth Verlag, Berlin, Germany
34. Fixson SK (2006) A road map for product architecture costing. In: Simpson TW, Siddique Z, Jiao J (eds) Product platform and product family design: methods and applications. New York, Springer, pp 305–334

Generating Assembly Instructions by Demonstration—System Implementation and Evaluation

Dennis Keiser, Christoph Petzoldt, Dario Niermann, Burak Vur, and Michael Freitag

Abstract Assembly assistance systems are widely used in industrial assembly to help manage complexity due to various product variants and to avoid quality issues. The effectiveness of these systems heavily depends on clear and precise assembly instructions. However, these instructions are often still created manually, which can lead to errors, and the setup process can be very time-consuming. To improve the setup process for informational assistance systems, this article proposes a novel system for the automated generation of assembly instructions using the concept of teaching by demonstration. The foundation of the developed system lies in the application of state trees, which interpret the assembly process captured by cameras and convert it into an organized assembly sequence. The system also includes a human–machine interface that enables additional input options for further enrichment of the assembly instructions with context-based information. The assembly instructions created can be transferred via an interface to various assembly assistance systems for assembly process execution. The developed system was evaluated by a user study, which revealed both its potential for reducing the time needed to set up assembly instructions and needs for further research.

Keywords Assembly · Assistance systems · Assembly instructions · Teaching by demonstration · Human activity recognition

1 Motivation

Shorter product life cycles as well as the increasing customization and individualization of products lead to greater product variety and smaller lot sizes, resulting in a high complexity of manual assembly processes [1]. To address this complexity,

D. Keiser (✉) · C. Petzoldt · D. Niermann · B. Vur · M. Freitag
BIBA—Bremer Institut Für Produktion Und Logistik GmbH at the University of Bremen, Bremen, Germany
e-mail: ked@biba.uni-bremen.de

B. Vur · M. Freitag
Faculty of Production Engineering, University of Bremen, Bremen, Germany

M. Manns (ed.), *New Paradigms for Anticipated Uncertainty*, Lecture Notes in Mechanical Engineering, https://doi.org/10.1007/978-3-032-16889-4_50

assembly assistance systems are used, which provide workers in these demanding environments with informational guidance by delivering the right information at the right time and in the desired format [2]. Currently, setting up these assembly assistance systems and creating assembly instructions is largely a manual process. The providers of assembly assistance systems offer input masks for this purpose, in which textual information, photos and graphic visualizations are entered and the control of connected systems (e.g. cameras, connected torque drivers, pick-by-light modules) is configured. This does not only require trained specialist personnel but is also a very time-consuming process. Initial approaches are attempting to simplify this set-up process using function block-based configuration approaches or recommendation systems for process completion; however, more intuitive and efficient methods are still required, particularly for assembly stations with frequently changing assembly products and processes. Inspired by robotics research, programming by demonstration [3] is a promising novel approach to make the creation of manual assembly processes and instructions more efficient and highly automated. Instead of entering the basic assembly system instructions in input masks, the core idea is to let the process planner demonstrate the assembly process, while the assistance system automatically generates both the assembly instructions.

Driven by this vision, this paper presents a system and its implementation for programming assembly instructions through demonstration. In this approach, the setup of assembly assistance systems is achieved by recording a demonstration of the desired assembly process. During the demonstration, assembly steps are identified and documented using text and images, enabling the automatic configuration of assistance systems. Our approach aims to simplify and accelerate the assembly setup process, making it faster, less error-prone, and easier to use.

We advocate for the programming by demonstration approach due to the following advantages: First, it reduces errors in process sequences since the assembly is demonstrated using actual parts. This hands-on method minimizes the risk of missing information, as the process planner physically performs the task. Additionally, the demonstration-based approach bridges the information gap between workers and planners, as planners gain practical insights while conducting the assembly process themselves. Unlike CAD-based methods, which often generate instructions that are difficult for humans to interpret, this approach aligns naturally with human cognitive processes and workflows. Finally, demonstrations are more efficient than traditional-based inputs, because essential process-relevant data is captured automatically. The data, such as positions, actions, images, and process times, does not require extensive manual input.

The remainder of this paper is structured as follows: In Sect. 2, the state of the art of assistance systems in industrial assembly as well as the basics and previous approaches to the automatic generation of assembly instructions are presented. Section 3 describes the implementation of our developed system, covering requirements and hardware setup, the software architecture, the methodological approach for interpreting and analyzing human demonstrations, as well as the developed user interface. Section 4 presents the results of the conducted user study to evaluate the system. Section 5 summarizes the findings, discusses limitations, and highlights potential directions for future research.

2 State of the Art

2.1 Informational Assistance Systems and Assembly Instructions

Informational assistance systems for industrial assembly provide cognitive support to workers by delivering task-relevant information, such as step-by-step instructions, in a timely and accessible manner. Extensive morphologies detailing the characteristics and variations of assembly assistance systems can be found, for example, in [4]. These systems aim to reduce errors, improve productivity, and manage the complexity of modern assembly processes [5]. Assistance functionalities can be broadly divided into product- and process-centered functions, such as quality control, progress recognition, and assembly instructions, and human-centered functions, which address aspects like ergonomics, motivation, and individualization [5]. While traditional product- and process-related features—such as Pick-by-Light systems for component picking or automated quality control using industrial cameras—are widely implemented in industrial practice. The human-centered functionalities addressing worker motivation and active ergonomics support as well as the process-related feature of a fast and intuitive system configuration remain underexplored [6].

2.2 Automated Generation of Assembly Instructions

The generation and preparation of assembly instructions is time- and knowledge-intensive due to the high variety and complexity of the products. Therefore, ongoing research addresses the topic of automatic generation of assembly instructions [7]. One widely discussed approach in research is the generation of assembly instructions based on CAD data [8]. These methods define a technical assembly sequence based on the component geometry. Moreover, product features such as tolerances or the orientation of components relative to each other can be extracted from CAD and used for the creation of assembly instructions [9]. However, more information about the assembly sequence are necessary for assembly instructions. This context-based knowledge includes specific guidance and for the assembly worker: For example, guidelines regarding the usage of tools during the assembly process or further instructions for manual quality control loops are part of typical assembly instructions. Furthermore, usually a short description of the assembly step is given. For the generation of this text-based information one new research direction focuses on the application of large language models (LLM) [7, 10]. The proposed systems using LLM have great potential, but there is still the necessity for further product data such as a bill of material. These inputs need to be well-organized and kept up to date, which can often be challenging in practice. Especially for small and medium size companies (SMEs) without a Product-Lifecycle-Management (PLM) system, gathering the right data for this approaches is not easy. With the increasing possibilities of process recordings in assembly using sensors and cameras and the usage of intelligent algorithms, the automatic generation of assembly instructions seems to be one promising research direction [11]. In contrast to the aforementioned methods, such a demonstration-based approach does not require data to be provided in advance.

3 System Implementation

3.1 Requirements and Hardware Set-Up

For the development, initial requirements were gathered. These requirements are divided into those related to the process of creating assembly instructions. For example, one requirement aims for the assembly instructions to be as easily customizable by the user as possible. These process requirements were subsequently categorized into sensors, human–machine interface (HMI), and process language, and translated into technical requirements. The full list of the detailed requirements gathered has already been published by the authors in [12]. Based on these requirements, the hardware depicted in Fig. 1 was assembled. The assembly station integrates all necessary hardware, including the cameras, positioning system, and displays for the HMI. Furthermore, material boxes for the supply of the assembly parts and a projector is part of the assembly station.

3.2 Software Implementation

Software Architecture. As depicted in Fig. 2, the overall system consists of an assembly station equipped with sensors, a software backend for processing assembly demonstrations, and a human–machine interface (HMI). The assembly station includes position sensors (Sarissa PositionBox) embedded in Smart Tools, such as a torque screwdriver, and depth cameras (two ZED 2 cameras) for tracking hand and object positions. The software backend processes the captured sensor data using data fusion (by merging the point clouds from both depth cameras into a unified 3D workspace representation and locating the tool position data into the workspace representation calibrated through Aruco-markers) as well as hand and object tracking techniques (Google MediaPipe) to create an inferred world state. Further details on the implementation and precision of the hand and object tracking as well as the calibration procedure can be found in our previous work [12]. This

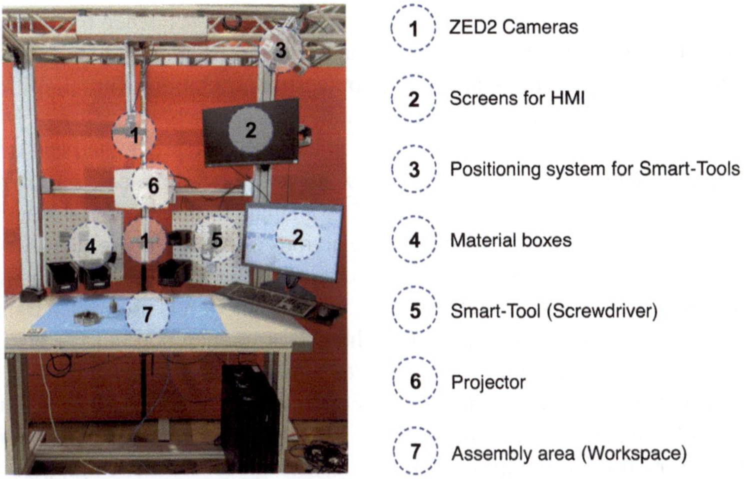

Fig. 1 Hardware set-up and integration: manual assembly work station with cameras and screens

information is structured within a state tree and is processed as described below to infer a world state and an associated process graph, which serves as the basis for generating assembly instructions. The HMI (consisting of an interactive touchscreen-interface and a projector-based worktop-projection system) visualizes a digital twin based on the inferred world state and enables users to modify both the detected process steps and the generated assembly instructions. Further, recorded assembly processes can be saved as JSON files to be loaded in worker assistance systems for assembly execution.

Processing Pipeline and Algorithm for Automatic Process Interpretation. The processing pipeline follows a structured approach to interpret demonstrated assembly actions. Based on the captured and preprocessed sensory data, the inferred world

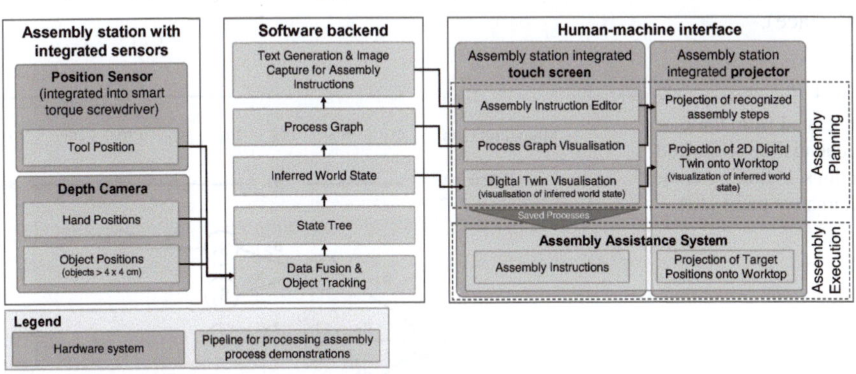

Fig. 2 System architecture and processing pipeline of the overall system for the automatic generation of assembly instructions

state is interpreted, and structured in terms of the process sequence by a state tree, as illustrated in Fig. 3. Each node s_x in the state tree represents a system state, while edges denote detected actions a_x such as object manipulation. The algorithm for interpreting assembly steps follows a probabilistic approach, comparing observed states to predefined action models. The available actions are predefined and include, for example, reaching into an assembly box to grasp a component, removing the hand, opening or closing the hand, or moving a smart tool out of its defined area. Different heuristics are used to recognize the actions (including a probability score), e.g. the action "grasping an object" is recognized based on the position overlap between hand contours and object positions. When an action is inferred, a transition occurs within the state tree by appending the action to each node of the node tree. This makes it possible to skip incorrectly recognized actions that did not actually represent a real assembly step: Either if conflicting or ambiguous states occur, or after a configurable number of measurements m_x, the system resolves inconsistencies by comparing transition probabilities. For this, generic process rules (e.g., predefined sub-process models for pick-and-place), as well as the detection reliability scores for the individual actions are utilized to compare the probabilities of the transitions between the individual states, e.g., from s_0 to s_1, from s_0 to s_2, or from s_0 to s_3, etc. Thus, paths in the state tree that are likely to contain incorrect actions (such as misclassified grasping actions or consecutive joining operations of the same object at the same location), are pruned (to reduce complexity of the state tree) if alternative state transitions exhibit higher likelihoods. This allows for fault tolerance and robust step recognition even under uncertain conditions.

When an action a_x is recognized, an RGB image is also captured with the integrated camera and saved in association with the action. Also, the assembly instructions are automatically generated for each process step by using the saved images on the one hand and a textual description based on predefined action-text-templates (e.g., "pick a perforated plate from component box 6") on the other. Here, the object recognition automatically identifies the corresponding component box from which the component was taken, while the assignment of components (with their names or IDs) and component boxes is predefined (user input in the digital twin of the user interface).

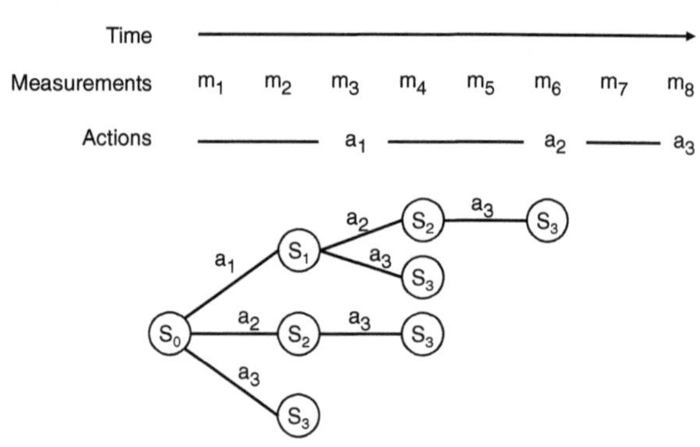

Fig. 3 Exemplary state tree generated from three sequentially recognized actions, including three alternative paths [12]

Human–Machine Interface. The implemented user interface is shown in Fig. 4. Figure 4a shows the user interface for the assembly planner, in which the digital twin based on the inferred world state, the inferred process sequence and the generated assembly instructions associated for the currently recognized process step are displayed. The most likely inferred process sequence is always displayed, which is calculated based on a cumulative score of the probabilities along the paths across all branches of the tree. If required, users can switch between the top-ranked inferred world states via a button or edit the textual description of the generated assembly instructions. An optional developer mode allows for debugging by visualizing the complete state tree, see Fig. 4b. Additionally, the integrated projector visualizes the teaching-relevant information directly onto the workspace by displaying recognized assembly steps and objects as well as the inferred human hand inventory (e.g. screwdriver or picked components assumed to be in hands) directly on the worktop, as shown in Fig. 4c. This enables the demonstrating planner to check directly whether the steps and the inferred world state have been recorded correctly. Thus, it enables a natural interaction and seamless workflow during the demonstration without shifting focus between the physical workspace and the digital interfaces.

4 Evaluation Study

Study Goals and Design. The evaluation study aimed to assess the user-friendliness and overall functionality of the developed system while comparing it to the manual creation of assembly instructions. To achieve these objectives, a multi-stage study design was employed: First, participants received a brief introduction to the software for demonstrating assembly instruction explained above, and a prepared spreadsheet (MS Excel) for the manual process of creating assembly instructions. This ensured that all participants had a basic understanding of the study process and the functionalities of the software as well as the prepared spreadsheet. Following the introduction, participants created assembly instructions for two assembly components—first using the developed system and then manually with the spreadsheet. The first component was rather simple (cubes), designed to be straightforward to provide an easy entry point. After completing this initial task, participants moved on to a more complex component. The used component was an oil pump out of the automotive industry and includes all typical assembly steps such as handling and screwing of sub-parts. Thus, this second part offers a realistic assembly scenario.

Study Results. Overall, this first evaluation study was conducted with eight participants. The demographic characteristics of the participants show that they are a relatively young group of scientists with a background in engineering. The majority of participants are between the ages of 30 and 34, and have prior experience with assembly processes as well as the support with assembly assistance systems.

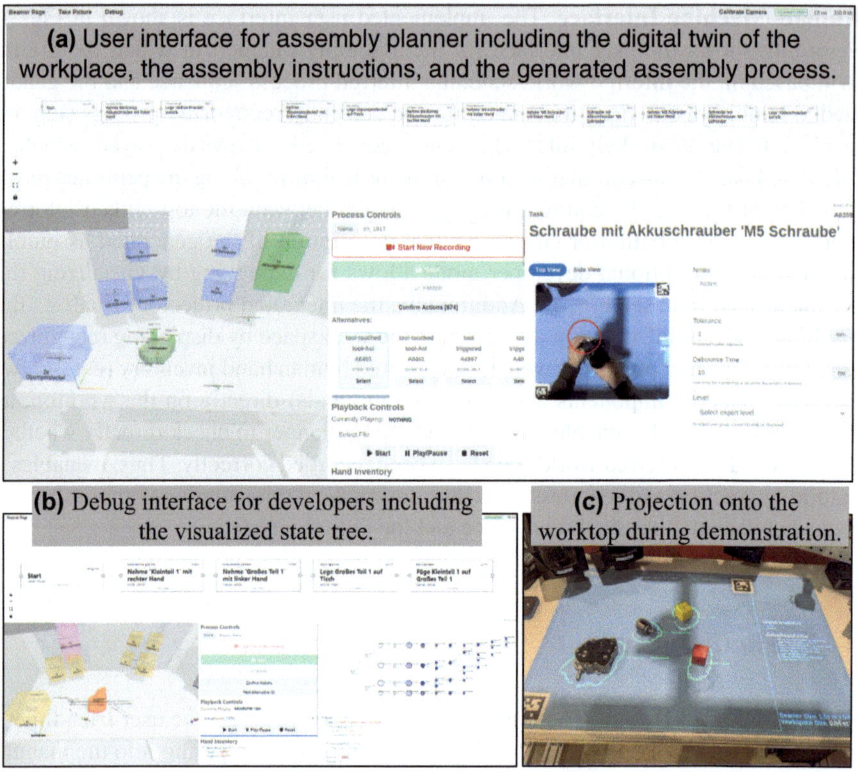

Fig. 4 Implemented graphical user interface (implemented in German): the figure is only showcasing the different functions and interfaces of the tool. It is not intended to show the entire GUI in all details

Regarding the analysis of the results in terms of productivity, Table 1 shows a significant difference in efficiency measured by the time required to create assembly instructions. The manual creation, which was represented in the study using the spreadsheet, takes significantly longer. Additionally, observations showed that the manual creation of the instructions lead to more errors.

In addition to the software efficiency, user-friendliness plays an important role, particularly in ensuring high acceptance during future implementation. As shown in Fig. 5, and categorized based on [13], the developed software achieved a system

Table 1 Results of user study

	Simple example (cubes)		Complex example (oil pump)	
	Developed system	Spreadsheet	Developed system	Spreadsheet
Time	00:47	06:31	00:51	05:47

Comparison of time to prepare instructions for simple and complex example. Time in (mm:ss)

usability score (SUS) of 77.81. Compared to the spreadsheet, which had a SUS score of 59.06, this indicates already a good level of user-friendliness. A SUS score above 70 is also classified as *good* and is particularly positive considering the prototype status of the application.

For measuring the perceived workload, we used the standardized NASA-TLX questionnaire [14]. The results are summarized in Fig. 6. Looking at the six dimensions, our developed software performed again well in comparison to the spreadsheet. In particular, the dimensions mental demand and frustration show a significantly better performance.

Furthermore, an open feedback session was answered by all participants. This qualitative feedback supported the quantitative study findings. Participants acknowledged that the developed software works effectively and enabled them to complete the

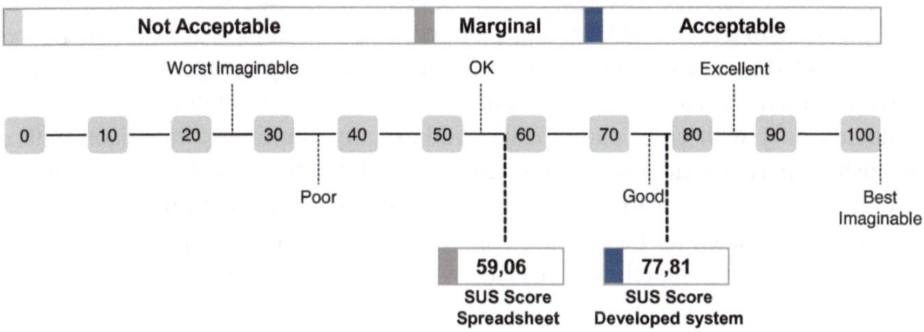

Fig. 5 Calculated SUS Score based on study results. Comparison of spreadsheet vs developed system (classification and graphical representation based on [13])

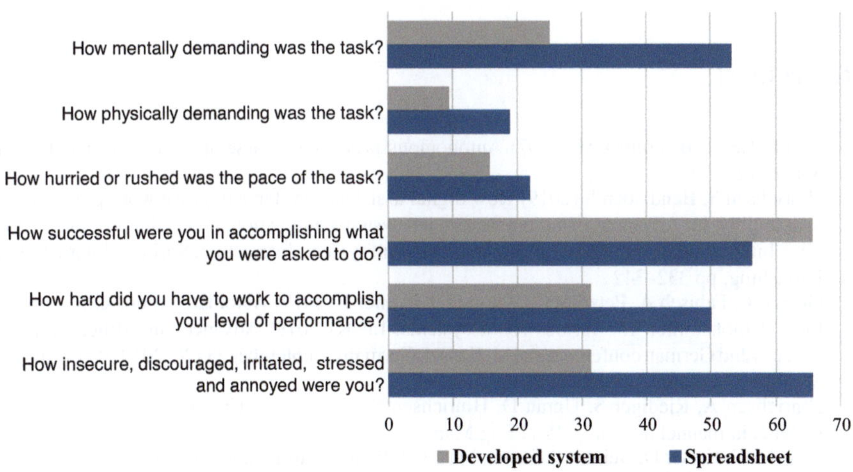

Fig. 6 NASA TLX results. Comparison of spreadsheet vs. developed system

assigned tasks. However, they highlighted the need for various HMI improvements in future development phases.

5 Discussion and Outlook

This paper presented a novel system for the automated generation of assembly instructions using the concept of teaching by demonstration. We proposed a system that captures the manual assembly demonstration through a camera system and a positioning system for smart tools. Both are integrated on a typical workstation for manual assembly. Core of the proposed system is the developed software solution and the methodological approach for the interpretation of the camera stream. For this, we presented a software architecture consisting of the integrated sensors, the software backend including a data processing pipeline, and the human–machine interface. The conducted user study shows promising first results regarding the usability and working principles. However, the evaluation has different limitations. First, it was conducted only with eight participants. Additionally, the time comparison with a spreadsheet may not fully reflect the real-world alternatives, such as input masks, for creating assembly instructions. Also concealment issues with the cameras are not fully investigated so far and a comparison with commercial planning tools is also needed. Future research will therefore focus on a broader evaluation outside the laboratory environment.

Acknowledgements This research has been funded by the Federal Ministry for Economic Affairs and Climate Action of Germany as part of the project AITeach (project number KK5082910SW2).

References

1. Scholz-Reiter B, Freitag M (2007) Autonomous processes in assembly systems. CIRP Ann 56(2):712–729
2. Hinrichsen S, Bendzioch S (2019) How digital assistance systems improve work productivity in assembly. In: Advances in human factors and systems interaction: proceedings of the AHFE 2018 International conference on human factors and systems interaction. Springer International Publishing, pp 332–342
3. Gutzeit L, Fabisch A, Petzoldt C, Wiese H, Kirchner F (2019) Automated robot skill learning from demonstration for various robot systems. In: KI 2019: Advances in artificial intelligence: 42nd German conference on AI, Kassel, Germany, September 23–26, 2019, Proceedings. Springer International Publishing, pp 168–181
4. Hinrichsen A, Riediger S, Unrau D, Hinrichsen S, Riediger D, Unrau A (2016) Assistance systems in manuel assembly. Prod Eng Manag
5. Petzoldt C, Keiser D, Beinke T, Freitag M (2020) Functionalities and implementation of future informational assistance systems for manual assembly. In: Subject-oriented business process management. The digital workplace–nucleus of transformation: 12th International conference, S-BPM ONE 2020, Bremen, Germany, December 2–3, 2020, Proceedings 12. Springer International Publishing, pp 88–109

6. Keiser D, Petzoldt C, Beinke T, Freitag M (2021) Mitarbeiterzentrierte Assistenzsysteme für die manuelle Montage - Systematische Evaluation von Montageassistenzsystemen. Industrie 4.0 Manag 37(3):11–15
7. Meyer F, Freitag L, Hinrichsen S, Niggemann O (2024) Potentials of large language models for generating assembly instructions. In: IEEE 29th International conference on emerging technologies and factory automation, pp 1–8
8. Neb A (2019) Review on approaches to generate assembly sequences by extraction of assembly features from 3d models. Procedia CIRP 856–861
9. Sanfilippo EM, Borgo S (2016) What are features? An ontology-based review of the literature. Comput-Aided Design 9–18
10. Herbort R, Green D, Hinrichsen S (2025) Automatic creation of assembly instructions by using retrieval augmented generation. In: Intelligent human systems integration (IHSI 2025): integrating people and intelligent systems, p 160
11. Burggräf P, Adlon T, Steinberg F, Bröhl F, Moriz A, Engeln C, Schütz M, Jaworek F (2024) Automatic generation of assembly instructions by analyzing process recordings—a concept overview. Procedia CIRP 126:775–780
12. Niermann D, Keiser D, Freitag M (2024) Towards a novel system for creating assembly instructions through demonstration. Procedia CIRP 130:1182–1187
13. Bangor A, Kortum P, Miller J (2009) Determining what individual SUS scores mean: adding an adjective rating scale. J Usability Stud 4(3):114–123
14. Hart SG (2006) NASA-task load index (NASA-TLX); 20 years later. In: Proceedings of the human factors and ergonomics society annual meeting. Sage Publications, CA, vol 50, no 9, pp 904–908

Development of a Methodology for the Semi-automated Determination of Optimized Workplace Design in Manual Assembly

Fabian Adler, Sebastian Flierl, and Rainer Müller

Abstract In many small and medium-sized enterprises (SMEs), assembly workplaces often evolve over time without structured planning. This frequently results in inefficient workflows and suboptimal workplace layouts. A primary reason is the complexity of existing optimization methods, which typically require extensive expert knowledge and are therefore challenging to apply in SMEs. This paper presents the development of a user-friendly methodology for the semi-automated determination of optimal assembly workplace configurations. The methodology integrates both assembly time- and ergonomics-based optimization criteria, allowing both to be considered simultaneously. By incorporating both aspects within the optimization process, the methodology identifies configurations that maximize efficiency while minimizing physical strain on employees. To support data collection in non-digitized workplaces, an augmented reality (AR) application was developed for the Microsoft HoloLens 2. This application assists workers in capturing critical workplace parameters, including component distances, movement sequences, and process flows. The methodology analyses the workplace using a structured combination of primary-secondary analysis, Methods-Time Measurement Universal Analyzer System (MTM-UAS), and a variant of the Rapid Upper Limb Assessment—NEPRA. Afterwards the workplace is optimized by systematically changing the arrangement of all parts containers. The methodology also guides the user through the optimization by providing visual and textual instructions based on the results of the analyses. The proposed methodology provides companies, especially SMEs, with a practical tool for systematic workplace optimization without requiring extensive prior expertise.

Keywords Assembly workplace · Optimization methodology · Ergonomics · Time-efficiency · Augmented reality

F. Adler (✉)
ZeMA gGmbH, Research Division Assembly Systems, Saarbrücken, Germany
e-mail: fabian.adler@zema.de

S. Flierl · R. Müller
Chair of Assembly Systems, Saarland University, Saarbrücken, Germany

M. Manns (ed.), *New Paradigms for Anticipated Uncertainty*, Lecture Notes in Mechanical Engineering, https://doi.org/10.1007/978-3-032-16889-4_51

1 Introduction

In large enterprises, the design and optimization of assembly workplaces are well-established processes [1]. High production volumes allow for investments in optimization strategies, enabling the exploitation of existing potential in terms of efficiency and cost-effectiveness. In contrast, small and medium-sized enterprises (SMEs) often operate under fluctuating order volumes and batch sizes, adapting to the market through competitive strategies such as high quality, customized solutions, and specialized services [2, 3]. As a result, their assembly processes tend to be manual, with frequently changing working conditions that necessitate regular workplace adjustments. This makes extensive workplace optimization economically unfeasible. Empirical evidence suggests that assembly workplaces in SMEs evolve over time without systematic planning, leading to unexploited optimization potential [1].

Only a small fraction of SMEs have established industrial engineering structures or possess the necessary expertise in workplace design and optimization [4]. Larger enterprises can economically justify the costs associated with workplace optimization and the employment of highly trained specialists. Existing optimization methods are often impractical for smaller companies with limited resources and are therefore not utilized, despite the presence of significant economic and ergonomic optimization potential [1].

Therefore, this paper investigates how a simplified and user-friendly methodology combining assembly time and ergonomics analysis can enable SMEs to identify time and ergonomic optimization potential through systematic workplace layout proposals. The approach begins with the assisted collection of all necessary workplace information using an Augmented Reality (AR) application. The methodology focuses on the semi-automated determination of various optimization recommendations for manual assembly workplaces, requiring little to no investment for implementation. Time-saving optimization measures can often be achieved through simple restructuring and reorganizing the assembly workplace, although these improvements depend on the specific products and processes involved. Furthermore, in highly flexible and dynamic assembly environments, employees represent the most critical resource [5], necessitating the protection of their well-being through ergonomically designed workplaces. Studies also show that ergonomic workplace design enhances productivity [6].

2 Related Work

Several methods exist for analyzing workplaces that can be applied without specialized training or extensive prior knowledge. Regarding ergonomics, the Key Indicator Method developed by the German Federal Institute for Occupational Safety and Health (BAuA) is explicitly designed for small enterprises [7]. For time-related analysis, the BasicMOST method offers a simple and easily learnable approach [8]. However, a combination of ergonomics and time efficiency is only found in the Ergonomic Assessment Worksheet (EAWS) and Methods-Time Measurement (MTM), both which require specialized training [9].

Optimization methods can be categorized into two distinct groups, *analysis-based methods* and *recommendation-based methods*. The first group of methods consists such as Failure Mode and Effects Analysis (FMEA) and Rapid Upper Limb Assessment (RULA). While these methods can identify problematic areas within a workplace, the actual problem-solving process relies on the expertise of the user. The second group comprises collections of expert-based recommendations, such as the principles for ergonomic workplace design proposed by Lotter et al. [10]. However, no existing method combines the advantages of both groups.

In general workplace analysis within an assembly context (e.g., arrangement of components, assembly positions, or distances between tasks), literature reviews suggest that these aspects are often assumed to be either given or insufficiently complex to warrant further investigation. As a result, only a limited number of studies address these aspects in a specific context [11]. Many approaches assume the availability of CAD files or focus on methods—especially in conjunction with AR—that are already integrated into the design and planning phases [12, 13].

3 Implementation of the Methodology for Optimized Workplaces

The methodological approach focuses on optimizing non-value-adding activities, as the optimization of value-adding activities is highly application-specific. According to Lotter et al. [10], assembly process efficiency is largely affected by (non-value-adding) handling movements. Therefore, the developed method optimizes the workplace by adjusting the arrangement of component containers. The structure and procedure for collecting essential workplace information (see Sect. 3.1) and optimizing an assembly workplace (see Sect. 3.2) are illustrated in Fig. 1.

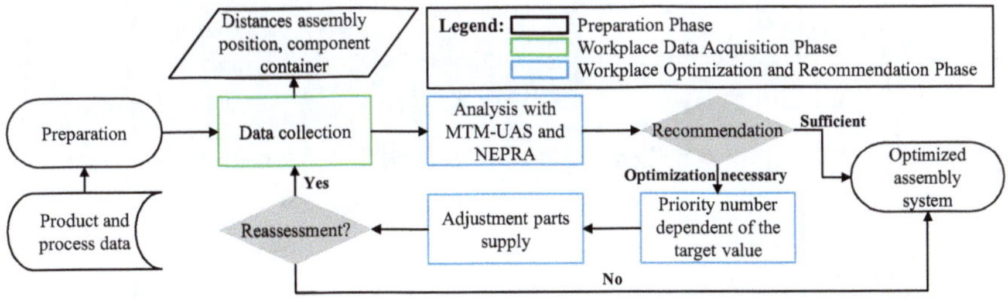

Fig. 1 Methodology sequence from information request to the derivation of optimization recommendations

The methodology consists of multiple phases (black, green, blue) and is applied iteratively. Step two and three are further explained in the following Sects. 3.1 and 3.2. The first phase (Fig. 1, black) includes all necessary preparations for applying the method. In this phase, general information about the assembly process and the required components is collected. This includes, for example, the bill of materials and the precedence graph, which define the scope of analysis.

3.1 Workplace Acquisition

The second phase of the methodology (Fig. 1, green) collects AR-based workplace data. As shown in Fig. 2, this phase is designed to allow the user manually define and collect essential data for create the "digital" workplace. For this purpose, an AR headset—*Microsoft HoloLens 2*—was used. In principle, other AR devices such as a smartphone or tablet could also be utilized. However, the advantage of an AR headset is the ability to freely interact with both hands in the virtual and real environments.

To interact with the AR application, a menu has been developed (Fig. 2, left side) to perform all the functions described hereafter. First, a physical anchor point in the form of a QR code is placed at the **assembly workplace (W)**, which is initially scanned using the **HoloLens (H)** before workplace data acquisition begins. The QR code also serves as a reference point when loading a previously saved workplace configuration. Second, one or more **assembly positions (M_i)** are defined in the AR headset according to the workplace, representing the locations where assembly operations take place. Third, based on the available components and component

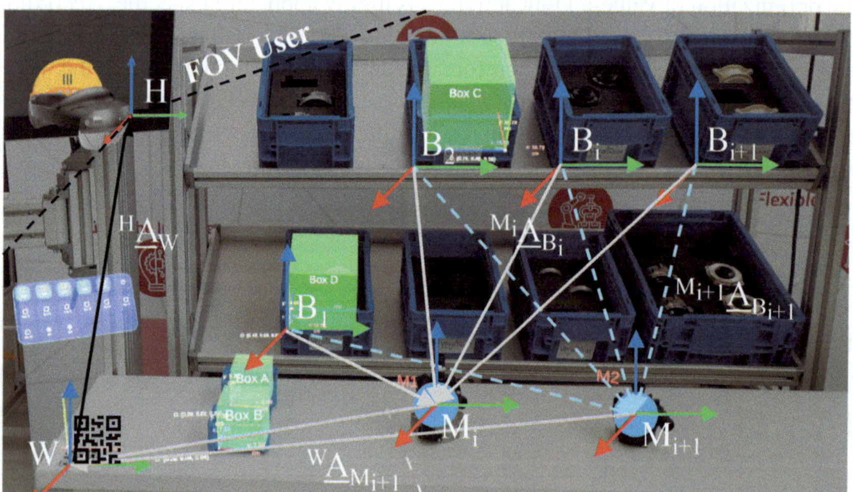

Fig. 2 Structure of the AR application for workplace acquisition and measurement

containers, any number of **component containers (B_i)** can be defined. These bins are named according to the components they contain and are adjusted in dimensions to match their real-world counterparts. Each virtual object has its own coordinate system. The component containers are positioned relative to all assembly positions, while the assembly positions are referenced to the QR code, which represents the assembly workplace. All information about the configuration—such as component names, container dimensions, relative positions, orientations, and distances—is sent as a JSON file via MQTT.

3.2 *Workplace Optimization*

The third phase (see Fig. 1, blue) comprises workplace optimization and adjustment recommendations. Initially, the positioning of component containers before and after optimization is processed using simplified versions of MTM-UAS and NEPRA. The movement duration is calculated using the MTM-UAS module "Get and Place." To assess and optimize ergonomic strain, ergonomic zones were identified, with each zone assigned a specific ergonomic score based on NEPRA. The score quantifies the ergonomic strain created by a movement to the corresponding zone. The differentiation of zones is based on variations in joint positions and body posture according to NEPRA and is illustrated in Fig. 3a. The zones are labeled with a letter followed by an index, where the letter indicates the position on the work surface and the index classifies the height above the work surface (1 for 0–20 cm, 2 for 20–40 cm, and 3 for 40–60 cm). Finally, a *Primary-Secondary Analysis* according to Lotter [10] is conducted, which distinguishes between value-adding (primary) and supporting (secondary) activities to identify optimization potential. This helps the user evaluate the effectiveness of the changes in terms of assembly time.

After data processing, a three-part optimization is performed. First, general recommendations guide the user in interpreting results and prioritizing optimization goals based on ergonomic strain and primary-secondary analysis. Second, a priority number (PN), is calculated for each component, considering criteria such as quantity,

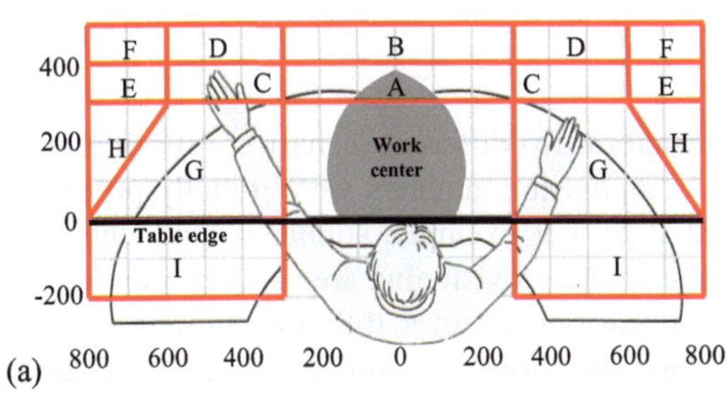

(b)

Rank	Zone	Rank	Zone	Rank	Zone
1	A1	10	C2	19	D3
2	B1	11	D2	20	G3
3	A2	12	I2	21	F1
4	B2	13	A3	22	E2
5	C1	14	B3	23	F2
6	D1	15	C3	24	I3
7	G1	16	H1	25	H3
8	I1	17	H2	26	E3
9	G2	18	E1	27	F3

Fig. 3 **a** Ergonomic zones based on joint positions or postures based on [14] and **b** recommendation for placement of component containers in each zone

weight, dimensions, and container design. The higher the priority number the more critical the optimal arrangement becomes. The calculation of the priority number is presented in Formula (1). The weighting factors a_1 to a_4 allow for an adjustable focus with the variable F between ergonomic optimization and optimization for assembly time. F can be set from − 1 to 1, where $F = 1$ represents a maximum focus on assembly time and $F = -1$ a maximum focus on ergonomics. The values for a_1 to a_4 were determined by assessing the relative importance of the criterion, choosing a suitable influence of the variable F.

$$PN_i = a_1 * n_i + a_2 * m_i - a_3 * d_i + a_4 * b_i \tag{1}$$

with $a_1 = 1.5 + (0.5*F)$; $a_2 = 1$; $a_3 = 0.5 + (0.25*F)$; $a_4 = 0.5 - (0.25*F)$.

The number of pieces n_i was selected as criteria, as the more frequently a component $i \in \{1, 2, \ldots, N\}$ is required for the product to be assembled, the greater is the achievable optimization. Heavier components m_i impose a higher ergonomic strain, thereby increasing the duration of the handling process. Additionally, larger components inevitably require more space in the parts provisioning area. To take that into account, the size of the component d_i is factored negatively into the priority number. The design of the component container b_i should be as space-efficient as possible to ensure an optimal arrangement of all component containers and facilitate ergonomic component retrieval. In practice, this ideal state is not always achieved, making differentiation within the priority number necessary.

Third, the methodology recommends an order for stocking zones with component containers (see Fig. 3). The basis for this recommendation is the movement distance from each respective zone to the assembly position. The priority number and ergonomic zone classification guide the user in positioning component containers. Placement follows a descending order, starting with the highest priority in the top-ranked zone. Within each zone, users arrange containers as needed, skipping unusable zones. For example, if a process module is set up in zone G1 (see Fig. 3), the user would skip the zone during the container placement. Likewise, if the table is narrower than 1.6 m, then Zones E, F, and H might not be usable.

4 Results

The developed methodology was tested in two different scenarios, differing in the size of the component containers. Component containers were initially positioned to reflect a realistic, non-optimized state. For example, similar components were grouped together, but their placement in the provisioning area was chosen without a specific systematic approach. In both scenarios, eleven different components were arranged, including circuit boards, fasteners, housing components, and mechanical parts. The methodology was applied as described in Sect. 3.2. For the calculation of priority numbers (Formula 1), a neutral balance between the optimization targets, the assembly time and the ergonomics was selected.

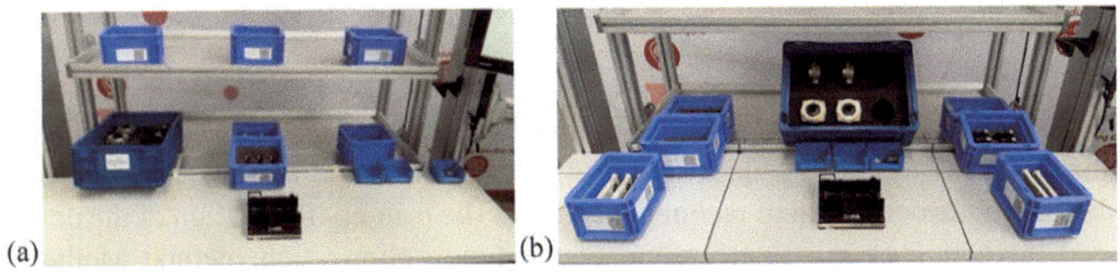

Fig. 4 Setup one: Workplace **a** before and **b** after performing and applying the methodology

The following results were obtained for the first setup (Fig. 4). Only a single component container was positioned per zone to simulate the use of larger containers. An exception was made for bulk materials such as bolts. The handling time was reduced from 36.9 to 29.5 s. Acceptable ergonomic strain decreased from 90.5 to 38%, while the proportion of low ergonomic strain increased from 9.5 to 62%. Figure 4b illustrates the differences in the arrangement. These improvements result from placing frequently used or heavy components in ergonomically favorable zones (e.g., A1–C1 in Fig. 3) near the assembly position. As ergonomic load increases with distance, this positioning shifts movements into low ergonomic strain zones and reduces physical effort (see Formula 1).

In the second setup, multiple containers were allowed per zone. Assembly time was reduced from 37.26 to 27.36 s. Acceptable ergonomic strain decreased from 33 to 28.5%, while low strain increased from 67 to 71.5%.

5 Discussion

The results demonstrate that the proposed methodology leads to a substantial improvement in workplace design. Handling time was reduced by 20–25%, accompanied by a consistent decrease in overall ergonomic strain. Although an expert in workplace design or extensive established methods might achieve superior outcomes, the developed methodology offers a practical alternative by delivering results quickly and without requiring prior training. Unlike related approaches, the methodology integrates elements from both analysis-based and recommendation-based methods. This hybrid structure enables context-sensitive, actionable recommendations while maintaining flexibility. A limitation identified during validation relates to the placement sequence of component containers (see Fig. 3b). It became evident that ease of refilling should be included as an additional decision criterion. Furthermore, the methodology should account for the spatial competition between containers, tools, and consumables when determining placement priorities.

6 Conclusion

This paper proposes a comprehensive semi-automated approach, encompassing the entire process from the acquisition of essential data to the optimization of workplaces and the derivation of recommendations. The primary innovation of the developed methodology is its integration of the time optimization method Methods-Time Measurement Universal Analyzer System (MTM-UAS) and the ergonomic analysis method NEPRA, a variant of the Rapid Upper Limb Assessment. This allows for a simultaneous consideration of key objectives—assembly time and ergonomics—while preventing optimization actions at the ergonomic expense of workers.

Currently, the display of the required actions and the integration of the optimized workplace into the AR application are still pending. Following these improvements, a user study on the entire methodology is planned to obtain more substantiated insights into its usability. For future identification of component containers in the workplace, ArUco markers will be used, which employees can place on the containers to enable automatic recognition via the AR application.

Nevertheless, the methodology provides a strong foundation for future extensions particularly for companies with limited investment capacity and technical expertise.

Acknowledgements This paper was written in the context of the project iSMAT ("*innovative Produktion von und mit Smarten Materialsystemen*") which is funded by the European Fonds for Regional Development (EFRE). The project is located in the federal state of Saarland and is executed by ZeMA—Centre for Mechatronics and Automation Technology gGmbH.

References

1. Gote H, Glatzel T (2019) Empirische Studie über die Arbeitsplatzgestaltung von manuellen Montagearbeitsplätzen in KMU. In: Augustin S (ed) Arbeit interdisziplinär analysieren - bewerten - gestalten. 65. Kongress der Gesellschaft für Arbeitswissenschaft e.V., Dresden, 27.02-01.03.2019. GfA-Press, Dortmund
2. Wettbewerbsstrategien im Mittelstand | KfW. https://www.kfw.de/Über-die-KfW/KfW-Research/Wettbewerbsstrategien.html. Accessed 23 Feb 2025
3. Schwartz M (2013) Mittelstand im rauen Wind: Wettbewerb wird intensiver, die drei „I" versprechen Erfolg. Fokus Volkswirtschaft. KfW
4. Junkind W, Könneker I, Pläster I, Reuber M (2019) Handbuch der Prozessoptimierung. Die richtigen Werkzeuge auswählen und zielsicher einsetzen, 1st edn. REFA-Kompendium Arbeitsorganisation, vol 2. Carl Hanser Verlag GmbH & Co. KG, München
5. Dombrowski U, Krenkel P (2021) Ganzheitliches Produktionsmanagement. Springer Berlin Heidelberg, Berlin, Heidelberg
6. Frank C (2019) Produktivitätsfaktor Ergonomie. Zeitschrift für wirtschaftlichen Fabrikbetrieb. https://doi.org/10.3139/104.112102
7. Bundesanstalt für Arbeitsschutz und Arbeitsmedizin (2019) BAuA - Handlungshilfen für die Gefährdungsbeurteilung. Gefährdungsbeurteilung mit Hilfe der Leitmerkmalmethode -. https://www.baua.de/SharedDocs/Handlungshilfen/DE/Gefaehrdungsbeurteilung/BAuA/Gefaehrdungsbeurteilung-mithilfe-der-Leitmerkmalmethode. Accessed 23 Feb 2025

8. Viharos ZJ, Ban B (2020) Comprehensive comparison of MTM and BasicMOST, as the most widely applied PMTS analysis methods. In: Global trends in testing, diagnostics & inspection for 2030. 17th IMEKO TC 10 and EUROLAB virtual conference, online, 20–22.10.2020
9. Schaub KG, Mühlstedt J, Illmann B, Bauer S, Fritzsche L, Wagner T, Hoffmann ACB, Bruder R (2012) Ergonomic assessment of automotive assembly tasks with digital human modelling and the 'ergonomics assessment worksheet' (EAWS). IJHFMS. https://doi.org/10.1504/IJHFMS.2012.051581
10. Lotter B, Deuse J, Lotter E (2016) Die Primäre Produktion. Springer Berlin Heidelberg, Berlin, Heidelberg
11. Steinbauer CMT (2012) Modell zur Konfiguration der Kleinserienmontage. EINE THEORETISCHE UND EMPIRISCHE ANALYSE. Dissertation, Technische Universität München
12. Ong SK, Pang Y, Nee A (2007) Augmented reality aided assembly design and planning. CIRP Ann. https://doi.org/10.1016/j.cirp.2007.05.014
13. Simonetto, M., Arena, S., Peron, M.: A methodological framework to integrate motion capture system and virtual reality for assembly system 4.0 workplace design. Safety Science (2022). https://doi.org/10.1016/j.ssci.2021.105561
14. Koether R, Meier K-J (2020) Lean Production für die variantenreiche Einzelfertigung. Springer Fachmedien Wiesbaden, Wiesbaden

Optimizing Finite State Machines for Process Time Analysis in Industrial Production

Dan Eisenkrämer and Bernd Lüdemann-Ravit

Abstract Efficient production planning, machine optimization, and bottleneck analysis are critical for reducing cycle times in manufacturing processes. This paper introduces an automated process mining (PM) approach designed to streamline these tasks by generating and optimizing finite state machines (FSMs) from event logs derived from signal changes in industrial plants. Unlike conventional methods, this approach requires no prior knowledge of the configuration or behavior of the target programmable logic controllers (PLCs). The synthesized FSMs facilitate the extraction of process-oriented insights, enabling the identification of underperforming manufacturing processes and detailed analysis of their durations. The proposed method is validated through a case study on a real industrial plant, demonstrating its efficacy in uncovering process inefficiencies and supporting decision-making. This work provides a novel, generalizable framework for process analysis in manufacturing environments, contributing to the broader field of automated process optimization.

Keywords Process control · Data mining · Manufacturing optimization · Finite state machine · Bottleneck analysis

1 Introduction

Reducing production cycle times while maintaining efficiency is vital for manufacturers to stay competitive in increasingly dynamic markets. Traditional methods depend on expert analysis and simulations, which are resource-intensive and poorly suited for real-time use. This paper addresses these challenges by presenting a novel automated process mining (PM) technique that leverages event logs to generate finite state machines (FSMs). These FSMs provide a structured representation of production processes, enabling manufacturers to identify bottlenecks, analyze pro-

D. Eisenkrämer (✉) · B. Lüdemann-Ravit
Institute of Production and Informatics, Sonthofen, Germany
e-mail: dan.eisenkraemer@hs-kempten.de
URL: https://www.hs-kempten.de/ipi

M. Manns (ed.), *New Paradigms for Anticipated Uncertainty*, Lecture Notes in Mechanical Engineering, https://doi.org/10.1007/978-3-032-16889-4_52

cess durations, and optimize operations without prior knowledge of the underlying system configurations.

Automating FSM synthesis and optimization simplifies complex manufacturing system analysis. The efficacy of the method is demonstrated through its application to a real-world production line, highlighting the practical benefits of automated process insights and actionable recommendations.

2 Related Work

PM merges business process management and data mining to optimize industrial operations. Recent research highlights its potential to enhance resource allocation and efficiency, particularly in manufacturing environments. However, challenges remain in integrating PM with FSM synthesis for real-time process analysis and optimization.

Ingemansson et al. [1] and Onggo et al. [2] discussed bottleneck detection and the substantial effort needed for data acquisition in simulation projects. Son et al. [3] emphasized preprocessing raw event data, and Gunz [4] noted the challenge of requiring expert knowledge to reprogram modular plant systems. Yurtay [5] reviewed PM's niche role in manufacturing optimization.

Efforts to generate state machine models from signals include Röder et al. [6], who proposed a semi-automated tool for hierarchical models from text-based event logs. However, this requires manual configuration and is unsuitable for low-level input/output (I/O) data. Meanwhile, Theis et al. [7] synthesized Petri nets from event logs using machine learning to approximate unknown programmable logic controller (PLC) logic. Koehler and Jing [8] forcibly reconfigure known PLCs to generate traces for dependency detection. Chivilikhin et al. [9] used a secondary PLC to collect Boolean signal traces from legacy systems. This method, reliant on asynchronous scan cycles, may introduce inaccuracies. Their SAT-based reconstruction targets controller migration, not performance evaluation.

Unlike these methods, the presented approach directly synthesizes FSMs from raw I/O signal changes to analyze throughput and performance. It operates without structured logs or system-specific knowledge, enabling real-time diagnostics in complex or undocumented industrial environments.

3 Preliminaries

Process time refers to the duration required to complete a manufacturing task. Key metrics include ***cycle time***, measuring the interval between repeated signal patterns, and ***throughput time***, which defines the total duration for completing a process cycle.

A ***finite state machine*** (FSM) models systems with finite states and input-driven transitions. The transitions between these states occur in response to inputs [10].

Formally, an FSM is a quintuple $A = (Q, \Sigma, \delta, q_0, F)$, where Q is the set of states, Σ the input alphabet, δ the transition function, q_0 the initial state, and F the accepting states [11]. The generally defined transition function δ allows a state to be transferred to several subsequent states when a character is read. In control engineering, however, each control assumes a very specific state at a specific point in time [12]. In such cases, we speak of a ***deterministic finite automaton*** (DFA) whose transition function describes a total function with $\delta : Q \times \Sigma \rightarrow Q$ [11].

4 Approach

4.1 Event Log Processing

Event logs are derived from time-stamped changes in binary I/O signals and represent sequences of unique events [7]. Stored in structured formats (e.g., JSON), they form the basis for FSM synthesis.

Each input or output signal is identified by participant ID id and signal position pos. The value of an input signal at time t is denoted by $I_{pos}^{id}(t)$, with rising and falling edges marked as $0 \rightarrow 1$ and $1 \rightarrow 0$, respectively.

4.2 FSM Synthesis

To generate a DFA from the event log, the set of all possible activities a represents the input alphabet Σ of the finite state machine G. The start state q_0 describes the state of the system before the first recorded signal change in the event log. The activity of each event in the log transfers the DFA to a new, previously unreached state whose index is greater than all previous indices. All states that are reached in this way form the set of accepting states F. For each pair $(q_m \in F, a \in \Sigma)$ for which no state transition is defined as a result, a transition to a common trap state q_t is defined. The set of all states Q thus consists of the accepting states and the trap state: $Q = F \cup \{q_t\}$. For reasons of clarity, we omit the trap state from transition diagrams. Since the trap state is the only state that is not contained in the set of accepting states, the end states are also not marked in this informal representation.

To illustrate this, an event log is given for a system in which there is only one bus participant that can receive two binary output signals from the PLC. The number of possible activities in this example is $3^{|I|} + 3^{|O|} - 2 = 3^0 + 3^2 - 2 = 8$. The transition diagram of the example DFA B in Fig. 1 results from the event log.

In the transition graph, the transitions resulting from the event log are labeled with the variables $t_i : i \in [0, 5]$. The specific activity that triggers the transition can be assigned to each t_i via the activity function $a(t_i)$. In addition, the time function $t(t_i)$ is used to assign each transition the number of times it occurs from the event

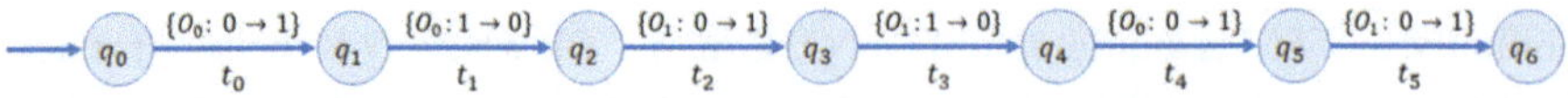

Fig. 1 Informal representation of the transition diagram

Fig. 2 Functiontable for C

i	$s(q_i)$	$a(t_i)$	$t(t_i)$
0	$\{O_0 : 0,\ O_1 : 0\}$	$\{O_0 : 0 \rightarrow 1\}$	{0,15s}
1	$\{O_0 : 1,\ O_1 : 0\}$	$\{O_0 : 1 \rightarrow 0\}$	{0,25s}
2	$\{O_0 : 0,\ O_1 : 0\}$	$\{O_1 : 0 \rightarrow 1\}$	{0,50s}
3	$\{O_0 : 0,\ O_1 : 1\}$	$\{O_1 : 1 \rightarrow 0\}$	{0,75s}
4	$\{O_0 : 0,\ O_1 : 0\}$	$\{O_0 : 0 \rightarrow 1\}$	{1,25s}
5	$\{O_0 : 1,\ O_1 : 0\}$	$\{O_1 : 0 \rightarrow 1\}$	{2,00s}
6	$\{O_0 : 1,\ O_1 : 1\}$	–	–

log. Furthermore, a signal value function s shall assign all states the signal value combination that is present in the event log when the state is reached. The initial signal values $s(q_0) = O_0 : 0,\ O_1 : 0$ are assigned to the start state of the DFA. The function $a(q_i, q_j)$ with $q_i, q_j \in Q$ outputs the input character $\alpha \in \Sigma$, which transfers the state q_i to q_j. The result of a concatenation $s \cdot a$ between the set of signal values s and the activity a is the set of signal values obtained by applying the signal changes contained in a to s. In cases where the original value of a signal change applied to a signal does not match the value of the signal from s, the concatenation is not defined. shows The functions for the given example are shown in Fig. 2.

The function s shows that the signal values assigned to the states q_0, q_2 and q_4 are identical. The same applies to the states q_1 and q_5. This knowledge is used to create a new DFA that combines all states with identical signal value assignments into one state. We will call this machine a combined DFA (CDFA). The indices of the states of the CDFA correspond to the order that results when the states are sorted according to the smallest index of their combined states. The CDFA C combined from the state

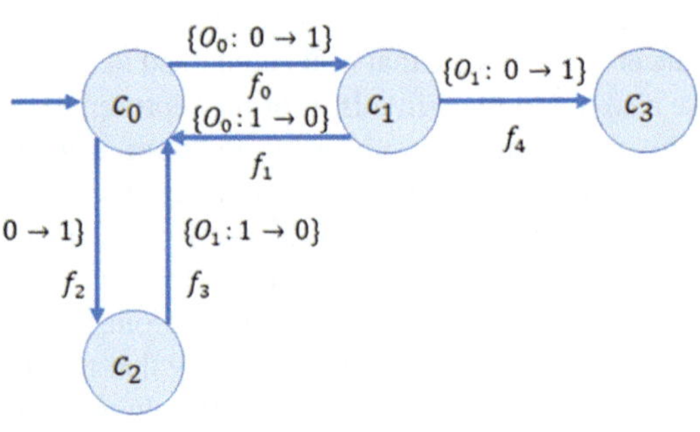

Fig. 3 Transition diagram of C

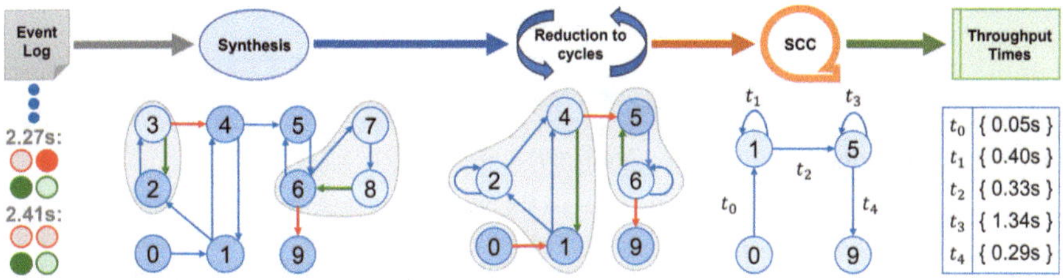

Fig. 4 Optimization procedure

machine B consists of the state set $Q(C) = \{c_0, c_1, c_2, c_3, c_t\}$ and the same input alphabet as B. c_0 is the start state and c_t is the trap state of C. Figure 3 shows the informal transition diagram of C.

While B models the sequence of signal changes of the unknown system recorded in a certain time window, C represents the possible subsequent signal changes for each recorded signal value combination. In contrast to B, C can contain cycles whose duration can be determined using the time function t. The set of times for each transition f of C is the union of the times of all transitions of B in which the signal values of the active state and the target state match the signal values of the corresponding states from f:

$$t(f_x) = \bigcup_{t_y \in T_x} t(t_y) \quad \text{with } f_x = (\delta(C)(c_i, \alpha) = c_j) \text{ and}$$

$$T_x = \{\delta(B)(b_m, \alpha) = b_n \mid b_m, b_n \in Q(B) : s(b_m) = s(c_i) \wedge s(b_n) = s(c_j)\}$$

4.3 *Optimization*

To optimize a CDFA $A = (Q, \Sigma, \delta, q_0, F)$ with regard to the cyclic process times of the industrial production plant, the cycles of the machine must be identified. In order to be able to use algorithms that require as little information as possible, A is abstracted to the essentials. This results in a digraph G with nodes $V(G) = F$ and edges $E(G) = \{[x, y] \mid x, y \in V, \alpha \in \Sigma : \delta(x, \alpha) = y\}$, which correspond to the states and transitions of the CDFA. The graph has the same cycles as A. To ensure that it is still possible to determine the cycle times, the edges of the graph are also assigned the set of timestamps of the corresponding transitions of the CDFA: $t([x, y]) = t(\delta(x, \alpha) = y)$. The corresponding digraph for the state machine C is shown in Fig. 5.

Each start node of a cycle in the graph is characterized by the fact that at least one returning edge leads to it. In the case of returning edges, the start node has a higher index than the destination node. The time stamps assigned to them represent the end times of a cycle and therefore the relevant information for determining the process

Fig. 5 Digraph G for C

times. In the first optimization step, the nodes and edges that have no corresponding information content are removed. The start node of the graph is always retained for the assignment of the initial signal values. In a second step, we combine the individual strongly connected components (SCCs) of the graph into a single node. An SCC groups all cycles that can be reached from each node within the SCC and can no longer be reached after leaving the SCC. The duration of the cycles that begin at the start node of an SCC thus describe throughput times of the greatest cyclic processes of the plant. Figure 4 shows the procedure of the optimization starting from the event log on a given example.

The procedure for reducing a graph G to its cycles runs through the start node of the graph and every node with an input degree greater than one. This node set contains the start nodes of all cycles, since a returning edge ends in them and they are either the start node or an edge from a node with a lower index leads to them. This node set represents that of a new, reduced graph $G' = (V', E')$. For each node $v \in V$ that is part of the set ($v \in V'$), each of its nodes $w \in V$ is examined to see whether it is also contained in the new graph. If this is the case, i.e. if $w \in V'$ is valid, the edge $[v, w] \in E$ is added to the edge set E'. If $w \notin V'$, its successor nodes $x \in V$ are examined in the same way. If $x \in V'$, the edge $[w, x] \in E$ with its associated timestamps is added to the graph G' as edge $[v, x] \in E'$: $t([w, x] \in E) = t([v, x] \in E')$. If x is also not part of the node set V', its successor nodes are examined using the same recursive procedure.

The end times of a cycle underlying the node are assigned to the loops of the reduced graph G' generated with the method. Different cycles can be summarized with the same start node, whereby the information is lost as to whether the sequence within a loop is the same or whether it is based on different sequences. This loss of information is not decisive for determining the cycle times, as the difference between a timestamp assigned to a loop and the previous transition time continues to describe the duration of a cycle.

Using Tarjan's algorithm [13] and a corresponding variant of Pearce [14], the SCCs can be identified and used to create a optimized graph. In this graph, the nodes of an SCC are combined into a common node and only contain the edges that lead out of an SCC or back to its start node. Each edge has the union of the timestamps that are assigned to the original edges from which the edge was created in the optimized graph. For all digraphs that can be created using the method presented in this paper, it is guaranteed by definition that a successor node with a smaller index is always part of the same SCC. If an edge $[v_j, v_i] \in E$ with $j < i$ exists, an edge path leading from v_i to v_j must exist due to the event log. In addition, it is ensured that a depth-first search, which starts with the start node of the graph, runs through every node.

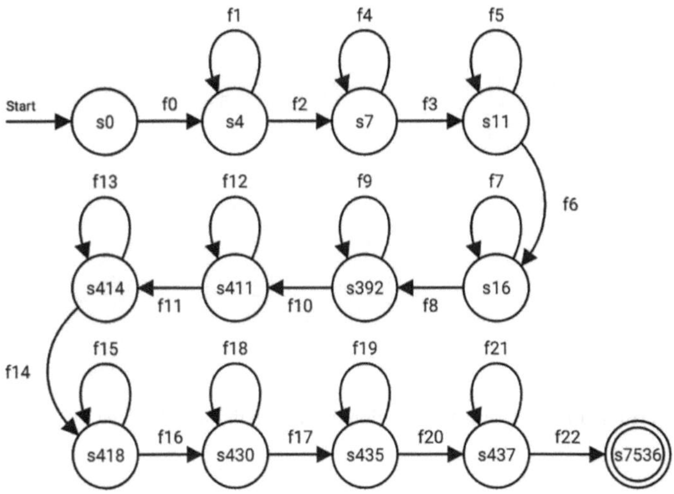

Fig. 6 Optimized transition diagram

5 Case Study

The proposed method was validated on a transfer system at the Institute of Production and Informatics in Sonthofen with two conveyor belts circulating four workpieces (WPs), each completing a cycle before returning to its start.

An event log of 8,723 signal transitions was used to generate and optimize the FSM.[1] The optimized CDFA is shown in Fig. 6, with selected transition times in Fig. 7.

Figure 8 visualizes the process using Gantt charts. Bars represent high signal values over time. The vertical lines in the diagram represent the time at which a state from the fully optimized automaton occurs for the first time. The red, solid

Fig. 7 Transition times

i	$t(f_i)$
7	$\{0.08s; 1.27s; 2.02s; 2.02s; 1.80s; 0.16s\}$
9	$\{3.44s\}$
21	$\{7.05s; 25.10s; 25.12s; 44.60s; 24.12s\}$
22	$\{17.91s\}$

[1] Event log and C++ implementation: www.github.com/Todesbengel0/CycleTimeFSM.

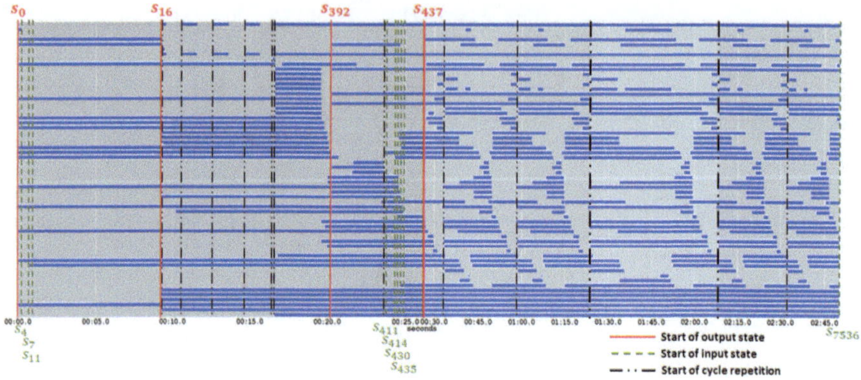

Fig. 8 Comparison of the determined transition times with the signal values

lines reference the start of an output state in the optimized automaton and the green, dashed lines the start of an input state. The respective cycle repetitions are marked with the black dashed two-dot lines. Using the Gantt chart, it can be validated that state s_{437} represents the main cycle of the plant, while the previous states are the start-up phase.

The calculated throughput times in the main cycle contain two discrepancies from an originally assumed 25 s. The first deviating cycle time of 7.05 s indicates an initialization phase before regular operation. At around 44.60 s, the low-performance cycle takes almost twice as long as the previous and subsequent cycles. Further analysis of state transitions identified two critical signal changes ($\{O_{43}^{0} : 1 \rightarrow 0\}$ and $\{O_{41}^{FFFE} : 1 \rightarrow 0\}$), correlating with extended belt activation times. The PLC configuration confirmed that reduced motor speeds led to longer transport times, explaining the performance fluctuation.

6 Conclusion

This work introduced an automated method for PM in industrial production through finite state machine synthesis and optimization. The proposed approach demonstrated its ability to analyze event logs, identify inefficient processes, and support decision-making without requiring prior knowledge of system configurations. By integrating state minimization and cycle detection, the method reduces manual effort and enhances process transparency.

Compared to traditional simulation-based approaches, the proposed method offers rapid deployment and minimal configuration effort, making it ideal for dynamic production environments. While simulation methods often require expert modeling and detailed system knowledge, our FSM-based approach operates purely on fieldbus data. However, a limitation is that process relevant actors and sensors cannot be discerned from signal changes used for meta communication. Without prior filtering of signals large plants might bloat the FSM synthesis to an unreasonable extend.

Future research should focus on extending the approach toward data-driven simulation models to enable real-time, adaptive analysis. Integrating computational intelligence can enhance predictive capabilities and support dynamic process improvements. Additionally, combining FSMs with probabilistic or timed models may better capture process variability. Finally, scaling the method to distributed or multi-line systems and developing adaptive FSMs that evolve with incoming data would improve long-term applicability.

References

1. Ingemansson A, Ylipää T, Bolmsjö GS (2005) Reducing bottle–necks in a manufacturing system with automatic data collection and discrete–event simulation. J Manuf Technol Manag 16(6):615–628
2. Onggo B, Hill J, Brooks R (2013) A pilot survey on data identification and collection in simulation projects. In: European simulation and modelling conference, ESM 2013
3. Son S, Yahya B, Song M, Choi S, Hyeon J, Lee B, Jang Y, Sung N (2014) Process mining for manufacturing process analysis: a case study. In: Ouyang C, Jung J-Y (eds) Asia pacific conference on business process management 2014. Brisbane, QLD. Springer, Cham
4. Gunz JM (2021) Process-Mining in der digitalisierten Produktion. Master's thesis, Fachhochschule Vorarlberg, Bodolz, Germany
5. Yurtay Y (2022) Process mining in manufacturing: a literature review. J Comput Inf Sci 5(3):341–355. Sakarya University
6. Röder A, Vasyutynskyy V, Kabitzsch K, Zarbock T, Luhn G (2005) Log-based state machine construction for analyzing internal logistics of semiconductor equipment. In: International conference on modeling and analysis of semiconductor manufacturing. Singapore, pp 54–60
7. Theis J, Mokhtarian I, Darabi H (2019) Process mining of programmable logic controllers: input/output event logs. In: International conference on automation science and engineering. IEEE, Vancouver, BC, pp 216–221
8. Koehler W, Jing Y (2020) Trace induction for complete manufacturing process model discovery. Int J Adv Manuf Technol. 110:29–43
9. Chivilikhin D, Patil S, Cordonnier A, Vyatkin V (2019) Towards automatic state machine reconstruction from legacy PLC using data collection. In: 17th international conference on industrial informatics. IEEE, Helsinki, Finland, pp 147–151
10. Hopcroft JE, Motwani R, Ullmann JD (2011) Einführung in die Automatentheorie, Formale Sprachen und Berechenbarkeit. Munich, ed. 3. Pearson Studium
11. Wagenknecht C, Hielscher M (2022) Formale Sprachen, abstrakte Automaten und Compiler. Wiesbaden, ed. 3. Springer
12. Wellenreuther G, Zastrow D (1995) Steuerungstechnik mit SPS. Braunschweig/ Wiesbaden, ed. 3. Friedr. Vieweg & Sohn Verlaggesellschaft mbH
13. Tarjan R (1971) Depth-first search and linear graph algorithms. In: Symposium on switching and automata theory. IEEE, East Lansing, MI, pp 114–121
14. Pearce DJ (2016) A space-efficient algorithm for finding strongly connected components. Inf Process Lett 116(1):47–52

Resistance Spot Welding: Quantitative Assessment of Its Impact on Cycle Time and Robotic Assembly Line Balancing

Dawid Stade and Martin Manns

Abstract Process time variance is frequently assessed in assembly line balancing research, a crucial production planning step. However, it has been largely overlooked in planning of highly automated car body construction assembly lines, owing to the deterministic nature of robots. Resistance spot welding, the most prevalent joining technique, employs feedback control systems for quality assurance, leading to varying process times. When multiple welds are done per cycle, these variations lead to fluctuations in overall cycle times. A quantitative analysis of 16 assembly lines reveals a 0.6 s median cycle time range. By incorporating process time variance in the planning of car body construction, this cycle time variance can be controlled to mitigate effects on assembly line performance. However, modelling of process times is required because these cannot be known in advance. As simulations on the balancing of car body construction assembly lines indicate, the model accuracy, however, needs to be high. Inaccurate distributional shape and spread predictions cause deficient planning assumptions, resulting in unpredictable outcomes.

Keywords Resistance spot welding · Robotic assembly line balancing · Process time variance · Car body construction · Closed-loop control

1 Introduction

A car body-in-white is made by joining components to assemblies that are manufactured in separate assembly lines that are arranged in a fish bone pattern. This arrangement enables a continuous and directed flow of materials to converge onto a central main assembly line for all intermediate goods [11]. A typical assembly line comprises standardized workstations. A workstation is equipped with robots designed for either material handling or joining processes. Each joining robot executes a sequence of joining operations. Due to the parallel operation of all robots, unloading can only commence upon completion of all joining operations. To meet

D. Stade (✉) · M. Manns
Universität Siegen, PROTECH-Institut für Produktionstechnik, Siegen, Germany
e-mail: dawid.stade@uni-siegen.de

M. Manns (ed.), *New Paradigms for Anticipated Uncertainty*, Lecture Notes in Mechanical Engineering, https://doi.org/10.1007/978-3-032-16889-4_53

stipulated production targets, the cycle time at each workstation must remain below a defined limit.

The division of the total workload for the fabrication of a body-in-white among multiple stations within an assembly line is a critical step of production planning. This challenge is formally recognized within the literature as the Robotic Assembly Line Balancing Problem (RALBP), first introduced by Rubinovitz and Bukchin [6]. Due to the deterministic character of robots, variable process times have received limited attention within the scientific community [1]. However, Resistance Spot Welding (RSW), the most common joining technique in car body construction, often utilizes Feedback Control Systems (FCS) to maintain weld quality under external disturbances. Controlling welding parameters, such as current and time, regulates weld nugget growth, thereby optimizing joint strength [13]. This necessitates the introduction of varying process times. Specifically, due to the interaction between robots and stations, these may significantly impact the efficiency of car body construction assembly lines. Hence, this work initially aims to analyze and quantify the effects of RSW process time variance on the performance of several real-world car body construction assembly lines.

In response to significant impacts, considering varying process times in the balancing of robotic assembly lines may help to mitigate these already in planning. Since process time distributions are initially unknown, modeling is necessary. To effectively control the impact of process time variations on assembly line performance, understanding the required accuracy for the distribution modeling is crucial. Hence, this work subsequently aims to quantify the effects of distribution model fidelity on robotic assembly line balancing outcomes. To this end, simulations of car body construction assembly line balancing are conducted, varying process time distribution shapes and spread widths. The balancing results are then compared to those obtained from empirical process time distributions. This simulation-based approach captures real-world data variability, thus obviating approximate distribution models.

The remainder of this work is organized as follows: Sect. 2 reviews the research on assembly line balancing, with a particular emphasis on car body construction and the consideration of variable process times. In Sect. 3, the influence of variability on actual assembly lines is quantified. Subsequently, Sect. 4 examines the impact of distribution characteristics on the RALBP. The results are discussed in Sect. 5. Section 6 concludes this study by summarizing the findings.

2 State of Research

Real-world robotic assembly line balancing complexity in car body construction often challenges optimal solutions. Recent studies address this through various approaches: Michalos et al. [4] addressed the complexity by decomposing the RALBP into line balancing and equipment selection, solved independently in a three-step method. Michels et al. [5] reduced the solution space by grouping operations based on joining techniques and linearizing cubic constraints. Hagemann and Stark

[3] achieved optimal solutions by exploiting the hierarchical structure of automotive body-in-white, enabling the decomposition of the RALBP into smaller, manageable units. However, process time variance has not been considered in the aforementioned publications. In assembly line balancing, process time variance can be modeled by stochastic, fuzzy, or robust approaches. The choice depends on the use case and the information available. Stochastic methods rely on probability distributions, typically assuming normal distributions [7], which may not be sufficient in practice. Stade and Manns [9] therefore present a genetic algorithm for the RALBP that is capable of considering arbitrary distributions. The risk of cycle time violations can be controlled by setting a probability limit. Fuzzy models utilize membership functions to map uncertainties more flexibly [12]. Robust methods minimize the influence of uncertainties through conservative modelling. Process times are modeled using intervals, and it is assumed that the number of tasks per workstation that necessitate the maximum process time is known [2].

3 Process Time Variance in Resistance Spot Welding

The RSW process under investigation has already been analyzed by Stade and Manns [8]. The process steps are: The robot positions and initiates closure of the welding gun. If a minimal gap remains, the welding controller proceeds closing until the clamping force remains within a preset range for a set duration. Subsequently, a feedback-controlled welding phase adjusts current and time to external influences. A holding and post-welding phase concludes the process, enabling gun opening. The closed-loop controlled gun-closing and welding exhibit process time variance. This work, therefore, focuses on these steps.

Process times for 2363 Mercedes-Benz car body weld joints are collected over one month for data analysis. Irregular times due to cleaning or electrode cap changes are excluded. Figure 1 presents histograms of the 5- to 95-percentile ranges of the process time distributions, referred to as the spread width. Both process steps and their respective combination are depicted. It is evident that the median spread width of 82 ms for welding is almost four times higher than that of 22 ms for closing. The value range is approximately 0.8 s, whereas the gun-closing step unfolds only in

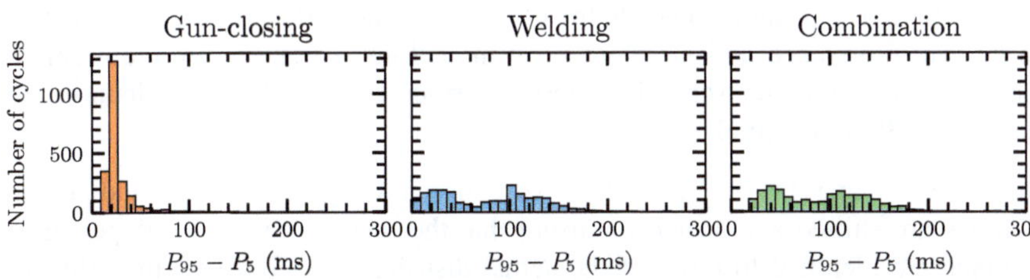

Fig. 1 The spread of weld time distributions is higher than that of gun-closing

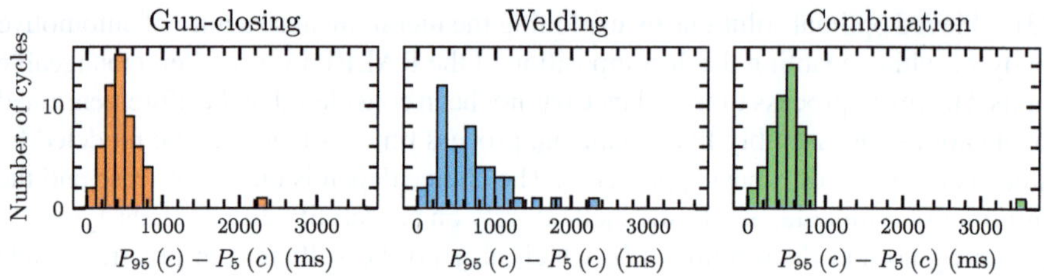

Fig. 2 As a result of more wide-spread weld times, these also lead to higher cycle time fluctuations than gun-closing

0.19 s. The combination results in a median spread width of 93 ms and a range of 0.77 s. The same observations can be applied to the distributions' shape, even if this cannot be clearly measured (Fig. 2).

Station cycle time variations exceed those of single joints due to the execution of multiple welds per robot and interdependencies within the stations. Figure 2 confirms this, showing cycle time spread widths of 55 stations across 16 assembly lines. The analysis distinguishes between gun-closing, welding, and their combination. In this case as well, the median spread width of the station cycle times for welding, at 0.46 s, is significantly higher than that for gun-closing, with a value of 0.14 s. The same applies to the range of values, with a significantly lower value of around 0.6 s for gun-closing, while neglecting the outlier at 2 s. Welding results in a time range of around 1.5 s. Because each robot produces multiple weld joints per cycle, the station cycle time distribution approximates a normal distribution in accordance with the central limit theorem. Consequently, the variations in the shape of the cycle time distributions are relatively minor. In combination, the varying process times of closing and welding in resistance spot welding result in a median spread width of approx. 0.6 s. The spread width fluctuates within a range of approx. 1.6 s and can even reach 3.6 s in exceptional cases.

4 Influence of Process Time Variance on the RALBP

Planning can control the impact of varying process times on cycle times in car body construction. This requires knowledge of process time distributions, which cannot be known in advance and therefore need modelling. To assess the requirements for model accuracy, two scenarios altering various aspects of the distributions are analyzed in RALBP simulations:

Distribution shape (DS): A normal distribution assumption is made, and process times are shifted and scaled to ensure that the median and 5- to 95-percentile range are identical to those of the actual distributions. To assess the influence of the distribution shape on the planning process, a comparison with the actual distribution results is conducted.

Spread width (SW): The actual distributions are preserved; however, the 5- to 95-percentile range is normalized to the average value of all joints. The influence of the spread width of process time distributions on the RALBP is determined by comparing the results to those obtained by actual distributions.

Four of the previously mentioned assembly lines (A-D) are selected for balancing in the experiments. The assemblies to be produced contain 90 to 169 primary and secondary weld joints, produced in four to six joining stages. A joining stage couples at least two components. All primary joints within a stage, ensuring structural integrity, must precede further welding. Secondary joints reinforce connections and can be produced later. A precedence matrix can be established that is adhered to by the RALBP.

The RALBP is solved using the genetic algorithm presented by Stade and Manns [9], which allows for consideration of arbitrary process time distributions. The algorithm allows control of the risk of cycle time violations using a probability limit α. Increasing the probability limit may free up capacity in rear stations, potentially reducing equipment costs. Therefore, the chosen probability limit should balance cost savings with controlled cycle time violation risk. To simulate this process, all experiments are repeated with $\alpha = 0\,\%, 5\,\%, 10\,\%$ and $50\,\%$. The genetic algorithm's parametrization includes a population size of 500, a crossover rate of 0.2, and a mutation rate of 0.02.

Two metrics presented by Stade et al. [10] are considered to quantify the effects on the RALBP: *Capacity savings* (CS) makes it possible to evaluate the extent to which the two scenarios affect the predicted capacity savings. This metric is important in practice. An incorrect estimation of capacity savings could result in an increased willingness of risk taking, without this paying off in practice. Since increasing the probability limit opens up capacity in the rear stations, according to Eq. 1 introduced by Stade et al. [10], savings CS are measured by the difference in median cycle times in the last workstation. The CS metric is determined by comparison of the predicted and actual capacity savings by increasing the limit from α_1 to α_2. The aim is to achieve a small positive value, which represents a slight underestimate of the capacity savings.

$$\text{CS}\,(\alpha_1, \alpha_2) = cs^{actual}_{\alpha_1 \to \alpha_2} - cs^{predicted}_{\alpha_1 \to \alpha_2} \tag{1}$$

The *probability limit violation* metric (PLV) indicates the probability of violating the preset limit α in practice. For this purpose, the balancing results are used to calculate the station cycle time distributions, but the process times are replaced by their actual counterparts. This is represented by function $F'(C)$ in Eq. 2 introduced by Stade et al. [10], that returns the probability of cycle times being below or equal to the predetermined cycle time C. Therefore, a positive PLV metric indicates a violation of the probability limit. Application of a balancing that results in probability limit violations could lead to an assembly line configuration that does not meet the expectations and increase costs.

$$\text{PLV}\,(\alpha) = 1 - F'(C) - \alpha \tag{2}$$

Table 1 Inadequate spread widths negatively affects the CS metric and results in potential overestimations of capacity savings due to increased probability limits

			CS (α_1, α_2)		
Line	Step	Scenario	0% → 5%	5% → 10%	10% → 50%
A	Welding	DS	0.0 s	0.0 s	9.02 s
		SW	0.0 s	−0.02 s	−2.96 s
	Both	DS	0.0 s	0.0 s	9.02 s
		SW	0.0 s	0.0 s	−2.98 s
B	Welding	DS	0.0 s	0.0 s	6.06 s
		SW	0.0 s	0.0 s	−5.91 s
	Both	DS	0.0 s	0.0 s	9.04 s
		SW	0.0 s	0.0 s	−5.92 s

4.1 Capacity Savings

Table 1 displays the results of the CS metric, which are filtered for values below or equal to −0.5 s to focus on major overestimations of capacity savings. This filtering shows that the CS metric only has negative examples in assembly lines A and B, as well as in the welding process step or combination of both. Furthermore, the savings are only overestimated if the probability limit is increased from 10 % to 50 %. It becomes apparent that the overestimations only occur when the spread width is normalized. On the other hand, the distributions' shape has only a minor negative impact on the CS metric, but in 12 cases it leads to a significant underestimation of at least 5 s of capacity savings. The findings demonstrate a significant impact of both the distributions' shape and spread width on the CS metric, even though only the spread width leads to significant overestimations.

4.2 Probability Limit Violation

The results of the PLV metric are shown in Fig. 3, which presents the results of both scenarios in histograms. In this instance, the outcomes have been reversed, and a correlation with the process steps cannot be discerned. While the normalized spread width violates the probability limit a few times, the normal distribution assumption results in a total of 63 violations. The majority of these violations are within the range of 40 % to 50 %. The findings indicate that the distributions' shape exerts a significant impact on the PLV metric.

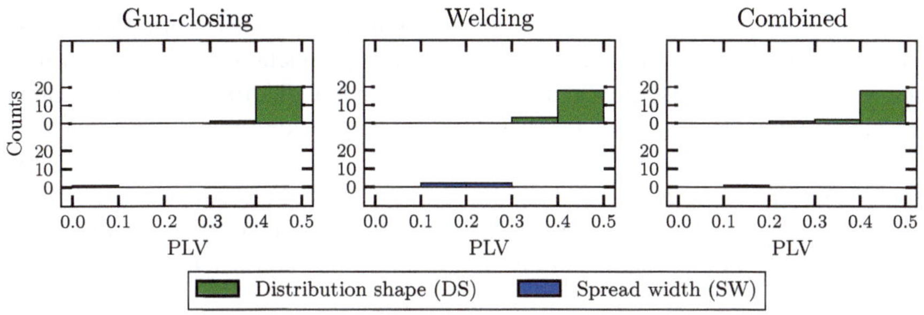

Fig. 3 False assumptions of normally distributed process times negatively impacts the PLV metric, with 63 violations concentrated between 40% and 50%

5 Discussion

Production data analysis indicates that variable process times affect car body construction performance by influencing station cycle times. Apart from other factors, investigations suggest that process time variations contribute to instability, potentially requiring buffers. Considering these in production planning can help reduce costs, as fewer compensatory measures are required. However, this requires adequate distribution models because process times cannot be known in advance. As the simulations demonstrate, both the distributions' shape and spread width must be estimated with high accuracy. A discrepancy between the predicted and actual spread widths of process time distributions has a negative impact on the estimation of capacity savings. On the other hand, an inappropriate distribution assumption causes frequent and significant violations of the predetermined probability limit. Hence, both aspects of the process time distributions directly impact the quality of balancing results. Therefore, modeling techniques, including statistical learning, must accurately reflect these attributes. Diverse joining techniques in car body construction necessitate significant effort. However, current results do not fully incorporate all technical restrictions and distributional characteristics. Future research should address technical constraints like weld gun incompatibilities, account for outlier process times, and consider irregular processes such as electrode cap changes to achieve more practically meaningful results.

6 Conclusion and Future Research

This study quantitatively investigates the impact of varying process times in resistance spot welding on the cycle times in car body construction. In addition, simulations are conducted to investigate the influence of the process time distributions on the RALBP. It has been demonstrated that the process time variance in resistance spot welding alone is responsible for significant cycle time variations in production. The simulations show the potential to predict and thus control these during production planning. In practice, this can contribute to a stable production output. Furthermore, the results also demonstrate the clear need for suitable distribution models. Otherwise, the planning results may be misleading, which could result in additional costs. Therefore, finding and parameterizing an appropriate distribution model is of central importance and requires a great deal of effort in practice. Future research may focus on reducing the effort involved by using suitable models and processes, considering the large number of joining techniques used in car body construction.

References

1. Chutima P (2022) A comprehensive review of robotic assembly line balancing problem. J Intell Manuf 33(1):1–34. https://doi.org/10.1007/s10845-020-01641-7
2. Gurevsky E, Hazır Ö, Battaïa O, Dolgui A (2013) Robust balancing of straight assembly lines with interval task times. J Oper Res Soc 64(11):1607–1613. https://doi.org/10.1057/jors.2012.139
3. Hagemann S, Stark R (2020) An optimal algorithm for the robotic assembly system design problem: an industrial case study. CIRP J Manuf Sci Technol 31:500–513. https://doi.org/10.1016/j.cirpj.2020.08.002
4. Michalos G, Fysikopoulos A, Makris S, Mourtzis D, Chryssolouris G (2015) Multi criteria assembly line design and configuration - an automotive case study. CIRP J Manuf Sci Technol 9:69–87. https://doi.org/10.1016/j.cirpj.2015.01.002
5. Michels AS, Lopes TC, Sikora CGS, Magatão L (2018) The Robotic Assembly Line Design (RALD) problem: model and case studies with practical extensions. Comput Ind Eng 120:320–333. https://doi.org/10.1016/j.cie.2018.04.010
6. Rubinovitz J, Bukchin J (1991) Design and balancing of robotic assembly lines. In: Proceedings of the fourth world conference on robotics research, Pittsburgh, pp 17–19
7. Şahin MC, Tural MK (2023) Robotic stochastic assembly line balancing. Flexible Serv Manuf J 35(4):1076–1115. https://doi.org/10.1007/s10696-023-09494-x
8. Stade D, Manns M (2023) Influence of time-variant robotic welding processes in car body construction. In: Galizia FG, Bortolini M (eds) Production processes and product evolution in the age of disruption. Springer, Bologna, pp 581–588, https://doi.org/10.1007/978-3-031-34821-1_63
9. Stade D, Manns M (2023) Robotic assembly line balancing with multimodal stochastic processing times. In: Advances in automotive production technology – towards software-defined manufacturing and resilient supply chains. Springer International Publishing, Stuttgart, pp 78–84, https://doi.org/10.1007/978-3-031-27933-1_8
10. Stade D, Spoor JM, Manns M, Ovtcharova J (2024) Process time distribution simulation in robotic assembly line balancing. Int J Prod Res pp 1–18, https://doi.org/10.1080/00207543.2024.2416570
11. Wemhöner N (2006) Flexibilitätsoptimierung zur Auslastungssteigerung im Automobilrohbau. PhD thesis, Aachen
12. Zacharia PT, Nearchou AC (2013) A meta-heuristic algorithm for the fuzzy assembly line balancing type-E problem. Comput Oper Res 40(12):3033–3044. https://doi.org/10.1016/j.cor.2013.07.012
13. Zhou K, Yao P (2019) Overview of recent advances of process analysis and quality control in resistance spot welding. Mech Syst Signal Process 124:170–198. https://doi.org/10.1016/j.ymssp.2019.01.041

A Dynamic Digital Twin Modelling Environment Integrating Maintenance Policy in RMS Operations' Management

Marco Bortolini, Francesco Gabriele Galizia, Ludovica Diletta Naldi, Giulia Cardelli, Michele Micci, Andrea Sanfilippo, and Alberto Regattieri

Abstract In Industry 4.0 and 5.0 eras, Reconfigurable Manufacturing Systems (RMSs) rose as effective solution to meet the highly dynamic customers' needs. RMSs deal with intelligent machines able to perform a large variety of tasks, by 'reshaping' themselves thanks to auxiliary modules which can be assembled and disassembled to/from their base structures. However, despite auxiliary modules increase reconfigurability level of these systems, their frequent assembly/disassembly and the highly different working parameters and conditions subject them to greater wear and tear risks. Hence, an appropriate maintenance policy must be planned to maximise the modules' lifespan, reducing downtimes and costs. In this context, embracing the progress of digital technologies, Digital Twins (DTs) are becoming a widespread tool in smart manufacturing for performance analysis, process simulation and the development of *what-if* DfX analysis. This paper presents a dynamic DT modelling environment, based on a commercial 3D scale simulation platform, to integrate maintenance policy in RMS operations' management. By focusing on the most relevant challenges posed by RMSs, the model allows simulating the impact of responsive and flexible maintenance alternatives, tracking their impact on the system uptime and productivity. The proposed DT modelling environment is preliminary tested through a simplified industrial case study inspired from the metalwork mechanical sector, showcasing its effectiveness in supporting operations' activity planning and management within complex and adaptable production environments.

Keywords Digital twin · Predictive maintenance · RMS simulation

M. Bortolini · F. G. Galizia (✉) · L. D. Naldi · G. Cardelli · M. Micci · A. Sanfilippo · A. Regattieri
Department of Industrial Engineering, Alma Mater Studiorum—University of Bologna, Bologna, Italy
e-mail: francesco.galizia3@unibo.it

M. Manns (ed.), *New Paradigms for Anticipated Uncertainty*, Lecture Notes in Mechanical Engineering, https://doi.org/10.1007/978-3-032-16889-4_54

1 Introduction and Background

The evolution of market dynamics increases demand variability, leading to a growing need for high-quality customized products [1–3]. Customers constantly seek innovative and personalized solutions, pushing companies to introduce new product variants more and more frequently. At the same time, product life cycles become shorter, due to rapid technological advancements and the shift of the customer preferences [4]. In this context, traditional production systems, mainly characterized by high volume production through standardized processes, are not able to guarantee high flexibility standards. Hence, flexible production systems are of help to respond efficiently and quickly to this evolving scenario while maintaining cost-effectiveness and operational efficiency [5]. Among them, Reconfigurable Manufacturing Systems (RMSs) emerged as an efficient solution [6]. These systems are able to modify their functionality and production capacity quickly and cost-effectively, in response to unpredictable demand changes. RMSs are characterized by a set of Reconfigurable Machine Tools (RMTs), which are modular machines that can be equipped with advanced tools, i.e., auxiliary modules, able to perform different manufacturing tasks [7, 8]. Furthermore, a reconfigurable system lies on six main characteristics, i.e., modularity, defined as the ability of the system to be made by a set of modular components devoted to a specific function; scalability, allowing the arrangement of the production capacity according to the market demand; customization, that allows product variation without significative system reconfigurations; convertibility, defined as the capacity to modify production functions to second different tasks; advanced diagnostics, used to monitor the system for preventive maintenance and downtime reduction; interoperability, which ensures effective communication between machines and digital systems in the production environment [9]. Thus, the adoption of RMSs represents a strategic solution for companies, enabling them to enhance competitiveness by improving responsiveness to market changes and optimizing the use of production resources [10].

Focusing on auxiliary modules, which are the key element of RMSs, due to frequent assembly and disassembly actions, such components face continuous risks of potential failures. Hence, an effective maintenance strategy is crucial to maximizing their operational lifespan, ensuring long-term reliability. Despite the industrial relevance of this topic, literature exploring maintenance strategies in RMS environments is still rare [11]. Among the existing contributions, Xia et al. [11] developed a novel dynamic maintenance strategy to make real time schedules for system-level opportunistic maintenance. However, in this study, the behavior of auxiliary modules is not analyzed. Maintenance strategies in RMSs need to include corrective and preventive maintenance policies. Corrective maintenance policy addresses failures after they occur, requiring fast reactive interventions that cause downtime and production delays. Preventive maintenance policy aims to reduce failure risks through scheduled inspections, component replacements, and predictive analytics [12]. By implementing preventive maintenance policies, companies enhance system performance, reduce operational disruptions, hence optimizing the lifecycle of the auxiliary modules.

During the last years governed by Industry 4.0 and, more recently, by Industry 5.0, Digital Twins (DTs) are widely used to simulate the behavior of complex systems and processes, monitoring their performance [13]. In this study, a dynamic DT model, based on a commercial 3D scale simulation platform, is proposed, to integrate maintenance policy in RMS operations' management. The impact of implementing a preventive maintenance strategy compared to a reactive approach, i.e., failure-based, is studied and a sensitivity analysis, varying the number of auxiliary modules, determines the influence of different maintenance policies on the overall system efficiency and operational stability. The DT model helps to answer the main research question lying behind this study: *'What is the impact of the adopted maintenance policy in the RMS and auxiliary modules' availability and performances?'*

Starting from this background, the remainder of this paper is structured as follows. Section 2 presents the dynamic DT model, while Sect. 3 discusses some results of industrial relevance. Finally, Sect. 4 concludes this paper with final remarks and future research opportunities.

2 Dynamic Digital Twin Model

To evaluate the impact of different maintenance policies in RMS operations' management, a dynamic DT of a small-scale production system is developed using AutoMod simulation software. The DT represents a simplified industrial case study inspired by the metalwork mechanical sector, allowing to evaluate the impact of the adoption of a statistical based preventive maintenance on the system uptime and productivity, compared with the adoption of a corrective failure policy.

The reference layout of the virtual production system is sketched in Fig. 1.

It is composed of a production area, consisting of 4 different RMTs served by a conveyor belt, and of an auxiliary module storage and repairing area, consisting of a modular warehouse served by an Automated Guided Vehicle (AGV) and an operator/resource in charge of the modules' maintenance. The conveyor belt handles 34 different types of parts from the input area on the left, i.e. *in_area*, to the RMTs, represented as blue cuboids on the belt corners, that perform the production operations. Finished parts exit the system reaching the output area, labeled as *out_area*, on the right. The use of 10 types of modules enables an RMT to perform up to 16 different operations. The AGV manages the handling of the different modules from the storage area, labeled as *mod_storage* in Fig. 1, to the RMTs. Each RMT owns a buffer area storing the modules waiting to be set up on the machine or retrieved by the AGV.

The parts' production cycles include between 1 and 8 operations. Each RMT can perform different operations, requiring a specific set of auxiliary modules, i.e., from a minimum of 1 to a maximum of 3 auxiliary modules. A feasible operation-RMT-modules combination results in different working times, while the assembly

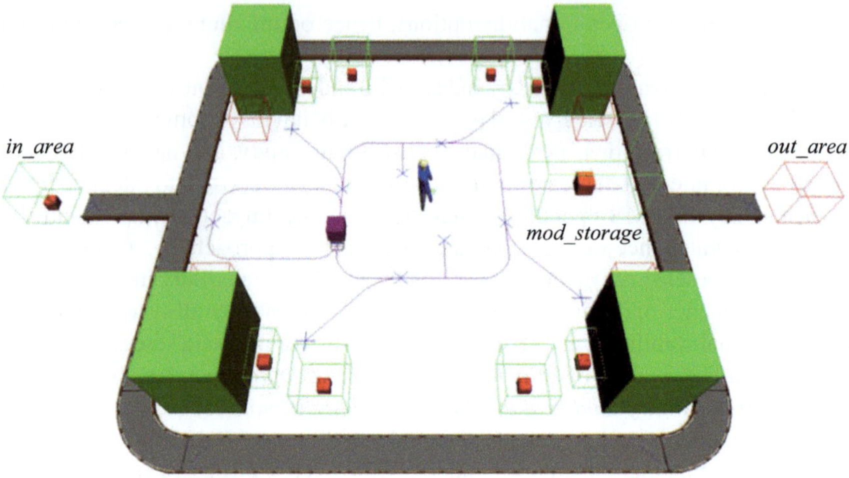

Fig. 1 Layout of the virtual production system on AutoMod

and disassembly time of a module is independent on the RMT that uses it. For the sake of brevity, details on parts and required operations, operation-RMT-modules compatibility as well as the other input data are omitted and can be found in Bortolini et al. [8].

The auxiliary modules are limited in quantity and, consequently, not always available when requested. A preliminary static sizing of their minimum number, given the target parts' production volumes, leads to 15 units (see the detail in Table 1, first row). Starting from this configuration, a multi-scenario analysis evaluates the system performance if the number of the modules increases up to 28 units (see the detail in Table 1, last row). At each iteration, to define the auxiliary module to add, the most used type is chosen.

Based on the presented layout and data, the adoption of a statistical based preventive maintenance policy for the auxiliary modules is simulated and compared to the use of corrective policy. Two conditions trigger each preventive maintenance activity, i.e., the auxiliary module cumulative number of working hours (set to 180 h in this case study) and the cumulative number of assembly and disassembly tasks (set to 250 in this case study). Additionally, two Weibull distribution functions model the

Table 1 Multi-scenario analysis varying the number of modules used for each module type. (MTx in the column headings)

Module type	MT1	MT2	MT3	MT4	MT5	MT6	MT7	MT8	MT9	MT10	TOT
Initial (static) number	1	1	2	2	2	2	1	1	2	1	15
Final (dynamic) number	1	3	7	2	4	3	1	2	4	1	28

auxiliary modules' failures due to the working time (shape parameter equal to 3, and scale parameter of 806,300 s in this case study) and to the number of assembly/ disassembly tasks (shape parameter equal to 3, and scale parameter of 315 in this case study). If a failure occurs, the module reaches the maintenance area, and it is repaired by a dedicated operator/resource. The proposed DT is based on the following hypotheses:

- Modules are always repairable and repaired *as good as new*;
- Failure detection and preventive maintenance activities on a module are performed at the end of each RMT operation.

Finally, the system's behavior and performances are simulated using the DT and considering three different parts' feeding policies, i.e., longest processing time (LPT), shortest processing time (SPT), and random (RND) scheduling.

Globally, 84 scenarios are simulated by completing 40 sequences of parts to process (approximately 900 working hours per each simulation run). Data about the system productivity, the number of failures and the average modules' use rate are collected as relevant KPIs to monitor and plot. Key results are discussed in Sect. 3.

3 Results and Discussion

Figure 2 plots the overall system productivity, in pcs/hour, varying the number of auxiliary modules. The six different lines are for the simulated combinations of maintenance and parts' feeding policy.

The benefit of adding auxiliary modules decreases as their total number increases. Between 15 and 17 auxiliary modules the hourly productivity increases of about $\nabla_1 =$ 15.5 pcs/hour per each added unit. This value decreases to $\nabla_2 = 3.6$ pcs/hour between 17 and 22 modules and is almost null beyond ($\nabla_3 = 0.33$), stating there is no reason to further invest in such costly resources. Additionally, SPT and LPT feeding policy outperform RND scheduling especially if the auxiliary modules are a scarce resource.

Figure 3 focuses on the impact of preventive maintenance policy on the number of failures the auxiliary modules experienced. The use of the DT suggests the adoption of this maintenance policy if the auxiliary modules are in a limited number, while for the scenarios with 20 or more modules the failures' number reduction is almost negligible between the preventive (in blue) and corrective (in black) scenarios. This is mainly due to the lower use of the auxiliary modules if their number is high, i.e., highly intermitted use, that makes the preventive maintenance policy less effective.

Finally, all the maintenance and parts' feeding policies present a similar and almost linear decreasing trend of the auxiliary modules' average use rate while the total number of units increases (Fig. 4). Analytically, the use rate quantifies the percentage of production time of the modules over the total available time. The scattered trend for 15 to 17 auxiliary modules is due to the higher impact of down times because of failures of modules.

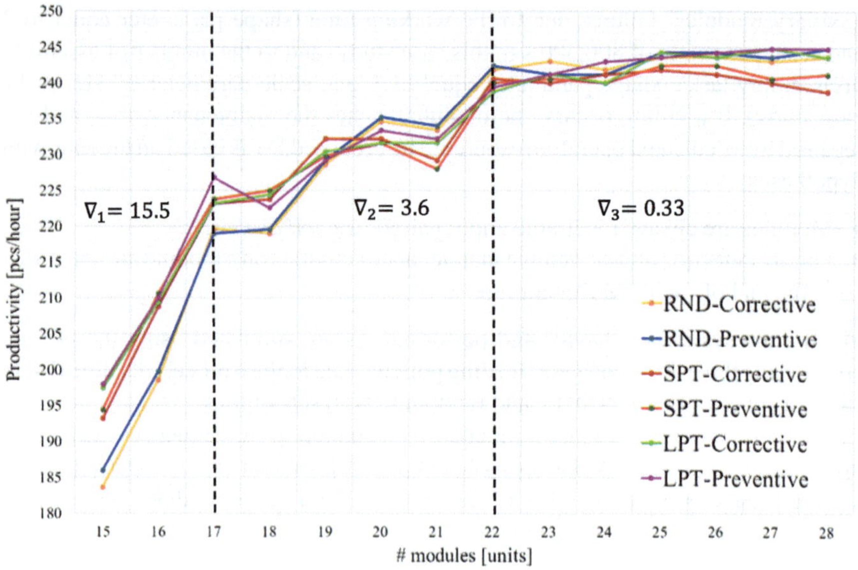

Fig. 2 Industrial case study: productivity analysis varying the number of auxiliary modules

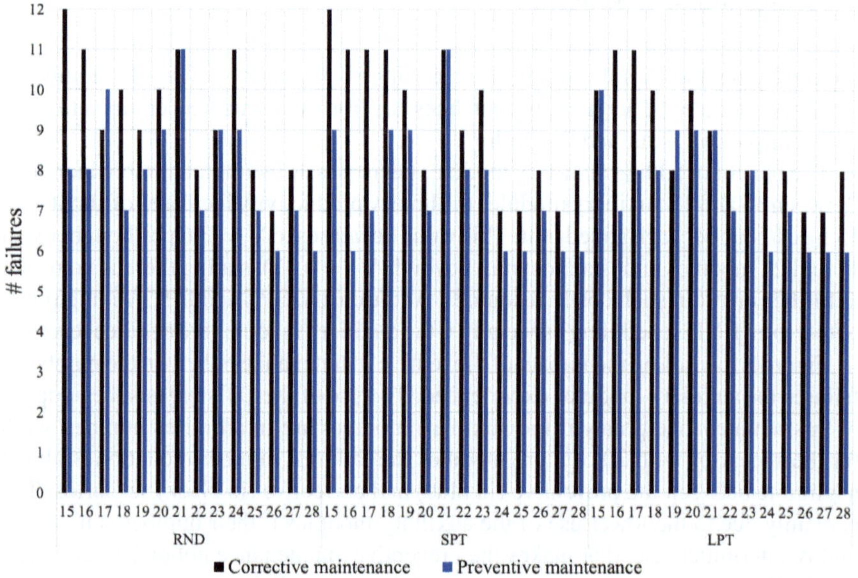

Fig. 3 Industrial case study: preventive maintenance impact on the auxiliary modules' failures

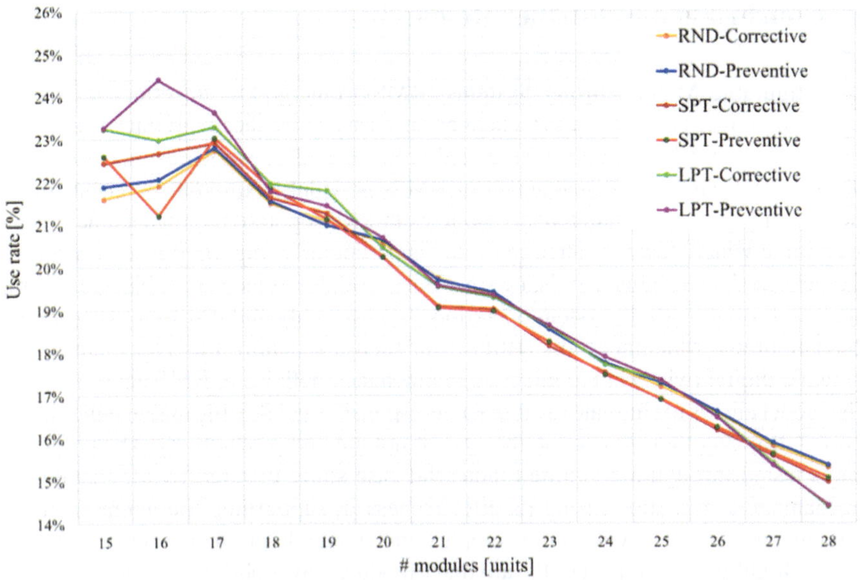

Fig. 4 Industrial case study: auxiliary modules' use rate analysis varying their number of units

The obtained results emphasize multiple relevant insights of potential help for practitioners and industry enhancing traditional methods and models for operations' management and production scheduling [9, 12]. Among them:

- The investment in preventive maintenance is more and more of interest as the auxiliary modules are scarce in number and frequent in their use, assembly and disassembly. Structured maintenance plans for 'lumpy' modules have a longer return on investment. This phenomenon is further emphasized in the case auxiliary modules are not always fully reparable;
- A deep failure root cause analysis is a cornerstone to tune the preventive maintenance plans. Both data on the auxiliary modules' production time and their assembly and disassembly from RMTs are needed as different and interconnected stresses that affect the auxiliary modules' working state and reliability;
- Designing the production plans not to overstress auxiliary modules with frequent assembly and disassembly activities is an interesting insight for production planners further enabling opportunistic maintenance activities;
- Transferring the DT performance back to real industrial systems is crucial to fully exploiting the potential of the proposed modelling environment, driving the change toward new approaches of managing production systems and operations plans.

4 Conclusion and Further Research

Reconfigurable Manufacturing Systems (RMSs) emerged as an effective answer to recent industrial and market changes in terms of reduced product lifecycles, increasing demand for highly customized products and dynamic customer demand. Despite the design and management of RMSs is widely explored by recent literature, the integration of maintenance policies in RMS operations' management is an open topic which deserves attention. This is particularly important since RMSs are characterized by advanced tools, i.e., auxiliary modules, which are subject to several stresses during the daily system operation, i.e., assembly and disassembly from the Reconfigurable Machine Tools (RMTs). In such a scenario, this paper presents a dynamic digital twin (DT) to integrate maintenance policies in RMS environments. The model allows to simulate the impact of responsive and flexible maintenance alternatives, tracking their impact on the system uptime and productivity. The proposed DT is tested through a simplified industrial case study inspired by the metalwork mechanical sector, showcasing its effectiveness in supporting operations' activity planning and management within complex and adaptable production environments. Results highlight the impact of the auxiliary modules' availability, the adopted maintenance and the parts' feeding policies on the system productivity, failure and use rate of the auxiliary modules, enabling insights of potential help for practitioners and industry.

Future research deals with the validation of the DT through its application to larger real-sized industrial instances, facing the challenges of scaling real-world large-scale implementations, and the development of optimization models to best size the maintenance effort and parameters varying the auxiliary modules' number, features and behavior. Additionally, simulating non fully repairable auxiliary modules by including their performance degradation profiles is expected to strengthen the results' reliability.

References

1. Bi ZM, Lang SYT, Shen W, Wang L (2008) Reconfigurable manufacturing systems: the state of the art. Int J Prod Res 46(4):967–992
2. Bortolini M, Galizia FG, Mora C (2018) Reconfigurable manufacturing systems: literature review and research trend. J Manuf Syst 49:93–106
3. Naldi LD, Galizia FG, Bortolini M (2025) Balancing storage cost and customization time in product platform design: a bi-objective optimization model. Int J Adv Manuf Technol 1-14
4. Koren Y, Shpitalni M (2010) Design of reconfigurable manufacturing systems. J Manuf Syst 29(4):130–141
5. Youssef AM, ElMaraghy HA (2007) Optimal configuration selection for reconfigurable manufacturing systems. Int J Flex Manuf Syst 19:67–106
6. Koren Y, Heisel U, Jovane F, Moriwaki T, Pritschow G, Ulsoy G, Van Brussel H (1999) Reconfigurable manufacturing systems. CIRP Ann 48(2):527–540

7. Eguia I, Molina JC, Lozano S, Racero J (2017) Cell design and multi-period machine loading in cellular reconfigurable manufacturing systems with alternative routing. Int J Prod Res 55(10):2775–2790
8. Bortolini M, Ferrari E, Galizia FG, Regattieri A (2021) An optimisation model for the dynamic management of cellular reconfigurable manufacturing systems under auxiliary module availability constraints. J Manuf Syst 58:442–451
9. Huang A, Badurdeen F, Jawahir IS (2018) Towards developing sustainable reconfigurable manufacturing systems. Procedia Manuf 17:1136–1143
10. Mehrabi MG, Ulsoy AG, Koren Y (2000) Reconfigurable manufacturing systems: key to future manufacturing. J Intell Manuf 11:403–419
11. Xia T, Xi L, Pan E, Ni J (2017) Reconfiguration-oriented opportunistic maintenance policy for reconfigurable manufacturing systems. Reliab Eng Syst Saf 166:87–98
12. Gackowiec P (2019) General overview of maintenance strategies–concepts and approaches. Multidiscip Asp Prod Eng 2
13. Javaid M, Haleem A, Suman R (2023) Digital twin applications toward industry 4.0: a review. Cogn Robot 3:71–92

Design of a Decision Logic of a Hybrid Matrix Assembly for Electric Vehicle Production

Stefanie Dechant and Hans-Christian Möhring

Abstract Automobile manufacturers face increasing final assembly complexity due to expanding model portfolios, customized configurations, and emerging powertrain technologies. Conventional assembly lines struggle to absorb process variability efficiently. In response, this study introduces a hybrid assembly concept that integrates cycle-based line segments with flexible matrix segments to combine efficiency with adaptability. A structured scoring-based decision methodology is developed to allocate process modules based on criteria such as process time variance, variant diversity, labor requirements, and precedence constraints. The approach is validated through a case study at a German automotive OEM. Results show that 11 out of 17 assembly clusters benefit from matrix integration, with reduced cycle time losses and improved flexibility. The findings demonstrate the practical applicability of the methodology and its potential to support the transition toward future-oriented, modular assembly systems.

Keywords Assembly · Modularization · Matrix production · Hybrid assembly · Efficiency

1 Introduction

Automobile manufacturers face increasing competitive pressure and growing demand for individualized vehicles. This necessitates a systematic expansion of the model portfolio and a greater diversification of optional features [1]. Many OEMs still rely on synchronized high-volume production [2], though increasing product variance lowers line efficiency, reinforcing the need for adaptable concepts

S. Dechant · H.-C. Möhring
Institute for Machine Tools, University of Stuttgart, Stuttgart, Germany

S. Dechant (✉)
Mercedes-Benz Group AG, Stuttgart, Germany
e-mail: Stefanie_ruth.dechant@mercedes-benz.com

M. Manns (ed.), *New Paradigms for Anticipated Uncertainty*, Lecture Notes in Mechanical Engineering, https://doi.org/10.1007/978-3-032-16889-4_55

[3]. Matrix production, a modular and flexible assembly concept based on decoupled workstations, has become a key topic of discussion in both academia and industry [4]. The fundamental goal is to reduce manufacturing costs for highly variant products while increasing flexibility and efficiency [3]. Therefore, the system requires a structured task allocation [4]. Greschke et al. [4] introduce a dynamic, sequential task allocation method that assigns tasks based on current system conditions, considering processing times, labor requirements, and average cycle time to balance workload adaptively. Kern et al. [5] propose eight deterministic assignment rules derived from the assembly precedence graph. These rules combine product-related aspects, including task dependencies and installation sequences, with process-oriented considerations like spatial requirements. Accordingly, Hottenrott and Grunow [6] compare matrix layouts with traditional mixed-model assembly lines. By using automated guided vehicles (AGV) and allowing vehicles to bypass unnecessary stations, they reduce idle times and improve station utilization. Their results show an average efficiency gain of 24.5%. A simulation study by the Boston Consulting Group demonstrates an 11% increase in overall worker utilization in matrix systems, achieved through balanced task allocation and reduced idle times [7]. The system architecture is supported by a flexible material flow and intelligent production planning, enabling demand-driven control of material flow while compensating disruptions via redundant cells [8].

However, the widespread adoption of modular concepts in industrial practice remains limited. High investment costs, control complexity, and a lack of standardization represent key barriers [3]. To address this, Kampker et al. [9] propose a hybrid assembly system, combining matrix and line segments. To allocate tasks to flexible workstations, they present a five-step methodology. Tasks are modularized focusing on those with high execution time variance. The allocation is guided by cycle-based capacity matching and considers worker utilization, entry/exit times, and buffers. Design principles include minimizing task dependencies, avoiding redundancy, grouping logically, and ensuring spatial proximity to enhance ergonomics and reduce overhead. While Kampker et al. [9] report a 19% increase in a simulation study, integrating hybrid systems into existing environments remains challenging. Göppert et al. [10] explore a comprehensive strategy that merges the benefits of both assembly structures by incorporating planning software and digital twins alongside a flexible layout and AGV transport routes. Bebersdorf and Huchzermeier [3] apply this concept to a cycle-based configuration, but their study focuses on predefined layouts and lacks data-driven evaluation. While existing approaches provide useful foundations, they remain limited in providing empirical decision support for task allocation for future Electric Vehicle (EV) assembly. This paper presents a scoring-based approach to support the systematic distribution of assembly processes between line and matrix segments. It aims to enhance reproducibility and reduce dependence on expert intuition in high-variance EV production. Section 2 introduces the decision methodology. Accordingly, Sect. 3 demonstrates and discusses its application through an automotive case study. Finally, Sect. 4 concludes the paper and outlines future research directions.

2 Decision Logic for the Design of Hybrid Assembly Systems

A structured method is required to allocate process modules in hybrid assembly systems under conditions of high variant diversity and limited process transparency. This section presents a scoring-based methodology based on a set of quantifiable, decision-relevant criteria. Scoring models provide a framework for translating complex qualitative factors into transparent, comparable measures [9]. The model determines, based on empirical data and quantifiable criteria, which assembly tasks are suited for sequential line production and which, due to high variant diversity and flexibility demands, are better integrated into matrix segments. Each vehicle passes through the complete production system, but the associated assembly operations are split between flexible matrix and synchronized line segments depending on their characteristics, as shown in Fig. 1.

Given the high process variability, there is no complete and detailed assembly priority diagram for each vehicle variant in practice (up to 10^{32} theoretical variants [11]). The allocation is therefore carried out at an aggregated level in the form of process modules p_i, each encompassing multiple task operations $j \in p_i$. While highly standardized processes facilitate seamless integration into line assembly, matrix production requires greater adaptability to accommodate variant-rich assembly scenarios. To quantify the flexibility requirements of individual process modules, the process time variance is incorporated as the first factor into the scoring model.

Process Time Variance (S) The process time variance F_S quantifies the relative spread of processing times within a process module p. It is calculated as the difference between the maximum and minimum processing time, normalized by the average time $\overline{T}_p$.

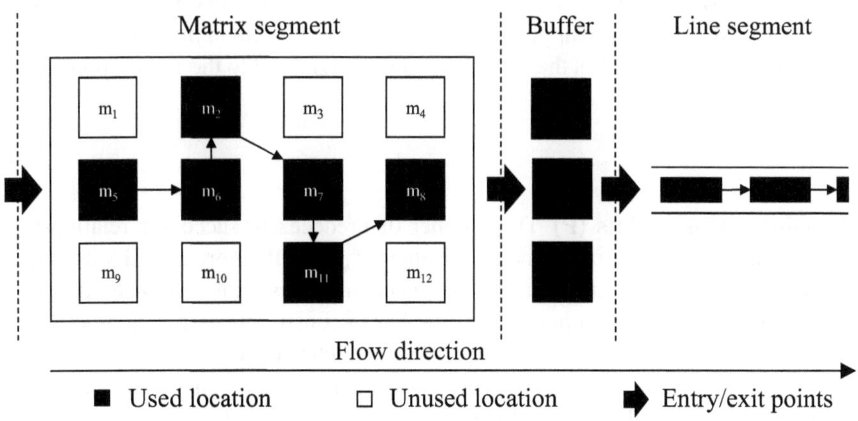

Fig. 1 Functional diagram of a hybrid matrix assembly system

$$F_S = \frac{T_{p,\max} - T_{p,\min}}{\overline{T}_p} \quad (1)$$

This formulation allows F_S to take on values greater than 1 when the range of processing times exceeds the average. In practical applications, F_S ranges from 0 to a maximum of 2, the latter occurring when $T_{p,\,min} = 0$.

Variant Diversity (V) Variant diversity quantifies the degree of adaptability required for a process and the necessity of individual process adjustments. The metric includes setup and sequence variability.

$$F_V = \frac{V_R + V_S}{2} \quad (2)$$

The setup time V_R reflects the need for adjustments in the process and is calculated as the ratio of the setup time for a module to the maximum setup time across all modules. Process sequence variability V_S considers the number of distinct process sequences required for different product variants. It is calculated as the ratio of the distinct process sequences for a module relative to the maximum number of sequences within the system. Accordingly, the overall indicator is calculated as the average of both factors. To ensure that the value of F_V falls within the range [0, 1], normalization is carried out. V_R is normalized by dividing the setup time for the module by the maximum setup time across all modules in the system. Similarly, V_S is normalized by dividing the number of distinct process sequences for the module by the maximum number of distinct process sequences in the system. A high variant index indicates a significant need for flexibility, justifying assignment to matrix production.

Labor Requirement (L) The labor requirement factor F_L represents the number of workers needed to perform a process. It is defined as a discrete normalized value, with $F_L = 1$ for single-worker tasks and $F_L = 2$ for two-worker tasks. The scale can be expanded for future applications involving multi-person processes. A higher labor requirement increases the score, favoring assignment to the matrix production system.

$$F_L = L \quad (3)$$

Precedence Constraints (P) The number of predecessor-successor relationships presents the complexity of a process. To quantify this, the normalized value F_P is defined, where P is the actual number of precedence constraints for the module, and P_{max} represents the maximum possible number of constraints. For each module, the number of predecessor and successor modules is determined. This value is normalized, and therefore $F_P \in [0, 1]$. A higher F_P-value reflects a higher need for flexible adjustments, making the module better suited for matrix production.

$$F_P = \frac{P}{P_{max}} \quad (4)$$

These four criteria are combined into an overall score F, incorporating weightings w_1 to w_4 to reflect their relative importance. The overall score is calculated as follows:

$$F = w_1{}^*F_S + w_2{}^*F_V + w_3{}^*F_L + w_4{}^*F_P \quad (5)$$

The weightings w_i are calibrated based on industrial experience or empirical data, reflecting the relevance of each factor. The combined score F thus reflects the process module's overall requirement for flexibility. A higher score indicates a higher need for variant handling, process adaptability, and labor complexity.

To determine whether a process module is assigned to line or matrix production, a threshold value T is defined. If the combined score F falls below T, the module is integrated into line assembly, as this indicates a high degree of standardization and low variant diversity. If F exceeds the threshold, this suggests a high flexibility requirement, leading to assignment to matrix production. The specific determination of T depends on the production environment. T may be defined via mean values or derived empirically. One pragmatic approach is to determine the score distribution using a representative sample of valid process modules and define T such that an approximate balance between line and matrix assignments is achieved. The 50/50 distribution serves as a neutral starting point in the absence of system-specific calibration data and facilitates comparative analysis of both assembly strategies. In practice, however, this ratio can be adapted to reflect varying production priorities, resource availability, or strategic focus, such as a preference for higher flexibility or throughput. During an expert workshop with assembly planners at Mercedes-Benz AG, overhead tasks have been defined as an exclusion criterion for matrix production. The need for individual, expensive lifting systems and vehicle lowering at each station causes substantial inefficiencies and delays, primarily due to limited AGV accessibility. Likewise, indivisible automation processes demand fixed physical and logical integration, making them unsuitable for flexible matrix layouts due to high duplication costs. These constraints are incorporated into the extended decision logic, as illustrated in Fig. 2. The design of the line segments is based on the findings of Eversheim [12].

3 Application of the Design Logic

3.1 Case Study from the German Automotive Industry

The decision methodology is applied to a case study from the German automotive industry. Based on the manufacturing structure outlined by Keckl [13], the assembly process for a modular EV is analyzed. The results are presented in Table 1.

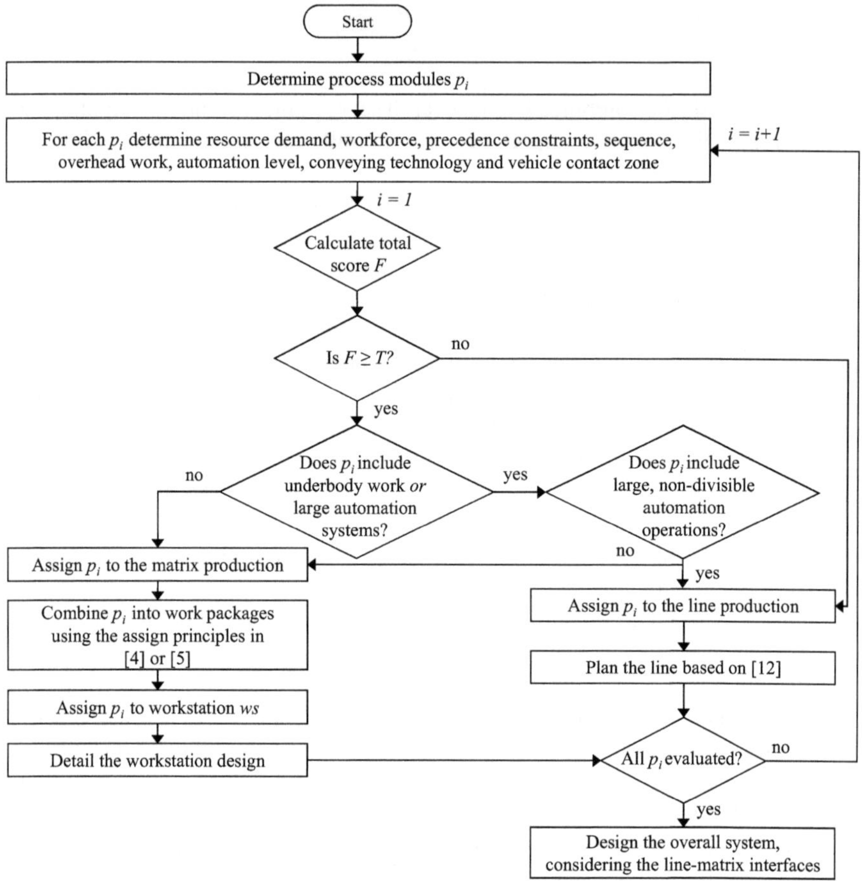

Fig. 2 Decision logic for the design of a hybrid assembly system

The process modules are structured into clusters $p_i \in CS$ to facilitate comparison with the conveyor segments of traditional line production. The cluster formation is based on spatial, technological, and organizational proximity. Each cluster represents a contiguous assembly section along the value stream with related process steps and similar infrastructure requirements. The process is split into 17 conveyor clusters, each comprising between 5 and 60 workstations. This includes a complete line balancing process, incorporating all tasks and their respective processing times, as well as a production program with the necessary orders and the resulting production rates. The clusters follow the sequential value stream from battery to final assembly, reflecting the chronological order of assembly activities. The analysis indicates that 11 out of 17 clusters can be implemented in matrix segments.

Table 1 Assignment of work contents to the line or matrix segments based on the decision logic

Cluster	Work content	Stations	Automation	Overhead	Assignment	
					Line	Matrix
1	Battery, rear axle	15	x	x	x	–
2	Underbody	15	–	x	x	–
3	Door removal, tunnel module	60	–	–	x	x
4	Cockpit, windows	15	x	–	x	–
5	Front exterior	15	–	–	x	x
6	Flashing	10	–	–	x	x
7	Front axle	15	x	x	x	–
8	Filling	5	–	–	x	x
9	Flashing	10	–	–	x	x
10	Interior equipment	25	–	–	x	x
11	Front module	40	–	–	x	x
12	Underbody	25	–	x	x	–
13	Tires	10	–	–	x	x
14	Doors	15	–	–	x	x
15	Commissioning	30	–	–	x	x
16	Roller test	–	–	–	x	x
17	Finish	–	–	–	x	–

3.2 Interface View of the Line-Matrix Assembly

Interface analysis ensures smooth transitions between matrix and line segments, avoiding inefficiencies. A key interface is identified between Cluster 2 and Cluster 3, where the vehicle transitions into the modular segment after overhead tasks. Door removal was shifted to the line segment to streamline flow and avoid redundant handling. Another critical transition occurs after Cluster 3, where the vehicle moves from the modular area into the fully automated line segment. To ensure smooth synchronization between both segments, a buffer is introduced at this transition point. Cluster 5 (front-end assembly) is initially assigned to the matrix system, but transitions back to a line layout after 25 stations. This requires adjustments to conveyors and buffer systems. Cluster 5 was combined with Cluster 6 in the line segment for stability and reduced complexity. Another matrix segment is assigned to Cluster 8, where vehicle filling takes place. The high process complexity and the integration of 80 stations support the efficient implementation of a modular structure in this section. Matrix segments should be large to minimize transitions and avoid workload imbalance. By systematically applying the decision logic, matrix and line segments are assigned in a way that optimizes production flow, enhances flexibility, and ensures a structured and efficient transition between different assembly processes.

3.3 Calculation and Evaluation of the F-Score for Matrix Segment

The design and evaluation of the first matrix segment form the basis for the transferability of the approach to other segments. In this initial segment, the tunnel module, along with various wiring harnesses, electrical, and electronic components, is assembled. For detailed planning, process modules with high time variability were selected. Four process modules were evaluated using F-scores against a threshold $T = 1$. A lower threshold could result in an excessive number of processes being unnecessarily assigned to the matrix, leading to higher costs and redundant resource allocation. For each process module, as shown in Table 2, the parameters variant diversity (V), process time variance (S), labor requirement (L), and precedence constraints (P) were determined. These values are based on industrial experience, process data analysis, and expert assessments. The labor requirement was modeled with either one or two operators per process module, reflecting standard practice. Scenarios requiring more than two workers are typically addressed through task division or automation and are therefore not considered in this evaluation. This modeling approach ensures practical relevance while maintaining conceptual simplicity. The detailed raw data and process module-specific values have been omitted for proprietary reasons. The weighting factors w_1, w_2, w_3 and w_4 were set based on the practical relevance of each parameter. In the current calculation, $w_1 = 0.35$, $w_2 = 0.4$, $w_3 = 0.3$ and $w_4 = 0.35$, as expert assessments indicated that all factors held similar significance. Based on this, the calculated F-scores for all four process modules exceeded the threshold T, justifying their inclusion in matrix production. This outcome was further confirmed through a validation session conducted with experts at Mercedes-Benz AG. A panel of six assembly planners and production system designers participated in the expert workshop, all with more than five years of experience in the planning of final assembly systems for EVs. The participants were selected based on their operational involvement in hybrid assembly planning projects and their familiarity with matrix-based production environments. The evaluation was conducted using a structured feedback format. Experts validated the assignments through structured feedback, confirming the scoring approach. Consensus was reached on the validity of the proposed allocations through facilitated discussion rounds. Quantitative scoring was not applied at this stage, as the aim was to validate applicability and identify implementation constraints. The experts emphasized that the high variant diversity, significant process time variance, and elevated labor requirements associated with these modules made a synchronized line production approach impractical, reinforcing the decision to assign them to matrix production.

Table 2 Calculation of the F-score for matrix segment

p_i	Process module	Process time variance S	Variant diversity V	Labor requirement L	Precedence P	Score F
p_1	Back wall and cushion frame *left*	0.4	0.7	2	0.2	1.16
p_2	Back wall and cushion frame *right*	0.4	0.7	2	0.2	1.16
p_3	Supports for window bag	0.5	0.8	1	0.7	1.04
p_4	Grab handle consoles *left*	0.6	0.8	1	0.7	1.01

4 Conclusion

The scoring-based logic supports process allocation in hybrid assembly systems by integrating key variability and resource factors. The methodology is particularly suitable for production environments characterized by high product variability, significant manual work content, and fluctuating demand structures, such as EV assembly. It provides structured support for early-stage system design and enables data-driven allocation decisions. However, its applicability may be limited in cases where high process standardization prevails, task modularization is infeasible, or reliable planning data is unavailable. A key limitation is the subjectivity involved in defining the weighting factors. Future work should therefore explore sensitivity analyses or learning-based weighting approaches. Additionally, it remains open in which scenarios hybrid assembly concepts provide significant advantages over conventional systems and how product designs can be specifically adapted to support greater modularization. Practical validation in industrial applications is necessary to assess the transferability of the decision logic and identify potential constraints in real production environments.

References

1. Kampker A, Heimes HH (2024) Elektromobilität: Grundlagen einer Fortschrittstechnologie. Springer, Berlin
2. Rossit D, Rossit D, Nesmachnow S (2024) Enhancing mass customization manufacturing: multiobjective metaheuristic algorithms for flow shop production in smart industry. SN Comput Sci 5(6):782
3. Bebersdorf P, Huchzermeier A (2021) Variabler Takt. Springer Books, Berlin
4. Greschke P, Schönemann M, Thiede S, Herrmann C (2014) Matrix structures for high volumes and flexibility in production systems. Procedia CIRP 17:160–165
5. Kern W, Rusitschka F, Bauernhansl T (2016) Planning of workstations in a modular automotive assembly system. Procedia CIRP 57:327–332

6. Hottenrott A, Grunow M (2019) Flexible layouts for the mixed-model assembly of heterogeneous vehicles. OR Spectr 41(4):943–979
7. Küpper D, Sieben C, Kuhlmann K, Ahmad J, Lim YH (2018) Will flexible-cell manufacturing revolutionize carmaking. Boston Consult Group Online 18
8. Hofmann C, Brakemeier N, Krahe C, Stricker N, Lanza G (2019) The impact of routing and operation flexibility on the performance of matrix production compared to a production line. In: Lanza G (ed) Advances in production research: proceedings of the 8th congress of the German academic Association for Production Technology (WGP), Aachen, November 19–20, 2018. Springer, Cham, pp 155–165
9. Kampker A, Kawollek S, Marquardt F, Krummhaar M (2020) Potential of hybrid assembly structures in automotive industry. Available SSRN 3848756. https://doi.org/10.2139/ssrn.3848756
10. Göppert A, Schukat E, Burggräf P, Schmitt RH (2021) Agile hybrid assembly systems: bridging the gap between line and matrix configurations. In: Schmitt RH (ed) Advances in automotive production technology—theory and application: Stuttgart conference on automotive production (SCAP2020). Springer, Cham, pp 3–11
11. Stautner U (2001) Kundenorientierte Lagerfertigung im Automobilvertrieb: Ergänzende Ansätze zum Supply Chain Management. Springer, Berlin
12. Eversheim W (1989) Arbeitsplanung Organisation in der Produktionstechnik: Arbeitsvorbereitung. Springer, Berlin, pp. 24–119
13. Keckl S(2017) Gestaltung einer Endmontage in der variantenreichen Serienfertigung am Beispiel der Automobilproduktion. Shaker Verlag, Aachen

Improving the Robustness of Reconfigurable Assembly Systems with an Agent-Based Control Architecture

Jasper Wilhelm, Artem Schurig, Aaron Heuermann, Burak Vur, and Michael Freitag

Abstract This paper introduces a novel approach for improving the adaptability and scalability of modular assembly systems by integrating decentralized, agent-based control with state-based planning. Traditional assembly systems struggle with adapting to varying production demands due to their reliance on fixed, centralized workflows. Our proposed concept leverages state-based descriptions of assembly processes and a multi-agent system to enable real-time reconfiguration and dynamic process planning. This approach facilitates seamless adaptation to changes in production volume and product types, reduces installation times, and extends the operational life of assembly equipment. The results demonstrate improved flexibility and resilience compared to conventional methods, offering significant implications for future research and application in manufacturing environments, particularly in industries requiring rapid responsiveness to market shifts.

Keywords Assembly · Reconfiguration · Reconfigurable assembly systems (RAS) · Reconfigurable manufacturing systems (RMS)

1 Introduction

The rise of Industry 4.0 and the emerging paradigm of Industry 5.0 are driving significant changes in manufacturing, emphasizing resilience, adaptability, and personalization [1]. With shorter product life cycles, smaller batch sizes, and an increasing demand for product diversity, especially in consumer electronics and automo-

A. Heuermann and B. Vur—Equal contribution.

J. Wilhelm · A. Schurig · A. Heuermann (✉) · M. Freitag
BIBA – Bremer Institut für Produktion und Logistik GmbH at the University of Bremen, Bremen, Germany
e-mail: her@biba.uni-bremen.de

B. Vur (✉) · M. Freitag
University of Bremen, Faculty of Production Engineering, Bremen, Germany
e-mail: bvur@uni-bremen.de

M. Manns (ed.), *New Paradigms for Anticipated Uncertainty*, Lecture Notes in Mechanical Engineering, https://doi.org/10.1007/978-3-032-16889-4_56

tive sectors, assembly systems face mounting pressure to become more flexible and responsive [2–4]. Current rigid assembly processes, designed for mass production, struggle to keep up with the frequent reconfigurations required by fluctuating market demands [5–7]. To meet these challenges, we propose a decentralized, agent-based control architecture for modular assembly systems, which supports dynamic scalability and leverages hardware-independent task descriptions to reduce planning and setup costs. A central component is an agent-based mediator system that enables real-time reconfiguration, fostering resilient and responsive production environments capable of handling high variability [8].

Implemented on the robotics middleware framework ROS 2, the system supports modularity, scalability, and fault tolerance in complex automation contexts. The architecture has been validated in a controlled laboratory environment through a representative assembly use case. Although no physical assembly was conducted, the system was evaluated based on task execution and coordination logic to verify its effectiveness and responsiveness. By combining hardware-independent task descriptions with an agent-based mediator framework, the architecture enables flexible, state-based assembly process planning [9], providing a robust technological foundation to meet the evolving demands of modern manufacturing.

2 Related Work

2.1 Reconfigurable Systems

rms balance productivity and adaptability, enabling rapid reconfiguration of hardware and software to meet changing production needs [10]. RAS extend this flexibility by incorporating modularity, allowing frequent assembly system adjustments for greater product variety and adjustable production capacity [5, 10, 11]. The plug-and-produce paradigm facilitates seamless module integration, reducing downtime and improving responsiveness [12]. A variety of Reconfigurable Assembly System (RAS) implementations focus on software interfaces and hierarchical capability descriptions for configuration-independent robot-assisted assembly [13–15]. Efforts to standardize interfaces and develop self-reconfiguring systems using autonomous guided vehicle (AGV) technology are ongoing [16]. A review of ten modular systems highlights varying levels of flexibility and robustness, with research gaps in dynamic scalability, multi-module coordination, and automated system configuration. Key areas for further exploration include:

- System-independent assembly process descriptions.
- Hardware-independent task descriptions with dynamic distribution.
- Dynamic process control managing incompatible constraints.
- Machine interaction for collaborative execution.
- Event-driven fault handling via adaptive automation.

2.2 Decentralized and State-Based Control

Decentralized, agent-based control addresses key limitations of traditional, rigid systems [17]. Unlike centralized approaches, it distributes decision-making, enabling real-time adjustments and greater system resilience. By treating modules as autonomous agents, decentralized control allows for dynamic task allocation, reducing planning complexity and enhancing flexibility [18–20]. State-based task descriptions further enhance adaptability by defining desired states before, during, and after process steps, enabling hardware-independent execution [21]. Hierarchical task decomposition aligns with *plug-and-produce* principles, allowing modular skills and process adjustments [18, 19, 22]. The product-process-resource(PPR) structure matches tasks to resources dynamically, improving system responsiveness [13, 21, 23, 24]. State-based descriptions excel in modular systems, where module positions and types vary. By focusing on desired states rather than fixed parameters, they enable flexible, real-time task execution [21]. Multi-agent architectures in decentralized RAS show promise for mass customization, allowing real-time adjustments based on environmental and system states [20, 25]. Research continues toward finer granularity, where individual modules function as independent agents.

3 Concept

The proposed concept introduces a decentralized, agent-based control architecture and state-based planning for hybrid assembly systems, offering a flexible and scalable solution to modular assembly. Unlike traditional systems that rely on linear, predefined workflows, this architecture dynamically adapts to changes such as varying batch sizes or frequent product variations. A core innovation is the decoupling of process logic from hardware-specific implementations, resulting in a system that is modular, hardware-agnostic, and highly adaptable to diverse production setups.

3.1 State-Based and Hierarchical Skill Parameterization

At the heart of this concept is the use of state-based descriptions for tasks. Rather than depending on fixed sequences, each process step is determined by the current state of the assembly components, available resources, and tasks. This dynamic state-driven approach enables real-time adjustments, eliminating the need for extensive reprogramming during reconfiguration. The system leverages a generalized ontology of tasks, independent of the hardware used, ensuring seamless adaptation across different assembly lines. This hardware-agnostic parameterization enhances flexibility, making it easier to respond quickly to production changes. Building on condition-based skills models [26], this concept incorporates an enhanced framework for skill-based task descriptions [21]. The skill descriptions structure assembly tasks by detailing the resources, parameters, and conditions required for successful execution. These descriptions include the following elements:

- *Resources*: Specifies the robots, sensors, and end effectors needed for task completion.
- *Input and Output Parameters*: Defines object-related data such as positions or physical properties relevant to the task.
- *Pre-Conditions and Post-Conditions*: Establishes the initial and final states, ensuring validation of task execution.
- *Process Conditions*: Outlines the operational constraints that need to be met for successful task execution.
- *Optional Information*: Includes metadata like task descriptions, cycle times, and log data, which are useful for future planning and analysis.

The described skills are based on simple primitives that can be combined hierarchically, akin to holons [18]. Fundamental primitives represent the most basic system actions, such as moving an end-effector or activating and deactivating signals. By combining these primitives, more complex skills are created, which can themselves be further combined to define increasingly sophisticated skills. Tasks can either be specified through existing skills or by directly defining the necessary primitives. Figure 1 illustrates two such skills—pick and place—constructed from primitives, along with a more complex pick-and-place skill, all designed for execution by a vacuum gripper. The conditions for these skills are described using variables, with parameterization providing the specific values. The PPR approach is applied to determine the appropriate position for grasping, based on the object being handled (a lid in the given example).

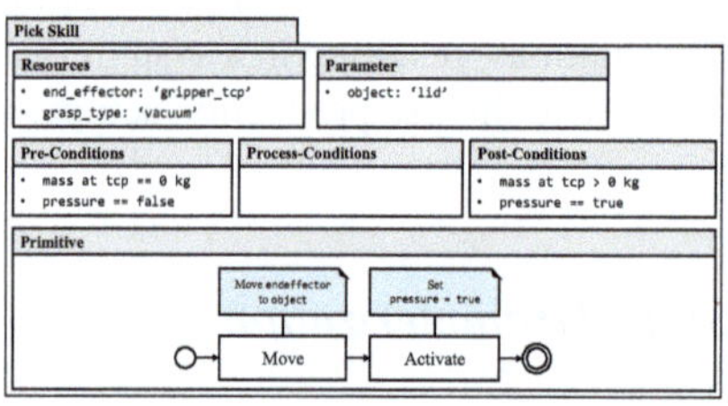

(a) Pick skill based on the primitives *move* and *activate* parameterized for picking the lid

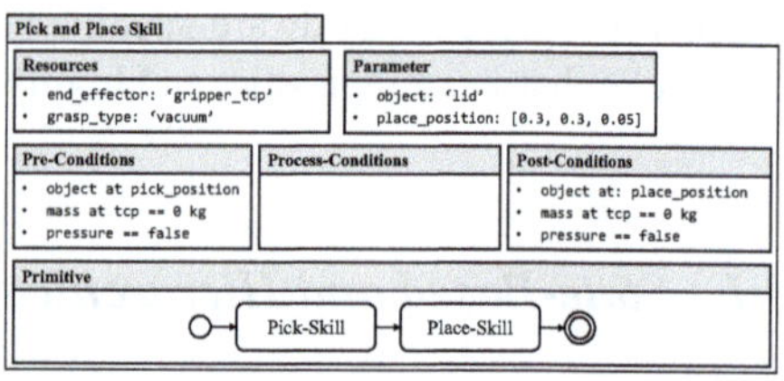

(b) Pick-and-Place skill based on the pick skill (left) and another skill *place*. Parameterized to place the lid at a given position.

Fig. 1 Skill architecture for skills made from primitives as well as more complex ones

3.2 Decentralized Architecture and Adaptive Mediator

The decentralized architecture is structured around a multi-agent system, where each assembly module operates autonomously. Agents collaborate using a mediator-based communication protocol, with the mediator role shifting dynamically between modules depending on the current task. This decentralized control distributes decision-making across the system, enabling self-organizing behavior and eliminating the bottlenecks associated with centralized control. By incorporating pre- and post-conditions for task execution, modules can negotiate in real time, selecting optimal resources based on current states and dynamically resolving any task conflicts. The adaptive mediator coordinates these interactions, managing task allocation based on resource availability and system needs, thus improving overall system resilience and scalability. This adaptive mechanism ensures continuous workflow even in the event of task failures, by automatically rerouting tasks to available modules. The system operates through the following control sequence:

1. Order entry via a Human-Machine Interface (HMI) module
2. HMI module assumes the initial mediator role
3. First assembly step is selected
4. Mediator requests resources from all modules
5. Modules calculate and submit resource bids based on their capabilities
6. Mediator evaluates bids and assigns tasks to the module with the best bid
7. Task execution by the selected module
8. The executing module becomes the new mediator
9. Steps 4–8 are repeated until the product assembly is complete

This results in the module with the best resource offer initiating the assembly process. Following this, the mediator module communicates the partial assembly plan to all other modules, ensuring that each module remains informed about the source, process, and destination at every stage of the workflow. This guarantees seamless coordination and transparency throughout the entire assembly sequence.

3.3 Control Architecture

The control architecture for hybrid modular assembly systems is based on a decentralized, agent-based mediator structure. Each module operates autonomously, while system-wide coordination is managed dynamically by a rotating *Task Manager Agent*. This role shifts between modules based on task execution, leveraging the multi-ContractNet protocol for multi-level assembly process planning and inter-agent negotiation. Communication between the *Task Manager Agent*, *Resource Agents*, and *Machine Agents* ensures efficient operation. *Resource Agents* manage sub-requests and provide cost-based metrics, which guide the mediator in optimizing resource allocation and process planning. This distributed architecture integrates successive

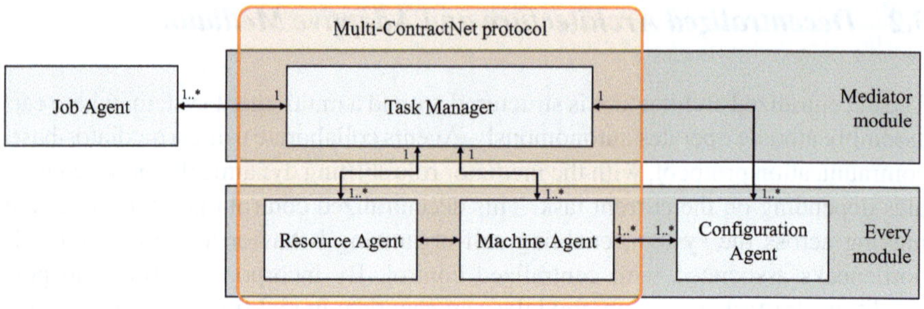

Fig. 2 Control architecture for modular hybrid assembly systems

assembly steps, enhancing flexibility and resolving task execution conflicts dynamically. The system includes multiple instances of *Resource*, *Machine*, and *Configuration Agents*, implemented across all modules with a many-to-many $(n : m)$ relationship. The *Task Manager Agent* maintains a one-to-many $(1 : n)$ relationship, ensuring coordinated management. A hardware-independent task description, based on semantic descriptions, supports state-based planning, enabling platform adaptability and system reconfigurability (Fig. 2).

4 Implementation

4.1 Prototypical Application of the Control Architecture

The prototypical implementation utilizes an agent-based control architecture to increase flexibility in hybrid modular assembly systems. The system runs on Ubuntu with ROS2 (Humble) as the software framework, using Raspberry Pi 4 devices as the hardware platform. These modules communicate over a shared network, with remote management via SSH. The system employs the Data Distribution Service (DDS) middleware for message-based communication, using the publisher-subscriber mechanism to coordinate system configuration and implement the multi-ContractNet protocol. A custom ROS2 package manages the user-defined message types crucial for agent communication and task execution. Key agents, implemented as Python classes within the ROS2 structure, drive the core system functionality.

- The *Task Manager Agent* oversees state transitions using the ROS2 YASMIN library, with state definitions and transitions governing assembly step progression.
- *Resource Agents* managed by the *ResourceManager* class, handle resource requests and generate sub-requests, maintaining distributed resource allocation across six DDS instances.
- *Configuration Agents* broadcast and process configuration messages, relying on statically defined XML files for system setup during this development phase.

Process conditions ensure that tasks meet predefined constraints, and optional metadata such as cycle times and log data enhance the system's usability and flexibility. This implementation demonstrates the system's decentralized control capabilities, with DDS and ROS2 ensuring robust communication, and the agent-based architecture enabling dynamic reconfiguration and scalability across multiple modules.

4.2 Validation

A functional test was conducted using a skateboarding assembly case study to validate the control system under real-world conditions. The assembly process, consisting of six stages—such as attaching the deck, trucks, and wheels—was modeled using the system's hardware-independent task description format, which was then converted into XML for integration. This approach underscores the system's modularity and adaptability for various applications. The validation focused on assessing the functionality and scalability of the agent-based architecture and the state-based control model. With DDS middleware and ROS2 ensuring reliable communication and coordination, the system demonstrated real-time reconfiguration and scalability across modules, validating its decentralized control capabilities. The test approach covered three key scenarios:

1. Automatic Configuration: Modules were initialized and subjected to simulated disconnections and reconnections. The system adapted in real-time, handling dynamic changes without disruption.
2. Network Propagation: The test verified the reliable transmission of task requests, sub-requests, and responses between agents. No errors occurred, and only relevant neighboring modules were contacted, confirming the robustness of network communication.
3. Full Process Execution: The complete assembly process was run through the state machine, successfully executing all predefined steps with no errors, demonstrating reliable state transitions.

These results confirmed the system's ability to manage configuration changes and network communication in a distributed environment, ensuring scalability and error-resilient process execution.

5 Discussion and Outlook

This research successfully developed and validated a control concept designed to enhance the flexibility of hybrid RAS. The core innovation of this work lies in integrating advanced software frameworks with established application-specific methods, resulting in a prototype that utilizes the multi-ContractNet protocol within a publisher-subscriber communication structure in the ROS2 framework. This novel implementation demonstrated the capability of the control architecture to adapt flexibly to changes, recover from failures, and autonomously configure itself. The condition-based description of assembly processes introduced in this study provides a semantic and machine readable format, which facilitates dynamic allocation of resources and efficient process planning. This approach enhances system resilience and scalability by allowing real-time adjustments and supporting seamless integration of new modules. While the research addressed the effectiveness of decentralized control systems, the comparative flexibility between decentralized and centralized approaches remains partially unexplored. Future research will focus on addressing this gap by developing a RMS demonstrator. This demonstrator will enable direct comparison of decentralized and hierarchical centralized control systems, providing insights into their respective advantages and limitations. The ongoing project PassForM2 is poised to build on this work by further refining the proposed control architecture, incorporating advanced state-based planning techniques, and exploring additional features such as enhanced modular integration and adaptive process management. These advancements will contribute significantly to the field of modular assembly systems, offering broader implications for industries that require adaptable and scalable manufacturing solutions.

5.1 Implications for Future Research

This concept lays the foundation for future research into dynamic, decentralized assembly systems. The state-based process description and agent-based control architecture open new avenues for exploring how modular systems can be further optimized for flexibility, scalability, and resilience. Future work could investigate how these principles can be applied in increasingly complex assembly environments, including the integration of advanced AI techniques for predictive maintenance and process optimization. The system-independent task descriptions, in particular, offer exciting potential for the development of industry-wide standards, enabling broader interoperability between different manufacturers and assembly systems. Similarly, the decentralized control approach could inspire new research into multi-agent collaboration in manufacturing environments, particularly in scenarios where real-time adaptability and fault tolerance are critical.

Acknowledgements This research has been funded by the Federal Ministry for Economic Affairs and Climate Action of Germany as part of the project PassForM2 (project number 01IF23333lN).

References

1. Breque M, de Nul L, Petridis A Industry 5.0: Towards a sustainable, human-centric and resilient European industry
2. Cenci MP, Scarazzato T, Munchen DD, Dartora PC, Veit HM, Bernardes AM, Dias PR (2022) Eco–friendly electronics—a comprehensive review. Adv Mater Technol 7(2):2001263
3. Sabadka D, Molnár V, Fedorko G (2019) Shortening of life cycle and complexity impact on the automotive industry. TEM J 8(4):1295–1301
4. Wilhelm J, Hoppe NH, Freitag M (2023) Requirements and concept development for a reconfigurable assembly system with individual and interchangeable modules. Proc CIRP 118:964–969
5. ElMaraghy HA (ed) (2009) Changeable and reconfigurable manufacturing systems. Springer series in advanced manufacturing. Springer, London
6. Krüger J, Wang L, Verl A, Bauernhansl T, Carpanzano E, Makris S, Fleischer J, Reinhart G, Franke J, Pellegrinelli S (2017) Innovative control of assembly systems and lines. CIRP Ann 66(2):707–730
7. Koren Y, Xi G, Guo W (2018) Reconfigurable manufacturing systems: Principles, design, and future trends. Front Mech Eng 13(2):121–136
8. Feldmann K, Schöppner V, Spur G (2014) Montagezellen. In: Feldmann K, Schöppner V, Spur G (eds) Handbuch Fügen. Handhaben, Montieren, Hanser eLibrary. Hanser, München, pp 433–588
9. Miguel Á (2023) González-Santamarta, Francisco J. Rodríguez-Lera, Vicente Matellán-Olivera, and Camino Fernández-Llamas. Yasmin: Yet another state machine. In: Tardioli D, Matellán V, Heredia G, Silva MF, Marques L (eds) ROBOT2022: Fifth Iberian Robotics Conference, vol 590. lecture notes in networks and systems. Springer International Publishing, Cham, pp 528–539
10. Lotter B, Wiendahl H-P (2009) Changeable and reconfigurable assembly systems. In: ElMaraghy HA (ed) Changeable and reconfigurable manufacturing systems. Springer series in advanced manufacturing. Springer, London, pp 127–142
11. Minhas SH, Lehmann C, Städter JP, Berger U (2011) Reconfigurable strategies for manufacturing setups to confront mass customization challenges. In: 21st international conference on production research (ICPR21)
12. Arai T, Aiyama Y, Maeda Y, Sugi M, Ota J (2000) Agile assembly system by "plug and produce." CIRP Ann 49(1):1–4
13. Onori M, Lohse N, Barata J, Hanisch C (2012) The ideas project: plug & produce at shop-floor level. Assem Autom 32(2):124–134
14. Siebert W, Nägele, Hansen MR Full release of rampup framework
15. Kroischke X (2018) 688642 – rampup deliverable d3.7: Revision of developers toolkit for modular automation components
16. Zhang D, Leng J, Xie M, Yan H, Liu Q (2022) Digital twin enabled optimal reconfiguration of the semi-automatic electronic assembly line with frequent changeovers. Robot Comput-Integrated Manufact 77:102343
17. Qin Z, Yuqian L (2021) Self-organizing manufacturing network: a paradigm towards smart manufacturing in mass personalization. J Manuf Syst 60:35–47
18. Arai T, Aiyama Y, Sugi M, Ota J (2001) Holonic assembly system with plug and produce. Comput Ind 46(3):289–299
19. Sugi M, Maeda Y, Aiyama Y, Harada T, Arai T (2003) A holonic architecture for easy reconfiguration of robotic assembly systems. IEEE Trans Robot Autom 19(3):457–464
20. Becker T, Weimer D, Pannek J (2015) Network structures and decentralized control in logistics: topology, interfaces, and dynamics. Int J Adv Logist 4(1):1–8
21. van de Ginste L, de Cock A, van Alboom A, Singh Y, Aghezzaf E-H, Cottyn J (2021) A formal skill model facilitating the design and operation of flexible assembly workstations. In: Dolgui A, Bernard A, Lemoine D, von Cieminski G, Romero D (eds) Advances in production management systems. artificial intelligence for sustainable and resilient production systems,

IFIP advances in information and communication technology, vol 634. Springer International Publishing, Cham, pp 108–116
22. Heuss L, Gonnermann C, Reinhart G (2022) An extendable framework for intelligent and easily configurable skills-based industrial robot applications. Int J Adv Manufact Technol 120(9–10):6269–6285
23. Danny P, Ferreira P, Lohse N, Dorofeev K (2018) An event-based automationml model for the process execution of plug-and-produce' assembly systems. In: 2018 IEEE 16th international conference on industrial informatics (INDIN), pp 49–54. IEEE
24. Perzylo A, Rickert M, Kahl B, Somani N, Lehmann C, Kuss A, Profanter S, Beck AB, Haage M, Hansen MR, Nibe MT(2019) Smerobotics: Smart robots for flexible manufacturing. IEEE Robot Autom Mag 26(1):78–90 (2019)
25. Liu P (2019) Decentralized Robust Capacity Control of Job Shop Systems with Reconfigurable Machine Tools. Dissertation, Universität Bremen, Bremen
26. Stenmark M (2013) Industrial robot skills. Twelfth Scandinavian Conference on Artificial Intelligence. Frontiers in artificial intelligence and applications. IOS Press, Amsterdam, pp 295–298

Smart Automation

Identification of Barriers and Opportunities for Adoption of Machine Vision for Small- and Medium-Sized Enterprises: A Stakeholder Investigation

Mikkel Graugaard Antonsen, Lasse Christiansen, and Christian Black Jørgensen

Abstract The digital transformation of industry relies on a range of technologies related to manufacturing, data processing and sensors. Among these, machine vision enables real-time quality inspection. However, small- and medium-sized enterprises (SMEs) struggle to adopt this technology. To support awareness and informed decision-making, this study examines the key barriers and opportunities for machine vision adoption. Drawing on stakeholder interviews, the analysis includes three enterprise categories: machine vision manufacturers, Danish SMEs in manufacturing and engineering services and production firms. Using the Gioia methodology, the findings are categorised into sensing, seizing, and transformation barriers. To deepen the analysis, the study also incorporates insights from automation engineering students based on their internship experiences. The results reveal that environmental and competency-related factors are widely recognised as barriers, while perceptions of transparency and other technical issues vary across vendors, builders and end users. These insights can inform practitioners and researchers working to advance machine vision implementation in industrial contexts.

Keywords Vision · Digital transformation · SME · Technology adoption · Barriers

M. G. Antonsen (✉) · L. Christiansen · C. B. Jørgensen
University College Northern Denmark, Sofiendalsvej 60, 9200, Aalborg SV, Denmark
e-mail: MIGA@ucn.dk

L. Christiansen
Aalborg University, Fibigerstrædet 16, 9220, Aalborg E, Denmark

M. Manns (ed.), *New Paradigms for Anticipated Uncertainty*, Lecture Notes in Mechanical Engineering, https://doi.org/10.1007/978-3-032-16889-4_57

1 Introduction

The recent changes in the industrial paradigm are characterised by interconnectivity, automation, machine learning, real-time data [1] and machine vision [2]. Large companies with strong technological capabilities tend to benefit most from this transition. However, these changes may negatively affect the competitiveness of small- and medium-sized enterprises (SMEs) because these firms often lack the resources to keep pace with larger competitors [3].

Vision-based inspection systems are particularly appealing to SMEs due to their potential for high reliability and performance at relatively low manufacturing costs [4]. Various vision systems have already been successfully developed [5–8], and improvements in accuracy have been shown to yield significant cost savings [7]. In Denmark, machine vision systems for quality assurance and operational efficiency have been demonstrated in commercial settings [9].

SMEs are often financially constrained and, as a result, tend to underperform in research and development. Compared to multinational enterprises, limited access to resources and awareness make it challenging for SMEs to upgrade and adopt advanced manufacturing technologies, thereby threatening their competitiveness and survival [10]. Furthermore, the lack of readily available technological infrastructure and IT integration hinders their ability to adopt innovative solutions [11].

Previous studies have explored barriers to the adoption of technologies [12, 13], and specific barriers faced by SMEs have also been examined [14]. Research has investigated obstacles to implementing virtual prototypes in design processes [15] and the barriers and opportunities associated with machine vision adoption in SMEs [16]. However, previous studies have not systematically analysed these barriers and opportunities across different stakeholder types. This study addresses that gap by posing the following research question: "What are the barriers and opportunities for different types of SMEs in adopting machine vision systems?" This article extends the findings of the short article presented in [16] by incorporating additional results, including the distribution of barriers and opportunities across various stakeholder categories. These results are significant because different barriers and opportunities correlate with specific types of stakeholders.

The remainder of this article is structured as follows: Sect. 2 presents the methodology, Sect. 3 details the results, and Sect. 4 offers the discussion and conclusion.

2 Methods

Different types of companies were investigated to understand the barriers and challenges regarding the implementation of machine vision systems for SMEs in Northern and Central Jutland. The companies were selected based on size (small- or medium-sized enterprises), location (Northern or Central Jutland) and professional category (vendors/distributors, engineering service providers and manufacturing companies). The participating companies include seven machine vision system vendors/distributors, 10 engineering service providers and nine manufacturing SMEs. The SMEs were selected using the Orbis database [17] and the authors' professional network. Agreements were made with all companies, and the participants held technical or managerial positions, such as senior engineers, engineering managers, chief technology officers (CTOs) and directors.

The mix of professions and positions provides insight into technological awareness and financial and human resources. This study is based on empirical data from 27 questionnaires, 27 semi-structured interviews with 26 businesses and four semi-structured interviews with automation engineering students during their internship courses.

2.1 Gioia Methodology

The Gioia methodology is a systematic approach to processing raw data [18]. In the first-order analysis, many concepts emerge, and there is little attempt to distil categories. The second-order analysis focuses on identifying similarities and differences among the concepts identified in the first-order analysis. This process helps organise the concepts into themes, thereby significantly reducing the number of categories. The second-order themes represent a more theoretical level compared to the first-order concepts. Finally, the themes are grouped under aggregate dimensions [18]. The three themes, sensing, seizing and transforming, originate from the dynamic capabilities framework [19]. Sensing refers to scanning, searching and exploring across technologies and markets. Seizing involves developing new products, processes or services. Lastly, transforming describes how a business can be reconfigured when the market and/or technology is inevitably transformed.

2.2 NVivo and Obsidian

NVivo was used to organise and structure the qualitative data from the transcripts of 27 semi-structured interviews, questionnaire responses and relevant literature. First, the data were coded into first-order concepts. These concepts were then categorised into themes by identifying similarities and differences during the second-order analysis.

Finally, the themes were grouped under aggregate dimensions, supported by the literature, and conclusions were drawn. Obsidian was used to organise and structure the four semi-structured interviews conducted with automation engineering students during their internship course. These interviews served to validate the emerging theory developed using the Gioia methodology.

3 Results

Table 1 presents the results of the analysis. Based on 197 statements, nine barriers were identified, while four opportunities were identified from 39 statements.

3.1 Distribution Among Types of Vendors

Figure 1 shows the distribution of the barriers and opportunities among the three stakeholder types in the initial interviews. The stakeholders perceive the barriers and opportunities differently, as illustrated below. For instance, builders did not identify a lack of transparency as a barrier, whereas the other stakeholder types did. Similarly, technology vendors did not emphasise quality requirements to the same extent as builders and producing enterprises. Furthermore, producing enterprises either have not recognised the opportunity presented by low implementation costs or do not acknowledge its significance. With the present sample size, barriers seem well identified, while opportunities can be investigated further.

Table 1 Identified barriers to machine vision applications

Aggregate dimension	Results based on analysis	
	Barrier	Opportunity
Sensing	Lack of customer demand	Increased quality requirements
	Lack of understanding of machine vision	
Seizing	Lack of funds to support machine vision projects	
	Lack of competencies	
	Mistrust in machine vision	
Transforming	Robustness issues	Deep learning
	Interoperability issues	Technological development rate
	Lack of performance transparency	Decrease in implementation cost
	Product and manufacturing environment limitations	

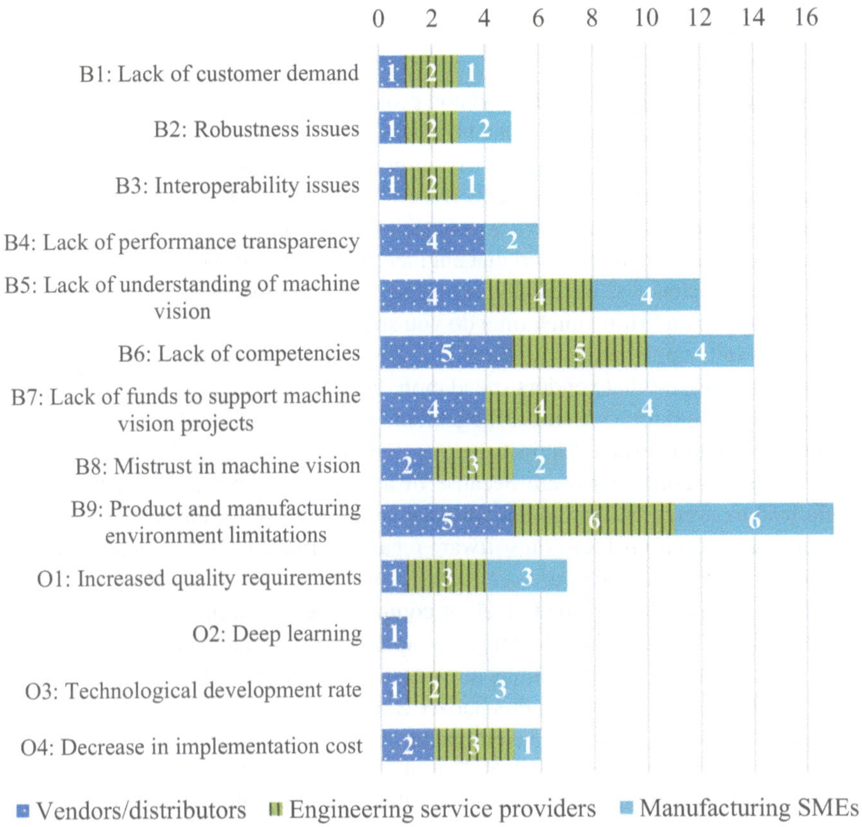

Fig. 1 Distribution of barriers and opportunities among types of stakeholders

3.2 Sensing

Twelve statements from seven companies referred to the barrier as "lack of customer demand." However, increasing quality requirements from customers and the need to optimise quality control processes are seen as necessary to become or remain competitive. Therefore, it can be concluded that SMEs recognise the potential of machine vision systems to meet these customer demands.

"Lack of understanding of machine vision" emerged as the most frequently mentioned barrier. It was referenced in 15 interview statements across 11 interviews and six questionnaire answers. Vendors and integrators noted that SMEs have unrealistic expectations regarding machine vision systems, and statements made by SMEs during the interviews supported this claim. For example, some SMEs expressed the belief that machine vision is incapable of handling certain quality control operations.

3.3 Seizing

"Lack of funds to support machine vision projects" and "lack of competencies" are two of the most frequently mentioned barriers in the interviews and questionnaires. "Lack of funds to support machine vision projects" was identified in seven questionnaire responses and 47 statements from 12 interviews. "Lack of competencies" appeared in seven questionnaire answers and 44 statements from 14 interviews.

Machine vision vendors stated that engineering service providers and SMEs lack technical competencies, especially regarding hardware setup. To investigate this claim, the question "How much time do you spend on physical setup vs. software?" was asked and supplemented with insights from in-depth interview conversations. The findings suggest that vendors spend more time selecting suitable equipment and arranging the setup for a specific arrangement. This leads to the software setup being relatively straightforward. Engineering service providers and SMEs tend to make hasty decisions about hardware because of a lack of competencies or time. This could lead to overcomplicating the software setup process. Vendors also noted that educational institutions focus on software regarding machine vision, leading to low competency regarding hardware.

Fifteen companies identified "lack of competencies" as a barrier. Some companies attempt to address this challenge by upskilling their employees and building competencies in-house, while others hire consultants or experts from machine vision vendors. However, overcoming this barrier can be time-consuming, particularly for companies pursuing in-house development. This is often due to a trial-and-error approach, as reflected in respondents' comments such as "We have not taken any course", "We learn by doing", and "We get the experience along the way".

Regarding the barrier "lack of funds to support machine vision," some SMEs lack the necessary financial resources. Contributing factors include the high salaries of experts and the significant implementation costs of turnkey solutions. Additionally, some companies seem to hold misconceptions about the actual costs involved.

3.4 Transforming

"Product and manufacturing environment limitations" is the most frequently recurring barrier in the aggregate dimension of transforming. This barrier was mentioned in 67 interview comments and encompasses challenges such as poor control of lighting conditions, speed of the manufacturing lines, harsh environment, spaces, production materials and so on. High-quality imaging, proper planning and an understanding of the physics behind imaging play a role in the performance of machine vision systems. Companies reported that using devices from different vendors can lead to "interoperability issues" and that the barrier of "lack of performance transparency" is related to the technological readiness of machine vision. Machine vision vendors have tried to break down these barriers by developing deep learning-based image analysis, improving cameras (e.g., light and lenses) and developing better image processing algorithms. The last barrier, "robustness issues," results from unrealistic expectations, such as expecting 100% accurate identification, or from performance problems encountered when replacing equipment.

3.5 A Deeper Dive Based on Questionnaires

The nine barriers presented in Table 1 were identified based on data collected from vision system manufacturers, engineering service providers and manufacturing companies. All companies were asked: "Do you use machine vision?" Of these, 10 reported that they did, 12 did not, and five were producers of vision technology. Vision system manufacturers were analysed separately, as they naturally use machine vision due to their role in developing such systems.

The companies that used machine vision equipment were asked to identify barriers to implementation. The most prominent barrier was the lack of an expert workforce (40%), followed by a lack of funds (30%). None of them indicated a lack of vision competencies among integrators. The vision manufacturers noted the lack of an expert workforce and funds to the same extent. However, they also pointed to a lack of knowledge regarding these opportunities.

The same groups were also asked to identify the main challenges in machine vision applications. Among the companies that used machine vision, lighting was a general issue among most enterprises, while half of them experienced trouble with product variations. Software, surface problems and environment are also mentioned, but to a lesser extent. Among the vision manufacturers, lighting, tight spaces, harsh environments, blocked views, and a lack of understanding of the application were challenges.

Four automation engineering students (A, B, C and D) assessed nine barriers during internships as a second measure to dive deeper into the field. In sensing, Student B noted a "lack of customer demand", while Student D highlighted a "lack of understanding of machine vision" due to internal resistance. In seizing, Students A, C and D observed that companies rely on a single dedicated employee for machine vision implementation, risking fragility. Student D also mentioned "mistrust in machine vision" due to management's lack of technical knowledge. In transforming, Student D cited "robustness issues" in harsh environments, while Student A encountered "interoperability issues" and "lack of performance transparency" when blind coding. Students also identified "product and manufacturing environment limitations", such as light interference and software misinterpretations in license plate recognition.

4 Discussion and Conclusion

As stated in the introduction, there is untapped potential to enhance competitiveness and productivity for SMEs through digitalisation and data. However, SMEs struggle with the implementation of machine vision systems. To explore this issue, the following research question was posed: "What are the barriers and opportunities for different types of SMEs in adopting machine vision systems?".

This study addressed the research question by identifying nine barriers and four opportunities across different stakeholder types. Raising awareness of these nine barriers could be a critical first step in supporting SMEs in the implementation of machine vision systems. Further work could focus on validating these barriers by expanding the sample size, which would help determine which barriers are most frequently encountered. This insight could guide more targeted support for SMEs in their adoption efforts.

Notably, the lack of an expert workforce is a concern shared by both manufacturers of vision equipment and the enterprises that use it. This aligns with the principles of the industrial paradigm, which highlights the need for competent employees. Therefore, a purely technological solution may not be sufficient; instead, long-term upskilling initiatives in this field could significantly contribute to value creation.

References

1. Epicor (2019) What is industry 4.0—the industrial internet of things (IIoT)?
2. Javaid M, Haleem A, Singh RP, Rab S, Suman R (2022) Exploring impact and features of machine vision for progressive industry 4.0 culture. Sensors International 3:100132
3. Kukharuk A, Gavrysh J (2019) Competitiveness of SMEs in terms of Industry 4.0. In: 2019 International conference on creative business for smart and sustainable growth (CREBUS), Sandanski, Bulgaria, pp 1–4. https://doi.org/10.1109/CREBUS.2019.8840103
4. Adamo F, Attivissimo F, Di Nisio A (2010) Calibration of an inspection system for online quality control of satin glass. IEEE Trans Instrum Meas 59(5):1035–1046
5. Chen T, Wang Y, Xiao C, Wu QMJ (2016) A machine vision apparatus and method for can-end inspection. IEEE Trans Instrum Meas 65(9):2055–2066. https://doi.org/10.1109/TIM.2016.2566442
6. Bjørlykhaug E, Egeland O (2019) Vision system for quality assessment of robotic cleaning of fish processing plants using CNN. IEEE Access 7:71675–71685. https://doi.org/10.1109/ACCESS.2019.2919656
7. Zhang H et al (2018) Automated machine vision system for liquid particle inspection of pharmaceutical injection. IEEE Trans Instrum Meas 67(6):1278–1297. https://doi.org/10.1109/TIM.2018.2800258
8. Ekwongmunkong W, Mittrapiyanuruk P, Kaewtrakulpong P (2016) Automated machine vision system for inspecting cutting quality of cubic zirconia. IEEE Trans Instrum Meas 65(9):2078–2087. https://doi.org/10.1109/TIM.2016.2566858
9. Intuitive machine vision inspection of food and packaging (2025) TriVision. Accessed: 20 Feb 2025. [Online]. Available: https://trivision.io/
10. Mittal S, Khan MA, Romero D, Wuest T (2018) A critical review of smart manufacturing & Industry 4.0 maturity models: Implications for small and medium-sized enterprises (SMEs). J Manuf Syst 49:194–214. https://doi.org/10.1016/j.jmsy.2018.10.005

11. Menon S, Shah, S (2020) Are SMEs ready for Industry 4.0 technologies: an exploratory study of I 4.0 technological impacts. In: 2020 International conference on computation, automation and knowledge management (ICCAKM), pp 203–208. https://doi.org/10.1109/ICCAKM46823.2020.9051550
12. Glass R, Meissner A, Gebauer C, Stürmer S, Metternich J (2018) Identifying the barriers to Industrie 4.0. Procedia CIRP 72:985–988. https://doi.org/10.1016/j.procir.2018.03.187
13. Kumar P, Bhamu J, Sangwan KS (2021) Analysis of barriers to Industry 4.0 adoption in manufacturing organizations: an ISM approach. Procedia CIRP 98:85–90. https://doi.org/10.1016/j.procir.2021.01.010
14. Larsen MSS, Magnusson M, Lassen AH (2022) Industry 4.0 holds a great potential for manufacturers, so why haven't they started? In: Andersen A-L, Andersen R, Brunoe TD, Larsen MSS, Nielsen K, Napoleone A, Kjeldgaard S (eds) Towards sustainable customization: bridging smart products and manufacturing systems, in Lecture Notes in Mechanical Engineering. Springer International Publishing, Cham, pp 721–729. https://doi.org/10.1007/978-3-030-90700-6_82
15. Christiansen L, Borregaard T, Antonsen MG, Laursen ES, Brunoe TD (2022) Implementing virtual prototyping for the production of customized products: an SME study. In: Andersen A-L, Andersen R, Brunoe TD, Larsen MSS, Nielsen K, Napoleone A, Kjeldgaard S (eds) Towards sustainable customization: bridging smart products and manufacturing systems, in Lecture Notes in Mechanical Engineering, pp 762–769. Springer International Publishing, Cham. https://doi.org/10.1007/978-3-030-90700-6_87
16. Basar MS, Christiansen L, Nannerup PD, Graugaard Antonsen M (2022s) Identification of barriers to and opportunities for adoption of machine vision for small and medium-sized enterprises. In: Presented at the 2022 IEEE 27th international conference on emerging technologies and factory automation (ETFA). IEEE
17. Orbis—BvD is now Moody's (2026) Accessed 20 Feb 2025. [Online]. Available: https://www.moodys.com/web/en/us/capabilities/company-reference-data/orbis.html
18. Gioia DA, Corley KG, Hamilton AL (2013) Seeking qualitative rigor in inductive research: notes on the Gioia methodology. Organ Res Methods 16(1):15–31. https://doi.org/10.1177/1094428112452151
19. Teece DJ (2007) Explicating dynamic capabilities: The nature and microfoundations of (sustainable) enterprise performance. Strateg Manag J 28(13):1319–1350. https://doi.org/10.1002/smj.640s

Comparison of Feature Extraction Methods for Time Series Data in Fault Classification of Screw Connections

Michelle Henkies, Nikolai West, and Jochen Deuse

Abstract High-frequency time series data in manufacturing creates computational challenges for machine learning applications. This study compares four feature extraction methods (PAA, PCA, catch22, tsfresh) for classifying surface-based defects in screw connections. Using 12,500 tightening operations across eight defect classes, we evaluate these methods based on classification performance, computational efficiency, and memory usage. Results show tsfresh variants achieve the highest accuracy (up to 11% improvement over benchmark), while PAA offers the best balance between performance and computational efficiency. For minimal resource usage, catch22 maintains performance within ± 6% of benchmark using only 24 features. These findings provide concrete guidelines for selecting feature extraction techniques based on application requirements.

Keywords Feature extraction · Tightening process · Time series data · Machine learning · Multiclass classification

1 Introduction

The proliferation of high-frequency time series data in manufacturing creates significant storage and processing challenges. While time series data collection has become widespread, efficient utilization of this data remains problematic. Feature extraction methods offer a solution by transforming high-dimensional data into compact representations while preserving critical information [1]. In our previous research [2], we successfully classified surface manipulations in screw connections using raw torque

M. Henkies (✉) · N. West · J. Deuse
RIF Institute of Research and Transfer, Dortmund, Germany
e-mail: michelle.henkies@tu-dortmund.de

Institute of Production Systems, Technical University Dortmund, Dortmund, Germany

J. Deuse
Centre for Advanced Manufacturing, University of Technology Sydney, Sydney, Australia

M. Manns (ed.), *New Paradigms for Anticipated Uncertainty*, Lecture Notes in Mechanical Engineering, https://doi.org/10.1007/978-3-032-16889-4_58

measurements. However, the high dimensionality of the data limited real-time applications. This raises a critical question: can we maintain or improve classification performance while reducing data dimensions?

This study compares four feature extraction methods—Piecewise Aggregate Approximation (PAA), Principal Component Analysis (PCA), tsfresh, and catch22—using a dataset of 12,500 screw connections with eight distinct surface properties. The research contribution lies in a structured empirical assessment of established feature extraction techniques. This offers comparative insights to provide practical guidelines for selecting feature extraction methods in industrial time series classification tasks.

2 Fundamentals and Methods

Machine learning for manufacturing defect detection has shown significant potential in supervised and unsupervised approaches [3]. Time series data, consisting of sequentially ordered values [4], requires appropriate feature extraction to transform it into suitable representations for analysis. Literature identifies four primary approaches to time series representation: (1) value-based using raw numeric data [4], (2) approximation-based capturing trends or shapes [5], (3) model-based employing mathematical models [4, 5], and (4) feature-based summarizing key characteristics [6]. Here, we compare popular methods for these categories: raw data, PAA as approximation-based, and three feature-based methods: PCA, tsfresh and catch22. These cover classical (PAA, PCA) and modern (catch22, tsfresh) methods designed for time series feature extraction:

- **PAA**: Divides time series into segments, representing each by its average value [7]. Its straightforward implementation divides the original time series into equal-sized segments, calculating the mean value for each segment. The number of segments is flexible, making PAA ideal for finding the minimum number of time series values while preserving information from the screw data.
- **PCA**: Transforms correlated variables into uncorrelated components while retaining variance [8]. It computes the covariance matrix of the data and finds its eigenvalues and eigenvectors, with the first component capturing most of the variance. Torque measurements often show correlation over time, and PCA identifies the main components of variation needed to distinguish normal data from anormal observations.
- **tsfresh**: Extracts statistical features based on scalable hypothesis tests. The extraction of statistical-based features is achieved through customizable dictionaries. Currently, the comprehensive feature set employs 783 distinct features, including statistical measures, correlation properties, and frequency characteristics, making it particularly suitable for complex screw data analysis [6].

- **catch22**: Comprises 22 features designed for time-series classification, selected from 4,791 candidates based on performance across 93 datasets [9]. These features capture key aspects of time-series dynamics, including distribution properties, linear and nonlinear temporal autocorrelation, and fluctuation scaling. Its compact yet effective feature set provides a comparison with fewer features than tsfresh.

We test all representations using four supervised models that West et al. also applied as a benchmark [2]. These methods were selected to represent different algorithm families, ensuring a comprehensive evaluation across diverse approaches:

- **k-Nearest Neighbors (KNN)** is a non-parametric instance-based method that classifies a sample based on the majority class of its k nearest neighbors [10]. It makes no assumptions about the underlying distribution and is effective for problems with irregular decision boundaries. The algorithm stores all training instances and makes predictions by calculating distances between the query point and training points.
- **Multi-Layer Perceptron (MLP)** is a feedforward artificial neural network that consists of multiple layers of nodes. The layers include an input layer, one or more hidden layers with nonlinear activation functions, and an output layer [11]. MLP learns complex patterns through backpropagation, making it well-suited for capturing intricate relationships in high-dimensional data like time series.
- **Naive Bayes (NB)** is a probabilistic classifier based on Bayes' theorem, assuming conditional independence between features [12]. Despite this simplifying assumption, NB often performs well even when the independence assumption is violated, making it computationally efficient and a good basis for comparison.
- **Random Forest (RF)** is an ensemble learning method that constructs multiple decision trees and outputs the mode of the classes for classification [13]. By combining many trees and introducing randomness in feature selection, RF reduces overfitting problems common to single decision trees while capturing nonlinear relationships.

This combination of algorithms represents different approaches to classification—distance-based (KNN), neural (MLP), probabilistic (NB), and ensemble tree-based (RF)—providing a robust starting point for comparing feature extraction methods.

3 Experiment Design

We utilize our open-source dataset [14, 15], which is particularly suitable for feature extraction comparison due to its high-dimensional time series data and diverse fault patterns. In this experiment, two thermoplastic workpieces are connected with two identical screw connections using Delta PT 40 $\times$ 12 screws from EJOT. Due to the limited number of workpieces and the investigation of multiple screw runs on thread-cutting connections [16], the dataset contains 25 screw runs for each component. For each run, torque, angle, and gradient measurements are recorded with timestamps.

For our analysis, we defined two scenarios that represent different practical applications in manufacturing quality control:

1. **First cycles** (S_{1st}): Only first cycles *(using only 500 screw runs total)*
2. **All cycles** (S_{all}): All 25 cycles *(using all 12,500 screw runs total)*

The surface manipulations were applied only for the first cycle, highlighting the importance of differentiating between these scenarios [2]. To evaluate the feature extraction methods under varying classification complexities, we implemented three different labeling strategies. The "Singles" strategy maintains all original class distinctions, presenting the most challenging classification task. The "Grouped" strategy combines similar error types, representing a moderate complexity level. The "Binary" strategy offers the simplest task by distinguishing only between normal and anormal. Table 1 summarizes the eight error classes and their organization under these labeling strategies.

This experimental design enables an assessment of feature extraction methods across different data volumes and classification challenges, allowing us to evaluate how each method performs under varying conditions of practical relevance to manufacturing.

Table 1 Error classes and their organization under different labeling strategies

Error class			Labeling strategies		
No.	Description	Size	"Singles"	"Grouped"	"Binary"
1	Baseline (normal screw runs with no manipulations)	2500	E_B	E_B	E_B
2	Repeated runs (top part used 25 times)	2500	E_{RR}	E_{RR}	E_{G1} (All errors in one class)
3	Lubricant water (surface wetted with water)	1250	E_{LW}	E_{G1} (Lubricant-based errors)	
4	Lubricant oil (surface wetted with oil)	1250	E_{LO}		
5	Coarse sanding (surfaces altered with 40 grid)	1250	E_{CS}	E_{G2} (Sanding-based errors)	
6	Fine sanding (surfaces altered with 400 grid)	1250	E_{FS}		
7	Plastic adhesive (dried adhesive on surfaces)	1250	E_{PA}	E_{G3} (Surface-alterations)	
8	Surface chipped (chip shaved by drill)	1250	E_{SC}		

4 Modeling

Data are loaded via the PyScrew library [14] using default get_data() parameters, except target_length=800. For model evaluation, we implemented a stratified fivefold cross-validation with an 80/20 train-test split to ensure balanced class distribution across folds. We adopted the optimized hyperparameters from [2] for all models and trained using torque values with scikit-learn version 1.6.1. For reference, all reported durations from experiments were conducted on a simple CPU (Intel Core i7-1355U).

To systematically evaluate dimensionality reduction performance, we configured each feature extraction method with parameters appropriate to its characteristics:

- **PAA and PCA**: Tested with 18 different feature dimensions (2, 3, 5, 8, 10, 15, 20, 25, 50, 75, 100, 125, 150, 175, 200, 300, 400, 500), covering both extreme reduction for computational efficiency and moderate reduction for maintaining accuracy. The upper limit of 500 corresponds to the number of samples in the S_{1st} scenario.
- **tsfresh** (three *predefined* sets): Three primary sets with 783 features for *comprehensive* cover of diverse time series characteristics, 777 features for *efficient* excluding computationally expensive calculations and 10 core statistical features for *minimal*. We use extract_relevant_features() to extract and then select the relevant ones. [6]
- **tsfresh** (two *specialized* sets): Two focused configurations—*Autocorrelation* set using time lags (10, 50, 100) based on anomaly patterns, and *FFT* set using Fast Fourier Transformation coefficients [6].
- **catch22**: Compact set with all 24 features, including the base 22 features plus mean and standard deviation, using default settings for consistent comparison [9].

This approach allows us to evaluate each extraction method´s performance across a range of dimensionality reduction levels, providing insights into the optimal balance between computational efficiency and classification accuracy for fault detection.

5 Results

The macro F1 score (here referred to as F1 score) is selected to demonstrate the performance of all scenarios and labeling strategies in **Fig. 1**. The mean of the F1 score is represented by the dot in the center of the bar, and the length of the bars indicates the standard deviation of the fivefold cross-validation. Figure 1a) shows the labeling strategy single with both scenarios and four machine learning (ML) methods with each of the eight extraction methods. The extraction method *None* uses the raw data as a reference, in which no features were extracted within the data set. The other two labeling strategies are shown in Figs. 1b) grouped and 1c) binary. The order of the extraction methods is consistent across the different ML models, scenarios and labeling strategies. For PAA and PCA Fig. 1 shows the highest F1 score of the 18 tested feature dimensions.

As illustrated in Fig. 1, certain extraction methods, in particular the efficient and comprehensive sets of tsfresh, show an improvement in performance across various scenarios and labeling strategies compared to the benchmark (None). PAA achieves similar F1 scores as the benchmark, regardless of the configuration and exceeds it by up to 11% (NB, binary, S_{all}). Additionally, PAA in combination with RF or MLP is often one of the top F1 scores of each scenario and labelling strategy. The combination of PCA with RF or MLP approximates the benchmark to a maximum deviation of about 7%. Using KNN with PCA, results in F1 scores that are analogous to the benchmark. Only the ML model NB outperforms the benchmark when using PCA up to 17% (grouped, S_{all}). catch22 is with a maximum of ± 6% close to the benchmark in every combination with only 24 features. The tsfresh_min shows acceptable performance with KNN and NB achieving about ± 5% in comparison

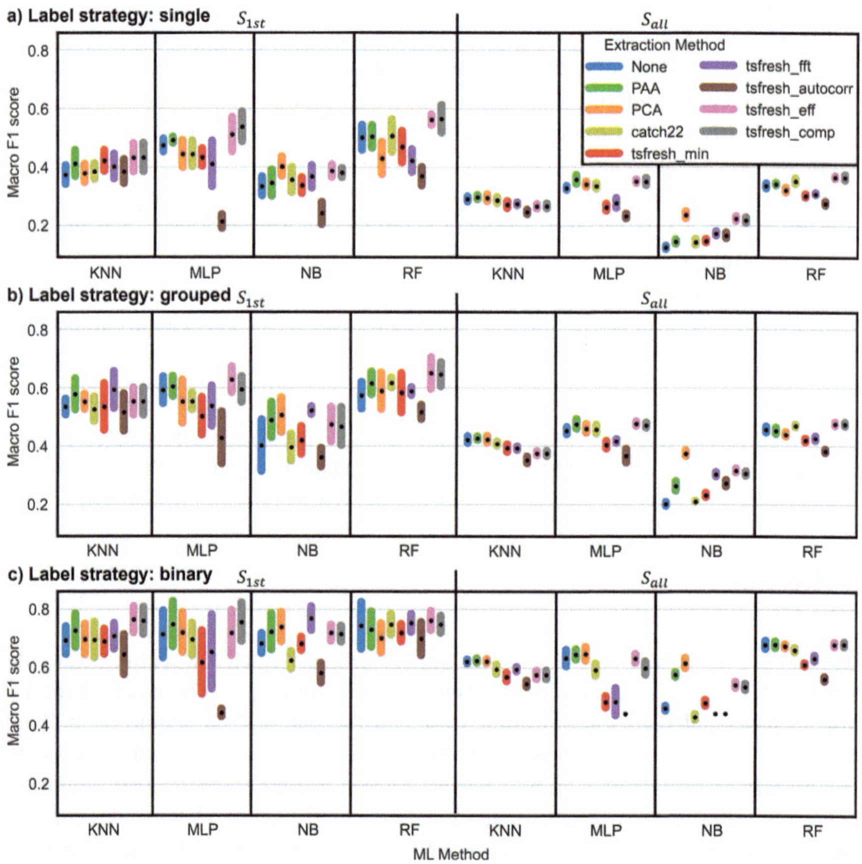

Fig. 1 Macro F1 scores of every feature extraction method, ML model, scenario (left: S_{1st}, right: S_{all}) and label strategy **a** single; **b** grouped, **c** binary

to the benchmark. When using MLP, this extraction method achieves around 10%-15% lower F1 scores than the benchmark. RF reaches with 3–7% lower F1 scores better results than MLP. tsfresh_fft behaves similarly to tsfresh_min in terms of ML methods, but tsfresh_fft often delivers slightly better F1 scores than tsfresh_min. The application of autocorrelation with tsfresh (tsfresh_autocorr) yields the worst F1 scores for most configurations and models. The efficient and comprehensive set of tsfresh achieved very similar results and often achieved optimal results with RF or MLP in each combination of label strategy and scenario. With the NB classifier, up to 11% better F1 scores are achieved with the efficient set compared to the benchmark. The performance of the RF is up to 8% better than the benchmark and the results with MLP exceed the benchmark with up to 6%.

Table 2 presents the training times, test times, feature extraction times and the number of features per model and extraction method for scenario S_{all} with single labelling strategy. It can be observed that KNN and NB generally have a very short runtime. It is noticeable that the test time of KNN for 2500 samples is longer than the training time of 10,000 samples. The KNN classifier only stores the data in the training phase. The test time includes the calculation of the distances to the training data, which leads to longer test times that train times. RF requires a training time of less than 5 s for this dataset. The MLP classifier trains for up to 24 s, but requires about ten times less test time than RF. Moreover, the train and test times are reduced by the extraction methods, with the exception of using the efficient and comprehensive sets of tsfresh. The feature extraction time increases with increasing number of features. The extraction methods with the best F1 scores also take the longest to calculate. In general, PAA and PCA are very fast for different numbers of features.

Table 2 Computation times depending on the extraction method (Single, S_{all})

Feature extraction method	Number of features	Feature extraction time (s)	Avg. train time (s) k =5; n= 10,000				Avg. test time (s) k =5; n= 2500			
			KNN	MLP	NB	RF	KNN	MLP	NB	RF
None	800	0.000	0.009	23.033	0.098	2.267	0.373	0.008	0.102	0.043
PAA	200	0.085	0.004	12.575	0.049	1.444	0.136	0.001	0.035	0.051
PCA	200	0.231	0.003	6.571	0.021	2.363	0.115	0.004	0.019	0.039
catch22	24	47.986	0.003	6.201	0.000	0.616	0.030	0.003	0.003	0.054
tsfresh_ autocorr	3	9.116	0.006	2.154	0.000	0.373	0.028	0.000	0.000	0.044
tsfresh_fft	4	8.775	0.004	3.360	0.007	0.512	0.034	0.003	0.000	0.054
tsfresh_ min	9	13.929	0.005	2.333	0.000	0.516	0.058	0.000	0.000	0.049
tsfresh_ eff	699	668.284	0.003	14.568	0.116	3.708	0.446	0.006	0.101	0.043
tsfresh_ comp	705	2031.349	0.002	23.649	0.088	4.245	0.437	0.008	0.118	0.052

Unlike randomly generated time series data, screw-driving processes exhibit distinct physical characteristics, including monotonic torque trends, identifiable process phases, and surface defect signatures that create structured signal patterns. The superior performance of domain-appropriate feature extraction methods (particularly tsfresh's comprehensive feature set) over simpler approaches demonstrates that manufacturing time series contain exploitable physical relationships generic dimensionality reduction cannot capture. These results suggest that for real-time monitoring of screw connections, PAA offers the best compromise between computational efficiency and classification performance, while offline analysis where computational time is less critical, tsfresh's comprehensive set provides the highest accuracy, with targeted feature selection critical for reliable fault detection and process optimization in production environments.

6 Conclusion

This study addresses a fundamental challenge in manufacturing analytics: selecting the optimal feature extraction method for screw tightening time series data based on specific application requirements. Our findings demonstrate that while dimensionality reduction is achievable across methods, the choice should be guided by clear priorities:

- **For accuracy-critical applications**: The efficient tsfresh delivers superior classification performance, improving F1 scores by up to 11% over raw data, albeit requiring longer feature extraction time and greater storage capacity. This makes it ideal for detailed quality analysis and process optimization in non-time-critical settings.
- **For real-time applications**: PAA or PCA provide the best balance, maintaining classification accuracy while reducing computational demands. This enables immediate fault detection in production lines without requiring specialized hardware.
- **For resource-constrained systems**: catch22 and the FFT or minimal set with tsfresh offer minimal memory footprint and processing requirements with performance within 6% of benchmark, suitable for embedded systems or rapid analysis scenarios.

While this study is focused on screw-driving, the time series characteristics, such as the monotonic trends, process phases, and sensor signal profiles, are also found in other industrial processes like welding, stamping, or forming. The insights derived from the presented use case may be generalizable to other applications, pending validation.

In summary, while each extraction method offers specific advantages, our research indicates a critical need for developing a universal, domain-specific feature set for screw-driving applications. Current general-purpose methods like tsfresh contain hundreds of potentially redundant features for this specific application. Future research should focus on identifying the most discriminative features across surface defect types to create a standardized approach that optimizes both performance and efficiency for manufacturing quality control.

References

1. West N, Schlegl T, Deuse J (2021) Feature extraction for time series classification using univariate descriptive statistics and dynamic time warping in a manufacturing environment. Int Conf Big Data, Artif Intell Internet Things Eng 2:762–768. https://doi.org/10.1109/ICBAIE52039.2021.9389954
2. West N, Trianni A, Deuse J (2024) Detection of surface-based anomalies for self-tapping screws in plastic housings using supervised machine learning. Int Conf Comput Ind Eng 51:1–10
3. West N, Schlegl T, Deuse J (2023) Unsupervised anomaly detection in unbalanced time series data from screw driving processes using k-means clustering. CIRP Conf Manuf Syst 120:1185–1190. https://doi.org/10.1016/j.procir.2023.09.146
4. Laurinec P, Lucká M (2016) Comparison of representations of time series for clustering smart meter data. In: Ao SI, Douglas C, Grundfest WS (eds) Proceedings of the world congress on engineering and computer science 2016, vol 1, pp 458–463. Newswood Limited, San Francisco. https://doi.org/10.48550/arXiv.2412.02026
5. Zhang M, Pi D (2017) A new time series representation model and corresponding similarity measure for fast and accurate similarity detection. IEEE Access 5:24503–24519. https://doi.org/10.1109/ACCESS.2017.2764633
6. Christ M, Braun N, Neuffer J, Kempa-Liehr AW (2018) Time series feature extraction on basis of scalable hypothesis tests (tsfresh—a python package). Neurocomputing 307:72–77. https://doi.org/10.1016/j.neucom.2018.03.067
7. Keogh E, Chakrabarti K, Pazzani M, Mehrotra S (2001) Dimensionality reduction for fast similarity search in large time series databases. Knowl Inf Syst 3:263–286. https://doi.org/10.1007/PL00011669
8. Jolliffe IT, Cadima J (2016) Principal component analysis: a review and recent developments. Philos Trans R Soc A: Math, Phys, Eng Sci, vol 374. https://doi.org/10.1098/rsta.2015.0202
9. Lubba CH, Sethi SS, Knaute P, Schultz SR, Fulcher BD, Jones NS (2019) Catch22: canonical time-series characteristics. Data Min Knowl Disc 33:1821–1852. https://doi.org/10.1007/s10618-019-00647-x
10. Cover TM, Hart PE (1967) Nearest neighbor pattern classification. IEEE Trans Inf Theory 13:21–27. https://doi.org/10.1109/TIT.1967.1053964
11. Rumelhart DE, Hinton GE, Williams RJ (1986) Learning representations by backpropagating errors. Nature 323:533–536. https://doi.org/10.1038/323533a0
12. Friedman N, Geiger D, Goldszmidt M (1997) Bayesian network classifiers. Mach Learn 29:131–163. https://doi.org/10.1023/A:1007465528199
13. Breiman L (2001) Random forests. Mach Learn 45:5–32. https://doi.org/10.1023/A:1010933404324
14. West N, Deuse J (2025) PyScrew: a comprehensive dataset collection from industrial screw Driving Exp. https://doi.org/10.48550/arXiv.2505.11925
15. West N, Deuse J (2025) Industrial screw driving dataset collection: Time series data for process monitoring and anomaly detection. v1.2.2 s02_variations-in-surface-friction. https://doi.org/10.5281/zenodo.15393134
16. West N, Deuse J (2024) A comparative study of machine learning approaches for anomaly detection in industrial screw driving data. In: Bui TX (ed) Proceedings of the 57th Hawaii international conference on system sciences, pp 1050–1059. HICSS Conference Office, Honolulu, Hawaii. https://doi.org/10.24251/HICSS.2024.127

A 3D Scanning-Based Method for Surface Defect Detection in Galvanized Scaffolding

Fengyun Shao, Tadele Belay Tuli, Florian Schreiber, and Martin Manns

Abstract Surface defects in galvanized scaffolding such as excess zinc on the metal surface can significantly affect assembly precision and durability. Traditional inspections based on manual visual assessment are limited in efficiency and cannot be easily automated. In this work, a 3D scanning-based method is proposed for automated defect detection in galvanized scaffolding. A 3D scanner is employed to acquire point cloud data of scaffolding components. The scanned point cloud is then aligned with an uncoated 3D Computer-Aided Design (CAD) reference model using the Iterative Closest Point (ICP) algorithm. Defects are identified by analyzing spatial deviations between the scanned data and the reference model. Deviations are considered as defects when exceeding a predefined threshold. To eliminate false positives defects and deviations at irrelevant areas, two region-specific filters are applied. Experimental validation demonstrates the improved automation of the proposed method compared to manual visual inspection.

Keywords Defect detection · Experimental design · Production engineering

1 Introduction

Galvanized scaffolding components are widely used in construction and industrial applications, requiring high geometric precision to ensure structural integrity and ease of assembly [8, 12]. Zinc coating process, which is vital for corrosion protection, often introduces surface irregularities, such as excess material accumulation [3, 6]. These defects may interfere with component assembly, which leads to increased manual rework and production delays [10].

In the current industrial scenario, defect detection is mainly carried out through manual inspection or data-driven methods [7]. The manual inspection is flexible

F. Shao (✉) · T. B. Tuli · F. Schreiber · M. Manns
PROTECH - Institut für Produktionstechnik, Universität Siegen, Paul-Bonatz 9-11, 57076 Siegen, Germany
e-mail: Fengyun.Shao@uni-siegen.de
URL: https://protech.mb.uni-siegen.de/fams

M. Manns (ed.), *New Paradigms for Anticipated Uncertainty*, Lecture Notes in Mechanical Engineering, https://doi.org/10.1007/978-3-032-16889-4_59

and capable of detecting various defect types. However, it is labor-intensive, time-consuming, and highly dependent on the experience and subjective assessment of the inspector [14]. Besides, variations in the detection results could occur due to differences in inspector skills, fatigue, and environmental changes. Those factors limit the repeatability and reliability of manual inspection. Although 3D scanning has been previously applied to detect surface defects in manufacturing contexts [11, 13], there has been limited application of such methods to galvanized scaffolding components. This paper proposes a novel deviation-based analysis pipeline tailored to this specific context. Data-driven approaches have become an alternative solution, which leverage machine learning-based [4, 15] or deep learning-based methods [2, 9] to automatically detect defects. These methods could essentially reduce reliance on human inspectors and increase consistency in the detection results. However, data-driven approaches require large amounts of high-quality datasets for training the model. Due to the lack of datasets, such methods are hard to deploy in industrial situations [16]. Unlike large-scale open source datasets used in research settings, industrial defect detection typically involves application-specific defect datasets that require extensive manual annotation, which is an expensive and time-consuming process [5].

2 Objective

To address these challenges, a 3D scanning-based defect detection method that overcomes the limitations of both manual and data-driven approaches is proposed. Instead of relying on human interpretation or large-scale labeled datasets, the proposed method utilizes geometric deviation analysis by comparing point cloud data of the scanned workpiece with 3D Computer-Aided Design (CAD) reference models. By applying thresholding and region-specific filtering, the method can systematically identify critical defects without requiring extensive training data or subjective human assessment.

3 Proposed Method

The proposed method follows a structured workflow for automated defect detection on the workpiece, as shown in Fig. 1. The process begins with 3D scanning, where a point cloud of the workpiece surface is acquired. The scanned data is then aligned with the 3D CAD model of the uncoated workpiece to establish a reference for deviation analysis. A thresholding and filtering step is applied to distill significant defects while removing irrelevant deviations. Finally, the identified defects are then visualized to clearly illustrate surface irregularities.

One type of galvanized vertical scaffolding tube is selected for measurement, representing a typical component in modular scaffolding systems. A 3D scanner captures

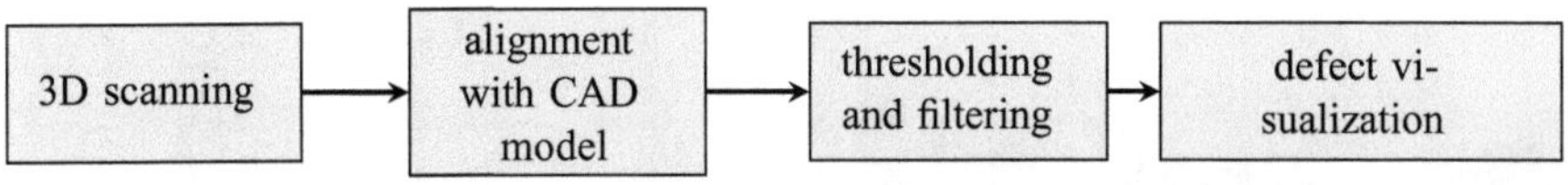

Fig. 1 Illustration of the proposed workflow of the defect detection process

high-resolution point cloud data, which is stored in .PLY format and processed using tools such as MeshLab [1]. The CAD model is aligned with the scan using a two-step process: coarse manual landmark matching followed by fine registration via the Iterative Closest Point (ICP) algorithm. ICP is chosen for its high performance in aligning partially overlapping point clouds when a good initial estimate is available.

The CAD model provided by the manufacturer does not include weld seams, although most modern 3D CAD tools support modeling them. These features can be added in future work to reduce false-positive deviations near welded areas.

After alignment, point-wise deviations are computed between the scan and CAD surface. To avoid misclassification due to small geometric variations or alignment imperfections, a deviation threshold is applied. Region-specific filtering further improves accuracy by excluding weld seams and ignoring irrelevant deviations near the pipe ends, focusing only on areas critical for defect detection.

4 Experiment and Results

4.1 Experimental Setup

To validate the proposed method, multiple galvanized scaffolding samples with surface defects were scanned using a Zeiss T-Scan Hawk II 3D scanner, with an example shown in Fig. 2. The T-Hawk 3D scanning system offers a spatial resolution of 20 μm and a measurement accuracy of $\pm 15\,\mu$m under standard conditions. The scanning process covered the whole workpiece surface, to ensure that all potential defect areas were included. The obtained point cloud data was then processed and aligned with the uncoated CAD reference model to enable deviation analysis.

To quantify surface deviations, each point in the 3D scan is projected onto the nearest surface point of the aligned CAD model using a point-to-surface distance calculation. The deviation is defined as the signed Euclidean distance along the normal vector of the CAD surface. No curve fitting or surface reconstruction is applied to the scan; instead, raw point cloud data is used to preserve measurement accuracy. This method avoids potential smoothing errors introduced by high-order fitting, and ensures that localized deviations such as burrs or zinc build-ups are retained in the deviation map.

Fig. 2 An example of surface defects on a galvanized scaffolding sample

4.2 Results and Evaluation

Based on the proposed method in Fig. 1, the results from 3D scanning using GOM (Gesellschaft für Optische Messtechnik) system is given in Fig. 3. On the illustration, the heatmap of the deviations using ZEISS INSPECT®which is a commercial tool is shown. Based on this, different threshold values were tested, and 1.7 mm was ultimately chosen as provides a balance between filtering out irrelevant deviations and ensuring that actual defects are not missed. This result is a preliminary finding that lead to new method development.

For testing the proposed method, first, the 3D scanning was performed. Then, the 3D scanned point clouds and original CAD designs are aligned as shown in Fig. 4 which compares the CAD and 3D scanned data by overlapping. Then a threshold and filters are applied to detect regions of defects e.g., a 1.2 mm threshold (left) and a 1.7 mm threshold (right). To exclude deviations at weld seams, defects detected within 50 mm of the workpiece tube axis are filtered out.

As shown in Fig. 5a, defects at weld seams are highlighted in yellow. Since these points are within 50 mm of the tube's center axis, they are removed after localized filtering (see Fig. 5b, c). For defects at the end part of the workpiece, Fig. 5d, f illustrate the filtering process. The initial detection using a 1.7 mm threshold is shown in Fig. 5e, where the red points represent false detections (False Positive) due to scanning variance, while the blue and yellow points correspond to defects on the inner and outer sides of the end part, respectively. A localized filter was then applied to remove points more than 20 mm away from the end part. Figure 5f presents the defects after applying both filtering steps.

After the initial filtering and localized filtering, it can be observed that defects mostly pop up at the rosette parts, outer side of end part and inner part of the end part. As illustrated in Fig. 6a, defects on the rosette, defects on the outer end part

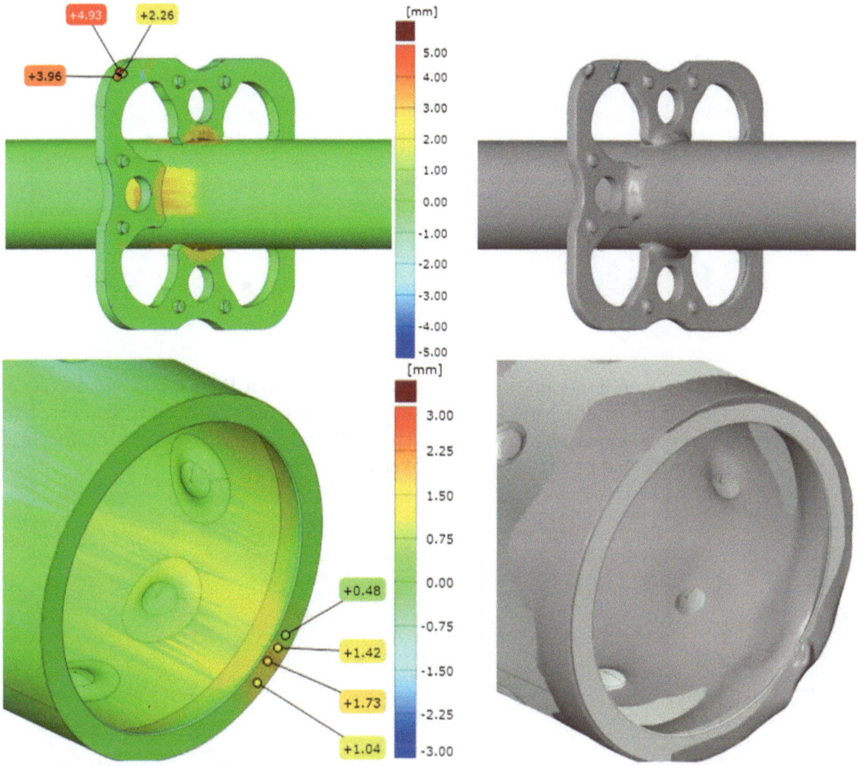

Fig. 3 Heatmap of deviations before filtering. Large deviations are detected in weld seam regions, which are not actual defects. These false positives are removed in the filtering process

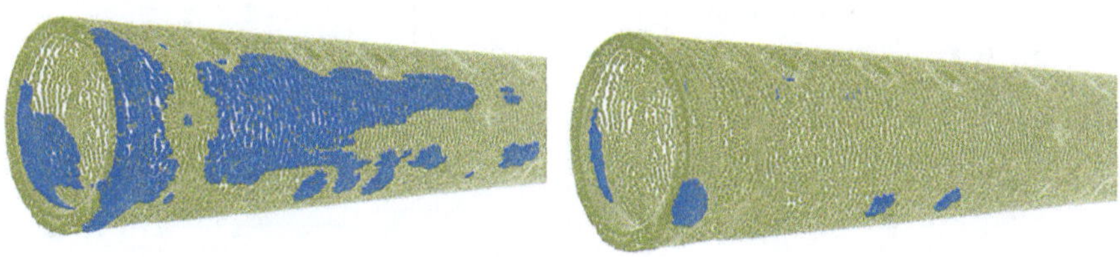

Fig. 4 Comparison of the reported defects using a 1.2 mm threshold (left) and a 1.7 mm threshold (right). Green indicates acceptable deviations within the defined threshold, while blue highlights areas with deviation higher than the threshold

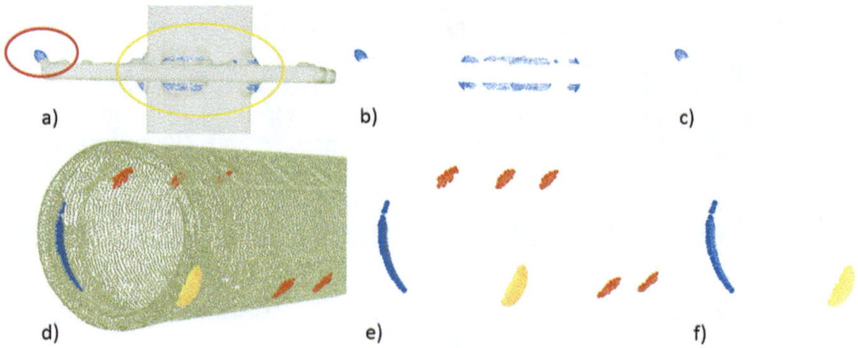

Fig. 5 Illustration of the filtering on the rosette part and end part. **a** Point cloud and defects on rosette part. **b** Defects on rosette part without localized filtering. **c** Defects on rosette part after localized filtering. **d** Point cloud and defects on end part **e** Defects on the end part before localized filtering. **f** Defects on the end part after localized filtering

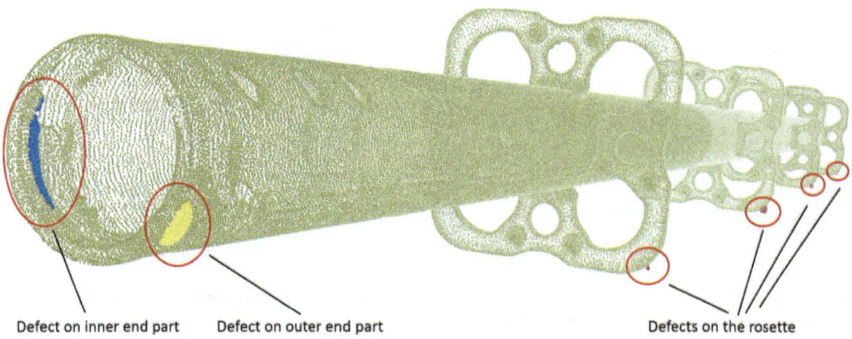

(i)) Visualization of the defects on the scaffolding.

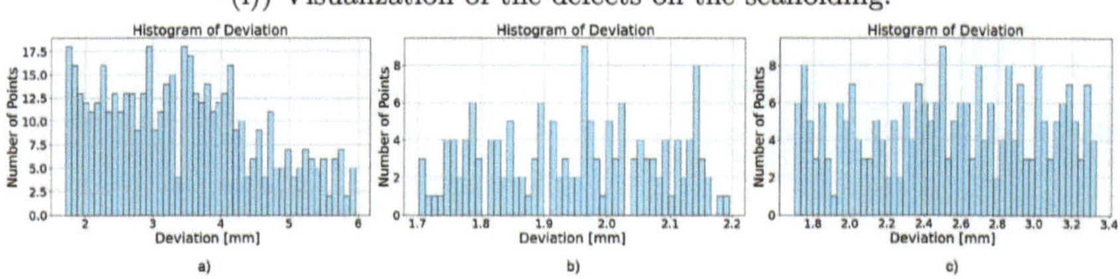

(ii)) Histogram of the defects surface deviation distribution on different parts of the workpiece. a) Defects on rosette part. b) Defects on inner end part. c) Defects on outer end part.

Fig. 6 Defect visualization and histogram analysis

and defects on the inner end part are marked in green, yellow and blue, respectively. Figure 6b shows the defects surface deviation distribution on different parts of the workpiece.

Table 1 Relative error in defect measurement at different locations

Defect location	Rosette	End part (outer side)	End part (inner side)
Average relative error	0.0389	0.164	0.022

$$\texttt{relative\_error} = \frac{|\texttt{measured\_deviation} - \texttt{real\_deviation}|}{\texttt{real\_deviation}} \quad (1)$$

Table 1 resents the relative error of the defect at different part of the workpiece.

5 Discussion

The proposed 3D scanning-based method overcomes the limitations of manual inspection and data-driven approaches. Manual checks are prone to inconsistency, while deep learning requires large labeled datasets, which are often unavailable in industrial settings. In contrast, our geometry-based method offers a reliable, repeatable solution without human interpretation or extensive data collection. However, scanning resolution and alignment accuracy still affect detection performance. Future work may explore adaptive thresholding to improve robustness under varying manufacturing conditions.

While manual inspection is faster for a single part, our method scales better for high-throughput production. Further optimization of alignment and filtering could reduce computational cost. Integrating robotic defect removal would also enable a fully automated inspection and correction pipeline, improving overall quality control.

This study uses the manufacturer's 3D CAD model as the reference for deviation analysis. However, we acknowledge that in industrial practice, 2D drawings with GD&T specifications define allowable tolerances. Since the current CAD model lacks such information, all deviations are calculated relative to nominal geometry. Incorporating GD&T data in future work will enable tolerance-aware defect classification and improve industrial applicability.

6 Conclusion

The proposed method detects defects on galvanized metal scaffolding surfaces by using a 3D scanner to scan the metal surfaces and converting the scanned results into a point cloud to compare with the CAD file of the metal scaffolding which has no defects. Two filters were added to exclude irrelevant defects and False True defects. The results show that the proposed method offers a higher degree of automation compared to manual visual inspection. In the future, adaptive thresholding techniques that dynamically adjust based on defect location and functional requirements of the component should be explored.

Acknowledgements The cooperation project "RoboPutz" (Grant number: KK5057119LF3) is funded by the Federal Ministry of Economics and Climate Protection (BMWK) as part of the Central Innovation Program for SMEs (ZIM) and is supervised by the Working Group of Industrial Research Associations (AiF). The 3D scanning result depicted in Fig. 3 has been conducted by the Chair for Forming Technology, University of Siegen.

References

1. Cignoni P, Callieri M, Corsini M, Dellepiane M, Ganovelli F, Ranzuglia G (2008) Meshlab: an open-source mesh processing tool. In: Scarano V, De Chiara R, Erra U (eds), Eurographics Italian chapter conference, pp 129–136. Eurographics
2. Dong H, Kechen Song Yu, He JX, Yan Y, Meng Q (2019) Pga-net: Pyramid feature fusion and global context attention network for automated surface defect detection. IEEE Trans Industr Inf 16(12):7448–7458
3. Gutoff EB, Cohen ED (2006) Coating and drying defects: troubleshooting operating problems. Wiley
4. He Yu, Song K, Meng Q, Yan Y (2019) An end-to-end steel surface defect detection approach via fusing multiple hierarchical features. IEEE Trans Instrum Meas 69(4):1493–1504
5. Hernavs J, Ficko M, Klančnik L, Rudolf R, Klančnik S (2018) Deep learning in industry 4.0–brief overview. J Prod Eng pp 1–5
6. Hong M-H, Saka H (2004) Fib and tem observations of defects in hot-dip zinc coatings. J Electron Microsc 53(5):545–552
7. Jeon YJ, Choi DC, Lee SJ, Yun JP, Kim SW (2015) Steel-surface defect detection using a switching-lighting scheme. Appl Opt 55(1):47–57
8. Ronny Kühne, Voelkel J, Bartsch H, Feldmann M, Pinger T, Sharma R, Reisgen U (2024) Fatigue strength of additively manufactured hot-dip galvanized steel coated with a zn–5al alloy. Progress in additive manufacturing, pp 1–13
9. Li J, Zhenfeng S, Geng J, Yin Y (2018) Real-time detection of steel strip surface defects based on improved yolo detection network. IFAC-PapersOnLine 51(21):76–81
10. Marder A, Goodwin F (2023) The metallurgy of zinc coated steels. Elsevier
11. Mariniuc AM, Cojocaru D, Abagiu MM (2024) Building surface defect detection using machine learning and 3d scanning techniques in the construction domain. Buildings 14(3):669
12. Murugesu J (2017) Design and analysis of frame scaffolding capacity using thickness and material as variables. Master's thesis, University of Malaya (Malaysia)
13. Rocha G, Mateus L, Fernández J, Ferreira V (2020) A scan-to-bim methodology applied to heritage buildings. Heritage 3(1):47–67
14. Dong X, Tong FX, Yang G (2023) A detection method of spangle defects on zinc-coated steel surfaces based on improved yolo-v5. Int J Adv Manuf Technol 128(1-2):937–951
15. Yun JP, Kim D, Kim K, Lee SJ, Park CH, Kim SW (2017) Vision-based surface defect inspection for thick steel plates. Opt Eng 56(5):053108–053108
16. Zohuri B, Moghaddam M (2020) Deep learning limitations and flaws. Mod Approaches Mater Sci 2:241–250

Anomaly Detection in Industrial Robotic Assembly with Variational Autoencoders

Niklas Grambow, Lisa-Marie Fenner, Dingyuan Wan, Jörg Krüger, and Kevin Haninger

Abstract Robots today still struggle with adaptation and generalization to changes in the task, a major barrier to deploying robots in semi- and unstructured tasks. If robots can detect novel situations that are out-of-distribution, defined as scenarios not represented in the training data, they can initiate a fail-safe or fallback action to recover or at least avoid damage. Anomaly detection identifies data patterns that deviate from expected behavior. We apply a variational autoencoder (VAE) approach to time series in robotics for an industrial cabling task. Inputs are force measurements and the robot's end-effector positions, from both nominal processes and various failure scenarios. In validation, the VAE model achieved an AUROC of 0.82 in detecting process-related failure. In the overall evaluation, two of three types of failures were reliably detected, while the third, which had smaller magnitude deviations in the force profile, proved challenging to identify robustly.

Keywords Anomaly detection · Event detection · Process monitoring · Variational autoencoder · Robotic assembly

1 Introduction

In the field of advanced robotic control, which is often required for complex industrial assembly tasks, sensor-based machine learning (ML)-based approaches allow robots to perform more complex tasks [1]. Integrating ML approaches into industrial applications remains challenging, primarily due to concerns about data quality and model robustness, particularly in out-of distribution scenarios [2]. ML is applied statistics, and thus cannot be inerrable. For factory automation, this raises the question of how to deal with the permanent risks of failures without constant human oversight. We propose a variational autoencoder-based anomaly detection (AD) approach to evaluate the assembly process in a realistic industrial setting, including contact with the environment and variance in the task execution, to determine whether the cabling

N. Grambow (✉) · L.-M. Fenner · D. Wan · J. Krüger · K. Haninger
Fraunhofer Institute for Construction Science and Design Technology IPK, Berlin, Germany
e-mail: niklas.grambow@ipk.fraunhofer.de

M. Manns (ed.), *New Paradigms for Anticipated Uncertainty*, Lecture Notes in Mechanical Engineering, https://doi.org/10.1007/978-3-032-16889-4_60

task was performed properly or if any errors occurred during the task execution. This work assesses whether AD is suitable for failure detection within complex robot industry tasks and discusses potential advantages and limitations. Integrating AD within industrial assembly processes, such as the given cabling task (Fig. 2a), can improve process reliability and contribute to autonomous behavior. In the end, our goal is to reduce scrap, accelerate commissioning, and enable early integration.

The paper is structured as follows: Sect. 2 reviews related work, Sect. 3 presents the method, Sect. 4 details the experimental setup, Sect. 5 discusses the results, and Sect. 6 concludes.

2 Related Work

Anomaly Detection (AD) methods can be categorized in various ways: by their core principle (e.g., statistical, distance-based, density-based, or cluster-based) [3, 4]; by method family (e.g., deep learning, classical machine learning, or signal analysis) [5]; or by their underlying modeling approach, distinguishing between data-driven, model-based, and knowledge-based methods [6]. Model-based approaches typically employ mathematical or logical descriptions of the underlying process or system, while knowledge-based methods rely on pre-defined fault initializations to subsequently recognize faults [6, 8]. Although data-driven methods often result in high computational effort to train, with decreasing costs for computation capacity, these approaches become more popular [6]. Furthermore, data-driven methods offer key advantages, such as not requiring process-specific knowledge [6]. Graebaek et al. [6] evaluated various anomaly detection methods using their proposed dataset and demonstrated that AD becomes significantly more challenging when robots interact with the environment. Furthermore, they emphasize the efficiency of the neural network based approaches compared to simple approaches like k-Nearest-Neighbors. Brockmann et al. [7] propose another dataset, where a conveyor is used to provide various positions of the object in a pick-and-place task. Beside common state-of-the-art approaches multivariate time series flow is proposed, that outperforms the other algorithms on the acquired data. Chen et al. [8] implement a variational autoencoder (VAE) including convolutional layers in the architecture of the en- and decoder. In a pick-and-place task the online AD on an industrial robot is then evaluated using joint angles and currents as inputs while processing the data with sliding windows. Unlike previous work [8], this analysis focuses on validating anomaly detection in a contact-rich assembly task involving various failure scenarios, including subtle ones. Furthermore, the evaluation accounts for complex control strategies that may introduce variability in task execution.

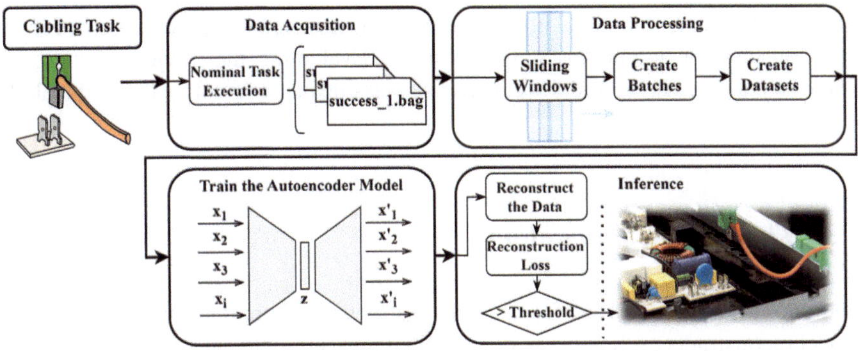

Fig. 1 Anomaly detection framework including data acquisition, training an autoencoder model and deploying within the inference

3 Method and Materials

3.1 Problem Statement

We consider AD in time series data, particularly from force/torque sensors. The time series data are given by $X = \{x_1, x_2, ..., x_m\}$, with the total number of time steps $m \in \mathbb{N}$ and $x_i \in \mathbb{R}^n$ being a single multidimensional sensor measurement. The goal is a model representing the time series data from nominal task execution that is capable to detect deviation from the expected nominal task data in incoming observations. Figure 1 illustrates the principal framework.

3.2 Machine-Learning Anomaly Detection Approach

Traditional failure detection approaches, such as time-synchronized methods like Dynamic Time Warping, map each data point to a specific model time step derived from nominal data [9]. However, these methods are sensitive to temporal variations, which are common in more complex robot control. The proposed semi-supervised approach trains an autoencoder-based model to represent nominal process behavior without requiring temporal alignment.

Sensor data is processed using a sliding window w of length T. A sequence w associated with the time step t can be expressed as $w_t = \{x_{t-T+1}, ..., x_t\} \subseteq X$ [8]. Each window w_t is used as input to an autoencoder, as visualized in Fig. 1. The autoencoder comprises an encoder-decoder pair, where the decoder mirrors the encoder. The encoder compresses input data into a low-dimensional latent vector, and the decoder reconstructs it [10, 11]. The network is trained to reconstruct the original data by minimizing $\|x_i - g_\phi(e_\theta(x_i))\|$, where $g_\phi : \mathbb{R}^l \to \mathbb{R}^{n \times T}$ is the decoder and $e_\theta : \mathbb{R}^{n \times T} \to \mathbb{R}^l$ the encoder with l defining the latent space dimension. After training the model utilizing nominal task executions, the expected reconstruction loss for anomalous states increases. Evaluating the model's reconstruction $\epsilon = |x_i - g_\phi(e_\theta(x_i))|$ during inference classifies an anomaly using a threshold $\epsilon > c_f$.

3.3 Threshold Computation

The threshold c_f is derived from the reconstruction of nominal process data. The model reconstructs each window of a nominal measurement, and the mean absolute reconstruction loss $\epsilon_k = \frac{1}{T}\sum_{i=0}^{T} |x_{k+i} - g_\phi(e_\theta(x_{k+i}))|$ is computed for each window k in the entire measurement. For each feature, the threshold is set to the maximum loss across all windows, representing the upper bound of normal behavior. A positive tolerance factor may be added to reduce false positives, at the cost of lower failure sensitivity.

4 Experimental Setup

4.1 Hardware Setup and Cabling Task

In the experiments, data is collected from an industrial use case involving electric stove assembly, where tasks such as plugging components are performed. The experimental setup according to Fig. 2a comprises a lightweight robot (Franka Research 3) with a parallel gripper at the flange. The robot uses individual gripper fingers to grasp the plug and perform the cabling task shown in Fig. 2a. The robot first picks up the connector, then moves toward the target socket on the stove surface, where it establishes contact by joining both connector ends. For the robot control an imitation learning approach is used according to [12]. By leveraging both camera inputs and robot parameters (TCP pose and trajectories) during training and inference, diffusion steps are integrated to learn effective visuomotor policies. Demonstrations are provided using a teleop-system comprising a joystick [13], enabling precise control of the robot's motion - particularly during the approach near the connector. Sensor information from demonstrations serve both control and AD training.

4.2 Collected Dataset

The dataset consists of measurements at a frequency of 500 Hz from both nominal task executions, defined by successful part assembly and introduced failures. Failures were intentionally induced to reflect realistic scenarios that lead to unsuccessful

(a) Robot automated cabling task. (b) Failure from human interference. (c) Incorrect part positioning leads to a tilted plug.

Fig. 2 Experimental setup: robot cabling task with process failures

assembly and exhibit varying absolute force differences compared to the nominal task execution. In industrial environments, such failures may arise from inaccuracies in image recognition, positional variances, external disturbances, worn or defect components. The failure scenarios include: (1) human interference with the robot's motion, (2) incorrect part positioning resulting in a tilted plug, and (3) absence of a plug in the gripper. While human disturbances introduce significant force variations, other failure types are more subtle and primarily related to the assembly process. In Fig. 2, the failure scenarios involving human interference (Fig. 2b) with the robot's motion and a mispositioned part leading to a tilted plug during insertion (Fig. 2c) are shown.

The experiments included ten nominal cases and five demonstrations for each failure type, with varying measurement lengths. Force data from the robot's wrench and joint states (position and velocity) were recorded. Forward kinematics are used to extract end-effector positions as features. To better evaluate the performance of the AD model, additional labeling of the data is carried out. While recording failure scenarios, an operator annotates the precise onset of anomalies. These additional labels enable a detailed, window-wise evaluation of the model's classification performance. While the model is trained solely on nominal task executions, the data labels serve only for validation purposes.

4.3 Autoencoder Initialization and Training

As AD model, we propose a variational autoencoder (VAE) architecture. VAEs, first introduced by [15], are generative models that use probabilistic methods to learn a compressed latent representation of the training data. At their core, VAEs employ random sampling to capture the underlying data distribution. The model is trained using a combined loss function that balances reconstruction performance with the quality of the learned distribution, measured using the Kullback-Leibler divergence. Representing the latent space as a distribution increases model complexity while reducing information loss. Consequently, VAEs not only assess whether a new data point aligns with the learned representation but also estimate its likelihood within the learned distribution. Since direct gradient computation on sampled distributions is infeasible, the reparameterization trick enables backpropagation during training. During training, latent variables are sampled accordingly, while inference uses only the mean for stable and deterministic reconstructions.

A preliminary hyperparameter search (not detailed here) yielded the following training settings: window size 100, batch size 25, latent dimension 15, and 500 epochs. The model's encoder has a quite simple structure, comprising three fully connected layers with 30 output features each, followed by a bottleneck layer that aligns with the specified latent dimension. The decoder mirrors the encoder's structure. Semi-supervised training is employed using all ten nominal bags to establish the AD model. The model is intended to represent the nominal process by being trained exclusively on nominal task executions. Consequently, overfitting has not been evaluated explicitly, and no failure bags were included in the training. The test set comprises fifteen previously unseen failure bags, covering three distinct failures types. However, the data for training and testing are processed per window as shown in Fig. 1.

4.4 Evaluation Metrics

To assess the model performance, standard classification metrics are used, following prior work [6–8, 14]. The Receiver Operating Characteristic (ROC) curve plots the true positive rate (TPR) against the false positive rate (FPR) across thresholds. Its area (AUROC) summarizes this trade-off as a scalar between 0 and 1, with 1 indicating perfect classification [7]. While the ROC and AUROC consider performance across multiple thresholds, the confusion matrix provides a detailed evaluation at a specific threshold, as visualized in Fig. 5. This should provide an indication of whether the approach is suitable for potential end users. The matrix visualizes correct classifications, true positive and true negative and wrong classifications false positive and false negative. In this evaluation, anomalies are treated as positive cases, whereas nominal behavior is treated as negative.

5 Results and Discussion

To evaluate the results, the VAE model was trained using the parameters outlined in Sect. 4.3. A baseline autoencoder (AE) was included for comparison, using similar training parameters to those proposed. The training data for the model included force measurements at the wrench (F_x, F_y, F_z) and Cartesian end-effector positions (x, y, z), which are derived from the joint angles. Sliding windows $w_t \in \mathbb{R}^{6\times T}$ with $T = 100$ time steps are used, where each input sequence is of the form $w_t = \{x_{t-T+1}, \ldots, x_t\}$, and $x_i \in \mathbb{R}^6$ represents the sensor observation at time step

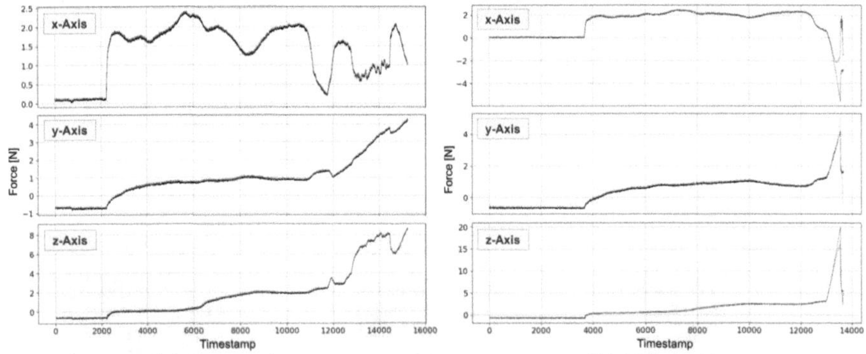

(a) Nominal. (b) Tilted plug.

Fig. 3 Force reconstruction from the VAE model. Measured forces are in black; reconstructions are shown in red, green, and purple for each axis

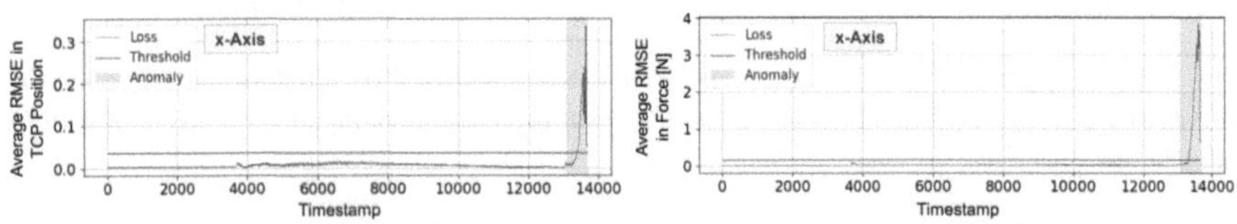

(a) Reconstruction loss in TCP position. (b) Reconstruction loss in forces.

Fig. 4 The average RMSE of the VAE reconstruction for the tilted plug failure is exemplified along the x-axis, with similar characteristics observed across all axes. Loss is shown in blue, the threshold in red, and the anomaly label in gray

i. A nominal task lasts approximately 27 s, while failure scenarios vary in duration. The threshold was set as described in Sect. 3.3.

5.1 Visualization of the Reconstruction

Figure 3 shows the model reconstructions compared to measured data for both a nominal and a failure case. The training included force and end-effector (TCP) position data. In the nominal case (Fig. 3a), the reconstructed forces closely match the reference. In the failure scenario (Fig. 3b), the model initially aligns with the reference but shows significant deviation in the final task phase.

Figure 4 shows the average RMSE per window for the TCP positions (Fig. 4a) and forces (Fig. 4b). Throughout most of the task, the reconstruction losses stay below the threshold but rise sharply close the labeled anomaly. This pattern is evident in both features, with the z-axis force showing a particularly strong deviation, increasing by several orders of magnitude.

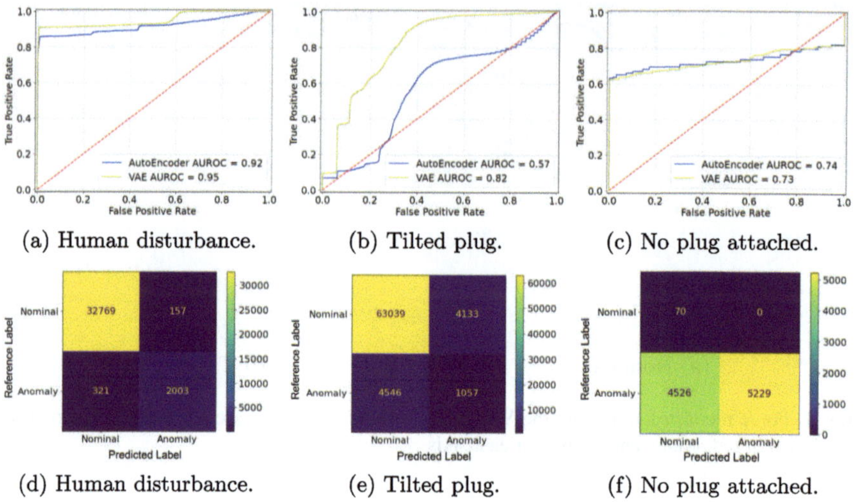

(a) Human disturbance. (b) Tilted plug. (c) No plug attached.

(d) Human disturbance. (e) Tilted plug. (f) No plug attached.

Fig. 5 Model evaluation across three failure scenarios comparing VAE with baseline AE: **a–c** ROC curves and **d–f** corresponding confusion matrices for the classifications of the VAE on the entire failure data

5.2 *Anomaly Detection Performance*

Figure 5a–c shows ROC curves comparing the baseline AE with the proposed VAE for three failure scenarios: human disturbance, tilted plug, and missing plug. The VAE outperforms the AE in the first two cases, achieving AUROC scores of 0.95 and 0.82, respectively, while the AE scores slightly lower in the first and significantly lower (0.57) in the second. Both models perform worse in the missing plug scenario, with the VAE reaching an AUROC of 0.73.

Figure 5d–f presents window-wise confusion matrices showing the VAE's classification performance across failure scenarios at a fixed threshold. The human disturbance scenario caused only 478 misclassifications, likely due to the distinct deviations typical of this failure type. The tilted plug scenario showed occasional errors but was largely detected. In contrast, the missing plug led to more false negatives, indicating worse detection. An evaluation on unseen nominal data yielded only true negatives, consistent with the accurate reconstruction in Fig. 3a. The results indicate that the VAE model outperforms the AE model, potentially due to its greater representational capacity and model complexity. Furthermore, the validation indicate that failure characteristics significantly impact the AD performance. The first two failures are detected with high robustness, while the tilted plug failure has some misclassifications, likely due to uncertainties in error labeling, as the exact onset of failure propagation is unclear. For the missing connector failure, the model lacks robustness, with many anomalies incorrectly classified as nominal (false negatives). In such failure cases, the model's reconstruction remains comparable to that under nominal conditions, rather than deteriorating as initially expected. As shown in Fig. 5f, the absence of false positives indicates no anomalies were detected. This could be due to subtle differences in the force profile or an insufficient threshold, which may require further adjustments for improved detection.

6 Conclusion

Within a real-world industrial cabling task, we conduct an experimental study integrating a semi-supervised variational autoencoder architecture, trained solely on nominal demonstrations. The collected multi-modal data set includes force measurements and robot parameters such as joint positions. Compared to a simple autoencoder, the VAE performed better on unseen failure data, achieving AUROC scores of 0.95 for human disturbance and 0.82 for the tilted plug, but struggled with missing plug detection, reaching only 0.73. This may be due to the lower force magnitudes involved, leading to subtler deviations that are harder to distinguish from the nominal task execution. Despite this, the visualization of the reconstructed input data highlights consistently lower reconstruction loss within the nominal cases compared to the failure scenarios, emphasizing the model's general effectiveness.

Acknowledgements This project has received funding from the European Union's Horizon 2020 research and innovation programme under grant agreement No 101058521– CONVERGING.

References

1. Billard A, Kragic D (2019) Trends and challenges in robot manipulation. Science **364**, eaat8414 (2019). https://doi.org/10.1126/science.aat8414
2. Mayr A et al (2019) Machine learning in production—potentials, challenges and exemplary applications. *Proc CIRP* 86:49–54 (2019)
3. Chandola V et al (2009) Anomaly detection: a survey. ACM Comput. Surv. **41**(3):15:1–15:58 (2009)
4. Wang H et al (2019) Progress in outlier detection techniques: a survey. IEEE Access 7:107964–108000. https://doi.org/10.1109/ACCESS.2019.2932769
5. Schmidl S et al (2022) Anomaly detection in time series: a comprehensive evaluation. Proc VLDB Endow 15:1779–1797
6. Graabæk SG et al (2023) Comparison of anomaly detection methods for collaborative robot manipulators. IEEE Access 11:65834–65848. https://doi.org/10.1109/ACCESS.2023.3289068
7. Brockmann JT et al (2023) The dvoraus-AD dataset for anomaly detection in robot applications. IEEE Trans Robot
8. Chen T et al (2020) Unsupervised anomaly detection of industrial robots using a VAE. IEEE Access (2020). https://doi.org/10.1109/ACCESS.2020.2977892
9. Kloska M et al (2023) Expert enhanced DTW-Based Anomaly detection. Expert Syst Appl 220:120030. https://doi.org/10.1016/j.eswa.2023.120030
10. Chen Z et al (2018) Autoencoder-based network anomaly detection. Proc WTS 1–5 (2018). https://doi.org/10.1109/WTS.2018.8363930
11. Michelucci U (2022) An introduction to autoencoders. arXiv:2201.03898

12. Chi C et al (2023) "Diffusion policy: visuomotor learning via Action Diffusion. Int J Robot Res. https://doi.org/10.1177/02783649241273668
13. Dass S et al (2024) Telemoma: a modular teleoperation system for mobile manipulation. arXiv:2403.07869
14. Park D et al (2017) A multimodal anomaly detector for robot-assisted feeding using an LSTM-VAE. Sci Syst Robot
15. Kingma DP, Welling M (2019) An introduction to variational autoencoders. Found Trends Mach Learn 12(4):307–392

A Force Controlled Pose Adaptation Method for Robot Based Defect Removal

Fengyun Shao, Tadele Belay Tuli, Florian Schreiber, and Martin Manns

Abstract In robotic surface finishing, ensuring feasible contact force during defect removal is critical for maintaining grinding quality and tool safety. This paper presents a force controlled pose adaptation method for robotic grinding. In this respect, success of grinding process, defect removal speed and surface quality are investigated to evaluate the proposed method. The control system employs real-time force feedback to adjust the TCP trajectory, moving backward when excessive force is detected to maintain contact force within a feasible range. Experiments of grinding tasks on zinc-coated scaffolding demonstrate that the proposed method enables successful defect removal while maintaining force levels.

Keywords Robot grinding · Force control · Experimental design · Production engineering

1 Introduction

Corrosion remains one of the key challenges in the manufacturing and maintenance of metallic components. As an electrochemical process, it leads to the gradual degradation of metal surfaces and causes economic losses of approximately \$ 1.8 trillion annually worldwide [1]. To mitigate this issue, various protective methods are employed–one of the most effective being hot-dip galvanizing. In this process, the component is fully immersed in molten zinc, resulting in the formation of a corrosion-resistant coating. Despite the high level of automation in galvanizing, unavoidable surface defects such as zinc run-offs or burrs occur during the cooling phase.

The subsequent post-processing of these defects—referred to as finishing—is currently performed manually. This is primarily due to the highly inhomogeneous distribution of defects and the often limited accessibility, particularly in complex geometries and hollow sections. Automated post-processing are not yet sufficient, as

F. Shao (✉) · T. B. Tuli · F. Schreiber · M. Manns
PROTECH - Institut für Produktionstechnik, Universität Siegen, Siegen, Germany
e-mail: Fengyun.Shao@uni-siegen.de
URL: https://protech.mb.uni-siegen.de/fams

M. Manns (ed.), *New Paradigms for Anticipated Uncertainty*, Lecture Notes in Mechanical Engineering, https://doi.org/10.1007/978-3-032-16889-4_61

they present significant technological challenges. In light of an increasing shortage of skilled labor and rising costs associated with inconsistent quality, innovative solutions to automate the finishing of hot-dip galvanized components are needed.

2 Objective

The objective of this work is to propose a method to remove hot dip galvanized surface defects occurring during the cooling phase based on robotic grinding system. In this regard, the proposed system aims to investigate if robotic grinding based on force controlled pose adaptation, the defect removal speed and surface quality are applicable. To achieve this objective, a real-time force controlled pose adaptation is implemented in ROS 2 [2], integrating both force and position feedback for adaptive surface interaction during defect removal. The system's performance is evaluated based on TCP position, force and the resulting surface quality of the polished parts. The outcome of this study is expected to contribute to more efficient and reliable surface finishing processes in hot dip galvanized component production lines.

3 Related Works

Robotic solution for surface quality improvement of construction materials such as scaffolding and galvanized steel is a growing area of interest due to the increasing demand for consistent, high-performance coatings in outdoor environments. Coatings such as zinc or Zn-Al alloys are widely used to improve corrosion resistance [10], but coating processes often introduce surface defects like spangles, inclusions, or thickness variations [4, 6, 12]. These defects not only compromise the protective properties of the coatings but also pose tolerance, aesthetic and structural issues in long-term deployments, particularly in scaffolding systems that experience frequent handling and environmental exposure [13].

Traditional inspection and defect correction processes in this domain are manual, labor-intensive, and subject to inconsistency [16]. Recent research has focused on integrating vision-based and deep-learning methods to detect surface anomalies with high precision. For example, Jeon et al. [8] use a switching-lighting scheme for defect detection. In contrast, Xiao et al. [17] and He et al. [5] apply enhanced YOLO and hierarchical feature fusion for identifying spangle and inclusion defects in zinc-coated steel. Despite their accuracy, these solutions are primarily limited to detection, leaving a gap in robotic defect removal.

To address this, robotic polishing and grinding systems have been investigated. Early methods employed hardcoded paths and static force parameters, such as in turbine vane finishing [9], which lack adaptability for irregular construction surfaces. More advanced systems adopted fuzzy logic [3] and proportional-integral (PI) force control [15] to ensure consistent grinding pressure, thereby improving surface quality.

Hybrid control strategies like the PI/PD system proposed by Huang et al. [7] further enhance adaptation to varying geometries, critical for the uneven surfaces seen in coated scaffolding.

Despite these advancements, current robotic systems still face challenges in responding dynamically to localized coating defects under variable force and contact conditions. Hierarchical control systems offer a promising path forward by integrating position and force feedback in a layered structure [11, 14], enabling finer control over surface interaction.

In this context, this work will propose a robotic grinding system that leverages force controlled pose adaptation control to remove coating defects on hot dip galvanized components, such as scaffolding selectively. This includes trajectory planning, force control and defect removal method in order to bridge the gap between precise defect detection and removal.

4 Proposed Method

To address the challenge of defect removal from hot dip galvanized components, the proposed robotic system mainly focus on maintaining controlled force while adjusting the position of the Tool Center Point (TCP). In this regard, a TCP Force-Guided Pose Adjustment method is proposed. This method adjusts the TCP's pose based on real-time force feedback to ensure that the applied force remains within a predefined threshold, thereby preventing excessive force and protecting the grinding tool. With the proposed method, the TCP moves towards the desired position, and if the force exceeds the threshold, it adjusts the TCP's position accordingly to maintain control. The specific details of this force-guided adjustment method can be found in Algorithm 1.

Algorithm 1 TCP Force-Guided Pose Adjustment

Input: TCP force, force threshold, target pose
Output: TCP pose
while TCP pose $\neq$ target pose **do**
    while TCP force $<$ force threshold **do**
        Move TCP step distance d_s forward towards target pose
    end while
    if TCP force $\geq$ force threshold **then**
        Move TCP backward by distance d
    end if
end while
Return TCP pose

Fig. 1 Experimental setup and TCP coordinate systems

5 Experimental Setup

In this experimental setup, an UR3e robot equipped with Dremel 3000 grinding tool is employed to perform a grinding task to remove defect on the zinc coated metal scaffolding. The robot end-effector tool follows the surface of the defective metal scaffolding in the Y-direction to remove the defects. Leveraging the ROS 2 [2], robot TCP force `wrench.force.y` is subscribed through Topic `/force_torque_sensor_broadcaster/wrench`. Based on current TCP force, robot adjust its TCP target pose, then transform TCP target pose into configuration space through Service `/compute_ik` in the type of `GetPositionIK`. This target position in configuration space is sent to the robot with a Publisher.

To evaluate the effectiveness of the proposed force control strategy, a comparative study is conducted between two approaches: grinding with and without the proposed force controller. The comparison focuses on three aspects: whether the grinding process successfully removes the defect; the grinding efficiency, measured by the material removal rate (MRR) defined in Eq. 1; and the surface quality after grinding, evaluated qualitatively. In the experiments involving grinding with the proposed force controller, two different tool rotation speeds are tested to observe the influence of tool speed on the performance of the control method. A physical experimental setup and the TCP coordinate system is shown in Fig. 1.

$$\text{MRR} = \frac{\text{Depth of Dressing}}{\text{Grinding Duration}}. \tag{1}$$

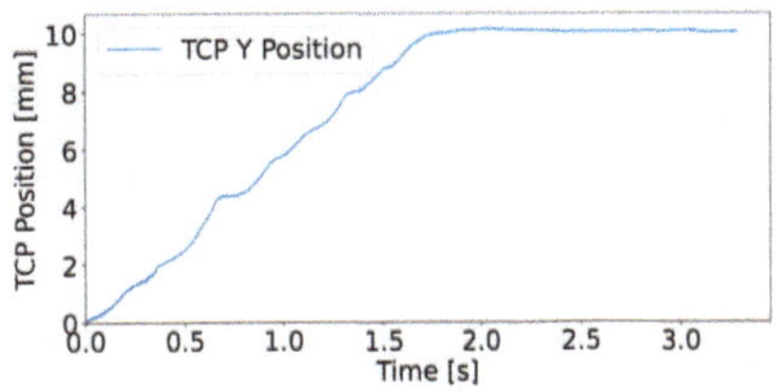

(a) TCP position during grinding process without the proposed controller.

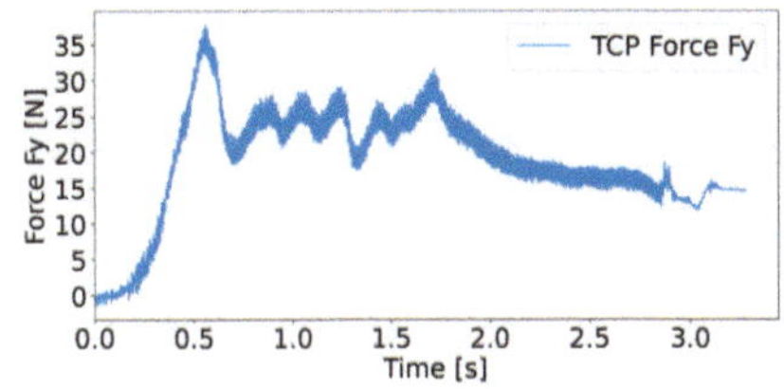

(b) Grinding process without the proposed controller reaches up to 35N.

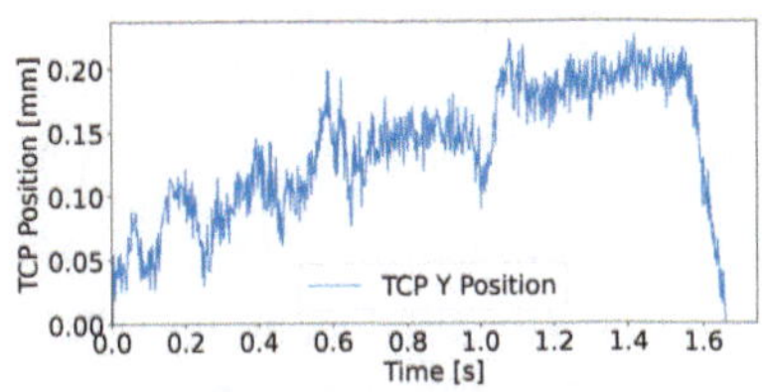

(c) TCP position threshold 2.8N with tool rotation speed 10000 RPM.

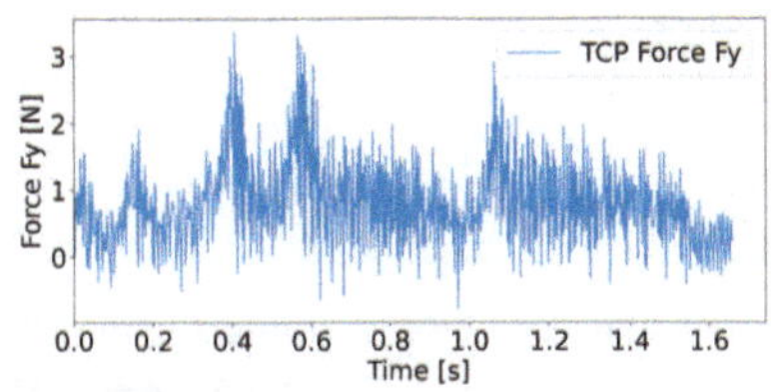

(d) Force signal with threshold 2.8N with tool rotation speed 10000 RPM.

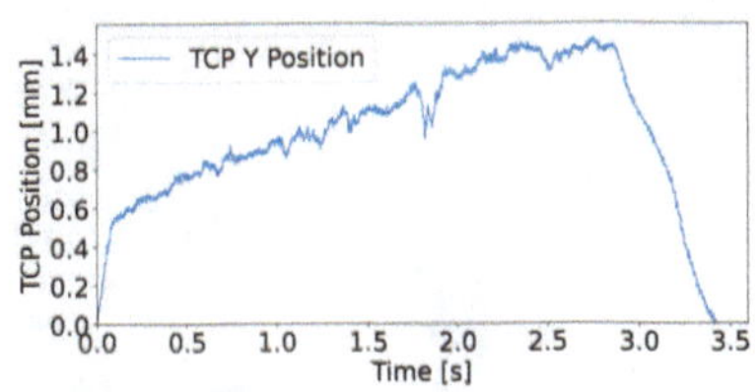

(e) TCP position threshold 2.8N with tool rotation speed 17667 RPM.

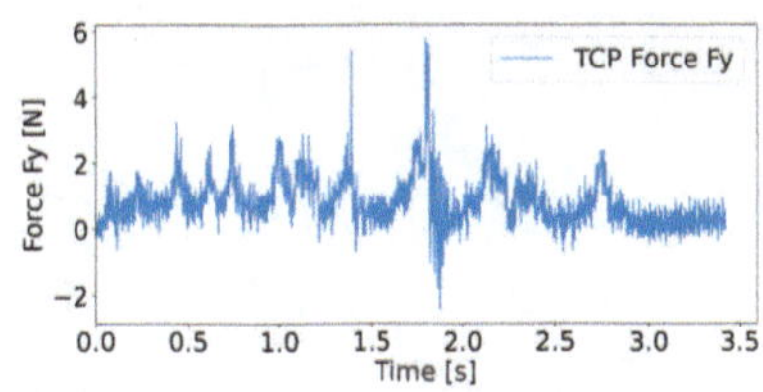

(f) Force signal with threshold 2.8N with tool rotation speed 17667 RPM.

Fig. 2 Grinding process visualization for TCP position (mm) verses time (s) and TCP force (N) verses time (s) using the proposed force controller (threshold = 2.8N, tool speed = 10000RPM and 17667RPM)

6 Result

Figures 2a, 2b illustrate the TCP position and TCP force during grinding process without proposed force controller. Since no force controller was employed to adjust the TCP force, the force applied during the grinding process reached a peak of approximately 35N, which exceed the maximum allowable force of the grinding tool. As a result, the tool head becomes stuck on the defect, leading to a failure of the grinding process.

Experimental results show that when the TCP force exceeds 28N, the grinding tool reaches its maximum force limit, causing the tool head to get stuck on the metal surface. Therefore, in the proposed control method, a force threshold of 2.8N (which is one-tenth of the maximum) is used as a validation example. In Fig. 2, the results

(a) defect before grinding in top view (b) defect after grinding in top view

(c) defect before grinding in front view (d) defect after grinding in front view

Fig. 3 A comparison of the surface quality before and after the grinding process with the proposed control method

of grinding process with proposed control method are presented. Figure 2c, d show the TCP position and force with force threshold value 2.8N and tool speed of 10000 Revolutions Per Minute (RPM), where the MRR is 0.136 mm/s and the TCP force is mostly within the pre-defined threshold (2.8N), with only 2 peaks slightly larger than 3N. Figure 2e, f present the corresponding results for a higher tool speed of 17667 RPM. In this case, the MRR increases to 0.432 mm/s, but the TCP force shows stronger oscillations, with peaks reaching around 6N. The surface quality before and after the grinding process with proposed control method is shown in Fig. 3.

7 Discussion

The results in Sect. 6 show that with proposed control method, the robot performs the grinding tasks successfully. While increasing the tool rotation speed presents higher MRR, it can also lead to higher oscillation of TCP forces and higher force above the threshold during the grinding process. This is primarily due to the fact that higher rotational speeds introduce greater dynamic oscillations of the grinding tool, which in turn cause larger oscillations in the measured TCP force.

Regarding the TCP force exceeding the pre-defined threshold, it should be noted that the threshold in the current control setup only serves as a trigger for the backward motion, rather than acting as a hard safety limit. Therefore, the threshold should be set lower than the actual maximum force the grinding tool can withstand. As shown in Fig. 2, the force overshoot beyond the threshold increases with higher tool rotation speeds. This indicates that when operating at higher tool speeds, the threshold should be set with a larger safety margin below the tool's mechanical force limit to avoid excessive contact force.

8 Conclusion and Outlook

This work presents a robotic grinding system for removing surface defects from hot-dip galvanized surfaces using real-time force-controlled pose adaptation in ROS 2. The system adjusts the TCP motion based on both force and position feedback to ensure safe and adaptive surface interaction. Experimental results demonstrate that, unlike the case without the proposed control method which fails due to excessive TCP force, the proposed method enables successful defect removal while maintaining force levels. Higher tool rotation speeds are shown to significantly improve the MRR. To address the limitation of force overshoot, future work will explore strategies to add hard constrains to limit the TCP force. This approach provides a promising foundation for intelligent and reliable robotic surface finishing in industrial applications.

Acknowledgements The cooperation project "RoboPutz" (Grant number: KK5057119LF3) is funded by the Federal Ministry of Economics and Climate Protection (BMWK) as part of the Central Innovation Program for SMEs (ZIM) and is supervised by the Working Group of Industrial Research Associations (AiF).

References

1. Bender R, Féron D, Mills D, Ritter S, Bäßler R, Bettge D, De Graeve I, Dugstad A, Grassini S, Hack T et al (2022) Corrosion challenges towards a sustainable society. Mater Corros 73(11):1730–1751
2. Chitta S (2016) Moveit!: an introduction. Robot Operating System (ROS) The Complete Reference (Volume 1), pp 3–27
3. Gao H, Li L, Wang G, Luo W, Wang X (2016) Force control based robotic grinding system and application. In: 2016 IEEE international conference on advanced intelligent mechatronics (AIM). IEEE, pp 1168–1173
4. Gutoff EB, Cohen ED (2006) Coating and drying defects: troubleshooting operating problems. Wiley
5. He Yu, Song K, Meng Q, Yan Y (2019) An end-to-end steel surface defect detection approach via fusing multiple hierarchical features. IEEE Trans Instrum Meas 69(4):1493–1504
6. Hong M-H, Saka H (2004) Fib and tem observations of defects in hot-dip zinc coatings. J Electron Microsc 53(5):545–552
7. Huang T, Jiang Y, Luo X, Zhu L, Peng H (2019) Robotic grinding of complex components: A step towards efficient and intelligent machining-challenges, solutions and applications. Chin J Aeronaut 32(1):256–268
8. Jeon YJ, Choi DC, Lee SJ, Yun JP, Kim SW (2015) Steel-surface defect detection using a switching-lighting scheme. Appl Opt 55(1):47–57
9. Kim H, Min BK, Lee D (2002) Robotic grinding and polishing for turbine-vane overhaul. Robot Comput-Integrated Manufact 18(3–4):211–218
10. Kühne R, Voelkel J, Bartsch H, Feldmann M, Pinger T, Sharma R, Reisgen U (2024) Fatigue strength of additively manufactured hot-dip galvanized steel coated with a zn–5al alloy. Prog Addit Manufact, pp 1–13
11. Lishchenko N, Gushchin A, Larshin V (2024) Hierarchical control in mechatronic technological systems. Machines 12(10):697
12. Marder A, Goodwin F (2023) The metallurgy of zinc coated steels. Elsevier
13. Murugesu J (2017) Design and analysis of frame scaffolding capacity using thickness and material as variables. Master's thesis, University of Malaya (Malaysia)
14. Tuli TB, Manns M (2019) Hierarchical motion control for real time simulation of industrial robots. Procedia CIRP 81:713–718
15. Wang L, Fan X, Zhao J, Lu B, Wang J (2018) Robotic grinding of a blisk with two degrees of freedom contact force control. Int J Adv Manufact Technol 99(9–12):2711–2721
16. Wen X, Yu JS, He, Song K (2022) Steel surface defect recognition: a survey. Coatings 13(1):17
17. Xiao D, Xie FT, Gao Y, Li ZN, Xie HF (2023) A detection method of spangle defects on zinc-coated steel surfaces based on improved yolo-v5. Int J Adv Manufact Technol 128(1-2):937–951

Rebound Reasoning: Enhancing Quantized Language Models Through Iterative Prompting as a Service for Production Systems

David Golchinfar, Daryoush Vaziri, Darius Hennekeuser, and Dirk Schreiber

Abstract This paper introduces Rebound Reasoning, an architecture leveraging iterative prompting and mixture of agents to improve the reasoning capabilities of quantized language models for efficient deployment as services in production systems. Our approach enables quantized models to narrow the performance gap with unquantized counterparts while requiring only a fraction of computational resources, making them suitable for manufacturing environments. We demonstrate the effectiveness of Rebound Reasoning through comprehensive experiments using the Alpaca Eval benchmark. The architecture integrates with vLLM and ollama infrastructure, generating multiple perspectives through iterations to reduce biases and synthesize more comprehensive answers. This approach enhances output diversity and quality while improving error tolerance and consistency—particularly valuable for decision support in production environments. Our experiments show Rebound Reasoning achieves comparable performance to unquantized models while offering significant improvements with quantized models. The approach's flexibility allows adaptation to various model configurations, making it scalable for different production applications. This work contributes to making large language models more efficient and accessible as services for production systems, broadening the application of AI reasoning capabilities in manufacturing environments.

Keywords LLM · Architectures · Generative AI · Reasoning

David Golchinfar, Daryoush Vaziri, Darius Hennekeuser: These authors contributed equally.

D. Golchinfar · D. Vaziri (✉) · D. Schreiber
University of Applied Sciences Bonn-Rhein-Sieg, Sankt Augustin, Germany
e-mail: daryoush.vaziri@h-brs.de

D. Hennekeuser
VAGO Solutions, Hennef, Germany

M. Manns (ed.), *New Paradigms for Anticipated Uncertainty*, Lecture Notes in Mechanical Engineering, https://doi.org/10.1007/978-3-032-16889-4_62

1 Introduction

Large Language Models (LLMs) have demonstrated remarkable capabilities in natural language processing tasks, but their increasing size and computational requirements pose significant challenges for deployment in production environments [1, 2]. Model quantization has emerged as a promising approach to reduce computational and memory footprints while maintaining acceptable performance [3, 4]. However, a performance gap often exists between quantized and full-precision models, particularly in complex reasoning tasks [5].

Prompt engineering, iterative refinement, and Mixture of Agents have shown promise in enhancing language model performance [6, 7]. Building on these insights, we propose Rebound Reasoning, a novel architecture that combines quantized models with iterative prompting and Mixture of Agents to enhance reasoning capabilities as a service for production systems.

Rebound Reasoning explores whether simplifying Mixture of Agents to using one prompt with one LLM could improve the reasoning capabilities of quantized LLMs when deployed as services. It introduces a multi-step approach that generates and synthesizes multiple perspectives from a quantized model, effectively bootstrapping its reasoning process.

Our key contributions include:

- The Rebound Reasoning architecture enabling quantized models to achieve reasoning capabilities comparable to unquantized counterparts, making them effective services for production systems.
- Seamless integration with vLLM and ollama infrastructure, demonstrating significant improvements in reasoning tasks for quantized models when deployed as services.
- Insights into the broader implications for production systems, including cost savings, improved user experience, and new avenues for research in prompt engineering and model behavior.

2 Related Work

Our work integrates several key research areas in natural language processing and machine learning.

2.1 *Foundation Technologies: Model Quantization and Infrastructure*

Model quantization has emerged as a crucial technique for reducing the computational and memory requirements of large language models. Lin et al. [4] provide a comprehensive survey of quantization methods for neural network inference. Frantar et al. [8] introduced GPTQ, an accurate post-training quantization method for generative pre-trained transformers, while Dettmers et al. [5] proposed LLM.int8(), an 8-bit matrix multiplication technique for efficient transformer scaling. GPT-Generated Unified Format (GGUF) and Activation-aware Weight Quantization (AWQ) represent two significant recent quantization methods that establish the state-of-the-art standards for GPU and CPU inference, respectively [9].

For efficient model deployment, Xiao et al. (2023) introduced vLLM, a high-throughput inference engine with PagedAttention mechanism that allocates non-contiguous memory blocks for attention states, reducing fragmentation and improving computational efficiency. Similarly, Ollama has emerged as another optimized inference engine, providing a lightweight container-based approach with strong support for quantized models. Kwon et al. [10] demonstrated how enhanced memory management techniques can substantially reduce operational costs while maintaining model quality in production settings.

For evaluation, Taori et al. [11] developed the Alpaca Eval benchmark, which provides a comprehensive framework emphasizing practical utility and real-world task performance rather than abstract academic metrics.

2.2 Prompt Engineering and Reasoning Techniques

Wei et al. [6] introduced chain-of-thought prompting to elicit step-by-step reasoning in large language models, showing significant improvements in complex reasoning tasks. Kojima et al. [12] demonstrated that large language models can act as zero-shot reasoners when prompted effectively, while Wei et al. [6] explored chain-of-thought prompting in zero-shot scenarios.

Shinn et al. [13] proposed Reflexion, a framework for language agents that iteratively improve responses through self-reflection. Madaan et al. [7] introduced Self-Refine for iterative self-refinement across various tasks. In ensemble approaches, Wang et al. (2022) showed that aggregating multiple reasoning paths improves few-shot learning performance, especially in arithmetic and commonsense reasoning.

Sumers et al. [14] developed the Cognitive Architectures for Language Agents (CALA) framework employing specialized cognitive modules to enhance language model capabilities. Rebound Reasoning extends these approaches by applying the mixture of agents paradigm specifically to quantized language models deployed as services in production systems. While previous research focused on combining distinct models or agents, we apply this concept to a single quantized model, treating each iteration as a distinct "agent" with its own perspective.

3 Rebound Reasoning: Architecture and Mechanisms

Rebound Reasoning is an iterative prompting architecture that improves the reasoning capabilities of quantized language models when deployed as services. The core concept involves generating multiple perspectives on a given task, synthesizing these perspectives, and producing a final, more comprehensive response. The architecture consists of three key components: (1) a quantized language model optimized for efficient inference; (2) a Response Synthesizer that combines and refines multiple outputs; and (3) vLLM integration for optimized performance.

The workflow builds on previous research in iterative prompting and mixture of agents, integrating them into a service architecture for efficient prompting strategies:

1. **Initial Prompt Generation**: The user's prompt is sent to the quantized LLM n-times, generating multiple answers.
2. **Parallel Processing**: The model processes prompts in parallel using ollama's or vLLM's efficient inference capabilities.
3. **Response Collection and Analysis**: Outputs from each prompt iteration are collected and analyzed.
4. **Synthesis and Refinement**: The Response Synthesizer identifies common themes, unique insights, and potential inconsistencies across responses.
5. **Final Response Generation**: A final prompt produces a comprehensive, refined answer incorporating multiple perspectives.

Example: A quantized model, for instance Qwen2-1.5B-Instruct-AWQ generates two answers to an alpaca eval 2.0 prompt. The same model then acts as the aggregator to conduct the final reasoning based on the two answers.

Rebound Reasoning integrates with ollama and vLLM, leveraging efficient memory management and optimized inference. This integration facilitates parallel processing of multiple prompts, effective memory utilization, and maximized throughput. The system is designed to be:

- **Model-agnostic**: Applicable to various model sizes and architectures
- **Task-flexible**: Suitable for diverse reasoning tasks
- **Resource-efficient**: Offering iteration control based on task complexity and resource constraints

A key strength of the architecture is its enhanced error tolerance through multiple mechanisms:

- Multiple perspectives reduce the impact of individual erroneous responses
- Cross-validation identifies potential hallucinations or inconsistencies
- Iterative refinement allows error correction in subsequent reasoning rounds.

By combining these elements, Rebound Reasoning creates a robust framework that enhances quantized models' capabilities for complex reasoning tasks when deployed as services in production environments. The architecture of Rebound Reasoning is schematically illustrated in Fig. 1.

Fig. 1 Rebound reasoning

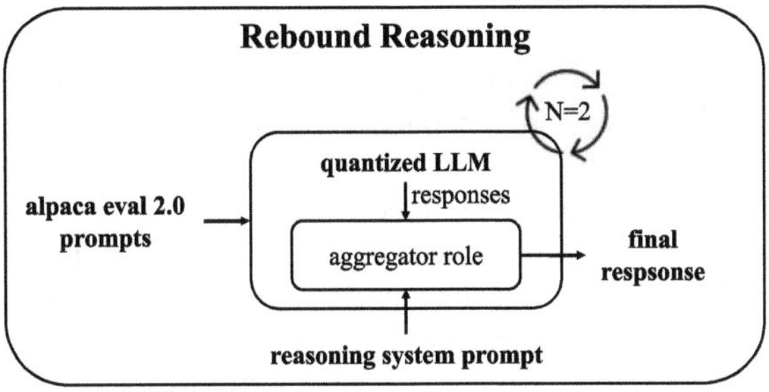

4 Experimental Setup and Results

4.1 Experimental Setup

We performed different experiments to compare the effects of Rebound Reasoning on different Model Architectures when deployed as services. For quantization, we used AWQ and GGUF for all experiments, as these are two of the most acknowledged and powerful quantizations available now. The models used for the experiments are listed in Table 1.

For Evaluation, we used the Alpaca Eval benchmark [11], which covers a wide range of tasks including reasoning, question-answering, and problem-solving. We performed the Alpaca Benchmark for the base model and its AWQ and GGUF counterpart to have a reliable measure for comparison of service performance. For our Rebound Reasoning experiments, we used the configuration presented in Table 2.

As user input, we provided the original alpaca eval 2.0 prompts to the unquantized and quantized models. The answers of each model were handed to the aggregator to perform the final reasoning.

Table 1 Model overview for experiments

Model name	Model size	Architecture	Model quantization
Qwen2-1.5B-Instruct	1.5 B parameters	Qwen2	AWQ, GGUF Q4_K_M
Qwen2-7B-Instruct	7 B parameters	Qwen2	AWQ, GGUF Q4_K_M
Meta-Llama-3-8B-Instruct	8 B parameters	Llama-3	AWQ, GGUF Q4_K_M
Hermes-2-Pro-Mistral-7B	7 B parameters	Mistral	AWQ, GGUF Q4_K_M

Table 2 Experiment configuration

Parameter	Characteristics
Number of iterations	2 answers + final synthesized answer
Prompt variations	1 per iteration
Temperature	0.7 for output generation

The aggregator used following reasoning system prompt, based on the original mixture of agents prompt: "You have been provided with a set of responses from various open-source models to the latest user query. Your task is to synthesize these responses into a single, high-quality response. It is crucial to critically evaluate the information provided in these responses, recognizing that some of it may be biased or incorrect. Your response should not simply replicate the given answers but should offer a refined, accurate, and comprehensive reply to the instruction. Ensure your response is well-structured, coherent, and adheres to the highest standards of accuracy and reliability.

It is important to mention, that the aggregator and the quantized LLM are the same model (see Fig. 1). Further, in our experiments we illustrate a minimal setup with only two prompts and the final aggregation. In practice, this can and should be increased to the domain's needs and requirements.

4.2 Results

Table 3 presents the overall performance of Rebound Reasoning compared to the baselines.

The results illustrate that Rebound Reasoning improved lc_winrate in all experiments, with exception to experiment 4. With respect to win_rate RR improved the results of all quantized models compared to their unquantized counterparts, demonstrating its effectiveness as a service approach. These improvements were consistently observed across different model architectures and quantization methods (see Table 3), underscoring the model-agnostic nature of the Rebound Reasoning approach.

With respect to memory usage rebound reasoning suggested noticeable advantages when compared to the unquantized version of the LLM. Memory usage was reduced by up to 80%. Even though we used 3 prompt iterations with rebound reasoning, inference time ranged between 62 and 73% of the unquantized version. Rebound Reasoning suggests that superior performance can be achieved, while significantly reducing memory usage and maintaining competitive inference times, which is particularly valuable for deploying these models as services in resource-constrained production environments.

5 Discussion

Our experiments demonstrate that Rebound Reasoning significantly improves the performance of quantized language models when deployed as services. We discuss the implications of our findings, analyze strengths and limitations, and consider applications in production service systems.

Table 3 Overall results of rebound reasoning

Model	Alpaca benchmark	
	lc_winrate	win_rate
Qwen/Qwen2-1.5B-Instruct-single (AWQ)	2.82	1.76
Qwen/Qwen2-1.5B-Instruct-RR (AWQ)	3.68	2.63
Qwen/Qwen2-1.5B-Instruct-single (GGUF4)	2.83	1.92
Qwen/Qwen2-1.5B-Instruct-RR (GGUF4)	3.44	2.53
Qwen2-1.5B-Instruct-bfloat16	**4.77**	**3.26**
Qwen/Qwen2-7B-Instruct-single (AWQ)	20.97	19.04
Qwen/Qwen2-7B-Instruct-RR (AWQ)	**25.22**	26.64
Qwen/Qwen2-7B-Instruct-single (GGUF4)	19.44	21.63
Qwen/Qwen2-7B-Instruct-RR (GGUF4)	23.36	**29.46**
Qwen2-7B-Instruct-bfloat16	23.57	21.62
Meta-Llama-3-8B-Instruct-single (AWQ)	17.48	16.96
Meta-Llama-3-8B-Instruct-RR (AWQ)	21.45	**23.61**
Meta-Llama-3-8B-Instruct-single (GGUF4)	15.88	15.44
Meta-Llama-3-8B-Instruct-RR (GGUF4)	22.84	23.46
Meta-Llama-3-8B-Instruct-bfoat16	**22.9**	22.6
Hermes-2-Pro-Mistral-7B-single (AWQ)	15.33	10.71
Hermes-2-Pro-Mistral-7B-RR (AWQ)	18.92	17.32
Hermes-2-Pro-Mistral-7B-single (GGUF4)	20.49	21.39
Hermes-2-Pro-Mistral-7B-RR (GGUF4)	**24.21**	**22.07**
Hermes-2-Pro-Mistral-7B-bfloat16	19.04	12.57

RR Rebound Reasoning

5.1 Theoretical Implications and Integration with Research

Rebound Reasoning extends current research in several key ways. In model quantization, our results show that performance losses in complex reasoning tasks can be significantly reduced through architectural innovations—expanding the spectrum of compression strategies discussed by Zhang et al. [15]. As an adaptation of the Mixture-of-Agents paradigm, our approach demonstrates that perspective diversification can be achieved within a single quantized model through iterative prompting, enhancing practical applicability as a service in production systems. Unlike specialized reasoning models that often require larger resources, Rebound Reasoning offers a model-agnostic, resource-efficient method for improving various reasoning tasks. Finally, our seamless integration with vLLM infrastructure directly addresses the deployment efficiency challenges identified by Kwon et al. [10].

5.2 Cognitive and Technical Implications

The success of multi-perspective reasoning in our approach aligns with cognitive theories emphasizing the importance of different viewpoints in problem-solving. Similar to Wang et al. [16], our results show that aggregating multiple reasoning paths leads to more robust results, suggesting that even quantized models have variable internal representations that can be activated through different prompting iterations.

From a technical perspective, Rebound Reasoning demonstrates significant efficiency gains. The combination of model quantization and optimized architecture results in significantly reduced memory usage (up to 82% less with 4-bit quantization) while simultaneously improving performance, directly addressing the challenges in scaling transformer models discussed by Korthikanti et al. [17]. Additionally, training and aligning native reasoning models like Deepseek-R1 can be very resource-intensive. Companies searching for cost-efficient ways to make use of reasoning capabilities might find a solution in Rebound Reasoning.

5.3 Limitations and Future Research Directions

Despite promising results, several limitations motivate future research. Our experiments used a minimal configuration with two answers plus synthesis; a systematic investigation of different iteration numbers and prompt variations would be valuable, particularly regarding potential saturation effects. Domain-specific optimization could further enhance Rebound Reasoning for particular reasoning types relevant to production environments. The interaction between quantization methods and Rebound Reasoning merits deeper analysis, as our experiments showed performance differences between AWQ and GGUF quantized models. Finally, our experiments rely solely on the alpaca evaluation framework. Future research should use additional benchmarks (such as GSM8K or MMLU) to further validate Rebound Reasoning's generalizability.

6 Conclusion

Rebound Reasoning represents an innovative approach in the field of language model infrastructure that extends the boundaries of what is possible with quantized models when deployed as services. By combining model quantization, iterative prompting, and efficient inference architecture, our approach directly addresses the challenge of providing advanced reasoning capabilities in resource-efficient configurations for production systems.

The results of our experiments demonstrate that Rebound Reasoning is not just a theoretical concept but delivers practical, measurable improvements across various model architectures. This opens new perspectives for the democratization of AI capabilities and the development of more sustainable AI services for production environments.

Future research can build on these foundations to deepen the theoretical foundations of the approach, develop domain-specific optimizations for production applications, and expand the practical service possibilities in manufacturing contexts. Rebound Reasoning thus contributes to the broader goal of making AI systems more accessible, efficient, and sustainable as services for production—a critical concern in a time of rapid AI development and adoption.

References

1. Bommasani R, et al (2021) On the opportunities and risks of foundation models. arXiv:2108.07258
2. Touvron H, et al (2023) LLaMA: open and efficient foundation language models. arXiv:2302.13971
3. Gholami A, et al (2021) A survey of quantization methods for efficient neural network inference. arXiv:2103.13630
4. Lin J, et al (2023) AWQ: activation-aware weight quantization for LLM compression and acceleration. arXiv:2306.00978
5. Dettmers T, et al (2022) LLM.int8(): 8-bit matrix multiplication for transformers at scale. arXiv:2208.07339
6. Wei J, et al (2022) Chain-of-thought prompting elicits reasoning in large language models. arXiv:2201.11903
7. Madaan A, et al (2023) Self-refine: iterative refinement with self-feedback. arXiv:2303.17651
8. Frantar E, Ashkboos S, Hoefler T (2023) GPTQ: accurate post-training quantization for generative pre-trained transformers. arXiv:2210.17323
9. Rajput S, Sharma T (2024). Benchmarking emerging deep learning quantization methods for energy efficiency. In: Proceedings of the 21st IEEE international conference on software architecture (ICSA-C 2024). https://doi.org/10.1109/ICSA-C63560.2024.00049.
10. Kwon W, Li Z, Zhuang S, Sheng Y, Zheng L, Yu CH, Gonzalez JE, Zhang H, Stoica I (2023) Efficient memory management for large language model serving with pagedattention. Proceedings of the 28th ACM symposium on operating systems principles (SOSP 2023). https://doi.org/10.48550/arXiv.2309.06180
11. Taori R, et al (2023) Alpaca: a strong, replicable instruction-following model. arXiv:2303.16199
12. Kojima T, et al (2022). Large language models are zero-shot reasoners. arXiv:2205.11916
13. Shinn N, et al (2023) Reflexion: language agents with verbal reinforcement learning. arXiv:2303.11366.
14. Sumers TR et al (2023) Cognitive architectures for language agents. arXiv:2309.02427
15. Zhang Z, et al (2022) A survey on model compression for large language models. arXiv:2308.07633
16. Wang H, et al (2023) Self-consistency improves chain of thought reasoning in language models. arXiv:2203.11171
17. Korthikanti VA, Casper J, Lym S, McAfee L, Andersch M, Shoeybi M, Catanzaro B (2023) Reducing activation recomputation in large transformer models. In: Proceedings of the 5th conference on machine learning and systems (MLSys 2023). https://doi.org/10.48550/arXiv.2205.05198
18. Chowdhery A, et al (2022) PaLM: scaling language modeling with pathways. arXiv:2204.02311

19. de Vries A (2023) The growing energy footprint of artificial intelligence, Joule, vol 7, Issue 10. https://doi.org/10.1016/j.joule.2023.09.004
20. Du N, et al (2023) GLaM: efficient scaling of language models with mixture-of-experts. In International conference on machine learning (ICML) 2023. PMLR, pp 8547–8569
21. Gu Y, et al (2023) MiniLLM: knowledge distillation of large language models. arXiv:2306.08543

Leveraging LLM-Based Reasoning for Human-Robot Cooperative Disassembly

Daniel Gusenburger, Marco Giangreco, Attique Bashir, and Rainer Mueller

Abstract In recent years, large language models (LLMs) have gained significant popularity, driving innovation across various domains. LLMs are widely used across fields, including general-purpose robotics. However, their use in industrial applications still has a lot of unexpressed potential. This paper presents a prototypical implementation of a collaborative robot system which uses a set of carefully instructed LLMs to plan a collaborative assembly/disassembly process. Using natural language descriptions, the system generates action plans, assigning tasks to robots or humans based on capabilities and tools. Central to this implementation are state-of-the-art object detection and a reasoning process to deduce action plans. In addition, we implemented a monitoring system employing sequential LLMs to check for and correct errors during task execution. We detail the integration of system components, including peripherals and prompt engineering methods to optimize task outputs. To validate the approach, we demonstrate its application on the assembly and disassembly of a Raspberry Pi-based product. This work explores how LLMs enhance human-robot collaboration (HRC) in industrial assembly and disassembly, improving task planning, real-time error detection, and adaptive correction.

Keywords Human-robot collaboration · Large language models · Assembly · Disassembly · Task planning

1 Introduction

The rise of LLMs and their increased usage especially in human-centric applications like service chat bots showcases the power of natural language processing and large

D. Gusenburger · R. Mueller
Chair of Assembly Systems, Department of Natural Sciences and Technology, Saarland University, Saarbrücken, Germany

M. Giangreco (✉) · A. Bashir
Department of Assembly Systems, Center for Mechatronics and Automation Technology gGmbH, Saarbrücken, Germany
e-mail: marco.giangreco@zema.de

M. Manns (ed.), *New Paradigms for Anticipated Uncertainty*, Lecture Notes in Mechanical Engineering, https://doi.org/10.1007/978-3-032-16889-4_63

language models. Not only do these models enable easy interaction with humans, their improved reasoning capabilities make it more and more interesting for use in research, especially in domains focusing on reasoning such as task planning for robotic applications or more specifically human-robot collaboration applications. HRC brings together human cognitive flexibility and a robots' meticulous accuracy, providing significant benefits in industrial settings, especially for complex assembly and disassembly tasks. HRC is usually divided into different categories, based on the level of interaction between human and robot. These categories range from basic interactions, where robots and humans perform distinct tasks in the same environment to working on the same part of the product simultaneously [1]. Despite the potential advantages of HRC, several challenges remain. One of these challenges is the modeling and/or planning of human-robot interactions including the distribution of tasks between human and robot which sometimes requires complex reasoning. Traditional reasoning approaches, e.g., PDDL[1]-based methods, often rely on predefined task descriptions and strict task modeling, which limit the adaptability of the system in dynamic environments [2]. Modularity and applicability to new products suffer from this rigidity as well since it takes expert knowledge to describe a new product, making it hard for laymen to reconfigure the system. Our work focuses on cooperative tasks, i.e., where human and robot simultaneously work in the same work area, but not on the same workpiece simultaneously.

This paper presents a novel approach for HRC in industrial robotics, utilizing the capabilities of LLMs to address these challenges. Our objective is to deploy a collaborative robotics (cobotics) assistance system that, based on a natural language description of a selected product and supported by an AI assistant, distributes tasks between human and robot while simultaneously evaluating the quality and correctness of the overall process. Such a system could significantly reduce the commissioning effort required for a collaborative robotics application. Therefore, we propose a cognitive assistance system that combines LLMs with a state-of-the-art object detection model to facilitate cooperative assembly and disassembly in real time. By deriving a product's logical structure from its natural language description, our system generates an online action plan that attributes tasks to either the robot or a human based on their respective capabilities and available tools. In addition to the planning module, the system integrates sequential LLMs for error detection and correction during the execution stage. We validate this approach by applying it to the assembly and disassembly of a prototypical product, demonstrating the potential of LLMs in industrial contexts and setting the foundation for future applications of LLM-driven task planning in cooperative assembly processes. By integrating LLMs into human-robot collaboration/cooperation, we enable more efficient, adaptive, and intelligent automation in industrial assembly and disassembly.

[1] Planning Domain Definition Language.

2 Related Work

Recently, we have been seeing a significant shift towards integrating machine learning methods, especially LLMs into robotics and manufacturing, particularly in human-robot collaboration (HRC) scenarios [3]. Rekik et al., e.g., apply machine learning and computer vision to track human intentions in collaborative assembly, improve task allocation, and reduce errors [4]. Overall, three major trends can be identified: first, the use of LLMs as natural language interfaces to bridge the communication gap between humans and robots, allowing more intuitive interactions; second, the development of sophisticated frameworks that combine LLMs with visual perception systems and digital twins to enhance environmental awareness and task execution; and third, the implementation of novel approaches to handle LLM limitations, such as hallucination management and the integration of human demonstration learning [5, 6].

Bashir et al. already showed that an LLM is able to explore assembly state spaces in real time [7]. We expand on this approach by enhancing task planning with advanced prompt engineering and product structure analysis from natural language descriptions. Unlike prior studies focused solely on assembly, our system enables bidirectional planning while optimizing task allocation between humans and robots based on their skills and tools. Traditional approaches for disassembly, e.g., Disassembly Sequence Structure Graphs (DSSG), use expert rules and constraints to enhance product design and minimize environmental impact [8]. In this context, LLMs advance reasoning by decomposing tasks, mitigating performance issues with large datasets, and enabling breakthroughs in table-based reasoning, even surpassing human performance [9]. Furthermore, LLMs can generate domain-specific models such as PDDL, improving reliability through automated consistency checks [10]. Other approaches use LLMs as reward functions for reinforcement learning on a relational graph convolution network to compute disassembly sequences [11].

LLMs have demonstrated significant potential to provide high-level semantic knowledge for robotic tasks, particularly in planning and executing complex, temporally extended instructions in natural language [12]. However, LLMs lack real-world experience, limiting their ability to make contextually grounded decisions for specific physical environments. Their effective implementation thus requires careful integration with other technologies and robust error handling mechanisms. Recent approaches address this by integrating LLMs with real-world grounding through pre-trained robotic skills and leveraging feedback from the environment [12]. These skills constrain the model to propose feasible and contextually appropriate actions, allowing the robot to act as the model's "hands and eyes" while the LLM supplies task-specific semantic guidance. Closed-loop feedback, such as success detection, scene description, and human interaction, further enables LLMs to reason dynamically and adapt to the evolving context of embodied tasks.

3 Implementation

Our approach utilizes LLMs for intelligent task sequencing, error detection, and corrective adaptations, ensuring seamless human-robot interaction during assembly and disassembly sequentially distributing tasks to the human and the robot. We validate our system using a single-person workstation with a cooperative robot and depth camera. The robot is mounted on a table and equipped with a two-finger gripper. The table is divided into three distinct areas: (1) the assembly area with two cushion boards for assembly and disassembly processes, (2) the assembly area for components left from each operation, and (3) the storage area for tools. For validation and to showcase our system's capabilities, we custom-designed a product with medium assembly complexity called the 'sensor box'. In addition to a Raspberry Pi and two sensors, it includes multiple joint types (sliding, click, and screw connections) to test different assembly/disassembly methods (see Fig. 1a).

We developed a system integrating object detection, LLMs, and cooperative robotics (see Fig. 1b). The system aims to reason and optimize assistance for human workers, including process guidance. As the cobot performs tasks, the camera system tracks the workspace, including parts, positions, and the current assembly step. The different parts of the product in the workspace are detected by a fine-tuned version of YOLOv8m [13] and then linked over consecutive frames using DeepSORT [14].

Similar to the approach used by Bashir et al. [7], we use GPT-4o to analyze natural language instructions enriched with relevant data from manuals in PDF-form describing the product and available tools and decompose them into sub-tasks. To extract relevant data from the manuals, we stored them in a vector database equipped with OpenAI's text embedding. Before publishing a query to the OpenAI API, we perform a similarity search on the vector store to extract parts of relevant documents and attach them to the prompt The prompt template was carefully engineered using

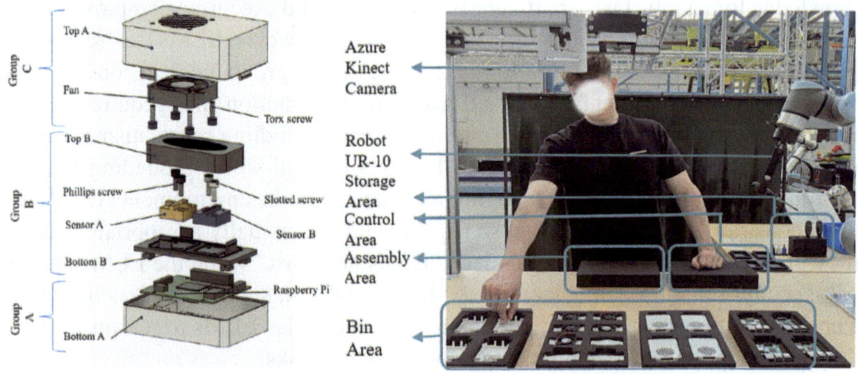

(a) Exploded view of the product. (b) Composition of the workstation.

Fig. 1 System components overview

chain-of-thought and examples for few-shot learning to coerce the language model into generating an action plan in form of a list of JSON objects describing each action. Each action object contains information about its action type (e.g. attaching an object to another object, moving an object from storage to the work area), responsible agent and the required tools and components. The model is prompted with a general rule set for the dis-/reassembly of the product as well as some examples for both cases respectively. In the disassembly scenario, the goal is to compute a plan to disassemble the product to a state where the specified component can be removed or replaced, e.g., remove the top part to expose the fan and subsequently remove the screws such that it can be replaced. In a similar fashion, the assembly/reassembly starts with the already disassembled product where the specified component has been replaced and aims to reverse the disassembly process. We provide 8 examples in total, four each for assembly and disassembly (see also Table 1), which covers half of all possible cases (two plans for each of the 8 components). A single example consists of the input prompt, e.g., "Compute a disassembly plan in order to replace the circuit board in the sensor box.", combined with the ground truth plan for this particular scenario in the specified JSON format. The structured response is then sent to both the robot controller and the human assistance module. Additionally, to reduce costs and inference times, our system employs a caching system based on the LangChain[2] framework.

Once the general plan is published, the LLM-based task execution monitor (see Fig. 2) continuously check for errors in task execution by monitoring the scene state, robot status, and planned LLM action details. The error monitor or error detection module compares the intended action with results from the object detection and the current state of the robot to determine whether the action was performed correctly. If an error is detected, e.g., the wrong component was detached from the product, this error is immediately displayed by the human assistance module and simultaneously sent to the error correction module to suggest a corrective action. The error correction module consists of an LLM, which is instructed similar to the action planner module. In addition to the incorrect action, a partial action plan is passed to the module to provide the necessary context. Based on the knowledge about the product and the assembly process, the model suggests an action correcting the mistake made by the worker. The corrective action is then passed to the LLM-based task execution monitor which updates the action plan and informs the worker via the human assistance module. To perform actions with the physical robot, we use the traditional approach of computing object locations by applying a series of coordinate transformations to obtain the location in robot base coordinates. Based on these locations, we use the ROS library MoveIt! to calculate a pick and place motion plan for the robot arm which is then sent to the robot controller.

[2] https://www.langchain.com/langchain.

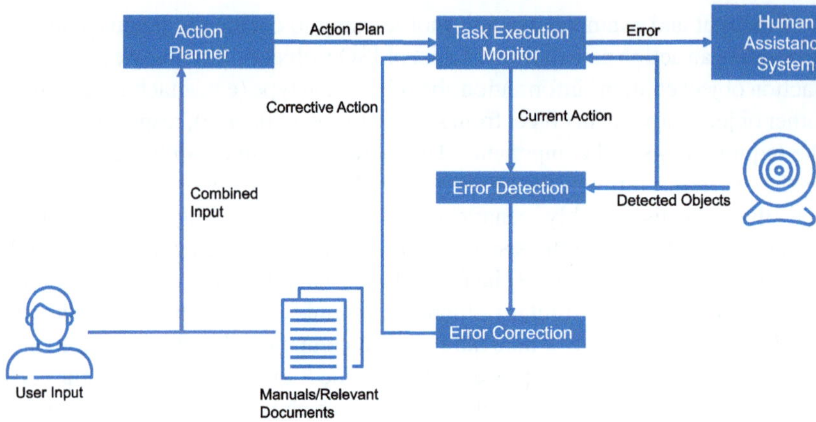

Fig. 2 Diagram of the system architecture

Table 1 Correctness evaluation results for the dis- and reassembly tasks. Cases marked with * were included in the few shot training. Component names were abbreviated, e.g., BotA stands for Bottom A as seen in Fig. 1a

	BotA	Pi	BotB	SenA	SenB	TopB	Fan	TopA	**Avg.**
Disassembly	1.00*	0.7	0.9*	1.00*	0.9	0.55	0.75	1.00*	**0.85**
Reassembly	1.00*	0.60*	1.00	0.85	1.00*	1.00	0.80*	0.35	**0.83**
Average	1.00	0.65	0.95	0.925	0.95	0.775	0.775	0.675	**0.84**

Table 2 Test and validation correctness. Test correctness includes only the tasks that were provided as examples in the prompt

	Test/training correctness	Validation correctness
Disassembly	0.975	0.725
Reassembly	0.85	0.8

4 Evaluation

We evaluated the action planner's correctness and consistency by comparing computed action plans against a human-defined ground truth. To this end, we repeatedly query the language model to compute new action plans and check if they are semantically equivalent to the ground truth. In total, we compute 20 action plans per task. As we are only focusing on the quality of the action plans, these computations were made in simulation. Semantic equivalence in this case means that we allow certain permutations of the ground truth as well, such as plans where manipulate actions occur several steps before action where the manipulated object is needed. It is important to note that the caching mechanism is disabled during this evaluation. We define the correctness/consistency of the planner as the average ratio of valid plans for each component (see Table 1), i.e., the percentage of computed plans that is semantically equivalent to the respective ground truth. As expected, plans for components which were included as examples in the few-shot-prompt achieve high accuracy most of the time. Most notably, the correctness of the reassembly plan after replacing the circuit board (Pi) is lower, even though the example was explicitly specified in the prompt. The comparatively high accuracy of plans for sensor A and sensor B, respectively, is explained by the similarity of their action plans, varying only in the type of screws used to fasten them to the underlying board. Overall, the correctness of the cases that were not included in the few-shot 'training', i.e., the validation correctness, is 72. 5% for disassembly and 80% for the reassembly cases (see Table 2).

5 Discussion

Our LLM-driven approach demonstrates reasonable generalization in deriving procedural sequences from natural language. However, it struggles with complex tasks like parallel processes due to limitations of few-shot prompting, which performs well only in narrowly defined scenarios. Designing broadly applicable prompts remains challenging. Future work should explore fine-tuning on domain-specific data for better scalability and robustness, and investigate how larger context windows and advanced models can improve interpretation of multi-step or concurrent operations, particularly given the small scope of current evaluations.

6 Conclusion and Future Work

This work introduces an LLM-driven human-robot cooperation system for industrial disassembly and assembly, integrating object detection and LLM-based task planning for efficient task distribution and real-time error correction. The system enhances automation flexibility through intelligent planning and adaptability. Future efforts will focus on improving model robustness via prompt engineering and exploring local or fine-tuned LLMs, expanding product compatibility, and incorporating multimodal feedback[3] to enhance scalability and applicability in[4] manufacturing.

Acknowledgements This summarizes the findings of two research projects: VADER, and RICAIP.

[3] Vernetzter digitaler Assistent für das datengetriebene Engineering von roboterbasierten Produktionsanlagen–Förderkennzeichen 13IK017B, funded by the German Federal Ministry for Economic Affairs and Climate.

[4] Research and Innovation Centre on Advanced Industrial Production—European Union Grant Agreement No. 857306.

References

1. Arents J et al (2021) Human robot collaboration trends and safety aspects: a systematic review. J Sens Actuator Netw 10(3). ISSN: 2224-2708. https://doi.org/10.3390/jsan10030048. https://www.mdpi.com/2224-2708/10/3/48
2. Alami R et al (2005) Task planning for human-robot interaction. In: Proceedings of the 2005 joint conference on smart objects and ambient intelligence: innovative context-aware services: usages and technologies. sOc-EUSAI'05. Association for Computing Machinery, Grenoble, France, pp 81–85. ISBN: 1595933042. https://doi.org/10.1145/1107548.1107574
3. Wang J et al (2024) Large language models for robotics: opportunities, challenges, and perspectives. arXiv:2401.04334 [cs.RO]
4. Rekik K et al (2024) Predictive intention recognition using deep learning for collaborative assembly. In: 2024 10th international conference on control, decision and information technologies (CoDIT), pp 1153–1158. https://doi.org/10.1109/CoDIT62066.2024.10708459
5. Wang R et al (2024) LLM-based robot task planning with exceptional handling for general purpose service robots. arXiv:2405.15646 [cs.RO]
6. Lim J et al (2024) Enhancing human-robot collaborative assembly in manufacturing systems using large language models. arXiv:2406.01915 [cs.RO]
7. Bashir A et al (2024) State space exploration with large language models for human-robot cooperation in mechanical assembly. In: 2024 7th Iberian robotics conference (ROBOT), pp 1–6. https://doi.org/10.1109/ROBOT61475.2024.10796918
8. Smith S, Smith G, Chen WH (2012) Disassembly sequence structure graphs: an optimal approach for multiple-target selective dis- assembly sequence planning. Adv Eng Inform 26(2). Knowledge based engineering to support complex product design, pp 306–316. ISSN: 1474-0346. https://doi.org/10.1016/j.aei.2011.11.003. https://www.sciencedirect.com/science/article/pii/S1474034611001091
9. Ye Y et al (2023) Large language models are versatile decomposers: decompose evidence and questions for table-based reasoning. arXiv: 2301.13808 [cs.CL]
10. Smirnov P et al (2024) Generating consistent PDDL domains with large language models. arXiv: 2404.07751 [cs.RO]
11. Xia L et al (2024) Large language models empower the reliability of dis- assembly in remanufacturing. Manuf Lett 41. 52nd SME North American manufacturing research conference (NAMRC 52), pp 1728–1733. ISSN: 2213-8463. https://doi.org/10.1016/j.mfglet.2024.09.200. https://www.sciencedirect.com/science/article/pii/S221384632400292X
12. Huang W et al (2022) Inner monologue: embodied reasoning through planning with language models. arXiv: 2207.05608 [cs.RO]
13. Jocher G et al (2022) ultralytics/yolov5: v7.0-YOLOv5 SOTA realtime instance segmentation. Version v7.0. https://doi.org/10.5281/zenodo.7347926
14. Wojke N, Bewley A, Paulus D (2017) Simple online and realtime tracking with a deep association metric. arXiv:1703.07402 [cs.CV]

Leveraging LLM for Assembly Instructions Using Controlled Natural Language

Michael Jonek and Martin Manns

Abstract With increasing product complexity and shorter product lifecycles, manufacturing companies—especially small and medium-sized enterprises (SMEs)—need to adapt their production planning processes quickly and reliably. A key step is the creation of accurate and easily interpretable work instructions for manual assembly tasks. Using Controlled Natural Language (CNL) for these instructions offers distinct advantages: it preserves the readability of natural language while enforcing a standardized syntax and vocabulary. In this context, recent advances in Large Language Models (LLMs) provide an opportunity to streamline the generation of CNL-based assembly instructions. This paper presents a proof of concept that investigates whether an LLM, coupled with a Retrieval-Augmented Generation (RAG) approach, can generate CNL-compliant instructions from Product-Process-Resource (PPR) models. Two representative use cases demonstrate the potential of this approach to improve structural consistency and reduce manual effort. However, the evaluation remains limited to a small number of examples and a single LLM, and current PPR models lack certain critical information such as precise positioning. These findings indicate that while LLMs show promise for supporting automated generation of CNL-based instructions, further work is needed to refine domain adaptation, improve data quality, and validate the approach with broader datasets and alternative language models. This work thus lays the groundwork for more flexible, automated production planning processes that exploit the synergy of LLMs and controlled linguistic frameworks.

Keywords Generative artificial intelligence · Large language models · Assembly planning · Retrieval-augmented generation model

M. Jonek (✉) · M. Manns
PROTECH—Institut für Produktionstechnik, Universität Siegen, Siegen, Germany
e-mail: michael.jonek@uni-siegen.de

M. Manns (ed.), *New Paradigms for Anticipated Uncertainty*, Lecture Notes in Mechanical Engineering, https://doi.org/10.1007/978-3-032-16889-4_64

1 Introduction

Modern manufacturing environments face increasing product complexity and rapidly shortening product lifecycles [1, 2]. These trends are particularly challenging for Small and Medium-sized Enterprises (SMEs), which must quickly adapt their production planning processes to remain competitive [3]. One essential step in production planning is generating clear and standardized work instructions for manual assembly tasks.

Recent advances in Large Language Models (LLMs) offer new avenues for automating and streamlining the generation of such instructions [4]. By leveraging methods like Retrieval-Augmented Generation (RAG), LLMs can incorporate domain-specific data and respond more accurately to manufacturing scenarios. However, instructions for manual assembly must balance human readability with strict adherence to domain-specific constraints. A Controlled Natural Language (CNL) meets these requirements by offering the familiarity of natural language while enforcing standardized syntax and terminology [5].

In this paper, we address the challenge of automatically generating work instructions in a CNL by leveraging a RAG approach with an LLM. Although CNL is intended to simplify text by enforcing standardized syntax and vocabulary, implementing it for complex assembly instructions isn't trivial [6]. CNL must accurately capture domain-specific constraints, avoid ambiguity in multi-step sequences, and remain flexible enough to incorporate variables like product variants and evolving process data [7]. These requirements create significant hurdles, especially when striving for automated solutions that do not rely on extensive manual editing [8]. By integrating these elements, we seek to bridge the gap between generative AI capabilities and the stringent demands of modern production planning. This work presents a proof of concept to investigate the fundamental feasibility of combining RAG with LLMs for generating CNL instructions.

2 Objective

The objective of this research is to investigate whether LLMs, that are supported by a RAG approach can automatically produce high-quality CNL-based work instructions for manual assembly. By leveraging comprehensive domain knowledge, enforcing precise CNL-specific constraints, and providing detailed input data on the assembly, we aim to determine if unambiguous and context-aware instructions can be generated with minimal human intervention. These instructions are analyzed for being both linguistically consistent and practically applicable in modern production environments.

3 State of the Art

Planning assembly sequences and creating assembly instructions are essential yet challenging tasks in manual assembly, as both demand substantial domain knowledge and detailed product data. Much of the existing research leverages data extracted directly from Computer-Aided Design (CAD) systems, whose 3D models encapsulate the geometric shapes and non-geometric features such as material type, weight, tolerances, and part orientations [e.g. 9, 10]. Due to the large amount of assembly information, CAD systems are an ideal source of information for knowledge-intensive assembly planning. There is a range of works that employ them in order to obtain automated assembly sequence planning [11].

However, assembly instructions contain more information that is not available in CAD data, such as the tools or manual activities required for assembly [12]. Therefore a standard for digital factory process and tool data structures has been established [13], that specifies product, process and resource as a foundation. The standard is established in common industrial planning and simulation tools such as Tecnomatix and DELMIA [14, 15]. Manufacturing-relevant information is added by the process planner when designing the process. This enables an automated export of corresponding assembly instructions for the worker. In the last years, several authors have published approaches to automate this. Examples are CAD geometry extraction [16] and ontology-based models that map domain-specific knowledge [12]. Further approaches rely on rule-based systems [17] or deep learning frameworks that use 3D geometries to determine assembly sequences [18]. Wang et al. [19] analyses the potential of ChatGPT for the future of manufacturing and identifies four central fields—uman–robot collaboration, knowledge management, design innovation and engineering education and training. There ChatGPT is found capable of making a valuable contribution. Although existing CAD or ontology-based methods already offer initial solutions for the semi-automated creation of assembly instructions, a literature review show that it is hardly used in industrial practice.

Meyer et al. [4] have gone one step further and researched generative AI in the form of an LLM for the automatic creation of assembly instructions. Experiments with different Models from OpenAI, different input data, and various prompting techniques reveal that while LLMs excel at generating textual quality advice, they struggle with planning accuracy of assembly sequences.

One application where the strength of LLMs, the generation of natural texts, can be used profitably is in the automatic generation of work instructions for manual assembly [4]. A challenge in the automated generation of assembly instructions is the format in which the work instructions are written. Depending on which LLM is used or which assembly is considered, a coherent text or an enumeration of the assembly steps can be output. Additional information that is only indirectly related to assembly, such as preparing the parts or checking or reworking, can also be added depending on the information in the knowledge database of the LLM. The wording of the texts can also vary. In [6], the researchers Manns et al. analyse differently worded work instructions in a company and quantified the effects in the form of a vagueness measure.

4 Method

To standardize work instructions generated automatically by LLMs, we propose a method for generating CNL-based instructions using a RAG approach. This method relies on a predefined sentence structure and vocabulary (Fig. 1), as developed for manual assembly planning [6, 20], ensuring that instructions are both human- and machine-readable.

Figure 2 illustrates the workflow of our proposed architecture. The input data includes a PPR model and a CAD rendering or image of the assembly. These are passed to a vector database, which retrieves the most relevant information fragments (top-k chunks) from a modular knowledge base. This base contains structured representations of CNL templates, company-specific vocabulary, safety guidelines, and domain knowledge about the assembly process.

A prompt builder module composes a structured prompt by combining the retrieved knowledge with the assembly-specific input. The final prompt is sent to an LLM (OpenAI GPT model), which generates work instructions in CNL. The output is validated and returned in machine-readable format (e.g., JSON).

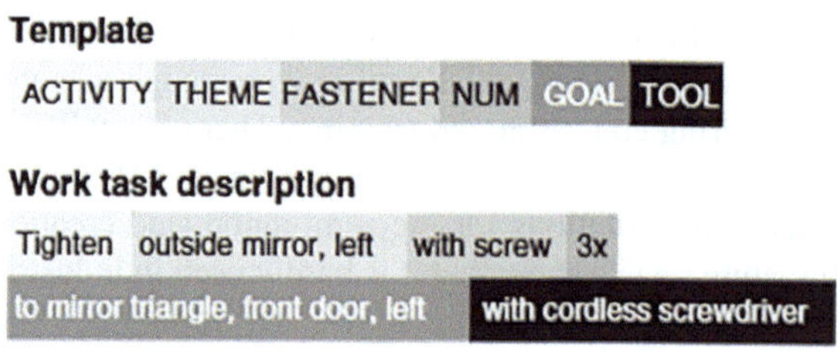

Fig. 1 Example of a work task description in the controlled natural language [5]

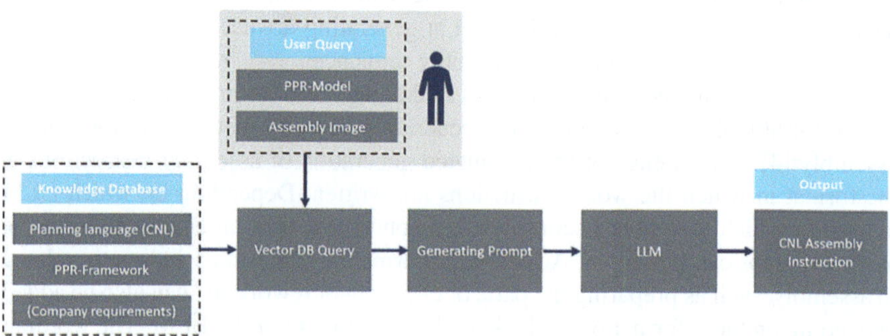

Fig. 2 Modell of a RAG for automatic generation of assembly instructions in CNL

Compared to conventional prompting approaches, our method allows the LLM to be dynamically guided by company-specific and contextual domain data, increasing the consistency of the generated output and its compliance with CNL syntax.

4.1 Implementation

The core architecture of our implementation follows a modular RAG pipeline that enables controlled and context-aware generation of assembly instructions. The process begins with the user providing input through a graphical user interface (GUI), where both a PPR model and a corresponding CAD image of the assembly are uploaded. These inputs form the query for a vector-based semantic retrieval system implemented using FAISS. This system performs similarity searches using cosine distance and retrieves the top-k most relevant information chunks from a dedicated knowledge base, ensuring that only the most contextually meaningful data is used for generation.

The knowledge base is composed of structured content, including previously validated CNL assembly instructions, definitions of PPR model elements, company-specific vocabulary constraints, and relevant safety or process regulations. Once relevant content is retrieved, a dedicated prompt builder assembles the selected fragments into a cohesive and structured prompt template. This prompt adheres to the formal sentence structure required by CNL standards while allowing limited flexibility to accommodate product-specific nuances.

The composed prompt is subsequently transmitted to the OpenAI ChatGPT model (o1-2024-12-17) via API. The language model processes this input and generates a draft of the assembly instructions in accordance with CNL syntax. The resulting output is then parsed and converted into a structured format, such as JSON, to facilitate further evaluation, validation, or integration into downstream systems like MES or quality control platforms.

4.2 Use Case

We evaluated our proposed method using two representative use cases. The first use case involved the manual assembly of a pneumatic cylinder and served as the baseline for comparison. As shown in Fig. 3, the instruction generation was initially performed using a direct prompt approach, where the PPR model and an image were provided to OpenAI's ChatGPT o1. The resulting instruction output lacked uniformity in phrasing, contained redundant preparatory steps, and used a variety of synonymous expressions, which led to inconsistencies with the target CNL format. In the second step, we applied our RAG-based approach with the exact same input. As depicted in Fig. 3, the resulting assembly instructions adhered closely to the CNL syntax template. The introductory preparation step was correctly omitted, and all process steps were uniformly structured using approved CNL patterns. Although the model inserted some terms that were not explicitly part of the defined CNL vocabulary, such as "secure" instead of "fasten", the lexical structure and clarity improved substantially. This highlights the system's ability to generate coherent and semantically accurate instructions, even when some vocabulary constraints are not fully enforced.

To further evaluate the scalability and robustness of our approach, we applied the same RAG-based method to a second, more complex use case involving the manual assembly of an angular gearbox (Fig. 4). This task represents a realistic industrial scenario with multiple components, including a gearbox housing, a drive shaft with flange, and several fastening screws. It requires sequential positioning, alignment of bolt patterns, and conditional dependencies between assembly actions. In this scenario, we again provided the PPR model and a corresponding CAD image as input via the GUI, triggering the RAG pipeline. The resulting instructions for the gearbox assembly were more detailed than the pneumatic cylinder case. For example, the output without the RAG approach unnecessarily included both preparation and

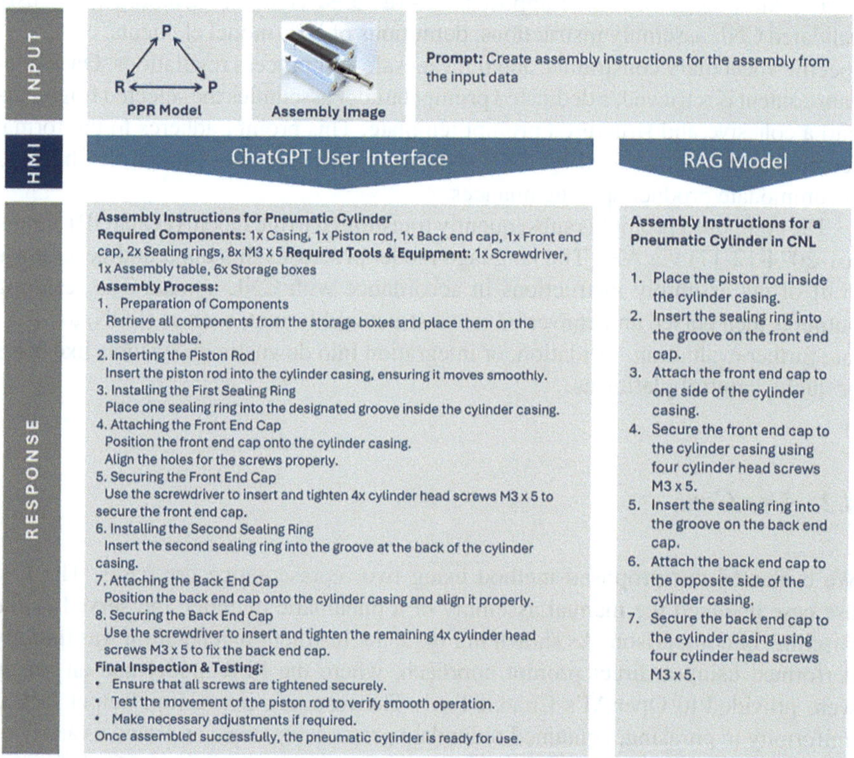

Fig. 3 Comparison of assembly instructions for a pneumatic cylinder: The upper section shows the input data (PPR model and assembly image), while the lower section presents the LLM-generated responses: on the left the free-form instructions, on the right the CNL-structured version

post-processing steps. The steps were presented, as in the first use case, with a short headline followed by a descriptive text for each action, which led to additional irrelevant content compared to the concise target CNL format. Compared to the free-form instructions generated without retrieval, the RAG-based approach clearly removed unnecessary preparation and post-processing steps and produced only the essential assembly actions. Apart from the title, no additional content has been added to the response by the LLM. It followed the defined CNL schema more strictly, resulting in concise, uniform step descriptions and improved readability, which reduced ambiguity and made the output more directly applicable in an industrial context.

We evaluated the output along three dimensions: completeness of the required actions, lexical accuracy (percentage of CNL-approved verbs), and the manual post-editing time estimated by an assembly expert. In this gearbox case, the instructions covered all required steps without omissions, and lexical accuracy was fully compliant, with correct use of verbs such as "fasten" according to the defined CNL schema. Manual editing time decreased significantly, from 20 min for free-form output to about 12 min for the RAG-based output. Quantitatively, across both use cases we observed a reduction in post-editing time of around 40%, and vocabulary compliance increased from 50% in the baseline case to 85% with the RAG-based

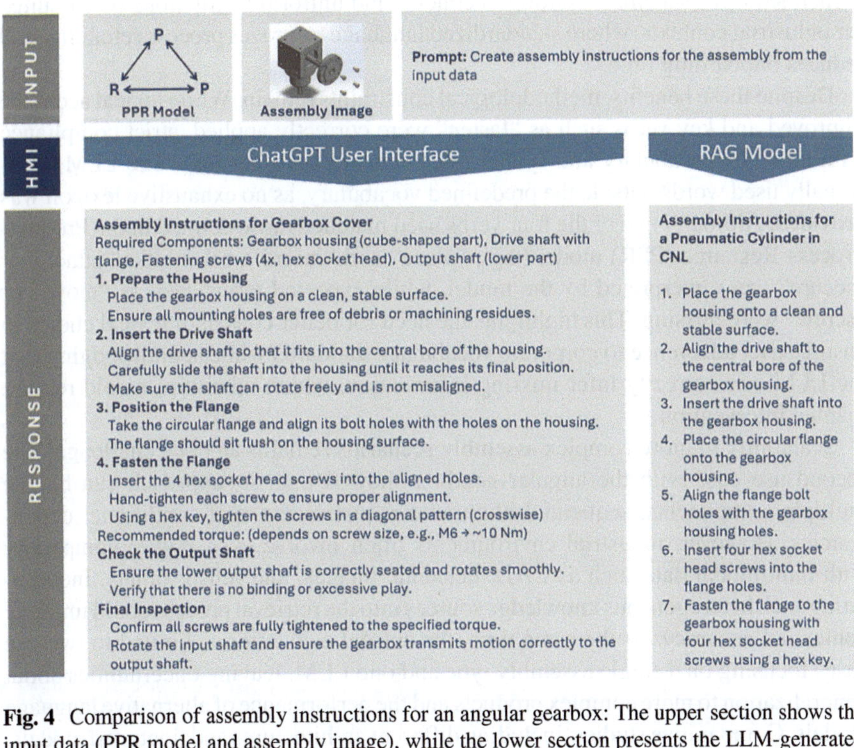

Fig. 4 Comparison of assembly instructions for an angular gearbox: The upper section shows the input data (PPR model and assembly image), while the lower section presents the LLM-generated responses: on the left the free-form instructions, on the right the CNL-structured version

approach. These results underline the potential of the proposed method to support scalable, semi-automated generation of high-quality work instructions suitable for deployment in real manufacturing environments.

5 Discussion

The results of the two evaluated use cases demonstrate both the potential and the current limitations of combining LLMs with RAG for generating CNL work instructions. In line with existing literature on automated assembly planning and text generation [4, 16–18], our results demonstrate that LLMs can reduce manual effort and achieve consistent instruction formatting, thus meeting a key objective identified in prior studies on AI driven assembly support. The RAG-based approach improved structural consistency and removed unnecessary preparatory and post-processing steps compared to free-form outputs generated without retrieval. By retrieving contextually relevant information from a knowledge base and applying strict syntactic rules, the model reduced well-known issues of "hallucination" [21, 22] and inconsistent phrasing commonly observed in unconstrained LLM outputs [23]. Due to domain-specific knowledge, the system adhered more closely to the predefined CNL schema, resulting in concise and uniform instructions better suited for industrial contexts, where standardized language improves process reliability and reduces onboarding time.

Despite these benefits, methodological constraints remain. While lexical accuracy improved and key verbs such as "fasten" were correctly applied, strict compliance with all CNL vocabulary and syntax rules was not fully ensured. The LLM occasionally used words outside the predefined vocabulary, as no exhaustive lexicon was provided. Although two of the four verbs used matched those defined in the Product-Process-Resource (PPR) model (e.g., "place" and "insert"), others like "attach" or "secure" were interpreted by the model, while expected terms such as "move" or "screw" were missing. This highlights the need for better control of lexical choice to ensure strict adherence to corporate vocabulary standards while acknowledging that the LLM can correctly infer missing information, which otherwise would require manual intervention.

Scalability to more complex assembly scenarios remains an open challenge. The second use case with the angular gearbox showed that the approach can handle multiple components, sequential alignment requirements, and conditional dependencies. However, industrial environments often involve even higher complexity with multimodal data such as CAD metadata, images, and sensor inputs. Incorporating such heterogeneous knowledge sources into the retrieval process could improve contextual accuracy. At the same time, the current evaluation is limited to two use cases focusing on a single assembly type and one LLM, leaving uncertainties about generalization to more complex products and the performance of alternative language models. Comparative studies, including different architectures and finetuned models, are necessary to assess robustness across various domains.

Another limitation concerns the input data. The PPR models used lacked critical information such as precise positioning and spatial relationships, which must either be inferred from additional sources like images or rely on the pre-trained knowledge of the LLM. While this level of input was sufficient to validate feasibility, a production-ready implementation would require higher data quality and completeness. Moreover, the effort to create and maintain PPR models for each product raises questions about how best to extract relevant data directly from CAD files or integrate automated assembly sequencing methods with retrieval-augmented generation.

To draw statistically significant conclusions, broader empirical studies are needed, combining expert validation and quantitative metrics such as CNL compliance rates or human post-editing time. Comparative analyses against alternative approaches, such as finetuned LLMs without retrieval or rule-based systems, could further clarify cost–benefit trade-offs. In addition, future research should refine domain adaptation strategies, integrate multimodal data for improved instruction depth and accuracy, and ensure that evolving product features are reliably captured.

In summary, the presented RAG-based proof of concept demonstrates that combining retrieval with LLMs is a viable first step toward automated CNL generation for industrial assembly instructions. To achieve production-level robustness, future work should focus on domain-specific language control, richer multimodal knowledge integration, improved data completeness, evaluation of different LLMs, and comprehensive empirical validation on larger and more diverse datasets.

6 Conclusion and Outlook

This work presents a proof of concept demonstrating that a RAG-based approach, leveraging an LLM guided by domain-specific data and strict syntactic rules, can generate CNL work instructions from PPR models. In two representative use cases, the approach improved structural consistency and reduced redundant steps compared to free-form LLM outputs, showing the potential to reduce manual effort in creating standardized assembly instructions. However, these findings remain preliminary. The evaluation is limited to two example assemblies and a single LLM model, and current PPR models lack essential details such as precise positioning information. These limitations mean that the present implementation serves primarily to demonstrate feasibility rather than provide a production-ready solution.

Future research should therefore extend the evaluation to more complex products and multiple LLMs, and focus on improving the completeness and quality of input data. In particular, more efficient techniques for generating and maintaining PPR models, potentially leveraging CAD data and existing automated sequence planning, are needed. Incorporating multimodal data such as CAD metadata, images, or sensor information could provide richer context and improve the accuracy and robustness of generated instructions. By addressing these challenges, RAG-based instruction generation could evolve from a feasibility study into a reliable component of advanced, flexible assembly planning, bridging the gap between automated processes and user-friendly documentation.

Acknowledgements The authors would like to acknowledge the financial support by the Federal Ministry for Economic Affairs and Climate Action (BMWK) within the program "Future Investments for Vehicle Manufacturers and Supplier Industry" (KoPa 35c) for the project SkaLaB (grant number: 13IK025B).

References

1. Bakker CA, Mugge R, Boks C et al (2021) Understanding and managing product lifetimes in support of a circular economy. J Clean Prod 279:123764. https://doi.org/10.1016/j.jclepro.2020.123764
2. ElMaraghy H, ElMaraghy W (2022) Adaptive cognitive manufacturing system (ACMS)—A new paradigm. Int J Prod Res 60:7436–7449. https://doi.org/10.1080/00207543.2022.2078248
3. Schuh G, Lukas G, Frey C (2023) Design model of an ecosystem for resilient and sustainable value creation of SMEs in single and small batch production. publish-Ing, Hannover
4. Meyer F, Freitag L, Hinrichsen S et al. (2024) Potentials of large language models for generating assembly instructions. In: 2024 IEEE 29th international conference on emerging technologies and factory automation: ETFA 2024: 10–13 September 2024, Padova, Italy. IEEE, Piscataway, NJ, pp 1–8
5. Manns M, Arteaga Martin NA (2014) A vagueness measure for concurrent engineering of manual assembly lines. Prod Eng Res Devel 8:225–231. https://doi.org/10.1007/s11740-013-0494-y
6. Manns M, Martín NAA (2013) Automated DHM modeling for integrated alpha-numeric and geometric assembly planning. In: Abramovici M, Stark R (eds) Smart product engineering. Springer, Berlin, Heidelberg, pp 325–334
7. Manns M, Wallis R, Deuse J (2015) Automatic proposal of assembly work plans with a controlled natural language. Procedia CIRP 33:345–350. https://doi.org/10.1016/j.procir.2015.06.079
8. Manns M, Fischer K, Du H et al (2018) A new approach to plan manual assembly. Int J Comput Integr Manuf 31:907–920. https://doi.org/10.1080/0951192X.2018.1466396
9. Pan C (2005) Integrating CAD files and automatic assembly sequence planning
10. Pan C, Smith SS, Smith GC (2006) Automatic assembly sequence planning from STEP CAD files. Int J Comput Integr Manuf 19:775–783. https://doi.org/10.1080/09511920500399425
11. Gu P, Yan X (1995) CAD-directed automatic assembly sequence planning. Int J Prod Res 33:3069–3100. https://doi.org/10.1080/00207549508904862
12. Rusch T, Steuer J, König M et al (2021) Tool-based automatic generation of digital assembly instructions. Procedia CIRP 99:454–459. https://doi.org/10.1016/J.PROCIR.2021.03.065
13. VDI—Verein Deutscher Ingenieure (2008) 4499 Blatt 1: digital factory—Fundamentals, 2008-02. DIN Media GmbH, Berlin
14. Lv N, Liu J, Ding X et al (2017) Assembly simulation of multi-branch cables. J Manuf Syst 45:201–211. https://doi.org/10.1016/J.JMSY.2017.09.007
15. Afizul NA, Selimin MA, Pagan NA et al (2024) Modelling an assembly line using tecnomatix plant simulation software. rmtb 5:1048–1055
16. Gors D, Put J, Vanherle B et al (2021) Semi-automatic extraction of digital work instructions from CAD models. Procedia CIRP 97:39–44. https://doi.org/10.1016/j.procir.2020.05.202
17. Meisels A, Davidov D (1989) A rule-based systems generator for industrial assembly planning. Eng Appl Artif Intell 2:65–71. https://doi.org/10.1016/0952-1976(89)90041-9

18. Neves M, Neto P (2022) Deep reinforcement learning applied to an assembly sequence planning problem with user preferences. Int J Adv Manuf Technol 122:4235–4245. https://doi.org/10.1007/s00170-022-09877-8
19. Wang X, Anwer N, Dai Y et al (2023) ChatGPT for design, manufacturing, and education. Procedia CIRP 119:7–14. https://doi.org/10.1016/j.procir.2023.04.001
20. Rychtyckyj N (2005) Intelligent manufacturing applications at Ford Motor Company. NAFIPS 2005—2005 Annual meeting of the North American fuzzy information processing society, pp 298–302
21. Friel R, Sanyal A (2023) Chainpoll: a high efficacy method for LLM hallucination detection
22. Martino A, Iannelli M, Truong C (2023) Knowledge injection to counter large language model (LLM) hallucination. In: Pesquita C, Skaf-Molli H, Efthymiou V et al (eds) The semantic web: ESWC 2023 satellite events: Hersonissos, Crete, Greece, May 28–June 1, 2023, proceedings. Springer, Cham, pp 182–185
23. Beurer-Kellner L, Fischer M, Vechev M (2024) Guiding LLMs the right way: fast, non-invasive constrained generation

Human Motion Modelling for Collaborative Handling of Rigid Objects

Raza Saeed, Tadele Belay Tuli, and Martin Manns

Abstract Human motion modelling is essential in understanding and enhancing human-human collaboration, particularly during the collaborative handling of rigid objects. This paper presents a data-driven approach to developing a new model for human motion in collaborative object-handling scenarios, aiming to enhance the effectiveness and adaptability of cooperative tasks. This study also addresses the accurate analysis of captured data from human-human collaboration. Through an experimental setup utilizing a motion capture system, human motion data is collected and preprocessed for a collaborative handling task to create a model that can generate motion styles. The proposed model is trained using experimental data based on Gaussian Mixture Models (GMM) and functional Principal Component Analysis (FPCA). The performance of the proposed model is evaluated in terms of its spatial accuracy. The results demonstrate that the model is capable of capturing the dynamics of human-human collaboration and motion styles, which makes it feasible for collaborative processes and potential applications in human-robot collaboration (HRC).

Keywords Human motion modelling · Human-human collaboration · Collaborative object handling · Motion capture · Data-driven modelling

1 Introduction

In manual assembly lines, some tasks require two humans to collaborate in handling objects, where high coordination and precise positioning are necessary due to factors such as size, weight, or geometric tolerance. Collaboration between two humans may also depend on the workplace setting that describes processes to lift, align, and position the part accurately. In handling tasks during assembly processes, where two

R. Saeed (✉) · T. B. Tuli · M. Manns
PROTECH-Institut für Produktionstechnik, Universität Siegen, Siegen, Germany
e-mail: raza.saeed@uni-siegen.de
URL: https://protech.mb.uni-siegen.de/fams/

M. Manns (ed.), *New Paradigms for Anticipated Uncertainty*, Lecture Notes in Mechanical Engineering, https://doi.org/10.1007/978-3-032-16889-4_65

humans work together to move a rigid object, they must coordinate their actions and agree on a movement strategy to transport the object from its initial position to its target position. Object manipulation depends on the collaboration between humans and the object, as well as communication, physical strains arising from pull-push [1], and speed mismatch [11] effects. Handling a car's windshield by two people during an automotive assembly operation is a real-world example where collaborative object handling is crucial.

In human-human collaboration, individuals must continuously adapt to each other's motion behaviour during joint object manipulation. This coordination requires both humans to anticipate the potential trajectory, sense the action and reaction forces, and carefully observe and respond to their co-worker's movements and intentions [11]. Therefore, to ensure successful collaboration in physical tasks such as joint object handling, it is necessary to investigate how two humans understand each other's actions and communicate effectively to coordinate their actions when lifting, aligning, and positioning a part during collaborative tasks. This includes exchanging verbal and nonverbal cues such as motion synchronization, mutual gaze, and force feedback. For example, during the lifting and positioning of a part, each individual must predict their partner's intentions and adjust their movements accordingly. Investigating these implicit communication channels can inform the design of human-robot systems that must operate under similar physical and temporal constraints. Exploring human-human collaboration may guide the development of human-robot collaboration (HRC) systems that operate under comparable physical and temporal constraints. Collaborative handling of an object is one of the current challenges regarding HRC's actual collaboration capabilities. HRC combines the extended capabilities of humans and robots to create a more inclusive and human-centred future in the production industry [4]. However, it remains unclear how robots and humans should interact to handle an object jointly and anticipate mutual care. Transfer of knowledge of human motion enables robots to work alongside humans more naturally and efficiently. One critical factor for achieving effective collaboration is the accurate modelling of human motion, which allows robots to anticipate, adapt, and respond more effectively to human actions. By leveraging human motion models, HRC systems may enhance speed, accuracy, agility, and controllability, thereby potentially narrowing the automation gap in product assembly.

This study presents a human motion modelling approach for human-human collaboration scenarios involving the joint handling of rigid objects, aiming to generate natural human motion that enhances collaboration effectiveness and adaptability. The model analyses motion samples by examining motion duration, canonical timelines, and cross-feature correlations. Additionally, the study employed an iterative feedback mechanism designed to improve human-human collaboration.

2 Related Work

In manual assembly environments, humans perform various tasks, including reaching, grasping, picking, moving, and placing product parts at specific locations. Understanding the full range of human body movements while performing these tasks is crucial. To study human motions, tasks performed by human workers are predetermined based on an analysis of human motion. The study of human motion involves generation models that are mainly based on data-driven approaches (e.g., [9]), which rely on reference and example datasets. Human motion is often too complex to be accurately captured by model-based approaches. These approaches, which rely on simplified function approximations, can simulate basic motion simulation but may not fully capture the complexity of natural human behaviour. Data-driven motion generation approaches are considered in various research and industrial applications, including film, sports, ergonomics, health, and production environments [3, 12]. Deep Learning-based motion modelling and synthesis is one of the data-driven approaches widely investigated for character animation. These methods are known for their high accuracy but are limited in their interpretability and challenging to enforce constraints. The relevance of constraint is to achieve plausible and realistic motion limited by starting and target points [5].

Comparatively, statistical models that contain an infinite number of motion variants, enabling the generation of a rich repertoire of motions based on a finite number of input datasets, are interpretable and controllable. Statistical models are based on the assumption that the human motion generator system comprises an infinite number of styles and postures. They assume that different executions of a single motion primitive exhibit an intrinsic relationship, which can be approximated using statistical models. This is because data-driven motion generation techniques depend on the quality of the motion model, which can be affected by several factors, including the accuracy of the motion capture system, sample size, modelling techniques, application environment, and the granularity of motions and segmentation processes [7]. With a particular focus on sample size, several works have reported the difficulty of handling high-dimensional data resulting from the many human body segments, each represented by position and orientation in both spatial and temporal domains. The work of Min et al. [9] presented natural human motion modelling using the Gaussian mixture model (GMM). This is a prominent approach for setting up Latent Space (LS) based on the maximum likelihood of expectation by fitting mixtures of Gaussian models to samples in high-dimensional spaces [8]. A probabilistic model is established that maximises the expectation by fitting mixtures of Gaussian models to samples in high-dimensional spaces. Samples in high-dimensional spaces may be considered latent spaces, in which the GMM helps cluster data into a set of kernels; each GMM kernel represents a motion strategy. More controllable models, such as probabilistic models, allow the planner to define constraints, which are essential to analysing human-machine interaction and planning the process efficiently. An extension of these models may offer valuable additional information for HRC, as currently, the latent spaces only consider a human and not a robot, and the generation of heavily constrained trajectories is not real-time capable.

3 Method

A proposed human motion model for jointly handling an object is represented in Fig. 2 using the IDEF0 method [6]. This method is built upon [5], which presents a motion model construction process that follows steps such as segmentation, time normalisation, dimensionality reduction, statistical motion model generation, and feature clustering. This work focuses on extending and testing the method for two humans collaborating to handle an object jointly while minimising the variance using a feedback mechanism. The process involves the following steps:

- *Segmentation*: An experimental setup for human motion capture using inertial measurement units (IMUs) is discussed in [13] and illustrated in Fig. 1. The segmentation function input consists of Biovision Hierarchy (BVH) motion capture files, which are segmented into discrete motion snippets. Metrics such as jerk

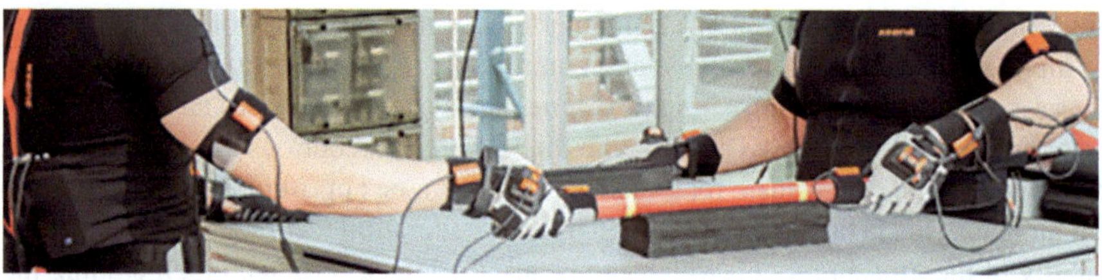

Fig. 1 Experimental setup for data collection for human-human collaboration

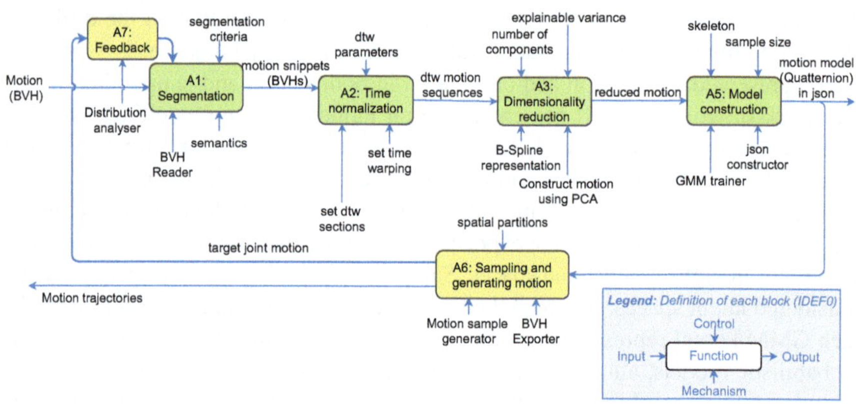

Fig. 2 IDEF0-based illustration of FPCA-based human motion modelling for collaborative object handling tasks. A1-A5 are developed based on state-of-the-art, while A6 and A7 are new contributions

peaks are proposed to detect transitions in motion dynamics and automate the segmentation process.
- *Time normalisation*: Each segmented motion snippet exhibits a different frame size (duration). At this step, spatially similar motions are also assumed to be similar temporally. The segmented motions undergo temporal alignment using Dynamic Time Warping (DTW) to ensure that motion sequences of varying durations are mapped onto a canonical timeline (c. [10]), enabling consistent comparison and synthesis.
- *Dimensionality reduction (spatial and temporal)*: Motion segments registered to the canonical timeline have similar frame sizes, each represented as a vector. This vector comprises the root position and all joint orientations. FPCA is employed to reduce motion data from 189 to 12 dimensions by capturing essential movement patterns. The initial 189 dimensions derive from 27 joints in the human skeleton, and each is represented by a 7D quaternion (4D orientation and 3D position). FPCA reduces motion data by extracting 9 Principal Components (PCs) for spatial and 3 PCs for temporal variations. Hence, each motion is approximated with a cubic B-spline in quaternion space to ensure singularity-free motions. At the same time, each motion segment is Z-Transformed to approximate the temporal deviation [8].
- *Motion model construction*: This step applies the GMM to train the motion model based on the outputs of aligned, reduced, and transitioned motion primitives. The output is a motion model in quaternion angles, a structured and reusable representation of motion. The motion models are represented using JSON files, which can be easily used for motion synthesis and analysis.
- *Sampling, analysis and feedback for optimisation*: The generated motion model has a high variance if the motion data shows significant variation. In this regard, verifying the model's fidelity in spatial and latent spaces is essential before using it. Then, based on space partitioning, position and orientation offset are computed. Then, these offsets are approximated and corrected iteratively. This optimisation step feeds back into improving the segmentation, ensuring that the segmentation and alignment process remains robust against artefacts introduced by dimensionality reduction.

As depicted in Fig. 2, the novel contribution of this work is to extend a human motion modelling method with a minimal variance feedback mechanism. This method is developed based on the morphable graphs library,[1] in which Gaussian Mixtures are employed to create a statistical model composed of multivariate Gaussian distributions (c. [2]). The motion model training configuration for GMM and the motion model parameters are described in Table 1.

[1] https://github.com/dfki-asr/morphablegraphs.

Table 1 Configuration settings and parameters

Input		Configuration	
Parameter	Value	Parameter	Value
No. of motion segments	200	Fraction	0.90
Sampled motion size	200	No. of spatial PCs	9
Segment frame size (μ, max, min)	314.28, 432, 240	No. of temporal PCs	3
Canonical frame line	315	Temporal precision	0.99

4 Results

The right-hand position is visualized in Fig. 3 for the captured motion data Fig. 3 (left) and snippets data Fig. 3 (middle) in the X-Z plane. The Z-axis shows upward, while X and Y represent the workspace plane where the participating human stands. The captured motion data in BVH is segmented into snippets, and then the duration histogram and the computed canonical frame line are shown Fig. 3 (right). The training configuration for the motion model using GMM and its parameters are detailed in Table 1. The trained human motion model has 9 PCs for spatial components, as shown in Table 1, which reduces the motion dimension by approximately 93.65% with an explainable variance fraction of 90%. The trained model is used to generate human movement trajectories in the Cartesian space, and the result is shown in Fig. 4 (left). The right wrist path is shown in red colour, and the path is slightly tightened due to retargeting to a new skeleton and human model. The Fig. 4 (middle) illustrates the correlation between the features (PCs) of human motion in latent space. The figure shows that some PCs have almost no correlation, e.g., PC_0 vs PC_1. A positive correlation (0.1494) exists between PC_1 vs PC_5, and 0.1971 between PC_5 vs PC_4. A negative correlation is also observed between PC_3 and PC_4, with a correlation coefficient of –0.1199. The model-generated motion sample is exported as a BVH file and simulated using skinned human mesh models for result verification, as shown in Fig. 4 (right).

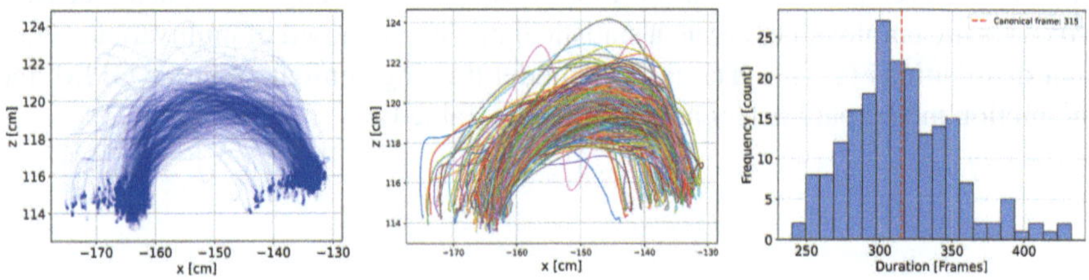

Fig. 3 Captured motion data visualization (left), segmentation (middle) on the XZ plane and histogram of motion snippets duration (right)

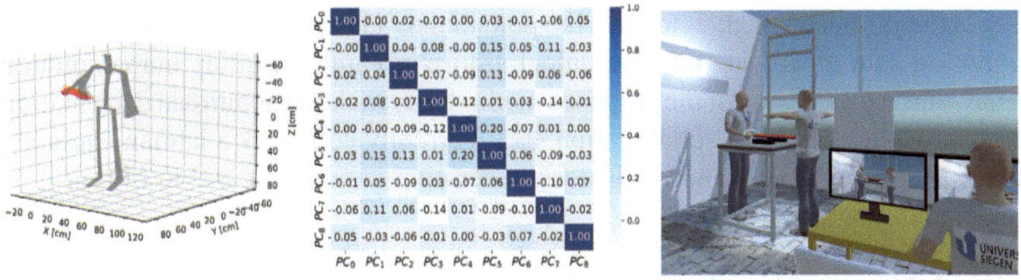

Fig. 4 Motion sample data visualization on the XZ plane (left), latent feature correlations of PCs (middle) and verification of the generated motion sample in simulation environment (right)

5 Discussion

The correlation matrix of the latent space features indicates that most PCs capture independent directions of variance, confirming the effectiveness of the proposed method. The 9 selected PCs reduced the motion dimension by approximately 95.23% while retaining 90% of the explainable variance, demonstrating the model's efficiency in representing human motion with significantly lower dimensionality. The correlation analysis (e.g., PC_5 vs PC_4) suggests underlying dependencies, and the slight negative correlation observed (PC_3 vs PC_4) may highlight compensatory motion behaviour in specific motion dimensions. Furthermore, the visualisation of the right wrist trajectory in the CS confirms that the motion paths remain consistent after retargeting to a new skeleton and human model. The model-generated motion samples were exported into BVH format and simulated with a human skin mesh model, aligning well with the expected motion behaviour and validating the model's applicability. By focusing on the PCs that explain the highest variance (PC_0, PC_1, PC_4), the model can effectively capture critical motion attributes such as speed and posture, mainly when dealing with noisy sensor data. These findings highlight the potential of this model in human-robot collaboration, where understanding human motion patterns enables robots to adapt their movements dynamically during collaboration. However, further investigation is needed to interpret the latent features more comprehensively and refine the model's ability to generalise across different motion scenarios.

6 Conclusion and Future Outlook

The proposed method for modelling human motion for collaborative handling of rigid objects yielded low variance in the generated motion samples, which is a promising result in terms of model performance. The canonical timeline and the average frame of motion snippets are closely aligned. Correlation analysis of each sample reveals a strong positive correlation between the motions, indicating that the motion data for the *RightWrist* joint across the BVH files are identical or move in unison. This allows the proposed method to have the potential to address not only key challenges in motion data processing but also unlock future research regarding human motion behaviour. An extension of the proposed method may make an additional contribution to adaptive human-robot collaboration, as it currently considers a human and not a robot. Moreover, constrained trajectory generation is not yet capable of real-time implementation.

Acknowledgements The authors would like to acknowledge the financial support by the DFG within the HiSMoT project (500490184) and the Horizon Europe FLEX4RES project (101091903).

References

1. Argubi-Wollesen A et al (2017) Human body mechanics of pushing and pulling: analyzing the factors of task-related strain on the musculoskeletal system. Saf Health Work 8(1):11–18
2. Bishop CM (2006) Pattern recognition and machine learning. Springer, New York
3. Field M et al (2011) Human motion capture sensors and analysis in robotics. Ind Robot 38(2):163–171
4. Glogowski P et al (2017) Task-based simulation tool for human-robot collaboration within assembly systems. In: Tagungsband des 2. Kongresses Montage Handhabung Industrieroboter. Springer, pp 155–163
5. Herrmann E et al (2017) Accelerating statistical human motion synthesis using space partitioning data structures. Comput Animat Virtual Worlds 28(3–4), e1780. https://doi.org/10.1002/cav.1780
6. Li Q, Chen YL (2009) Idef0 function modeling. In: Modeling and analysis of enterprise and information system. Springer, pp 98–122
7. Manns M, Otto M, Mauer M (2016) Measuring motion capture data quality for data driven human motion synthesis. Procedia CIRP 41:945–950. https://doi.org/10.1016/j.procir.2015.12.068
8. Manns M et al (2018) A new approach to plan manual assembly. Int J Comput Integr Manuf 31(9):907–920
9. Min J, Chai J (2012) Motion graphs++ a compact generative model for semantic motion analysis and synthesis. ACM Trans Graph (TOG) 31(6):1–12
10. Myers DE (1981) Dynamic time warping. IEEE Trans Acoust Speech Signal Process 29(6):1372–1379. https://doi.org/10.1109/TASSP.1981.1163455
11. Noohi E, Žefran M, Patton J (2016) A model for human- human collaborative object manipulation and its application to human- robot interaction. IEEE Trans Robot 32(4):880–896
12. Pons-Moll G et al (2010) Multisensor-fusion for 3D full-body human motion capture. In: 2010 IEEE computer society conference on computer vision and pattern recognition. IEEE, pp 663–670
13. Saeed R, Tuli TB, Manns M (2025) Experimental setup for motion capturing two humans jointly handling an object. Procedia CIRP 134:401–406

AI-Driven Analysis of 2D Technical Drawings for Agile and Reconfigurable Manufacturing Systems

Marco Fries and Thomas Ludwig

Abstract The automotive industry faces major challenges due to shifting market demands and technological advances, particularly in balancing innovation, cost efficiency, and growing product variability. Flexible manufacturing techniques (FMTs) with automation support adaptability to diverse component designs. Despite progress in 3D modeling, 2D technical drawings remain essential due to their compatibility with existing workflows. Automating their processing is therefore crucial for their seamless integration into modern manufacturing systems. This paper presents an AI-based approach for analyzing 2D technical drawings by detecting geometric features (e.g., reinforcements, embossments) and extracting manufacturing-relevant text (e.g., dimensions, tolerances). The prototype uses Faster R-CNN for object detection and a Keras-OCR-based pipeline for text recognition. A dataset of 204 labeled drawings covering eight feature types was created. Results confirm the system's ability to identify complex patterns and extract critical data for downstream automation. The approach provides a scalable solution, especially for legacy components lacking CAD data and relying on scanned 2D documents.

Keywords Artificial intelligence · Technical drawings · Object detection · Optical character recognition · Reconfigurable production systems

1 Introduction

The automotive industry is undergoing a transformation driven by changing market conditions and technological innovations [1, 2]. These developments impact not only original equipment manufacturers (OEMs), but also suppliers, who face challenges in cost-efficient manufacturing [2, 3]. Those challenges arise from low production volumes of single parts and an increasing variety of bodywork designs, making it difficult to maintain profitability while meeting customer demands [4]. Consequently,

M. Fries (✉) · T. Ludwig
University of Hagen, Hagen, Germany
e-mail: marco.fries@fernuni-hagen.de

M. Manns (ed.), *New Paradigms for Anticipated Uncertainty*, Lecture Notes in Mechanical Engineering, https://doi.org/10.1007/978-3-032-16889-4_66

FMTs are gaining importance due to their ability to offer variant and capacity flexibility [3, 5, 6]. At the same time, FMT demand digital, machine-readable, information, like computer aided design (CAD) data for automated processing to unveiling its full potential. Despite progress in 3D modeling, 2D technical drawings remain necessary in the manufacturing industry [7]. This is partly because 3D models are not always accessible in day-to-day operations, whereas 2D drawings are easier to handle, portable, and compatible with a wide range of workflows [8–10]. They can be used without specialized software and are often the default format in documentation and communication. Moreover, in many real-world scenarios, especially in the context of legacy parts, only 2D technical drawings are available, sometimes as scans or prints, while the corresponding CAD data is either lost or not provided. In some cases, 2D documentation is shared exclusively due to licensing restrictions, confidentiality concerns, or software incompatibilities. This makes 2D drawings not just a legacy artifact, but a necessary and practical foundation for many automated manufacturing processes today.

These factors highlight the continued relevance and usability of technical drawings as information providers in the manufacturing industry for decades. Dillenhöfer et al. [11] demonstrate that an AI-driven automated detection (information extraction) in technical drawings is technically feasible achieving a detection accuracy of 89% for geometric features. Whilst this example shows the potential in object detection, they focused specifically on gear boxes. A more general perspective for flexible and reconfigurable manufacturing that addresses a wealth of diversity in possible components is missing. Additionally, for further processing dimension information as well as tolerances are needed. Haar et al. [7] and Villena Toro et al. [12] demonstrate good results using OCR technologies to extract text from technical drawings. Their solutions in turn miss a detection of geometric features.

The objective of this study is to develop and evaluate an AI module that can detect key features and relevant textual information for manufacturing, enabling further automated processing. In this paper, we first review the latest open-source libraries and frameworks, identify existing research gaps, and present our approach. We then describe the steps for data preparation, model training, and feature selection. Using our trained models, we present a framework that integrates them, combining object and text detection. We evaluate our framework and AI models with example parts (Sect. 5). Finally, we conclude with a brief discussion, conclusion and outlook (Sect. 6).

2 Related Work

In what follows, we present the current advancements around object detection and OCR on 2D technical drawings.

2.1 *Object Detection with Artificial Intelligence on Technical Drawings*

Technical drawings are essential in mechanical engineering, capturing design, manufacturing, and assembly specifications. They typically include dimensions, processes, surface finishes, and tolerances, summarized in an information block containing general tolerances, material data, and other details. Multiple views ensure geometric clarity, with tolerances defined via a Feature Control Frame (FCF). Recent advances in object recognition have increased interest in extracting such information automatically [13]. Object recognition, a computer vision technique, detects objects in images, marks their coordinates, and assigns confidence scores [14]. It relies on neural networks like Convolutional Neural Networks (CNNs) to learn visual features [15, 16]. Object detectors are classified as single-stage or two-stage models. Single-stage models predict classes and bounding boxes directly, without intermediate steps. Examples include YOLO (You Only Look Once) [17] and SSD (Single Shot MultiBox Detector) [18]. Both offer high speed and competitive accuracy, making them suitable for real-time applications such as autonomous driving [19]. Two-stage models, such as R-CNN (Region Convolutional Neural Network), Faster R-CNN, Mask R-CNN, and G-RCNN (Granulated Recognition Convolutional Neural Network), first generate region proposals before classification [15, 20, 21]. Pretrained models, e.g. from the TensorFlow Detection Model Zoo, can be fine-tuned for specific applications. The most prominent detectors are Faster R-CNN, SSD, and YOLO [19]. Faster R-CNN is a widely used and highly versatile object recognition model [13]. In technical drawing contexts, Faster R-CNN performs well, achieving 88.8% average precision in [22]. Other works [7, 23–25] highlight YOLO for its speed and accuracy. As summarized by [13], YOLO and Faster R-CNN represent state-of-the-art solutions for object detection in technical drawings.

2.2 *Character Recognition with Artificial Intelligence on Technical Drawings*

For text extraction in technical drawings, most approaches rely on OCR [26]. OCR automatically identifies characters from images and converts printed or handwritten text and symbols into machine-readable formats [27–29]. Typically, OCR systems consist of two models or sequential steps: a first model for feature extraction and a second for reading text sequences [30]. These technologies are widely used in domains like autonomous vehicles, retail analytics, and assistive tools for the visually impaired [31]. However, generic OCR is not well-suited for technical drawings, as recognition accuracy drops with increasing document complexity [26]. Drawings often include varied text formats and symbols (e.g., dimensions, tolerances, title block data), which standard OCR tools cannot reliably handle [7, 23, 28]. For instance, OCR typically reads from left to right, making it incapable of distinguishing between upper and lower deviations in nominal sizes [23]. Although generic OCR fails in this context [32], alternative methods exist for processing complex documents [33]. Recent research, as highlighted by Haar et al. [7] and Villena Toro et al. [12], focuses on combining OCR with machine or deep learning to enhance performance on engineering drawings. A particularly effective solution is eDOCr, presented in [12]. Based on Keras-OCR, it targets challenges such as recognizing dimensions, tolerances, and other domain-specific symbols. eDOCr also supports customization, allowing users to adapt the tool by training it on their own character sets.

3 Research Gap and Approach

Previous approaches effectively leverage advanced object recognition and OCR frameworks, demonstrating robust detection of shapes, text, and symbols in technical drawings [12]. These solutions can identify geometric elements or retrieve textual data, forming a solid basis for further automation in manufacturing. However, the gap in existing research lies in the lack of an integrated AI system capable of detecting both geometric features and relevant textual information from 2D technical drawings, which is critical for automated production processes. While deep learning models have shown promise in text and object recognition, most existing approaches lack the ability to combine these two aspects in the context of manufacturing-specific applications (e.g. automotive production). This integration is particularly important when aiming for flexible production processes that rely on technical drawings for controlling manufacturing steps. Our research question therefore constitutes as follows:

RQ: 'How can AI models detect both geometric features and manufacturing-relevant textual information in 2D technical drawings, and how can these outputs be integrated to support automated production processes?

To address our research question, we used a set of technical drawings collected in collaboration with industry partners. Based on industry-relevant requirements for agile and reconfigurable manufacturing, we identified key features to detect and labeled them for training. This enabled the development of an AI model tailored to these tasks. In parallel, we evaluated the eDOCr framework and trained models for text recognition. By combining outputs from geometric feature detection and OCR, we linked each geometric element with its corresponding descriptive text. The result is a structured JSON file that contains detailed information on detected features (e.g., bends, holes, embossments) and associated parameters such as dimensions and tolerances. This data can be directly used by manufacturing machines to identify required operations and configure process parameters. For example, a bending machine can extract angles and lengths needed for the task, reducing manual input to material placement, or potentially enabling full automation via robotic systems. The JSON file thus serves as a bridge between visual interpretation of 2D drawings and executable manufacturing actions in flexible production environments.

4 Design and Introduction of the Technology

In what follows we present our feature selection, the process of data preparation (technical drawings) and how we trained our AI models on the data.

4.1 Feature Selection, Data Preparation

For data preparation, we leveraged a raw dataset comprising 51 technical drawings provided by industry partners. The original files were vector-based PDF exports of classical technical drawings showing three projection views (front, top, side). These were rasterized and converted to JPEGs, then downscaled to 2048 pixels on the longer side to reduce memory load and enable GPU-based training. This trade-off maintained sufficient visual detail for detection tasks but posed challenges for identifying small-scale features. We facilitated collaborative discussions under domain experts within a workshop series to identify and prioritize essential geometric features that our AI should detect and potential manufacturing techniques. For instance, detecting holes would imply the need for drilling or laser cutting techniques, while identifying bending edges would require machinery capable of handling such tasks. Following these discussions, the labeling process was carried out using Label Studio (https://labelstud.io). The labeling criteria were informed by the insights gained from the workshop, ensuring that an augmented dataset of 204 labeled technical drawings (rotation and flipping [34] of the initial 51 drawings for enhanced model robustness) accurately reflected the eight geometric features (Fig. 1) of interest [3dBending (1), BoxCorner (2), BendingEdge (3), Notch (4), Reinforcement (5), ThroughHole (6), InnerContour (7), Embossment (8)]. We differentiated technically further labels (13), but sub summarize them into eight distinct features. This comprehensive approach ensures that our AI model is trained to recognize and interpret features accurately.

In terms of training, Lin et al. [8] and Vabalas et al. [35] highlight the significant role of data partitioning in deep learning. Babu et al. [36] suggest an 80–20 split of the

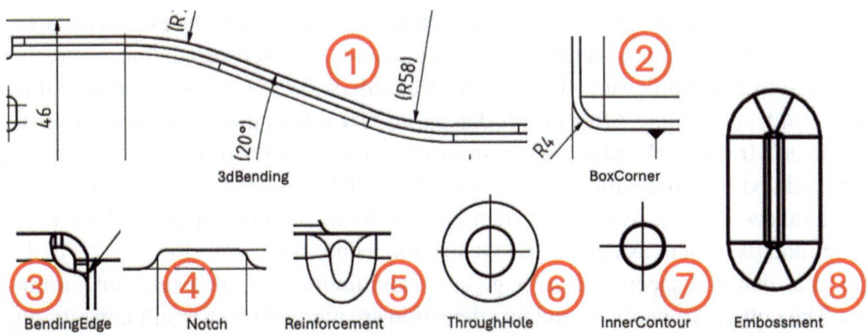

Fig. 1 Sample representations of the eight geometric features

data, with 80% used as training data and 20% as test data, where the remaining 20% should be split into 10% validation and 10% test data. In this study, the challenge of data partitioning and the associated limitations arise from the dataset's size. Due to legal restrictions, we were not able to gather more technical drawings from industry partners. Thus, implementing the recommended percentage splits is deemed impractical with the available dataset of 204 images. Consequently, we made the decision to initially use 90% of the dataset for training and the remaining 10% for testing. The trained model is evaluated based on this test dataset. A separate validation dataset to assess model performance during training is omitted due to the aforementioned limitations. This step is largely replaced by analyzing the training process using TensorBoard. By handling the data in this manner, the problem of overfitting is slightly higher and intentionally accepted to maintain a dataset size that enables an efficient model training [34]. We used a Resnet 50—faster-RCNN pretrained model from the Tensorflow Model Zoo, to make use of the transfer learning technique and not to start from scratch [37].

4.2 Model Training and Performance Test

For the training process, we utilized a high-performance notebook equipped with an RTX 4080 GPU featuring 12 GB VRAM, 32 GB RAM, and a Core i7 processor. Both text recognition and object detection models were trained using this hardware setup. Python along with Docker served as our primary environment for training the AI models and executing data processing tasks.

4.3 Framework Conception

We developed the framework as a Python program comprising multiple classes to manage training, testing, and execution of AI models for detecting geometric features and extracting relevant text from technical drawings. It processes JPEG or PDF files, which are analyzed using an object detection model. Detected features are visualized with bounding boxes in a second version of the drawing. Users can define a detection threshold to adjust precision. The system stores all detected features, their positions, and confidence scores. After object detection, text recognition is performed using the eDOCr framework, which we customized with our trained models (Fig. 2—1). Detected text is stored with positional data. If available, additional information such as tolerances is extracted. The system checks the info block for general tolerance values and uses them as default unless specific values are found (Fig. 2—2). It also identifies material specifications (e.g., steel, aluminum) and other relevant parameters. The outputs from object and text detection are merged into a result table linking features with associated text, then exported as a structured JSON file for further use (Fig. 2—3). The entire framework runs in a Docker container with a basic web

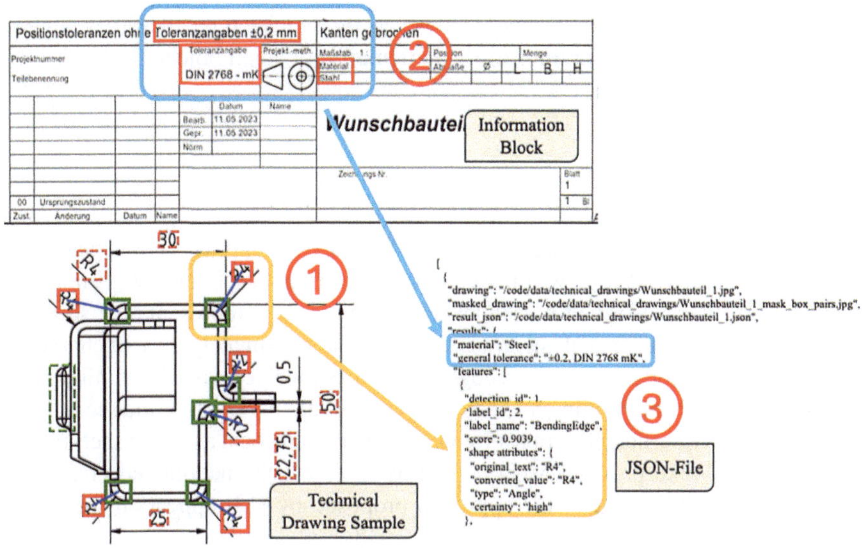

Fig. 2 Example of the proposed representation pipeline: (1) detected geometric annotations in the technical drawing, (2) extracted title block information, and (3) the corresponding structured JSON representation

interface that allows users to upload drawings and execute detection. Results are presented as JSON files along with annotated images showing the detected features.

5 Results and Discussion

In this section, we present and discuss the evaluation results of our AI-based framework. To reflect the dual focus of our system, the results are structured into three parts: object detection (Sect. 5.1), text recognition (Sect. 5.2), and the integration of both types of information into a unified feature table (Sect. 5.3).

5.1 Object Detection

When evaluating the overall accuracy of our object detection model, we achieved a mean Average Precision (mAP) of 0.5012 and a mean Average Recall (mAR) of 0.5799, resulting in an F1-score of 0.5377. Analyzing performance across different object sizes in the technical drawings, we obtained that large objects are detected with a mAP = 0.6639 and Recall = 0.6779, medium-sized objects with a mAP = 0.5254, and Recall = 0.5993 while small objects showed a mAP = 0.1612 and Recall = 0.2935. The overall loss decreased from 3.6875 to 0.2735, indicating a significant improvement in model convergence. However, the precision/recall values suggest that there is still room for optimization, particularly for small objects.

5.2 Text Recognition

For text recognition, we leveraged eDOCr, which does not allow for an exact determination of accuracy. While we did not conduct a formal benchmark with annotated ground truth, we performed a manual assessment across representative samples. Based on this qualitative analysis, we estimate that approximately 97% of characters were recognized correctly. Most recognition errors involved visually similar characters, such as misinterpreting a '3' as ')' or an 'm' as 'n'. Overall, the recognition performance was sufficiently robust for extracting dimensional and tolerance-related information. Unlike previous works, such as [26], which encountered challenges with vertically oriented text, the approach proposed by [12] appears to be highly promising.

5.3 Combined Feature Table

Figure 3 shows an example detection of geometric features and text, followed by detailed information in Table 1, combining both recognitions.

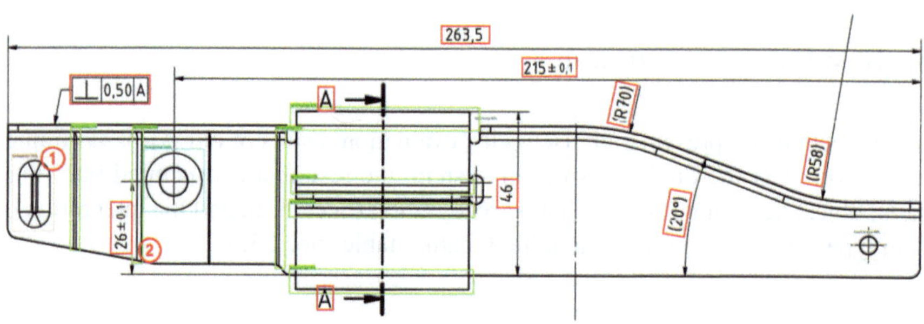

Fig. 3 Sample section of a technical drawing with recognized features and text

Table 1 Excerpt of the final feature table with the combination of geometric and textual information

ID	Feature	Score (%)	Assigned text	Tolerance	Tolerance upper	Tolerance lower	Text type
1	Embossment	99.69		General			
2	BendingEdge	82.91	26 ± 0.1	Specific	+ 0.1	– 0.1	Length

The labels exhibit varying detection probabilities, highlighting the need to improve detection of small objects. This is due to the limited number of training examples, the large scale of technical drawings, and their inherent complexity. Additionally, class imbalance persists, as only 51 drawings were available [38]. It is evident that the current dataset is insufficient to cover all labels adequately. To address this, synthetic data could expand the training base [11, 37]. As it stands, the limited dataset increases the risk of overfitting [34]. A more granular subdivision of geometric features would further support learning by enabling the model to better differentiate between similar elements. Despite these limitations, our system effectively detects both geometric features [11] and associated manufacturing-relevant text [12, 26], thus answering our research question. The Docker-based implementation integrates these components successfully. However, associating features with the correct textual information remains challenging and highly dependent on the quality of the technical drawings. In cases where dimension annotations are incomplete or imprecise, text-to-feature assignment becomes unreliable. Moreover, identifying relevant information, such as diameters or angles, is still suboptimal. Our current rule-based approach, which links text to features based on spatial proximity, requires refinement. Alternative strategies should be explored. The system currently does not distinguish between different projection views within a drawing, limiting its ability to associate measurements from multiple perspectives with the same feature. This is especially relevant for complex parts represented through front, top, and side views. To address this, we are experimenting with rule-based heuristics and spatial segmentation to detect and classify projection areas. Initial results are promising but require further refinement and validation to ensure robust implementation. At this stage, human intervention remains essential—not only to review and interpret results, but also to adjust where the AI struggles to distinguish relevant from irrelevant information. This human–AI interplay is critical to ensure high-quality outputs and reliable decision-making in production environments.

Nonetheless, our results show that combining object detection and text recognition provides essential data for downstream automated FMTs, which are increasingly valuable due to their variant and capacity flexibility as well as cost efficiency [5, 6]. The proposed approach enables the use of 2D technical drawings for legacy components (e.g., spare parts), allowing production without expensive tooling or large spare parts inventories. This capability supports the reproduction of components even decades after a product's lifecycle has ended, based solely on the original drawings.

6 Conclusion and Outlook

FMTs are gaining relevance due to their capacity and variant flexibility [5, 6]. While 3D models are increasingly used, 2D technical drawings remain widespread in practice [9, 10]. Most existing research focuses either on detecting geometric features [11, 13] or extracting text [12, 26], but automated production requires their integration.

This paper introduced a Docker-based application that combines object detection (Faster R-CNN) and text recognition (Keras-OCR) to analyze 2D technical drawings. The system processes PDF and JPEG files and outputs structured JSON files for downstream automation. Our object detection model was trained on 51 technical drawings (204 images) and performed well overall, though with limitations for small features. The OCR component reached ~ 97% character-level accuracy, with typical errors from visually similar characters [e.g., '3' vs. ')']. As the model is trained on German data, it cannot recognize characters outside the German alphabet. Linking geometric and textual information remains a challenge, particularly across different projection views. Our current rule-based assignment requires further refinement. Future work may explore layout analysis and machine learning to enhance this mapping. Multimodal models such as LLaVA [39] could support more advanced interpretation and interactive use cases [40, 41]. Overall, our integrated approach shows strong potential for extracting actionable data from legacy 2D drawings, especially in remanufacturing and spare parts production where no CAD data exists. We hope this work stimulates further research in combining vision and language models for production automation.

Acknowledgements The authors acknowledge the financial support by the Federal Ministry for Economic Affairs and Climate Action in the project SkaLaB (project number 13IK025B).

References

1. Túry G (2019) Electromobility in the automotive industry. What role does technology change play in the geographic pattern of production? 7:112–120
2. Pichler M, Krenmayr N, Schneider E, Brand U (2021) EU industrial policy: between modernization and transformation of the automotive industry. Environ Innov Soc Transit. 38:140–152. https://doi.org/10.1016/j.eist.2020.12.002
3. Demlehner Q, Schoemer D, Laumer S (2021) How can artificial intelligence enhance car manufacturing? A Delphi study-based identification and assessment of general use cases. Int J Inf Manage 58. https://doi.org/10.1016/j.ijinfomgt.2021.102317
4. Gupta P, Vardhan S (2016) Optimizing OEE, productivity and production cost for improving sales volume in an automobile industry through TPM: a case study. Int J Prod Res 54:2976–2988. https://doi.org/10.1080/00207543.2016.1145817
5. El-Khalil R, Darwish Z (2019) Flexible manufacturing systems performance in U.S. automotive manufacturing plants: a case study. Prod Plan Control 30:48–59. https://doi.org/10.1080/09537287.2018.1520318.
6. Geschonke M, Hecht B, Rössinger M, Zastera J (2023) New concepts for a flexible car body shop of hang-on parts: from lightweight engineering fixture to modular flexible cell concept. In: Galizia FG, Bortolini M (eds) Production processes and product evolution in the age of disruption. CARV 2023. Lecture Notes in Mechanical Engineering. Springer, Cham. https://doi.org/10.1007/978-3-031-34821-1_39
7. Haar C, Kim H, Koberg L (2023) AI-based engineering and production drawing information extraction. In: Kim KY, Monplaisir L, Rickli J (eds) Flexible automation and intelligent manufacturing: The human-data-technology nexus. FAIM 2022. Lecture Notes in Mechanical Engineering. Springer, Cham. https://doi.org/10.1007/978-3-031-18326-3_36

8. Lin Y-H, Ting Y-H, Huang Y-C, Cheng K-L, Jong W-R (2023) Integration of deep learning for automatic recognition of 2D engineering drawings. Machines 11:802. https://doi.org/10.3390/machines11080802
9. Engelke T, Keil J, Rojtberg P, Wientapper F, Schmitt M, Bockholt U (2015) Content first. In: Proceedings of the 6th ACM multimedia systems conference. ACM, New York, NY, USA, pp 105–111. https://doi.org/10.1145/2713168.2713169
10. Gattullo M, Scurati GW, Fiorentino M, Uva AE, Ferrise F, Bordegoni M (2019) Towards augmented reality manuals for industry 4.0: a methodology. Robot Comput Integr Manuf 56:276–286. https://doi.org/10.1016/j.rcim.2018.10.001
11. Dillenhöfer F, Doksanbir A, Kraske JL, Künne B (2023) Automatic generation of training data for AI object detection in terms of technical drawings in engineering. In: Auer ME, Langmann R, Tsiatsos T (eds) Open science in engineering. REV 2023. Lecture Notes in Networks and Systems, vol 763. Springer, Cham. https://doi.org/10.1007/978-3-031-42467-0_60
12. Villena Toro J, Wiberg A, Tarkian M (2023) Optical character recognition on engineering drawings to achieve automation in production quality control. Front Manuf Technol 3. https://doi.org/10.3389/fmtec.2023.1154132
13. Bhanbhro H, Hooi YK, Hassan Z (2022) Modern approaches towards object detection of complex engineering drawings. https://doi.org/10.1109/ICDI57181.2022.10007400
14. Andreopoulos A, Tsotsos JK (2013) 50 Years of object recognition: directions forward. Comput Vis Image Underst 117:827–891. https://doi.org/10.1016/j.cviu.2013.04.005
15. Girshick R (2015) Fast R-CNN. In: Proceedings of the IEEE international conference on computer vision (ICCV) (2015)
16. Mani S, Haddad MA, Constantini D, Douhard W, Li Q, Poirier L (2020) Automatic digitization of engineering diagrams using deep learning and graph search In: Proceedings of the IEEE/CVF Conference on Computer Vision and Pattern Recognition Workshops (CVPRW), pp. 176–177
17. Redmon J, Divvala S, Girshick R, Farhadi A (2016) You only look once: unified, real-time object detection. In: Proceedings of the IEEE conference on computer vision and pattern recognition, pp 779–788
18. Liu W, Anguelov D, Erhan D, Szegedy C, Reed S, Fu CY, Berg AC (2016) SSD: Single shot multibox detector. In: Leibe B, Matas J, Sebe N, Welling M (eds) computer vision – ECCV 2016. Lecture Notes in Computer Science, vol. 9905, pp. 21–37. Springer, Cham. https://doi.org/10.1007/978-3-319-46448-0_2
19. Kim JA, Sung JY, Park SH (2020) Comparison of faster-RCNN, YOLO, and SSD for real-time vehicle type recognition. In: 2020 IEEE international conference on consumer electronics—Asia, ICCE-Asia 2020. Institute of Electrical and Electronics Engineers Inc. https://doi.org/10.1109/ICCE-Asia49877.2020.9277040
20. Carranza-García M, Torres-Mateo J, Lara-Benítez P, García-Gutiérrez J (2021) On the performance of one-stage and two-stage object detectors in autonomous vehicles using camera data. Remote Sens (Basel) 13:1–23. https://doi.org/10.3390/rs13010089
21. Xavier AI, Villavicencio C, Macrohon JJ, Jeng JH, Hsieh JG (2022) Object detection via gradient-based mask R-CNN using machine learning algorithms. Machines 10. https://doi.org/10.3390/machines10050340
22. Nguyen T, Van Pham L, Nguyen C, Van Nguyen V (2021) Object detection and text recognition in large-scale technical drawings. In: Proceedings of the 10th international conference on pattern recognition applications and methods. SCITEPRESS—Science and technology publications, pp. 612–619. https://doi.org/10.5220/0010314406120619
23. Ackermann M, Haase C (2023) Machine learning-based identification of interpretable process-structure linkages in metal additive manufacturing. Addit Manuf 71:103585. https://doi.org/10.1016/j.addma.2023.103585
24. Elyan E, Jamieson L, Ali-Gombe A (2020) Deep learning for symbols detection and classification in engineering drawings. Neural Netw 129:91–102. https://doi.org/10.1016/j.neunet.2020.05.025
25. Yao R, Cheng X, Hui Z, Li Z, Min Z (2022) Intelligent reading for multi-scale engineering drawings based on adaptive object detection and optical character recognition. In: Proceedings

of WCSE 2022 Spring event: 2022 9th international conference on industrial engineering and applications. WCSE. https://doi.org/10.18178/wcse.2022.04.171
26. Schlagenhauf T, Netzer M, Hillinger J (2023) Text detection on technical drawings for the digitization of brown-field processes. Procedia CIRP 118:372–377. https://doi.org/10.1016/j.procir.2023.06.064
27. Awel MA, Abidi AI (2008) Review on optical character recognition. Int Res J Eng Technol 3666
28. Khallouli W, Pamie-George R, Kovacic S, Sousa-Poza A, Canan M, Li J (2022) Leveraging transfer learning and GAN models for OCR from engineering documents. In: 2022 IEEE world AI IoT congress (AIIoT). IEEE, pp 015–021. https://doi.org/10.1109/AIIoT54504.2022.9817319.
29. Francois M, Eglin V, Biou M (2022) Text detection and post-OCR correction in engineering documents. In: document analysis systems, pp. 726–740. Springer, Cham. https://doi.org/10.1007/978-3-031-06555-2_49
30. Du Y, Chen Z, Jia C, Yin X, Zheng T, Li C, Du Y, Jiang Y-G (2022) SVTR: scene text recognition with a single visual model
31. Wang K, Babenko B, Belongie S (2011) End-to-end scene text recognition. In: 2011 International conference on computer vision. IEEE, pp 1457–1464. https://doi.org/10.1109/ICCV.2011.6126402.
32. Joshua M, Ragav V, Ibrahim S (2024) Advanced knowledge extraction of physical design drawings, translation and conversion to CAD formats using deep learning. In: Janmanee P (eds) advanced in creative technology- added value innovations in engineering, Materials and Manufacturing. ICCT 2023. Lecture Notes in Networks and Systems, vol 979. Springer, Cham. https://doi.org/10.1007/978-3-031-59164-8_28
33. Jamieson L, Francisco Moreno-García C, Elyan E (2024) A review of deep learning methods for digitisation of complex documents and engineering diagrams. Artif Intell Rev 57. https://doi.org/10.1007/s10462-024-10779-2
34. Mutasa S, Sun S, Ha R (2020) Understanding artificial intelligence based radiology studies: what is overfitting? https://doi.org/10.1016/j.clinimag.2020.04.025
35. Vabalas A, Gowen E, Poliakoff E, Casson AJ (2019) Machine learning algorithm validation with a limited sample size. PLoS One 14. https://doi.org/10.1371/journal.pone.0224365
36. Babu MAA, Pandey SK, Durisic D, Koppisetty AC, Staron M (2024) Impact of image data splitting on the performance of automotive perception systems. In: Bludau P, Ramler R, Winkler, D, Bergsmann, J (eds) Software Quality as a Foundation for Security. SWQD 2024. Lecture Notes in Business Information Processing, vol 505. Springer, Cham. https://doi.org/10.1007/978-3-031-56281-5_6
37. Talukdar J, Gupta S, Rajpura PS, Hegde RS (2018) Transfer learning for object detection using state-of-the-art deep neural networks. In: 2018 5th international conference on signal processing and integrated networks (SPIN). IEEE, pp 78–83. https://doi.org/10.1109/SPIN.2018.8474198
38. Xu Y, Goodacre R (2018) On splitting training and validation set: a comparative study of cross-validation, bootstrap and systematic sampling for estimating the generalization performance of supervised learning. J Anal Test 2:249–262. https://doi.org/10.1007/s41664-018-0068-2
39. Liu H, Li C, Wu Q, Lee YJ (2023) Visual instruction tuning. In: advances in neural information processing systems (NeurIPS), vol. 36
40. Khan MT, Chen L, Ng YH, Feng W, Tan NYJ, Moon SK (2026) Fine-tuning vision-language model for automated engineering drawing information extraction. In: Guan S (eds) Proceedings of the 9th International Conference on Innovation in Artificial Intelligence. ICIAI 2025. Lecture Notes in Electrical Engineering, vol 1458. Springer, Singapore. https://doi.org/10.1007/978-981-95-0887-7_3
41. Khan MT, Chen L, Ng YH, Feng W, Tan NYJ, Moon SK (2024) Leveraging vision-language models for manufacturing feature recognition in CAD Designs. arXiv preprint arXiv. https://arxiv.org/abs/2411.02810

Combining Virtual Commissioning and Discrete-Event Simulation in Digital Manufacturing: A Literature Review and Future Research Directions

Martin Birtic, Enrique Ruiz Zúñiga, and Anna Syberfeldt

Abstract Advanced digital technologies are developed and implemented to address challenges in complex manufacturing systems. Two well-known technologies in this domain are virtual commissioning and discrete-event modeling and simulation. These simulation tools are supported by well-substantiated theoretical frameworks and proven in industrial applications. However, the two approaches target different aspects of production systems and their combination is not extensively researched, thereby motivating this study. This paper reviews existing literature on the combination of these simulation approaches, aiming to identify current opportunities, challenges, and research gaps, while proposing innovative directions for future exploration informed by the findings. The results suggest future research should explore combining VC and DES to enable test-driven development, reduce modeling effort through joint model design, and extend simulation model usability across the production system life cycle.

Keywords Virtual commissioning and discrete-event simulation · Hybrid simulation · Model-based system engineering · Test-driven development · Digital twin

1 Introduction

Production systems are facing growing demands, as evidenced by the increasing complexity of production and logistics systems driven by short product life cycles, rising product variants, and higher product complexity [1]. Digital manufacturing attempts to address the complexity issues of modern production systems by leveraging advances in simulation technology, which enables experimentation, validation, and testing of product, process, and system designs in increasingly sophisticated and decentralized manufacturing environments [2]. In industry, Virtual Commis-

M. Birtic (✉) · E. R. Zúñiga · A. Syberfeldt
Intelligent Automation Department, University of Skövde, Högskolevägen, Sweden
e-mail: Martin.Birtic@his.se

M. Manns (ed.), *New Paradigms for Anticipated Uncertainty*, Lecture Notes in Mechanical Engineering, https://doi.org/10.1007/978-3-032-16889-4_67

sioning (VC) and Discrete Event Simulation (DES) are proven simulation methods. While both are based on digital models, they have fundamental differences. VC is the simulation-based testing and validation of automation and control systems at the level of logic and signals, ensuring functionality and integration before physical installation and deployment. DES is modeling system-level performance by simulating the flow of events and resources at a higher abstraction level than VC. While VC focuses on validating low-level control logic, DES begins at the operational level, analyzing workflows, resource allocation, and overall system efficiency without addressing lower level details such as control logic. Recognizing that different simulation approaches address different aspects of production systems, hybrid simulation methods (HSM) have emerged, exemplified by the integration of DES and System Dynamics (SD), which leverage their complementary strengths, DES for detailed process modeling and SD for understanding systemic behaviors; such methods are able to capture a broader range of system dynamics and complexities [3].

Although hybrid simulation is studied extensively in operations research, the majority of this research focuses on combining DES, SD, and Agent-Based Simulation (ABS) within the model development and implementation stages of simulation studies, as evidenced by examples such as DES-ABS-SD models and ABS-DES hybrids, which are widely considered to be standard hybrid methods [4]. To the best of the authors' current knowledge, the explicit combination of VC and DES in production system development and operation remains underexplored. This apparent gap, combined with the broader goal of supporting the advancement of digital tools in an increasingly digitalized manufacturing domain, motivates the present effort to clarify the current state of research and identify possible paths forward. Accordingly, the main objective of this review is to map the existing research on combining VC and DES, and the secondary objective is to propose novel research directions informed by the findings of the review. Using Cooper's taxonomy [5] to position this review, it is a selective, thematically organized review aimed at identifying research gaps and guiding future work. The focus is on exploring the current state of combining VC and DES in manufacturing-specific applications to support further research and development. For the purpose of structuring the research, the following research questions are outlined.

RQ1 What are the current opportunities, challenges, and future research directions identified in relation to VC, DES, and their combination within production system modeling and simulation?

RQ2 What additional novel research directions could emerge from answering RQ1 to further advance the understanding and application of merging VC and DES?

2 Literature Survey

The literature selection process includes structured database searches, application of inclusion and exclusion criteria, and stepwise selection of titles, abstracts, and full texts. The number of items identified, screened, excluded, and included is summarized in a PRISMA diagram depicted in Fig. 1, describing the key steps in the selection process [6]. Figure 2 shows the distribution of papers over year.

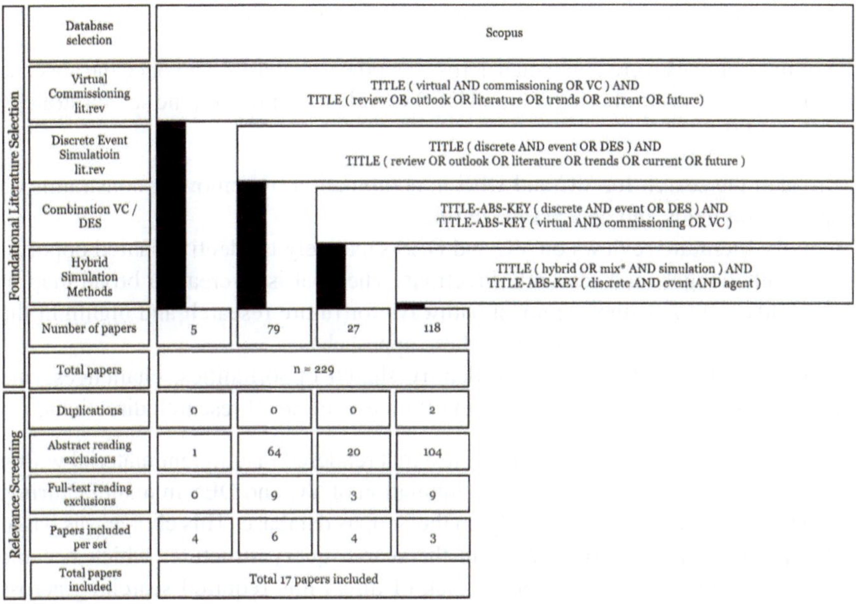

Fig. 1 The literature survey process initially began with 229 papers and was refined to include 17 papers after filtering

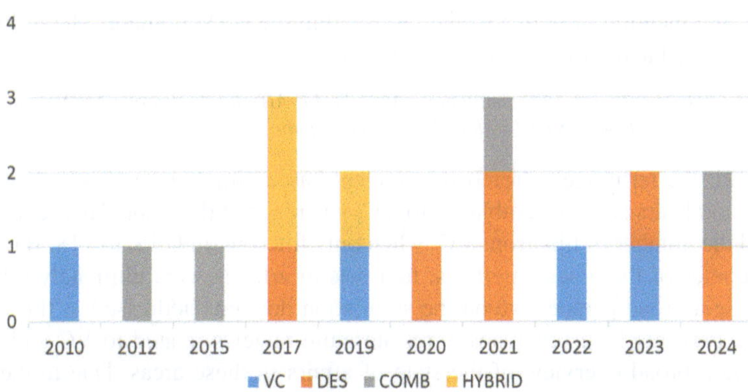

Fig. 2 Distribution of papers over publication year

Initial search experiments were performed on several databases, including Scopus, IEEE Xplore, Web of Science, ScienceDirect, ACM Digital Library and Google Scholar. Below is one of the search queries used in the early experiments to identify papers that combine VC and DES. Both terms and their appropriate abbreviations are required in the title to reflect the main theme. To narrow the scope to relevant application areas, the query also requires that manufacturing, production or industry be mentioned in the title, abstract, or keywords. This search returned a single, non-relevant result on Scopus:

TITLE (virtual AND commissioning OR VC) AND TITLE (discrete AND event OR DES) AND TITLE-ABS-KEY (manufacturing OR production OR industry)

Similar searches were conducted across the other databases, and once the absence of explicit research combining VC and DES was clearly apparent, the search strategy was revised to:

- Broaden the search for VC and DES in combination by removing constraints on application fields and appearance in title.
- Identify literature reviews on VC and DES separately to identify shared opportunities, challenges, and research directions. The goal is to create a broad map of the fields and to synthesize novel pathways for future research and highlight the potential advantages of combining these approaches.
- Broad search into HSM also to identify shared opportunities, challenges, and research directions for inclusion in the synthesis of novel research directions.

A further consequence of this initial test that rendered insufficient material within the primary search area, the specific combination of VC and DES in manufacturing, was the decision to focus exclusively on the Scopus database. This choice was based on the prioritization of consistency in the search query structure, which became increasingly important given that the lack of direct hits required searching across several related topics. The explicit rationale is that using a single consistent database promotes methodological transparency and replicability. Also, one note about the search query related to HMS. In this context, an initial search was conducted to identify any mention of both VC and DES within the HSM domain. The following query was used and returned only a single result:

TITLE (hybrid OR mix AND simulation) AND TITLE-ABS-KEY (discrete AND event OR DES) AND TITLE-ABS-KEY (virtual AND commissioning OR VC)*

Instead, the term agent (referring to agent-based simulation) was included, as this method bears some resemblance to VC, whereas SD does not. This refinement reduced the number of hits from 565, when only filtering on DES, to 118. Based on the outcomes of these search efforts, four sets of articles were ultimately selected to provide a broad perspective on the simulation domains addressed in this study. The first two sets focus exclusively on literature reviews related to VC and DES, providing a broad overview of the state of affairs in these areas. Due to the lack of existing literature that explicitly addresses the combination of VC and DES, the third and fourth sets were compiled to include articles that explore the mixing of

Table 1 Inclusion and exclusion criteria

Inclusion	Exclusion
Relevant use of simulation method	Non-relevant usage
Focus on industrial manufacturing	Other application field
Journal or conference article	Other source
Published in English	Non-english publication
Published from 2000 onward	Older publication
Non-duplicate	Duplicate
Accessible	Non-Accessible

simulation paradigms in different contexts making them relevant to this study. After completing the foundational literature selection process abstracts and full-texts were read applying the inclusion and exclusion criteria listed in Table 1. Thematic analysis was conducted systematically by reading each article in the final selection and identifying and coding recurring themes and insights. The process focused on identifying specific opportunities, challenges, and suggested future research directions. Similar or overlapping themes were consolidated where appropriate to ensure clarity and reduce redundancy. The results are presented in separate tables in the following section.

3 Opportunities, Challenges, and Future Research from Literature

3.1 Opportunities of HSM, VC, and DES

The main opportunities are listed in Table 2 and can be summarized in the following points. Simulation approaches, particularly in DES and VC, are *gaining widespread adoption*, reinforcing their importance in modern engineering and industrial prac-

Table 2 Opportunities

Theme	HSM	VC	DES
Growing interest in simulation	[19]	[9]	[11, 12]
Early-stage error recognition	[17, 18]	[7, 10]	
Accelerated commissioning and re-commissioning	[20]	[8, 9]	
Decision making in complex systems	[21, 22]		[14, 15]
Operator safety	[23]	[7, 8]	
Operational safety	[23]	[9]	

tices. Both HSM and VC approaches have significant potential for *recognizing errors in the early stages* of system development, highlighting their role in reducing rework and ensuring system reliability. VC methods are particularly effective at *speeding up commissioning and re-commissioning processes*, enabling quicker system deployment and adaptability. The DES approach is particularly strong in *aiding decision-making for complex systems*, making it invaluable for planning and optimization. Both HSM and VC prioritize safety, with a focus on *operator and operational safety*, ensuring risk minimization in system design and implementation.

3.2 Challenges in HMS, VC, and DES

The main challenges are listed in Table 3 and summarized in the following points. *Resource consumption and effort* in building and simulating models are prevalent challenges across all domains, suggesting this is a significant and critical barrier to effective implementation. Both HSM and DES face *challenges in acquiring and utilizing data* which is an important input to models. This highlights the need for better data management practices and integration capabilities. *Complex integration with IT infrastructure* affects HSM and DES, indicating the importance of robust infrastructure planning and alignment with digital systems. *Limited standardization* is a common challenge, particularly in HSM and VC, emphasizing the need for unified frameworks and protocols. *Specific optimization domains*, such as layout design and multi-objective problems, are particularly complex and present challenges across

Table 3 Challenges

Themes	HSM	VC	DES
Resource consumption and effort	[21]	[7, 8],	[12, 13]
	[22, 23]	[9, 10]	[14]
Challenges in acquiring and utilizing data	[17]		[13, 15, 16]
Complex integration with IT infrastructure	[18]		[13, 14]
Limited standardization	[17, 21]	[10]	
Complexities in optimization domains	[20]	[8]	[11–13]
Challenges in testing and validation of assumptions	[18]		[11, 13, 16]
Few comprehensive and transparent sucessful examples	[23]		[13, 14, 16]
Overlooking critical factors in simulations	[22]	[7]	[11, 15]
Prolonging model lifetime	[17]	[8, 9]	

all domains. This suggests that certain simulation areas demand more advanced methods or tools to address their inherent complexities effectively. *Challenges in testing and validating assumptions*, especially in HSM and DES, point to the need for more reliable validation tools and frameworks. *Few comprehensive and transparent examples* are a significant barrier, particularly in VC and HSM, underlining the need for better documentation and case studies. *Overlooking critical factors in simulations* is a major issue across all domains, suggesting a need for more comprehensive simulation practices and attention to detail. *Prolonging the life cycle of simulation models* is crucial for maximizing their value and making these technologies more appealing to both academia and industry.

3.3 Suggested Future Research in the Literature

Table 4 highlights the following key directions for future research found in the literature. *Standardization and interoperability* are suggested areas for future research, with efforts focused, among others, on unifying standards for simulation components, exchange formats, frameworks, methods, and communication protocols. *Incorporating additional dynamics into simulation models* aims to broaden the scope of simulation models, enabling the simulation of more complex and comprehensive systems by introducing new dimensions, such as the human factor. *Real-time simulation and synchronization* also have the potential to enable expanded uses of simulation, for example by enabling advanced bi-directional digital twin models that are often referenced in the literature. It is critical to *reduce the significant simulation demands* to minimize resources and effort consumed by modeling and simulation activities. *Extending the life cycle of simulation* models by maximizing their usability can reduce long-term costs while providing a clearer understanding of return on investment opportunities. One way to make simulation models live longer is to *increase their testing coverage* expanding the variety of experiments and tests they can be

Table 4 Future research suggestions

Themes	HSM	VC	DES
Standardization of simulation components	[17, 21]	[7–10]	
Incorporating additional dynamics	[17, 18]	[7, 22, 23]	[15]
Real-time simulation and synchronization	[17, 18, 21, 22]		[14]
Minimize resource consumption and effort	[17, 21]	[7, 9, 10]	[11, 13]
Prolonging model life cycle	[17, 20]		[11–13]
Hybrid simulation framework development	[17, 20]		
Multi-level digital twins	[17]		
Increasing testing coverage	[20]		

used for. They may then be used at various points during the life cycle of the production system. *Developing hybrid simulation frameworks* for VC and DES is crucial to systematically work with the combination of the simulation paradigms.

4 Novel Research Directions

Based on the literature review, three future research directions derived from the previous chapter are presented here.

4.1 Combining VC and DES in Test-Driven Production System Development

Both VC and DES offer opportunities to support system design in different ways at its early stages. By providing simulation-based evidence at different levels, errors and inefficiencies can be identified and addressed. The combination also has the potential to expand testing coverage by addressing areas that were previously overlooked or excluded using only one simulation technique. Future research could investigate how the combination of VC and DES can be integrated into a specific test-driven system development methodology. A test-driven approach is a development strategy that emphasizes creating and executing tests early and continuously during system development to guide design decisions and verify system functionality throughout the development life cycle. In such a method, DES would be used to simulate and validate overall system workflows and resource allocation, while VC would support the testing and refinement of low-level control logic. Research should focus on how this type of iterative testing and optimization, applied throughout the system life cycle, can help detect and correct design errors while accelerating system commissioning and reconfiguration. The intended outcome is a system design and operational methodology that continuously integrates simulation-based empirical insights to support decision-making, rather than relying solely on human judgment. This shift toward data-driven design is critical, as exclusive dependence on human intuition in complex and dynamic manufacturing environments can result in overlooked errors, inefficiencies, and suboptimal outcomes. *Project proposal*: Adaptation of test-driven development methods in software development to the design of software-intensive production systems. The method initially involves developing test methodologies against and with the help of simulation models, followed by iterative testing on cyber-physical production systems to ensure functional and robust development of system behavior.

4.2 Co-Development of VC and DES Models

The literature emphasizes that all forms of model building and simulation require significant resources and efforts. One potential way to reduce this burden is through the co-development of VC and DES models. By establishing shared data structures, modular components, and efficient data exchange mechanisms, it becomes possible to coordinate data collection and feasibility studies in a way that informs the creation and emergence of both simulation models. This integrated development process can enable a lower total development cost where subsequently the use of both models can contribute to synergies that potentially generate value that exceeds the cost of creating and maintaining the models. Ultimately, a simulation framework for the combination can be developed that further reduces development costs while clarifying how the benefits of the combination can be realized. *Project proposal*: Investigating the potential to reduce total development costs by establishing a common simulation framework for the co-development of VC and DES models, leveraging joint prestudies, shared data structures and standardized exchange mechanisms to maximize the value generated through their coordinated use.

4.3 Combining VC and DES to Extend Model Life Cycle

As previously noted, developing simulation models requires considerable resources and effort, which makes it desirable to extend their usable lifespan once the initial creation is done. This highlights the importance of exploring how VC and DES models can be designed to complement each other while also being adaptable to repeated use across multiple stages of the production system's life cycle. The literature suggests several opportunities in this regard. For instance, model components can be standardized and structured to meet the requirements of both VC and DES. By integrating extended dynamics, enabled through the combination of both approaches, digital testing capabilities can be expanded, allowing for more detailed predictions and scenario evaluations. Moreover, since VC models are typically applied in the early stages of a system's life cycle and DES models in later stages, there may be untapped synergies where each model type can serve as a supporting or auxiliary model in the phase where the other is more dominant. *Project Proposal*: Investigate how VC and DES models can be co-developed for reuse across multiple stages of the production system life cycle, extending their value beyond initial deployment. The project focuses on identifying new opportunities to leverage the complementary strengths of VC and DES to enable broader testing scope, more detailed scenario evaluation, and increased life cycle coverage.

5 Limitations

This literature-based study has several limitations that may affect the validity of its findings. First, there may be alternative terminology for VC or DES that was not captured by the selected search queries. As a result, some relevant studies may have been overlooked, potentially limiting the completeness and resolution of the research space presented in this review. Second, cross-disciplinary research involving the combination of VC and DES was intentionally excluded to maintain a clear focus on manufacturing-specific applications and avoid obscuring domain relevance. While this may have resulted in potentially useful insights from adjacent fields being omitted, the exploratory initial searches conducted suggest that such cross-domain studies are very limited. Third, the final literature set was retrieved from a single database to ensure methodological consistency and transparency, supported by preliminary searches in other databases that yielded few additional relevant results. Fourth, the sets of articles included in the thematic analysis varied in type. Two sets consisted of literature reviews focused on individual simulation methods, while the other two consisted of other types of studies on the mixing of simulation paradigms. This variation introduced differences in the type and depth of reported insights: literature reviews tend to offer broader summaries, while primary studies provide more detailed, implementation-specific accounts. As a result, it may be difficult to make direct comparisons across all four sets, which may have affected the consistency of theme extraction and interpretation. Finally, the majority of the thematic analysis was conducted by a single researcher. This increases the risk of subjectivity in how themes are identified, categorized, and interpreted. Although care was taken to apply consistent criteria and critically reflect on the process, the lack of multiple opinions limits the robustness and reproducibility of the results.

6 Conclusion

As digitalization accelerates in manufacturing, it is important to understand and develop the tools and methods that support this transformation. Among them, VC and DES are established industrial methods. However, they are typically applied independently and at different stages of the system life cycle. This study investigates their combined use by reviewing literature on VC, DES, and HSM, framed around a research question concerning current opportunities, challenges, and suggested directions for future work. While some overlapping areas were identified, the analysis revealed that VC and DES largely address different aspects of production systems, highlighting a gap, and potential, for further research into how they might complement one another if combined. However, the idea that their combination could produce a synergetic effect, mitigating the limitations of each individual method, is compelling but currently lacks empirical validation and remains speculative. A second research question was stated in regards to feasible research directions to further investigate the combining of the two simulation methods. To initiate research to explore whether any synergy is achievable, the study contributes three novel research directions based on the literature review results and which specifically focus on the fusion of VC and DES. Future research should explore how the combination of VC and DES can support the development of test-driven methodologies in production system design, where DES validates high-level workflows and VC refines control logic through continuous simulation-based testing. Another promising direction is the joint development of VC and DES models using shared data structures, modular components, and standardized data exchange mechanisms to reduce overall development costs and enable coordinated use. In addition, research can be conducted to investigate whether merging these methodologies enables the extension of the lifetime of simulation models across the entire production system life cycle by improving reusability, expanding test coverage, and exploiting their complementary roles at different system stages.

Acknowledgements This work was conducted within the framework of the REFUSE project, which provided the funding and research context for the study.

References

1. Böse F, Windt K (2007) Catalogue of criteria for autonomous control in logistics. In: Understanding autonomous cooperation and control in logistics: the impact of autonomy on management, information, communication and material flow. Springer, Berlin, Heidelberg, pp 57–72
2. Mourtzis D, Doukas M, Bernidaki D (2014) Simulation in manufacturing: review and challenges. Procedia Cirp 25:213–229
3. Morgan JS, Howick S, Belton V (2017) A toolkit of designs for mixing discrete event simulation and system dynamics. Eur J Oper Res **257**(3):907–918
4. Mustafee N, Powell JH (2018) From hybrid simulation to hybrid systems modelling. In: 2018 winter simulation conference (WSC). IEEE, pp 1430–1439
5. Cooper HM (1988) Organizing knowledge syntheses: a taxonomy of literature reviews. Knowl Soc 1(1):104 Mar
6. Page MJ, McKenzie JE, Bossuyt PM, Boutron I, Hoffmann TC, Mulrow CD, Shamseer L, Tetzlaff JM, Akl EA, Brennan SE, Chou R (2021) The PRISMA 2020 statement: an updated guideline for reporting systematic reviews. bmj. 2021 Mar 29;372
7. Lechler T, Fischer E, Metzner M, Mayr A, Franke J (2019) Virtual Commissioning-Scientific review and exploratory use cases in advanced production systems. Procedia CIRP 81:1125–1130
8. Lidell A, Ericson S, Ng AHC (2022) The current and future challenges for virtual commissioning and digital twins of production lines. In: SPS2022. IOS Press, pp 508–519
9. Striffler N, Voigt T (2023) Concepts and trends of virtual commissioning–A comprehensive review. J Manuf Syst 71:664–680
10. Hoffmann P, Schumann R, Maksoud TMA, Premier GC (2010) Virtual commissioning of manufacturing systems a review and new approaches for simplification. ECMS 1(2010):175–181
11. Qiao D, Wang Y (2021) A review of the application of discrete event simulation in manufacturing. J Phys: Conf Ser 1802(2):022066. IOP Publishing
12. Prajapat N, Tiwari A (2017) A review of assembly optimisation applications using discrete event simulation. Int J Comput Integr Manuf 30(2–3):215–228

13. Greasley A, Edwards JS (2021) Enhancing discrete-event simulation with big data analytics: a review. J Oper Res Soc 72(2):247–267
14. Agalianos K, Ponis ST, Aretoulaki E, Plakas G, Efthymiou O (2020) Discrete event simulation and digital twins: review and challenges for logistics. Procedia Manuf 51:1636–1641
15. Kranz M, Nitsch V, Mütze-Niewöhner S (2023) Linking discrete-event simulation with artificial intelligence: a literature-based analysis of existing approaches in the context of manufacturing planning and control. In: 2023 IEEE international conference on industrial engineering and engineering management (IEEM). IEEE, pp 1194–1198
16. Aretoulaki E, Ponis ST, Plakas G, Tzanetou D (2024) (DT4Smart) discrete event simulation and Digital Twins in warehouse logistics: a bibliometric and content analysis-based systematic literature review. Int J Comput Integ Manuf (2024):1–28
17. Kattenstroth F, Disselkamp J-P, Lick J, Dumitrescu R (2024) Challenges in the implementation of simulation models for the digital factory twin-a systematic literature review. Procedia CIRP 128:442–447
18. Mykoniatis K, Harris GA (2021) A digital twin emulator of a modular production system using a data-driven hybrid modeling and simulation approach. J Intell Manuf 32(7):1899–1911
19. Bangsow S, Günther U (2012) Creating a model for virtual commissioning of a line head control using discrete event simulation. In: Use cases of discrete event simulation: appliance and research. Springer Berlin, Heidelberg, pp 117–130
20. Schmidt PP, Fay A (2015) Transformation of continuous simulation models of automated manufacturing systems into discrete event models on different levels of detail. In: 2015 IEEE 20th conference on emerging technologies and factory automation (ETFA). IEEE, pp 1–5
21. Barbosa C, Azevedo A (2017) Hybrid simulation for complex manufacturing value-chain environments. Procedia Manuf 11:1404–1412
22. Roemer AC, Strassburger S (2019) Hybrid system modeling approach for the depiction of the energy consumption in production simulations. In: 2019 winter simulation conference (WSC). IEEE, pp 1366–1377
23. Oliveira MLM, Montevechi JAB, Pinho AF, Miranda RC (2017) Using hybrid simulation to represent the human factor in production systems. Int J Simul Model 16(2):263–274

Implementing AI-Driven Augmented Reality: Insights from an Industrial Use Case Implementation

Katharina Schmenn, Lukas Baeck, Peter Burggräf, Carl René Sauer, Alexander Becher, and Maximilian Lutz

Abstract Augmented reality (AR) and artificial intelligence (AI) are increasingly applied to improve accuracy and efficiency in industrial workflows. Their combination offers potential for supporting complex manual tasks such as visual quality inspection. Although AR has shown promise in applications like assembly guidance and training, most implementations have been evaluated in controlled environments. Empirical evidence on AI-assisted AR under real production conditions remains limited. This case study presents the implementation of an AI-supported AR system for weld seam inspection in a manufacturing environment. The system integrates CAD-based AR visualization with AI-driven image classification to replace a manual, paper-based inspection process. The approach improved consistency and traceability of inspections, particularly for less experienced operators. Challenges included lighting sensitivity, hardware limitations, and the need for structured image data management to ensure classification reliability. Acceptance among experienced staff was cautious, highlighting the importance of transparent decision logic and user involvement. The results indicate that AI-assisted AR can enhance industrial inspection processes when technical feasibility is paired with human-centered system integration. The approach is transferable to other tasks relying on CAD models and visual criteria and provides practical insights into deploying digital assistance systems in operational contexts.

Keywords Cyber production management · Artificial intelligence · Smart manufacturing · Vuforia · Case study

K. Schmenn (✉) · L. Baeck · P. Burggräf · C. R. Sauer · A. Becher · M. Lutz
International Production Engineering and Management (IPEM), Universität Siegen, Kreuztal, Germany
e-mail: katharina.schmenn@uni-siegen.de

M. Manns (ed.), *New Paradigms for Anticipated Uncertainty*, Lecture Notes in Mechanical Engineering, https://doi.org/10.1007/978-3-032-16889-4_68

1 Introduction

Increasing competitive pressure, cost reduction demands, and the need for higher efficiency are driving the digitalization of industrial processes and the use of advanced technologies to optimize workflows and minimize errors [1]. Among these, augmented reality (AR) has emerged as a promising tool in manufacturing, maintenance, and quality assurance [2]. In combination with artificial intelligence (AI), AR can support decision-making, automate tasks, and improve operational efficiency [3].

AR adoption has advanced in assembly assistance, defect detection, and workforce training [4]. Studies have demonstrated that AR-based systems can lead to significant efficiency gains, such as reduced process times [5], lower error rates in quality assurance [6], and improved worker performance through real-time guidance and contextualized digital information mechanisms [7]. These benefits make AR an attractive solution in modern industrial workflows.

However, many of these applications remain limited to laboratory settings or conceptual prototypes [8]. There is a lack of empirical evidence regarding the practical integration of AR systems into production environments, especially when combined with AI-based assistance [9]. This paper addresses this gap by presenting a practical case study on the implementation of an AI-driven AR system for weld seam inspection in a real industrial context. The system combines CAD-based AR visualization with AI-driven image classification using the Vuforia platform [10]. It evaluates technical, procedural, and human factors to identify success conditions, practical challenges, and the potential for transfer to other use cases.

2 Related Work

AR has been increasingly applied across manufacturing domains to enhance process monitoring, worker assistance, and quality control. Various studies have demonstrated the feasibility of AR for industrial inspection tasks. For example, Baron and Braune [11] presented an early AR concept for process supervision using industrial UI standards, while Alves et al. [12] developed an AR-supported step-by-step verification tool to assist operators during manual quality checks. Seeliger et al. [13] and Szajna et al. [14] empirically showed that AR systems, particularly head-mounted displays, can significantly improve accuracy and reduce cognitive load during complex inspections in production environments.

Specific to weld seam quality assurance, Havlíková et al. [15] proposed an AR-based inspection tool using Vuforia to project interactive instructions and 3D models onto real welded components. Their user study highlighted efficiency gains and reduced mental workload, especially for less experienced inspectors. While such studies provide promising results, they are often conducted in pilot environments or early-stage setups. A systematic review by Ho et al. [16] confirms that the majority of

AR applications in quality control still reside in experimental or laboratory settings, with limited exploration of real industrial integration.

Beyond inspection, AR has also been widely explored for training purposes, as shown by Papakostas et al. [17], who investigated the acceptance of AR-based welding simulators in engineering education. Etonam et al. [18] further emphasize the potential of AR in maintenance and shop-floor applications, but also point out technical and organizational challenges such as high initial costs and limited interoperability.

In contrast to most prior work, this paper focuses on the real-world deployment of an AI-driven AR system for weld seam inspection in an actual industrial environment. By identifying technical and organizational barriers and analyzing factors that influence accuracy, usability, and effectiveness, our contribution addresses the current research gap between AR prototypes and practical applications in full-scale production settings.

3 Use Case: Quality Assurance for Welding Components

The successful implementation of AI-driven AR systems in industrial applications requires development platforms that enable accurate spatial tracking, flexible workflow design, and integration of digital content into physical environments [19]. In quality assurance tasks such as weld seam inspection, these systems help verify compliance with defined standards and detect defects such as cracks or porosity that could compromise weld integrity. Traditionally, such inspections rely on visual assessments and paper-based instructions [20].

While traditional inspection methods are well-established, they are time-consuming, rely heavily on skilled personnel, and are prone to human error. Operator fatigue and differences in expertise can lead to inconsistencies, increasing the risk of undetected defects [21]. In this context, AI-driven AR systems offer an opportunity to increase consistency, reduce inspection time, and support both experienced and novice users [22].

This paper presents the implementation of such a system in a real industrial setting, focusing on the inspection of a cantilever arm within a semi-automated robotic welding process. The component consists of two end plates, a footplate, and an I-beam, assembled manually prior to welding. Although welding is automated, the quality inspection of eleven weld seams is still performed manually using paper-based documentation. The goal of this paper is to replace this process with a paperless, more consistent, and efficient solution using AI-driven AR.

3.1 Software Specifications

Vuforia was selected as the AR development platform based on its support for CAD-based model tracking and its compatibility with mobile devices commonly used in industrial environments. These features are essential for the accurate spatial localization of weld seams. In contrast to platforms like ARKit (Apple) and ARCore (Google), which are optimized for consumer applications, Vuforia allows the direct use of CAD data and supports structured inspection workflows. Compared to Unity, it also enables faster implementation with fewer programming requirements for the use case presented.

Vuforia offers two primary implementation methods: CAD model integration for precise visualization and localization, and room scanning for flexible digital overlays. For the weld seam inspection task, the CAD-based method was chosen to ensure high alignment accuracy between digital content and the physical component.

The process begins with importing the CAD model, defining inspection steps, and placing visual markers at each weld seam. The evaluation criteria were developed in collaboration with quality assurance staff and include both ideal weld characteristics and typical defect types, such as binding errors, pores, cracks, burn-through, and slag inclusion. Corresponding digital reference images are embedded in the AR workflow to support visual comparison. Operators follow the guided procedure on their device. The system recognizes the cantilever arm and overlays the relevant markers at the weld seam positions. The operator inspects the weld, captures an image, classifies the result, and may add a comment.

The inspection process is supported by the Step Check feature, which introduces AI-based visual inspection. The module uses images captured during inspections as training data to build a classification model. While the specific architecture and training process are proprietary, it can be assumed that supervised learning methods from the field of image classification are employed. Step Check evaluates new images in real time, providing classification feedback such as "Success" (flawless) or "Error" (defective) based on the defined criteria.

The effectiveness of the model depends on the quality and consistency of the training data provided by the users. To improve classification accuracy, images from previous inspections can be reused in other procedures based on the same CAD model. However, using identical images across tasks with different evaluation goals may lead to misclassifications. This risk can be reduced through careful dataset management and consistent labeling.

Vuforia's configuration interface allows for efficient integration of reference models and inspection logic. While basic functionalities can be used with limited prior knowledge, configuring advanced features such as Step Check requires a deeper understanding of the Vuforia toolset and the specific inspection process. This includes specifying relevant image regions, defining classification criteria, and ensuring consistent capture conditions. Proper dataset structuring also enhances traceability, supports the identification of recurring defects, and contributes to continuous process improvement.

3.2 Hardware Specifications

Alongside software configuration, hardware selection is critical for AR-based quality assurance. Devices must offer reliable spatial tracking, sufficient image quality, and compatibility with the AR platform. This paper evaluated two commercially available options: the RealWear Navigator 520 and the Apple iPad Pro (12.9″, 6th generation, 128 GB) with integrated LiDAR.

The RealWear Navigator 520 supports hands-free operation via voice control, which can be useful in industrial settings. However, it does not support 3D model rendering required for CAD-based tracking in Vuforia. Its fixed-focus 13 MP camera also delivered limited image quality, especially under varying lighting conditions. The absence of depth sensing further restricted its applicability for this use case.

The iPad Pro, by contrast, includes LiDAR and a high-resolution camera, enabling accurate spatial mapping and consistent image capture. Its handheld form factor, however, limits ergonomic comfort during prolonged inspections. Despite this, the device fulfilled all technical requirements for the Vuforia-based inspection workflow.

While Vuforia supports a broader range of AR-capable devices, including Android tablets, the RealWear Navigator 500, Microsoft HoloLens 2, and Trimble XR10, these were not included in this evaluation. The selection was based on practical factors: the iPad Pro was already available, and the RealWear device could be tested temporarily through an external loan. Acquiring and integrating further devices exceeded the scope and resources of this project. The evaluation is therefore limited to hardware that was accessible during implementation.

3.3 Implementation in Production Environment

Following the selection and implementation of appropriate hardware and software, the AR-based quality assurance system was integrated into a live production environment. Several test runs were conducted under real operating conditions to evaluate functionality, usability, and inspection accuracy. These trials provided valuable feedback that informed system refinements.

The system guided operators through each weld seam inspection step using CAD-based visual markers and reference images. Real-time AI-based classification provided additional support, particularly in borderline cases. Although the AI component helped reduce subjective variation, final decisions remained with the human operator. Contextual information and standardized evaluation criteria were used throughout the process to improve decision consistency.

A total of 48 inspection cycles were carried out across six operators with varying experience levels. The average inspection time per cantilever arm did decrease from using the paper-based method (4.5 min) to using the AR-based system (4 min). But the Operator agreement on weld quality classification increased from 75 to 93% when using the system's guided workflow and Step Check feedback. All inspection results were automatically logged within the system, supporting documentation requirements and traceability. While the system was positively received by both novice and experienced users, further testing is required to assess its long-term effects on process stability and training efficiency.

4 Lessons Learned and Potential for Further Research

Following the implementation of the AR-based quality assurance system, several key insights emerged regarding its strengths and limitations. A notable advantage was its ability to support consistent and standardized inspections. By guiding attention to each step, the system promoted more uniform evaluations, which is particularly beneficial under time pressure or during extended shifts, when fatigue may reduce concentration. The system also facilitated the onboarding of less experienced operators by reducing the need for constant supervision and making training less dependent on language skills or prior knowledge. However, the effectiveness of AI-assisted assessments proved sensitive to image quality, especially under inconsistent lighting conditions. Additional lighting or more controlled environments are recommended to improve reliability.

In terms of hardware, neither of the two tested devices proved ideal. The RealWear Navigator enabled hands-free use but lacked camera quality and spatial tracking ability. The iPad Pro offered better image capture and tracking but was less ergonomic for prolonged use. Hardware selection should therefore be based on specific task and environmental requirements. Structured decision-making tools such as morphological boxes can assist in evaluating trade-offs between factors like tracking precision, field of view, usability, and cost.

Integrating AI through the Step Check module required more technical effort than for the basic AR workflow, but the effort remained manageable without extensive programming knowledge. Reliable classification depended on high-quality training datasets and clearly defined inspection procedures. Misclassifications occurred when image data were reused across different tasks without proper labeling. Ensuring consistent lighting and accurate annotation improved overall system performance. In addition, effective digital asset management was identified as a critical factor. Assets such as 3D models, reference images, and training data must be versioned, structured, and backed up. Since deleted assets in Vuforia cannot be restored, clear asset management practices are essential to support long-term usability and future adaptation.

Beyond the immediate use case, the systems structure and methods are transferable to other visual inspection tasks. These include surface defect identification, assembly verification, or part orientation checks. Given the availability of CAD data and clear evaluation criteria, the approach could be applied across domains such as automotive, aerospace, or electronics manufacturing.

Operator feedback highlighted differing perceptions of the system. While AI-based suggestions were seen as helpful in ambiguous cases, they were not always fully trusted. Experienced users appreciated the structured documentation but sometimes viewed the system as unnecessary for routine tasks. To support long-term acceptance, it is important to involve experienced operators early, maintain manual decision authority, and clearly communicate the supportive, non-substitutive role of the system.

One specific benefit for experienced personnel lies in the value of systematic documentation. By collecting labeled images and inspection histories, the system helps capture and preserve expert knowledge. This supports the development of high-quality training materials and enables experienced users to contribute their expertise without additional workload. Rather than replacing their skills, the system offers a platform to formalize and transfer them.

Long-term integration of AI-driven AR systems requires transparency, trust, and regular user engagement. Operators should be able to interpret AI-based suggestions and override them when needed. Incorporating structured feedback and updating training datasets as procedures evolve will help maintain the system's effectiveness and relevance in practice.

5 Conclusion

This paper demonstrated the practical implementation of an AI-driven AR system for weld seam inspection in a real industrial setting. The approach enabled more consistent inspections by combining step-by-step visual guidance with AI-based image classification. The system reduced reliance on paper-based documentation and supported less experienced operators through structured workflows and digital reference materials.

Despite these benefits, several challenges emerged during implementation. Effective use of AI-supported inspection required well-prepared image datasets and clearly defined procedures. Misclassifications occurred when images were reused across inspection tasks without consistent labeling or contextual separation. Structured data management and thoughtful dataset design proved essential for maintaining classification accuracy. Hardware-related factors also influenced system performance, particularly image quality and spatial tracking under varying lighting conditions. Although a comprehensive hardware comparison was beyond the scope of this case study due to limited device availability, preliminary evaluations indicated that camera resolution, depth sensing, and ergonomics significantly affect practical usability.

While this paper focused on a single use case, the underlying approach is transferable to other inspection tasks where CAD data and visual evaluation criteria are available.

In summary, AI-driven AR show potential to improve inspection quality, consistency, and documentation in industrial settings. Their successful application depends not only on technical feasibility but also on human-centered integration and continuous refinement.

References

1. Topic M, Biedermann H (2020) Increasing resource efficiency through digitalization – chances and challenges for manufacturing industries. In: Anisic Z, Lalic B, Gracanin D (eds) Proceedings on 25th international joint conference on industrial engineering and operations management—IJCIEOM. The next generation of production and service systems. Lecture notes on multidisciplinary industrial engineering, 1st edn. Springer, Cham, pp 1–11 (Imprint)
2. Eswaran M, Bahubalendruni MVAR (2022) Challenges and opportunities on AR/VR technologies for manufacturing systems in the context of industry 4.0: a state of the art review. J Manuf Syst. https://doi.org/10.1016/j.jmsy.2022.09.016
3. Kästner L, Eversberg L, Mursa M, Lambrecht J (2020) Integrative object and pose to task detection for an augmented-reality-based human assistance system using neural networks. http://arxiv.org/pdf/2008.13419
4. Moya A, Bastida L, Aguirrezabal P, Pantano M, Abril-Jiménez P (2023) Augmented reality for supporting workers in human-robot collaboration. MTI. https://doi.org/10.3390/mti7040040
5. Ababsa F (2020) Augmented reality application in manufacturing industry: maintenance and non-destructive testing (NDT) use cases. In: Paolis LT, de Bourdot P (eds) Augmented reality, virtual reality, and computer graphics. 7th International conference, AVR 2020, Lecce, Italy, September 7–10, 2020, Proceedings, Part II, vol 12243. Springer eBook collection, vol 12243, 1st edn. Springer, Cham, pp. 333–344 (Imprint)
6. Eswaran M, Raju Bahubalendruni M (2023) Augmented reality aided object mapping for worker assistance/training in an industrial assembly context: Exploration of affordance with existing guidance techniques. Comput Ind Eng. https://doi.org/10.1016/j.cie.2023.109663
7. Sahu CK, Young C, Rai R (2021) Artificial intelligence (AI) in augmented reality (AR)-assisted manufacturing applications: a review. Int J Prod Res. https://doi.org/10.1080/00207543.2020.1859636
8. Husár J, Knapčíková L (2024) Implementation of augmented reality in smart engineering manufacturing: literature review. Mobile Netw Appl. https://doi.org/10.1007/s11036-023-02121-x
9. Marino E, Barbieri L, Bruno F, Muzzupappa M (2024) Assessing user performance in augmented reality assembly guidance for industry 4.0 operators. Comput Ind. https://doi.org/10.1016/j.compind.2024.104085
10. Vuforia Enterprise Augmented Reality (AR) Software (2025) PTC. https://www.ptc.com/en/products/vuforia. Accessed 3 Feb 2025
11. Baron L, Braune A (2016) Case study on applying augmented reality for process supervision in industrial use cases. In: 2016 IEEE 21st international conference on emerging technologies and factory automation (ETFA). 2016 IEEE 21st international conference on emerging technologies and factory automation (ETFA), Berlin, Germany, 9/6/2016–9/9/2016. IEEE, pp 1–4. https://doi.org/10.1109/ETFA.2016.7733639
12. Alves J, Marques B, Dias P, Santos BS (2021) Using augmented reality and step by step verification in industrial quality control. In: Karwowski W, Ahram T, Etinger D, Tanković N, Taiar R (eds) Human systems engineering and design III, vol 1269. Advances in intelligent systems and computing, vol 1269. Springer, Cham, pp 350–355
13. Seeliger A, Cheng L, Netland T (2023) Augmented reality for industrial quality inspection: an experiment assessing task performance and human factors. Comput Ind. https://doi.org/10.1016/j.compind.2023.103985
14. Szajna A, Stryjski R, Woźniak W, Chamier-Gliszczyński N, Królikowski T (2020) The production quality control process, enhanced with augmented reality glasses and the new generation computing support system. Procedia Comput Sci. https://doi.org/10.1016/j.procs.2020.09.024
15. Havlíková K, Hořejší P, Kopeček P (2023) Effect of augmented reality support on quality inspection of welded structures. Appl Sci. https://doi.org/10.3390/app132111655
16. Ho PT, Albajez JA, Santolaria J, Yagüe-Fabra JA (2022) Study of augmented reality based manufacturing for further integration of quality control 4.0: a systematic literature review. Appl Sci. https://doi.org/10.3390/app12041961

17. Papakostas C, Troussas C, Krouska A, Sgouropoulou C (2022) User acceptance of augmented reality welding simulator in engineering training. Educ Inf Technol. https://doi.org/10.1007/s10639-020-10418-7
18. Etonam AK, Di Gravio G, Kuloba PW, Njiri JG (2019) Augmented reality (AR) application in manufacturing encompassing quality control and maintenance. IJEAT. https://doi.org/10.35940/ijeat.A1120.109119
19. Blanco-Novoa O, Fernandez-Carames TM, Fraga-Lamas P, Vilar-Montesinos MA (2018) A practical evaluation of commercial industrial augmented reality systems in an industry 4.0 shipyard. IEEE Access. https://doi.org/10.1109/ACCESS.2018.2802699
20. Zou YY, Zhao MY, Liu LZ (2010) Post-weld quality inspection for CO_2 laser tailored blank welding with vision sensor. AMR. https://doi.org/10.4028/www.scientific.net/AMR.97-101.4337
21. Li Y, Hu M, Wang T (2020) Visual inspection of weld surface quality. IFS. https://doi.org/10.3233/JIFS-179993
22. Zhang B, Wang X, Cui J, Wu J, Xiong Z, Zhang W, Yu X (2024) Enhancing weld inspection through comparative analysis of traditional algorithms and deep learning approaches. J Nondestruct Eval. https://doi.org/10.1007/s10921-024-01047-y

Ergonomics-Oriented Parameter Analysis Based on Digital Human Models with a Focus on Manual Work

Aydin Ünlü, Pascal Petri, and Karsten Kluth

Abstract A Digital Human Model (DHM) enables computer-aided analysis of production processes and supports the ergonomic design of workflows. In particular, DHMs are often used to analyze manual work and evaluate ergonomic parameters. Previous studies show that the hand models for ergonomic evaluation are still limited due to inaccurate movement patterns. This study focuses on the simulation of manual assembly work using an anthropometric model. It illustrates with motion measurements how large the angular deviation of the hand postures is. To investigate ergonomic parameters, a cordless screwdriver was used to assemble screws on differently angled mounting plates, which caused different hand-arm postures. The motion capture was done with IMU sensors. Fifteen male subjects aged between 20 and 30 took part in these motion measurements. The results show the effect of the inclined position of the mounting plate on the wrist angles of the respective test subjects. Finally, the deviations between simulation and measurement are discussed and the limitations are shown.

Keywords Digital ergonomics · Wrist angles · Parameter study · Anthropometric human model · Motion capture

1 Introduction

Digital human models (DHM) are computer-based representations of the human body that are used to analyze and optimize workplaces and to evaluate products in terms of their ergonomics. They enable detailed simulation and evaluation of motion sequences, force effects and stresses in virtual environments. In the field of ergonomics in particular, DHMs offer a valuable opportunity for the early analysis and design of workflows because they enable the rapid evaluation of different scenarios [1].

A. Ünlü (✉) · P. Petri · K. Kluth
University of Siegen, Siegen, Germany
e-mail: aydin.uenlue@uni-siegen.de

M. Manns (ed.), *New Paradigms for Anticipated Uncertainty*, Lecture Notes in Mechanical Engineering, https://doi.org/10.1007/978-3-032-16889-4_69

A key objective of scientific work on DHMs is to develop realistic motion simulations in order to conduct more precise ergonomic assessments. Simulation models are often validated by experimental measurements in order to minimize deviations between real and simulated movements. For example, studies show that motion data is used to improve DHMs by serving as a basis for the realistic reproduction of complex motion sequences, such as getting in and out of vehicles [2].

Ergonomic analyses increasingly require detailed hand models, since hand movements and hand postures play a central role in manual work. Nevertheless, there are only a few studies that explicitly deal with the simulation of wrist angles. An approach was presented by Romero et al. [3], who developed the MANO model. This model is based on high-resolution 3D scans of hands and enables realistic simulation of hand movements. Further research shows that digital hand models are increasingly being used in ergonomic studies to evaluate gripping comfort and hand strain. For example, Ma et al. [4] developed a methodology for integrating digital human models into ergonomic evaluation procedures to analyze physical fatigue in certain work tasks. Sharma et al. [5] showed that proactively integrating DHMs into the design process can reduce strain. Gust and Ünlü [6] and Ünlü et al. [7] have already shown that ergonomic product design can be carried out with the help of digital hand models, particularly with regard to hand posture or the design of work equipment such as the shape of handles.

To validate and improve DHM simulations, motion capture systems are increasingly employed to provide real-world reference data. In this context, inertial measurement units (IMUs) are particularly suitable for capturing dynamic wrist movements in manual tasks due to their lightweight, wireless configuration and ability to measure joint angles in real-time without restricting natural motion. However, despite advances in motion capture and simulation, the integration of digital hand models into ergonomic product design for parameter-based optimization such as reducing wrist strain remains limited. Moreover, reliably simulating realistic hand movements is still challenging due to the complexity of musculoskeletal dynamics and variability in human behavior [8]. Therefore, the aim of this study is to compare simulated wrist angles with motion data captured during manual assembly tasks using IMUs and to carry out a parameter analysis to evaluate ergonomic implications.

2 Method

To examine the wrist angle simulation in comparison to real wrist angle measurement, an experimental test stand was developed according to Fig. 1. The test stand consists of a stable worktop with evenly distributed internally threaded sleeves. Hexagon socket screws were screwed into these sleeves using a cordless screwdriver. The torque limiter of the screwdriver was set to a medium level to avoid an abrupt stop and thus a jerky torque at the wrist. The screw positions were set in a horizontal and a vertical arrangement. The distance between the screw positions was 100 mm both horizontally and vertically, forming a regular grid. The cordless screwdriver

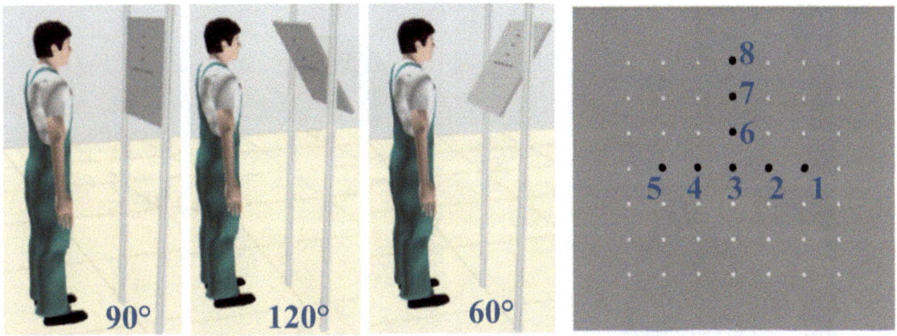

Fig. 1 Experimental setup in the simulation environment using emaWD [9]. Depiction of the three configurations of the screw plate at shoulder height and screw arrangement on the plate from 1 to 8

used had a total length of 200 mm, a handle diameter of 40 mm, and a trigger offset of 30 mm from the handle axis. The numbering of the positions indicates the order of the screwing process. The hand posture remained constant throughout the entire test, with the handle of the cordless screwdriver held horizontally. This arrangement allows for a systematic variation of hand postures while keeping the body posture and distance to the test stand constant. Each screw was tightened to a screw-in depth of 1 cm per cycle, with a uniform tightening time of 3 s being specified. A clock standardized the screw change to 5 s.

A total of 15 male subjects aged between 25 and 38 years took part in the study. Before the measurements began, individual body measurements were taken. Inertial measurement units (IMU sensors) from the Xsens motion capture system were attached to the entire body of the subjects to collect the motion data. The height of the worktop was individually adjusted: the minimum height corresponded to the elbow height of the respective test subject, while the maximum height was set to shoulder height. The screw plate was brought into the three constellations 90°, 120° and 60° for both heights and tilted around the horizontal axis of symmetry. The test person monitored the movements and documented them on video. To minimize fatigue effects, five-minute breaks were taken between the runs. The study focused in particular on the wrist angles in palmar/dorsal and radial/ulnar directions. In addition, arm elevation and forearm rotation (pronation/supination) were recorded and analyzed as further measurement and simulation parameters.

The digital human model emaWD (ema Work Designer, [9]) was used to simulate the hand movements. The digital strain analysis was carried out with an anthropometric human model that simulated the work process under identical conditions. A digital model of the cordless screwdriver was created, with lengths and handle angles corresponding exactly to the real dimensions. The simulation was adapted so that the body height and position of the digital human model corresponded to those of the test subjects, in order to ensure that the hand and arm postures were reproduced as realistically as possible. In line with the real measurement, both the screw-in depth and

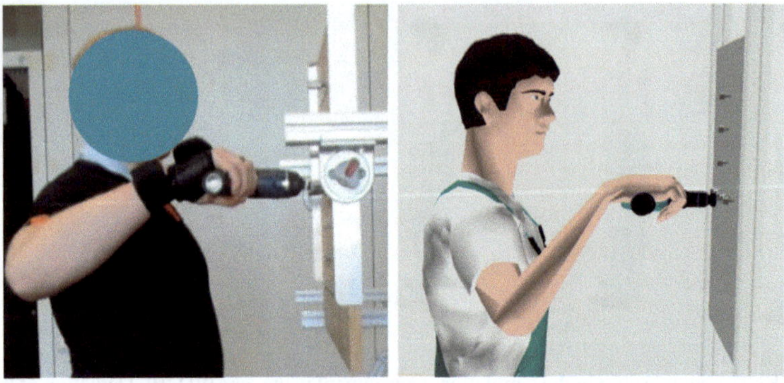

Fig. 2 Test and simulation procedure of the plate at 90° (shoulder height)

the screw-in time per screw were taken into account in the simulation. This ensured comparability between the experimental measurement and the digital simulation.

A descriptive statistical analysis was performed, including mean values and standard deviations of measured and simulated wrist angles. The root mean square deviation (RMSD) was calculated to quantify the average discrepancy per screw position.

Figure 2 shows an example of a simulation and test procedure for a tilt angle of 90° (plate at shoulder height). It can be seen that the simulation model does not fully correspond to the measurement data with regard to the elbow position. The wrist angles (ulnar/radial abduction and palmar/dorsal flexion) show a dependency on the arm angles, which is reflected in the measurements of the different constellations.

3 Results

The analysis in Fig. 3 illustrates the variance between the simulation and the measured values. Within the measurement, the scattering can be attributed to the fact that the adapted human models show more uniform movement behavior, while the test subjects show individual variations in their movement patterns. The increased variability in the measured wrist angles compared to the simulation can be attributed to subject-specific movement strategies and compensatory actions not captured in the digital model, especially in extreme positions.

For further parameter analysis, the body sizes of all test subjects were averaged and a standardized human model was defined. Such a standardized human model should cover all test subjects. The plate was also brought to the corresponding height on the model, with slight deviations arising between the models. It can be assumed that this has no effect on the analysis and simplifies further analysis. The simulation model was then adapted to the average hand-arm angle measurements so that the elbow movement corresponded to the realistic movements. This reduced the deviations of

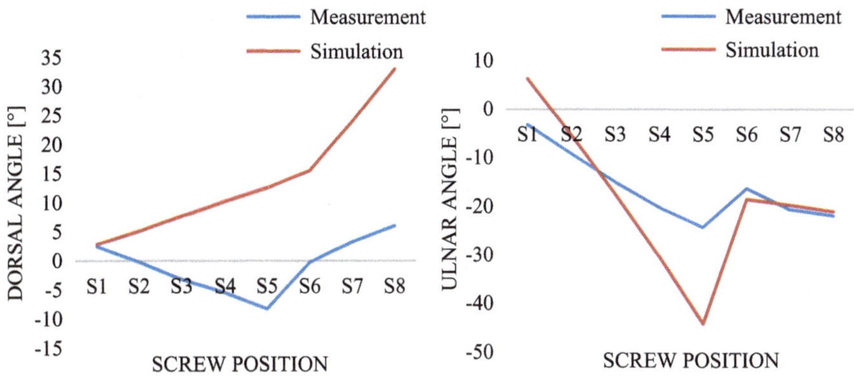

Fig. 3 Exemplary mean values from the simulated and measured wrist angle along the individual screws of the plate at 90° (shoulder height)

the wrist angle. The analysis shows the tilt angle and screw position at which the wrist angle is at its lowest. To do this, the simulation was run again with tilt angles in 5-degree increments and the wrist angle was evaluated in relation to the screw position. A quantitative comparison between simulation and measurement, such as a statistical test or RMSD calculation, was not conducted in this exploratory study. Instead, the analysis focused on identifying trends and deviations in wrist angles across screw positions and tilt angles. The visual comparison suggests systematic discrepancies, particularly in ulnar abduction at extreme positions.

At the elbow height, tilt angles between 90° and 100° appear to be particularly favorable for the ulnar hand posture, while tilt angles between 85° and 95° are advantageous for the dorsal hand posture. It is noticeable that screw position S2 is particularly favorable for both the ulnar and dorsal hand postures at 85°. The ulnar hand posture here is 1.7°, and the dorsal hand posture is – 1.3°. When examining the shoulder height, it can be seen that most screw positions have the smallest wrist angles at a tilt angle of 105°. Screw position S2 is particularly advantageous here, with an angle of – 4.3° in the ulnar hand posture. For the dorsal hand posture, the deflection for S2 is approximately – 14°. However, since S2 has the second-highest angle for the dorsal hand posture, it seems more useful to consider the 100° tilt angle. Although this results in a slightly greater ulnar hand posture of – 4.8° for S2, it also results in a significantly lower dorsal hand posture of – 7.6°.

The mean wrist angles are shown for both heights and for all screws in Fig. 4. The average wrist angles for all screw configurations were analyzed as a function of the tilt angle change. Regardless of the screw position, a tilt angle range between 95° and 100° results in small wrist angles for the wrist.

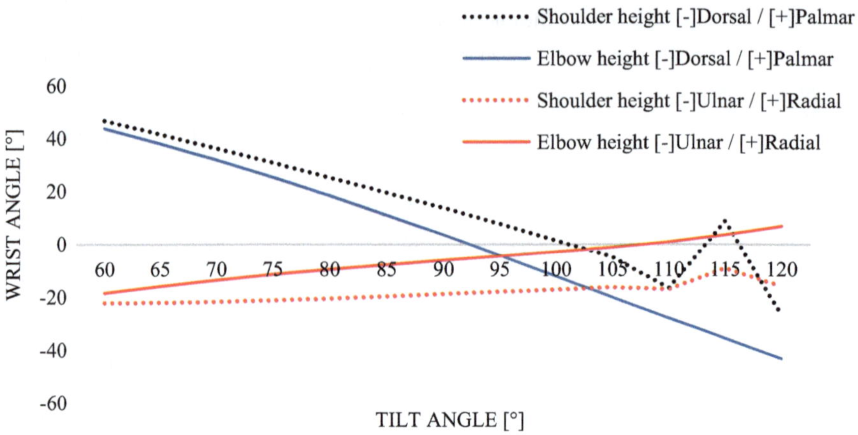

Fig. 4 Average wrist angles for all screw configurations as a function of tilt angle. Dashed line: shoulder height, solid line: elbow height. Blue: dorsal flexion, red: ulnar abduction

4 Discussion

The present study shows that digital human models can be used to analyze ergonomic parameters of hand postures. However, it became clear that the quality of the results of the exemplary simulated wrist angles varies depending on the inclination of the plate and the screw position. Therefore, deviations between the simulation results and the real measured values could be determined. The simulation results do however confirm that the tilt angle of the work surface has a significant influence on hand posture and that a targeted adjustment of the screw positions can help to reduce unfavorable wrist angles. Overall, it can be concluded that an ergonomic parameter analysis with anthropometric human models is fundamentally possible.

However, such models have limitations, especially when compared to finite element models (FEM), which can simulate localized tissue deformation and detailed biomechanical interactions. In contrast, DHMs simplify contact mechanics and joint behavior, which reduces their suitability for simulating fine motor interactions or soft tissue dynamics. Digital human models are therefore less suitable for the precise simulation of fine motor interactions with products. In addition, the forearm and upper arm angles significantly influence the wrist angles, so that unrealistically simulated body or arm postures can also lead to unrealistic hand postures. An improved fitting function that adapts the simulation model to the experimental measurement values could reduce this problem. However, its advantage lies in the comparatively short computing time, which enables quick analyses.

A further limiting factor of this study is the failure to take into account external and internal forces, such as muscle forces or gripping forces. In this study, no forces in the sense of a multi-body simulation were determined and no deformations were calculated.

The focus was exclusively on the analysis of the hand posture as a function of varying working conditions. Future research could supplement the analysis with further parameters, such as muscle activity and external stresses. For example, the effects of masses and the resulting torques exerting strain upon the joints could be studied. The integration of aspects of this kind could enable a more comprehensive assessment of existing stresses and indicate further scope for optimization.

Another relevant research aspect is the use of digital human and hand models for the ergonomic design of work equipment. In earlier work, e.g., Kluth et al. [10], the application of such models was studied for the optimization of screwdrivers. In the present study, a simplified product example was used to show how an optimal plate angle can be determined between the determined hand postures. Further analysis could investigate a systematic parameter optimization of the screw positions over the entire plate as a function of the tilt angle. This could be done, for example, by a design of experiments (DOE) approach. Such an approach could provide insights for ergonomic workplace design and contribute to further improving digital human models in ergonomic analysis. Variability in the measurement data can also stem from sensor inaccuracies, skin movement artifacts, and differences in how subjects executed the screwing motion. Addressing these sources of error would further improve the validation of simulation models.

5 Conclusion

This study demonstrates that digital human models, when calibrated with real motion data, can be effectively used to identify ergonomic wrist postures in manual assembly tasks. The combination of IMU-based wrist angle measurements and simulation revealed systematic deviations, especially at specific plate tilt angles and screw positions. Optimal configurations to minimize wrist stress were identified. Future research should aim to enhance simulation fidelity by including muscle activity data and individual motor control strategies to better reflect real-world human motion variability.

References

1. Chaffin DB (2001) Digital human modeling for vehicle and workplace design. SAE International
2. Mühlstedt J (2016) Digitale Menschmodelle. In: Homo Sapiens Digitalis-Virtuelle Ergonomie und digitale Menschmodelle, pp 73–182
3. Romero J, Tzionas D, Black MJ (2022) Embodied hands: modeling and capturing hands and bodies together. arXiv:2201.02610
4. Ma L, Chablat D, Bennis F, Hu B, Zhang W (2010) Integrating digital human modeling into virtual environment for ergonomic oriented design. In: ASME world conference on innovative virtual reality, vol 49088, pp 203–210.

5. Sharma NK, Tiwari M, Thakur A, Ganguli AK (2021) Biomechanical simulation and a detailed analysis of the roadside cleaning activity. In: Congress of the international ergonomics association. Springer, Cham, pp 183–190
6. Gust P, Ünlü A (2022) Comfort optimization method for work equipment based on a digital hand model. In: Advances in ergonomics in design, usability & special populations: Part II, pp 263
7. Ünlü A, Gust P, Mersch F (2018) Analysis for objective evaluation of the stress of the hand. In: Anatomy, posture, prevalence, pain, treatment and interventions of musculoskeletal disorders
8. Luo Y, Xie K, Andrews S, Kry P (2021) Catching and throwing control of a physically simulated hand. In: Proceedings of the 14th ACM SIGGRAPH conference on motion, interaction and games, pp 1–7.
9. imk Industrial Intelligence GmbH.: ema Software Suite/ema work designer (version 2.4.1) [computer-software]. https://imk-ema.com/. Last accessed 31 Mar 2025
10. Kluth K, Chung HC, Strasser H (2004) Verfahren und Methoden zur Prüfung der ergonomischen Qualität von handgeführten Arbeitsmitteln: professionelle Schraubendreher im Test. Haefner

Analysis of Cooperative Activities Between Two People Including Motion and Muscle Activity Measurements

Aydin Ünlü, Saeed Raza, Martin Manns, and Karsten Kluth

Abstract The aim of this study is to examine the cooperative activity of two people who work together to transport a heavy box in the same way. To do this, an experiment was set up with different positions at and around a table using a heuristic approach. The aim was to find out what significant differences in posture and muscular strain exist between the two subjects. The movement and muscle activity measurements focus on the level of differences in the stress on the right hand-arm system. Finally, the causes of these differences are discussed and a hypothesis is formulated for further research into the relationship between muscle activity and human support behavior.

Keywords Muscle activity · Hand-arm movement · Hand position · Surface electromyography (sEMG) · Inertial measurement unit (IMU)

1 Introduction

When two people work together, the question often arises as to how the physical strain is distributed between them and who bears the main burden. One example of this is when two people are carrying a heavy object together, with one person taking on a larger share of the load while the other is less strained. Studies such as those by [1, 2] show that the distribution of the load in such scenarios is not always even. These studies mainly focus on the upper extremities, especially the back, and measure muscle activity (surface electromyography, sEMG) without taking movement dynamics into account. The results of the electromyographic measurements confirm that unequal load distribution often occurs when heavy loads are carried cooperatively. The study by [3] uses Inertial Measurement Unit (IMU) based motion capture to analyze the movements between two people and to understand how they

A. Ünlü (✉) · K. Kluth
Work Science/Ergonomics Division, University of Siegen, Siegen, Germany
e-mail: aydin.uenlue@uni-siegen.de

S. Raza · M. Manns
Chair of Production Automation and Assembly, University of Siegen, Siegen, Germany

M. Manns (ed.), *New Paradigms for Anticipated Uncertainty*, Lecture Notes in Mechanical Engineering, https://doi.org/10.1007/978-3-032-16889-4_70

share or change roles during collaboration. Various parameters such as total time, distance, and acceleration in repetitive tasks are considered.

This study examines right hand–arm movement and muscle activity in a two-person cooperative lift, combining IMU-based joint-angle tracking with continuous EMG recordings. For the first time, we show how different hand-placement and bracing strategies in an industrially realistic scenario alter joint kinematics and muscle stress.

2 Experiment

A heuristic approach was used to analyze the cooperative work processes between two people. Two male subjects aged 25 years participated in this pilot study (body weight: approx. 75 kg and 80 kg; height: 171 cm and 175 cm) and each performed five repetitions under every lift condition, resulting in ten trials in total. Given this small sample size in this pilot study and the absence of true repeated-measures statistics, the findings must be regarded as preliminary and interpreted with caution. In each trial, the participants cooperatively moved a 10 kg box across a table to evoke variations in posture and muscle strain, while joint angles were recorded via IMUs and muscle activity via sEMG.

The schematic test set-up is shown in Fig. 1. The movement was performed by both subjects with their right hand. The position on the table and the type of grip (power grip, back of the hand facing upwards) were maintained. In order to make it easier to concentrate on the sequence of movements, no fixed pace was set. Instead, the subjects performed 5 repetitions of the movement sequence in each of two rounds. The subjects performed the box movement (lift—carry—put down) together in two rounds from a defined starting position to a target position. The following positions were used for the movements: A-B-C and C-B on the table, B-Fl and Fl-B (Fl = floor), B-Rl and Rl-B (Rl = rack level), B-Rl-bh (bh = both hands).

To strengthen statistical validity, future studies should involve a larger cohort and substantially increase the number of repetitions per condition. Employing a repeated-measures design and appropriate statistical analyses will allow robust assessment of both intra- and inter-individual strategy differences and support generalizable ergonomic recommendations.

The movement and muscle activity data were recorded using IMU and EMG sensors. The following muscle groups were selected: biceps brachii (flexion and supination of the arm), triceps brachii (extension and stabilization of the elbow), extensor carpi ulnaris (ulnar abduction and dorsal flexion at the wrist) and flexor carpi radialis (radial abduction and palmar flexion) (cp.) [4]. These muscle parts were each normalized with resting activity and maximum muscle activity (MVC method).

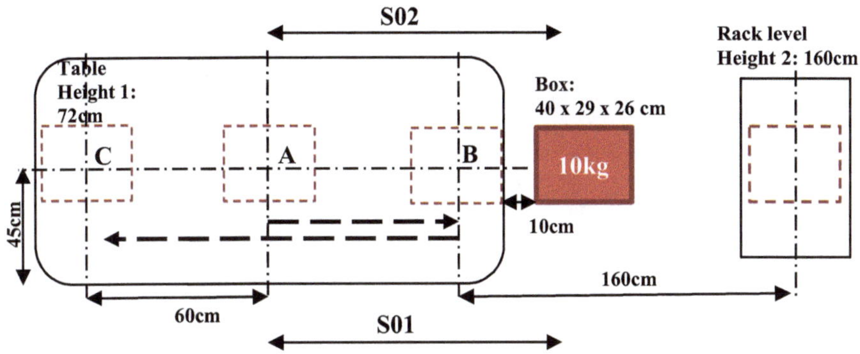

Fig. 1 Box positioning: one example of a motion sequence is moving the box from the center of the table (position A) to the right edge of the table (position B) and then to the opposite edge of the table (position C)

3 Results

The measurements of movement and muscle activity for both test subjects (S01 and S02) throughout the entire movement sequence showed that the percentage strain in terms of muscle activity is on average 5% higher for S01 than for S02. With regard to the analysis of the total body load using the key indicator method for manual work processes according to [5], S01 shows an 8% higher load. The greatest deviation can be seen in the placement of the box on the rack (B-Rl or Rl-B). The t-scores of the EMG evaluation show a difference of 8%, while the key indicator method for manual work processes shows a difference of 18% between S01 and S02.

A detailed examination of the right arm and hand area in Fig. 2 shows that the joint angles of the two subjects differ despite the same specifications. The angle measurements are reported in degrees. EMG amplitudes are normalized to each participant's MVC and expressed as percentage values to enable between-subject comparison. The greatest angle difference of 66° to 81° occurs due to the radial and ulnar abduction of the hand, especially when placing the box on the floor and in the rack.

A reflection of the movement data, in which the movement of the right hand was analyzed either towards or away from the center of the body, also shows differences. Here, S02 tends to perform ulnar abduction, while S01 performs radial abduction more frequently. However, both test subjects show palmar flexion in all sequences, with S01 bending the right hand more towards the palm than S02. This occurs in particular when placing the box on the rack or on the floor. The smallest angle differences of a maximum of 19.5 degrees occur when placing the box on the table (movement sequence A-B-C or C-B).

Muscle activity analysis also shows significant differences (see Fig. 3). The greatest differences of 5.8% to 8.5% were observed in the biceps brachii during all movement sequences. For the flexor carpi radialis, the maximum difference was

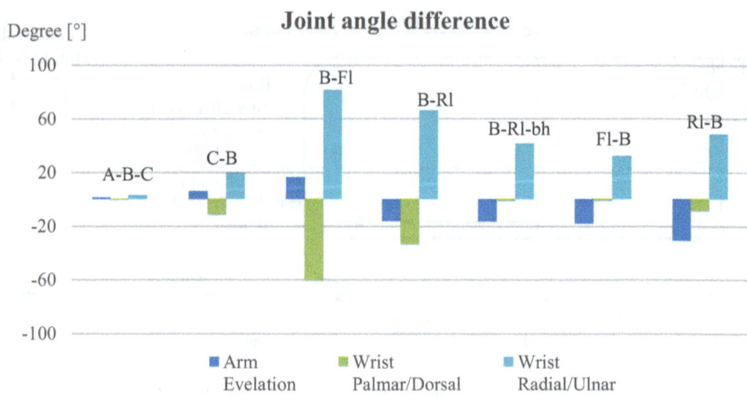

Fig. 2 Difference in joint angle from the normal position when moving the box. The calculation shows the difference between S01 and S02

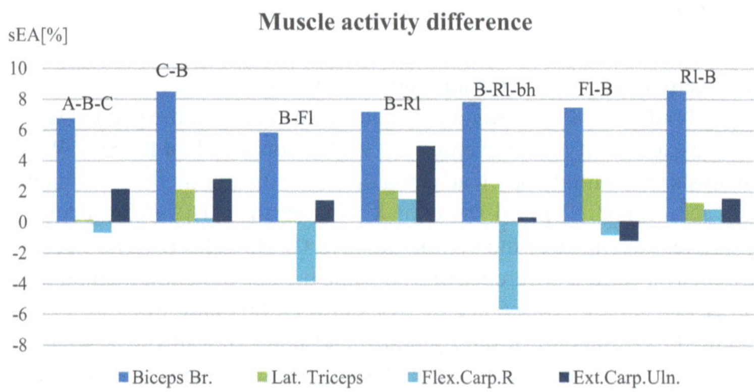

Fig. 3 Muscle activity differences of the right hand-arm system when moving the box. The determination shows the difference between S01 and S02

5.7% (B-Rl and Rl-B), while for the extensor carpi ulnaris, a maximum difference of 4.9% (B-Rl) was observed.

4 Discussion

The analysis of the movement and muscle activity data clarifies the differences in the demands on the two test subjects. The causes of these differences are discussed using the example of the movement sequences from position B to the floor and from position B to the rack. In general, the two subjects' postures differ when lifting the box. Interestingly, the observed inter-subject variation in lift strategy, such as one participant bracing with the free hand on the knee, was not anticipated. This finding underscores the substantial effect that even minor individual technique differences can have on both joint mechanics and muscle load.

Our results suggest that simple modifications in lift technique (e.g., free-hand bracing, grip width) can yield significant changes in joint-angle profiles and EMG activation patterns. A systematic comparison of these techniques in future work would be essential to derive concrete ergonomic guidelines.

In Fig. 4, left, S01 bends his upper body less than S02 and thus lifts the box from a different height. These differences in the movement sequence could have led to different muscular strain in the arm area, particularly in the biceps brachii. Since the biceps plays a crucial role in pulling movements [6], the strain varies depending on the posture and movement execution. In addition, it has been shown that S02 turns the upper body more strongly towards the ground and supports himself with his left hand on his knee. This movement behavior could not only have influenced the hand movement, but also led to differences in muscle activity.

As shown in Fig. 4 on the right, S02 lifts his elbow higher and positions his right hand differently than S01 when placing the box on the rack. These differences in movement execution influence muscle tension and movement patterns in the arm area [7]. In contrast, in S01, the biceps brachii takes on a larger part of the tensile load due to the smaller elbow movement. This could explain the higher biceps activity measured for S01 compared to S02. However, it should be emphasized that the

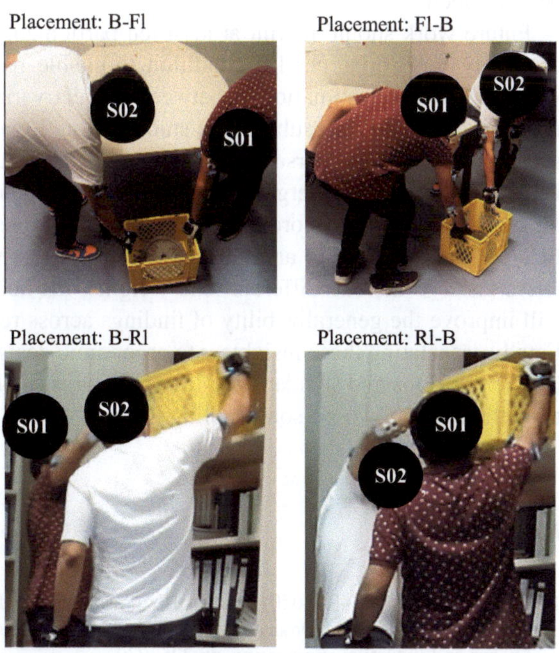

Fig. 4 Selection of experiments and detection of different postures. Placement of the heavy box on the floor (B-Fl and Fl-B) and on the rack (B-Rl and Rl-B)

maximum biceps muscle tension for S01 was 36% lower than for S02. By comparison, the triceps showed maximum differences of up to 2.7%.

These findings may inform the ergonomic design of collaborative workstations and human–machine interfaces, where optimal placement of handles and lift, surfaces can help reduce joint stress during cooperative tasks. Integrating real-time joint-angle and muscle-activity feedback into assistive devices could further enhance safety and efficiency in shared-load operations.

5 Conclusion

This pilot investigation uniquely combined IMU-derived joint-angle tracking with continuous sEMG monitoring in a two-person cooperative lift. The movements and muscle activities of the two test subjects were measured separately, without synchronization. For this reason, no direct comparison could be made between posture or movement and muscle activity. However, it was found that the selected muscles as well as hand movements showed considerable differences between the test subjects.

One hypothesis, for example, is that the degree of assistance can be determined based on the slope of the muscle activity: the steeper the slope, the higher the degree of assistance. Building on the method of [3], which used hand accelerations to predict movement behavior, a similar approach using muscle activity could also be developed.

Future work should recruit at least ten participant pairs and collect a minimum of twenty repetitions per lift condition to enable robust repeated-measures analyses and accurate estimation of between-subject variance. Instead of testing many task variations at once, subsequent studies should manipulate one factor at a time, such as hand-bracing versus free-hand lifts, and group participants by strategy to isolate strategy-specific ergonomic effects. In addition, future studies should focus on synchronizing the recording of movement and muscle activity to enable precise linkage between posture and muscular response. Expanding the range of collaborative tasks to include different load weights, lift heights, and grip configurations will improve the generalizability of findings across real-world industrial scenarios. Finally, translating these insights into ergonomic design guidelines, such as optimal handle positions and standardized lift procedures, will directly enhance both safety and efficiency in two-person handling tasks.

References

1. Lee CL, Lou SZ, Su YR (2020) Load distribution and lifting duration effects on upper limb strain for manual handling. In: Proceedings of the 2020 2nd International conference on management science and industrial engineering, pp 225–230

2. Schiefer C, Hermanns I, Schuster D, Brandt K, Ditchen D, Göbel F, Reichert T (2019) Untersuchung der physischen belastungen von rettungskräften beim patiententransport in treppenhäusern. DGUV/IFA
3. Tuli TB, Manns M, Jonek M (2022) Understanding shared autonomy of collaborative humans using motion capture system for simulating team assembly. Springer International Publishing, Aalborg, Denmark, October/November 2021 8, pp 527–534
4. Valerius KP, Frank A, Kolster BC, Hamilton C, Alejandre-Lafont E, Kreutzer R (2023) Das muskelbuch: anatomie–untersuchung–bewegung. KVM-Der Medizinverlag
5. Steinberg U, Behrendt S, Caffier G, Schultz K, Jakob M (2007) Leitmerkmalmethode manuelle arbeitsprozesse. In: Schriftenreihe der bundesanstalt für arbeitsschutz und arbeitsmedizin, F1994. Forschung Projekt F1994. Dortmund/Berlin/Dresden
6. Hermens HJ, Freriks B, Disselhorst-Klug C, Rau G (2000) Development of recommendations for SEMG sensors and sensor placement procedures. J Electromyogr Kinesiol 10(5):361–374. https://doi.org/10.1016/S1050-6411(00)00027-4
7. Kumar S (2001) Theories of musculoskeletal injury causation. Ergonomics 44(1):17–47. https://doi.org/10.1080/00140130120716

Modular Cognitive Robotics: Enabling Flexible Handling and Inspection in Automated Manufacturing

Martin Naumann, Leutrim Gjakova, Rico Löser, and Martin Dix

Abstract Flexibility and individuality in product portfolios are increasingly important in manufacturing, necessitating highly adaptable production systems. Industrial robots, despite their potential, are often underutilized due to limited autonomy and programming challenges. This paper presents a modular software toolbox designed for cognitive robotics while addressing the research question of how a modular software toolbox can enhance the cognitive capabilities and autonomy of industrial robots to improve adaptability and efficiency in automated manufacturing processes, particularly in the context of small to medium batch inspections, alongside the increase in efficiency during commissioning and application. The toolbox features reconfigurable algorithms for object recognition, handling, and quality inspections using machine learning methods such as CNNs and SVMs. Optimized for standard industrial computers, the system ensures efficient resource use, achieving object recognition at 20 frames per second. The setup includes a six-axis robot with a CNC controller and high-precision sensors, offering significant cost and time savings for small- to medium-batch inspections. Validation demonstrates the toolbox's potential to enhance flexibility and intelligence in manufacturing systems.

Keywords Cognitive robotics · Modular software toolbox · Machine learning in manufacturing · Flexible quality inspection · Industrial automation optimization

1 Introduction

The manufacturing industry is transforming, driven by a growing demand for flexibility and customization in product offerings. This shift requires production systems that are adaptable and intelligent, moving from large-series production to small batch

M. Naumann (✉) · L. Gjakova · R. Löser · M. Dix
Fraunhofer Institute for Machine Tools and Forming Technology IWU, Chemnitz, Germany
e-mail: martin.naumann@iwu.fraunhofer.de

M. Dix
Institute for Machine Tools and Production Processes, Chemnitz University of Technology, Chemnitz, Germany

M. Manns (ed.), *New Paradigms for Anticipated Uncertainty*, Lecture Notes in Mechanical Engineering, https://doi.org/10.1007/978-3-032-16889-4_71

or single-item manufacturing. Addressing the research question of how a modular software toolbox can enhance the cognitive capabilities and autonomy of industrial robots to improve adaptability and efficiency in automated manufacturing processes, particularly in the context of small to medium batch inspections, this paper introduces a modular software toolbox designed to enhance cognitive robotic functionalities. It integrates algorithms for object recognition, flexible handling, and precision quality inspection, leveraging machine learning techniques such as convolutional neural networks (CNNs), Support Vector Machines (SVMs), and transfer learning to enhance robot intelligence and autonomy. Optimizing these algorithms for standard industrial computers ensures efficient resource utilization while meeting stringent cycle time and data transmission constraints.

The significance of this research lies in its potential to improve manufacturing efficiency, particularly through the enhancement of cognitive robotic capabilities, which also leads to increased efficiency during commissioning and application. The demonstrator setup showcases the transformative potential of cognitive robotics, utilizing a six-axis industrial robot with advanced sensors and controllers. This approach promises significant cost and time savings, redefining automated inspection systems.

Key scientific challenges include recognizing diverse parts, autonomous path planning, and inline optical quality monitoring. Developing fast and resource-efficient machine learning methods for real-time data processing on limited hardware is crucial. Through detailed analysis of methodology, experimental setup, results, and implications for future systems, this investigation aims to contribute significantly to cognitive robotics, fostering the evolution of more flexible, intelligent, and efficient automated manufacturing solutions.

2 State of the Art

Industrial robotics has seen significant advancements, particularly in flexibility and autonomy, transitioning from traditional repetitive tasks to more dynamic applications [1]. While industrial robots excel in precision and efficiency [2], their potential is often constrained by complex programming and adaptability challenges in dynamic environments. Cognitive robotics offers a transformative approach by integrating machine learning and AI to enhance robotic systems' adaptability and autonomy, enabling robots to perceive, learn, and interact with environments akin to human cognition [3, 4]. However, while the literature predominantly addresses the complete autonomy of robots, the current approach aims to focus on specific aspects to ensure the necessary robustness of industrial systems.

Machine learning techniques such as CNNs, SVMs, and Transfer Learning have become pivotal in advancing robotic capabilities, particularly in object recognition tasks [5–7]. Numerous techniques exist for object recognition, including traditional methods like sliding window approaches, template matching, feature-based methods, texture classification and edge detection [8, 9]. Recently, CNNs have demonstrated state-of-the-art performance, with frameworks like YOLO (You Only Look Once) leading in real-time object detection with high precision, making them ideal for applications needing rapid and accurate identification [10, 11].

Quality inspection technologies have innovated with 2D [12] and 3D [13] inspection methods, leveraging advanced sensors and algorithms to ensure product quality and compliance in automated systems and robotic applications [14, 15]. Robot path planning continues to evolve, focusing on precision and reliability, especially in handling densely packed components without collision [16–18]. Strategies like landmark-based path planning are being employed for robust industrial applicability and maintenance.

Modular software development facilitates flexible and reconfigurable robotic systems, enabling rapid adaptation to customer demands and dynamic environments. Recent advancements allow for easy setup and programming, enabling non-experts to configure workcells for diverse applications, thereby enhancing efficiency and fostering innovation in automated manufacturing [19, 20].

Despite these developments, challenges remain in seamlessly integrating cognitive capabilities and addressing variability in manufacturing environments. Future research aims to refine machine learning models, improve sensor technologies, and develop robust strategies for dynamic operations [1, 2]. These advancements in the state of the art form the foundation for proposed solutions in this paper, highlighting the potential for cognitive robotics to revolutionize automated manufacturing with flexible, intelligent, and efficient systems.

3 Methodology

The modular software toolbox for cognitive robots consists of a collection of modules designed for object and component recognition, path generation for robotic movement, component handling, control of sensors and actuators, data acquisition and processing, and result evaluation. In a base application, these modules can be interconnected to create a cohesive system. They are available in a library and can be logically linked through a sequence description. A notable application of this modular toolbox is in flexible inspection robotics, which is particularly suited for small to medium batch sizes. For this specific application, the system primarily operates in two distinct modes: teaching new components and inspecting previously known components.

3.1 Teaching New Components

The process of teaching new components is structured into several phases. Initially, object recognition training is conducted by placing multiple objects of the same type within the robot's workspace. Training and evaluation datasets are generated through automated processes, which involve generating random poses for the robot and capturing images with the robot camera. Automated object recognition and labeling are performed using techniques such as Segment Anything algorithm (SAM) or binary large object analysis (Blob). Subsequently, comprehensive training and evaluation datasets are created to expand the base model and dataset with new classes and images. The object recognition model is trained using methods such as YOLO or feature analysis with OpenCV, coupled with SVM classification, before integrating the newly trained model into the application.

In configuring quality inspections, the system offers two options. The first option allows the selection of existing inspection approaches from the module catalog for quality inspection, such as actual versus target comparisons, which involves recording reference components to establish a model for the representation of the acceptable condition. The second option involves adding new inspection approaches, which may include developing specific image processing modules using the Xeidana® Framework [21] or python in combination with OpenCV [22].

3.2 Inspecting Components

The inspection process comprises several phases. Initially, object recognition is performed by capturing an overview image and identifying the object using a CNN like the YOLO network. This is followed by calculating the object's position to generate robot programs and deriving grip angles from Principal Component Analysis (PCA). The inspection process is then generated based on the stored configuration for the component. Two approaches are utilized: grasping the object and positioning it towards the appropriate sensor, such as a 3D scanner, or utilizing the sensor mounted on the robot to generate an inspection program to scan the component, such as with a 2D camera. Execution of the inspection process involves recording sensor data, specifically 2D or 3D images, and evaluating these datasets, typically through actual versus target comparisons. The results are presented in a user interface, and a robot program is generated for placement in acceptable or unacceptable storage.

3.3 Modules

The toolbox includes several essential modules for path planning and quality inspection. Path planning involves generating NC code for robot control and incorporates modules for path generation, collision checking, and hardware control, including grippers and sensors. Modules for handling components cover a range of actions, including grasping, positioning, placing, scanning, and reorienting components. Quality inspection is divided into 2D and 3D inspection. In 2D inspection, techniques such as contour comparison for anomaly detection and surface inspection for homogeneity are employed, alongside individual inspection through the Xeidana interface.

In 3D inspection, methods include actual versus target comparison using CAD or reference scans, potentially supported by Iterative Closest Point (ICP) algorithms.

The existing modules are managed in a library within a base application and can be logically interconnected for specific applications, such as the flexible inspection robot. To facilitate universal applicability, defined interfaces and various approaches are available for integrating new modules and expanding functionalities. These approaches include Reflection and Assembly Loading in C#, dynamic libraries (DLLs/Shared Libraries), which both allow the straightforward integration of Xeidana networks and modules, console calls with file transfer primarily utilized for Python modules, and prospective microservices via Google Remote Procedure Call (gRPC) to enable interaction with decentralized modules. This last approach is scalable and may allow for the offloading of resource-intensive algorithms.

This methodology underscores the modular and adaptable nature of the system, facilitating efficient and flexible inspection processes in automated manufacturing environments. The modules can be expanded as needed to integrate other robot controls, connect additional sensors or add new evaluation methods.

4 Experimental Setup

The experimental setup (Fig. 1) evaluates the automation capabilities of cognitive robotics for quality inspection and object handling within an integrated hardware and software system by utilizing the functions of the modular toolbox. It demonstrates an exemplary configuration of the modular software modules for the application of a flexible inspection robot. The overall functionality is achieved through the logical interconnection of the individual modules, which can be intuitively and individually designed by the user or system integrator. The setup includes a robot with NC control and a two-finger gripper, complemented by several grayscale 2D matrix cameras. An overview camera captures broad components, while a camera on the robot's end effector aids in precise location correction. A separate 2D camera serves as an independent inspection station and a 3D snapshot sensor is dedicated to quality inspection tasks. Conveyor belts facilitate the transport of objects, and a homogeneous black surface ensures consistent imaging with optional lighting available for enhanced illumination.

The software component is centered around a C# application that serves as the primary control and visualization interface. Tasks such as path planning and quality inspection are executed on a standard computer. The application integrates functionalities including object recognition, classification and path generation using the trained YOLO model. It also manages sensor data for quality inspection, interfacing with control systems for robot, sensor and actuator management. A user interface supports operations and visualization.

The setup follows a structured workflow, beginning with the training phase implemented in Python. Data generation involves placing homogeneous components within the robot's workspace and generating a robot program that randomly varies

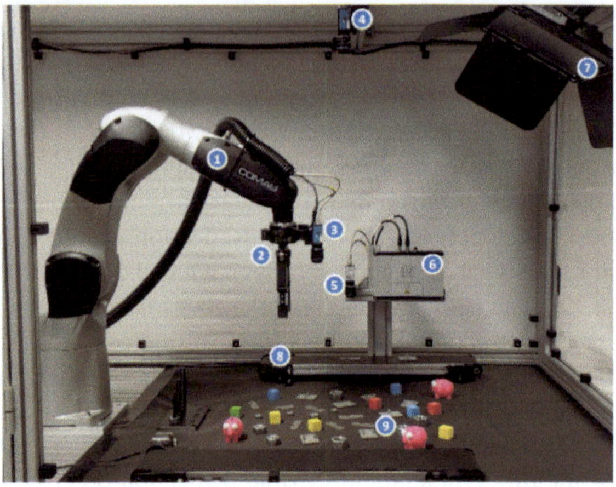

Fig. 1 The experimental setup comprises a six-axis industrial robot (1) and an electric two-finger gripper (2). It includes several 2D matrix cameras (3–5), a 3D snapshot sensor (6), additional illumination (7), a conveyor belt (8), and a workspace containing various test objects (9)

positions across all six degrees of freedom, triggering camera captures. For path planning, a support point-based NC code generation is utilized, whereby the fundamental movement capabilities are predefined and complemented by flexible segments based on the path planning modules for image capture and component handling. This approach achieves high robustness while simultaneously ensuring good maintainability and interpretability. Concurrently, collision detection is performed based on real-time coupling with a digital twin and envelope-based collision monitoring. Training data is recorded using the camera on the robot's end effector, with optional variations in lighting and shadow effects. Labeling and data preparation apply object recognition using Blob analysis and FAST SAM, deriving bounding boxes and applying type labels, followed by data preparation for CNN training. The training process uses transfer learning with YOLO V11 Nano, utilizing labeled data divided into training and validation datasets. AI training is conducted using Google Colab. Validation of object recognition quality is performed by placing mixed components in the workspace and applying the processing network to both the overview and robot cameras.

During the inspection phase, one or more objects are placed in the workspace and the inspection routine is initiated. Objects are captured using the overview camera, and object recognition is performed using the YOLO network. Inspection steps and methods for the identified component are retrieved from the component database. The robot program is generated for handling the object, e.g. placing it on the conveyor belt for transport under the 3D sensor, triggering the 3D sensor, and processing the result data against the target model. The object is then gripped and handled to either the acceptable or unacceptable storage.

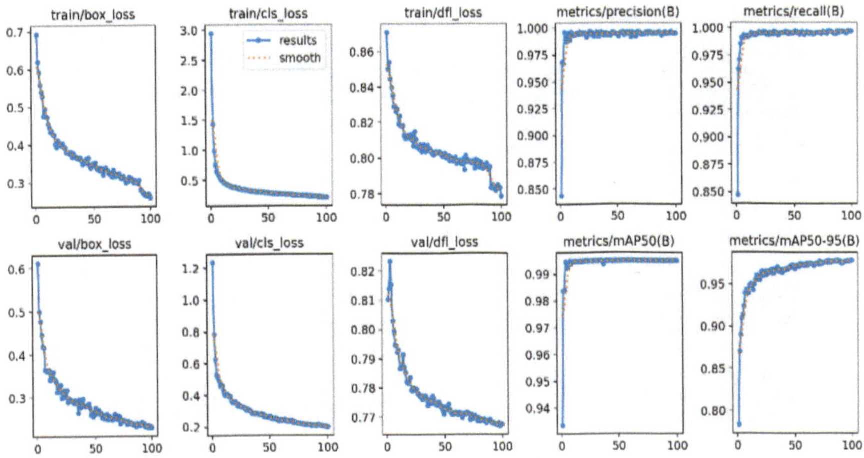

Fig. 2 Loss and metric curves illustrating the YOLO model's performance during the training and validation phases. The curves demonstrate the model's optimization across various metrics, including precision, recall, and mean average precision, reflecting strong predictive accuracy and effective generalization

5 Results

5.1 Object Detection Evaluation (YOLO Model)

In our setup, the YOLO object detection model (yolo11n.pt) was trained using a dataset with six classes over 100 epochs, leveraging pre-trained weights for efficiency. Training occurred on a CPU with a batch size of 2 and image dimensions of 640 × 640 pixels, utilizing data augmentation strategies such as translation, scaling, and mosaic augmentation. The model's performance, depicted in Fig. 2, showed consistent downward trends in training and validation loss curves, indicating effective optimization. Metrics like precision and recall approached 1.0, demonstrating high accuracy and robustness across varied intersection-over-union (IoU) thresholds. Adequate model performance was reached by the 17th epoch (approx. 49 min after start), though practical applications revealed dependencies on viewing angles and lighting conditions (Fig. 3).

5.2 Quality Inspection Evaluation

Two modules were tested: 2D contour inspection using the Enhanced Correlation Coefficient (ECC) algorithm for position adjustment, and 3D actual versus target comparison using ICP with CAD models or scans. Both methods proved effective in defect detection under consistent positional and lighting conditions, aided

Fig. 3 Validation of object recognition on the real demonstrator indicates that a variety of objects can be classified; however, external lighting conditions still exert influence

by robust orientation recognition and precise robot programming. Figure 4 illustrates these processes with 3D evaluations showing discrepancies highlighted in red and 2D defect identification under varied conditions. The image underscores the model's capability in defect detection using both 3D point cloud analysis and image processing techniques.

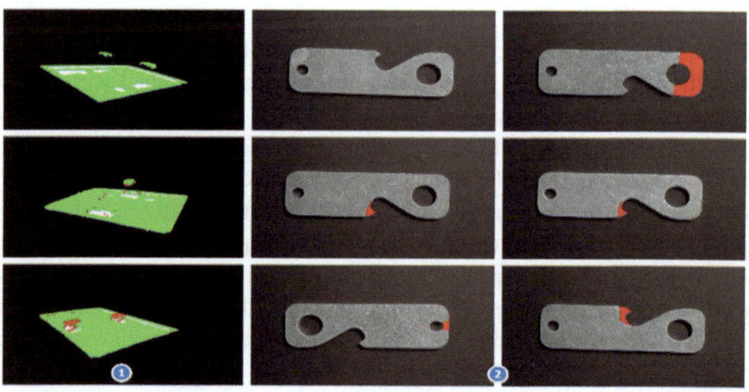

Fig. 4 The image illustrates two defect detection processes. The left side (1) presents a 3D evaluation using point clouds with ICP. The right side (2) displays images of 2D contour defects under varying angles and lighting conditions

6 Discussion

The implementation of a modular software toolbox has significantly enhanced the flexibility and adaptability of manufacturing systems, allowing for seamless integration and reconfiguration of robotic functionalities without extensive reprogramming. This approach supports incremental updates and expansions, optimizing resource allocation and reducing downtime, thereby improving operational efficiency. Object recognition was reliable under specific conditions, such as homogeneous backgrounds and non-overlapping components, despite limited training data and using the YOLO-Nano model. For each newly introduced object, a complete retraining of the model is required. Fortunately, this is feasible due to the currently short training times and the achieved accuracy. Furthermore, through the later integration of a microservices interface, this aspect can be offloaded to high-performance computing units, allowing for parallel execution alongside continued utilization. Challenges with overlapping components and lighting variations suggest the need for free-move strategies, utilizing the robot's end-effector camera to enhance recognition by adjusting position and orientation. To further address these challenges, additional approaches for synthetic data generation can be employed, such as leveraging Sim2Real techniques. These methods can be used to expand the current data augmentation strategies and achieve greater robustness in object recognition. However, it is essential to consider the existing performance resources to ensure that these enhancements do not compromise system efficiency or operational capacity. The system accurately detected component orientation for reliable gripping but faced challenges with closely positioned components, indicating a need for strategies to reorient or separate components for effective handling. 2D contour inspection was robust with consistent lighting, and optimization can be achieved through training under varied lighting conditions and dynamic thresholding. The 3D actual versus target comparison works well but requires smoothing techniques and improved object-background separation. Enhancing visibility through multi-angle scanning and stitching could enable comprehensive 3D capture. Overall, the system shows promising capabilities in object recognition, path planning, and quality inspection with identified areas for optimization to enhance robustness and reliability in dynamic manufacturing environments.

7 Conclusion and Future Work

This paper investigated the integration of cognitive robotic functionalities in automated manufacturing systems, focusing on object recognition, path planning, and quality inspection, under the utilization of the developed modular software toolbox. The YOLO-Nano model proved effective under controlled conditions but faced challenges with overlapping components and lighting variations. Path planning was reliable but highlighted the need for strategies in densely packed scenarios, while quality inspection showed robustness with potential for optimization in dynamic lighting and 3D scanning. Key areas for improvement include dynamic thresholding and multi-perspective scanning, which can enhance system capabilities. Future work will refine these approaches, aiming for complete workflow automation from data generation to quality evaluation. Simplifying the addition of new objects for sequential training can save resources, while automating validation and exploring color cameras may boost recognition capabilities. Implementing one-shot 2D inspection methodologies [23] and developing methods for separating mixed objects will further enhance system efficiency and adaptability in diverse manufacturing environments.

Acknowledgements The work is part of the project Kogni-Bot which is supported by the German Federal Ministry of Education and Research (BMBF).

References

1. Arents J, Greitans M (2022) Smart industrial robot control trends, challenges and opportunities within manufacturing. Appl Sci. https://doi.org/10.3390/app12020937
2. Licardo JT, Domjan M, Orehovački T (2024) Intelligent robotics—a systematic review of emerging technologies and trends. Electronics. https://doi.org/10.3390/electronics13030542
3. Chun WH (2024) Foundations of artificial intelligence and robotics. Chapman and Hall/CRC, Boca Raton
4. Gaurav A, Arya V (2025) AI technologies in robotics. In: Giannoccaro I, Gupta BB, Colace F (eds) AI developments for industrial robotics and intelligent drones. Advances in computational intelligence and robotics. IGI Global, pp 17–32
5. Bai Q, Li S, Yang J, Song Q, Li Z, Zhang X (2020) Object detection recognition and robot grasping based on machine learning: a survey. IEEE Access. https://doi.org/10.1109/ACCESS.2020.3028740
6. Manakitsa N, Maraslidis GS, Moysis L, Fragulis GF (2024) A review of machine learning and deep learning for object detection, semantic segmentation, and human action recognition in machine and robotic vision. Technologies. https://doi.org/10.3390/technologies12020015
7. Jozdani SE, Johnson BA, Chen D (2019) Comparing deep neural networks, ensemble classifiers, and support vector machine algorithms for object-based urban land use/land cover classification. Remote Sens. https://doi.org/10.3390/rs11141713
8. Sackewitz M (ed) (2016) Leitfaden zur inspektion und charakterisierung von oberflächen mit bildverarbeitung. Vision-Leitfaden, vol 16. Fraunhofer Verlag, Stuttgart
9. Ushma A (2014) Object detection in image processing using edge detection techniques. IOSRJEN. https://doi.org/10.9790/3021-04311013
10. Dhillon A, Verma GK (2020) Convolutional neural network: a review of models, methodologies and applications to object detection. Prog Artif Intell. https://doi.org/10.1007/s13748-019-00203-0
11. Redmon J, Divvala S, Girshick R, Farhadi A (2015) You only look once: unified, real-time object detection
12. Pierer A, Hauser M, Hoffmann M, Naumann M, Wiener T, León MAL, de Mende M, Koziorek J, Dix M (2022) Inline quality monitoring of reverse extruded aluminum parts with cathodic dip-paint coating (KTL). Sensors (Basel, Switzerland). https://doi.org/10.3390/s22249646
13. Huo L, Liu Y, Yang Y, Zhuang Z, Sun M (2023) Review: research on product surface quality inspection technology based on 3D point cloud. Adv Mech Eng. https://doi.org/10.1177/16878132231159523

14. Burski B, Garbacz P, Mężyk J, Czajka P, Mizak W (2015) The use of 2D/3D sensors for robotic manipulation for quality inspection tasks. SSP. https://doi.org/10.4028/www.scientific.net/SSP.237.77
15. Herakovic N (2010) Robot vision in industrial assembly and quality control processes. In: Ude A (ed) Robot Vision. InTech
16. Billard A, Kragic D (2019) Trends and challenges in robot manipulation. Science (New York, N.Y.). https://doi.org/10.1126/science.aat8414
17. Marturi N, Kopicki M, Rastegarpanah A, Rajasekaran V, Adjigble M, Stolkin R, Leonardis A, Bekiroglu Y (2019) Dynamic grasp and trajectory planning for moving objects. Auton Robot. https://doi.org/10.1007/s10514-018-9799-1
18. Gu Y, Otto RM, Naumann M, Gjakova L, Löser R, Dix M (2024) AI-driven force torque control strategies for further automate flexible high-precision, contact-intensive assemblies. Procedia CIRP. https://doi.org/10.1016/j.procir.2024.10.099
19. Bilancia P, Schmidt J, Raffaeli R, Peruzzini M, Pellicciari M (2023) An overview of industrial robots control and programming approaches. Appl Sci. https://doi.org/10.3390/app13042582
20. Simonic M, Pahic R, Gaspar T, Abdolshah S, Haddadin S, Catalano MG, Worgotter F, Ude A (2021) Modular ROS-based software architecture for reconfigurable, industry 4.0 compatible robotic workcells. In: 20th International conference on advanced robotics (ICAR), Ljubljana, Slovenia, pp 44–51. https://doi.org/10.1109/ICAR53236.2021.9659378
21. Putz M, Wiener T, Pierer A, Hoffmann M (2018) A multi-sensor approach for failure identification during production enabled by parallel data monitoring. CIRP Ann. https://doi.org/10.1016/j.cirp.2018.04.079
22. OpenCV Homepage. https://opencv.org/. Accessed on 01 Apr 2025
23. Wiener T Optische inline-überwachung umgeformter blechteile mittels robuster anomaliedetektion. Dissertation, Technische Universität Chemnitz

Timing Analysis of Low-Cost Edge Vision for Object Detection for Smart Factory Systems

Christopher Tofaute, Tadele Belay Tuli, Florian Schreiber, and Martin Manns

Abstract Modern manufacturing environments require adaptable, low-cost vision systems, particularly for robotic end-effectors that must handle a diverse range of workpieces. This study evaluates the ESP32-S3 microcontroller integrated with the OV2640 camera as a compact, wireless vision module compatible with ROS2. The system transmits JPEG-compressed images over Wi-Fi using the micro-ROS communication stack. Performance was analyzed across various image resolutions, and the video stream was processed on a remote ROS2 host using OpenCV and YOLOv8n. To identify performance bottlenecks, high-resolution timing measurements were conducted along the data pipeline. Results show that image acquisition and message preparation contribute negligible delay (< 0.1 ms), while most latency arises during image encoding, serialization, and wireless transmission, with a median delay of approximately 78 ms. This limits the achievable frame rate between 18.3 and 40 frames per second. The findings demonstrate the ESP32-S3's suitability for motion detection and basic object recognition in low-cost, real-time robotic applications.

Keywords Edge computing · Smart factory automation · ESP32-S3 · ROS2 · Time analysis · Object detection

C. Tofaute (✉) · T. B. Tuli · F. Schreiber · M. Manns
PROTECH—Institut für Produktionstechnik, Universität Siegen, Siegen, Germany
e-mail: christopher.tofaute@student.uni-siegen.de

T. B. Tuli
e-mail: tadele-belay.tuli@uni-siegen.de

F. Schreiber
e-mail: florian.schreiber@uni-siegen.de

M. Manns
e-mail: martin.manns@uni.siegen.de

M. Manns (ed.), *New Paradigms for Anticipated Uncertainty*, Lecture Notes in Mechanical Engineering, https://doi.org/10.1007/978-3-032-16889-4_72

1 Introduction

Modern manufacturing systems are undergoing significant transformation. Product life cycles are becoming increasingly shorter, while the number of product variants and the speed of new product development are rising rapidly (cf. [14]). This trend presents a significant challenge for production companies, which must continually adapt their equipment and tools to meet the evolving requirements of new products. A key component in assembly and handling operations is robotic grippers, which must be capable of flexibly grasping and manipulating a wide variety of workpieces [2]. To achieve this flexibility, the integration of intelligent sensor systems is essential. Among these, camera-based solutions offer a highly versatile approach for capturing environmental data and dynamically adjusting gripping strategies in response to changing object shapes or positions.

To enable scalable and retrofittable deployment, such camera systems must be compact, lightweight, and cost-efficient, particularly when mounted directly on the end-effector. At the same time, they must be capable of transmitting live video data with sufficient quality and minimal latency to a central processing unit in order to support reliable real-time image analysis. In this context, low-cost, microcontroller-based camera modules represent a promising alternative. Specifically, the ESP32-S3, which features integrated camera support and wireless communication via Wi-Fi, appears to be a suitable platform for developing lightweight, ROS2-compatible video solutions.

2 State of the Art

A variety of technical solutions are available for integrating visual sensing into robotic end-effectors, each differing in performance, complexity, and cost. Industrial cameras combined with industrial grade computers represent the most capable setup. These systems offer high-resolution image capture, low latency, and support for a wide range of industrial interfaces (cf. [6, 13]). Coupled with the computing power of modern industrial grade computers, they enable flexible deployment and seamless integration into ROS environments [4]. However, the associated costs are relatively high, typically exceeding €1000 for the camera and processing unit combined. Cameras that integrate imaging sensors and processing units, such as CPUs, GPUs, or FPGAs, into a single device, often referred to as smart cameras, offer simple operation and intelligence [15]. According to [15], these cameras support standards such as GigE Vision, USB3 Vision, or Camera Link, and are sometimes natively compatible with ROS. As they perform image processing onboard, there is no need for an external computer. Nonetheless, their prices are typically above €300 (e.g., see [1]), making them relatively expensive for large-scale deployment.

An alternative solution is an embedded AI platform, such as the Raspberry Pi, combined with a camera module, that strikes a balance between cost and functionality

[8]. These systems support basic image processing tasks using frameworks such as OpenCV and can be extended for neural inference workloads. With moderate power consumption (~ 5 W) and a cost of around €100 (see [11]), they offer a feasible solution for non-critical applications. However, they lack the robustness and deterministic performance found in industrial systems.

For a smart factory where cost is crucial and resources are limited, a low-cost edge computing device, such as an ESP32-S3 microcontroller equipped with a camera module, is presented as a potential candidate for real-time object detection tasks (cf. [10, 17]). Priced at under €20 (see [5]) and featuring onboard Wi-Fi connectivity, it offers the potential for scalable deployment in robotic applications. While ESP32-based platforms are widely used for control tasks and telemetry, their suitability for real-time video streaming and integration into ROS2 environments remains largely unexplored. In the context of industrial quality control, the ESP32-CAM has been effectively used for bolt detection on steel assemblies, achieving 95% accuracy with *Faster R-CNN*, which confirms the module's viability for structured inspection tasks in manufacturing [10]. Similarly, the integration of ESP32-CAM in surveillance robots for autonomous tracking using OpenCV and MobileNet models has shown that low-cost, embedded vision systems can deliver real-time detection capabilities while maintaining wireless communication and motor control, enabling flexible deployment in factory layouts [10]. For agricultural automation, ESP32-CAM combined with machine learning models has demonstrated 97% weed detection accuracy and successful robotic manipulation, suggesting that similar detection-then-actuation systems could be adapted for sorting or inspection tasks in production lines [17]. Additionally, the ESP32-S3 has been applied to object detection and environmental sensing tasks, such as pothole detection, utilizing ultrasonic sensors and YOLOv8n, which demonstrates reliable image classification, GPS-based reporting, and real-time inference with mobile and web integration [9]. These findings collectively indicate that ESP32-based platforms, when equipped with suitable compression, communication, and lightweight inference tools, have the potential to provide a scalable and cost-effective foundation for distributed object detection in smart factory environments.

Within the ROS2 ecosystem, communication between nodes is typically handled by the Data Distribution Service (DDS) [12]. For resource-constrained devices, the lightweight DDS-XRCE (eXtremely Resource Constrained Environment) protocol is used, via the micro-ROS framework (cf. [3]). Micro-ROS is shown to be feasible for transmitting control and status data from microcontrollers, which becomes effective in handling high-bandwidth video streams, but its extended application for real-time object detection has not been comprehensively evaluated. ROS2-based classical computer vision methods, such as frame differencing, as well as modern deep learning models, including You Only Look Once (YOLOv8) [5], are among the commonly used methods for object detection and classification. While the latter offer robust detection performance, they also demand higher image quality and real-time processing capabilities.

3 Objectives

The primary objective of this study is to develop and systematically evaluate a video transmission pipeline based on the ESP32-S3 microcontroller, integrated into a ROS2 network using the micro-ROS communication stack. The goal is to assess whether compressed live video data can be transmitted with sufficient quality and low latency to enable real-time processing for motion and object detection.

To validate the proposed system, the following research questions are addressed:

- Where are the bottlenecks in encoding, serialization, and transmission of video data on microcontroller-based platforms?
- How does the selected image resolution affect the achievable frame rate and overall real-time performance?
- Is it feasible to apply such edge devices for the detection and classification of objects using algorithms such as YOLOv8 [7] using the transmitted image stream in a real-time context?

By answering these questions, this study aims to assess the suitability of the ESP32-S3 as a cost-effective image sensor for ROS2-based industrial vision systems and its potential for integration in flexible robotic grippers.

4 Proposed Method

Consider a robotic packing application in smart factory where the robot's gripper is desired to be integrated with intelligent system for detecting an object for handling while detecting its motion and classifying it to its category (see Fig. 1). In this aspect, cost and performance is identified as a crucial factor for designing the intelligent system using vision edge device. To investigate the real-time video streaming performance of a low-cost ESP32-S3 microcontroller equipped with an integrated camera and Wi-Fi interface, functional capability and performance limitations must be analyzed. Hence, time performance analysis as well as suitability for motion detection and object recognition under constrained computational resources are the primary focus.

The experimental system comprises the following hardware and software components (see Table 1):

4.1 Image Frame Acquisition and Transmission Pipeline

Image frame acquisition and transmission from the edge device to the remote computer are necessary to implement an object detection application. The image frame acquisition from the onboard OV2640 camera is the first step implemented

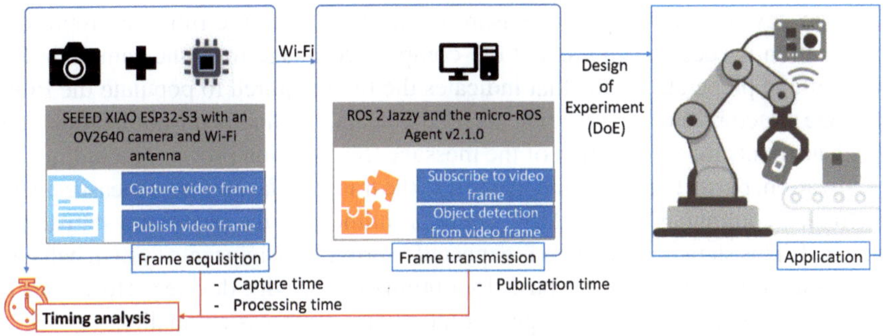

Fig. 1 Proposed method for time analysis of edge vision-based object detection and classification for robotic applications

Table 1 System requirement specification based on the proposed method in Fig. 1

Requirement	Specification
Microcontroller	SEEED XIAO ESP32-S3 sense, equipped with 8 MB PSRAM and an OV2640 image sensor
Camera module	OV2640 CMOS sensor connected via onboard interface
Communication	Wi-fi (2.4 GHz), using an external 5 dBi antenna
Firmware	Custom micro-ROS firmware configured to capture and publish compressed image frames
Host system	A remote Linux computer running ROS2 jazzy, serving as the micro-ROS agent and performing downstream image analysis

in the ESP32-S3 microcontroller. The image frame is represented in JPEG format, which is a compressed image format. The captured image frame is encoded and serialized as a compressed image message within the ROS2 framework. The process involves metadata such as timestamps and encoding information to prepare the image for network transmission and time analysis. Finally, the constructed message is transmitted over Wi-Fi using the micro-ROS middleware. This involves passing the data through the DDS-XRCE transport protocol, where the image is serialized into smaller packets and sent via UDP to a ROS2 agent running on a remote host. Each step is performed sequentially, and due to the congestion nature of the data transmission protocol, new frames are acquired once the previous frame is delivered.

4.2 *Timing Instrumentation*

A time analysis is proposed to identify bottlenecks affecting ESP32-S3 performance for image frame transmission and object detection, and also to quantify system limitations. In this regard, microsecond-resolution timestamps are applied directly to the edge device. This helps to measure image frame capture time that represents the duration needed to acquire a JPEG-compressed image from the camera buffer, the message preparation time that indicates the time required to populate the ROS2 message structure and assign the necessary metadata, and finally the publication time, which accounts for the duration of the message transmission process, including data serialization, buffer handling, and handoff to the transport layer. These measurements are logged continuously for a thirty-minute runtime, and the results are statistically analyzed. This statistical analysis is expected to provide insights into system behavior under load as well as serve as a basis for optimizing the edge device performance by adjusting parameters such as image resolution and compression quality.

4.3 Design of Experiment (DoE)

Test scenario—To investigate the feasibility of vision edge devices for intelligent automation and robotics, time performance analysis is conducted to assess their compatibility with production cycle times and responsiveness requirements. In this study, a detailed microsecond-resolution timing analysis is performed on the ESP32-S3 camera processing pipeline. Computation time is measured continuously for 30-minute intervals for frame rates configured at 5, 10, 15, and 60 frames per second. Frame rate performance is evaluated across four image resolutions: QVGA (320 × 240), VGA (640 × 480), SVGA (800 × 600), and UXGA (1600 × 1200). Each configuration is tested under stable Wi-Fi conditions (using a dedicated Wi-Fi router), and the average frame rate is measured over a sustained runtime. In parallel, a ROS2 node is implemented to process the incoming image stream using OpenCV, assessing the system's suitability for downstream vision tasks.

Methodological Constraints—Performance measurements are designed based on sequential frame handling, as concurrent or pipelined processing is not feasible on the ESP32-S3 under the given conditions. Second, image resolution and compression settings are adjusted to minimize the transmission payload size, given the limited throughput.

4.4 Feasibility Analysis for Machine Vision Application

Tests are conducted to evaluate frame rate performance across multiple image resolutions, including QVGA, VGA, SVGA, and UXGA. Each configuration is executed continuously for thirty minutes under stable Wi-Fi conditions, and the average frame rate is recorded for comparison. In parallel, a ROS2 node that analyzes images is implemented to transmit image frames using the OpenCV library, with the aim of assessing the feasibility of real-time applications. For motion detection evaluation, frame difference and contour analysis are applied (cf. [16]). Similarly, for object detection evaluation. Finally, object classification has been implemented using YOLOv8n, a lightweight deep learning model designed for multi-class inference [7].

5 Results

The *image capture time*, defined as the duration required by the OV2640 driver to acquire and deliver a JPEG-encoded image from the framebuffer, exhibits stable behavior across all frame rates. The median duration is approximately 0.058 ms, with values ranging from 0.014 to 0.265 ms. The message *processing time*, covering message construction and metadata assignment, is also consistent, with a median of 0.007 ms and values extending up to 0.024 ms.

Dummy-image is applied as a benchmark to compare the camera image frames. Accordingly, the dummy image confirms that these two stages are unaffected by actual image content, as their averages remain nearly constant across configurations. This is visualized by the overlaid dummy mean lines in Fig. 2.

In contrast, the *publication time*, encompassing message serialization, fragmentation into 1024-byte UDP packets, and wireless transmission via the DDS-XRCE protocol, has a median duration of approximately 49.6 ms, with observed values ranging from 3.2 ms to 136 ms. This stage dominates overall latency and reflects the cost of transmitting high-volume data from constrained embedded systems. Publication time shows the most spread, especially under higher FPS, indicating transmission overhead.

Figure 3 illustrates that the frame rate decreases as the resolution increases. QVGA exceeds 40 FPS, while UXGA drops below 5 FPS. For VGA and SVGA, the Dummy data and the camera system yielded nearly similar frame rates. Both exceed the throughput threshold of the ESP32-S3 pipeline, and further resolution increases may not add to already saturated bandwidth and processing time.

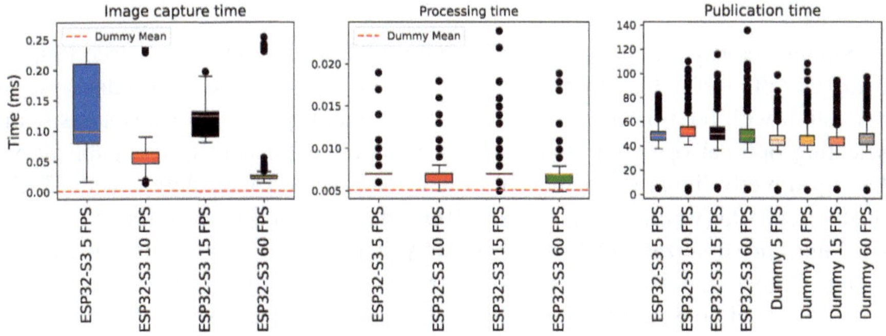

Fig. 2 Box plots for image capture, processing, and publication time

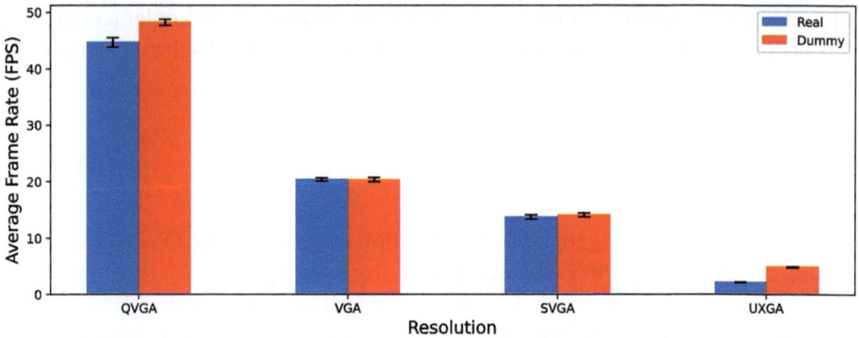

Fig. 3 Average frame rate comparison across sensor resolutions (QVGA, VGA, SVGA, UXGA) for real image acquisition and dummy image input. Error bars indicate standard deviation over a 30 min interval

6 Discussion

The timing analysis confirms that the ESP32-S3 camera module meets key requirements for integration into flexible, low-cost robotic systems envisioned in modern reconfigurable production environments. With stable image capture and message processing times in the microsecond range, the device supports fast and predictable local sensing, critical for real-time adaptation in robotic grippers handling diverse workpieces. While the publication time introduces measurable latency, with a median of 49.6 ms and peaks exceeding 130 ms, which is acceptable for many object detection tasks that do not require hard real-time constraints. These findings align with the demands outlined in the introduction: scalable, lightweight, ROS2-compatible vision modules that can support the agility and modularity required in smart factories facing rapid product evolution. The ESP32-S3 offers a promising path for embedding edge vision into robotic end-effectors, enabling dynamic, cost-efficient adaptation to product variation without the complexity or expense of traditional industrial vision systems.

Simultaneously, the image stream is processed on the ROS2 node (see Fig. 4) using three OpenCV-based pipelines, and YOLOv8n is utilized to correlate image resolution with both frame rate and detection capability. This provides a basis for selecting optimal operating conditions in practical applications. The publication latency directly limits the achievable frame rate. At VGA resolution with JPEG quality set to 25 (where zero is low and 100 is the best quality), each frame requires encoding and serialization into approximately 70 packets. These overhead limits the system to a maximum sustainable frame rate of approximately 12–18 FPS, depending on Wi-Fi conditions and the microcontroller's load. Higher resolutions result in a bottleneck due to increased payload size, while lower resolutions (e.g., QVGA) can reduce serialization and improve throughput.

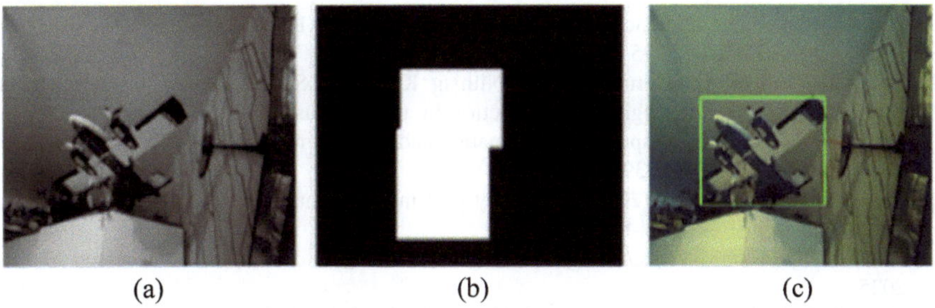

Fig. 4 Object detection and classification application: a gray scale, b frame difference, and c YOLOv8n

7 Conclusion

These findings confirm that the primary performance limitation is not due to sensor speed or local image processing, but rather to the encoding, serialization, and transmission load imposed by the DDS-XRCE protocol on constrained hardware. Future work will focus on system improvement by simulating the sensor mounted on a robot arm to investigate how the proposed approach enables robots to intelligently manipulate objects.

References

1. Advantech (2025) Industrial smart camera. https: //www.advantech.com/industrial-camera
2. Ali H, Zainur MI, Elshaikh M, Aman AM (2022) Development of vision based smart gripper for material handling using internet of things. J Phys Conf Ser 2312:012040. https://doi.org/10.1088/1742-6596/2312/1/012040
3. Azumi T, Maruyama Y, Kato S (2020) ROS-lite: ROS framework for NoC-based embedded many-core platform. In: 2020 IEEE/RSJ international conference on intelligent robots and systems (IROS), pp 4375–4382. 10.1109/IROS45743.2020.9340977
4. D'Avella S, Avizzano CA, Tripicchio P (2023) ROS-industrial based robotic cell for industry 4.0: eye-in-hand stereo camera and visual servoing for flexible, fast, and accurate picking and hooking in the production line. Robot Comput-Integr Manuf 80:102453. https://doi.org/10.1016/j.rcim.2022.102453
5. Espressif Systems (2025) ESP32-S3 overview. https://www.espressif.com/en/products/socs/esp32-s3
6. Golnabi H, Asadpour A (2007) Design and application of industrial machine vision systems. Robot Comput-Integr Manuf 23:630–637. https://doi.org/10.1016/j.rcim.2007.02.005
7. Jocher G, Qiu J, Chaurasia A (2023) Ultralytics YOLO. https://github.com/ultralytics/ultralytics
8. Mazzia V, Khaliq A, Salvetti F, Chiaberge M (2020) Real-time apple detection system using embedded systems with hardware accelerators: an edge AI application. IEEE Access 8:9102–9114. https://doi.org/10.1109/ACCESS.2020.2964608
9. Mishra M, Shukla A, Patel K, Varshney H (2024) ESP32-S3 DevKit-C integration for enhanced road safety for pothole detection and nighttime driving assistance. In: 2024

2nd international conference on disruptive technologies (ICDT), pp 238–244. 10.1109/ICDT61202.2024.10488995
10. Pesha JV, Jotham G, Thirumalai CK, Nithisg Kumar V, Sasirekha M (2024) Enhanced surveillance abilities through object detection in autonomous robots. In: 2024 international conference on system, computation, automation and networking (ICSCAN), pp 1–5. 10.1109/ICSCAN62807.2024.10893951
11. Raspberry Pi Foundation (2025) Raspberry pi 4 model B. https://www.raspberrypi.com/products/raspberry-pi-4-model-b/
12. ROS on DDS (2025). https://design.ros2.org/articles/ros_on_dds.html. Last accessed 27 July 2025
13. Sága M, Bulej V, Čuboňova N, Kuric I, Virgala I, Eberth M (2020) Case study: performance analysis and development of robotized screwing application with integrated vision sensing system for automotive industry. Int J Adv Robot Syst 17:1729881420923997. https://doi.org/10.1177/1729881420923997
14. Stauder L, Yang F, Grünert G, Niemietz P, Bergs T (2025) Economic evaluation of manufacturing process sequences under consideration of potential component variants. Prod Eng. https://doi.org/10.1007/s11740-025-01339-w
15. Stiefenhöfer P, Keppler P (2018) Revolution in industrial image processing. Opt Photonik 13:45–47. https://doi.org/10.1002/opph.201800030
16. Suzuki S, be K (1985) Topological structural analysis of digitized binary images by border following. Comput Vis Graph Image Process 30:32–46. https://doi.org/10.1016/0734-189X(85)90016-7
17. Vinod DN, Bhaskar B, Teja BR, Kamalesh A (2024) Riddance of weed using an agri-robot employed by arduino and ESP32 CAM. In: 2024 15th international conference on computing communication and networking technologies (ICCCNT), pp 1–6. 10.1109/ICCCNT61001.2024.10724013

Collaborative Robotics

Robust Automated Unscrewing with Cost-Effective Robot Control System

Sara Menetrey, Manuel Möckel, Holger Schlegel, Matthias Rehm, and Martin Dix

Abstract Research in the field of automated disassembly, in relation to end-of-life electric car batteries, has focused on the removal of screws over the last ten years. The most common detection system is based on visual recognition, which is unable to recognize the pitch or length of the screw. The unscrewing process therefore relies on knowledge from databases, human cooperation, or force sensors to regulate the movement of the robot in the direction of the screw axis during unscrewing. This movement is crucial during the screwing process to maintain the screwdriver's engagement with the screw drive and ensure the successful removal of the screw. In this paper, a new cost-efficient solution is presented in which a spring system and a foil potentiometer are embedded in a conventional industrial fixed screwdriver to measure the spring travel caused by contact between tool tip and screw head. Tests were carried out to determine the accuracy and precision of the spring travel measurement. The control of the robot's movements via flexible G-code programming is explained and mathematically proven to be safe, depending on the robot's feed rate and its processing time. Finally, practical experiments were performed to verify the correct execution of the process. The presented technology is intended to be easily reproduced and integrated into any CNC-controlled robot equipped with a conventional screwdriver to make the disassembly technology more accessible.

Keywords Disassembly · Spring travel sensing · Adaptability · Automated unscrewing · Robot control

S. Menetrey (✉) · M. Möckel · H. Schlegel · M. Rehm · M. Dix
Professorship Production Systems and Processes, Chemnitz Technical University, Chemnitz, Germany
e-mail: sara.menetrey@mb.tu-chemnitz.de

M. Manns (ed.), *New Paradigms for Anticipated Uncertainty*, Lecture Notes in Mechanical Engineering, https://doi.org/10.1007/978-3-032-16889-4_73

1 Introduction

An important part of the circular economy is disassembling end-of-life (EOL) products. It is divided into two kinds of processes: destructive and non-destructive. The use of non-destructive processes enables the recovery of intact parts for reuse, repair, or remanufacture. However, these processes tend to be more complicated and very specific to the product itself, whereas destructive processes mostly just cut through any materials that need to be separated.

For a product to be disassembled without being further damaged, it must be assembled using detachable fasteners. Those can involve magnetic fields, piezoelectric materials, light radiation, or temperature changes to engage or loosen a joint [2]. The most widespread detachable fastener is the screw (83 published standards and projects from the International Organization for Standardization (ISO) found under the category 'bolts, screws, studs' [8]). It is mechanically removable with a manual, electric, or pneumatic screwdriver. Automated disassembly systems are being developed in several sectors to make the dismantling process more efficient and less tiring or even dangerous for the operator.

Even if the principle of a screw remains the same, its drive, and size can vary considerably. For example, to assemble an electric vehicle battery (EVB) pack, different sizes of M4 to M8 screws are used [9]. These differences in size are crucial to the control of the robot, which must withdraw as the screw is removed to stay engaged with it.

As part of the KaDoTE project, a new automated system to disassemble EOL EVB packs is being developed. This includes an adaptable automated unscrewing system, which does not need information about the screw pitch and length [1]. This paper presents the current screw removal systems and the recurring problem of detection and adaptability, followed by a detailed description of the mechanical design and the control algorithm of the system.

2 State of the Art

The subject of automated disassembly of EOL products has been around since the early 2000s, when the focus lay on electronic devices [11]. With the mass marketing of EVs, research has devoted more and more time to developing robotic systems for dismantling EVBs since the early 2010s. In the literature, the topics can be separated into different sub-themes [10]:

- the detection, using a vision system or other sensors to determine the current condition of the various EOL products;
- the artificial intelligence, which assigns the recognized objects to processes and creates a disassembly sequence;
- the tooling designed to perform the disassembly and
- the degree of autonomy, either fully or semi-automatic, using human-robot collaboration.

The focus here is on fully automated tools and control technologies to loosen the widest possible variation of screws.

Industrial fixtured screwdrivers available on the market are not specifically designed for unscrewing. Notable manufacturers are Cleco, Onrobot, Stöger, Weber. Stöger and Onrobot mention that their tools are meant to be used with collaborative robots. No manufacturer specifies the installation of force sensors, which would enable robot control, in their screwdrivers. Most of them present automatic feeding systems used for screwing, but none of them presents an automatic removal system, which indicates that the tools' main purpose is not to remove but to mount screws. Adaptive systems are usually superfluous during assembly as the condition of the parts is known and the size, shape, and positions of the screws are predefined.

Ferre et al. [3] used a UR5e robot to perform a force analysis during screwing. To hold the correct position, they used the inbuilt impedance control of the robot, combining position and force control. Huang et al. [4] used a collaborative robot with an inbuilt impedance control (here called the compliant mode) to unfasten screws. To determine the length of the screw, they monitored the position of the robot along the screw axis. Kaarlela et al. [5] reviewed 63 articles focusing on robotized EVB disassembly from 2014 to 2023. They stated that "none of the presented publications presented a solution to detect the thread pitch of the screws".

Overall, the only methods that have been used to control and adapt the speed of the robot to the screw pitch are based on force-sensing or prior knowledge of the screw type. A lot of previous research relies on the compliant modes of the robots. So-called collaborative robots have inbuilt force sensors and impedance control modes that enable force-controlled motion. These robots often have a smaller working area and a low load capacity.

The aim of this paper is to propose a tool and a control algorithm solution that enable any robot equipped with a conventional fixtured screwdriver to unscrew. Especially robots with computerized numerical control (CNC) are getting more popular due to their higher precision compared with conventional robot controls and their affordability compared with CNC machines. Such robots can also be used for milling operations. The idea is to enable switching between destructive and non-destructive tools while keeping it simple for the operator by being programmable in G-code.

3 Design

The construction was briefly mentioned in [6]. A more detailed explanation is provided here. The screwdriver is a standard fixtured industrial screwdriver equipped with a transducer to measure the rotational angle and torque (Cleco 1BD53T1S-A) [6]. Its maximum torque is 53 Nm, thus being able to remove every kind of screw up to M8 (DIN 912).

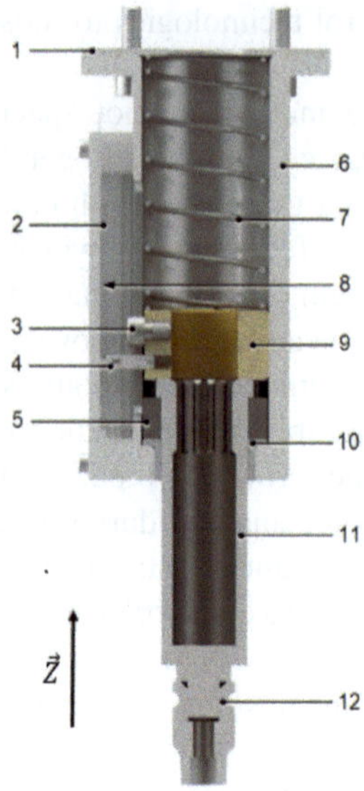

1 Spring limiter
2 Sensor housing
3 Guiding screw
4 Resilient contact element
5 Spacer sleeve
6 Spring housing with mounting flange
7 Spring
8 Foil potentiometer
9 Slider
10 Slide bearing
11 Square drive with wedge profile
12 Socket wrench

Fig. 1 CAD section of the screwdriver tip with spring system and foil potentiometer [6]

It has a controller, which is connected to the robot controller. This controller enables the selection between predefined screwing modes. For each mode, several elements must be specified in advance: the direction of the rotation, the rotational speed, the stopping angle, the stopping torque, the maximum angle, and the maximum torque. If the maximum values are exceeded, the operation is classified as *not okay*. A mode can be selected and started by variables sent between the controllers using Profinet.

The used robot is a COMAU NJ130-2.05 with a Sinumerik 840D SL CNC from Siemens. The tool tip is equipped with a 50 mm spring-loaded adapter. To control the robot's movement while unscrewing, the spring travel of this adapter is measured using a foil potentiometer. In this way, there is no risk of the equipment being exposed to contact forces greater than those caused by the stiffness of the spring itself. The construction of the adapter assembly is shown in Fig. 1.

The screwdriver guide (not represented here) rotates the square drive with wedge profile and the socket wrench. The rest of the construction stays static and is attached to the outer housing of the screwdriver. When the tool tip comes into contact with an object, the wedge profile is pressed into the spring housing and takes the slider and the spring-loaded contact element with it. The rotation of the slider is prevented by the guiding screw and ensures that the contact element always remains in touch with the foil potentiometer. The output voltage of the potentiometer is proportional to the spring travel, which can therefore be deduced from it. The chosen foil potentiometer has a measurement range of 0 to 50 mm, and its specifications indicate a service life of a minimum of one million cycles.

4 Control

The developed unscrewing sequence is shown in Fig. 2. The preconditions are: the screwdriver is already equipped with the correct bit, and the position and orientation of the targeted screw head are known.

First, the robot moves to this transmitted position and then moves along the screw axis (Z negative, Fig. 1) to achieve contact until a suspension of 3 mm with an error of 0.3 mm is reached. Then the *screw search* mode is started. This means the screwdriver rotates clockwise until a resistance of at least 1 Nm is found. If no resistance is found, the screwdriver stops turning after five turns, and an error is displayed. The screwing process is interrupted.

Once the screw is found, the robot pushes against the screw until reaching the suspension control value C (here 10 mm) and the *loosening* mode is started. The screwdriver rotates counterclockwise for a predefined number of turns (in this case, 10). During this process, the spring travel is measured repeatedly. If it deviates more than 0.3 mm from the specified 10 mm, the robot's position is corrected in the direction of the screw axis by the amount of the measured deviation. If not, a counter is incremented to determine when the screw is completely unscrewed. That counter is reset as soon as a position correction is required. When the tool has reached the number of revolutions, either the counter is less than or equal to two and the *loosening*

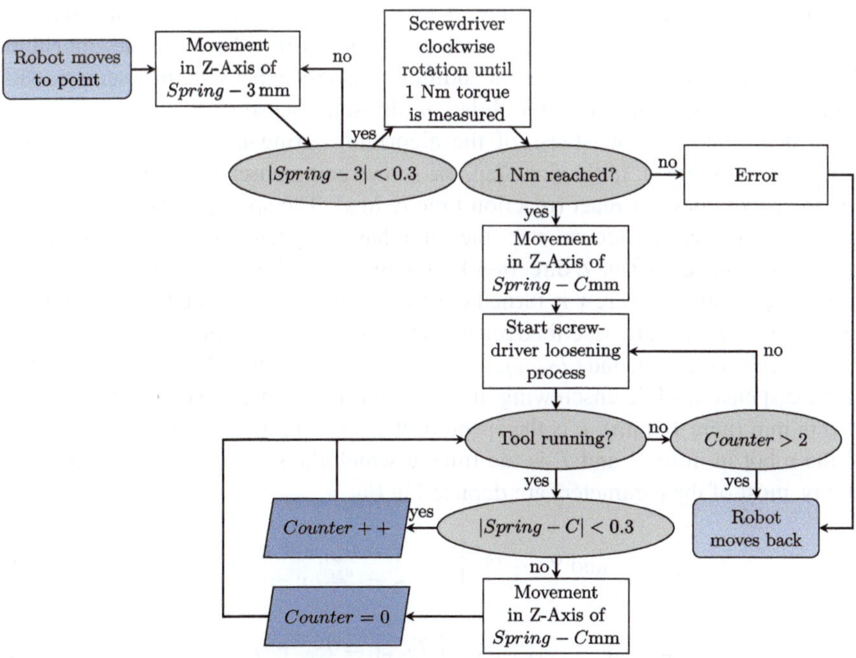

Fig. 2 Flowchart of the robot's control algorithm using the spring travel measurement

mode is restarted, or the counter is higher and the process is completed. The robot then moves back in the direction of the screw axis (Z positive). By constantly monitoring the spring suspension, this control system does not need to know the screw pitch or length to remove a screw.

5 Experimental Analysis

First, the accuracy and precision of the system are evaluated to assess its suitability for the control system. As mentioned before, the range of the foil is 50 mm. The value range of a Siemens analog input, to which it is connected, is 0 to 27648. Thus, the system has a theoretical resolution of 1.81 μm. To test its precision and accuracy, the screwdriver tip (without bit) was brought into contact with a flat surface. The robot was then moved further down in steps of 5 mm until a total distance of 30 mm was traveled. The same is executed going upwards until arriving back to a suspension of 0 mm. This was repeated five times. Between every step, the analog input value is read, and the exact position of the robot is measured using a laser interferometric displacement measurement system (Renishaw XL 80). The accuracy was then defined as the maximum difference between the measured position and the actual position and was found to be under 0.34 mm. The precision was defined as the maximum standard deviation between the ten potentiometer measurements and was found to be under 0.05 mm. The accuracy being much smaller than the total range and the precision being smaller than the permitted error in the algorithm, the system is considered suitable. In addition, it is less expensive than sensors with similar functions, such as infrared, lidar, or ultrasonic sensors.

To determine the robustness of the algorithm during unscrewing, a two-step sequence was derived. In the first step, the screw, being unscrewed, pushes the tool, and the robot does not react (reaction time r_t in s). The spring is then compressed at a constant rate. In a second step, the robot has recognized the state of the spring and starts a movement in Z direction by the length of this measured spring depth. At the same time, the screw is turned out further at the same speed. The sequences $(u_n)_{n\in\mathbb{N}}$ and $(T_n)_{n\in\mathbb{N}}$ are described recursively using two sub-sequences (respectively $(u_{2n+1})_{n\in\mathbb{N}}$, $(u_{2n+2})_{n\in\mathbb{N}}$ and $(T_{2n+1})_{n\in\mathbb{N}}$, $(T_{2n+2})_{n\in\mathbb{N}}$) in (1) and (2). Here, u is the state of the suspension while unscrewing in mm, minus the predefined controlled value C also in mm (here 10 mm), v is the speed of the screw in mm s^{-1}; F is the feed rate of the robot in mm s^{-1} and T is the time at which the step n takes place in s. For clarity, most of the parameters are depicted in Fig. 3.

$$u_0 = 0\text{ , and } \forall n \in \mathbb{N}, \begin{cases} u_{2n+1} = u_{2n} + r_t v \\ u_{2n+2} = \frac{u_{2n+1}}{F} v = \frac{u_{2n}+r_t v}{F} v \end{cases} \tag{1}$$

$$T_0 = 0\text{ , and } \forall n \in \mathbb{N}, \begin{cases} T_{2n+1} = T_{2n} + r_t \\ T_{2n+2} = T_{2n+1} + \frac{u_{2n+1}}{F} \end{cases} \tag{2}$$

Fig. 3 Picture of the screwdriver and an M6 Allen screw with annotations

The following explicit formulas can be derived from them, $\forall n \in \mathbb{N}$:

$$u_{2n} = \frac{r_t v^2}{F}\left(\frac{1-\left(\frac{v}{F}\right)^n}{1-\frac{v}{F}}\right); \qquad u_{2n+1} = \frac{r_t v^2}{F}\left(\frac{1-\left(\frac{v}{F}\right)^n}{1-\frac{v}{F}}\right) + r_t v \tag{3}$$

$$T_{2n} = \frac{\sum_{k=1}^{n} u_{2k-1}}{F} + n r_t; \qquad T_{2n+1} = \frac{\sum_{k=1}^{n} u_{2k-1}}{F} + (n+1) r_t \tag{4}$$

(u_{2n}) and (u_{2n+1}) therefore only converge if $v < F$. If this is the case, their convergence values can be determined, and the overall sequence $(u_n)_{n\in\mathbb{N}}$ will always fluctuate.

$$\lim_{n\to\infty} u_{2n} = \frac{r_t v^2}{F-v}; \qquad \lim_{n\to\infty} u_{2n+1} = \frac{r_t v^2}{F-v} + r_t v \tag{5}$$

The difference between both is $r_t v$. For the system to be safe, the larger value plus the control value should not exceed the spring range; otherwise collision will occur. When the range is noted R in mm, this rule results in (9).

$$\frac{r_t v^2}{F-v} + r_t v + C < R \tag{6}$$

$$\Leftrightarrow \quad \frac{vF}{F-v} < \frac{R-C}{r_t} \tag{7}$$

as $F - v > 0$

$$\Leftrightarrow \quad vF < (F-v)\frac{R-C}{r_t} \tag{8}$$

$$\Leftrightarrow \quad v < \frac{F\frac{R-C}{r_t}}{F + \frac{R-C}{r_t}} \tag{9}$$

Of course, u only represents the evolution of the suspension in the hypothetical case of an endless screw. In reality, at some point, the screw is completely unscrewed and no longer pushes the screwdriver. The curve would then fall back to zero. The time δt (in s) after which this happens is defined by (10) with L the screw length in mm.

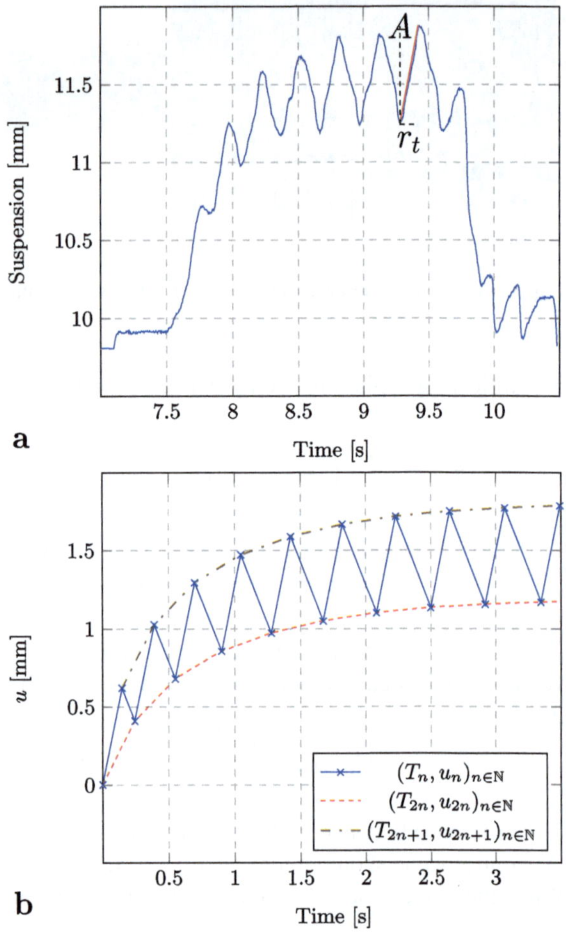

Fig. 4 Spring travel during unscrewing **a** real measurement **b** theoretical suspension for $v = 4.2\,\text{s}^{-1}$, $r_t = 0.147\,\text{s}$ and $F = 6.37\,\text{mm}\,\text{s}^{-1}$

$$\delta t = \frac{L}{v} \tag{10}$$

To verify these formulas in practice, the real position of the screwdriver and the suspension value were recorded during eight unscrewing processes with a sampling rate of 4 ms. All the tests were performed with Allen M6 screws of similar length and a robot feed rate of 400 mm/min. The eight resulting curves are all very similar, especially during the unscrewing phase. The major differences are found in the search phase, which is more or less long depending on the test. This is since the relative position of the tool in relation to the screw head is always different. Figure 4 shows an example of the resulting curves zoomed on the unscrewing phase and the calculated theoretical spring curve.

As predicted, the suspension curve oscillates, converging towards two extremes of oscillation, then falls back to the set value once the screw has been completely removed. The amplitude of the fluctuation (A in mm) is equal to the processing time of the program (r_t in s) times the speed of the screw (v in mm/s), which in turn depends on the screw pitch (S in mm) and the rotational speed of the screwdriver (ω in s^1).

$$\begin{aligned} v &= S \times \omega \\ A &= r_t \times v = r_t \times S \times \omega \end{aligned} \tag{11}$$

The measurement shows an amplitude of 0.63 mm and a reaction time of 0.12 s. So the screw speed had to be 5.25 mm s^{-1}. The rotational speed of the screwdriver was set to 5 s^{-1}. According to (11), the screw pitch should be about 1.0 mm, which is actually the standardized pitch of an M6 screw. The unscrewing process begins at 7.5 s and ends around 9.8 s. Using (10), the screw length can also be deduced, which would be 9.7 mm. The screw used for the tests was indeed an M6, length 10 mm. This algorithm, together with the measurement of the spring path, enables the user to deduce the pitch and length of the removed screw, knowing only the screwdriver's rotational speed.

6 Discussion and Conclusion

This paper presented a cost-efficient way to adapt existing tools for automated unscrewing. State-of-the-art research has indicated that there is currently no way of identifying the pitch and length of a mounted screw without using force sensors. However, the presented technology using a foil potentiometer can be adapted to any type of screw. It has already been successfully tested on star, Allen and hexagonal screws, from size up to M8, as these are the screw heads that are most frequently used in EVBs.

Its speed limitation is described by (9), which depends on the robots' reaction time, feed rate, as well as the range and control value of the spring system. This means that if the screwdriver system is set as mentioned above ($C = 10$ mm, $R = 50$ mm, $r_t = 0.12$ s and $r_t = 5$ s^{-1}) and if it is known that the robot reaches a speed of 10 mm s^{-1}, all the screws that are to be removed using this method need to have a smaller pitch than 1.94 mm. Otherwise, the spring system could get overloaded, and the screw could damage the tool. If larger screws are to be removed, the screwdriver rotation speed can be reduced to enable the unscrewing without risk. Nevertheless, the maximum jerk of the robot must be considered when determining the maximum speed. As the robot does not move smoothly but stops repeatedly, a low jerk setting can lead to a significantly lower average feed rate than programmed.

The formulas presented in this paper can be used to determine the pitch and the length of the screw. This information could then be stored in a database describing the product and reused when the same product has to be disassembled again. This can also be useful for further treatment of the dismantled parts. The screws could be sorted automatically without having to install additional measuring devices.

Subsequent experiments will combine this new tool with automated screw detection using visual data processing and artificial intelligence to recognize the screws and give their position. The goal is to develop a flexible, fully automated system for unscrewing screws of EVBs.

Acknowledgements This Project is supported by the German Federal Ministry for Economic Affairs and Climate Action (Bundesministerium für Wirtschaft und Klimaschutz, BMWK).

References

1. Bräunig G, Menetrey S, Hertel D (2024) Construction of a demonstrator for artificial intelligence-supported, automated dismantling of battery systems. In: IFIP international conference on advances in production management systems. Springer, pp 18–32
2. Duflou J, Willems B, Dewulf W (2006) Towards self-disassembling products. In: Innovation in life cycle engineering and sustainable development. Springer, Grenoble
3. Ferre M, Peralta PE, Gallego VR, Barbosa N, Sánchez-Urán MÁ (2023) Forces analysis on robotics screwing tasks. In: Iberian robotics conference. Springer, pp 553–563
4. Huang J, Pham DT, Li R, Jiang K, Lyu D, Ji C (2020) Strategies for dealing with problems in robotised unscrewing operations. In: International precision assembly seminar. Springer, pp 93–107
5. Kaarlela T, Villagrossi E, Rastegarpanah A, San-Miguel-Tello A, Pitkäaho T (2024) Robotised disassembly of electric vehicle batteries: a systematic literature review. J Manuf Syst 74:901–921
6. Menetrey S, Möckel M, Schlegel H, Dix M (2024) Design and validation of an automated removal system for screws during disassembly
7. Quattrucci L, Ceccarelli M, Santoro M, Russo M (2024) Problems and requirements for the robotic disassembly of lithium-ion batteries in the automotive industry. In: The international conference of IFToMM ITALY. Springer, pp 137–145
8. The International Organization for Standardization (ISO) (2024) Iso 21.060.10 bolts, screws, studs. https://www.iso.org/ics/21.060.10/x/p/1/u/0/w/0/d/0
9. Villagrossi E, Dinon T (2023) Robotics for electric vehicles battery packs disassembly towards sustainable remanufacturing. J Remanufacturing 13(3):355–379
10. Vongbunyong S, Chen WH, Vongbunyong S, Chen WH (2015) Disassembly automation. Springer
11. Weigl-Seitz A, Hohm K, Seitz M, Tolle H (2006) On strategies and solutions for automated disassembly of electronic devices. Int J Adv Manuf Technol 30:561–573

Role-Based Human-Robot Interaction for Symbiotic Collaboration

Tadele Belay Tuli, Raza Saeed, and Martin Manns

Abstract Interpreting human motion during physical human-robot interaction is critical for enhancing robot adaptability and acceptance in collaborative tasks such as joint handling of rigid objects. Effective modeling of the trajectories of human joints from initial to target positions allows robots to adapt their motion accordingly. However, the dynamics of role shifts between human and robot during such adaptation remain insufficiently understood. This work presents a trajectory-based approach for modeling role transitions in human-robot collaboration (HRC). This method analyzes situations that trigger role-switching and interaction conflicts by evaluating force and velocity vectors at the end-effector level. A finite state machine (FSM) is implemented to classify interaction states, such as following, leading, or resisting, based on spatial deviation and motion alignment. In addition, this method monitors force unit within a desired threshold while planning robot motion along a generated human motion corridor. Unlike traditional pre-programmed robotic systems, this role-adaptive approach enables flexible and responsive behavior, supporting more natural and robust collaboration in dynamic production environments.

Keywords Motion adaptation · Mutual care · Anticipatory model · Human motion · Human-robot collaboration

1 Introduction

Recent studies have shown that modeling human motion behavior is critical for enabling robots to understand and interpret human actions in collaborative tasks [5]. When robots can interpret human movement, they can adapt their actions accordingly, enhancing their acceptance in collaborative environments. Such robots require human motion interfaces that provide relevant information and control mechanisms

T. B. Tuli (✉) · R. Saeed · M. Manns
PROTECH - Institut für Produktionstechnik, Universität Siegen, Siegen, Germany
e-mail: tadele-belay.tuli@uni-siegen.de
URL: https://protech.mb.uni-siegen.de/fams

M. Manns (ed.), *New Paradigms for Anticipated Uncertainty*, Lecture Notes in Mechanical Engineering, https://doi.org/10.1007/978-3-032-16889-4_74

for seamless human-robot interaction [12]. In collaborative tasks like joint object handling, interactions typically involve shared workspaces, direct contact, a common goal, simultaneous actions, resource sharing, and varying autonomy levels, leading to a form of *symbiotic collaboration* [2].

In symbiotic collaboration, humans and robots are mutually dependent, characterized by action, guidance, and protection. Here, "action" refers to movements by both agents, "guidance" involves collaborative planning and trajectory adaptation, and "protection" includes strategies like obstacle avoidance.

The success of symbiotic collaboration relies on the quality of information exchanged between humans and robots (c.[13]). This raises research questions regarding interaction dynamics and trajectory adaptation in response to human motion styles.

For joint object handling, a role-based human-robot interaction (HRI) model capable of recognizing role-switching is essential for real-time adaptation. We propose a finite state machine (FSM) that determines roles based on calculated forces from motion capture data of two humans handling an object. For example, if a human accelerates toward a robot, the force direction indicates that the human is leading and the robot is following; balanced force implies co-manipulation [2].

In human-robot collaboration, recognizing and labeling each agent's role (e.g., leader, follower) is critical for responsive motion adaptation. Labeled roles serve as cues for the robot to lead, follow, or synchronize its motion. Without this labeling, the robot lacks context to adjust its behavior appropriately. This work uses role state identification as the basis for transitioning between motion modes in a finite state machine. For instance, if a robot detects a human transitioning into a leading role, it can proactively adjust its trajectory to avoid resistance or instability.

2 Objective

This study presents a simulation-based method for enabling role-based human-robot interaction within symbiotic collaboration scenarios. The objective is to develop a trajectory-level motion generation and adaptation method that responds to human behavioral cues in real time and supports dynamic role-switching between humans and robots. Instead of relying on fixed pre-planned paths, the robot's trajectory dynamically adapts based on interaction states inferred from motion patterns and inferred forces, allowing it to transition smoothly between leading, following, or co-manipulating. The method may enable transfer learning principles and domain adaptation to enable intuitive and efficient collaborative object handling by coordinating robot motion strategies with human intent and role behavior.

3 Related Works

Interpreting human motions during human-robot interaction (HRI) for collaborative tasks such as simultaneous object handling presents several key challenges due to heterogeneity in human behaviors that are diverse and unpredictable. Therefore, it is difficult for robots to consistently interpret dynamic gestures and motions [3]. The variability in how robots perceive human actions makes developing robust handling approaches hard.

According to [11], overcoming uncertainty in human intentions and communication barriers between humans and robots, including interpreting verbal and non-verbal commands, is necessary for seamless collaboration. However, this often depends on data from various sensor types (e.g., motion sensors, force/torque sensors). Tracking and interpreting human motions add additional challenges for effective HRI [8, 15].

The study in [4] presents an approach for extraction of interactive skills of human-robot collaboration from observation of task to identify interaction phases, which involves modeling primitive interactions using partial trajectory. [1] investigates interaction phases to identify whether a human and a robot harmoniously manipulate objects or engage in conflict using haptic data from force and torque sensors. The concept of roles for human-robot interaction requires the understanding of human-human interaction, which takes into account mechanisms of human intention inference (cf. [10]), role shifts and learning effects (cf. [14]).

In general, these studies identify roles during human-robot interaction through behavior analysis (e.g., identifying dominant body joint such as hand for grasping actions), interaction patterns (e.g., analysing total time, distance, and acceleration variance [14]), conflict detection (e.g., measuring interaction force [1]), and by developing models such as predictive [4] or adaptive control [10] by anticipating future trajectories and proactively following when there is a leader or proactively contributing when interactions are not harmonious.

Currently, role-based human-robot interaction (HRI) faces several limitations, one example of which is how robots dynamically adjust their behavior based on the states of human intention and the interaction context. This requires advanced cognitive architectures that are still in the early stages of development [6]. However, this work proposes a simplified method that derives role states from motion trajectories with an assumption that robots naturally follow a unique path unless the interaction force from the human partner deviates from its planned path.

4 Method

We propose a trajectory-based method for analyzing physical human-robot interaction by integrating motion normalization, trajectory warping, and finite state classification. First, recorded human hand trajectories are normalized using a 3D alignment algorithm that translates and rotates each motion segment to a consistent canonical orientation. The normalized trajectories are then warped to a predefined spatial segment using scaled directional mapping, ensuring uniform representation across trials. A circular trajectory is defined to represent the robot's ideal motion path, parameterized by start, stop, and peak positions. This trajectory is projected onto a directional plane fixed at a specified Y-coordinate, and intersection points with the human trajectory are used to extract role-relevant spatial features.

To characterize interaction dynamics, we implement a finite state machine (FSM) that classifies each frame of the trajectory into one of three discrete roles: *following*, *leading*, or *dragging*. This classification is based on force direction derived from the spatial deviation between the human and robot trajectories, as well as the alignment of their motion directions. Interaction forces are estimated using second-order derivatives of position, and transitions between states are tracked using a transition matrix. This method enables frame-level interpretation of role allocation, effort distribution, and control exchange. Figure 1a illustrates the whole pipeline, including trajectory generation, digital human-robot-object modeling, sensor model integration, FSM-based decision-making, and simulation-based execution evaluation.

The IDEF0 diagram in Fig. 1b depicts a digital twin-based finite state machine (FSM) architecture for role-based human-robot interaction. The process starts with

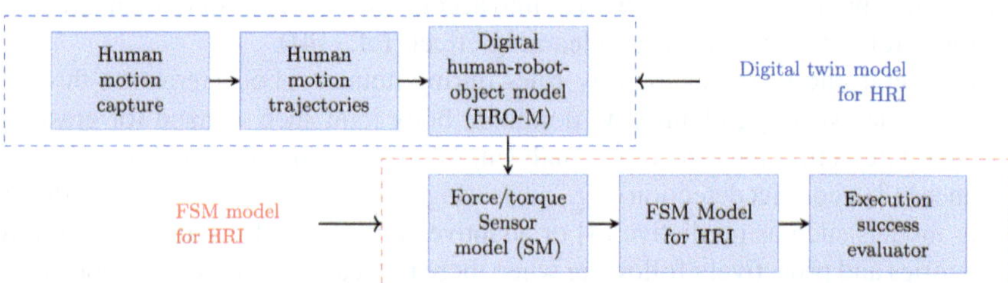

(a) A high-level illustration of a role-based simulation workflow. The novel contribution of the current work falls within FSM model for HRI.

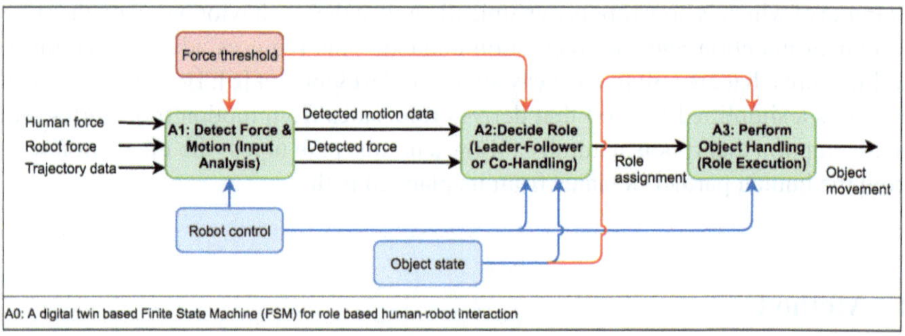

(b) IDEF0 (cf. [7]) representation of finite state machine (FSM) model for HRI. Connectors: control(red), mechanisms and resources (blue), and input and output (black).

Fig. 1 Illustration of the proposed method

force and motion input analysis (A1), where human and robot motions are processed to extract features like velocities and applied forces. In the next stage (A2), these features are assessed against thresholds to determine the interaction role, whether the human or robot is leading, following, or co-handling, via a rule-based FSM.

This inferred role state guides the robot's object-handling behavior (A3), adjusting its actions based on real-time updates and force feedback. A feedback loop between A3 and the robot controller to A1 and A2 facilitates continuous behavior monitoring and adjustment. This digital twin setup enables adaptive motion generation to enhance natural human-robot collaboration, allowing the system to respond dynamically to changes in applied force during co-manipulation scenarios.

5 Results

The recorded human motion data from the experimental tests in the lab setup during joint object handling are analyzed to evaluate the proposed method. In this study, 200 cycles are recorded for picking and placing tasks (see Fig. 2). Focusing on hand movement, warped canonical circular paths are generated with the initial and final hand positions as references. The resulting canonicalized 2D and 3D representations are shown in Fig. 2a,b, illustrating consistent warping to a common motion structure while preserving motion variability across repetitions.

In order to study the human motion behavior, we defined a hypothetical robot path using circular motion, which is overlayed with the human hand path (see Fig. 2c). Comparing hand path with the ideal robot, interaction dynamics are identified between *following*, *leading*, and *dragging* states throughout the task. As shown in Fig. 3a, the human spends approximately 21.7% of the time *following* the robot's path, 26.7% *leading* the interaction, and 51.6% in a *dragging* state, indicating delayed response or passive resistance. Force estimation derived from second-order kinematic derivatives reveals a clear correlation between effort and role. As illustrated in Fig. 3b, the average force is lowest during *following* (18.16 units), increased during *dragging*

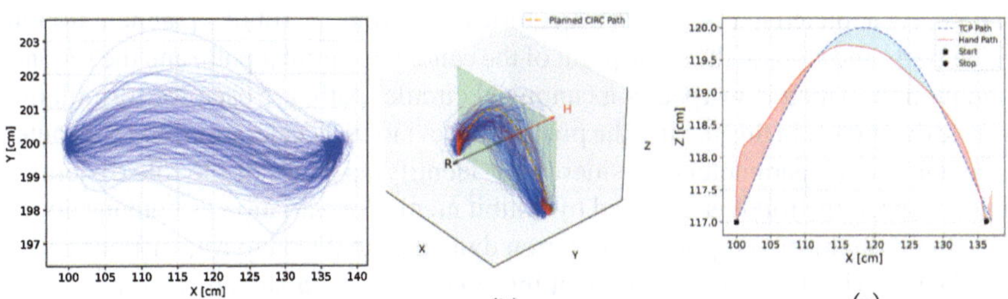

(a) (b) (c)

Fig. 2 Visualization of the 2D and 3D trajectories after canonical time scaling for the starting and end points

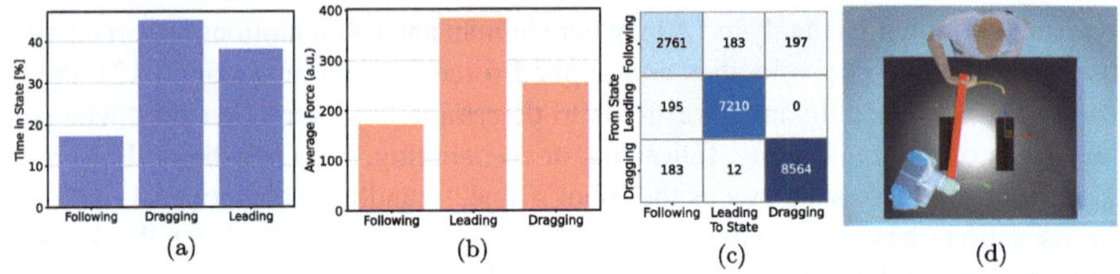

Fig. 3 Quantitative results of **a** role time in state, **b** estimated average force unit per state, **c** states transition count and **d** simulation demo top view

(29.22 units), and peaked during *leading* (45.87 units), reflecting higher physical effort in assertive or resisting roles. The FSM transition matrix in Fig. 3c further supports the presence of persistent behavioral phases, with dominant self-transitions indicating low switching frequency. Specifically, *dragging* and *leading* states exhibit high persistence ratios of 97.8% and 97.4%, respectively, while *following* shows a moderately stable pattern at 87.9%. These findings suggest that once a role is adopted, it tends to be maintained over sustained interaction periods.

The simulation environment for digital twins shown in the Fig. 3 illustrates a role-based human-robot interaction. A jointly handled rigid object by the human and the robot enabled the simulation of collaborative object handling. The trajectories of the human hand and the robot's TCP are calculated using inverse kinematics and visualized as point clouds using a particle system. This setup allows real-time verification of motion coordination, spatial alignment, and role transitions in a virtual environment.

6 Discussion

This study introduces an FSM-based method for modeling role dynamics in human-robot interaction using human motion capture for collaborative object-handling tasks. Human motion trajectories are first normalized and then warped to a canonical circular path. As depicted in Fig. 2, two trajectories, one from the robot's planned motion, which starts and stops at the same point of the canonical circular path, and the second-hand motion, which is warped to a canonical circular path, are used for formulating the role dynamics. In this regard, the positional deviation, acceleration, and computed force unit are the parameters considered for identifying states of the role dynamics. In this respect, the robot is assumed to exhibit no motion variation as humans do (cf. [9]). Based on this assumption, interaction dynamics that lead to role differentiation have been derived and tested by mapping trajectories to a shared reference frame and projecting them within a defined spatial plane, which we call a threshold plane. Using a digital twin Fig. 3, the FSM-based classification model enabled real-time labeling of control roles, identifying whether the human was leading, following, or resisting the robot's intended motion.

Furthermore, the results demonstrated distinct behavioral phases characterized by different physical roles. The prolonged duration of the *dragging* state, as well as the high force levels observed in *leading* and *dragging*, indicate that role switching is often accompanied by increased physical demand. These findings may lead to the importance of adaptive role recognition and motion planning in collaborative robotics. The FSM approach enables an interpretable segmentation of interaction dynamics and may be usable for simulation and real-world use.

7 Conclusion and Future Outlooks

This work analyzed role states in human-robot interaction within symbiotic collaboration scenarios, focusing on shared control during object handling. Accurate tracking of role dynamics over time was enabled by combining 3D motion normalization, spatial trajectory warping, and finite state classification. A canonical reference trajectory can be used for consistent comparison across trials, while the FSM provided interpretable segmentation of control strategies based on spatial deviation and directional alignment. Results from the simulation demonstrated the method's capability to distinguish leading, following, and resisting behaviours, supported by force trends and transition stability.

The method shows promise for scaling for various manufacturing and assembly tasks, particularly those requiring coordinated co-manipulation. Further enhancements could include adaptive role negotiation strategies, transitioning to physical validation using motion capture from human trials and force sensors with a collaborative robot to confirm the transferability of the proposed method beyond simulation.

Acknowledgements The authors would like to acknowledge the financial support by the German Research Foundation (DFG) within the HiSMoT project (500490184) and Horizon Europe within the Flex4Res (101091903).

References

1. Al-Saadi Z, Hamad YM, Aydin Y, Kucukyilmaz A, Basdogan C (2023) Resolving conflicts during human-robot Co-Manipulation. In: Proceedings of the 2023 ACM/IEEE international conference on human-robot interaction, ACM, Stockholm Sweden, pp 243–251. https://doi.org/10.1145/3568162.3576969
2. Barravecchia F, Bartolomei M, Mastrogiacomo L, Franceschini F (2023) Redefining Human-Robot Symbiosis: a bio-inspired approach to collaborative assembly. Int J Adv Manuf Technol 128(5):2043–2058. https://doi.org/10.1007/s00170-023-11920-1
3. Chen R, Shek A, Liu C (2023) Robust and context-aware real-time collaborative robot handling via dynamic gesture commands. IEEE Robot Autom Lett 8(6):3510–3517. https://doi.org/10.1109/LRA.2023.3268586

4. Cui Y, Poon J, Miro JV, Yamazaki K, Sugimoto K, Matsubara T (2019) Environment-adaptive interaction primitives through visual context for human-robot motor skill learning. Auton Robot 43(5):1225–1240. https://doi.org/10.1007/s10514-018-9798-2
5. Edmonds M, Gao F, Liu H, Xie X, Qi S, Rothrock B, Zhu Y, Wu YN, Lu H, Zhu SC (2019) A tale of two explanations: enhancing human trust by explaining robot behavior. Sci Robot 4(37):eaay4663. https://doi.org/10.1126/scirobotics.aay4663
6. Kim SK, Kirchner EA, Schloßmüller L, Kirchner F (2020) Errors in human-robot interactions and their effects on robot learning. Front Robot AI 7. https://doi.org/10.3389/frobt.2020.558531
7. Knowledge Based Systems, Inc (1993) Integration definition for function modeling (idef0). Technical report FIPS PUB 183, US Air Force, Texas, USA. https://nvlpubs.nist.gov/nistpubs/Legacy/FIPS/fipspub183.pdf, federal Information Processing Standards Publication
8. Kyrkjebø E (2018) Inertial human motion estimation for physical human-robot interaction using an interaction velocity update to reduce drift. In: Companion of the 2018 ACM/IEEE international conference on human-robot interaction, association for computing machinery, New York, NY, USA, HRI '18, pp 163–164. https://doi.org/10.1145/3173386.3176955
9. Manns M, Fischer K, Du H, Slusallek P, Alexopoulos K (2018) A new approach to plan manual assembly. Int J Comput Integ Manuf 31(9):907–920. https://doi.org/10.1080/0951192X.2018.1466396
10. Mielke E, Townsend E, Wingate D, Salmon JL, Killpack MD (2024) Human-robot planar co-manipulation of extended objects: Data-driven models and control from human-human dyads. Front Neurorobot 18. https://doi.org/10.3389/fnbot.2024.1291694
11. Muratore L, Laurenzi A, Tsagarakis NG (2019) A self-modulated impedance multimodal interaction framework for human-robot collaboration. In: 2019 international conference on robotics and automation (ICRA), pp 4998–5004. https://doi.org/10.1109/ICRA.2019.8794168
12. Ötting SK, Masjutin L, Steil JJ, Maier GW (2022) Let's Work together: a meta-analysis on robot design features that enable successful human-robot interaction at work. Hum Fact 64(6):1027–1050. https://doi.org/10.1177/0018720820966433
13. Tuli TB (2024) An approach to human behavior adaptable collaborative robot workplace design. Fortschritt-Berichte VDI, Reihe 02, Nr 713. VDI Verlag, Düsseldorf
14. Tuli TB, Manns M, Jonek M (2022a) Understanding shared autonomy of collaborative humans using motion capture system for simulating team assembly. Springer International Publishing, Cham, Lecture Notes in Mechanical Engineering, pp 527–534. https://doi.org/10.1007/978-3-030-90700-6_59
15. Tuli TB, Manns M, Zeller S (2022) Human motion quality and accuracy measuring method for human-robot physical interactions. Intel Serv Robot 15(4):503–512. https://doi.org/10.1007/s11370-022-00432-8

Scenario-Based Collaborative Robot Evaluation (SCORE) Framework

Ali Al-Yacoub, Alina Peter, Dennis Guck, Felix Fleschhut, and Michael Keckeisen

Abstract Human-Robot Collaboration Systems (HRCSs) are integral to the automation industry. Robots work alongside humans to enhance productivity and efficiency across an extensive range of applications. In each case, ensuring the safety of HRCSs is crucial, necessitating a robust safety verification process—ideally while maintaining efficiency and optimizing cost. This imposes contradicting requirements. For example, an efficient assembly task might require high robot speed, but high speed can endanger humans. Both the complexity of the, typically manual, safety validation process and the resulting restraints on efficiency can hinder widespread adoption of HRCSs. Optimization in the face of the combinatorial explosion of possible HRCSs implementations and scenarios will require a new approach. Here, we propose a scenario-based safety verification framework for collaborative robots that can systematically generate and analyze scenarios and test simulated control strategies to ensure that human safety is maintained while achieving optimal productivity. The proposed approach consists of a text-based HRCSs operational scenario that describes the scene, hardware setup, and dynamical elements in the scene. Unlike existing methods, the proposed approach is intended to enable automated scenario generation and adaptive risk assessment, providing a scalable and repeatable framework for systematically designing and evaluating collaborative robot safety in dynamic operational settings.

Keywords Collaborative robots · Safety verification · Scenario-based simulation · Human-robot interaction · Risk assessment

---

A. Al-Yacoub (✉) · A. Peter · D. Guck · F. Fleschhut · M. Keckeisen
TWT GmbH Science & Innovation, Stuttgart, Germany
e-mail: ali.alyacoub@twt-gmbh.de
URL: https://twt-innovation.de

M. Manns (ed.), *New Paradigms for Anticipated Uncertainty*, Lecture Notes in Mechanical Engineering, https://doi.org/10.1007/978-3-032-16889-4_75

1 Introduction

Collaborative robots have emerged as pivotal assets in the automation industry, working alongside humans to amplify productivity, efficiency, and cost-effectiveness. They promise greater flexibility, versatility and human-centric work compared to traditional robot installations, and have to potential to be more adaptable to rapid changes in supply chains, requirements, or tasks. However, integrating collaborative robots introduces safety concerns, requiring meticulous safety verification not only at the manufacturing stage but also once installed on the shop floor [1]. This makes deployments of such systems costly, lengthy procedures requiring a lot of expertise, manual, iterative procedures and ad hoc adaptations to a single situation at hand. There is no established standardized, automated method for risk assessment. Moreover, to ensure human safety, available standards and risk assessment procedures impose strong constraints on collaborative robots, limiting their performance. For example, in collaborative mode, robots are often limited to speeds around 1.6 to 2.0 m/s depending on the application and risk assessment [2], reducing their efficiency significantly. On top of that, physical testing, using prototypes or actual collaborative robots, is limited to controlled environments, to evaluate their behavior under various scenarios, including human-robot interaction (HRI), collision detection, and emergency stop functionality [3]. However, such controlled testing does not account for the unpredictability of real-world environments.

Ensuring the safety of Human-Robot Collaborations Systems (HRCSs) in a scalable manner and for ever more capable robots requires a structured and automated verification framework that effectively balances productivity and risk mitigation in dynamic, unstructured environments.

We propose a scenario-based simulation framework for evaluating the safety features of collaborative robots in automated systems. Such an approach will provide manufacturers, integrators and end-users with a design tool that allows affordable, systematic, reproducible testing and validations of HRCSs. The proposed scenario-based safety validation and verification approach draws inspiration from the scenario-accompanied, text-based, iterative Evaluation of automated driving Functions (stiEF) [4] approach used for autonomous driving. Our approach leverages text-based scenario descriptions to systematically simulate various HRIs and assess safety mechanisms. By enabling iterative refinement based on simulation results, including numerical benchmarks, the proposed method ensures that potential risks are identified and mitigated before deploying the physical robot or setup, achieving an optimal balance between safety and efficiency.

The paper is structured as follows: Sect. 2 presents a comprehensive literature review. The Scenario-based Collaborative Robot Evaluation (SCORE) framework is introduced in Sect. 3, detailing its descriptive language, automated scenario generation, and risk assessment mechanisms. Finally, Sect. 4 concludes the paper, summarizing key contributions and outlining future research directions.

2 Literature Review

The increasing deployment of HRCSs in industrial and human-centric environments has necessitated rigorous safety validation methods. Over the past decade, researchers have explored various approaches to assessing and mitigating risks in HRCS. This section reviews key contributions to HRCSs safety standards, risk assessment methodologies, and validation techniques, highlighting gaps and emerging trends.

HRCSs safety standards: The safety assurance of collaborative robots is governed by multiple international standards, with ISO/TS 15066 playing a pivotal role in defining safety parameters such as power and force limitations as well as speed and separation monitoring [5] to ensure human safety during direct interactions [6]. Chemweno et al. [2] reviewed the practical application of these standards, proposing a framework based on ISO 31000 and ISO/TS 15066 to align design safeguards with hazard analysis and risk assessment outcomes. Rosenstrauch et al. [1] critiques ISO/TS 15066, highlighting that adherence to its force and pressure limits does not guarantee safety in HRC, especially with heavier robots. Both studies demonstrate the challenges in implementing these standards in real-world applications, showing that adherence alone neither guarantees safety nor efficiency.

Sensor-enabled safety mechanisms: A major strand of research investigates sensor-enabled safety mechanisms that dynamically adjust robot behavior based on real-time human motion detection. Scholz et al. [7] provide a brief review of sensor technologies for HRC safety, emphasizing the need for advanced proximity sensors with greater accuracy and range, multisensor integration for enhanced perception, and standardized benchmarks like a robotic intelligence index for consistent evaluation. They also highlights the importance of supporting design and decision-making processes for sensor integration. The integration of multisensor fusion systems allows for robust perception, enabling SSM in compliance with ISO/TS 15066 standards [8]. Recent developments in safe learning algorithms provide additional means for enhancing safety in robotics beyond sensor-based approaches [9].

Dynamic safety planning: Recent advances in risk assessment focus on real-time constraint-based safety planning, where robots dynamically adapt to human movement constraints [10]. Explicit reference governors (ERG) and constrained motion planning enhance stability and collision avoidance, with Merckaert et al. integrating Rapidly-exploring Random Trees (RRT) and ERG to improve real-time safety performance while maintaining high-speed operation. However, traditional RRT-based planners struggle with real-time adaptability due to computational overhead and lack of dynamic constraint handling [7, 11]. In addition, Hamad et al. [3] propose a multi-layered safety architecture integrating physical and cognitive aspects for effective HRI, enabling safety certification across diverse scenarios. Brunke et al. [12] review reinforcement learning techniques for safety but emphasize the need for integrating learning-based control with formal safety assurances. Ensuring safety in dynamic, unstructured environments remains a challenge, requiring adaptive measures.

Systematic risk assessment: Several studies highlight the role of systematic risk assessment in early hazard identification. Tian et al. [13] provide a comprehensive survey of risk assessment methods for HRCSs, covering standards and structured techniques like FMEA, HAZOP, and Bow-Tie analysis. Existing methodologies primarily rely on empirical testing and heuristic safety measures, which may not fully capture the complexity of real-world HRCSs [14]. Haggy et al. [15] propose a structured safety assessment framework that combines Failure Mode and Effects Analysis with Bow-Tie Analysis, offering systematic hazard identification and mitigation in collaborative assembly systems. Hanna et al. [16] introduced a deliberative safety approach, enabling dynamic switching between safety measures to balance flexibility, efficiency, and risk mitigation. Rathmair et al. [17] further advocate for symbolic model-checking to verify HRC safety compliance while addressing system variability. Despite substantial progress, a major research gap remains in automating the validation and verification of HRC safety in dynamic conditions [13]. Existing methodologies primarily rely on empirical testing and heuristic safety measures, which may not fully capture the complexity of real-world HRIs [14, 18].

A potential solution for designing and validating HRCSs lies in scenario-based simulation validation, where predefined safety-critical scenarios are systematically evaluated within digital environments [19]. However, this approach does not explicitly integrate standardized safety requirements, limiting its applicability for regulatory compliance. Using Natural Language Processing (NLP) techniques, researchers can extract requirements from industrial manuals and regulatory documents, automatically integrating them into simulation scenarios [20]. Moreover, Alenjareghi et al. [14] highlight how AI-enhanced risk assessment methods can support standard-compliant safety evaluations in HRC. AI-assisted formalization of the requirements, such as that presented in Copilot [21], further enhances this process. Users can then generate a rich set of scenarios (as introduced in Sect. 3) that can be evaluated against all desired standards and constraints. In addition to pre-deployment simulations, the integration of digital twins with real-time simulation environments offers a promising direction for continuous safety validation.

Despite substantial research in formal verification and risk assessment [22], to date, and to the best of the authors' knowledge, there are no generic design tools for automatic validation and verification of dynamic HRCSs, nor are there existing frameworks for automatic risk assessment generation [2]. Although tools like Tecnomatix and Imk-Ima support static risk analysis aligned with standards, they lack support for automated scenario generation and validation of dynamic human-robot interactions. This gap motivates the development of SCORE, a structured, repeatable, and scalable simulation-based safety verification framework, which will be described in detail in the next section. In this context, the proposed SCORE framework supports decision-makers in developing/evaluating HRCSs.

3 Methodology: Conceptual Scenario-Based Safety Verification for Human-Robot Collaboration

Inspired by scenario-based validation approaches in automated driving systems (as in [22], and [4]), this conceptual methodology introduces a descriptive language to formally define HRCSs components. New here is the automatic generation of safety-critical simulation scenarios, and evaluation of user-defined control algorithms under real-world constraints. By systematically varying HRIs in a simulated environment leveraging Artificial Intelligence (AI) tools, this framework is intended to enable robust testing across multiple collaboration modes. Unlike existing robotics validation frameworks, SCORE is not only designed to automate the generation of simulation trials and results collection but also to evaluate the effectiveness of user-defined safety and control strategies through structured risk quantification. By systematically analyzing the robot's response to varying human-robot interaction conditions, the framework will generate a comprehensive safety score and provide actionable design recommendations based on multimodal AI trained on relevant scenarios to enhance HRCSs efficiency and ensure compliance with safety standards. This end-to-end automation makes SCORE a novel approach to HRCSs validation. It is worth mentioning here that the framework has not yet been validated in a physical setup and is not designed for real-time monitoring but rather for offline evaluation using collected data.

Figure 1 illustrates the key elements and interactions of the SCORE framework. First, the SCORE pipeline provides systematic scenery and trial generators, based on a textual scenario provided by the user in the SCORE language. SCORE generators interface with established simulation platforms, making full use of ongoing advanced in realistic physical simulations. The output is then evaluated by a dedicated module, evaluating trials and generating risk assessments.

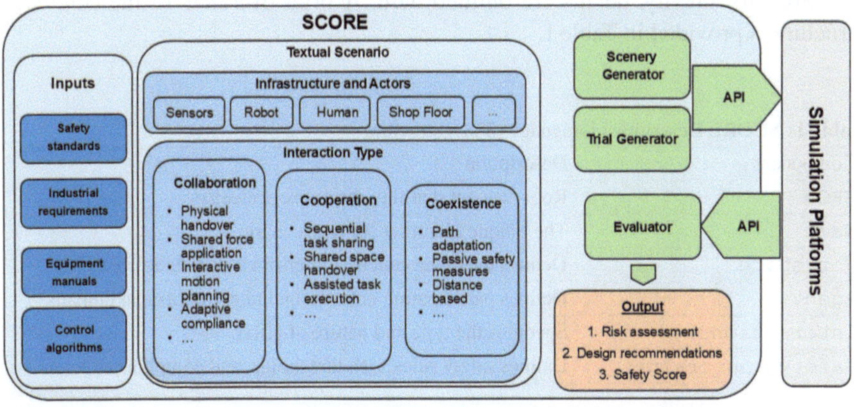

Fig. 1 Overview of the SCORE framework, with inputs in dark blue, and key AI-driven components in green and orange

The textual scenario model defines HRCSs components, including infrastructure (shopfloor, robots, sensors, and equipment) and actors (robots, humans, and equipment). After user input (blue boxes), including safety/control algorithms, standards, requirements, and manuals, the framework leverages language models and AI-driven analysis to extract relevant Key Performance Indicators (KPIs), under human oversight and in alignment with AI ethics and regulatory principles. These extracted KPIs might include proximity violations, collision events, time-to-complete deviations, and KPI-based safety thresholds. By comparing simulation results from systematically generated HRCS scenarios against these metrics, SCORE generates structured risk assessments, design recommendations, and a quantified safety score that balances operational efficiency with safety constraints. By integrating structured scenario-based safety verification, SCORE is intended to provide a structured, repeatable, and scalable simulation-based safety verification workflow for dynamic HRCSs. SCORE does not introduce new standards, but extracts requirements from existing standards using emerging AI tools. The framework enables users to select relevant standards according to their targeted application and evaluates the setup accordingly. By integrating structured scenario generation, dynamic simulation, and risk assessment, SCORE complements existing tools and aims at future interoperability with platforms like Gazebo and OPC UA-compliant systems.

The key elements of this proposed approach are described in more detail in the following paragraphs.

SCORE Descriptive Language (SDL) for HRCSs: The SDL provides a structured syntax for defining HRCSs in a standardized way. It enables the generation of simulation scenarios, automated safety validation, and risk assessments by allowing to define the workspace, robot configurations, and HRI zones.

SDL defines an HRCS through a hierarchical structure comprising Workspaces, Agents, Interactions, and Safety Constraints. Each component has attributes and parameters that can be further specified and that influence the simulation behavior. However, for brevity, details are omitted here. A brief overview of the core SDL structure is provided in Table 1.

Table 1 SCORE Descriptive Language (SDL) structure

Component	Description
`HRCS`	Root element defining the entire scenario
`Name`	The unique identifier for the scenario
`Workspace`	Defines the environment, dimensions, and static obstacles
`Agents`	Defines participating entities, including robots and humans
`Interactions`	Specifies the type and nature of HRIs
`Safety_Constraints`	Defines safety rules, risk thresholds, and compliance factors

Automated Scenario Generation: After the scenery is generated, edge cases that could impact human safety are systematically introduced into the simulation for hazard assessment. The system will automate the generation of simulation trials, tailoring them to the interaction types and validation goals. Users can input their robot safety and control algorithms, which are then systematically tested against safety-critical scenarios. Additionally, dynamic variations of human movement trajectories are incorporated to evaluate the robot's safety responses.

Simulation Interface: Automatically generated scenarios are simulated by generating XML descriptions of the scenery based on the SDL and corresponding CSV files for simulation parameters via a dedicated API. Trials reults will be then systematically collected; as CSV files, and saved to be evaluated.

Dynamic Risk Assessment (Evaluator): Automated evaluation of safety compliance will be conducted based on requirements extracted from regulatory documents and industry manuals and the simulation results. Using large language models, safety requirements are identified, along with KPIs to assess simulation results for specific applications using multimodal AI approaches. Additionally, a scenario-driven risk quantification approach is employed, adapting HRCS parameters to dynamically evaluate safety risks.

Using the SDL, users can define structured scenarios that enable automated risk assessment and KPI evaluation. Unlike traditional methods, SCORE supports early-stage virtual validation, improving reproducibility and scalability across platforms and standards. While not intended for real-time monitoring, it supports both simulation and real-world data, enabling efficient, adaptable, and transparent safety verification.

4 Conclusion and Future Work

This paper presents SCORE, a conceptual scenario-based safety verification framework for HRCS. Inspired by approaches like stiEF [4], it enables structured, simulation-driven validation through text-based scenario descriptions, automated generation of safety-critical trials, and AI-assisted risk evaluation. SCORE aims to reduce reliance on costly physical testing, support early-stage prototyping, and enhance reproducibility across dynamic industrial settings.

While promising, the framework is still conceptual. One of the main challenges ahead lies in identifying and sampling edge-case scenarios that effectively trigger safety-relevant events. Future work will focus on completing the simulation pipeline, integrating with platforms, and developing a graphical user interface backed by a Python-based SDL parser and XML/CSV generation tools. This will facilitate seamless scenario generation, enhancing the efficiency and accessibility of simulation-based safety validation.

Acknowledgements The authors acknowledge the financial support by the Federal Ministry for Economic Affairs and Climate Action in the project SkaLaB (project number 13IK025).

References

1. M. J. Rosenstrauch and J. Krüger, "Safe human-robot-collaboration-introduction and experiment using iso/ts 15066," in *3rd ICCAR*, pp. 740–744, IEEE, 2017
2. P. Chemweno and et al., "Orienting safety assurance with outcomes of hazard analysis and risk assessment: A review of the iso 15066 standard for collaborative robot system," 2020
3. M. Hamad and et al., "A concise overview of safety aspects in human-robot interaction," in *International Workshop on Human-Friendly Robotics*, pp. 1–18, Springer, 2023
4. F. Bock and et al., "Advantageous usage of textual domain-specific languages for scenario-driven development of automated driving functions," in *2019 SysCon*, pp. 1–8, IEEE, 2019
5. W. Li, "Safe human–robot collaboration for industrial settings: a survey," 2024
6. A. Ghanbarzadeh and E. Najafi, "Safe physical human–robot interaction through variable impedance control based on iso/ts 15066," *IJIDeM*, pp. 1–18, 2024
7. C. Scholz, H.-L. Cao, *et al.*, "Sensor-enabled safety systems for human-robot collaboration: A review," IEEE, 2025
8. Zhu C et al (2024) Task-oriented safety field for robot control in human-robot collaborative assembly based on residual learning. Expert Syst Appl 238:121946
9. S. Gu and et al., "Safe and balanced: A framework for constrained multi-objective reinforcement learning," 2024
10. Merckaert K et al (2024) Real-time constraint-based planning and control of robotic manipulators for safe human-robot collaboration. RCIM 87:102711
11. Gleirscher M et al (2022) Verified synthesis of optimal safety controllers for human-robot collaboration. Sci Comput Program 218:102809
12. Brunke L et al (2022) Safe learning in robotics: From learning-based control to safe reinforcement learning. ARCRAS 5(1):411–444
13. M. Tian and Z. Zou, "Safety-oriented human–robot collaboration in construction through human preference alignment," *Journal of Intelligent Construction*, 2025
14. M. J. Alenjareghi, "Safe human-robot collaboration: a systematic review of risk assessment methods with ai integration and standardization considerations," 2024
15. A. Haggy and et al., "Safety assessment of human-robot collaborations using failure mode and effects analysis and bow-tie analysis," in *Proceedings of the 21st ICINCO, November 18-20, 2024, Volume 1*, SCITEPRESS, 2024
16. A. Hanna and et al., "Deliberative safety for industrial intelligent human–robot collaboration: Regulatory challenges and solutions for taking the next step towards industry 4.0," *RCIM*, vol. 78, 2022
17. M. Rathmair, "Formal verification of safety properties of collaborative robotic applications including variability," in *2021 30th RO-MAN*, IEEE, 2021
18. Gualtieri L et al (2022) Development and validation of guidelines for safety in human-robot collaborative assembly systems. Computers & Industrial Engineering 163:107801 Jan.
19. T. P. Huck and et al., "Simulation-based testing for early safety-validation of robot systems," in *2020 IEEE Symposium on Product Compliance Engineering*, pp. 1–6, IEEE, 2020
20. Sajwan M, Singh S (2023) A review on the effectiveness of machine learning and deep learning algorithms for collaborative robot. ACM in Engineering 30(6):3489–3508
21. M. Obstbaum and et al., "Efficient formalization of requirements using artificial intelligence in the development of mbos," in *ELIV 2024*, VDI Verlag, 2024
22. C. Sippl and et al., "Scenario-based systems engineering: An approach towards automated driving function development," in *2019 SysCon*, pp. 1–8, IEEE, 2019

Energy Consumption in Robotics: A Simplified Modeling Approach

Lisa Lokstein, Kevin Haninger, Valentyn Petrichenko, Sophie Matthews, Miriam Schleipen, and Gregor Thiele

Abstract The energy consumption of a robot depends, among other factors, on its trajectory. Current methods for trajectory optimization with fixed start and end points can reduce the used energy by up to 30%. However, their use in trajectory planning for industrial robots is still restricted due to model complexity and lack of integration with planning tools that address other concerns (e.g. collision avoidance). We propose an approach to estimate the energy consumption for given trajectories. It includes differentiable inertial and kinematic models from standard open-source tools, integrating standard ROS planning methods. An inverse dynamics-based energy model is optionally extended with a single-parameter electrical model. By comparing the inertial and electrical models of a collaborative robot, we show that simplified models provide competitive accuracy while being easy to deploy in practice.

Keywords Energy efficiency · Industrial robots · Collaborative robots · Dynamic modeling · Optimization

1 Introduction

Gaining transparency and saving energy are key objectives of industrial energy management. This creates a demand for energy metering for industrial processes, including robots, and estimations of the impact of adaptations to the process to determine the most efficient settings. These estimations can be made using simulations to replicate real-world scenarios, as it can be difficult to test measures in ongoing production.

As there are over four million operating industrial robots worldwide, with an increase of 10% in 2023 [1], they represent a promising opportunity for energy optimization. A common approach to save energy in robotics is optimizing its trajectory

L. Lokstein (✉) · K. Haninger · V. Petrichenko · S. Matthews · G. Thiele
Fraunhofer Institute for Production Systems and Design Technology IPK, Berlin, Germany
e-mail: lisa.lokstein@ipk.fraunhofer.de

M. Schleipen
EKS InTec GmbH, Weingarten, Germany
e-mail: miriam.schleipen@eks-intec.de

M. Manns (ed.), *New Paradigms for Anticipated Uncertainty*, Lecture Notes in Mechanical Engineering, https://doi.org/10.1007/978-3-032-16889-4_76

and velocity [2]. Trajectory optimization is an important aspect of robotics that aims to find the most efficient path to follow in order to complete a given task. It has demonstrated significant effectiveness in designing dynamic movements for linear and nonlinear dynamical systems and takes into account the physical constraints imposed by the specific maneuvers, the surrounding environment, and the capabilities of the hardware involved [3]. This also involves minimizing energy consumption, time, or other resources. To determine which trajectory and movement speed result in the lowest energy consumption, dynamic robot models can be used to estimate the consumed energy for different motions [4]. Then, the settings with the lowest energy consumption can be determined without the need for a real robot to traverse the trajectory.

We propose two simplified robot energy models for collaborative robots that can be integrated into the design and commissioning process, offering new opportunities for improvement. By utilizing common data formats, these models can support gradient-based optimization, impact robot cell design, and guide kinematic decisions so that the gap between theoretical optimization and real-world deployment can be reduced. Gradient-based optimization methods allow a more computationally efficient minimization of the energy use, especially when hard constraints are involved. To reduce the modeling work for the energy optimization of a robot, the proposed energy models only use inverse inertial dynamics and either a constant overhead or a single electrical parameter. Compared with data-driven approaches, these models only require energy measurements in a handful of static conditions to identify the electrical parameter. We propose a robot energy modeling pipeline (see Fig. 1) where the model is integrated into open-source planning frameworks (MoveIt [5]) using differentiable robot dynamics (Pinocchio [6]) to support numerical optimization.

In this paper, an overview of the related work is provided in Sect. 2. Then, the mechanical and electrical power models are described in Sect. 3. In Sect. 4, these

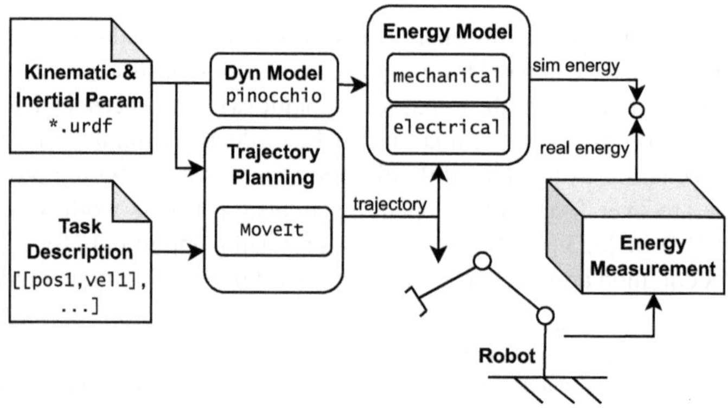

Fig. 1 Overview of the proposed energy modeling pipeline, using standardized data formats for robot model and task, open-source trajectory planning methods, and validation through energy measurements during robot execution

models are then used to determine the total energy consumption for given trajectories, including overhead power usage, which covers the power electronics, support computers, ventilation, end-effector, and other non-motor power usage of the robot. The proposed model is then validated on a low-payload, collaborative robot (Franka Emika), and the accuracy of the mechanical versus the mechanical-electrical model is evaluated within Sect. 5.

2 Related Work

For industrial robots, dynamic simulations are an established tool to plan processes and virtual commissioning [7]. By modifying the robot's motion interpolator parameters, like acceleration and velocity, these simulations can be used to optimize energy consumption [8]. Optimizing the trajectory (i.e. including the path) between fixed start and end points is well established and can save 15–30% of energy [9, 10]. However, the widespread implementation of existing trajectory optimization techniques in industrial robots faces several challenges. The majority of planning takes place in in-cell instruction or simulation environments, both of which have a limited ability to integrate data-driven models [11]. Additionally, ensuring compatibility with existing infrastructure requires the use of standard motion commands like joint or Cartesian moves, which complicates the use of methods that require the execution of an arbitrary optimized path.

There is a lot of research on energy consumption and energy modeling of industrial robots using different approaches. While some focus on analytical models [4, 12], others use data-driven methods, where the model of the robot is determined through energy measurements [13]. In [14], an approach mixing analytical and data-driven models is studied. Straczynski et al. analyzed the effects of velocity and payload on the energy consumption, separating the joints [15]. Heredia et al. focus on lightweight robots and presented the *EcBot*, a MATLAB-based tool for data-driven generation of dynamic energy models [16]. They also developed a benchmarking framework to evaluate the energy consumption of lightweight robots that requires a procedure for labeling the measured signals [17].

So while dynamic robot simulation is not entirely new, there is significant potential in further developing and utilizing inertial and kinematic models derived from standard formats [12]. These differentiable models open up new opportunities for applying gradient-based optimization techniques for both energy reduction and parameter identification. By focusing on these models, energy optimization can be more effectively integrated into the robot design, planning, and control stages, thereby advancing their application toward more efficient and sustainable robotic operations.

3 Dynamic Robot Models

This section discusses the inertial and electrical models, which are used for deriving the energy model. While several methods in the literature use detailed modeling for precise energy prediction along robotic trajectories, our approach prioritizes simplicity by minimizing the number of parameters that need to be identified. Then, the model can be easily scaled to other lightweight collaborative robots without extensive parameter identification.

3.1 *Mechanical Model*

The mechanical energy consumption of a robot can be calculated from the system dynamics. We assume the inertial dynamics are available in the form [18]:

$$M(q)\ddot{q} + C(\dot{q}, q) + G(q) = \tau, \tag{1}$$

with the joint positions $q \in \mathbb{R}^n$, velocities $\dot{q}$ and accelerations $\ddot{q}$ related via motor torque τ, inertial matrix M, Coriolis and centrifugal matrix C and gravitational terms G.

The inertial dynamics in Eq. (1) do not include friction effects, because we aim for a simplified model that can be calculated without extensive system identification. For lightweight collaborative robots like the Franka Emika (payload 3 kg) the static friction values are two orders of magnitude smaller than the gravitational torques [19], so the primary focus should be on the inertial dynamics represented by the system's mass and torque inputs.

3.2 *Electrical Model*

Robotic systems often utilize DC or permanent magnet synchronous motors [10], which require accurate electrical modeling. We employ a standard DC motor model [20], which describes the electrical characteristics of these motors effectively. According to [10], the motor electrical model is given by:

$$i = K_t^{-1}\tau, \tag{2}$$

$$v = K_{emf}\dot{q} + L\frac{di}{dt} + Ri, \tag{3}$$

with voltage v and current $i \in \mathbb{R}^n$, torque constant K_t, back-electromotive force (EMF) constant K_{emf}, motor electrical inductance L and resistance R. For many applications, the inductance L of the motor can be considered negligible due to its minimal impact on the overall dynamics [20]. This simplification reduces the model to

$$v = K_{emf}\dot{q} + RK_t^{-1}\tau. \quad (4)$$

The constants K_{emf} and K_t are crucial for determining the efficient conversion of electrical to mechanical energy. While theoretically equal, they often differ slightly in practice due to manufacturing tolerances, magnetic losses, and temperature variations.

4 Estimation of the Consumed Energy

This section derives the total energy required to follow a trajectory by analyzing the instantaneous power models of the robotic system.

4.1 *Dynamic Power Model*

Applying (2) and (4) to find electrical power results in

$$p = i^T v = \tau^T K_t^{-T} K_{emf}\dot{q} + \tau^T K_t^{-T} RK_t^{-1}\tau, \quad (5)$$

where the term $\tau^T K_t^{-T} K_{\text{emf}}\dot{q}$ represents the mechanical power required to generate mechanical motion, which directly relates to the velocity of the robot's joints. The term $\tau^T K_t^{-T} RK_t^{-1}\tau$ represents electrical losses due to the resistance in the motor windings, commonly referred to as Joule heating.

To simplify the modeling process, we assume that all resistances R are identical and that the back EMF constant K_{emf} equals the torque constant K_t under ideal conditions. This leads to the following equation for calculating the energy consumption:

$$p = \tau^T\dot{q} + RK_t^{-2}\tau^T\tau, \quad (6)$$

where $RK_t^{-2} \in \mathbb{R}^1$ represents the electrical characteristics of the motor model, which are assumed to be uniform across all robot motors.

4.2 *Complete Power Model*

The power Eq. (6) from the previous section does not consider the energy consumed by overhead systems. To address this, we added a term for this aspect:

$$p = \tau^T \dot{q} + RK_t^{-2} \tau^T \tau + p_{\text{overhead}}, \tag{7}$$

where p_{overhead} is assumed to be constant.

The torque τ and joint velocity $\dot{q}$ are determined from inverse dynamics, while the terms RK_t^{-2} and p_{overhead} need to be identified. This equation represents the proposed modeling approach and will be utilized for the validation of two potential modeling methods.

Method 1: When assuming the resistance $R = 0$, Eq. (7) simplifies to

$$p = \tau^T \dot{q} + p_{\text{overhead}}, \tag{8}$$

where only the mechanical power from the simulation and the mean overhead power are used to predict the energy consumption for a robot's trajectory. For this purpose, several robot poses need to be measured to calculate the mean power consumption of the overhead.

Method 2: In this case, Eq. (7) includes a torque-dependent term $RK_t^{-2} \tau^T \tau$, which accounts for pose-dependent power usage. The parameter RK_t^{-2} and the overhead p_{overhead} can be estimated from multiple experiments with the least square method.

4.3 Trajectory

The trajectory is represented as a list of trajectory points, defined as

$$\mathcal{T} = [(q_1, \dot{q}_1, \ddot{q}_1, t_1), (q_2, \dot{q}_2, \ddot{q}_2, t_2), \ldots], \tag{9}$$

with joint position, velocity, and acceleration $q_1, \dot{q}_1, \ddot{q}_1$ realized at the time t_1.

From the inverse dynamic model, this sequence can be transformed to a list of instantaneous motor torques via the inverse dynamics (1):

$$\tau_i = M(q_i)\ddot{q}_i + C(\dot{q}_i, q_i) + G(q_i). \tag{10}$$

With the calculated torques τ_i at each trajectory point, we can evaluate the instantaneous power consumption using the derived power model (7) or also reduced power model (8) assuming $R = 0$:

$$p_i = \tau_i^T \dot{q}_i + \tau_i^T \tau_i RK_t^{-2} + p_{\text{overhead}}. \tag{11}$$

The total energy consumption for the entire trajectory is computed by integrating the power over the time duration of the trajectory.

By applying this methodology, we can use the models to simulate the energy used for different trajectories and compare these results to actual measured energy consumption during robot operation.

5 Validation

This section compares the simulated and measured energy to evaluate the accuracy of the two models for a lightweight collaborative robot. The setup includes a Franka Emika Panda, a stationary computer for controlling the Franka Emika, and a ZES ZIMMER power analyzer LMG 671. To determine the unknown parameters of the models, the overhead was quantified in several static poses. Then, the unknown parameters of modeling Method 1 in Eq. (8) and Method 2 in Eq. (7) were estimated. In Method 1, averaging the power measurements across all poses yields an overhead power of 92.3 W. Alternatively, Method 2 employs a regression model, resulting in $RK_t^{-2} = 0.0036 \frac{\mathrm{W}}{\mathrm{N^2\,m^2}}$ and an overhead power of $p_{\text{overhead}} = 88.04\,\mathrm{W}$.

To support the validity of the simulation results, a horizontal, a vertical, and a diagonal movement were planned and executed, each at three different speeds without load. Since the considered collaborative robot has only a low payload, it is assumed that the load would not strongly influence the energy consumption in this case. The start and end pose coordinates, along with the corresponding joint configurations, are documented in [21].

Table 1 presents the measured and simulated energy consumption of the robot for different velocities and types of movement. Both simulation methods generally provide reliable estimates of energy consumption, with Method 1 being slightly closer to the measured values, although they tend to produce lower values compared to the measured data. The average deviations of 3.46% (Meth. 1) and 3.97% (Meth. 2) indicate that these methods approximate real values reasonably well for collaborative robots.

The comparison of simulated and measured power for a representative movement using the two proposed modeling approaches is shown in Fig. 2a and b. Comparing the simulated and the measured power shows that both methods achieve similar

Table 1 Comparison of simulated and measured energy consumption for different movements and scaled velocities on Franka Emika robot

Type of movement	Scaled velocity	Movement duration (s)	Measured energy (J)	Method 1 (J)	Deviation Method 1 (%)	Method 2 (J)	Deviation Method 2 (%)
Horizontal	1	8.58	814.13	794.96	2.35	802.20	1.47
Horizontal	5	4.26	408.98	395.88	3.2	397.31	2.85
Horizontal	10	3.22	311.98	299.89	3.88	300.02	3.83
Diagonal	1	9.66	935.92	902.74	3.55	900.01	3.84
Diagonal	5	4.82	476.54	455.88	4.34	452.52	5.04
Diagonal	10	3.58	358.10	341.33	4.68	338.22	5.55
Vertical	1	10.92	1049.08	1024.09	2.38	1012.82	3.46
Vertical	5	5.54	543.99	527.26	3.08	519.41	4.52
Vertical	10	3.92	392.23	377.62	3.72	372.05	5.14

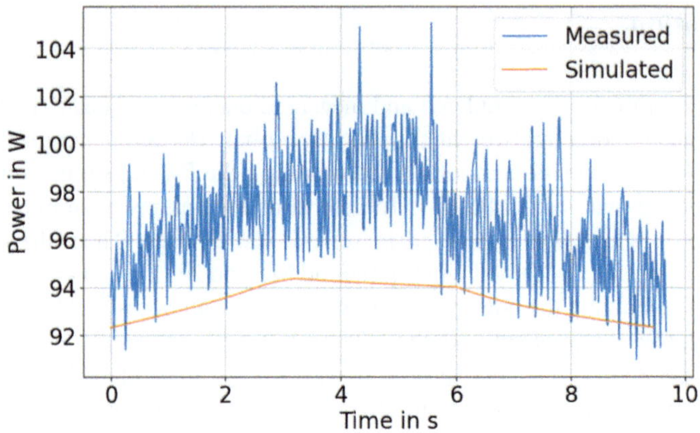

(a) Power for Diagonal Movement, Modeling Method 1

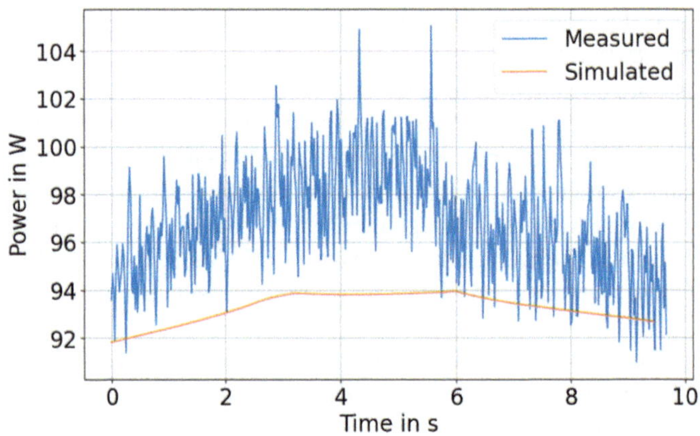

(b) Power for Diagonal Movement, Modeling Method 2

Fig. 2 Comparison of the power estimated by the two modeling methods and the measured power

results. It can be seen that the deviation between simulation and measurement varies for different phases of the movement. The deviation is smaller at the end pose than at the start pose of the robot. This can be attributed to the pose-dependent term in the simulation, as the influence of gravitational forces varies depending on the degree of extension of the robot arm. Both methods yield a similar course of the power as the measurements and correspond to the phases of the movement. In an acceleration phase, the power increases, before remaining constant in a phase with constant speed. Afterwards, the power decreases as the robot decelerates.

For the considered collaborative robot, the overhead accounts for approx. 95% of the active power consumption of the robot during a movement and is proportional to the duration of the movement. The total energy consumption is therefore largely determined by the duration of the movement, which is significantly lower at higher velocities than at lower velocities, despite the higher instantaneous power.

6 Conclusions

This study introduces a simplified approach to model the energy consumption of a collaborative robot that strikes a balance between accuracy and ease of integration with existing standard planning tools. By combining differentiable inertial and kinematic models with an adaptable single-parameter electrical model, the complexity of the identification process can be significantly reduced. This is particularly useful in scenarios where data is limited, implementation effort needs to be low, and quick deployment is essential.

The proposed model has limitations with respect to modeling and validation, particularly due to the use of a collaborative robot with low speed and payload. In this case, the overhead accounts for the largest part of the consumed energy. However, the results of the two proposed models approximate the actual energy consumption well for collaborative robots, exhibiting minimal deviation between measured and simulated energy consumption and supporting their suitability for practical use. Future work will focus on improving the model to better reflect the dynamic behavior and on testing the proposed model for applications where the movement exerts a greater influence on the consumed energy. This includes extending our modeling approach to industrial robots with higher payload capacities and validating its effectiveness across a broader range of robotic systems.

Acknowledgements This research work has been conducted within the EnEffNet project funded by the German Federal Ministry for Economic Affairs and Climate Action (grant no. 03 EN4023A-B) and the framework of the MODAPTO project funded by the European Union's Horizon Europe research and innovation program (grant agreement no. 101091996).

References

1. International Federation of Robotics (2024) World robotics 2024
2. Bukata L et al (2016) Energy optimization of robotic cells. IEEE Trans Ind Inform 13(1):92–102
3. Ratiu M, Prichici M (2017) Industrial robot trajectory optimization—a review. In: MATEC web of conferences, vol 126, p 02005
4. Carabin G, Wehrle E, Vidoni R (2017) A review on energy-saving optimization methods for robotic and automatic systems. Robotics 6(4):39
5. Coleman D et al (2014) Reducing the barrier to entry of complex robotic software: a Moveit! Case study. arXiv preprint arXiv:1404.3785
6. Carpentier J et al (2019) The Pinocchio C++ library: a fast and flexible implementation of rigid body dynamics algorithms and their analytical derivatives. In: IEEE/SICE international symposium on system integration (SII). IEEE, pp 614–619
7. Souflas T et al (2024) On the optimization of robot machining: a simulation-based process planning approach. Machines (Basel) 12(8):521
8. Torayev A et al (2024) Online and modular energy consumption optimization of industrial robots. IEEE Trans Ind Inform 20(2):1198–1207
9. Li X et al (2022) An efficient computation for energy optimization of robot trajectory. IEEE Trans Ind Electron 69(11):11436–11446

10. Gadaleta M, Pellicciari M, Berselli G (2019) Optimization of the energy consumption of industrial robots for automatic code generation. Robot Comput-Integr Manuf 57:452–464
11. Soori M, Arezoo B, Dastres R (2023) Optimization of energy consumption in industrial robots, a review. Cognit Robot 3:142–157
12. Ruscelli F et al (2022) Horizon: a trajectory optimization framework for robotic systems. Front Robot AI 9
13. Chang Q et al (2024) A data-driven method for predicting and optimizing industrial robot energy consumption under unknown load conditions. Actuators 13:12
14. Jiang P et al (2025) Industrial robot energy consumption model identification: a coupling model-driven and data-driven paradigm. Expert Syst Appl 262:125604
15. Straczynski P (2024) Energy consumption analysis for a 6-axis industrial robot in the packing process. PRZEGLAD ELEKTROTECHNICZNY
16. Heredia J, Schlette C, Kjærgaard MB (2023) EcBot: data-driven energy consumption open-source MATLAB library for manipulators. In: 21st international conference on advanced robotics (ICAR), pp 340–347
17. Heredia J et al (2023) Labelling lightweight robot energy consumption: a mechatronics-based benchmarking metric set. In: IEEE/RSJ international conference on intelligent robots and systems (IROS), pp 1789–1796
18. Spong MW, Hutchinson S, Vidyasagar M (2020) Robot modeling and control. Wiley
19. Gaz C et al (2019) Dynamic identification of the Franka Emika Panda robot with retrieval of feasible parameters using penalty-based optimization. IEEE Robot Autom Lett 4(4):4147–4154
20. Verstraten T et al (2016) Series and parallel elastic actuation: impact of natural dynamics on power and energy consumption. Mech Mach Theory 102:232–246
21. Lokstein L et al (2025) Start and end poses for energy modeling of a collaborative robot. https://fordatis.fraunhofer.de/handle/fordatis/445

A Methodology for Applying HRC on Remanufacturing Workstations: A Use Case from Household Appliances

Dionisis Vallianatos, Dimitrios Kaliakatsos-Georgopoulos, Dimosthenis Dimosthenopoulos, Christos Gkournelos, George Michalos, and Sotiris Makris

Abstract Remanufacturing transforms End-of-Life (EOL) products into high-quality, sustainable alternatives, contributing to significant economic and environmental benefits. However, due to the unpredictable state of EOL products, remanufacturing remains a complex and uncertain process. Human–Robot Collaboration (HRC) offers a human-centric, flexible approach to address these challenges while enhancing operator's overall well-being. This study proposes a methodology for designing and evaluating collaborative remanufacturing workstations by identifying optimal HRC scenarios for different tasks. Various production modes are assessed based on process quality, efficiency, and sustainability. A novel evaluation framework incorporating a Quality Index (QI) and a Sustainability Efficiency Index (SEI) is introduced to compare alternative workstation configurations. The approach is validated through a case study inspired by the household appliances industry, specifically focusing on domestic refrigerator disassembly for remanufacturing. The results highlight the superiority of HRC-based setups, demonstrating advantages over manual and fully automated processes in terms of quality, flexibility, ergonomics, and sustainability.

Keywords Remanufacturing · Workstation evaluation · Human robot collaboration · Sustainability · Household appliances

1 Introduction

Remanufacturing is defined as the "steps required to change the product into as new with at least equivalent performance and warranty of new" [1] and offers a guarantee of the same or higher-level performance than new products [2]. Among EOL

D. Vallianatos · D. Dimosthenopoulos · C. Gkournelos · G. Michalos · S. Makris (✉)
Laboratory for Manufacturing Systems and Automation (LMS), Department of Mechanical Engineering and Aeronautics, University of Patras, Patras, Greece
e-mail: makris@lms.mech.upatras.gr

D. Kaliakatsos-Georgopoulos
Teaching Factory Competence Center, Patras, Greece

M. Manns (ed.), *New Paradigms for Anticipated Uncertainty*, Lecture Notes in Mechanical Engineering, https://doi.org/10.1007/978-3-032-16889-4_77

disposal strategies, remanufacturing was claimed to have the greatest environmental benefits as it extends the life cycle of decommissioned products [3]. Remanufacturing includes logistics, disassembly, cleaning, inspecting and reassembling which are currently executed manually thanks to human flexibility on handling the uncertainties related to EOL products' condition. Disassembly is an inevitable process to retrieve high value cores within a product for remanufacturing. However, disassembling a product manually may not be effective as it often results in higher labour cost and ergonomic hazards [4].

Therefore, research is focused on the integration of Human–Robot Collaboration (HRC) in remanufacturing, promoting Circular Economy and Sustainability [5]. So far, there have been some studies on HRC disassembly, primarily on low payload operations. A method for press-fitted components evaluated a two-collaborative-robots-and-a-human-operator cell in the disassembly of an automotive water pump [6]. A case study investigating the disassembly of a personal computer and a power inverter focused on an HRC disassembly cell for extracting electrical components [7]. Considering the complete remanufacturing chain, a methodology has been proposed to aid in the selection and planning of the reconditioning processes based on the conditions of the products [8]. Another important aspect of enabling remanufacturing is to definition of the skills needed and their correlation with the capabilities of the resources [9].

In recent years, significant effort has been focused also on the design of HRC cells. Advances in Human–Robot Interaction (HRI) draw from cognitive engineering, psychology and sociology [10]. Human-centred robotic cell design is affected by factors, such as workload, vigilance, situation awareness, errors etc. as well as by cognitive engineering related aspects and ergonomics [11]. A key design element is the task allocation strategy, which defines how responsibilities are divided between human operators and robots influencing both ergonomics and system efficiency [12]. According to ISO 10218-1:2011, a "collaborative workspace" is one where humans and robots perform tasks simultaneously within a shared safeguarded area [13]. Current manufacturing practices require complete physical separation between people and robots to ensure safety [14]. An approach for introducing a fenceless hybrid workstation consists of a workflow for the analysis, design, implementation and commissioning activities that form a closed-loop process towards the achievement of all design specifications [15].

Considering the above, this work aims to propose a methodology for designing HRC remanufacturing workstations. Next follows a detailed presentation of the proposed framework and how it is implemented to evaluate different station configurations. Finally, the methodology is applied on a use case inspired by the refrigerators industry.

2 Approach

Designing an effective HRC workstation for remanufacturing involves several steps to optimise process efficiency and operator safety (Fig. 1). A key starting point is defining the remanufacturing process by identifying the core components that must be restored to like-new condition. These components often vary in wear, damage, or obsolescence, making their assessment crucial for planning the appropriate remanufacturing steps. Once identified, the required tasks can be defined accordingly.

2.1 Required Skills and Pool of Resources Specification

Once the process is defined, the set of skills required to handle the tasks within the process needs to be identified. A mapping between the tasks and skill requirements is performed to correlate with the capabilities of the available resources, e.g. limit parameters (payload, reach, etc.) and available tools. Thus, the resources suitable for each task emerge (Fig. 2). This procedure is repeated for each task forming alternative scenarios as different resources assigned to each task based on their capabilities.

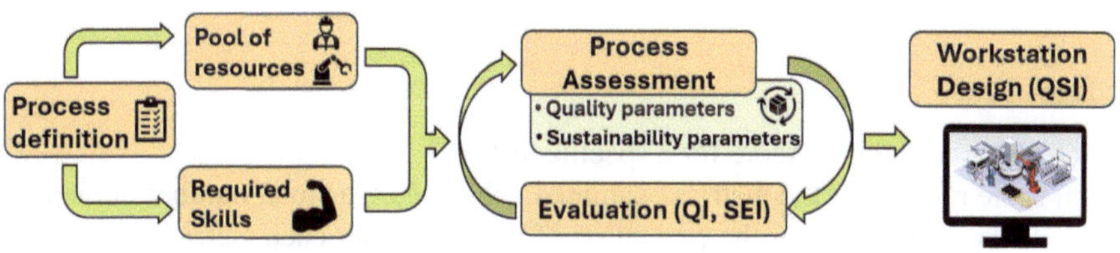

Fig. 1 Remanufacturing workstation design flowchart

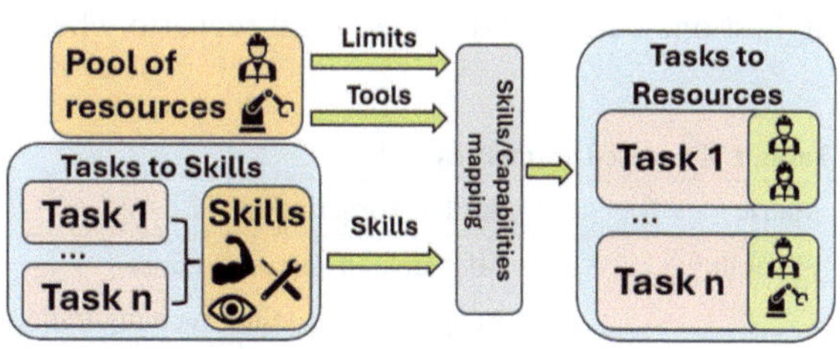

Fig. 2 Resources suitability definition framework

2.2 Process Assessment and Evaluation

For assessing these alternative scenarios, an analysis based on a novel yet simple approach is followed. More specifically, it aims to provide measurable indices calculated based on individual indicators which reflects performance, cost, quality, and flexibility aspects of the HRC application [16] as well as sustainability-related aspects. These indices are the Quality Index (QI) and the Sustainability Efficiency Index (SEI).

Quality Index (QI). Various metrics are taken into consideration under the QI calculation to evaluate the quality of an HRC process. These metrics are linked with factors of the human operators' work (human safety, ergonomics, etc.) and the robots' use (robot autonomy, manipulation ability, etc.). However, this index is limited in evaluating HRC schemes. Thus, under this paper, the QI has been expanded into evaluating also automated and manual scenarios with the introduction of three more metrics (a) Flexibility (process adaptability and reconfiguration effort needed to handle production variations), (b) Number of Operators and (c) Human Availability (Table 1).

Equation (1) provides the calculation of Q_t.

$$Q_t = \frac{\sum_{m=1}^{M}\left(\frac{v_m - v_m^{\min}}{v_m^{\max} - v_m^{\min}}\right)}{M} \tag{1}$$

where:

- Q_t is the cumulative score for a specific task,
- m is the number of the metrics examined, while M is the total number of metrics,
- v_m is the value set after the evaluation of the task under each metric,
- $v_m^{\min}$, $v_m^{\max}$ are the min and max values of each metric according to the index scale.

The lowest value implies a non-efficient human robot collaboration, while the highest one corresponds to the ideal scenario where the robot behaves as a second

Table 1 Proposed QI metrics scale

Metric	Scenario description	Value
Flexibility	Adaptable in any uncertainty	3
	Adaptable after minor reconfiguration effort by an operator	2
	Adapted after major hardware reconfiguration	1
No. of operators	0 Operators	3
	1–3 Operators	2
	4+ Operators	1
Human availability	Fully automated	2
	Human involvement	1

human being. After calculating the Qt for each specific task, the QI for the whole process is calculated considering the various alternatives based on Eq. (2).

$$QI = \frac{\sum_{t=1}^{T} Q_t}{T} \tag{2}$$

where:

- QI is the total quality score for a specific process,
- Q_t is the value calculated from Eq. 1 for each task
- T is the total number of tasks considered

Sustainability Efficiency Index (SEI). Considering the limitation in the literature for methods to evaluate remanufacturing processes, a dedicated index is being proposed (SEI) to evaluate remanufacturing processes in terms of sustainability (Table 2).

Equation (3) provides the calculation of SEt:

$$SE_t = \frac{\sum_{m=1}^{M} \left(\frac{v_m - v_m^{\min}}{v_m^{\max} - v_m^{\min}} \right)}{M} \tag{3}$$

where:

- SE_t is the cumulative SEI score for a specific task,
- m is the number of the metrics examined, while M is the total number of metrics,

Table 2 Proposed SEI metrics' scale

Metric	Scenario description	Value
Energy and environmental impact	80–100% below the industry baseline	5
	60–79% below the industry baseline	4
	40–59% below the industry baseline	3
	20–39% below the industry baseline	2
	≤ 20% less/beyond the industry baseline	1
Raw material consumption	80–100% less raw material than baseline consumption	5
	60–79% less raw material than baseline	4
	40–59% less raw material than baseline	3
	20–39% less raw material than baseline	2
	≤ 20% less/beyond baseline consumption	1
Useful material recovery	80–100% of materials are recovered	5
	60–79% of materials are recovered	4
	40–59% of materials are recovered	3
	20–39% of materials are recovered	2
	≤ 20% of materials are recovered	1

- v_m is the value which is set after the evaluation of the task under each metric
- v_m^{min}, v_m^{max} are the min and max values of each metric according to the index scale.

Respectively with QI, the SEI for the whole process is calculated based on Eq. (4).

$$SEI = \frac{\sum_{t=1}^{T} SE_t}{T} \quad (4)$$

where:

- SEI is the total score for a specific process,
- SE_t is the value calculated from Eq. 3 for each task
- T is the total number of tasks considered

2.3 Workstation Design Selection

Upon calculating the QI and SEI for each alternative process scenario, a method is needed to combine these results to be able to compare the alternative scenarios. Thus, a linear combination of these indices is utilized, as depicted in Eq. 5.

$$QSI = w_1 * QI + w_2 * SEI \quad (5)$$

where:

- QSI (Quality and Sustainability Index) is the cumulative score for a specific process,
- QI and SEI are the values calculated from Eqs. (2) and (4)
- w_1 and w_2 are the weights allocated to each index (the values selected are $w_1 = 0.7$ and $w_2 = 0.3$ to highlight the significance of the human-related metrics).

3 Implementation

Upon defining the process and therefore the respective tasks to be performed under this process, the required skills are described. Consequently, a skills library is created and stored in an internal database containing the skills offered by the available resources and the resources that can realise each skill based on their capabilities. A matching operation is then performed in between the defined required skills and the skills offered by the library to identify the resources that can perform each task. Through this process, the alternative scenarios are formed.

The evaluation process is based on the enhanced QI and SEI, as described above to compare the different production scenarios. The proposed design framework utilises

Table 3 QI, SEI and QSI calculated values for each scenario

Scenario type	QI	SEI	QSI
Fully manual	0.333	0.583	0.408
Fully automated	0.629	0.250	0.515
HRC 1	0.607	0.333	0.525
HRC 2	0.680	0.416	0.601

these Indices to evaluate different disassembly workstation configurations considering three different types of resources, (a) High-payload industrial robots, (b) Low-payload collaborative robots and (c) Human operators. Considering these types of resources, and their capabilities, the processes that need to be performed from the designed workstation are evaluated. For each process, four different scenarios are considered:

- **Manual**: A human takes over the whole process and no robots are involved.
- **Automated**: A Robot takes over the process and no human operators are involved.
- **HRC 1**: A human and a robot are working in the same station but on different areas of the product performing different tasks.
- **HRC 2**: A human and a robot are collaborating handling the same part.

For each of these scenarios, the metrics related to the QI and SEI are selected to calculate its value and select the optimal remanufacturing process scenario (Table 3).

It is evident that HRC is superior to automated and manual scenarios. By diving deeper into the HRC scenarios, under the HRC 1 scenario the operator and the robot work in the same workspace and deal with tasks sequentially but they do not work on the same part simultaneously. Contrariwise, in HRC 2 scenario the operator and the robot share tasks. Comparing these, it is evident that the second scenario is superior ($QSI_1 = 0.5$ while $QSI_2 = 0.6$) derived from the closer human–robot collaboration.

4 Case Study

This methodology is applied in a case study inspired by the household appliances industry dealing with the disassembly of refrigerators towards retrieving specific components for remanufacturing. The cores to be retrieved are the door and other mechanical (the compressor and the fan) and electrical components (PCB), as depicted Fig. 3.

The disassembly process involves handling heavy components (e.g., doors and compressors over 20 kg), along with lighter, detailed tasks such as unscrewing and tube cutting. Due to the variety in component geometries, the process demands high dexterity and adaptable tooling. Heavy tasks necessitate automation to reduce ergonomic strain from repetitive handling. These ergonomic considerations are addressed during process assessment and workstation design. An overview of tasks and required skills is provided in Table 4.

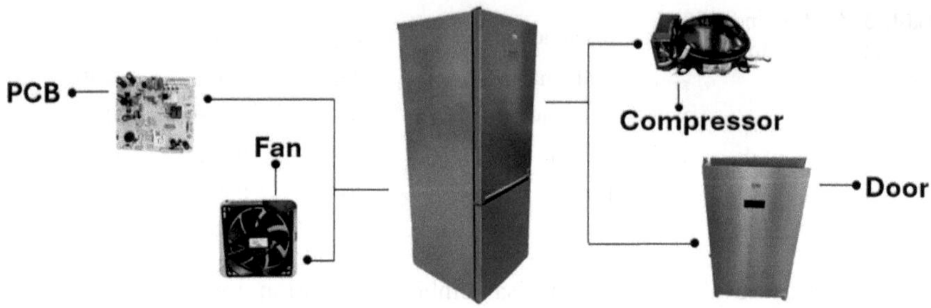

Fig. 3 Core refrigerator components considered

Table 4 Process overview

Process	Task	Part	Required skills
Refrigerator disassembly	Door removal	Door	20 kg payload, grasping, unscrewing
	Compressor removal	Compressor	20 kg payload, grasping, unscrewing, cutting
	PCB removal	PCB	3.5 kg payload, unscrewing, grasping
	Fan removal	Fan	3.5 kg payload, grasping

For the design of the remanufacturing workstation that will handle the disassembly of the refrigerator, the resource types selected need to have the required capabilities to realise the required skills for each task as described under Table 4. Considering the required skills, the resource types need to be selected to meet these skills. The capabilities of the different resource types are presented in Table 5.

Table 5 Required resources

Resource type	Capability	Capability value
Human operator	Payload	10 kg (ergonomics directives [17])
	Accessibility	High
	Tools	Hands, screwdriver, cutter
High-payload robot (HPR)	Payload	70 kg
	Accessibility	Low in tight spaces
	Tools	Suction gripper
Low-payload cobot	Payload	10 kg
	Accessibility	High
	Tools	Screwdriver, cutter

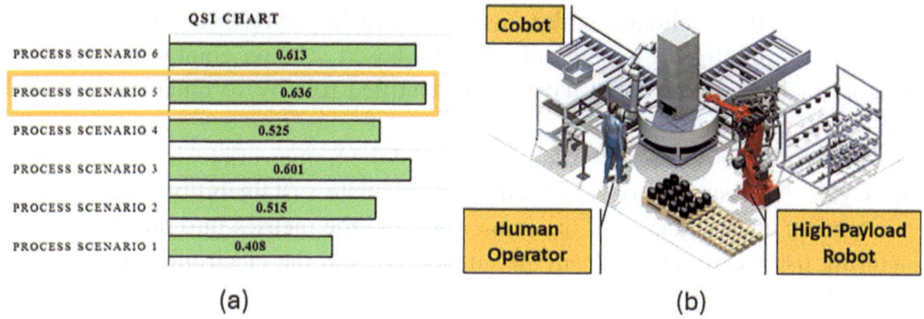

Fig. 4 **a** Workstation alternatives evaluation, **b** optimal workstation visualisation

4.1 Evaluation Results

Six different scenarios have been identified and analysed to identify the optimal one based on the proposed scenario:

- Process Scenario 1: Fully Manual (Two operators)
- Process Scenario 2: Fully Automated (Two robots)
- Process Scenario 3: Human Operator and Cobot
- Process Scenario 4: Human Operator and HPR
- Process Scenario 5: Human Operator, Cobot and HPR
- Process Scenario 6: Operator and Two Cobots

Based on the outcome of the analysis (Fig. 4a), the highest value corresponds to the scenario 5. The superiority of this scenario comes from the flexibility accomplished by the co-existence of the human operator with robots providing different capabilities, as they can dynamically adapt to different production requirements. Furthermore, the utilisation of the high-payload robot to handle heavy components as well as the cobot to assist in light, more flexible tasks, like disassembly, unscrewing and cutting removes strain from the human operator positively influencing ergonomics and throughput.

5 Conclusions and Future Work

Under this work, a framework for generating and evaluating (re)manufacturing workstations has been proposed. This framework is based on a) the selection of the resource types based on matching between the required skills and their capabilities and b) the evaluation of the emerged alternatives based on the QI and SEI. The QI has been proposed by the literature for evaluating HRC scenarios but under this work it has been expanded to meet the requirements of broader concepts, including manual, automated and HRC-based (re)manufacturing workstations. On the other hand, SEI has been introduced to evaluate production scenarios in terms of sustainability. The framework has been validated under a use case focusing on refrigerator disassembly workstations. Significant benefits of an HRC-based schema combining a human operator, an HPR and a cobot have been identified relative to other human/robot configurations scenarios.

Future work encompasses the enhancement of the framework by integrating LCA (Life Cycle Assessment) tools to accurately calculate sustainability metrics. Also, the extension and validation of the framework and the metrics into broader remanufacturing use cases derived from the automotive and the electronics industries is considered.

Acknowledgements This research is partially funded by the European project "RENÉE—Flexible Remanufacturing using AI and advanced Robotics for Circular Value Chains in EU Industry" (Grand Agreement: 101138415) (https://www.renee-project.eu/) funded by the EC.

References

1. Kerin M, Pham DT (2020) Smart remanufacturing: a review and research framework. J Manuf Technol Manag 31:1205–1235
2. Priyono A, Ijomah W, Bititci US (2016) Disassembly for remanufacturing: a systematic literature review, new model development and future research needs. J Indus Eng Manage (JIEM) 9:899–932
3. Peng S, Ping J, Li T, Wang F, Zhang H, Liu C (2022) Environmental benefits of remanufacturing mechanical products: a harmonized meta-analysis of comparative life cycle assessment studies. J Environ Manage 306:114479
4. Soh SL, Ong SK, Nee AYC (2014) Design for disassembly for remanufacturing: methodology and technology. Procedia CIRP 15:407–412
5. Chryssolouris G (2006) Manufacturing systems: theory and practice, 2nd edn. Springer, New York
6. Huang J, Pham DT, Wang Y, Qu M, Ji C, Su S et al (2020) A case study in human–robot collaboration in the disassembly of press-fitted components. Proc Instit Mech Eng Part B J Eng Manuf 234:654–664
7. Parthasarathi S, Eibar U, Alix T, Chavanne R, Cherif M, Perry N (2024) Smart disassembly cell for circularity: turn Industry 4.0 technologies for disassembly and recovery of components. Procedia CIRP 122:557–562
8. Kin STM, Ong SK, Nee AYC (2014) Remanufacturing process planning. Procedia CIRP 15:189–194
9. Van De Ginste L, De Cock A, Van Alboom A, Huysentruyt S, Aghezzaf E-H, Cottyn J (2023) A formal skill model to enable reconfigurable assembly systems. Int J Prod Res 61:6451–6466
10. Michalos G, Makris S, Tsarouchi P, Guasch T, Kontovrakis D, Chryssolouris G (2015) Design considerations for safe human-robot collaborative workplaces. Procedia CIRP 37:248–253
11. Ogorodnikova O (2008) Human weaknesses and strengths in collaboration with robots. Periodica Polytechnica Mech Eng 52:25
12. Evangelou G, Dimitropoulos N, Michalos G, Makris S (2021) An approach for task and action planning in human-robot collaborative cells using AI. Procedia CIRP 97:476–481
13. Andronas D, Argyrou A, Fourtakas K, Paraskevopoulos P, Makris S (2020) Design of human robot collaboration workstations—two automotive case studies. Procedia Manuf 52:283–288
14. Makris S (2021) Cooperating robots for flexible manufacturing. Springer International Publishing, Cham

15. Andronas D, Kampourakis E, Papadopoulos G, Bakopoulou K, Kotsaris PS, Michalos G et al (2023) Towards seamless collaboration of humans and high-payload robots: an automotive case study. Robot Comput Integr Manuf 83:102544
16. Kokotinis G, Michalos G, Arkouli Z, Makris S (2023) On the quantification of human-robot collaboration quality. Int J Comput Integr Manuf 36:1431–1448
17. Tirupachuri Y, Ramadoss P, Rapetti L, Latella C, Darvish K, Traversaro S et al (2021) Online non-collocated estimation of payload and articular stress for real-time human ergonomy assessment. IEEE Access 9:123260–123279

15. Andronas D, Kampourakis E, Papadopoulos G, Bakopoulou K, Kotsaris PS, Michalos G et al (2023) Towards seamless collaboration of humans and high-payload robots: an automotive case study. Robot Comput Integr Manuf 83:102544
16. Kokotinis G, Michalos G, Arkouli Z, Makris S (2023) On the quantification of human-robot collaboration quality. Int J Comput Integr Manuf [illegible]
17. [illegible] (2021) [illegible] assessment. IEEE Access [illegible]

Open Access This chapter is licensed under the terms of the Creative Commons Attribution 4.0 International License (http://creativecommons.org/licenses/by/4.0/), which permits use, sharing, adaptation, distribution and reproduction in any medium or format, as long as you give appropriate credit to the original author(s) and the source, provide a link to the Creative Commons license and indicate if changes were made.

The images or other third party material in this chapter are included in the chapter's Creative Commons license, unless indicated otherwise in a credit line to the material. If material is not included in the chapter's Creative Commons license and your intended use is not permitted by statutory regulation or exceeds the permitted use, you will need to obtain permission directly from the copyright holder.

Additive Manufacturing and Smart Tools

Benchmarking Additive Manufactured Silicone Actuators: A Trajectory and Bending Analysis

Florian Schreiber, Lukas Gugel, Philipp Vogelgsang, Jörg Franke, and Martin Manns

Abstract In soft robotics, pneumatic network actuators (pneunets) enhance safety in human–robot collaboration by reducing injury risks from rigid grippers. This study investigates the additive manufacturability and bending behavior of pneunets that are fabricated using various additive manufacturing (AM) processes, focusing on the Liquid Additive Manufacturing (LAM) process for real silicone. Compared to TPU-based pneunets manufactured via Fused Filament Fabrication (FFF) and Selective Laser Sintering (SLS), LAM pneunets achieve a larger bending angle (104°) without structural failure and exhibit superior resilience compared to PolyJet (PJ) pneunets. Trajectory analysis reveals that softer materials experience pre-bending due to gravity. However, the LAM pneunets demonstrate high elasticity and return to their original shape after actuation, making them promising for soft robotic applications such as delicate object handling. The findings highlight the potential of LAM manufactured silicone to improve pneunet performance. Future work focuses on optimizing the LAM process, evaluating long-term durability, and refining trajectory control to enhance functionality, for more flexible and efficient soft robotic systems in industrial settings.

Keywords Soft robotics · Pneunets · Additive manufacturing · Liquid additive manufacturing · Silicone

1 Introduction

A key challenge in human–robot collaboration is ensuring human safety. Although protective measures have been proposed [1], rigid gripping systems remain a potential injury risk. Soft robotics offers promising solutions in this regard: compliant grippers,

F. Schreiber (✉) · M. Manns
PROTECH—Institut für Produktionstechnik, Universität Siegen, Siegen, Germany
e-mail: florian.schreiber@uni-siegen.de

L. Gugel · P. Vogelgsang · J. Franke
Institute for Factory Automation and Production Systems (FAPS),
Friedrich-Alexander-Universität Erlangen-Nürnberg (FAU), Erlangen, Germany

M. Manns (ed.), *New Paradigms for Anticipated Uncertainty*, Lecture Notes in Mechanical Engineering, https://doi.org/10.1007/978-3-032-16889-4_78

such as those made of silicone, naturally adapt to objects and rely less on complex control systems to ensure safe operation [2]. However, their adoption in industrial manufacturing remains limited.

One type of soft gripper is the so-called pneumatic network, or pneunet, which generally consists of at least one air chamber [3]. When the pressure inside the chambers increases, they expand. By arranging the chambers in a specific pattern, controlled movements, such as bending motions, can be achieved. Pneunets are often manufactured using molding techniques. Recent advancements in additive manufacturing (AM) enable direct manufacturing of soft materials. This eliminates the need for conventional molds and casting processes, expands design possibilities and simplifies production.

2 State of the Art

In our previously published studies [4–6], the additive manufacturability of pneunets is analyzed. The pneunet design by Polygerinos [7] is selected as a reference and adapted to AM processes to enable the fabrication of functional pneunets. The fabricated pneunets are then analyzed for airtightness and bending behavior. Specifically, the bending angle is experimentally determined as a function of the applied pressure. Four different AM processes are evaluated. While the pneunets manufactured using *Fused Filament Fabrication* (FFF) and *Selective Laser Sintering* (SLS) with TPU are airtight and functional, the significantly softer pneunets made from silicone-like materials using *Stereolithography* (SLA) and *PolyJetting* (PJ) leak after a single actuation, rendering them non-functional. Due to the differences in Shore hardness, the silicone-like pneunets achieve greater bending angles despite material failure. Given the high bending angles observed, the additive manufacturing of functional pneunets from silicone is desirable.

Various processes can be utilized for AM of silicone (cf. [8]). Among these, *Direct Ink Writing* (DIW) enables extrusion-based deposition but demands precise rheological control, with curing methods including UV, thermal, or support-bath-assisted curing [9]. The latter, such as *Freeform Reversible Embedding* (FRE) or *Rapid Liquid Printing* (RLP) allow for large-scale, high-speed printing using standard silicones but necessitates complex toolpath calculations [10]. *Vat Photopolymerization* (VPP), including *Digital Light Processing* (DLP), offers high resolution but requires low-viscosity resins, often involving volatile or toxic reagents, and is constrained by high equipment costs and small build volumes [11]. Meanwhile, *Fused Filament Fabrication* (FFF) has seen limited success with elastomers due to interlayer adhesion challenges and nozzle interactions with the filament, often necessitating additional support structures [12]. Additionally, *Liquid Additive Manufacturing* (LAM) is an emerging 3D printing technique that enables the fabrication of highly flexible and durable silicone components by directly extruding and thermally curing viscous silicone materials [13].

In contrast to other AM processes, LAM enables the use of industrial-grade, highly viscous silicones without requiring photoinitiators or custom resin formulations. Unlike SLA, DLP, and PJ, which rely on modified, often brittle silicone-like materials, LAM produces elastomers with mechanical properties comparable to injection-molded silicone. Unlike DIW, which requires precise curing control, LAM uses thermal crosslinking for easier processing and consistent material performance in pneunet production.

3 Objectives

This work builds upon our previous studies on the additive manufacturability of pneunets [4–6]. The objective is to leverage the potential of silicone-based pneunets identified in these studies by fabricating pneunets from real silicone using the *Liquid Additive Manufacturing* (LAM) process. For this purpose, the pneunet reference model by Polygerinos [7] will be adapted and subsequently analyzed regarding its bending behavior. In addition to examining the bending angle, the trajectory of the LAM as well as the for FFF, SLA, SLS, and PJ-manufactured pneunets will also be investigated.

4 Methodology

4.1 Manufacturing

The silicone pneunet is manufactured at the FAPS institute in Erlangen using the LAM process on a LiQ 5 printer [14]. The system enables additive manufacturing with thermally cross-linking 2k silicones. The extrusion-based process is similar to the popular FFF process, whereby in LAM a reactive silicone rubber mixture is deposited as a liquid strand on the build platform. There, the strand remains in shape due to the high rubber viscosity and is thermally cross-linked into a silicone elastomer after each layer is applied using an infrared heating lamp.

Due to its geometry, the production of the pneunet is a challenge for LAM. The high viscosity silicone rubber is liquified by a curing lamp as the temperature rises, causing it to flow at this point. This poses a problem, particularly for the fabrication of overhangs and bridges. If not considered, segments can collapse and production can fail.

A second challenge is that the pneunet must be airtight. Small openings can render a component unusable. The transition between the infill and the side wall has proven to be particularly critical.

In order to enable the manufacturing of the component, the printing parameters of the LAM process—initially based on the default printing parameters provided by the

Table 1 Slicing properties for the manufacturing of the proposed silicone pneunet

Property	Parameter
Layer height (normal/round edges)	0.3/0.2 mm
Extrusion multiplier	1.1
Infill to perimeter overlap	35%
Perimeters/solid layers (top and bottom)	5/4
Infill density/infill pattern	100%/rectilinear

printer manufacturer—are specially adapted to the production of pneunets as well as to geometric adjustments to the part itself. Manufacturing parameters are shown in Table 1. The production of the critical round edges with overhang from 0 to 3 mm and bridges from 17 to 20 mm is enabled by a reduced layer height. In turn, airtight manufacturing is achieved by increasing the extrusion rate to 1.1, a high perimeter count of 5 and an increased infill to perimeter overlap of 35%.

4.2 Design

The pneunets investigated in this study are based on the reference design by Polygerinos et al. [7]. The pneunet consists of 14 air chambers connected via a central air channel. As the air pressure increases, the chambers expand and repel each other, resulting in a finger-like bending motion. Previous studies [4–6] have demonstrated that modifications to the design are necessary in order to fabricate the pneunets using different AM processes. A particular challenge in this context is the fabrication of airtight chambers. In general, the walls must neither be too thin, which would result in leakage, nor too thick, which would excessively limit the bending motion. For the LAM-manufactured pneunets, a wall thickness of 2 mm is chosen as a compromise to balance these opposing requirements.

In addition to the process adjustments described in Sect. 4.1, further design changes are made for manufacturing the pneunet using LAM. Figure 1 shows the modified part in section, with the cut edges colored green. The channels between the air chambers are sloped at 45° so that no bridges are necessary (Fig. 1a). Slopes (Fig. 1b) are created within each air chamber so that only small bridges (Fig. 1c) remain to close the ceilings, which can be handled by the process.

4.3 Material

The materials examined in this paper can be compared in terms of their Shore hardness and elongation at break. This allows comparing pneunets behavior. The softer the material and the greater its elongation at break, the higher the expected bending of the pneunets. An overview of the materials and their properties is provided in Table 2.

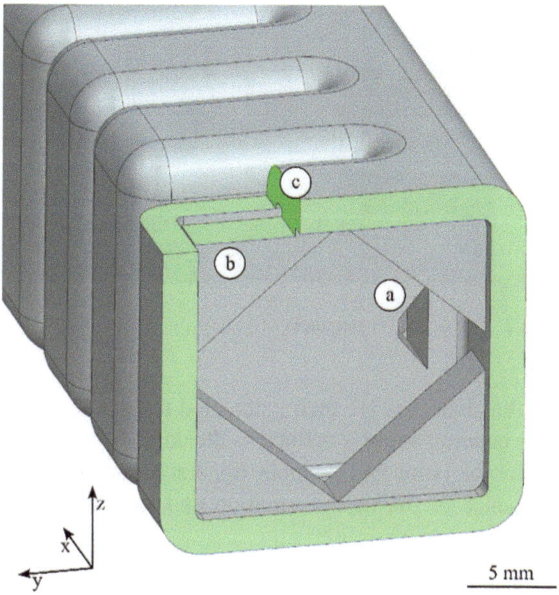

Fig. 1 CAD model of the additively manufactured pneunet with cutouts; **a** air channel with slopes, **b** wedge to support the air chamber ceiling, **c** remaining bridge

Table 2 AM materials and their key parameters

Material	AM process	Shore A hardness (A)	Elongation (%)
NinjaTek Cheetah filament	FFF	95	580
Luvosint TPU powder	SLS	92	520
Formlabs Elastic 50A resin V1	SLA	50	160
AR-G1L	PJ	35	160
Shin-Etsu KEG 2003H-30 Silicone	LAM	30	830

4.4 Experimental Setup

To evaluate the bending behavior of the pneunets, a testing setup that is based on the procedure outlined in [5] is employed. The pneunets are mounted horizontally on a fixed base. The internal pressure is gradually increased in 10 kPa increments, up to a maximum of 500 kPa, causing the pneunet to bend in a positive rotational direction. If the pneunet touches the flange before reaching the maximum pressure or is damaged, the test is terminated. To minimize the effects of relaxation, the pressure is held steady for 5 s after each increment. After each pressure step, an image is taken to track the position of the pneunet's tip. Figure 2 shows an overview of the test setup.

For accurate measurement of position changes, circular markers are placed on each chamber of the pneunet and on the fixed base. These markers are tracked using the HoughCircles function from the OpenCV library [15]. The positions of the markers

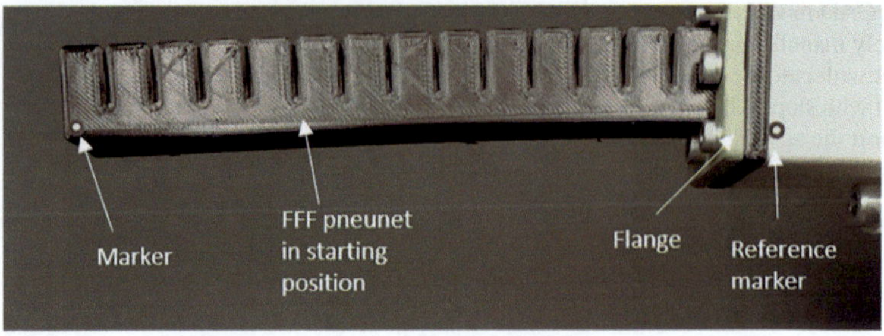

Fig. 2 Test setup for pneunets

on the chambers are then calculated relative to a reference point on the fixed base. The bending angle is defined as the angle between the x-axis and the vector between reference point and pneunet tip. Each pneunet is tested only once to prevent the influence of plastic deformation and to ensure precise evaluation of the bending behavior.

5 Results

When pressurizing the LAM pneunet, it shows no pressure loss, indicating airtightness. At a pressure of 60 kPa, the pneunet reaches the flange, achieving a bending angle of 104°. After testing, the pneunet shows no external damage and remains fully functional. A comparison with the results of the reference study [4] is given in Fig. 3.

The trajectories of the pneunets' tips are shown in Fig. 4. The initial position of each pneunet is represented by the leftmost marker, as the pneunets were clamped in a left-oriented horizontal starting position. Each subsequent marker represents a pressure increase of 10 kPa. While the FFF and SLS pneunets remain relatively close to the horizontal in their initial position due to the stiffer TPU material, gravity significantly affects the softer materials, causing the pneunets to sag and bend before actuation.

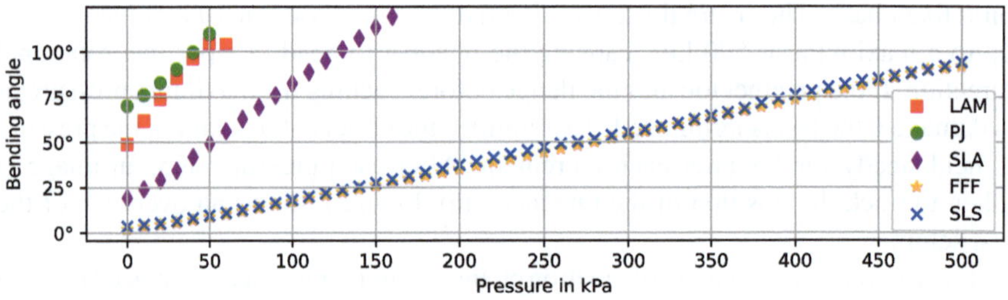

Fig. 3 Bending angle versus pressure

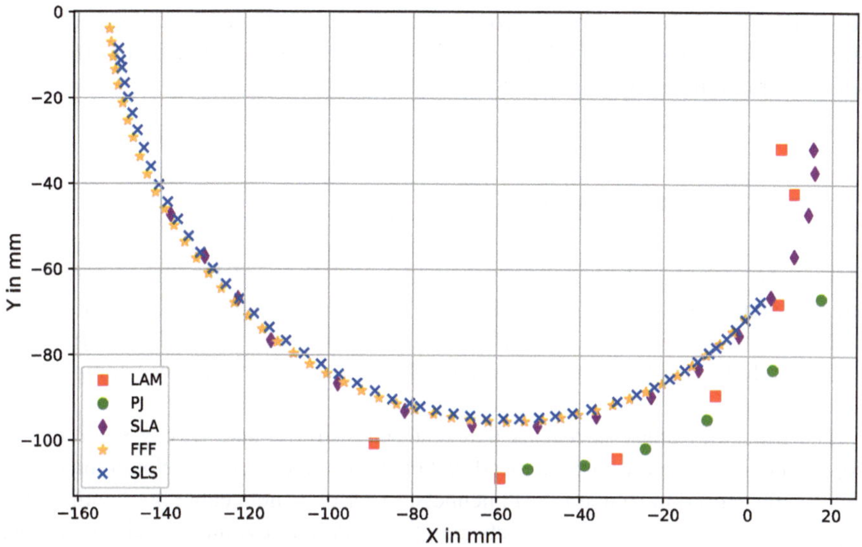

Fig. 4 Bending motion of tested pneunet from horizontal starting position

6 Discussion

The investigation of bending angles in pneunets shows that the behavior of the LAM pneunet is similar to the PJ pneunet. Due to gravity, the pneunets initially sag significantly, starting with a bending angle of 50° (LAM) and 70° (PJ), respectively. As the pressure increases, the chambers expand notably, resulting in a relatively large bending deformation. While the PJ pneunet ruptures during the experiment, the LAM pneunet reaches the flange. After the pressure is released, the pneunet returns back to its initial position. In contrast to TPU pneunets, the LAM pneunet does not exhibit time-dependent residual deformation and return to their original position. Consequently, the LAM pneunets can be classified as functional and, due to their large bending deformation, may be suitable for various applications, such as grippers for handling delicate objects.

When analyzing the trajectory, it is noticeable that the FFF, SLS, and SLA pneunets follow a common path, with specific points along this trajectory being reached at different pressure levels. Future studies should investigate the relationship between the pressure-dependent position along the common trajectory and the pneunets' material/design parameters used.

Although all investigated pneunets have the same length, the tips of the PJ and LAM pneunets lie below the common trajectory of the FFF, SLS, and SLA pneunets. This is due to initial gravity-induced bending. PJ and LAM pneunets bend significantly between the first and second air chambers, while the others remain unchanged and point downward. In contrast, FFF, SLS, and SLA pneunets achieve similar bending angles only when pressurized, causing deformation along the entire structure. As a result, the tips of PJ and LAM pneunets are positioned farther from the reference point than those of FFF, SLS, and SLA.

7 Conclusion and Outlook

This study investigates the additive manufacturability and bending behavior of pneunet actuators, introducing the LAM process for real silicone fabrication. The LAM pneunets achieve a 104° bending angle without structural failure, surpassing TPU-based FFF and SLS pneunets while exhibiting superior resilience compared to PJ pneunets.

The trajectory analysis reveals that material stiffness and gravity significantly impact the initial positioning and bending behavior of the pneunets. Softer materials, such as those used in PJ and LAM pneunets, experience noticeable pre-bending due to gravity, while TPU-based pneunets maintain a more horizontal initial orientation. Despite this, the LAM pneunets demonstrate high elasticity and return to their original shape after pressure release, making them promising candidates for applications requiring repeated actuation without permanent deformation.

The findings highlight the potential of real silicone for soft robotic actuators, particularly for grippers handling delicate objects. Future work focuses on optimizing the LAM process to improve manufacturability, investigating the long-term durability of AM silicone pneunets under cyclic loading, and refining trajectory control through material and design modifications. Additionally, expanding the study to include more complex actuator geometries further advances the integration of additively manufactured soft robotics in industrial applications.

Acknowledgements The authors would like to acknowledge the financial support of the Bavarian Ministry of Economic Affairs, Regional Development and Energy within the project "ProLAM".

References

1. Vasic M, Billard A (2013) Safety issues in human-robot interactions. In: 2013 IEEE International conference on robotics and automation. IEEE, pp 197–204
2. Wang Y, Wang G, Ge W et al (2024) Perceived safety assessment of interactive motions in human-soft robot interaction. Biomimetics 9:58. https://doi.org/10.3390/biomimetics9010058
3. Mosadegh B, Polygerinos P, Keplinger C et al (2014) Pneumatic networks for soft robotics that actuate rapidly. Adv Funct Mater 24:2163–2170. https://doi.org/10.1002/adfm.201303288
4. Schreiber F, Frohn-Sörensen P, Mouratidis M et al (2023) Case study on the additive manufacturability of printed soft-robotic bending actuators. Procedia CIRP 120:720–725. https://doi.org/10.1016/j.procir.2023.09.065
5. Frohn-Sörensen P, Schreiber F, Manns M et al (2022) Additive manufacturing of TPU Pneu-Nets as soft robotic actuators. In: Andersen A-L, Andersen R, Brunoe TD et al (eds) Towards

sustainable customization: bridging smart products and manufacturing systems. Springer International Publishing, Cham, pp 269–276
6. Manns M, Morales J, Frohn P (2018) Additive manufacturing of silicon based PneuNets as soft robotic actuators. Procedia CIRP 72:328–333. https://doi.org/10.1016/j.procir.2018.03.186
7. Polygerinos P, Lyne S, Wang Z et al (2013) Towards a soft pneumatic glove for hand rehabilitation. In: 2013 IEEE/RSJ International conference on intelligent robots and systems. IEEE, pp 1512–1517
8. Nethani H, Jangitwar A, Gupta S et al (2024) Silicone-based additive manufacturing. Polymer-Plastics Technol Mater 1–21. https://doi.org/10.1080/25740881.2024.2445610
9. Cheng Y, Chan KH, Wang X-Q et al (2019) Direct-Ink-Write 3D printing of hydrogels into biomimetic soft robots. ACS Nano 13:13176–13184. https://doi.org/10.1021/acsnano.9b06144
10. Sparrman B, Du Pasquier C, Thomsen C et al (2021) Printed silicone pneumatic actuators for soft robotics. Addit Manuf 40:101860. https://doi.org/10.1016/j.addma.2021.101860
11. Zhao T, Yu R, Li S et al (2019) Superstretchable and processable silicone elastomers by digital light processing 3D printing. ACS Appl Mater Interfaces 11:14391–14398. https://doi.org/10.1021/acsami.9b03156
12. Herzberger J, Sirrine JM, Williams CB et al (2019) Polymer design for 3D printing elastomers: recent advances in structure, properties, and printing. Prog Polym Sci 97:101144. https://doi.org/10.1016/j.progpolymsci.2019.101144
13. Da Brusa Costa Linn L, Danas K, Bodelot L (2022) Towards 4D printing of very soft heterogeneous magnetoactive layers for morphing surface applications via liquid additive manufacturing. Polymers (Basel) 14. https://doi.org/10.3390/polym14091684
14. InnovatiQ GmbH + Co KG (2025) LiQ 5. https://www.innovatiq.com/produkte/3d-drucker/liq-5/. Accessed 27 Mar 2025
15. Bradski G (2000) The OpenCV library. Dr. Dobb's J Software Tools 120–125

Fabrication of Cranioplasty Implants from Sheet Metal Plates in a Rapid Tooling Approach

Peter Frohn-Sörensen and Michael Schiller

Abstract Cranioplasty implants are applied for medical cure of human cranial diseases, e.g. skull cancer, malformation or cranial trauma after accident. Particular crucial for long term curing is the acceptance of the implant by the human body and its long-term integrity, which is generally achieved using biocompatible materials such as grade 5 titanium. As each medical implant is an individual product, flexible and adaptive manufacturing processes are in need to allow for an economic and automated fabrication. In the past, metallic implants have been fabricated directly by subtractive or additive processes. In an experimental approach, these implants were fabricated by forming of sheet metal plates in 3D-printed tools from polymer. In the light of this feasibility study, a rapid tooling process from cheap, recyclable materials, which can be applicated to directly manufacture molds from spatial scan data is demonstrated. Based on sheet metal forming, the presented process minimizes material effort of highly expensive titanium alloys while at the same time allows for the economical and rapid fabrication of individual single plate implant parts.

Keywords Rapid tooling · Customization · Cranioplasty · Medical implants · Biomedical engineering

1 Introduction

Cranioplasty implants are commonly applied for the reconstruction of a defected skull. The implantation is necessary for medical treatment of a variety of defects or injuries such as head trauma, raised intracranial pressure (ICP), removal of bone-invading tumors or infected bone flaps [1]. Amongst other methods of treatments, for example synthetic bone substitutes from polymers or ceramics, the implants are made as a single piece titanium plate or meshes [2]. These implants are particularly difficult to be manufactured because of diverse economic factors such as high priced metallic alloys and requirement of individualized shapes. From a manufacturing

P. Frohn-Sörensen (✉) · M. Schiller
Forming Technology Siegen, Institute of Production Technologies, Siegen, Germany
e-mail: Peter.Frohn@Uni-Siegen.de

M. Manns (ed.), *New Paradigms for Anticipated Uncertainty*, Lecture Notes in Mechanical Engineering, https://doi.org/10.1007/978-3-032-16889-4_79

perspective, a single piece implant might be desirable to fabricate from sheet metal, since it resembles a hollow shell with relatively moderate curvature. Classically, a sheet metal forming mold is made by machining so that it may be used for stamping processes [3]. However, the manufacturing of such forming tools requires extensive efforts, material wastage and as a consequence is only economically feasible for higher production volumes. Yet, implants are almost always individual geometries and cannot be standardizes in most cases. Therefore, finding novel, fast and economic production alternatives for manufacturing such molds is desirable.

In metal forming processes, an incremental sheet metal forming (ISF) process has been suggested in literature as an appropriate forming process for forming molds for cranioplasty implants. ISF has also been applied successfully in practical procedures [4]. The ISF process allows manufacturing of unique shapes with substantially high strain rates during sheet metal forming. Moreover, this can also be achieved while using non-specific tools. In this realm of manufacturing, there is a new technique of forming molds using additive manufacture techniques. Recently, Luo et al. presented a process flow to mold a single piece of titanium plate into a 3D-printed tool of ABS polymer by using hydrostatic pressing and subsequent manual finishing. This kind of titanium implant has recently been implanted successfully by cranial surgery on a 63-year-old patient [5].

However, it has been demonstrated by the authors, that available 3D-printing polymers vary significantly with regard to mechanical behavior and that careful selection of process parameters highly influences tool resilience [6]. In addition, particularly ABS is only of limited applicability for additive tooling [7]. Hence, the present study aims to improve the usage of additive tooling for cranioplasty implant manufacture by selecting suitable material and parameters for additive manufacture and analyzing the sheet metal forming process. Doing so, qualitative and quantitative results on the expectable manufacturing precision and necessary improvement steps for are identified.

2 Methodology

In this study, a novel manufacturing method to provide cranioplasty implants from single-piece sheet metal plates is analyzed with specific focus on the fabrication methods. The suggested methodology comprises.

i. obtaining spatial data from 3D-scans,
ii. manufacture of the corresponding tool molds by using direct polymer additive manufacturing and
iii. sheet metal forming in the rubber pad forming process as well as
iv. implant finishing.

While other studies on this topic have exploited the pre- and post-processing of such implants from computer tomography, e.g. De La Peña et al. [8], this paper is focused in detail on the peculiarities of both manufacturing processes, i.e. additive manufacture of tools and sheet metal forming.

2.1 Digital Preparation

In this study, a variation of different regions of a realistic human skull is regarded for the fabrication of implants. To this aim, an open access anatomy model of a human skull which has been composed from computer tomography scan data is utilized [9]. Regions which are relevant for cranioplasty applications are identified on this skull model from literature references. Three distinct exemplary regions of the skull model are identified. Namely, these are:

1. A top facing portion which is typical for treatment of brain issues such as tumors [1]. From an implant manufacturing perspective, this region features moderate and uniform spherical curvature resulting in a moderate manufacturing effort.
2. A part of the back of the head, where less uniform curvature and smaller geometrical features are present at the skull.
3. The forehead including the eye socket (orbit), which is, by manufacturing terms, a particularly challenging geometry due to the strong curvature and curvature gradients at the orbit. Such parts are typical for surgery in the upper face region, where bone fracture due to severe impacts requires treatment—and stabilization by an implant [1].

2.2 Additive Tooling

The aforementioned three geometry regions are derived from the open access skull model and implemented to a 3D-printing software (PRUSA SLICER 2.9.0) in order to additively fabricate molds for sheet metal forming from polymer by a 3D-printer. In detail, the fused filament fabrication FFF process is operated on a commercial 3D-printer, type PRUSA MK3S+. The graphical interface of the slicing procedure is shown in Fig. 1.

Polylactide (PLA) filament is used as raw material because, compared to other common low-cost 3D-printing polymers, the material provides a substantially high compressive elastic modulus. For additive tooling, feasible parameter sets have been identified in a preceding study [6], which allow to obtain a rigid and stiff polymer bulk material additive manufacturing. The 3D-printing parameters used for the molds are presented in Table 1. A solid (100% infill) crisscross hatch pattern (± 45°) is adjusted.

These molds are ready to be used without further post-processing after 3D-printing.

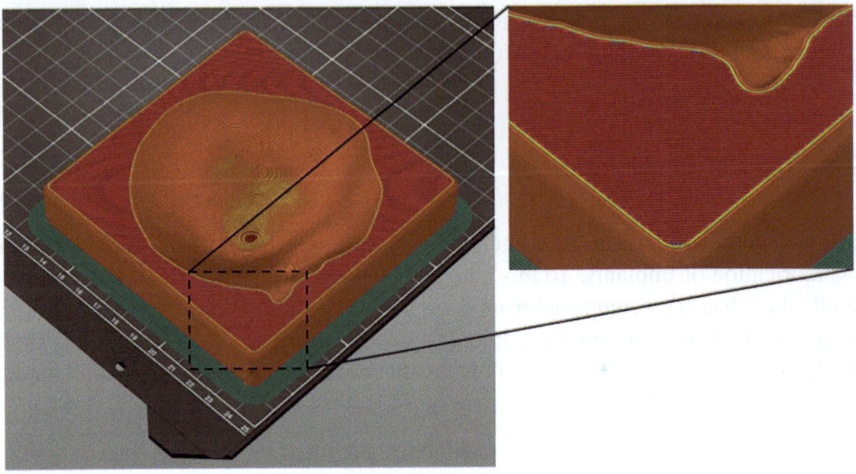

Fig. 1 Slicing algorithm showing tool paths of the 3D-printer applied to the tooling mold for a cranioplasty implant for case study #3 (forehead including upper eye socket)

Table 1 Printing parameters for the molds from polylactide acid PLA

Printing speed (mm/s)	Nozzle temperature (°C)	Bed temperature (°C)	Layer thickness (mm)	Number of perimeters (−)
20	215	65	0.15	3

2.3 Sheet Metal Materials

For medical applications, grade 5 titanium sheet metal type Ti6Al4V is commonly selected primarily due to its high biocompatibility. Here, sheet metal plates from this material are utilized with regard to the desired cranioplasty implant application, e.g. Refs. [2, 5]. Since this titanium alloy is of rather limited formability, as well as high strength and, compared to steel, considerably low elastic modulus, challenges for sheet metal forming arise with respect to shaping complex geometries. Therefore, a standard sheet metal forming grade steel 1.0338 with high ductility is introduced to the process as a baseline from a sheet metal forming perspective.

The relevant mechanical characteristics of these sheet metals are summarized in Table 2. The material parameters are obtained from the manufacturers' data sheets and the detailed material study conducted by Losertová et al. [10].

110 × 110 mm square blanks are cut from these materials and introduced to the forming process to obtain the desired shape according to the selected skull regions.

Table 2 Sheet metal materials and mechanical parameters

Material	Yield strength (MPa)	Ultimate tensile strength (MPa)	Tensile modulus (GPa)	Elongation (%)	Sheet thickness (mm)
1.0338	210	270–350	215	38	0.7
Ti6Al4V	852	925	106	5	0.8

2.4 Sheet Metal Forming Process

For sheet metal forming, the rubber pad forming (RPF) process is conducted on the rigid 3D-printed molds. To this aim, a fixture for a uniaxial tensile testing machine type ZWICK/ROELL Z250 from a preceding study is utilized [7]. The testing machine allows a precise process operation and an intrinsic evaluation of force over displacement curves. Thereby, the required press forces for different materials and geometries are evaluated and the required displacement until full formation of the geometry are obtained.

In the RPF process, a rubber cushion is compressed by a press load and pushed against a plane sheet metal lying on a mold. In detail, here, a polyurethane (PU) rubber cushion with a shore hardness of 40 units is enclosed in a solid steel restrainer with the same outer dimensions of the molds, i.e. nominal 115 × 115 mm^2, see Fig. 2a. On the opposite direction of the rubber cushion lies the 3D-printed mold with the sheet metal plate in between. During the process, the machine presses the rubber cushion against the sheet metal plate while the mold bears the load. After overcoming the sheet metal material´s yield strength, the rubber pad would plastically expand the sheet metal into the mold. The process ends as the sheet metal is in contact to the bottom of the cavity of the forming mold and the formed sheet metal plate is obtained.

For stretch forming of the initially planar sheet metal blanks into the cavities of the forming tools, nominal press forces of 17.5 and 22.5 metric tons are applied for conventional steel and titanium, respectively.

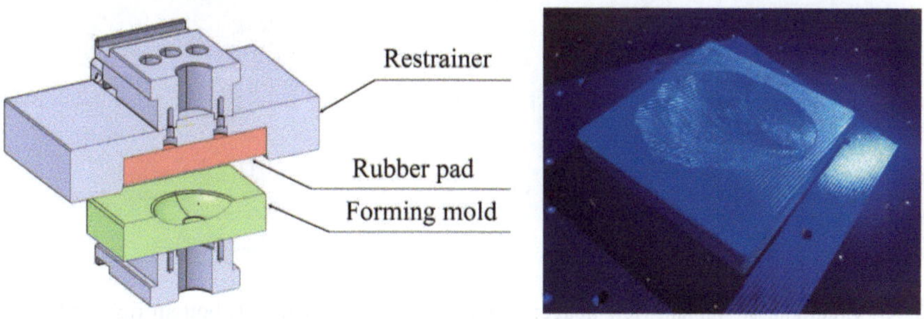

(a) Rubber pad forming assembly (b) 3D-line scanner

Fig. 2 Schematic principle of the rubber pad forming assembly and line scanning procedure of the 3D-printed forming mold for implant geometry #2

2.5 3D-Scanning

As a result, the manufactured sheet metal specimens are captured by a 3D-scanner type GOM Atos Core, which works on the laser triangulation principle. The scanning process is illustrated exemplarily for the mold of implant geometry #2 in Fig. 2b. After scanning, the formed geometries are digitally correlated to the intended skull implant geometry for quality inspection. This procedure is conducted in the software solution GOM Correlate and based on a best fit approach.

3 Results and Discussion

In the rubber pad forming process, the maximum force applied to the fixture is successively increased until full geometry formation is achieved in the molds of each of the selected implant geometries and for both studied sheet metal materials—i.e. the soft deep drawing alloy 1.0338 and the titanium medical grade Ti6Al4V.

Visually, the accuracy of fit is demonstrated by placing the herein fabricated implants on a 3D printed model of the human skull in Fig. 3. Subsequent to the forming process, these implants were manually trimmed at the edges.

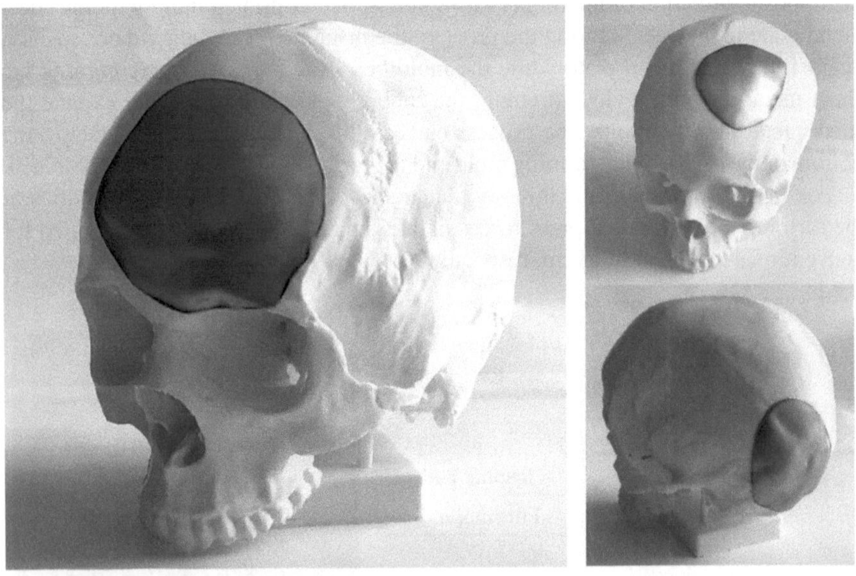

Fig. 3 Cranioplasty implant (left image: geometry #3, upper right: #1, bottom right: #2) from a single-piece metallic plate fabricated on an additively manufactured polymer mold. Illustrative positioning on a 3D-printed skull model. In particular, good fit is achieved for geometry #3 in the geometrically complex area of the eye socket (orbit)

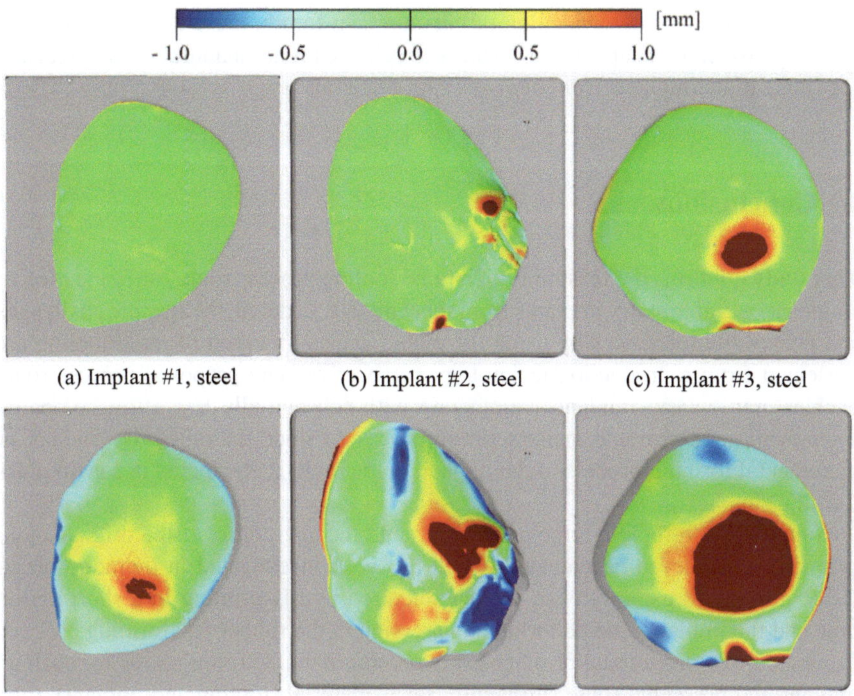

(a) Implant #1, steel (b) Implant #2, steel (c) Implant #3, steel

(d) Implant #1, titanium (e) Implant #2, titanium (f) Implant #3, titanium

Fig. 4 Surface deviations of the manufactured single plate implants

The forming results of the implants are evaluated by qualitative means using a 3D-laser line scanner. These measurements are correlated to the original geometry in a local best fit approach by fitting the surfaces without flanges into the mold geometries. The obtained deviations are plotted in Fig. 4.

The implant sections made from steel (a–c) align reasonably well to the desired skull geometry in areas where large global curvatures are demanded. For instance, implant geometry #1, which is taken from the top of the head, overall shows good fit. Geometry #2, which resembles a portion of the lower back of the head includes some features with high local curvature, which lead to smaller deviations in these areas. For geometry 3, which is the forehead including the eye socket (orbit), shows deviations in the central area. Interestingly, the geometrically complex feature of the edge of the eye socket is formed sufficiently, but the central deviation indicates that using 17.5 metric tons did not fully form the implant due to the higher drawing depth of this geometry.

For all Implants made from titanium (d–f), a considerably higher press force was applied (22.5 tons), which is required due to the material's higher strength (Table 2). The deviation plots globally indicate that elastic springback influenced the results which agrees to the lower modulus of elasticity of titanium compared to steel. All plots show positive deviations in the center and negative at the outer area which is explained by a widening of curvature. In future, these deviations need to be compensated by adapted tool surfaces to achieve higher manufacturing precision for single plate implants from high-strength materials.

4 Conclusions

This study confirms the feasibility of a rapid tooling process route for fabrication of single piece cranioplasty implants from sheet metals. In detail, the specific challenges and achievable tolerances by using the manufacturing method of rapid tooling are presented from a production technology perspective. The proposed process route enables cranioplasty implant manufacture with substantially less effort compared to erstwhile conventional manufacturing processes since the otherwise manual or incremental forming process of the implant in the mold is conducted in a swift press forming step herein.

Supposed, a sufficient tomography scan and reconstruction of the original skull geometry is achieved beforehand, which has been investigated in literature previously, a ready to use mold can be manufactured by 3D-printing in a short period time and the implants can be formed by pressing in the rubber pad forming process. Compared to conventional or other experimental manufacturing routes for such an implant, the presented route sufficiently saves material (tooling and implant), time and therefore cost. In addition, this route allows to fabricate individual implants in a cost-effective and flexible way. In future, the medical application will be addressed in deeper detail.

The overall vision of this research is to provide a manufacturing approach to enable a medical institution to fabricate a single plate implant in a rapid and straightforward approach during treatment. The required equipment would, besides medical apparatus, incorporate computing equipment for tomography computations, a 3D-printer for the mold, a small sheet metal press as well as metal cutting and finishing equipment.

Acknowledgements We highly appreciate the help and support of FabLab Siegen Team, particularly Marios Mouratidis, in 3D-printing equipment and advice.

References

1. Kolias AG, Kirkpatrick PJ, Hutchinson PJ (2013) Decompressive craniectomy: past, present and future. Nat Rev Neurol 9:405–415. https://doi.org/10.1038/nrneurol.2013.106
2. Bonda DJ, Manjila S, Selman WR, Dean D (2015) The Recent Revolution in the design and manufacture of cranial implants: modern advancements and future directions. Neurosurgery 77:814–824. https://doi.org/10.1227/NEU.0000000000000899
3. Rudman K, Hoekzema C, Rhee J (2011) Computer-assisted innovations in craniofacial surgery. Facial Plast Surg 27:358–365. https://doi.org/10.1055/s-0031-1283054

4. Ambrogio G, Luigi DN, Filice L, Gagliardi F, Muzzupappa M (2005) Application of incremental forming process for high customised medical product manufacturing. J Mater Process Technol 162:156–162. https://doi.org/10.1016/j.jmatprotec.2005.02.148
5. Luo J, Morrison DA, Hayes AJ, Bala A, Watts G (2018) Single-piece titanium plate cranioplasty reconstruction of complex defects. J Craniofac Surg 29:839–842. https://doi.org/10.1097/scs.0000000000004311
6. Frohn-Sörensen P, Geueke M, Engel B, Löffler B, Bickendorf P, Asimi A, Bergweiler G, Schuh G (2022) Design for 3D printed tools: mechanical material properties for direct polymer additive tooling. Polymers 14:1694. https://doi.org/10.3390/polym14091694
7. Frohn-Sörensen P, Geueke M, Belay TT, Kuhnhen C, Manns M, Engel B (2020) 3D printed prototyping tools for flexible sheet metal drawing. engrXiv (2020). https://doi.org/10.31224/osf.io/w8rgb
8. De La Peña A, De La Peña-Brambila J, Pérez-De La Torre J, Ochoa M, Gallardo GJ (2018) Low-cost customized cranioplasty using a 3D digital printing model: a case report. 3D Print Med 4(4). https://doi.org/10.1186/s41205-018-0026-7
9. Newey H. The anatomy of the human skull. https://sketchfab.com/models/baf6ac7b781a46218dca2b59dee58817/embed?autostart=1. Last accessed 12 May 2025
10. Losertová M, Štamborská M, Lapin J, Mareš V (2016) Comparison of deformation behavior of 316L stainless steel and Ti6Al4V alloy applied in traumatology. Metalurgija 55:667–670

4. Ambrogio G, [illegible] Filice L, Gagliardi F, Muzzupappa M (2005) Application of incremental forming process for high customised medical product manufacturing. J Mater Process Technol [illegible]
5. [illegible]
[illegible]

[illegible]

A Prototype System Integration Solution for Limp Fabric Handling Utilizing Reconfigurable and Scalable Grippers

Morteza Alebooyeh, Jill Urbanic, and Thanh Dat Vo

Abstract Fiber composite materials balance strength and mass criteria. Current practices rely on multi-stage manual operations for hand layup solutions. Alternative methods need to be developed to enable effective automation to be introduced in industrial domains. Specialty flexible grippers have been designed for dry, limp material gripping leveraging origami principles, and in this research, the proposed robotics systems integrated is presented. Design guidelines are presented along with solutions for gripping and wrinkle elimination. Modular experimental validation activities were performed for the grippers, different mold surfaces, and a prototype bladder based strategy for smoothing out the positioned material. The findings of the research pave the way for rapid, efficient, mechanically simple, and cost-effective, integration of origami based compliant mechanisms in fabric composite manufacturing settings.

Keywords Limp flexible materials · Material handling · Flexible grippers · Pick and place · Freeform surfaces · System layout

1 Introduction

Fiber composite materials, renowned for their stiffness and strength, have become integral in various industries, including automotive, aviation, and consumer goods. Light-weighting strategies, which balance strength to mass, are critical for many applications. However, the adoption of these materials presents challenges related to high production volumes and handling automation. Material handling of limp flexible materials is difficult. Careful handling is required to avoid damage and ensure accurate, wrinkle-free placement. Draping, a manual technique used to position textile patterns onto molds in composite manufacturing, relies heavily on the fabric's properties [1], including in-plane shear, bending stiffness, and structural stability, as well as the mold's geometric features. Fabrics constructed from heavy, coarse yarns with dense structures tend to drape poorly, whereas those with long floats and filament

M. Alebooyeh · J. Urbanic (✉) · T. D. Vo
University of Windsor, Windsor, ON, Canada
e-mail: jurbanic@uwindsor.ca

M. Manns (ed.), *New Paradigms for Anticipated Uncertainty*, Lecture Notes in Mechanical Engineering, https://doi.org/10.1007/978-3-032-16889-4_80

yarns with minimal twist drape more effectively. The flexibility of these textiles allows them to conform to varying curvatures, such as mold surfaces. However, this compliance, combined with factors like surface texture, construction, and geometric variability (e.g., shape, size, presence of cut-outs and slits), poses significant challenges for automation in terms of reliable handling without damaging the material [2–4].

Handling solutions include needles, clamps, vacuum systems, and adhesives, each with drawbacks like potential damage, high cost, or residue [4–7]. *Currently, there are no comprehensive, efficient, and rapid solutions, whether semi-automatic or automatic, for managing these flexible components* [8]. Manual layup operations are labor-intensive (creating process bottlenecks), and may pose health and safety risks, since layup tasks often result in awkward postures, leading to work-related musculoskeletal disorders and increased absenteeism as well as productivity losses. Another approach is required. Compliant mechanisms, which use flexible elements for movement and gripping, offer a promising solution [9]. Previous testing indicated that high surface area, compliant grippers can sustain high gripping forces from a robotic clamping mechanism without damaging the fibers [7]. Inspired by Miura-fold and origami designs [10–13], compliant grippers were introduced conform to complex shapes that handle delicate or irregularly shaped fabrics without damage. The Miura fold approach exhibits auxetic behavior, which is based on the geometry, not the material. When the Miura fold is stretched (or collapsed) in one direction, the folds open/close up, causing the material to expand/contract in the perpendicular direction.

The goal of this paper is to introduce an integrated system for handling limp fabric material utilizing reconfigurable and scalable grippers, leveraging previous research identifying potential wrinkle areas and quantifying the relationship between mold surface geometry and gripper design [8, 14]. Figure 1 illustrates where this paper is positioned among the overall research flow. A case study to illustrate an approach to automate handling of various fabric materials with compliant mechanism grippers designed for different non-planar mold surfaces is presented.

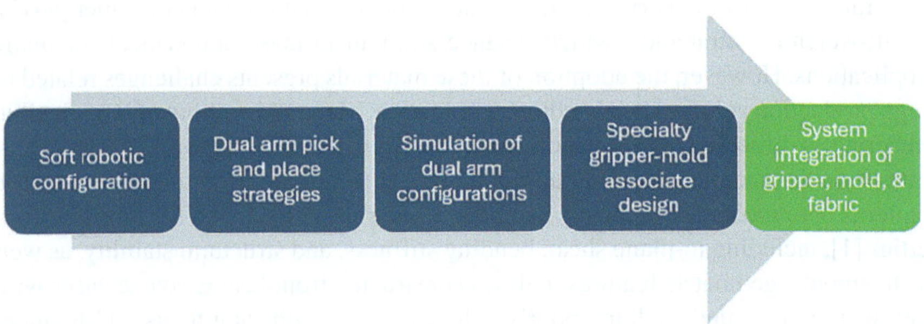

Fig. 1 Visual representation of the entire research flow highlighting the current study in the research progression (green background)

2 Methodology

The functional requirements (FRs) for the material handling system are determined to establish a baseline for the compliant gripper design rules. A summary of the system related functional requirements are:

- **Scalability**: The gripper should be designed with scalability in mind, allowing for potential future adaptations to handle different fabric sizes, weights, and materials and different mold surface profiles/contours.
- **Speed and Efficiency**: The gripper should be designed for fast and efficient pick and place operations, minimizing cycle times (target 75 pcs/h). The gripper system needs to be simple in design for closing and opening operations with a minimal number of sub-components (tessellations) needed for fabric manipulation.
- **Mechanical Simplicity**: The mechanism needs to be mechanically simple to minimize the risk of mechanical failures, but able to conform to mold surfaces.
- **Safety**: The gripper design should prevent accidental contact or entanglement during operation.
- **Integration**: The gripper should be designed for easy integration into an existing automation setup, including compatibility with robotic arms or other motion control systems. In this study a Hand-E robot arm attached to a UR5e Universal Robot is used for system's automation.

For the gripper, the design rules are:

- Utilize the geometry features of a quadrilateral to design parallelograms (Fig. 2).
- Use a Miura-fold gripper for mild non-flat surfaces regions. Limit the length of parallelogram to 30% to 50% of closing distance of the end effector (Fig. 3).
- Maintain acute angles between 35° and 45° in parallelograms.
- Apply fillets to external corners.
- Enhance the surface friction using built-in roughness fabrication methods or by adding coatings (this is limp fabric material dependent).
- Use a Miura-living hinges style gripper with Ori-living hinges connectors (Fig. 4) for moderate to steep non-flat surface regions. The number of connectors depends on mold surface geometry, and the width is dependent on the material thickness Note: the detailed design considerations for the gripper with respect to the mold geometry can be found in [14]. High curvature introduces stress zones.
- Apply grooves, slots, or gaps to the mid-region of living hinges for more compliance.
- *Use a rigid material for the tessellations (i.e., PLA), and a flexible material for the hinge connections (i.e. TPU)—whether a 'flat' connection, or a living hinge variant.*

Leveraging these design rules, a prototype frame and gripper design variants were fabricated using a material extrusion additive manufacturing (MEX) process, and tested using a Hand-E UR5E Universal collaborative robot (cobot) [15] and different materials (Table 1). The multi-material gripper prototypes were manufactured using

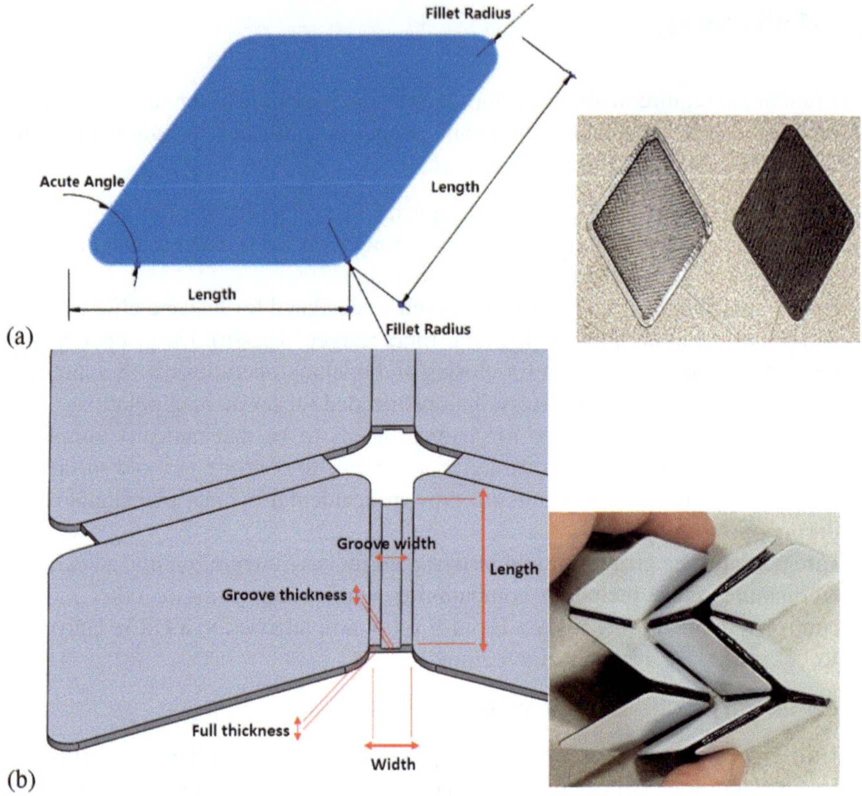

Fig. 2 **a** A tessellation 'panel' and parameters, and **b** a Miura fold gripper segment showing the living hinge used in the current study. Fabric is grasped in between panels

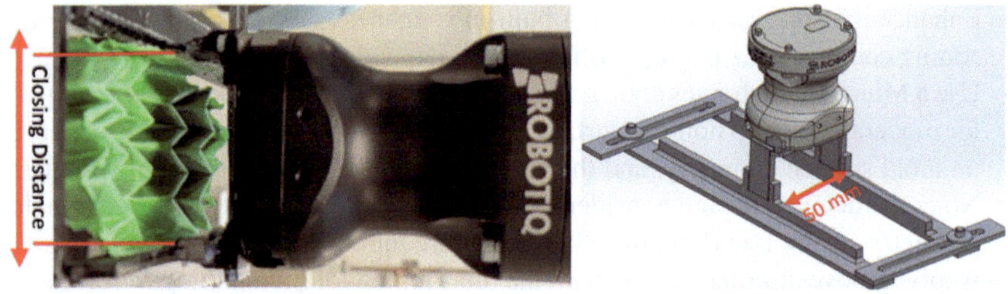

Fig. 3 Closing distance of robot end-effector for a Hand-E UR5e Universal Robot arm, including a schematic of the mounting frame

Fig. 4 The compliance of Miura-living hinge grippers linked with Ori-living hinge connectors (enclosed box) for curved and steep regions of a test mold

Table 1 Material summary table

Fabric type	Area weight (g/m^2)	Thickness (mm)
Carbon fabric	255	0.20
Cotton	170	0.18
Microfiber	95	0.11
Polyester	110	0.14

PLA (polylactic acid) and TPU (thermoplastic polyurethane) materials using a Prusa i3 MK3S + desktop machine. To address observed wrinkling issues, a standalone 'pneumatic press' was designed, fabricated, and tested. This flattening tool consists of a pneumatic air bladder made of a rectangular polyurethane envelope, restricting top and sides movement. The bladder was connected to compressed air supply (15 psi/ 103 kPa) and controlled by a manual valve. The operational parameters are: Inflation time: ~ 1.2 s; contact area: ~ 180 cm^2, and the pressure range during contact: 1.5–2.1 N/cm^2.

3 Results and Discussion

For the Miura fold gripper configuration used in this research, (Fig. 2) the parallelogram side lengths = 18 mm (short side) and 32 mm (long side), the internal acute angle = 37°, the panel thickness (PLA) = 2.4 mm, the hinge width (TPU) = 1.2 mm with variable groove cutouts to ensure directional compliance. The end-effector closing distance is 52 mm from fully open to fully collapsed. The gripper weight is 310 g (without the frame), with a fabric contact area per gripper face: ~ 38 cm^2. These values were determined through parametric iterations to balance grip force, surface contact, and manufacturing limits (e.g., the 3D printer resolution and TPU printability).

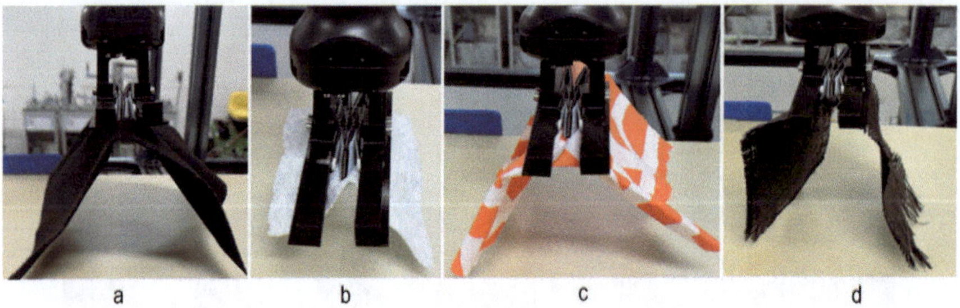

Fig. 5 Gripping and picking different fabric materials: **a** microfiber, **b** cotton, **c** polyester and **d** carbon fabric

A customized structural frame was designed and built to integrate a Miura-living hinge gripper to a Hand-E UR5e Universal cobot and tested with different materials (Fig. 5). This frame was designed to allow the gripper sides to freely slide when closing/opening, and to resist bending and twisting during the gripper fold/unfold activities. The average cycle time (grip, lift 25 cm, place, open) using the UR5e robot was 38–41 s per fabric. Multiple pick and place experiments were conducted for the microfiber, cotton, polyester, and carbon fiber materials and mold surfaces.

Shown in Fig. 5 are pictures from the carbon fibre material handling. No damage was observed, but the wrinkling is fabric stiffness dependent. It was observed that wrinkles were primarily formed at the inner concave zones and sharp corner transitions.

In Fig. 6, it can be seen that there are some wrinkles which need to be flattened, although there was no evidence of damaged fibres. A mechanism that can readily integrated into an automation cell was designed and fabricated to serve as an air bladder [16] with the capability to inflate and release air to introduce a flattening solution. This air bladder is deliberately enclosed on all sides and the upper portion, allowing air intake exclusively from the top while enabling air expulsion from the bottom. When the lower surface of the bladder comes into contact with fabric areas requiring de-wrinkling, this mechanism exerts pressure, compressing the fabric against the mold surface, thereby facilitating its conformity to the contours of the mold.

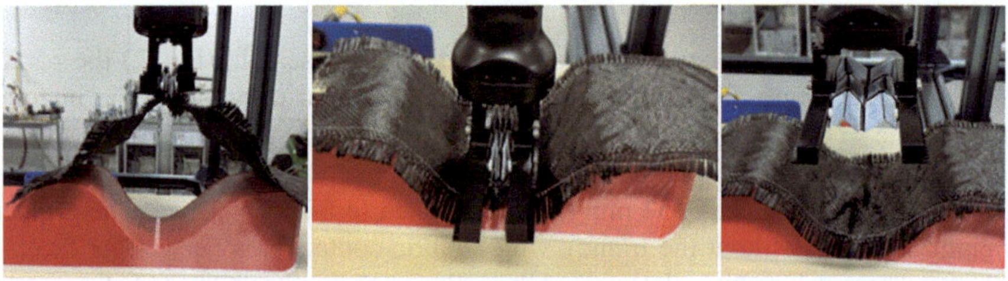

Fig. 6 Select pick and place experiments for a mold surface with a convex and concave surface

(a)

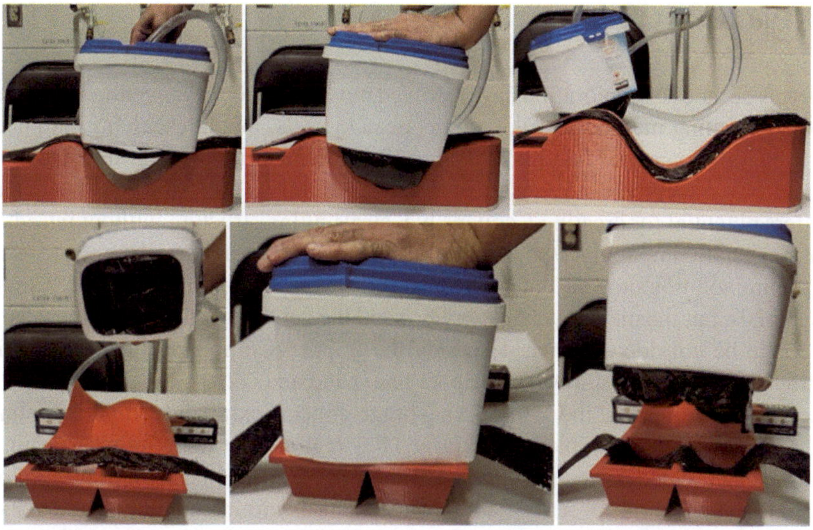

(b)

Fig. 7 Select placement refinement and dewrinkling experiments for a mold surface with convex and concave surfaces

Figure 7a depicts the fabric flattening process as it was applied to the carbon fibre fabric positioned over the flat-curvy mold. Notably, the carbon fabric situated over the concave region of the mold surface underwent a transformation as the air bladder exerts pressure, causing it to seamlessly adhere to the contours of the mold surface, and wrinkles were eliminated. However, springback occurred for the pockets with sharp internal corners (Fig. 7b). This prompts the question: Can a hybrid specialty-flattening mechanism be integrated into a single end effector? There are potential avenues for such a mechanism.

One concept involves crafting the specialty gripper tessellations with a thick, interconnected hollow structure, incorporating touch surfaces made of soft bladder-like materials. This design would enable air to pass through the gripper's inner wall, inflating each tessellation's end surface to assist in fabric un-wrinkling during gripper opening or immediately after placement. Another approach entails incorporating collapsed airbags at multiple intersection regions of the gripper's touch surface. These airbags can inflate simultaneously or independently after the gripper opens and placement is completed, aiding in fabric flattening. These initial concepts require further refinement and validation in future studies to ensure successful implementation.

4 Summary and Conclusions

Just as decades of exploration have resolved most obstacles in conventional pick-and-place handling methods of rigid components/objects, much research is still needed for effective material handling of limp, flexible materials, and compliant mechanism based methods introduce exciting opportunities. The Miura fold kinematics offers potential for creating a collapsible, curve fitting structure without introducing mechanical complexity associated with pneumatic or tendon driven systems. The gripping solution showed that this gripper style is capable of gripping a wide variety of materials, but the material characteristics and the geometry of the placement surface need to be considered when designing the gripper. Combining the gripper with a structural frame that allows the gripper to expand and collapse bi-directionally, and testing has shown that the Miura-living hinge grippers exhibited remarkable compliancy, bending over sharp edges and steep flex regions, but wrinkling can occur. The air-actuated bladder mechanism efficiently aided in fabric de-wrinkling. Exploring an integrated hybrid specialty-flattening mechanism within a single end effector solution or introducing a separate flattening operation should be considered in context of the desired process cycle times.

This solution can be tailored for fabrics with various shapes and slits. Ongoing research has validated this solution with different fabric geometries, including W-shaped fabrics with slits. These solutions mark significant progress in fabric material handling automation, but they are not mature. There is still much room for investigation. Alternative gripper materials, durability testing, alternative soft gripper—link styles, and systems integration etc. needs to be explored and a knowledge base of solutions developed. It is important to note that this solution may not be suitable for impregnated fabrics containing adhesive materials such as epoxy resin, as it relies on mechanical grip and release. Similarly, it is not recommended for fiber reinforced polymer composites (FRPCs) in solid states given its emphasis on flexibility and adaptability. While adjustments could be made for thinner and lighter FRPCs, other conventional methods may be more impactful and cost-effective for handling such materials.

Acknowledgements Funding by the NSERC CREATE grant and UWindsor Outstanding Scholars program.

References

1. Elder H-M, El-Tawashi M (1977) Some bending measurements on Yarns. J Text Inst 68(5):188–189. https://doi.org/10.1080/00405007708631378
2. Siqi Z (2015) An automated method for the layup of fiberglass fabric. Ph.D. Thesis, Iowa State University, Ames, Iowa. https://doi.org/10.31274/etd-180810-4504

3. Alebooyeh M, Wang B, Urbanic R-J (2020) Performance study of an innovative collaborative robot gripper design on different fabric pick and place scenarios. SAE Technical Paper. https://doi.org/10.4271/2020-01-1304
4. Raghavan L-K (2014) Industrial looks at ways of manufacturing defects of fiber reinforced polymer composites. M.Sc. Thesis, Iowa State University, Ames, Iowa. https://doi.org/10.31274/etd-180810-4054
5. Ebraheem Y, Emilie D, Dominique C-A (2021) Universal gripper for fabrics—design, validation and integration. Int J Cloth Sci Technol 33(4):643–663. https://doi.org/10.1108/IJCST-11-2019-0180
6. Krogh C, Sherwood J-A, Jakobsen J (2019) Generation of feasible gripper trajectories in automated composite draping by means of optimization. Adv Manuf Polym Compos Sci 5(4):234–249. https://doi.org/10.1080/20550340.2019.1699691
7. Alebooyeh M, Wang B, Urbanic R-J, Djuric A (2019) Investigating collaborative robot gripper configurations for simple fabric pick and place tasks. SAE Technical Paper
8. Hoffer J (2020) Development of a draping algorithm for non-structural aerospace composites. M.Sc. Thesis, University of Ottawa, Ottawa, ON. https://doi.org/10.20381/ruor-24878
9. Strelkova D, Urbanic RJ (2022) Design and analysis of dimensional compliant living hinges for automotive related pick and place applications and end of arm tooling, integrating origami into robotics via additive manufacturing, WCX conference presentation. SAE International, Detroit, Michigan
10. Mitani J (2018) The folded shape restoration and the rendering method of origami from the crease pattern. In: Proceedings of the International conference on geometry and graphics, pp 1–7
11. Turner N, Goodwine B, Sen M (2016) A review of origami applications in mechanical engineering. Proc Inst Mech Eng C J Mech Eng Sci 230(14):2345–2362
12. Shuguang L, Stampfli J-J, Xu H-J, Malkin E, Diaz E-V, Rus D, Wood R-J (2019) A vacuum-driven origami 'Magic-Ball' soft gripper. In: International conference on robotics and automation (ICRA). https://doi.org/10.1109/icra.2019.8794068
13. Strelkova D, Urbanic R-J (2021) Design and fabrication of novel compliant mechanisms and origami structures for specialty grippers. In: Proceedings of the CHAM, CARV 2021, MCPC 2021: towards sustainable customization: bridging smart products and manufacturing systems, Part of the Lecture Notes in Mechanical Engineering. Springer Link, pp 55–62. https://doi.org/10.1007/978-3-030-90700-6_5
14. Alebooyeh M, Urbanic R-J (2022) Mold geometry analysis tools for fabric material handling system applications, CAD & applications, CAD conferences and exhibition. Beijing, China
15. Hand-E UR5E Universal collaborative robot, Product specification of IRB 14000. ABB Robotics, Sweden (2015)
16. Krogh C, Jakobsen J, Wilm J (2020) Will it crease or Cease? A study of debulking of air pockets in automated prepreg composite layup. Compos Part A Appl Sci Manuf 138:106052. https://doi.org/10.1016/j.compositesa.2020.106052

A Framework for Medium Volume Production Strategies for Directed Energy Deposition Additive Manufacturing

Jill Urbanic

Abstract Additive Manufacturing (AM) processes enable the validation of design variants, and the manufacturing of low volume specialty components. A production on-demand solution can be established close to a customer, but the economies of scale need to be considered. Slow fabrication times are an issue for larger production volumes, but for the directed energy deposition (DED) and hybrid manufacturing (where additive and machining operations are interwoven), new process planning scenarios can be explored for both low and medium volume production levels, which aligns well with addressing on-demand service and out of production components. Prior to exploring multi-function machines and dynamic layouts, the precedence diagrams need to be determined as well as the process summary matrices. However, with DED and hybrid AM, there are unique scenarios for the dependencies and thermal cycling conditions. This preliminary research focuses on defining a framework for DED AM precedence diagrams, process summary data, and insights for systematically decomposing components for macro level process planning.

Keywords Additive manufacturing · Directed energy deposition · Process planning · Framework · Medium volume

1 Introduction

Additive Manufacturing (AM) processes enable the validation of design variants, and the manufacturing of low volume specialty components. A direct approach can be taken, where the component is fabricated from the CAD model, or an indirect approach can be taken [1, 2], where rapid tooling is designed and manufactured to facilitate production of the desired components. Both strategies assist production companies to adapt to on-demand production where individual or customized solutions are required, and address the on-going pressures for shorter delivery times.

J. Urbanic (✉)
University of Windsor, Windsor, ON, Canada
e-mail: jurbanic@uwindsor.ca

M. Manns (ed.), *New Paradigms for Anticipated Uncertainty*, Lecture Notes in Mechanical Engineering, https://doi.org/10.1007/978-3-032-16889-4_81

There are seven AM process families defined by ASTM [3], and interesting variants being introduced into the commercial domain such as Additive Friction Stir welding. A summary of the main processes and their characteristics is presented in Table 1. The focus of this research is on the directed energy deposition (DED) AM process family.

AM systems lend themselves to distributed manufacturing systems (DMS), as a production on-demand solution than can be established close to a customer. Rauch et al. [4] assessed the potential of AM technologies and identifies potential industry sectors for its application using the NACE (Nomenclature Statistique des Activités Économiques dans la Communauté Européenne) classification of industrial sectors.

Highlighted was the importance of leveraging new manufacturing strategies and technologies to meet the specific requirements of low volume production of customized artifacts. Medical and dental instruments and supplies were the top applications, with applications in aerospace and the automotive domain in the top eight. This is also reflected in research related to inventory and supply chain management [5, 6].

When developing process plans, two perspectives need to be considered: (i) micro levels and (ii) macro levels. At the micro level, the detailed information on the machine type, tools, operation parameters, and operation procedures are determined; whereas, at the macro level, the overall strategy and workflows are determined, such as the process sequences and workload balancing. Variant and generative process planning are two approaches employed in the machining and assembly domains;

Table 1 Summary of the main commercial AM processes

Process name	Characteristics
Stereolithography (SLA), Digital light processing (DLP)	Liquid polymer resin + UV light curing Laser beam is traced, or a digital projector is utilized
Binder jet printing (BJP)	Solid powder placed layer by layer, binder material is applied in the regions to be solid; no heat for this process
Material jet printing (MJP)	Solid powder placed layer by layer, polymer resin is applied in the regions to be solid; combined with UV light curing
Powder bed fusion (PBF)	Solid powder placed layer by layer fused / melted into an object; laser or electron beam is used as a heat source
Material extrusion (MEX) or fused fiber filament (FFF)	Plastic filament melted to a semi-viscous state and extruded through a nozzle for bead to bead placement
Layered object manufacturing (LOM)	Solid sheets cut to shape joined with adhesive
Directed energy deposition (DED)	Metal powder or filament melted using a laser, electron beam, plasma arc, or wire arc for bead to bead placement

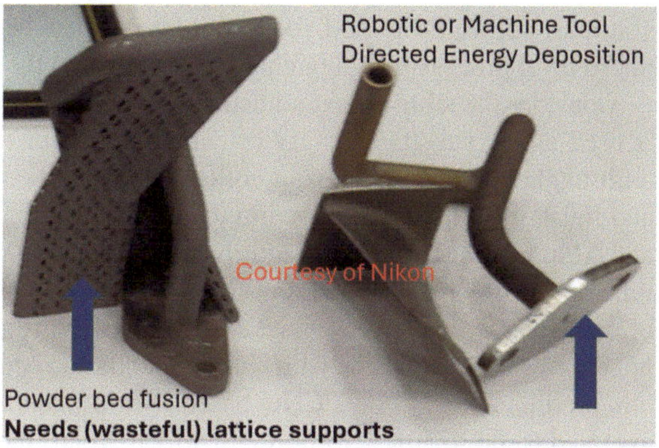

Fig. 1 A comparison of a PBF solution, and a DED multi-axis solution

however, the knowledge base does not exist for DED AM, and is outside of the scope of this work.

1.1 Micro Level Process Planning

The microlevel process planning for material extrusion and resin based systems may be trivial—after positioning and orienting a component, a slicer (i.e. CURA and Chitubox) will develop build information using preset values in the slicer. Hence the term 3D Printing, as the decision making is similar to printing a document. For developing effective DED AM build strategies, extensive knowledge is required as there is no core knowledge base to exploit. For DED AM, operation parameters, tool path generation, and simulation activities are sophisticated as there are multi-axis build strategies that can be utilized to eliminate support structures (Fig. 1), and heat management is a critical concern. The heat cycling introduces variations in the bead geometry during the deposition process, induced residual stresses as well as novel (and unpredictable) mechanical properties. Micro process planning is a challenge, but once process settings and effective tool paths are generated and validated, build strategies can be established, providing a foundation for variant process planning. However, introducing effective higher production solutions requires macro level process planning assessments.

1.2 Macro Level Process Planning

Macro level planning considers the time-task sequencing for component fabrication, the machine types and configurations, fixtures and setup configurations, the work flow, and so forth. Optimization typically focuses on minimizing non-value added operations, such as material handling or workload balancing for higher production volumes, but no DED AM focused approach has been developed to date. Unique micro level process planning challenges flow into the macro level planning.

Consider the table height adjustment block shown in Fig. 2 [7], where a cylinder is used as the substrate, rotary orientation is required to build up the component, a tilt angle is required to avoid collisions, and the component is built in segments, and finish machined where required. The tool paths were created using Mastercam and APlus for machining and additive manufacturing respectively. The process build time is 80 min or 0.75 pcs/h. For a 1000 part production run for a planning horizon of one week, 5 days per week, 2 shifts, 8 h per day, and a realistic OEE (80%), the hourly rate is 15.6 pcs/h, which corresponds to requiring 21 hybrid machines for this production run. For a CNC based hybrid manufacturing system, the costs are assumed to be $750K (US dollars), which would require a significant investment: $15,750K. When reviewing the process type and time elements, it can be seen that the AM portion is disproportionately longer than the machining activities (Fig. 2b), and when assessing the build approach, each operation requires 2½ or 3 axis once an orientation is established. As well, when analyzing the operation sets, it can be determined that there is approximately a 10:1 ratio of additive deposition time compared to the machining time. There is significant idle time associated with the machining aspects. Consequently, the long term research goal is to develop manufacturing processes that will augment the capabilities of DED and hybrid manufacturing, and provide cost and time competitive production capabilities for low to medium volume production systems. The goal of this research paper is to provide a foundation for process planning and systems design for DED and hybrid manufacturing for low volume production (100–2000 pcs) for short planning horizons (1 week to 1 month), i.e., inventory management, as there is significant variability in the components and the required production volumes. This aligns to 'medium volume' production levels.

2 Methodology

A systems approach is being taken to create a process planning roadmap. Detailed case study process plans are developed to fabricate components that represent unique geometric configurations. The specialty block (Fig. 2) and wheel (Fig. 3) are presented here. Precedence diagrams are generated reflecting the process planning. With machining and assembly process planning optimization, tables for summarizing the features, setup, tool path type, and axis/axis configuration are used for structured input data for optimization or near-optimization methods or heuristics, such as genetic algorithms (GA). The DED AM, there are additional considerations: the impact of the heat input/heat cycling, and the unique issues associated with segmenting geometry.

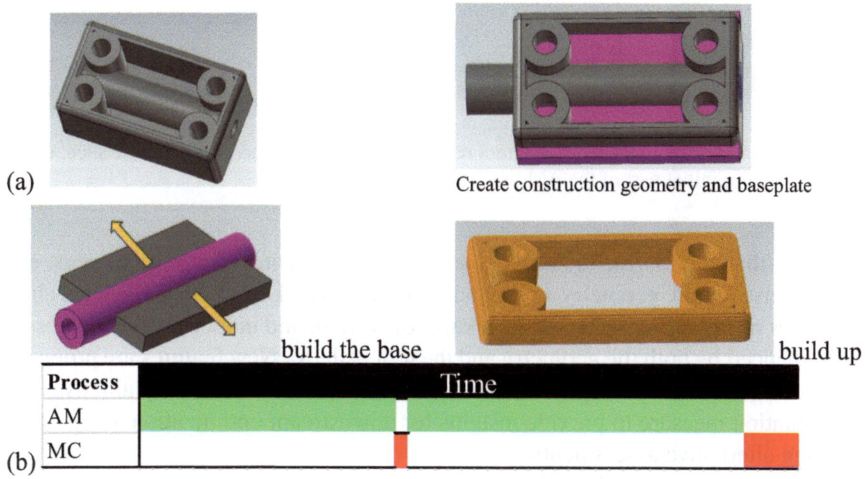

Fig. 2 **a** Micro process planning sequence (adapted from [7]), **b** a general manufacturing timeline. The CAD model is courtesy of PlusMfg

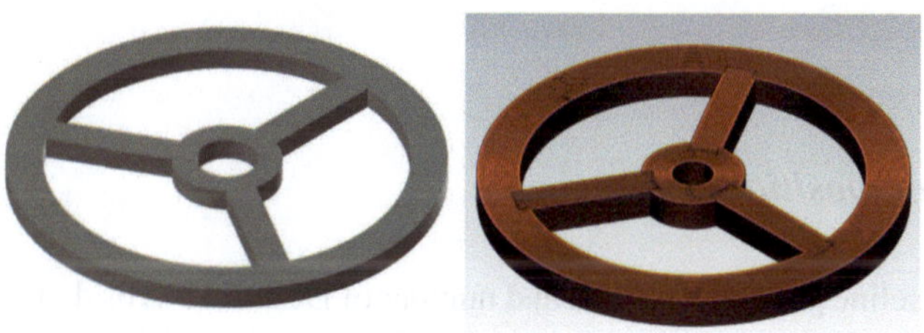

Fig. 3 Wheel to be built by the DED AM process, with localized tool paths for the spokes, centre hub and the outer ring

2.1 Precedence Diagrams and Process Summary Tables

Precedence diagrams are used to illustrate the planned sequence of activities. Nodes and arrows represent the task order and dependencies. Novel build scenarios exist for DED AM. They are: (i) introducing layer merging to interweave build conditions as there may be interference conditions or different materials and/or process settings. This is shown in Fig. 4a, where the merged operations are identified in a columnar table, with a thick outer boundary box; (ii) introducing variable process settings that create variable bead geometry, such as the spline slice shown in Fig. 4b, which is indicated by a variable symbol; and (iii), the potential of high levels of heat input. Heat accumulation will occur, but the rate will be higher for short travel paths, where an example is shown in Fig. 4c, and is shown as high heat at the end or beginning of the tool path.

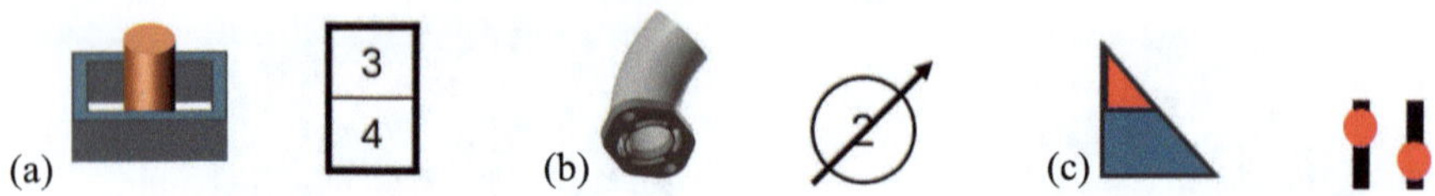

Fig. 4 DED AM novel precedence diagram representation: **a** layer merge, **b** variable slice, **c** heat build up (final or initial part of the AM tool path)

Tables that summarize the process setups, tooling data, and machine types, etc. typically use feature geometry as a foundation. However, DED AM heat related characteristics need to be captured as shrinkage, distortion, and induced residual stresses are influenced by both the geometry and the tool path [8]. New features are proposed to be collected: 'contact area', and surface area to volume ratio, and a qualitative heat accumulation measure to provide insights for downstream optimization and process planning alternative assessments.

These data sets provide insights with respect to build alternatives for a single component. This may not be cost effective for higher volumes. Therefore, a segmentation strategy is proposed for the additive manufacturing operations, interweaving machining operations to create in-process datum surfaces for a 'modular' fabrication approach.

2.2 *Decomposition Strategies*

First, the baseline production rates, and number of DED AM/hybrid machines needs to be determined for a given production scenario and OEE, and the related production strategies developed, which include, the (i) operation lists (s), precedence diagram(s), value stream maps, and (ii) process plan summary tables at the component and feature levels. Based on the initial single component production solutions, the number of segments can be determined. This is an iterative process, where feature based and time based segments need to be considered due to the opportunity to introduce dynamic process settings in conjunction with multi-axis motion. When assessing the segmentation strategies, the build feasibility (i.e. size and number and type of axes), orientation, and heat accumulation are considered. Due to heat cycling/ shrinkage, additional stock is required to be deposited at the segmentation interface, and intermediary datum planes need to be machined in place. When introducing the segmentation strategy, less axes may be required for the AM system(s) to build the features. However, orienting and positioning need to be considered, but this reframes the problem into a machine selection, cell design, and adaptive automation research problem. From this information, preliminary machine selection sets, fixturing strategies, layouts, etc. can be determined and utilized for downstream optimization.

3 Results

For the block, although the shape is prismatic, the substrate is rotary as the central geometric base is cylindrical. Therefore, bar stock can be employed, and clamped in a chuck, and the features built up (Table 2 and Fig. 5a). It is assumed that the exterior fillets between the top or bottom and side faces are not required to be finished machined. For the wheel, the circular geometry is to be built with concentric boundary passes, and the interior ribs built with the tool paths oriented axially. Therefore, there are 5 build segments, and there are issues with potential deposition head—deposited material collisions if each segment is built individually. Therefore, the standard precedence diagram with the enclosed additive operations, shown in Fig. 5b, is an impossible build case 'as is'. However, incrementally depositing the material, i.e., 1 layer for segment 1–5, with the localized tool paths, from layers 1 to *i* addresses the collision conditions. The representative presence diagram is shown in Fig. 5b.

It can be argued that using a layer merge strategy for the block would mitigate the potential distortion issues, i.e., perform layer merge on operations 2 and 3, and after the base plane is built, perform layer merge on operations 4 and 5 (Fig. 6). Then the machining operations can be performed.

The process summary, including the only DED AM elements, is presented in Table 3 due to space constraints. As all the block tool paths are planar, and the longest operations are op. 4 and 6 (building up the block body), a cell of 3 axis machines along with automation for material handling and orientation is required for a medium volume production line. Consider a scenario for the block, where machine 1 is a 3 axis machining centre which interacts with DED AM systems 2–4 (system 2 performs op. 2 and 3, system 3 performs op. 4, and system 4 performs op 6).

This four machine segmented approach introduces a production rate of 2.31 parts per hour; therefore, 7 cells are required. Assuming $350K for a 3 axis DED AM system, $300K for a machining centre, and $250K for automation integration, each cell requires an investment of $1,600K; therefore, 7 cells requires $11,200K investment—29% less.

Table 2 Process plan configurations for the block and wheel

Number	Description (block)	Number	Description (segmented wheel)
1	MC: tube substrate	1	Prepare substrate
2	AM: base segment 90°	2	AM: segment 1
3	AM: base segment -90°	3	AM: segment 2
4	AM: top geometry	4	AM: segment 3
5	MC: top face	5	AM: segment 4
6	AM: bottom geometry	6	AM: segment 5
7	MC: drill holes	7	MC: contour milling
8	MC: bottom face	8	MC: top face
9	MC: contour milling	9	MC: bottom face (if req'd)
10	MC: cutoff		
11	MC: cutoff face		

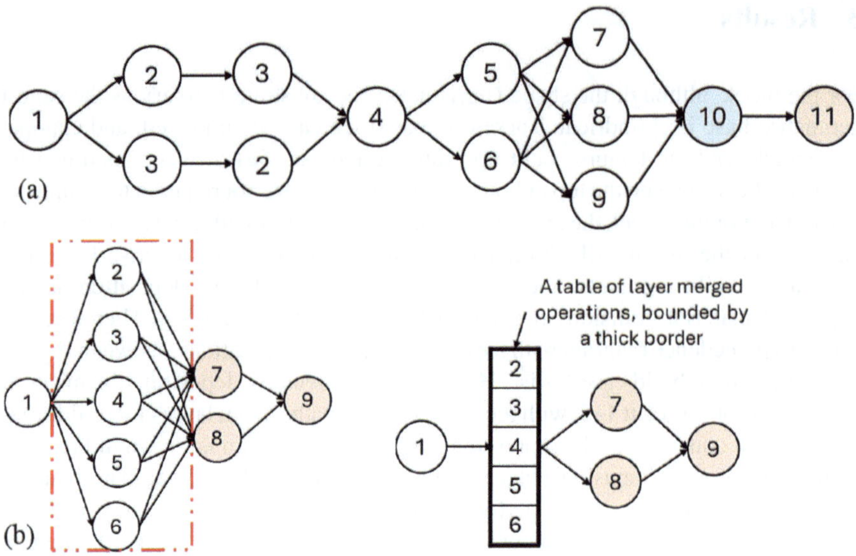

Fig. 5 Precedence diagrams for two build options—the blue indicates a cutoff operation and the orange represents a unique set up and final finish machining op

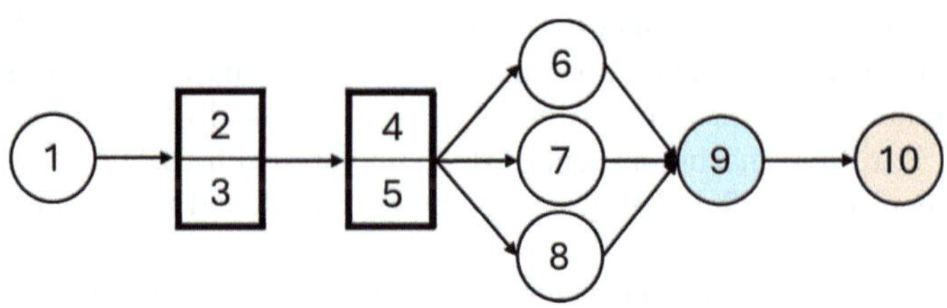

Fig. 6 The precedence diagram for the block with merged operations

Table 3 Process summary tables for the block (green) and wheel (blue)

				Axes no.									
No.	Description	Tool path type	Build plane	3	4	5	8	Axis stack	Cycle time (min)	Volume (cm^3)	Surf. Area (cm^2)	Contact a. (cm2)	Heat Acc.
2	AM: base segment 90°	Planar	XZ	1				XYZ	5	4	3.2	0.2	
3	AM: base segment -90°	Planar	-XZ	1				XYZ	5	4	3.2	0.2	
4	AM: Top geometry	Planar	XY + X rot		1	1	1	XYZA	25	17	14	3	
6	AM: bottom geometry	Planar	XY - X rot		1	1	1	XYZA	25	17	14	3	
2	AM: segment 1 - inner	Merged Planar	XY	1				XYZ	38	17	58	12	
3	AM: segment 2 - arm		XY	1				XYZ	30	13	41	9	
4	AM: segment 3 - arm		XY	1				XYZ	30	13	41	9	
5	AM: segment 4 - arm		XY	1				XYZ	30	13	41	9	
6	AM: segment 5 - outer		XY	1				XYZ	329	139	358	99	

The wheel production system should also be a set of 3 axis AM and machining systems. Unlike the block there are no off axis orientations or intermediary machining operations, so the material handling challenges are reduced, reducing investment costs. However, heat management will be an issue due to the large contact area (99 cm^2).

4 Discussion

With DED AM systems, build strategies are expanded beyond planar '2 1/2 D' approaches. When incorporating hybrid manufacturing stratagems with DED, the micro and macro process planning challenges have expanded. Consequently, new precedence diagram nomenclature for DED AM operations, layer merge operations, operations with variable process settings, and heat build up are introduced, as well as DED AM based process summary sheets. This allows researchers to effectively represent DED specific process information and provides a foundation to explore systems requirements alternatives. More sophisticated components need to be decomposed to evaluate the robustness of these new rules and to expand them as necessary. These solutions will form the basis for downstream macro level process optimization, which is in its preliminary stages of research. The targeted application is inventory management, as there is significant variability in the components' shape and the required production volumes.

References

1. Zaragoza VG, Rane K, Strano M, Monno M (2021) Manufacturing and performance of 3D printed plastic tools for air bending applications. J Manuf Process 66:460–469. https://doi.org/10.1016/j.jmapro.2021.04.045
2. Kalami H, Urbanic RJ (2019) Design and fabrication of a low-volume, high-temperature injection mould leveraging a 'rapid tooling' approach. Int J Adv Manuf Technol 105(9):3797–3813
3. ISO/ASTM 52900:2015 (2015) Additive manufacturing—General principles—Terminology
4. Rauch E, Unterhofer M, Nakkiew W, Baisukhan A, Matt DT (2021) Potential of the application of additive manufacturing technology in European SMEs. Chiang Mai Univ J Nat Sci 20(2):1–14. Article e2021023. https://doi.org/10.12982/CMUJNS.2021.023
5. Knofius N, van der Heijden MC, Zijm WHM (2019) Moving to additive manufacturing for spare parts supply. Comput Ind 113:103134. https://doi.org/10.1016/j.compind.2019.103134
6. Peron M, Sgarbossa F (2021) Additive manufacturing and spare parts: literature review and future perspectives. In Wang Y, Martinsen K, Yu T, Wang K (eds) Advanced manufacturing and automation X: IWAMA 2020. Lecture Notes in Electrical Engineering, vol 737. https://doi.org/10.1007/978-981-33-6318-2_78
7. Urbanic RJ, Hedrick B, Ramezani H, El Moghazi S, Saghafi M (2022) Hybrid manufacturing decomposition rules and programming strategies for service parts. In: Proceedings of the ASME 2022 International mechanical engineering congress and exposition. https://doi.org/10.1115/IMECE2022-95560

8. Mirazimzadeh SE, Pazireh S, Urbanic J et al (2022) Investigation of effects of different moving heat source scanning patterns on thermo-mechanical behavior in direct energy deposition manufacturing. Int J Adv Manuf Technol 120(11–12):4737–4753

Investigation of the Long-Term Durability of Soft Pneumatic Actuators

Florian Schreiber and Martin Manns

Abstract Considering increasing demands for adaptable manufacturing systems, soft pneumatic grippers—particularly those based on pneumatic networks (pneunets)—offer promising potential due to their intrinsic compliance and compatibility with additive manufacturing. However, their limited industrial adoption is due to concerns regarding long-term durability, which remains underexplored. This study investigates the long-term bending behavior of 3D-printed TPU pneunets with varying wall thicknesses. Using cyclic bending, the deformation trajectories of the pneunet tips are tracked and analyzed. Results reveal a consistent trajectory shift during initial cycles, which diminishes over time and can be effectively described using a logarithmic function. The model's accuracy is validated through high coefficients of determination ($R^2 > 0.95$), and the pressure-dependent change is further approximated via a quadratic fit. Findings indicate that wall thickness has no significant impact on long-term deformation. This work underscores the need for standardized testing protocols to enable reliable cross-study comparisons and support the broader application of soft robotic grippers in industrial contexts.

Keywords Soft robotics · Pneunets · Additive manufacturing · Cycle test · Long term durability

1 Introduction

In times of geopolitical tension and anticipated uncertainties, the demand for flexible manufacturing systems that can be rapidly adapted in terms of product range, production capacity, and location is increasing [1]. Conventional robots and their components are often rigid in design and tailored to specific tasks. A promising approach to enhance the flexibility of robotic grippers lies in the use of soft robotic grippers. In recent years, the soft robotics community has introduced a wide range of such

F. Schreiber (✉) · M. Manns
PROTECH—Institut Für Produktionstechnik, Universität Siegen, Siegen, Germany
e-mail: florian.schreiber@uni-siegen.de

M. Manns (ed.), *New Paradigms for Anticipated Uncertainty*, Lecture Notes in Mechanical Engineering, https://doi.org/10.1007/978-3-032-16889-4_82

grippers [2]. Among the various actuation principles, pneumatically driven systems have become particularly well established due to their simplicity and effectiveness.

One specific type of pneumatic gripper is based on so-called pneumatic networks, or "pneunets". These consist of at least one air chamber that expands when internal pressure is increased. Through the targeted arrangement of chambers and the incorporation of non-expandable elements, the motion can be directed and transformed into translational or rotational movement. For gripping applications, pneunets that produce a bending motion are commonly used. The key advantage of pneunet-based grippers lies in their intrinsic compliance, which allows them to adapt to unknown or varying object geometries, thereby increasing the range of objects that can be handled.

Moreover, advancements in additive manufacturing have enabled the processing of flexible materials, allowing pneunets to be fabricated directly through 3D printing. This supports decentralized, on-demand production of grippers. Despite these advantages, soft grippers have seen limited adoption in industrial settings. One major reason is the insufficient durability of pneunets, a factor that has rarely been systematically investigated in scientific research [3]. In industrial flexible production, grippers are expected to perform over many actuation cycles, often beyond what is typically tested in academic studies. However, most research examines behavior over relatively few cycles, limiting comparability. This highlights the importance of systematic durability studies like the present work.

2 Long-Term Durability of Pneunets

In a study on the research focus of publications in the field of soft robotics, Miriyev concludes that while many works explore new structures and designs, the long-term durability of the presented actuators is often not tested [3]. However, long-term durability is a fundamental requirement for the industrial application of pneunets. The few studies that do investigate durability report highly divergent results. These discrepancies can be attributed not only to the inherently different behaviors of the actuators examined but also to the lack of standardized testing procedures.

In the investigations by Bui et al. [4], fabric-reinforced pneunets made from Smooth-On DragonSkin 30 silicone are analyzed. During the investigation of the trajectory of the pneunet tip, the arc lengths of the trajectory increase by several millimeters. This change is especially evident during initial actuations. Bui et al. [4] attribute this effect to the so-called "fading memory" phenomenon [5], which enables the host material to preserve the effects of a recent input sequence within the system, facilitating the temporal integration of stimulus information. Beyond a threshold of 160 actuations, the trajectory exhibits no additional changes. A similar number of actuation cycles is observed in the investigation of additively manufactured TPU (thermoplastic polyurethane elastomers) pneunets by Yap et al. [6]. In contrast to the study by Bui et al. [4], however, the trajectory remained virtually unchanged, with deviations of less than 1 mm after 310 cycles. While many studies investigate actuator behavior over merely a few hundred cycles, Zhang et al. [7] have demonstrated performance over 2.27 million actuation repetitions.

These diverging findings illustrate the current lack of consensus regarding the long-term behavior of pneunets, particularly in the context of additive manufacturing. Considering the increasing relevance of soft pneumatic systems for industrial use, a systematic investigation into their durability under repeated actuation is essential. In this context, the present work aims to systematically analyze the long-term bending behavior of 3D-printed TPU pneunets under repeated actuation.

3 Objectives

In this study, the long-term bending behavior of additively manufactured TPU-pneunets is investigated. For this purpose, pneunets with different wall thicknesses are additively manufactured and then tested in cyclic bending tests. During the tests, the tip of the pneunets is tracked using markers and the change of the trajectory is examined.

Two research questions are pursued regarding the investigations:

1. Does the trajectory of a pneunet vary with an increasing number of actuation cycles (qualitative analyses)?
2. If a change is observed, how can the altered trajectory be approximated (quantitative analyses)?

4 Trajectory Analysis of a Pneunet

4.1 Pneunet Design and Manufacturing

The pneunet design is derived from the reference configuration proposed by Polygerinos et al. [8]. As illustrated in the cross-sectional view in Fig. 1, the structure comprises several air chambers linked by a central air channel. Upon increasing the internal pressure, the chambers expand and repel each other, inducing a bending motion. The pneunets measure 150.7 mm in length and have a cross-sectional dimension of 22 × 28 mm (width × height).

A Prusa MK4 3D printer, which operates using the Fused Filament Fabrication (FFF) technology, is used to additively manufacture the pneunets. NinjaTek Cheetah filament with a diameter of 1.75 mm is employed, using a nozzle with a diameter of 0.3 mm. To facilitate production via FFF, the reference design by Polygerinos et al. [8] is modified according to the results presented in [9]. A total of 6 pneunets with different wall thicknesses of 1.2, 1.5, and 1.8 mm (two pneunets each) are produced.

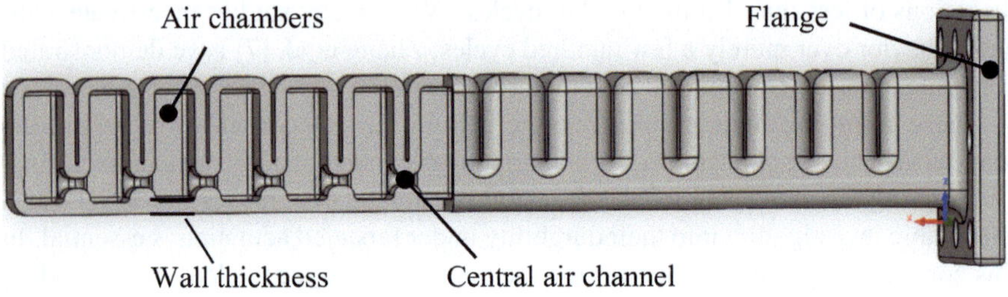

Fig. 1 Cross-section of a pneunet

The Pneunets are alphanumerically coded according to wall thickness (e.g., 1.2A, 1.2B, 1.5A, etc.).

4.2 Cyclic Deflection Tests

To evaluate the bending angle of the pneunets, a test setup is implemented following the methodology outlined in [10]. In this configuration, the pneunets are mounted horizontally, oriented to the left, and affixed to a rigid base. The internal pressure is incrementally increased in steps of 10 kPa up to a maximum of 500 kPa, inducing a bending motion in the positive rotational direction. After the maximum pressure is reached, the pressure is reduced in 10 kPa steps to 0 kPa. The cycle then starts again from the beginning and is repeated 200 times. To mitigate the effects of material relaxation, pressure is held constant for 5 s after each increment. Following each pressure step, an image is captured to determine the position of the pneunet tip.

For accurate tracking of positional changes, circular markers are placed on the pneunets' tip as well as on the fixed base. These markers are detected using the HoughCircles function from the OpenCV library [11]. The position of the tip marker is calculated relative to the reference marker on the fixed base. Since the pneunets extends to the left while the reference marker is positioned on the right, the resulting Cartesian coordinates are negative. The Cartesian coordinates (x, y) are then transformed into polar coordinates (r, θ). The angle θ is used to determine the bending angle t, whereby the initial angle θ_0 is subtracted from the actual bending angle θ.

4.3 Results

The trajectories of the pneunets are analyzed by comparing the coordinates of the actuator tip at defined pressure levels. For this purpose, the tip positions at a given internal pressure are evaluated across different actuation cycles. The analysis reveals a change in the bending behavior over time. Compared to the trajectory observed during the first actuation, subsequent tip positions shift along the initial trajectory in the direction of positive rotation.

Figure 2 illustrates this displacement for all start and end points of one representative pneunets' trajectory at 0 kPa (circular markers) and 500 kPa (cross markers), respectively. The progression along the original trajectory (grey line) is depicted by a color gradient ranging from blue (initial actuation) to yellow (150th and final actuation), highlighting the temporal evolution of the deformation. Since the pneunets move along the original path even with an increasing number of actuations, the change in trajectory can be described by the variation in the bending angle t.

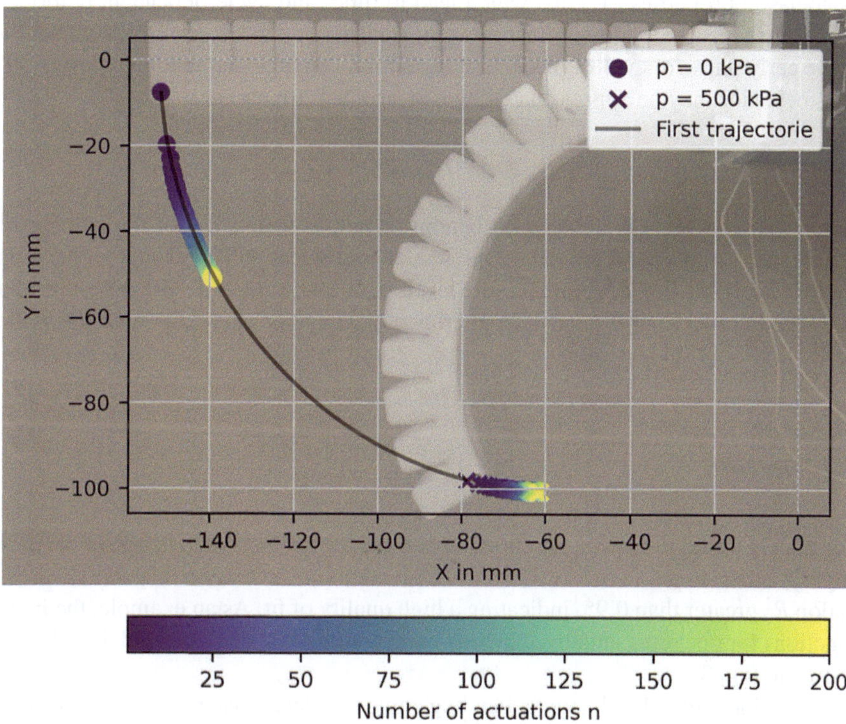

Fig. 2 Change in the start and end points of a pneunet's trajectory

5 Approximation of Pneunets' Long-Term Behavior

5.1 Methodology

Based on the results of the previous section, the change in trajectory can be attributed to a variation in the bending angle. It is evident that the pneunets exhibit a significant change in bending angle during the initial actuations, with the rate of change decreasing as the number of actuations increases. This behavior may be approximated by a logarithmic function. The observed trend is likely influenced by the viscoelastic properties of TPU, which lead to time- and cycle-dependent deformation under repeated loading. Stress relaxation and creep effects occur predominantly in the early phase of cycling, causing a rapid initial change that gradually levels off. This behavior may be approximated by a logarithmic function:

$$t_{diff}(n) = aln(n) \tag{1}$$

Here, t_{diff} represents the change in bending angle as a function of the number of actuations $n > 0$, characterized by the model coefficient a. The change in bending angle is analyzed for different pressure levels p: 0, 100, 170, 250, 330, 400, 450, and 500 kPa. In order to fit Eq. 1 to the experimental data, the "curve_fit" function of the SciPy library [12] is utilized.

5.2 Results

For the six tested pneunets, function (1) is fitted to the observed change in bending angle at different pressure levels. All fitted functions achieve a coefficient of determination R^2 greater than 0.95, indicating a high quality of fit. As an example, the fitted functions for one representative pneunet for pressure of 0, 250, 400, and 500 kPa are shown in Fig. 3.

It is evident that the change in bending angle t_{diff} decreases with increasing pressure p, which is reflected in lower values of the model coefficient a. In Fig. 2, this effect is further highlighted by the variation in arc length associated with the displacement of the trajectory's start and end points. In order to investigate the dependence of the model coefficient on the applied pressure, the fitted coefficients were compiled in a scatter plot (see Fig. 4).

Figure 4 shows that the dependence of the model coefficient a can be reasonably approximated by a second-degree polynomial, as indicated by a coefficient of determination R^2 of 0.97. The different wall thicknesses of the pneunets appear to have no significant influence on the model coefficient. The following equation can be used to approximate the model coefficients a as a function of pressure p:

$$a(p) = -1.98 \text{ x } 10^{-7}p^2 + 3.998x10^{-4}p + 0.055 \tag{2}$$

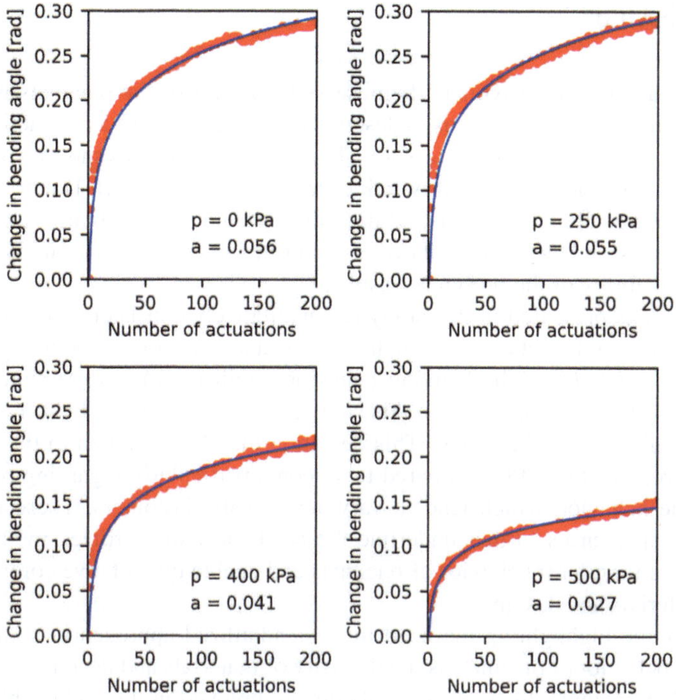

Fig. 3 Fitted functions for one representative pneunet

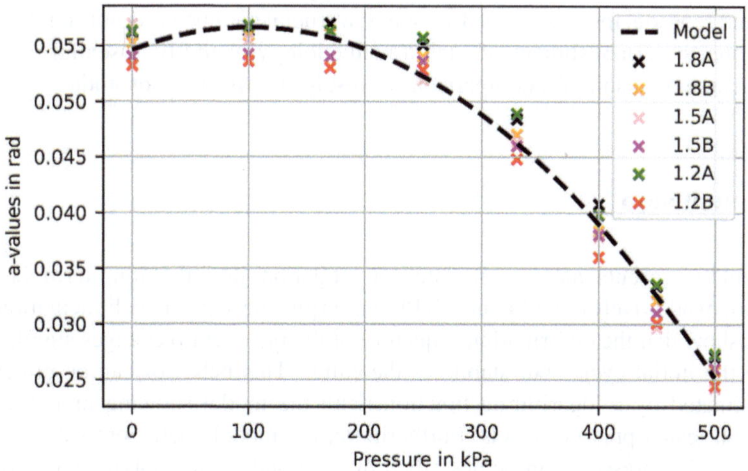

Fig. 4 Approximation of a-value depending on pressure for 6 different pneunets

6 Discussion

The fitted functions approximate the observed behavior of the pneunets with high accuracy, thus providing a means to describe the trajectory shift as a function of the actuation count and the applied pressure. The experimental results show no indication that the wall thickness significantly influences the trajectory shift. Due to the time-intensive nature of long-term tests, future studies should increase both the number of tested pneunets, and the number of cycles performed to improve the statistical power and reproducibility of the presented results.

The findings presented in this study are qualitatively consistent with the results from the study by Bui et al. [4], which describe a similar trajectory change. While Bui et al. observed that the bending behavior stabilized after approximately 160 cycles, the TPU-based pneunets in this study continue to exhibit slight increases in bending angle beyond that point. This deviation may be attributed to the different material properties of TPU compared to silicone, particularly regarding relaxation and fatigue behavior, which tend to vary across soft elastomers. These material-dependent differences further underline the need for a standardized methodology to assess the long-term behavior of pneunets and enable quantitative comparability across materials and designs.

A key issue highlighting the need for a standardized approach to studying the long-term behavior of pneunets is the diversity of materials and designs used in soft robotics. For example, the soft materials often employed in this field exhibit time-dependent, viscoelastic behavior, which is strongly influenced by the frequency of actuation. In the absence of a unified methodology, it is challenging to achieve comparable and reproducible results, as frequency, a crucial parameter, is not standardized. Therefore, the establishment of such a methodology would be essential to enable consistent and meaningful comparisons of results across different studies.

7 Conclusion

This study presents an experimental investigation into the long-term bending behavior of additively manufactured TPU-based pneunets under cyclic actuation. The results show that the deformation trajectory of the pneunet tip changes significantly during the initial cycles but stabilizes thereafter. This behavior can be effectively approximated by a logarithmic function, with the model showing high accuracy across all tested pressure levels. Furthermore, the model coefficient's dependency on the applied pressure can be described by a second-degree polynomial. Notably, variations in wall thickness within the tested range show no significant influence on the long-term trajectory shift. These findings contribute to a deeper understanding of the mechanical behavior of soft pneumatic actuators over time and underscore the importance of including durability assessments in the evaluation of soft robotic systems. To facilitate comparability across studies and designs, future research should focus on establishing standardized testing procedures. Such standards would improve the reliability of data and support broader adoption of soft robotic technologies in industry.

References

1. Tallman S, Koza MP (2024) Global strategy in our age of chaos. Cambridge University Press
2. Zhou L, Ren L, Chen Y et al (2021) Bio-inspired soft grippers based on impactive gripping. Adv Sci 8:2002017. https://doi.org/10.1002/advs.202002017
3. Miriyev A (2019) A focus on soft actuation. Actuators 8:74. https://doi.org/10.3390/act8040074
4. Bui PDH, Prugh B, Padilla AME et al. (2023) Endurance tests for a fabric_reinforced inflatable soft actuator. Front Mater 10. https://doi.org/10.3389/fmats.2023.1112540
5. Nakajima K, Li T, Hauser H et al (2014) Exploiting short-term memory in soft body dynamics as a computational resource. J R Soc Interface 11:20140437. https://doi.org/10.1098/rsif.2014.0437
6. Yap HK, Ng HY, Yeow C-H (2016) High-force soft printable pneumatics for soft robotic applications. Soft Robot 3:144–158. https://doi.org/10.1089/soro.2016.0030
7. Zhang Y, Kutani K, Wang Z (2024) Durability testing of soft pneumatic actuators with a portable device. In: 2024 International conference on advanced mechatronic systems (ICAMechS). IEEE, pp 49–54
8. Polygerinos P, Lyne S, Wang Z et al. (2013) Towards a soft pneumatic glove for hand rehabilitation. In: 2013 IEEE/RSJ International conference on intelligent robots and systems. IEEE, pp 1512–1517
9. Schreiber F, Frohn-Sörensen P, Mouratidis M et al (2023) Case study on the additive manufacturability of printed soft-robotic bending actuators. Procedia CIRP 120:720–725. https://doi.org/10.1016/j.procir.2023.09.065
10. Frohn-Sörensen P, Schreiber F, Manns M et al (2022) Additive manufacturing of TPU Pneu-Nets as soft robotic actuators. In: Andersen A-L, Andersen R, Brunoe TD et al (eds) Towards sustainable customization: bridging smart products and manufacturing systems. Springer International Publishing, Cham, pp 269–276
11. Bradski G (2000) The OpenCV library. Dr. Dobb's Journal of Software Tools:120–125
12. Virtanen P, Gommers R, Oliphant TE et al (2020) SciPy 1.0: fundamental algorithms for scientific computing in Python. Nat Methods 17:261–272. https://doi.org/10.1038/s41592-019-0686-2

Optimized Data Preprocessing for Design of Passive Compliance via Solid Generation from G-Code

Richard Matthias Hartisch, Danny Huang, Patrick Kahl, Linus Schneider, and Jörg Krüger

Abstract To support high-speed mechanical search and self-aligning insertion of parts with tight tolerances and sensitive components, compliance directly proximal to the gripped object has been proposed in previous work, provided by structured compliance in additively manufactured grippers. The grippers are designed parametrically via the slicer software, allowing a quick iteration over the design parameters. Due to the design process, each combination has to be remodeled manually in CAD in order to be analyzed in simulations. This work focuses on the current modeling problem, by proposing a python script to directly abstract a solid from G-Code. The object reconstruction runtime and quality are compared to other state-of-the-art approaches achieving reconstruction up to 97.9% times faster for objects with a higher complexity by also achieving a high reconstruction quality. The reconstruction time can be accelerated further in future work, by implementing a script-based modeling.

Keywords Additive manufacturing · CAD reconstruction · Compliant structures · G-Code processing · Passive compliance

1 Introduction

Passive, structured compliant grippers for rapid and sensitive contact-rich manipulation in dexterous tasks, such as electrical assembly, have been established as a means of error compensation over pose uncertainties. This principle works as an alternative to active compliance, e.g. force, impedance, or admittance control, which can adapt the relative pose in response to external forces during contact [1].

In previous work, compliant structures have been designed by creating an initial solid part in CAD. In this design step the outer geometry is created, as further described in [2, 3]. In the next step, the structured compliance is generated in the

R. M. Hartisch (✉) · D. Huang · P. Kahl · L. Schneider · J. Krüger
Department of Industrial Automation Technology, TU Berlin, Berlin, Germany
e-mail: r.hartisch@tu-berlin.de

M. Manns (ed.), *New Paradigms for Anticipated Uncertainty*, Lecture Notes in Mechanical Engineering, https://doi.org/10.1007/978-3-032-16889-4_83

slicer software, following a similar design principle as in [4]. After the design process in the slicer software, the manufacturing instructions are generated in a G-Code for an arbitrary 3D printer and the compliant structures can be manufactured. Thus, the final design is set in the G-Code and not in a solid. Efficient evaluation during the optimization process of these structures requires rapid iteration. Simulation offers a powerful tool for optimization. However, current workflows require manual CAD remodeling of each gripper, introducing time expenditure and potential inaccuracies in key geometric parameters. As additionally pointed out by [5], in many cases, only the G-Code file used for additive manufacturing is available, while the original .stl or CAD model are not existent anymore, providing a challenge if modifications are needed before reprinting. Thus, making a direct solid abstraction from G-Code crucial for streamlining optimization.

Existing work on solid abstraction from g-code employs distinct approaches. Sobieski and Kiński [5] introduce a reverse-engineering method where a custom Fortran script extracts the extruder's x-, y-, and z- coordinates from the g-code. Linear interpolation fills in extended paths, generating a dense point cloud that is then converted into a continuous surface via Poisson Surface Reconstruction. However, this method struggles with accurately capturing surfaces, curved and fine features as found by [5]. In contrast, in [6] a CAD-based technique is proposed. Their method uses MATLAB to extract coordinate data from the g-code and automatically generates a python script that creates layer-wise sketches and performs sweep operations in Abaqus/CAE. This reconstructs a CAD model that accurately reflects the deposited filament profiles and is suitable for finite element analysis (FEA). In [7] a CAD-based approach is also adopted. They extract toolpath coordinates using MATLAB, then employ a python script to filter and group continuous extrusion segments into a structured sweep list. The 3D model is reconstructed in FreeCAD via the sweep function and exported as a STEP file, enabling integration with CAD and FEA tools. In summary, while the point-cloud method proposed by [5] is innovative, it lacks precision in complex areas. Both [6, 7] offer automated, CAD-based workflows that yield high-quality models. The primary differences lie in their software choices. Vega et al. [6] utilizes Abaqus/CAE, whereas [7] uses FreeCAD as an open-source solution for 3D model reconstruction. However, the approach proposed in [7] necessitates MATLAB to extract the toolpath coordinates.

Here, the research gap can be identified; the proposed approach in this work aims to fill the gap from [6, 7] by providing a similar CAD-based workflow. However, solely relying on a python script to extract the coordinates, without the need of MATLAB, as in [6, 7], as well without the necessity of a dense point cloud combined with a surface conversion via Poisson Surface Reconstruction, as in [5]. As demonstrated later, the proposed python script offers runtime benefits through silent operation, process priority management, and parallelization, making it particularly effective in reconstructing parts involving multiple layers. The performance analysis in [7] indicates that their reconstruction process becomes increasingly time consuming with an increasing number of sweeps generated, i.e. in complex geometries or dense infill patterns. A desirable outcome of the solid abstraction would be to achieve a CAD independent generation of the final gripper design for the design process with a highest level of independence possible and generalizability from CAD and simulation tools.

This paper is organized as follows: In Sect. 2.1 the workflow of the python script for G-Code abstraction is described. Section 2.2 further describes the implemented measures to optimize efficiency. In Sect. 3 the performance of the python script is compared to a previously published works, as well as an overview of existing errors is given and potential improvement noted before the outlook and conclusion in Sect. 4. The in this paper used files have been uploaded to our github repository, which can be found here: https://github.com/richardhartisch/structuredcompliance.

2 Methodology

2.1 Workflow

The proposed python code implements a workflow for converting G-Code files into 3D solid models by interfacing with Autodesk Inventor's API. For this, the G-Code is parsed to extract layer and feature data, generating corresponding sketches and features through sweep and extrusion operations. The individual layers are finally combined into an assembly and converted to a solid. The code was developed following an iterative approach inspired by the V-model, described in [8].

The workflow is visualized in Fig. 1 and starts by reading a JSON configuration file to initialize parameters such as filament diameter, maximum parallel instances, and folder paths. The function *read_gcode_variables* extracts key parameters, i.e. the layer count and layer height from the G-Code file.

extract_layer_coordinates parses the G-Code to compute the 2D coordinates and corresponding extrusion widths for each layer. This gives a list of paths and a list of the corresponding path widths, used later in the function *generate_one_layer*.

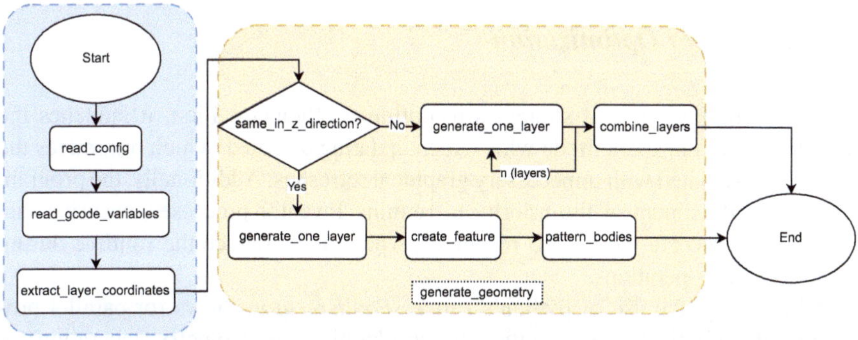

Fig. 1 Here the function workflow of the python script is demonstrated

In the top-level function *generate_geometry*, highlighted in Fig. 1, these G-Code parameters are read, selecting the appropriate processing method (single layer with vertical patterning or multiple layers), and the 3D model is ultimately assembled. If the model has the same dimensions and geometries in the z-direction, the function *generate_one_layer* is executed once, and the single generated layer is copied in an Inventor instance in correspondence to the amount of layers obtained from *read_gcode_variables*. The geometry abstraction is handled by the function *generate_one_layer*, which processes a single layer by utilizing the prior gathered coordinates from the G-Code using the function *extract_layer_coordinates*. Here, an *Inventor* object is instantiated, iterating over each feature to generate the corresponding solid layer. The *Inventor* class launches a silent, non-updating Inventor session. In this instance, profile sketches are created, based on the gathered coordinate data. The class then utilizes these sketches to form solid features via sweep or extrude operations. If a sweep operation fails due to complex geometry or unsolved self-intersections, it falls back to an extrusion method. The correct determination of movement direction of the extruder is essential for the sweep and extrusion features.

Once the geometry for a layer is created, *generate_one_layer* attempts to fill in any small gaps. The solid body is then created in *create_feature*. Depending on the configuration, the geometry is patterned along the z- direction to replicate the layer multiple times. However, if the model varies in the z- direction, a multiprocessing pool is created, and each layer is generated separately and by parallel execution of the aforementioned function *generate_one_layer* in correspondence to the amount of layers. For this, the *InventorAssembly* class is needed. This manages the creation of an assembly part by importing the individual layer parts created in *generate_one_layer*. The *combine_layers* function assembles the individual parts into an assembly, using a transformation matrix to position each part along the z-axis correctly, thereby enabling the combination of multiple layers into a single assembly. The assembly is then converted back into a single part file. Finally, the *main* function terminates existing Inventor processes for a new start, prompting the user for one or more G-Code filenames, and processing each file while reporting execution times.

2.2 Efficiency Optimization

The code employs several strategies to optimize efficiency. First, it launches the Inventor session in silent mode with screen updating disabled, which minimizes the overhead associated with unnecessary graphical refreshes. Additionally, the program allows the adjustment of the priority of running Inventor processes, ensuring that critical tasks receive more CPU resources, which helps reduce the runtime during intensive CAD operations.

Additionally, the use of python's *ProcessPoolExecutor* allows for parallel processing of multiple layers, significantly accelerating the geometry generation for multi-layer models. Finally, the *init_globals* function efficiently initializes necessary configuration parameters in each subprocess, avoiding redundant setup and further enhancing overall performance. The impact of these measures are evaluated in the following section.

3 Performance Evaluation

To evaluate the performance of the proposed script, a direct comparison of the computation times of various objects with different parameters is performed. The results are compared to the computation times obtained in [7]. Additionally, the reconstruction quality performed with the proposed python script is compared to the proposed workflow from [5].

Table 1 presents a comparison of computation times for two different models (Cube 10×10×10 and Dog-bone specimen) using various infill densities and types. The results are measured both with the method described in [7] and with the proposed program. At 20% Lines infill for the Cube 10×10×10 model, the computation time of the program (205.70 s) is 202.5% slower than the method in [7] (68.00 s). A similar trend is observed for 20% Gyroid infill, where the proposed program requires 328.72 s compared to 78.60 s, resulting in a slowdown of 318.2%. These slower times are attributed to the smaller number of operations per layer, making the overhead of creating Inventor instances and combining layers into assemblies more significant. For higher infill densities (50% and 100% lines), the proposed program demonstrates faster runtimes. For instance, at 100% Lines infill, the proposed program achieves a computation time of 263.53 s, which is 81.4% faster than the 1419.30 s required by the method in [7]. Regarding the Dog-bone specimen model, the proposed program outperforms the method in [7] across all tested infill densities and types. For example, at 100% Lines infill, the proposed program is 97.9% faster (1281.32 s vs. 60815.30 s). Similarly, for the Pyramid infill at 20%, the proposed program reduces the computation time to 1004.87 s from 1540.00 s.

Table 1 Computation time comparison with different infill types and densities

Part	Infill density [%]	Infill type [-]	Time [s] [7]	Time [s] (own program)	Difference [%]
Cube 10x10x10	20	Lines	68.00	205.70	–202.5
	50	Lines	361.30	248.46	+31.2
	100	Lines	1419.30	263.53	+81.4
	20	Gyroid	78.6	328.72	–318.2
Dog-bone specimen	10	Lines	693.00	607.59	+12.3
	20	Lines	1666.80	794.29	+52.3
	50	Lines	7705.10	1998.45	+74.1
	100	Lines	60815.30	1281.32	+97.9
	20	Pyramid	1540.00	1004.87	+34.8

The different computational times derive from the varying solution approaches. In the proposed python script, performance is enhanced by several key strategies. For instance, the Inventor session is launched in silent mode with screen updating disabled, and process priorities are adjusted, ensuring that the system devotes maximum resources to the CAD operations. Moreover, by leveraging python's parallel processing via the *ProcessPoolExecutor*, the script can process multiple layers in parallel, significantly reducing overall runtime for multi-layer models.

In contrast, the methodology described in [7] utilizes FreeCAD to perform sweep operations for geometry reconstruction. Their performance analysis indicates that the computational time grows with a power law relative to the number of sweeps generated. This means that for complex geometries or dense infill patterns, the reconstruction process becomes computationally expensive. Thus, potentially limiting its applicability for larger models.

Additionally, the reconstruction quality of the python script is compared to the Poisson Surface Reconstruction from [5]. In Fig. 2 the reconstructed cube (Fig. 2a and b) and dog-bone specimen (Fig. 2c) are visualized, demonstrating both efficient

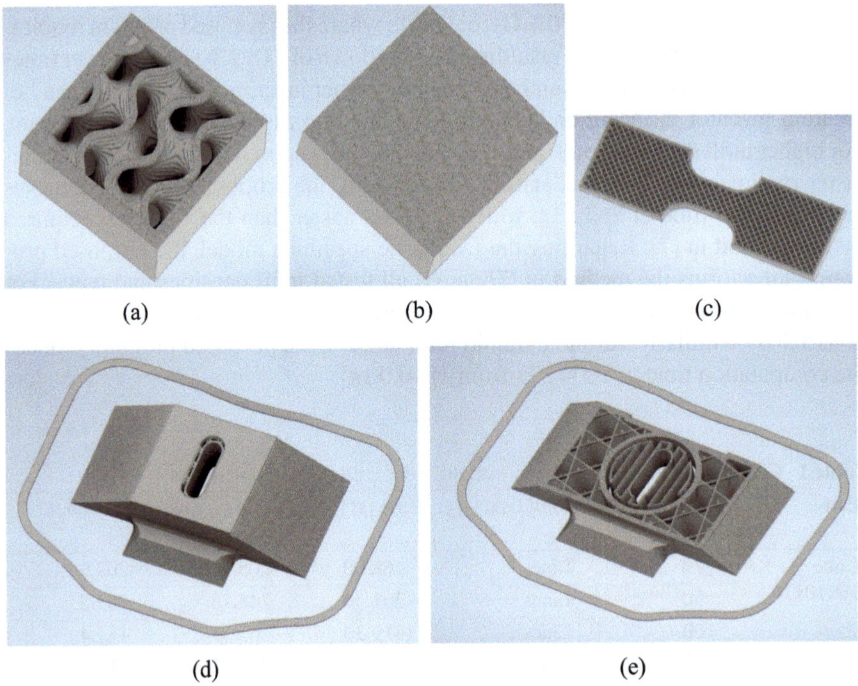

Fig. 2 Here, the reconstructed models can be seen, demonstrating feasibility. **a** and **b** visualize the reconstructed cube, **a** shows the infill type "gyroid" and **b** the closed cube. In (**c**), the dog-bone specimen with the infill type "lines" is shown. **d** and **e** visualize parts with deviations along the z-axis with a skirt on the print ped. In (**d**) the individual layers can be seen. In (**e**), the support structures supporting the overhang along the slotted hole is shown

surface and infill reconstruction, especially compared to the results in Fig. 8–10 from [5]. As concluded by [5], the Poisson Surface Reconstruction is not suitable to reconstruct the 3D objects in a satisfactory manner. However, it can be seen, that the outer walls in our work, as seen in Fig. 2 do not exactly represent the stacked layers in the z- direction. The individual layers, however, can be seen better in objects with deviating layers along the z- axis, as in Fig. 2d and e. It becomes apparent, that the extrusion during manufacturing is not represented ideally. The profile is simplified as a rectangular shape, instead of the circular profiles, as seen in Fig. 8 of [7]. However, it is unclear if this level of detail is relevant for simulation. Figure 2d and e additionally demonstrates the suitability of the proposed python script to reconstruct print bed adhesion and support structures, if necessary.

Further comparing the reconstruction quality of the proposed python script to the results from [7], it can be seen in Fig. 9 of [7], that the layers of the outer walls are wrongly arranged vertically. However, Fig. 8 in [7] demonstrates more suitable results, overall demonstrating a solid reconstruction quality.

However, while the proposed python script shows promising results, occurring failure cases have to be investigated. In Fig. 3a, a closer examination reveals that edges are not merged correctly in some areas. This occurs due to paths not being identified correctly as closed curves in the code, as they are not traversed continuously within a single path. Additionally, Fig. 3b shows gaps in certain parts of the model. It is assumed that they occur due to the flow of material not being represented accurately in the model.

(a) (b)

Fig. 3 In (**a**) and (**b**) resulting errors are demonstrated from the G-Code abstraction. These can be attributed to an incorrect identification of the paths of the extruder for (**a**) and an incorrect modeling of the material flow during extrusion in (**b**)

4 Conclusion and Outlook

This work proposes a python script to obtain CAD-models of parts from G-Code created in an arbitrary slicer software. Thus enhancing the foundation for the precise evaluation and optimization of additively manufactured components and enabling the analysis of a wide range of G-Code files in simulation. The python script enables this without the necessity of remodeling parts manually with various parameter combinations.

In direct comparison to existing work, as in [7], it is observed, that the performance gains achieved by the proposed program become more pronounced as the model complexity increases. While smaller models with fewer operations per layer show limited improvement, larger models demonstrate the scalability and efficiency of the proposed approach. This improvement is attributed to the better runtime complexity of the script, which becomes increasingly advantageous as model size and operation count grow. Focusing on the reconstruction quality, the proposed python script performs well in comparison with other solutions focusing on open-source solutions, as in [5]. However, there remain errors in the solid generation, as mentioned earlier. It is currently unknown, if this level of detail is relevant for the simulation, which is why differing levels will be investigated in future work.

As described earlier, the general scope of the G-Code abstraction is to achieve a CAD independent generation of the final gripper design for a future design-pipeline with a highest level of independence possible and generalizability from CAD and simulation tools. Other than existing work, as in [6, 7], the coordinate extraction is performed without the use of MATLAB. However, Autodesk Inventor is currently still needed for the solid abstraction. Thus, in future work the program will be improved by enabling a generation of a solid body directly in python, without relying on external software like Inventor. This could be made possible using libraries such as *cadquery* in future work. Since this library enables script-based modeling, it can be assumed that this approach would lead to significantly better results, further enhancing the runtime. Additionally, an approach as in [7] is possible, utilizing FreeCAD as an open-source solution.

References

1. Li R, Qiao H (2019) A survey of methods and strategies for high-precision robotic grasping and assembly tasks-some new trends. IEEE/ASME Trans Mechatron 24(6):2718–2732
2. Hartisch RM, Haninger K (2024) High-speed electrical connector assembly by structured compliance in a finray-effect gripper. IEEE/ASME Trans Mechatron 29(2):810–819
3. Hartisch R, Haninger K (2023) Compliant finray-effect gripper for high-speed robotic assembly of electrical components. In: 2023 IEEE/ASME international conference on advanced intelligent mechatronics (AIM). IEEE
4. Elgeneidy K, Lightbody P, Pearson S, Neumann G (2019) Characterising 3D-printed soft fin ray robotic fingers with layer jamming capability for delicate grasping. In: 2019 2nd IEEE international conference on soft robotics (RoboSoft). IEEE, pp 143–148

5. Sobieski W, Kiński W (2020) Geometry extraction from GCODE files destined for 3D printers. Tech Sci 23(2):115–130
6. Vega G, Paz R, Gleadall A, Monzón M, Alemán-Domínguez ME (2021) Comparison of cad and voxel-based modelling methodologies for the mechanical simulation of extrusion-based 3d printed scaffolds. Materials 14(19):5670
7. Bacciaglia A, Falcetelli F, Troiani E, Di Sante R, Liverani A, Ceruti A (2023) Geometry reconstruction for additive manufacturing: from G-CODE to 3D CAD model. Mater Today Proc 75:16–22
8. Mathur S, Malik S (2010) Advancements in the v-model. Int J Comput Appl 1(12):29–34

MIX
Papier aus verantwortungsvollen Quellen
Paper from responsible sources
FSC® C105338

If you have any concerns about our products,
you can contact us on
ProductSafety@springernature.com

In case Publisher is established outside the EU,
the EU authorized representative is:
Springer Nature Customer Service Center GmbH
Europaplatz 3, 69115 Heidelberg, Germany

Printed by Libri Plureos GmbH
in Hamburg, Germany